LELAND SAYLOR

Cost Estimating • Construction Management • Feasibility Reports

CURRENT CONSTRUCTION COSTS 2015

Founder & Author:
Leland Saylor

Managing Editor:
Brad Saylor

Contributing Editors:
Natalie Saylor
Jeff Saylor

Layout Editor:
Edward Savio

Associate Editors:
Dean Clevenger
Seth Danquah
Mike Kritscher
Warren Miller
Ian Slight

QUALITY COST DATA SINCE 1958

$99.95
Retail per copy. Price subject to change.

Copyright © 2015 by Leland Saylor.
All rights reserved.

No part of this publication may be reproduced, stored in a retrieval system, or transmitted in any form or by any means, without the prior written permission of the Publisher, except as otherwise permitted by law.

Requests for reprint permission should be addressed to:
Leland Saylor
Saylor Communications, Inc.
101 Montgomery Street, Suite 800,
San Francisco, CA 94104
(415) 291-3200.

Limit of Liability and Disclaimer of Warranty: While the Publisher attempts to make accurate and reliable data available in its publications, no warranty or guarantee, expressed or implied, is made for the content herein, including cost and statistics data, availability and legality of use of any components, materials, means or methods of construction listed, or the applicability its use. Leland Saylor shall have no liability to any purchaser, recipient or any other user of the content by any means, for any purpose, for any loss or damage, including but not limited to, lost profit or lost revenue, or incidental, special, or punitive damages, caused either directly or indirectly by or from use of the following content, or any content connected to this and/or other related publications, by any error or omission arising from the availability of its data past, present and future.

Saylor® is registered a trademark of Lee Saylor.
Saylor book logo is a trademark of Saylor Communications, Inc.

Saylor Communications, Inc.
Current Construction Costs, Edition 52

1 2 3 4 5 6 7 8 9 10
Printed in the United States of America

For Over 50 Years, Saylor Has Meant Precision.

Leland Saylor Associates
A Certified SDVOSB

We Wrote The Book.

Leland Saylor created the Current Construction Cost manual in 1963 to give architects, engineers, owners and contractors reliable cost information to help them produce accurate cost estimates. In 1967, he developed the Lee Saylor Material/Labor and Subcontractor Indices, published in ENR, to track construction costs over time. Today, Mr. Saylor continues his role as chief economist and editor-in-chief of the publications division.

For Over 50 Years, Saylor Has Been A Part Of Turning Vision Into Reality.

Services
- Design Phase Cost Estimating
- Value Engineering
- Construction Phase Cost/Change Management
- Project Controls
- Construction Management
- Construction Inspection
- Construction Market Research

Leland Saylor Associates
A Certified SDVOSB

For Over 50 Years, Saylor Has Been Helping To Build The Future.

Leland Saylor Associates (LSA) is a certified Service Disabled Veteran Owned Small Business (SDVOSB) construction consulting firm providing expertise in the areas of cost estimating, scheduling, value engineering, and claims analysis. Founded in 1958, LSA is one of the largest cost consulting firms in the United States with 5 offices throughout the country estimating over 400 projects per year, from multi-billion-dollar civic programs to small office tenant improvements. Each year, the staff is called on to estimate over twenty billion dollars of construction projects.

Leland Saylor Associates
A Certified SDVOSB

Main Office:
Leland Saylor Associates
101 Montgomery St., #800
San Francisco, CA 94104
415.291.3200 TEL
415.291.3201 FAX
info@lelandsaylor.com

Oakland Office:
Leland Saylor Associates
1629 Telegraph Avenue
Oakland, CA 94612
510.986.1212 TEL
510.444.0279 FAX

Los Angeles Office:
Leland Saylor Associates
2046 Armacost Avenue
Los Angeles, CA 90025
310.207.6900 TEL
310.207.6906 FAX

Washington DC:
Leland Saylor Associates
1050 Connecticut Ave. NW
10th Floor
Washington, DC 20036
202.550.3801 TEL
202.772.1046 DIRECT

Chicago Office:
Leland Saylor Associates
155 N. Wacker Drive
Suite 4250
Chicago, IL 60606
312.324.4566 TEL

Table of Contents

Foreword..I
Wage Rate Recap...VI
Cost Indices..VII
Major Cities Cost Relationship Index...XI
Seismic Zones..XV
Symbols & Abbreviations..XVI

Division #	Division	Page
01.0000 000	**GENERAL REQUIREMENTS**	**1**
01.1000 000	General Conditions	1
01.1010 000	Mobilization, On & Off	2
01.1020 000	Non-Distributable Labor	2
01.1030 000	Permits, Licenses & Fees	2
01.1040 000	Temporary Utilities, Structures & Fences	3
01.1100 000	Equipment Rental	3
01.1800 000	Other General Conditions	23
01.1900 000	Non-Manual Labor Distributables	25
01.2000 000	Overhead & Profit, Bonds	25
01.3000 000	Escalation	25
01.4000 000	Contingencies	25
01.5000 000	Geographical Differences	25
02.0000 000	**SITE WORK**	**26**
02.1000 000	Demolition	26
02.1100 000	Site Demolition	26
02.1200 000	General Building Demolition	26
02.2000 000	Excavation, Fill & Grading	30
02.3000 000	Piling	31
02.3500 000	Caissons & Drilling	32
02.4000 000	Shoring & Bulkheading	33
02.5000 000	Site Utilities	34
02.5100 000	Storm Drainage & Sanitary Sewer Pipe	35
02.5300 000	Water, Steam & Gas Distribution Piping	37
02.5400 000	Valves & Specialties	40
02.5500 000	Mechanical Utilities, Accessories	41
02.5600 000	Miscellaneous Site Equipment	41
02.5700 000	Electrical Distribution, Underground	42
02.5800 000	Electrical Distribution, Overhead	44
02.6000 000	General Site Work, Paving & Walks	45
02.7000 000	Miscellaneous Site Improvements	48
02.7400 000	Irrigation, Sprinkler Head Systems	48
02.7600 000	Landscaping	50
02.9100 000	Railroad Work	51
02.9500 000	Marine Work	51
03.0000 000	**CONCRETE**	**52**
03.0500 000	Concrete, In-Place	52
03.0600 000	Precase Concrete	53
03.0700 000	Specialty Concrete	53
03.0800 000	Excavation & Backfill	54
03.1000 000	Concrete Forms	55
03.1100 000	Foundation Forms	55
03.1200 000	Footing Forms	55

Table of Contents

03.1300 000	Forms, Slab on Grade	55
03.1400 000	Construction Forms, Vertical	55
03.1600 000	Column Forms	56
03.1700 000	Construction Forms, Horizontal	57
03.1800 000	Construction Forms, Horizontal, Heavy Duty	57
03.1900 000	Misc. Concrete Forms & Form Specialties	58
03.2000 000	Reinforcing Steel	59
03.2100 000	Reinforcing Steel, In-Place	59
03.2200 000	Reinforcing Steel, Built/up Cost	59
03.3000 000	Readymix Concrete	60
03.3100 000	Concrete Placement	60
03.3600 000	Slab Finishes	61
03.3700 000	Vertical Surface Finishes	61
03.3800 000	Miscellaneous Concrete Finishes	62
03.3900 000	Miscellaneous Concrete Items & Accessories	62
03.4000 000	Tilt Up Construction	62
03.4500 000	Compilation of In-Place Cost	63
03.5100 000	Insulating Concrete, Interior	63
03.5200 000	Concrete Deck, Exterior	64
03.5300 000	Insulating Decks	64
03.5400 000	Fiber Deck	64
03.6000 000	Epoxy Injection, Repair	64
04.0000 000	**MASONRY**	**65**
04.1000 000	Brick Masonry	65
04.2000 000	Concrete Unit Masonry	65
04.3000 000	Architectural Stonework	67
04.4000 000	Masonry Accessories & Miscellaneous Work	68
04.5000 000	Fireplaces	68
04.6000 000	Pargeting	68
05.0000 000	**METALS**	**69**
05.1000 000	Structural Steel	69
05.1100 000	Structural Steel Specialties	69
05.3000 000	Decking & Siding	70
05.5000 000	Miscellaneous Iron	72
06.0000 000	**CARPENTRY**	**75**
06.1000 000	Rough Carpentry	75
06.1100 000	Vertical Framing, Walls, Per 1,000 Board Feet	75
06.1200 000	Horizontal Framing, Per 1,000 Board Feet	76
06.1300 000	Miscellaneous Framing & Materials	77
06.1400 000	Sheathing	77
06.1500 000	Carpentry Specialties	78
06.2000 000	Finish Carpentry	79
06.2100 000	Carpentry, Installation Only	80
06.3000 000	Glu-lam Beams, Trusses & Heavy Timber	81
06.5000 000	Stairs, Wood	81
06.6000 000	Rough Hardware	81
07.0000 000	**THERMAL & MOISTURE PROTECTION**	**84**
07.1100 000	Waterproofing	84
07.2000 000	Thermal & Sound Insulation	85
07.3000 000	Roofing	87
07.3100 000	Composite Building Panels	88

Table of Contents

07.3200 000	Stone Panels, Manufactured	89
07.3300 000	Mineral Fiber Panels, Curtain Walls	89
07.4000 000	Exterior Insulation Finish Systems (EIFS)	90
07.6000 000	Sheet Metal & Fabricated Skylights	90
07.7000 000	Architectural Sheet Metal	93
07.9000 000	Caulking & Sealants	94

08.0000 000 DOORS, WINDOWS & GLASS ... 96

08.1000 000	Hollow Metal Doors & Frames	96
08.2000 000	Wood Doors & Frames	98
08.2100 000	Wood Garage Doors	100
08.2200 000	Wood Door Specialties	100
08.3000 000	Special Doors	100
08.4000 000	Vinyl, Windows and Doors	102
08.5000 000	Aluminum, Windows and Doors	103
08.6000 000	Wood, Windows and Doors	104
08.7000 000	Finish Hardware	105
08.8000 000	Glass & Glazing	107
08.9000 000	Curtain Wall & Storefront Systems	108
08.9100 000	Curtain Walls & Exterior Panel Systems	109

09.0000 000 FINISHES ... 112

09.1000 000	Lath, Plaster, Studding & Furring	112
09.1100 000	Studs	112
09.1200 000	Furring	113
09.1300 000	Lathing	113
09.1400 000	Plaster & Lath	114
09.2000 000	Gypsum Wall Board, Studding & Furring	115
09.2100 000	Gypsum Wall Board, Specialties	117
09.3000 000	Ceramic Tile	118
09.4000 000	Terrazzo	118
09.5000 000	Acoustic Treatment	119
09.6000 000	Wood Flooring	120
09.7000 000	Resilient Flooring	121
09.8000 000	Painting & Wall Covering	122
09.9000 000	Plastic & Factory Finish Wall Surfaces	124

10.0000 000 SPECIALTIES ... 125

10.1000 000	Chalk & Tack Boards	125
10.1500 000	Toilet Partitions & Compartments	125
10.2000 000	Partitions Folding, Relocatable & Demountable	126
10.4000 000	Toilet Accessories	127
10.5000 000	Miscellaneous Building Specialties	129

11.0000 000 EQUIPMENT ... 133

11.1100 000	Bank Equipment	133
11.1200 000	Ecclesiastical Equipment	133
11.1300 000	Educational Equipment	135
11.1400 000	Observatories & Planetariums	136
11.1500 000	Vocational Shop Equipment	136
11.1600 000	Food Service Equipment	136
11.1700 000	Gymnasium & Playground Equipment	140
11.1800 000	Industrial Equipment	141
11.1900 000	Parking Lot Equipment	142
11.2000 000	Material Handling Equipment	142

Table of Contents

11.2100 000	Laboratory Equipment	142
11.2200 000	Library Equipment	143
11.2300 000	Hospital Equipment	143
11.2400 000	Dental Equipment	146
11.2500 000	Mortuary Equipment	147
11.2600 000	Prison Equipment	147
11.2700 000	Central Vacuum System	147
11.2800 000	Stage Equipment	147
11.2900 000	Garbage Compactors	149
11.3000 000	Window Washing Equipment, Powered	149
11.4000 000	Vocational Equipment	149
12.0000 000	**FURNISHINGS**	**150**
12.1100 000	Blinds & Shades	150
12.3000 000	Cabinets & Laminated Plastic Tops	150
12.3500 000	Laminated Plastic & Simulated Marble Tops	151
12.4000 000	Carpets	151
12.5000 000	Draperies & Curtains	152
12.8000 000	Office Landscape, Furniture by Station	152
13.0000 000	**SPECIAL CONSTRUCTION**	**154**
13.1000 000	Special Construction	154
13.1100 000	Prefabricated Structures	154
13.1201 000	Radiation Protection	155
13.1203 000	Swimming Pools	156
14.0000 000	**CONVEYING SYSTEMS**	**157**
14.1000 000	Conveying Systems	157
15.0000 000	**MECHANICAL WORK - PLUMBING**	**161**
15.1000 000	Equipment	161
15.1200 000	Fixtures	162
15.1300 000	Piping	164
15.1400 000	Valves & Specialties	168
15.1500 000	Insulation, Piping	171
15.1600 000	Miscellaneous Plumbing Specialties	172
15.1700 000	Medical & Laboratory Equipment & Pipe	174
15.1800 000	Fees, Permits & Sterilization	176
15.1900 000	Industrial Piping	176
15.2000 000	Gate, Globe & Check Valves, Cast Steel	178
15.2100 000	Industrial Piping Insulation	179
15.3000 000	**MECHANICAL WORK - HVAC**	**182**
15.3100 000	Equipment, Furnaces	182
15.3200 000	Equipment, Hot Water & Steam Boilers	183
15.3300 000	Equipment, Cooling	184
15.3400 000	Equipment, Heating & Cooling Combinations	186
15.3500 000	Auxiliary Heating & Cooling Combinations	187
15.3600 000	Air Handling Equipment, Primary	188
15.3700 000	Distribution, Terminal Equipment	190
15.3800 000	Miscellaneous Equipment	192
15.3900 000	Controls	193
15.4000 000	Duct Work, Grills & Registers	193
15.4100 000	Piping & Insulation	195
15.4200 000	Fittings	197

Table of Contents

15.4300 000	Valves & Specialties	198
15.4400 000	Insulation, Piping	200
15.5500 000	Fire Protection Systems	201

16.0000 000 ELECTRICAL WORK 205

16.0100 000	Total Electrical Work, Buildings	205
16.1000 000	Electrical Costs, In-Place, Preliminary Estimates	205
16.1100 000	Main Switchboards, 600V, Service & Distribution	206
16.1200 000	Distribution Panels to 600V	206
16.1300 000	Transformers	207
16.1400 000	Raceway & Wire, Combined	208
16.1500 000	Underfloor Distribution Systems	208
16.1600 000	Lighting Fixtures, In-Place	209
16.1700 000	Branch Circuit Runs, Special Purpose Conduit & Wire	210
16.1800 000	Signal & Communications Systems	210
16.1900 000	Branch Circuit Outlets & Devices	211
16.2000 000	Equipment, Unit Substations	211
16.2100 000	Equipment, Switchgear & Transformers	212
16.2200 000	Equipment, High Voltage Transformers	212
16.2300 000	Service Sections	213
16.2400 000	Combination Service & Distribution Switchboards	215
16.3000 000	Motor Control Centers	218
16.4000 000	Panelboards, 600V Max, Bolt-on Breakers	220
16.4100 000	Transformers, Dry, Low Voltage	220
16.4200 000	Panelboards for Bolt-on Breakers, 120/240V, 1PH, 3W	221
16.4300 000	Load Centers, Main Lug & Circuit Breaker Types, 240V Max	221
16.4400 000	Plug-in Circuit Breakers, Type QO	222
16.4500 000	Special Gear	223
16.5000 000	PVC, RSC, IMC & Aluminum Raceway	223
16.5100 000	PVC, RSC, IMC & Alum Conduit Terminals, Elbows & Fittings	226
16.5200 000	EMT Raceway, Terminations & Elbows	227
16.5300 000	EMT, MI Cable & Terminations	229
16.5400 000	Specialty Fittings, Explosion Proof	229
16.5500 000	Underfloor & Flush Trench Duct, Cable Tray	230
16.5600 000	Steel Gutters, Pull Boxes, Unistrut Hangers	231
16.5700 000	Special Raceway Assembly Systems	231
16.5800 000	Conductor Only	232
16.5900 000	Busways	235
16.5950 000	Raceway & Wire Combined	237
16.6000 000	Lighting Fixtures	238
16.7000 000	Electric & Signal Devices	242
16.7100 000	Communication, Intercom, Public Address	244
16.7200 000	Special Hospital Systems	246
16.7300 000	Prison Cell Door Control Systems	247
16.7500 000	Soft Wire Systems, 3 Wire	247
16.7600 000	Energy & Building Management Systems	248
16.7700 000	Testing	248

Assembly Costs 249
   Commercial Square Foot Building Costs 316
   Index 330

Foreword

Please read and absorb this section in its entirety before using the technical sections of this manual.

INTRODUCTION

The prices contained herein are intended to be used as guidelines and may not be representative of any single project. They are compiled from a large number of projects, interviews with material suppliers and subcontractors, and costs worked out with general contractors for labor. Many of the costs in this volume have been averaged not only from actual project costs, but from opinions rendered by others involved in the construction industry as well. When multiple line items are used together, the user will find their sum an accurate portrayal of construction costs. This book is ideal for preparing conceptual, schematic, preliminary, and final estimates for architects, engineers, and estimators in all trades.

ARRANGEMENT

The general arrangement of Current Construction Costs uses the Construction Specifications Institute (CSI) division format. Within each division, the data is further organized into major sub-trade categories. The trades are formatted according to the Uniform System for Construction Specifications, Data Filing and Cost Accounting. The arrangement is further refined for estimating purposes. The basic alphanumeric system has been replaced with a numeric system. We believe that the user will find that the numeric system makes it easier for the estimator to locate components which must be collected together. This is especially true in the mechanical and electrical trades, where the combination of possible systems is infinite. Some deviations from the standard CSI format may occur as well. The reasons for the deviations are as follows:

1. Where the standard format causes a split between disciplines involved in a single trade, the identity of the trade has been preserved. For instance: light steel framing, lathing, plaster and fireproofing are collected under the title "Lath, Plaster, Studding & Furring." Although the trade discipline may be split in time or sequence, the sections are grouped for two basic reasons: 1) the trades normally work together, and, 2) for the estimator's convenience in taking off the whole work that his firm may perform.

2. Where more separations are needed than allowed for in the Uniform Accounting System, some liberty has been taken, as in General Conditions.

3. Where divisions under the CSI format do not fit a logical trade pattern or flow of materials, we have deviated so that the patterns will better fit the trades. Specifically for the Mechanical and Electrical divisions we have divided the patterns as we felt appropriate. We divide Mechanical into Plumbing, HVAC, and Fire Protection. Electrical is divided into a pattern as well.

4. Where the trades are so broken down that the sections become too small to conveniently handle in a cost book, as in the Specialties section, items used frequently are combined into one section called Building Specialties, Miscellaneous.

BUDGET ESTIMATES VS. BID ESTIMATES

For complicated trades in which a number of separate materials make up the final item prices, we are now showing the combination of material as a single item under a section called In-Place Costs. The example below represents a typical item seen in the In-Place Costs section.

ITEM: (IN-PLACE COSTS SECTION)
Asphalt paving (in place)

DETAIL: (BREAKDOWN)
Rough grade
Fine grade
Sub-base
Base
Binder
Asphalt concrete
Sealer

The above breakdowns were necessary for the following reasons:

1. For conceptual, schematic, and preliminary estimates, most of the detail is not necessary.

2. Where accurate final estimates are to be performed, extensive detail is shown on the plans and is necessary for in-depth quantity surveys. Contractors and subcontractors need the accuracy of extensive detail, while others performing budget type estimates may not find full detail necessary.

Foreword

DETAIL GROUPINGS

Within the detail groupings, the items are generally broken down in the logical order of performance. An example of one such breakdown is concrete, which is split into excavation for concrete, forms, pour, and finish. Where no special construction sequence is required, the individual trade is broken down into its logical components, such as electrical work into equipment, fixtures, conduit and wire, and miscellaneous electrical devices. Groupings are valuable in helping the estimator who is not familiar with all trades become aware of the various divisions within a given trade (almost all trades have them), as well as in ensuring that the non-obvious items are included in the estimate.

UNIT PRICE COMPILATION

The unit prices of the various trade items are broken down into three columns: material cost, labor cost, and total cost. Material cost means equipment, equipment rental, material, and factory fabrication, combined with applied profit and overhead. Labor cost is defined as field installation and erection, applied subcontractor's overhead and profit, all applicable fringe benefits, payroll taxes, and insurance. The only exception to this rule applies to trade items not normally subcontracted by the general contractor.

Almost all the general contractor items are contained in sections 1 (General Requirements), 3 (Concrete) and 6 (Carpentry). All of the above costs are included, in addition to profit for the trade involved. We have done this because most general contractors are now treating these sections much like a subcontract, regardless of the fact that the work is being performed by their own teams. See the section on overhead and profit for more detail on how overhead and profit are included for the purposes of this book. Divisions and sections which do not include subcontractor overhead and profit are labeled to further clarify cost breakdowns.

VARIATIONS IN THE QUANTITY SURVEY

Each trade has formulas that are unique to the trade, usually developed over a period of years to account for any variations that have not otherwise been considered in the standard formulas. For waste materials, job conditions, and non-detailed trim items, the following quantities should be added:

Minimum of 10%
Sand (10-30% for compaction)
Crushed rock (10-20% for compaction)
Compacted fill (10-30% for compaction)
Foundation concrete (for waste and over pours)
Slab membranes (for laps)
Reinforcing (for laps and bends)
Plaster (for waste and clean-up)
Carpet (for waste and fitting)
Linoleum and sheet vinyl (for fitting and covers)
Tongue-and-groove lumber (10-33% size dependent)
Rough lumber (miscellaneous backing and blocking)
Wire (for waste and pigtails)
Excavation, mass (10-30% for compaction)

Minimum of 5%
Concrete slabs on grade
Piping
Conduit
Mesh
Lath
Metal studding and furring
Structural steel, for details and clips (7%)
Ceramic tile
Resilient flooring

In general, openings are deducted in the rough trades and not deducted in the finish trades.

OVERHEAD AND PROFIT

For the purposes of this book, the subcontractor's overhead and profit is included in the trade prices for trades normally subcontracted. If this is not true for a particular section, it is noted at the beginning of that section.

For trades in which the general contractor does his own work, no home office overhead or profit is included. In general, these trades are concrete, building specialties and carpenter-installed items of other trades.

All field overhead is combined in a section called General Conditions (1.1000), including all permits, temporary work, supervision, payroll taxes, union fringes for the non-distributable labor, and bonds.

Depending on the project, the General Conditions section may comprise between five and fifteen percent of the contract. Since general conditions represent such a large percentage of the project, it is imperative that it be given its own section of the estimate and not be distributed into the trades. In addition, some of the composition of field overhead is technically not distributable into the various trades. Fringes and

Foreword

payroll insurance for trade labor may be included in the applicable trade section, or in the General Conditions section of the estimate, at the discretion of the estimator. After all other items have been determined, profit is added to the estimate summary. In the construction industry, it is customary to absorb home office overhead in the profit item. The section entitled Overhead & Profit, Bonds (1.2000) details the typical general contractor profit structure and overhead relationships for various types and sizes of projects.

ESCALATION

The subject of commodity inflation has crept into the daily conversations of construction project planners in recent years. News headlines often announce increasing prices for metals, energy and other raw materials. Shifting global trade patterns, increased demand for new infrastructure and other factors partially explain this phenomenon.

For years, Leland Saylor Associates (LSA) has been closely monitoring how material and labor price affect construction costs. Due to the recent heightened volatility in the commodity market, we have introduced a series of measures to keep material prices closely aligned with their fundamentals.

Since the user will also need to understand inflation trends, LSA offers two indices found in the preface. The LSA Material/Labor Index surveys twenty-three materials and nine construction trades in major metropolitan areas. When combined, the result is a powerful indication of current construction cost trends. The Subcontract Index diverges by introducing contractor and subcontractor pricing for the twenty-one most often quoted in-place construction items.

The user will find these indices most helpful in gauging trends related to inflation. Quarterly updating of these indices is available in the LSA Construction Costs Quarterly Newsletter and Update Service, which may be purchased through LSA.

MAJOR CITIES COST RELATIONSHIP INDEX

No country-wide manual would be complete without a geographical index. Our Major Cities Cost Relationship Index (MCCRI) is in the preface to this manual. The index uses San Francisco as base 100, while all other cities are represented relative to that base. The metropolitan statistical group of 395 U.S. and Canadian cities, are listed in a straightforward index number format. The system works well for budget estimates. Naturally, contractors working in a given area will receive sub-bids with local units and local accuracy.

LABOR HOURS

Construction costs vary between cities due to changes in productivity, access to resources and labor, and transportation costs. For this reason, LSA has generated an extensive list of 395 U.S. and Canadian cities' cost index values to help you adjust costs found in this book. The cities are grouped by region and state, and their numbers adjusted to reflect costs relative to San Francisco, which is base 100.

WAGES

Wage rates used in this book are prevailing averages for urban areas where the majority of construction is done.

Leland Saylor Associates' wages reflect the labor costs paid to construction employees by an employer. The base wage and total fringe of twenty-nine select trades are sourced from the Department of Industrial Relations' Prevailing Wage Determinates. Furthermore, similar to how the Major Cities Cost Relationship Index is based on San Francisco costs, wages included in this volume are based primarily on Northern California. From there, a series of calculations account for federal and state taxes, payroll, supervision, trade specific workers' compensation, overhead and profit. This final wage rate is then multiplied by the Labor Hours to complete a job, resulting in Labor Cost.

It should be noted that reduced wages in certain areas are related to lowered productivity. However, this statement is not applicable to fringe benefits. Care should be taken to compare productivity factors when possible. The user should note that a comparison of those projects bid on a unit price basis throughout the U.S. indicates that despite wage rates varying by almost 100% from low to high, the in-place bids for such units bear a much closer similarity than do the wage components included in the makeup of the unit prices. For this reason, the Major Cities Cost Relationship Index numbers appearing in this manual are closer than wages alone would indicate.

Foreword

ESTIMATE PREPARATION

It is always necessary to prepare a complete summary for the estimate. The summary should contain all the trades included in the Table of Contents in this book. Naturally, the estimator may not use every trade for the project in question. Items not used are blanked out. The summary then serves as a checklist to reassure the estimator that no trade has been left out. The estimate is then prepared from the ground up, beginning with the estimator surveying the quantities and building the project in his mind just as the project would be built in the field.

Each of the trades in the summary are then relisted and, within each trade, the appropriate quantities are listed. A unit price is then applied. The trade is then totaled and inserted into the summary. In the case of general contractor, general conditions, concrete, and building specialties only, the balance of trade prices is provided by the subcontractors. General conditions, escalation, and profit are then added, completing the estimate. The estimate is then reviewed for items that may have been left out and decimal point errors. A beautifully prepared estimate can easily cause a loss because of a misplaced decimal point!

BUDGET OR ENGINEERING ESTIMATES

See section 01. 4000 000 for contingencies.

The prices in this manual are based on normal bid practices, which average 4-5 bids for a project. The importance of the bid cannot be overstated. The bidding general contractor will invite bids from several subcontractors in each of the 63 trades or more. Since subcontractors on a typical project account for 60-80% of the total dollar volume, the most important determinant of the total bid price is the subcontractor. More general contractor bidding induces more subcontractor bidding.

Examination of a large number of bids received would indicate that the deviation from engineering estimates produced from complete drawings, using the pricing in this book, is as follows.

```
1 bid ................................. +15% to +40%
2-3 bids ........................... +8% to +12%
4-5 bids ........................... -4 to +4%
6-7 bids ........................... -7 to -5%
8 or more bids ................. -12% to -8%
```

It is not unusual for subcontract bids to vary as much as 100% for an individual trade.

COMBINATION OF MATERIAL

Below is a typical example of how the costs in this manual are combined.

This manual In-Place Costs section:

Concrete, Structural Slab, 1 story: 245-282 CY.

Actual estimators have broken down and detailed sections of this manual:

DESCRIPTION	MATERIAL	LABOR	TOTAL
Forming for Concrete Slab	2.250	3,820	6,070
Beams, Girders, Shoring	1,650	4,530	6,180
Shoring/Reshore	1,050	810	1,860
Reinforcing	4,155	1,475	5.630
Accessories	350	313	663
Pour Concrete	3,350	1,100	4,450
Finish Concrete	50	1,200	1,250
Strip & Store Forms	92	590	682
Patch & Sack	92	1,006	1,098
Concrete Pumping	375		375
Total Per 100 CY	**13,414**	**14,844**	**28,258**

ASSEMBLIES

Leland Saylor Associates also includes an assemblies section with this edition. Similar to the Combination of Material description above, the assemblies section takes commonly used finished items and combines all of their individual components into one price. For example, an interior wall will often include gypsum boards, studs, insulation, paint, finish, and a rubber base. Rather than finding the cost of each individual component and totaling their values, the assemblies section will allow the user to easily view the cost of the entire wall summed into a single price.

Foreword

COMMERCIAL SQUARE FOOT BUILDING COSTS

Preceding the index is LSA's Commercial Square Foot Building Costs section. The table outlines the square foot building costs of 65 different buildings, ranging from simple apartments to complex hospitals and airport terminals. Each building type has its per square foot cost listed respective to various exterior construction options, as well as in which seismic zone it will be built.

CONCLUSION

Our intention was to keep Current Construction Costs as brief as possible, while also including as many of the costs and materials needed to assemble estimates or reference materials. Should you come across any items that have not been included, please let us know.

We invite you to contact us for quotes on complete or partial estimates for projects of major significance, including complete or partial military schools, hospitals, or commercial or institutional structures.

Thank you for supporting the ongoing publication of Leland Saylor Associates' Current Construction Costs.

QUICK START GUIDE

Though much more is happening behind the scenes, the following diagram reveals the basic formula that we use to maximize the accuracy of the cost estimates. The entries in this book are not only detemined through our team's industry expert knowledge, but also by numerous price indices to reflect the latest market trends.

$$\text{Labor Hours} \times \text{Labor Cost} + \text{Material Cost} = \text{Total Cost}$$

Labor Hours
- Individual time
- Crew time

Labor Cost
- Base wage
- Fringe benefits
- Taxes
  - Federal
  - State
  - Payroll
- Supervision
- Workers' comp
- Overhead & Profit

Material Cost
- Prices
  - Material
  - Commodity
- Indices
  - Material
  - Commodity
  - Other

For example, consider a carpenter and painter crew spending one **labor hour** building and painting a wooden desk. Only after accounting for various costs, such as taxes and overhead, are we able to determine average hourly **labor cost**.

Next, add in the desk's **material costs**, including wood, nails, and paint. We have incorporated the latest prices.

The cost of labor and material are summated, resulting in the **total cost**.

Labor Wage Rate Recap

The base wage and total fringe of 29 select trades are sourced from the Department of Industrial Relations' Prevailing Wage Determinates. Total wage accounts for various federal and state taxes, payroll, supervision, workers' compensation, and overhead and profit rates, as paid by the employer. Inside the main Current Construction Costs section of this book, this total wage rate is multiplied by column Labor Hours to obtain column Labor Cost.

Trade	Base Wage $	Total Fringe $	Payroll Tax $	Supervision $	Workers' Comp %	Workers' Comp $	Subtotal $	Overhead, Profit %	Overhead, Profit $	Total Wage Rate $
Asbestos Worker	59.48	19.84	5.86	2.97	11.59%	6.89	95.04	22%	20.91	115.95
Bricklayer	36.56	28.44	3.76	1.83	14.55%	5.32	75.91	27%	20.49	96.40
Carpenter, Drywall/Lather	40.35	28.50	4.11	2.02	13.71%	5.53	80.51	33%	26.57	107.07
Carpenter, General	40.20	27.87	4.09	2.01	26.81%	10.78	84.95	27%	22.94	107.89
Carpenter, Hardwood Floorer	40.35	27.72	4.11	2.02	26.81%	10.82	85.01	27%	22.95	107.96
Cement Mason	30.00	23.66	3.16	1.50	14.55%	4.37	62.68	27%	16.92	79.61
Electrician	49.56	31.05	4.95	2.48	10.20%	5.06	93.09	20%	18.62	111.71
Elevator Constructor	59.19	30.64	5.83	2.96	5.77%	3.42	102.03	33%	33.67	135.71
Glazier	41.83	22.72	4.24	2.09	18.55%	7.76	78.64	24%	18.87	97.52
Hod Carrier	32.84	19.69	3.42	1.64	14.55%	4.78	62.37	27%	16.84	79.21
Laborer, Demolition	28.14	20.13	2.99	1.41	12.72%	3.58	56.25	15%	8.44	64.68
Laborer, General	29.09	20.13	3.08	1.45	28.28%	8.23	61.98	15%	9.30	71.27
Operating Engineer, General	38.43	28.38	3.93	1.92	17.80%	6.84	79.50	15%	11.93	91.43
Operating Engineer, Oiler	34.07	28.28	3.53	1.70	17.80%	6.06	73.65	15%	11.05	84.70
Painter, Drywall (Taper)	40.32	20.41	4.10	2.02	18.12%	7.31	74.64	25%	18.66	93.29
Painter, General	38.45	21.17	3.93	1.92	18.12%	6.97	72.44	25%	18.11	90.55
Pile Driver	39.59	28.38	4.04	1.98	17.83%	7.06	81.04	15%	12.16	93.20
Plasterer	37.48	25.11	3.84	1.87	19.29%	7.23	75.54	33%	24.93	100.46
Plumber	54.40	35.80	5.39	2.72	13.43%	7.31	105.62	15%	15.84	121.46
Resilient Floorer	29.55	18.37	3.12	1.48	10.99%	3.25	53.96	27%	14.57	68.53
Roofer	31.66	17.76	3.31	1.58	54.51%	17.26	71.57	22%	15.75	87.32
Sheet Metal Worker, Deck & Siding	33.86	31.83	3.51	1.69	14.44%	4.89	75.78	15%	11.37	87.15
Sheet Metal Worker, Mechanical	54.85	36.79	5.43	2.74	14.44%	7.92	107.74	22%	23.70	131.44
Sprinkler Fitter	52.42	25.77	5.21	2.62	9.19%	4.82	90.84	15%	13.63	104.46
Structural Iron Worker	33.50	28.20	3.48	1.68	32.87%	11.01	77.87	21%	16.35	94.22
Teamster, 4 yard	27.96	24.81	2.97	1.40	28.28%	7.91	65.05	15%	9.76	74.80
Teamster, 9.5 yard	28.91	24.81	3.06	1.45	28.28%	8.18	66.40	15%	9.96	76.36
Terrazzo Mechanic	40.42	25.58	4.11	2.02	7.74%	3.13	75.26	27%	20.32	95.58
Tile Setter	38.24	17.82	3.91	1.91	11.24%	4.30	66.18	25%	16.55	82.73

Cost Indices

MATERIAL COST INDEX

23 Selected Materials	$/Unit	2008	2009	2010	2011	2012	2013	2014	YOY Change
Aluminum Sheet 48"x96"	CWT	190.73	189.00	183.00	175.00	202.25	193.26	201.19	4.10%
Ashpalt Felt, 4 squares	ROLL	21.12	21.51	21.00	23.85	24.90	24.68	24.07	-2.47%
Block, Concrete 8x8x16	EA	1.43	1.42	1.47	1.47	1.55	1.57	1.62	3.18%
Brick, Standard Modular	M	348.97	351.00	368.00	348.00	356.00	362.21	367.50	1.46%
Cement, Portland	TON	101.76	102.00	98.95	98.09	101.24	102.99	107.34	4.22%
Concrete, Readymix 3000 PSI	CY	92.26	93.80	85.00	84.15	85.46	90.37	94.00	4.02%
Copper Pipe, 1/2" type L	MLF	2,130.62	2,217.00	1,638.00	1,766.00	1,799.00	1,760.00	1,677.89	-4.67%
Glass, 1/4"	SF	5.15	5.31	5.09	5.25	5.52	5.58	5.75	3.05%
Gypsum Wall Board, 1/2" 4x8	MSF	253.03	238.34	234.37	251.03	255.00	265.85	290.96	9.45%
Insulation, Fiberglass batts, r13	MSF	274.80	279.80	307.00	312.50	351.00	363.90	354.28	-2.64%
Insulation, rigid, 1", poly iso	MSF	303.98	310.72	461.23	478.75	481.00	495.91	532.54	7.39%
Lath, Metal, 3.4#, Galvanized	CSY	210.13	215.74	208.80	222.23	236.00	232.85	263.74	13.26%
Lumber, framing, 2x4, 2x6	MBF	427.74	411.13	388.75	341.90	444.00	457.20	469.23	2.63%
Mason's Lime, 50# bags	TON	205.46	203.36	215.00	210.53	219.00	229.00	233.04	1.76%
PG 64 liquid bulk shipped	TON	274.79	286.04	477.50	557.00	570.00	570.00	558.59	-2.00%
Pipe, 8" c-900 PVC water	LF	8.85	8.82	8.93	9.55	9.26	9.15	9.30	1.59%
Pipe, Reinforced Concrete, 24"	LF	24.96	26.60	22.00	22.99	24.70	25.82	26.95	4.37%
Plywood 19/32" 4x8 CD exterior T+G	MSF	597.32	578.65	499.90	504.28	643.38	610.00	628.75	3.07%
Steel sheets, Stainless, 304, 4x8	CWT	198.17	179.74	176.00	183.95	144.50	157.25	159.95	1.72%
Steel, Reinforcing	CWT	40.90	40.36	40.30	42.06	36.20	35.80	36.79	2.77%
Steel, Structural Shapes	CWT	44.67	43.21	42.10	43.39	42.55	42.50	44.71	5.21%
Coal Tar Pitch, modified	TON	643.50	644.29	445.00	488.54	486.22	480.25	483.56	0.69%
Titanium Dioxide pigment	CWT	126.09	126.09	137.00	154.21	162.50	162.24	168.75	4.01%
Total Average Change (Non-weighted)									**2.88%**

Cost Indices

LABOR COST INDEX

9 Selected Trades, $/hour	2008	2009	2010	2011	2012	2013	2014	YoY Change
Carpenters	47.13	50.89	52.87	53.75	55.08	56.77	58.39	2.85%
Bricklayers	47.66	51.17	53.27	54.10	55.07	56.18	57.93	3.11%
Iron Workers (Structural)	51.97	55.75	58.81	59.68	60.80	61.41	62.76	2.19%
Laborers (Building)	37.84	41.85	43.64	44.49	44.99	47.02	49.68	5.67%
Operating Engineers (Crane Operators)	51.06	54.92	57.01	57.85	58.89	60.27	64.55	7.10%
Plasterers	44.18	48.15	50.10	50.83	51.26	52.85	54.68	3.47%
Plumbers	53.78	59.14	61.09	62.16	62.97	64.65	67.66	4.65%
Electricians	54.89	57.21	59.29	59.68	61.00	62.07	64.50	3.91%
Teamsters	40.89	41.51	43.02	43.76	44.97	45.80	48.00	4.80%
Total Average Wage (Non-weighted)	**47.71**	**51.18**	**53.23**	**54.03**	**55.00**	**56.34**	**58.68**	**4.20%**

SUBCONTRACT COST INDEX

21 Basic In-Place Materials	$/Minimum Quantity	Unit	2010	2011	2012	2013	2014	Change
Acoustic Ceiling, incl grid, tiles	20,000 & up	SF	3.31	3.46	3.62	3.75	3.92	4.53%
Brick veneer, common, commercial	2,000 & up	SF	14.61	14.87	15.15	15.29	15.44	1.00%
Ceramic Tile 4x4, Mortar set	1,000 & up	SF	10.89	11.02	11.31	11.52	11.81	2.52%
Copper pipe, 1/2" Type L	2,000 & up	LF	12.15	11.89	12.31	12.55	11.98	-4.54%
Ductwork, Galvanized iron, 16 ga	10,000 & up	#	8.27	8.31	8.79	8.89	8.93	0.45%
Flooring, Terrazzo, Cementitious	3,000 & up	SF	15.75	16.15	16.79	17.01	17.26	1.47%
Glass, Float, 1/4"	1,000 & up	SF	9.87	9.97	10.15	10.40	10.76	3.46%
Glue Laminated beams, installed	10,000 & up	MBF	2,880.00	2,895.00	3,355.00	3,489.00	3,583.20	2.70%
Gypsum Wall Board, 5/8" 4x8	20,000 & up	SF	2.12	2.16	2.29	2.36	2.52	6.78%
Insulation, 1 1/2" rigid, poly iso	5,000 & up	SF	1.55	1.60	1.67	1.85	1.96	5.95%
Metal Roof Deck 1 1/2", 20 Ga., painted	10,000 & up	SF	3.15	3.16	3.35	3.45	3.58	3.77%
Paint, Interior, 3 coat over GWB, roll	10,000 & up	SF	0.71	0.78	0.81	0.82	0.83	1.10%
Piles, concrete, precast, 12"	2,000 & up	LF	43.59	43.15	43.41	44.56	45.91	3.03%
Plywood 19/32", 4x8, cd exterior, T&G, floor	10,000 & up	SF	2.25	2.19	2.28	2.45	2.58	5.31%
Roofing, 4 ply, built up, 20 year	5,000 & up	SF	3.89	4.16	4.24	4.14	4.20	1.45%
Steel, Reinforcing	20,000 & up	#	0.82	0.80	0.82	0.84	0.87	3.57%
Steel, Structural Shapes	250,000 & up	#	1.54	1.49	1.41	1.55	1.65	6.45%
Stucco, Exterior, Production Residential	2,000 & up	SY	36.51	36.12	36.27	36.49	36.87	1.04%
Vinyl Composition Tile, 3/32"	10,000 & up	SF	2.89	3.05	3.09	3.15	3.22	2.22%
Wire #12, THHN, Pulled in conduit	10,000 & up	LF	0.94	0.92	0.88	0.89	0.91	2.25%
Wood Stud walls, 2x4	10,000 & up	BF	2.85	2.82	2.98	3.15	3.34	6.00%
Total Average Change (Non-weighted)								**2.88%**

Cost Indices

LELAND SAYLOR ANNUAL COST INDICES 1967 - 2014

Year	Subcontract Index	%YoY	Material/Labor Index	%YoY
1967	100.00		100.00	
1968	108.50	8.5%	108.50	8.5%
1969	116.30	7.2%	116.30	7.2%
1970	122.70	5.5%	126.20	8.5%
1971	132.90	8.3%	139.20	10.3%
1972	143.30	7.8%	151.50	8.8%
1973	161.60	12.8%	163.70	8.1%
1974	188.30	16.5%	194.30	18.7%
1975	166.90	-11.4%	208.50	7.3%
1976	182.30	9.2%	221.20	6.1%
1977	209.30	14.8%	242.00	9.4%
1978	241.30	15.3%	263.00	8.7%
1979	281.30	16.6%	285.40	8.5%
1980	296.40	5.4%	310.30	8.7%
1981	331.40	11.8%	335.00	8.0%
1982	343.30	3.6%	354.20	5.7%
1983	350.80	2.2%	370.60	4.6%
1984	370.00	5.5%	386.50	4.3%
1985	367.10	-0.8%	395.30	2.3%
1986	376.20	2.5%	402.00	1.7%
1987	380.20	1.1%	409.70	1.9%
1988	390.90	2.8%	421.20	2.8%
1989	397.20	1.6%	433.60	2.9%
1990	415.30	4.6%	440.90	1.7%
1991	423.10	1.9%	444.30	0.8%
1992	434.70	2.7%	452.10	1.8%
1993	428.40	-1.4%	468.90	3.7%
1994	442.80	3.4%	492.30	5.0%
1995	459.40	3.7%	503.20	2.2%
1996	482.80	5.1%	511.90	1.7%
1997	499.90	3.5%	522.00	2.0%
1998	522.60	4.5%	533.00	2.1%
1999	534.30	2.2%	552.10	3.6%
2000	585.00	9.5%	575.50	4.2%
2001	602.70	3.0%	587.10	2.0%
2002	645.40	7.1%	609.90	3.9%
2003	651.60	1.0%	639.00	4.8%
2004	743.20	14.1%	685.70	7.3%
2005	811.80	9.2%	724.90	5.7%
2006	864.30	6.5%	769.50	6.2%
2007	910.40	5.3%	790.20	2.7%
2008	926.70	1.8%	816.10	3.3%
2009	908.23	-2.0%	823.52	0.9%
2010	852.80	-6.1%	835.98	1.5%
2011	862.58	1.1%	862.68	3.2%
2012	890.01	3.2%	893.99	3.6%
2013	917.47	3.1%	906.32	1.4%
2014	**944.07**	**2.9%**	**939.74**	**3.7%**

Cost Index Summary

The **Material Cost Index** includes 23 key construction material, measured at wholesale prices. The index does not include contractor overhead and profit. Prices represent year end values, while the average is calculated using the year on year percentage point change in prices.

The **Labor Cost Index** represents the wages paid to nine trades. The index does not include contractor overhead and profit. Each figure is derived as the average of the 16 highest paid metropolitan areas in the USA and Canada. These wages are used as industry approximates, and are not used in calculating this book's labor cost column. The final average value is calculated using the year on year percentage point change in prices.

The 21 listed prices in the **Subcontracts Cost Index** reflect material costs plus installation labor costs, with the total average change row determined using the year on year percentage point change in prices. As a market-basket index of contractor and subcontractor prices for the twenty-one most often quoted in-place construction items, it can help the reader gauge conditions in the contractor market. The measure is indexed to base year 1967.

The **Material/Labor Index** is comprised of two components, Labor Cost (54% weighting) and Material Cost (46% weighting). The measure is indexed to base year 1967.

Cost Indices

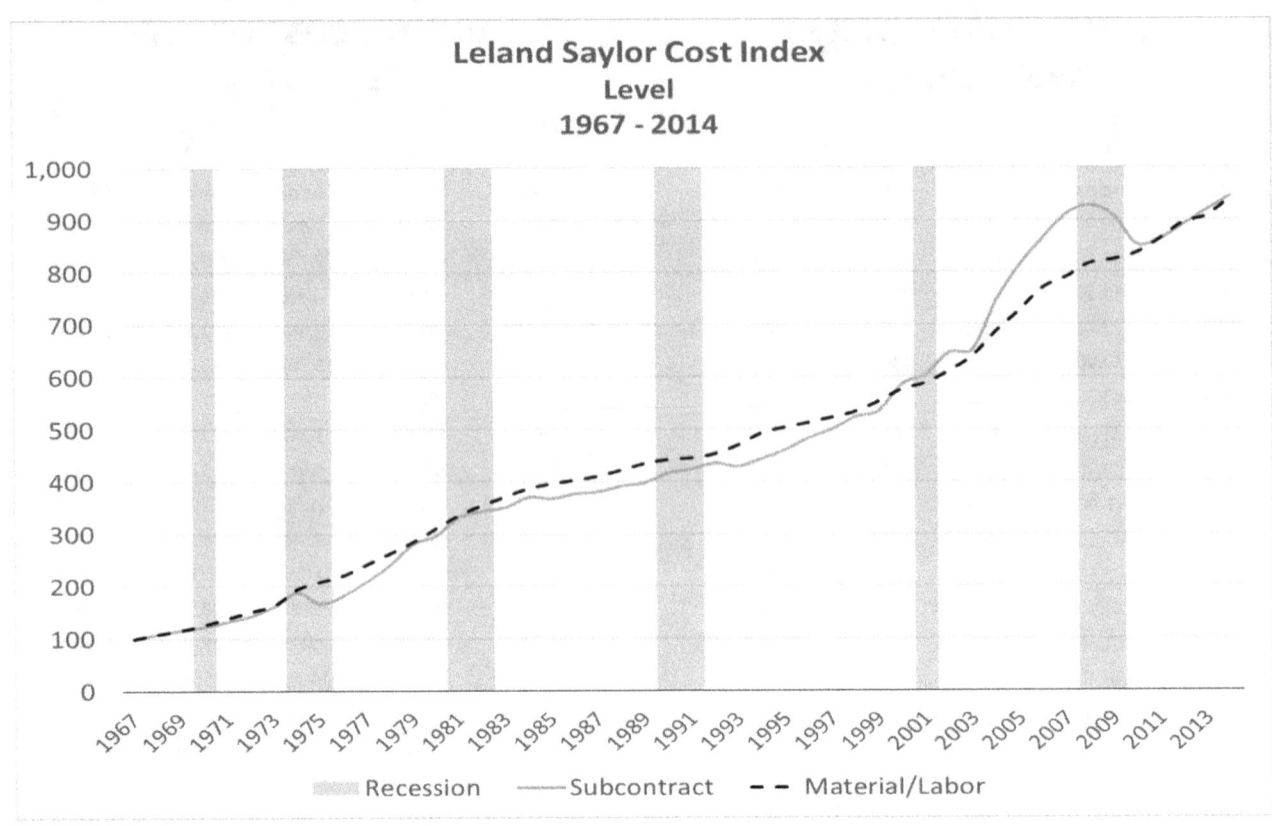

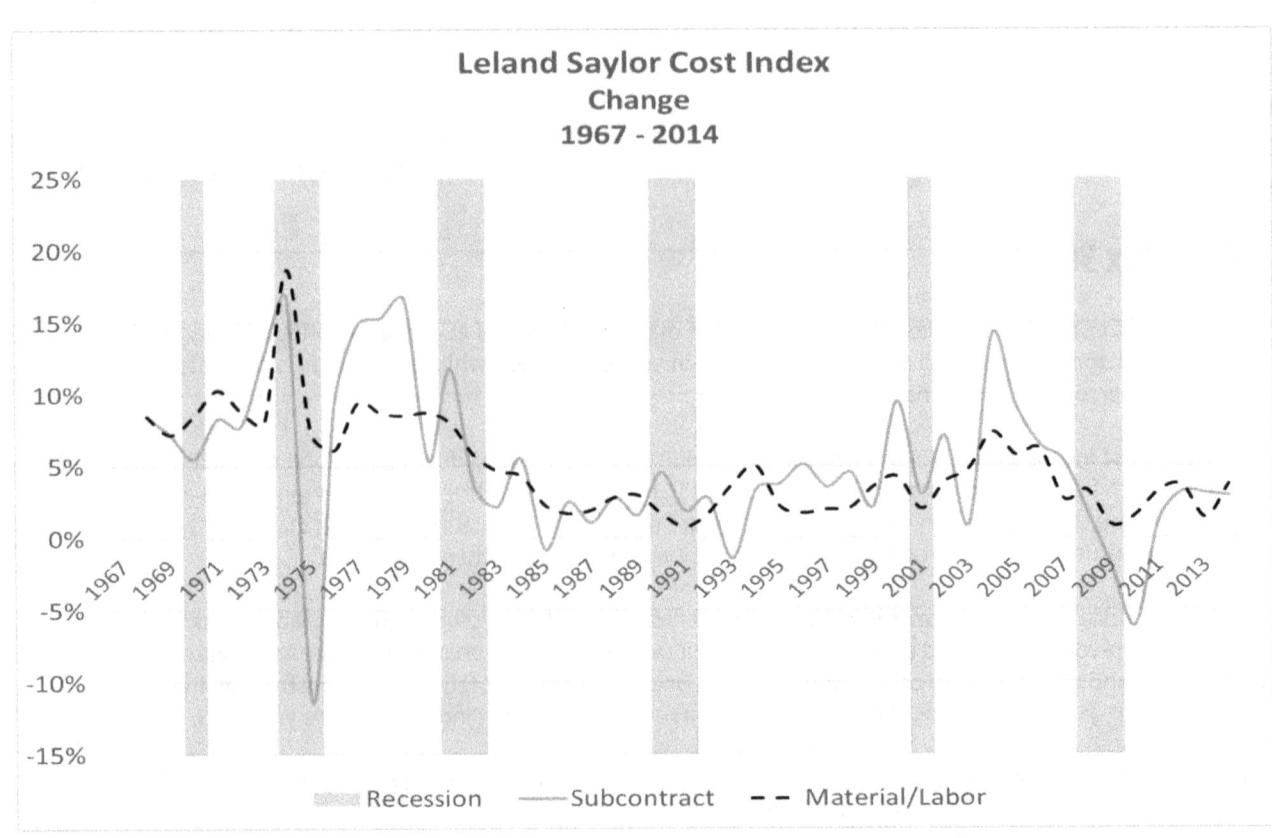

Major Cities Cost Index

2015 Cities Index for Current Construction Costs

ALASKA	Anchorage	121
	Fairbanks	119
	Juneau	119
	State Average	120
ARIZONA	Flagstaff	84
	Phoenix	82
	Tucson	80
	State Average	82
CALIFORNIA	Anaheim	93
	Bakersfield	89
	Eureka	88
	Fresno	90
	Long Beach	93
	Los Angeles	92
	Oakland	97
	Ontario	91
	Oxnard/Ventura	91
	Palo Alto	93
	Redding	88
	Riverside/San Bern.	92
	Sacramento	91
	Salinas/Monterey	90
	San Diego	92
	San Francisco	100
	San Jose	96
	Santa Ana/Irvine	94
	Santa Barbara	96
	Santa Cruz	93
	Santa Rosa	92
	Stockton/Modesto	88
	Vallejo	91
	State Average	92
COLORADO	Boulder	74
	Colorado Springs	77
	Denver	78
	Durango	73
	Grand Junction	75
	State Average	75
HAWAII	Hilo	115
	Honolulu	120
	State Average	120
IDAHO	Boise	74
	Coeur d'Alene	78
	Pocatello	75
	Twin Falls	72
	State Average	75
MONTANA	Billings	78
	Butte	75
	Great Falls	79
	Helena	77
	Missoula	78
	State Average	77
NEVADA	Carson City	82
	Elko	79
	Ely	85
	Las Vegas	87
	Reno	82
	State Average	83
NEW MEXICO	Albuquerque	76
	Clovis	75
	Gallup	76
	Las Cruces	74
	Roswell	76
	Santa Fe	77
	State Average	76
OREGON	Bend	86
	Eugene	86
	Klamath Falls	86
	Portland	87
	Salem	85
	State Average	86
UTAH	Logan	75
	Ogden	74
	Provo	75
	Salt Lake City	76
	State Average	75
WASHINGTON	Everett	86
	Olympia	86
	Richland	83
	Seattle	88
	Spokane	81
	Tacoma	87
	Vancouver	86
	State Average	85
WYOMING	Casper	69
	Cheyenne	73
	Newcastle	69
	Riverton	70
	Yellowstone	70
	State Average	70

Major Cities Cost Index

2015 Cities Index for Current Construction Costs

MIDWEST

ILLINOIS
City	Index
Bloomington	86
Carbondale	79
Champaign	86
Chicago	98
Galesburg	87
Joliet	96
La Salle	89
Peoria	86
Quincy	81
Rockford	90
Springfield	83
State Average	87

INDIANA
City	Index
Bloomington	78
Columbus	75
Fort Wayne	75
Gary	85
Indianapolis	79
Kokomo	75
Lafayette	76
Lawrenceburg	74
Muncie	77
South Bend	77
Terre Haute	79
State Average	77

IOWA
City	Index
Cedar Rapids	78
Council Bluffs	75
Davenport	80
Des Moines	79
Dubuque	75
Fort Dodge	70
Sioux City	72
Spencer	70
Waterloo	72
State Average	75

KANSAS
City	Index
Emporia	72
Fort Scott	73
Hutchinson	70
Kansas City	84
Salina	73
Topeka	75
Wichita	71
State Average	74

MICHIGAN
City	Index
Battle Creek	82
Detroit	84
Flint	80
Grand Rapids	73
Lansing	81
Muskegon	82
Traverse City	71
State Average	79

MINNESOTA
City	Index
Bemidji	86
Brainerd	88
Duluth	89
Mankato	85
Minneapolis	93
Rochester	87
St. Cloud	90
St. Paul	92
Windom	82
State Average	88

MISSOURI
City	Index
Bowling Green	82
Chillicothe	73
Flat River	81
Hannibal	80
Jefferson City	81
Joplin	74
Kansas City	85
St. Louis	85
State Average	80

NEBRASKA
City	Index
Alliance	74
Columbus	75
Grand Island	78
Lincoln	75
North Platte	78
Omaha	79
State Average	77

NORTH DAKOTA
City	Index
Bismark	73
Devils Lake	71
Dickinson	72
Fargo	74
Grand Forks	70
Minot	76
Williston	71
State Average	72

OHIO
City	Index
Akron	83
Canton	79
Cincinnati	79
Cleveland	86
Columbus	81
Dayton	78
Hamilton	78
Mansfield	80
Marion	76
Toledo	84
Youngstown	81
State Average	80

SOUTH DAKOTA
City	Index
Aberdeen	69
Mobridge	68
Rapid City	70
Sioux Falls	71
State Average	70

WISCONSIN
City	Index
Beloit	85
Eau Claire	84
Green Bay	84
Kenosha	87
La Cross	83
Lancaster	82
Madison	84
Milwaukee	88
New Richmond	83
Oshkosh	79
Racine	87
Wausau	82
State Average	84

Major Cities Cost Index

2015 Cities Index for Current Construction Costs

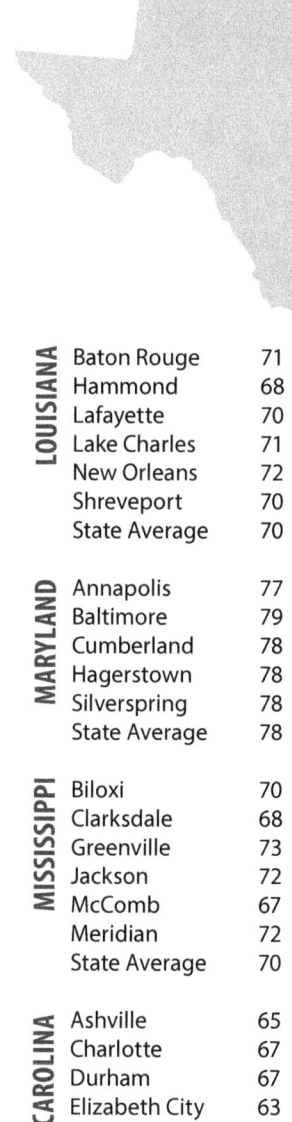

ALABAMA		
Birmingham	71	
Decatur	67	
Gadsden	69	
Huntsville	71	
Mobile	74	
Montgomery	70	
Selma	67	
State Average	70	

ARKANSAS		
Fayetteville	69	
Fort Smith	81	
Little Rock	72	
Texarkana	76	
State Average	75	

DELAWARE		
Newark	85	
Wilmington	88	
State Average	87	

FLORIDA		
Datona Beach	81	
Fort Lauderdale	78	
Fort Meyers	75	
Gainsville	76	
Jacksonville	73	
Lakeland	74	
Miami	79	
Orlando	76	
Pensacola	77	
Sarasota	79	
St. Petersburg	76	
Tallahassee	78	
Tampa	74	
West Palm Beach	75	
State Average	77	

GEORGIA		
Agusta	68	
Albany	69	
Athens	70	
Atlanta	73	
Columbus	70	
Dalton	69	
Gainesville	70	
Macon	68	
Savannah	68	
Waycross	68	
State Average	69	

KENTUCKY		
Bowling Green	76	
Campton	76	
Frankfort	74	
Lexington	74	
Louisville	77	
Owensboro	75	
Paducah	76	
Somerset	70	
State Average	75	

LOUISIANA		
Baton Rouge	71	
Hammond	68	
Lafayette	70	
Lake Charles	71	
New Orleans	72	
Shreveport	70	
State Average	70	

MARYLAND		
Annapolis	77	
Baltimore	79	
Cumberland	78	
Hagerstown	78	
Silverspring	78	
State Average	78	

MISSISSIPPI		
Biloxi	70	
Clarksdale	68	
Greenville	73	
Jackson	72	
McComb	67	
Meridian	72	
State Average	70	

NORTH CAROLINA		
Ashville	65	
Charlotte	67	
Durham	67	
Elizabeth City	63	
Fayetteville	68	
Gastonia	65	
Greensboro	65	
Raleigh	67	
Winston-Salem	66	
State Average	66	

OKLAHOMA		
Ardmore	70	
Clinton	70	
Durant	68	
Enid	70	
Guymon	59	
Lawton	69	
Muskogee	63	
Oklahoma City	71	
Tulsa	68	
State Average	68	

PUERTO RICO		
San Juan	68	

SOUTH CAROLINA		
Aiken	74	
Charleston	70	
Columbia	67	
Greenville	66	
Spartanburg	66	
State Average	69	

TENNESSEE		
Chattanooga	73	
Cookeville	70	
Johnson City	69	
Knoxville	70	
Nashville	74	
State Average	71	

TEXAS		
Abilene	69	
Amarillo	71	
Austin	70	
Beaumont	72	
Corpus Christi	67	
Dallas	74	
El Paso	67	
Fort Worth	72	
Galveston	74	
Laredo	67	
Lubbock	70	
McAllen	66	
Midland	68	
San Antonio	71	
Temple	66	
Waco	69	
Wichita Falls	66	
State Average	69	

VIRGINIA		
Alexandria	81	
Arlington	81	
Charlottesville	81	
Fairfax	79	
Fredericksburg	75	
Lynchburg	73	
Newport News	74	
Norfolk	75	
Richmond	74	
Roanoke	72	
State Average	77	

WEST VIRGINIA		
Bluefield	76	
Charleston	82	
Huntington	83	
Morgantown	82	
Parkersburg	82	
Wheeling	82	
State Average	81	

Major Cities Cost Index

2015 Cities Index for Current Construction Costs

NORTHEAST

CONNECTICUT		
	Bridgeport	88
	Hartford	90
	Meriden	88
	New Haven	88
	Norwalk	87
	Stamford	89
	State Average	88

D.C.		
	Washington	87

MAINE		
	Augusta	79
	Bangor	77
	Bath	77
	Portland	78
	Rockland	76
	State Average	77

MASSACHUSETTS		
	Boston	99
	Buzzards Bay	92
	Fitchberg	88
	Framingham	95
	Lowell	93
	New Bedford	91
	Reading	91
	Springfield	87
	Three Rivers	87
	Worcester	89
	State Average	91

NEW HAMPSHIRE		
	Claremont	79
	Concord	81
	Manchester	80
	Nashua	81
	Portsmouth	81
	State Average	80

NEW JERSEY		
	Atlantic City	92
	Camden	91
	Dover	95
	Elizabeth	95
	Hackensack	95
	Newark	96
	Trenton	93
	State Average	94

NEW YORK		
	Albany	84
	Bronx	106
	Brooklyn	110
	Buffalo	87
	Hicksville	103
	Jamaica	109
	Kingston	93
	Long Island City	109
	New Rochelle	99
	New York	98
	Niagra Falls	84
	Poughkeepsie	97
	Queens	109
	Rochester	84
	Syracuse	105
	Utica	84
	White Plains	98
	Yonkers	101
	State Average	96

PENNSYLVANIA		
	Altoona	76
	Allentown	84
	Bedford	79
	Butler	83
	Bradford	78
	Dubois	83
	Erie	86
	Greensberg	85
	Kittanning	85
	Johnstown	83
	Harrisburg	78
	Lancaster	79
	Lehigh Valley	89
	New Castle	82
	Philadelphia	92
	Pittsburgh	87
	Scranton	85
	York	82
	State Average	83

RHODE ISLAND		
	Newport	87
	Providence	91
	State Average	89

VERMONT		
	Brattleboro	75
	Burlington	74
	Montpilier	73
	Rutland	74
	State Average	74

NATIONAL TOTAL
Average	80
Minimum	59
Maximum	121

ALBERTA
Calgary	84
Edmonton	84

BRITISH COLUMBIA
Vancouver	87

MANITOBA
Winnipeg	82

NOVA SCOTIA
Halifax	83

ONTARIO
Ottawa	90
Toronto	93

QUEBEC
Montreal	83
Quebec City	83

SASKATCHEWAN
Regina	83

NATIONAL TOTAL
Average	85
Minimum	82
Maximum	93

Seismic Zones

In 1988, the Uniform Building Code divided the United States into five seismic hazard zones, as illustrated in the map below. Each hazard zone describes the potential damage and likely occurrence of earthquakes, and carries with it unique local construction practices and applicable codes, which are generally attributed to the structural components of the building. In the higher hazard zones, this means greater design emphasis on shear walls and reinforcing. In these areas, designers also tend to add extra safety features to the buildings.

The Commercial Square Foot Building Costs section of this resource outlines how these seismic zones affect the costs of various building types.

SEISMIC ZONE MAP OF THE UNITED STATES

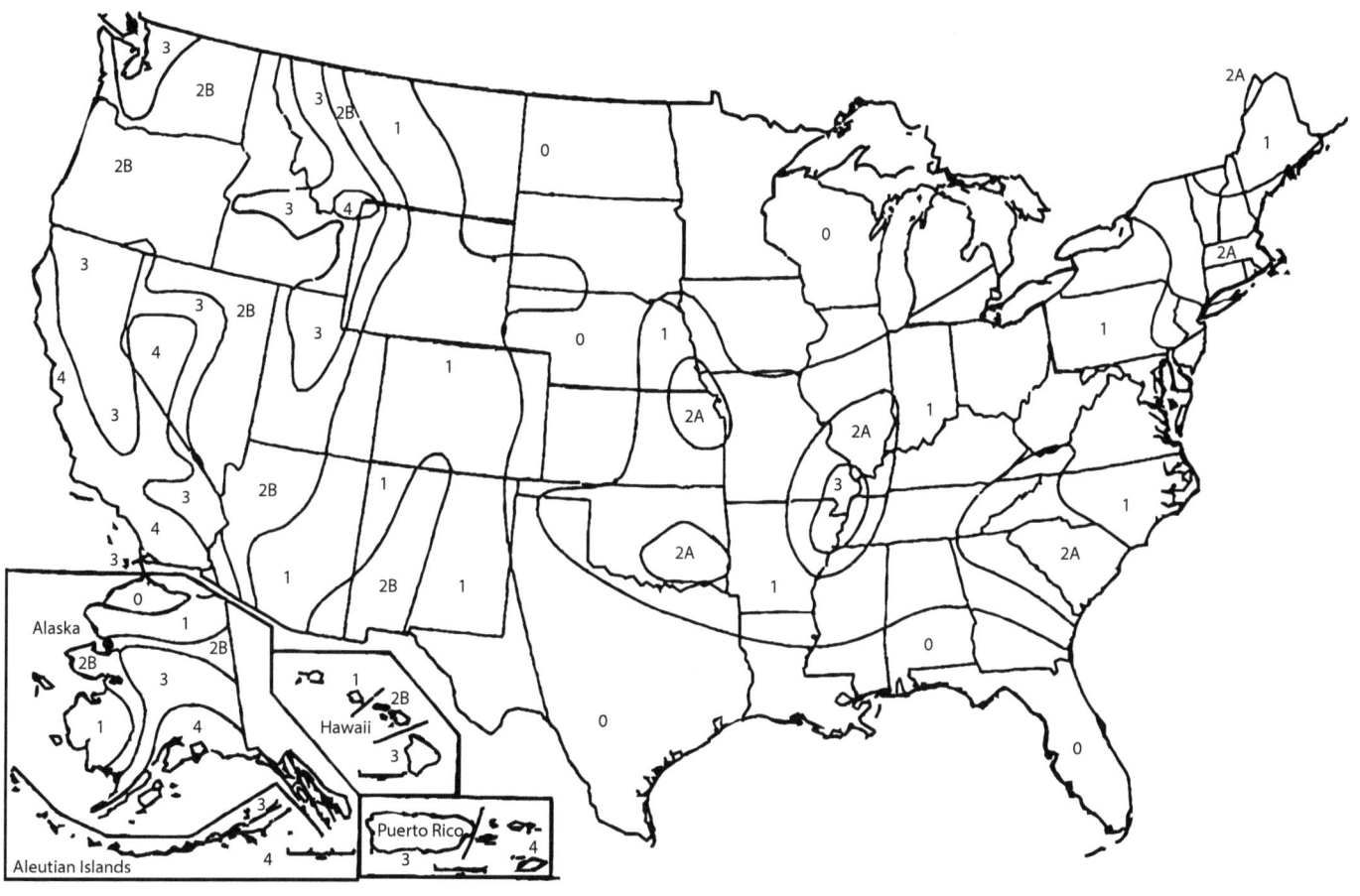

Reproduced from the 1988 edition of the Uniform Building Code, copyright © 1988, with permission of the publishers, the International Conference of Building Officials.

Symbols & Abbreviations

"	Inches
'	Feet
#	Pounds
A	Amperes
A/V	Audio Visual
AASHO	Am Assn Of State Hwy Off
Abras	Abrasive
Abs	Absorption
	Acrylonitrile Butadiene Styrene
AC	Asphaltic Concrete
	Alternating Current
Acb	Air Circuit Breaker
Acc	Accordion
Acp	Asbestos Cement Pipe
Acry	Acrylic
Acu	Air Conditioning Unit
Adj	Adjacent
	Adjustable
Agg	Aggregate
Aic	Ampere Interrupted Capacity
AJ	Aluminum Jacket
Al	Aluminum
Amb	Amber
Amp	Ampere
Anod	Anodized
Appd	Approved
Appl	Application
Arch	Architectural
Artic	Articulated
Asb	Asbestos
Asph	Asphalt
Assem	Assembly
Ass't	Assistant
ASTM	Am Soc For Testing Materials
AT	Air Terminal
Atu	Air Terminal Unit
Av	Average
Avb	Atmospheric Vacuum Breaker
B-O	Bolt-On
B-U	Built-Up
B&B	Board & Batten
B&G	Bolt & Gasket
B&W	Black & White

Barr	Barricade
Batt	Batt
	Batten
Bck	Back
Bckstp	Backstop
Bd	Board
Bel	Below
Bf	Board Foot
Bfly	Butterfly
Bfp	Backflow Preventer
Bhp	Brake Horsepower
Bitum	Bituminous
Bk	Back
Bldg	Building
Blk	Black
	Block
Bllws	Bellows
Bm	Beam
Bp	Blood Pressure
Bps	Bolted Pressure Switch
Brch	Birch
Brk	Break
	Brick
Brkr	Breaker
Brlap	Burlap
Brz	Bronze
Bsbd	Baseboard
Bsn	Basin
Bsp	Black Steel Pipe
Bttn	Batten
Bttr	Better
Btu	British Thermal Units
C/B	Circuit Breaker
CA	California
Calsil	Calcium Silicate
Cath	Cathedral
Cb	Catch Basin
Cc	Center To Center
Ccss	Copper Clad Stainless Steel
Ccu	Coronary Care Unit
Cdr	Cedar
Cem	Cement
Cent	Centrifugal

Symbols & Abbreviations

Cf	Cubic Foot	Const	Constant
Cfm	Cubic Feet Per Minute		Construction
Cfrd	Coffered	Conv	Conveyor
Cg	Construction Grade	Corr	Corrugated
Chem	Chemical	Cplg	Coupling
Chk	Check	Cpp	Corrugated Polyethylene Pipe
Chnnl	Channel	Cr-Moly	Chrome-Molybdenum
Ci	Cast Iron	Crb	Curb
Cip	Cast In Place	Crk	Cork
	Cast Iron Pipe	Cryo	Cryogenic
Circ	Circular	Csmnt	Casement
	Circulating	Ct	Coat
Ckt	Circuit	Ctd	Coated
	Cricket	Ctrl	Central
Cl	Coil	Cu	Copper
	Class	Cult	Cultured
Clg	Ceiling	Curr	Current
Clnr	Cleaner	Cust	Custom
Clr	Clear	Cvr	Cover
Clsd	Closed	Cw	Cold Water
Cmp	Corrugated Metal Pipe	Cy	Cubic Yard
Cmplx	Complex	D	Deep
Cmprsr	Compressor	Db Hng	Double Hung
Cmu	Concrete Masonry Unit	Dblwl	Double Wall
Cndr	Cinder	DC	Direct Current
Cnt	Center	Decon	Decontamination
Cntr	Counter	Decor	Decorative
Cntrl	Control	Defib	Defibrillator
Col	Column	Deg	Degree
Comb	Combination	Desc	Descending
Comm	Commercial	Det	Detector
Comp	Computer	Dev	Developing
	Composition		Device
	Component	Df	Douglas Fir
	Compartment	Di	Drop Inlet
	Compact	Dia	Diameter
Compl	Complete	Diaph	Diaphragm
Conc	Concrete	Dim	Dimension
	Concealed	Dip	Ductile Iron Pipe
Cond	Conduit	Dir	Direct
	Conditioning	Disc	Disconnect
Congl	Conglomerate	Disp	Dispose
Conn	Connection		Disposal

Symbols & Abbreviations

	Dispenser
Dist	Distribution
Div	Divider
Dr	Door
Drwr	Drawer
Dsl	Diesel
Dumbwtr	Dumbwaiter
Dwl	Dowel
DWV	Drainage, Waste & Vent
Dys	Days
Ea	Each
Econ	Economy
Ej	Expansion Joint
	Ejector
Elc	Extra Low Carbon
Elec	Electric
Elev	Elevated
	Elevator
Ellip	Elliptical
Emb	Embedded
Emerg	Emergency
Emt	Electrometallic Tubing
En	Enamel
Enc	Enclosure
Eq	Equal
Ep	Explosion Proof
Equip	Equipment
Erw	Electric Resistance Welded
Esc	Escalator
Exc	Excavate
Exh	Exhaust
Exp	Exposed
	Expansion
Ext	Exterior
F&S	Fittings & Supports
Fab	Fabricated
	Fabric
Fbgls	Fiberglass
Fbr	Fiber
Fbrbd	Fiberboard
Fbrgls	Fiberglass
Fg	Foundation Grade
Fibrebd	Fibreboard

Fin	Finished
Fiss	Fissured
Fit	Fittings
Fix	Fixture
Fl	Floor
	Float
Fld	Field
Fll	Fill
Flng	Flange
Flot	Flotation
Flt	Float
	Flight
Fltr	Filter
Fluor	Fluorescent
Fnd	Foundation
Fob	Free On Board
Fount	Fountain
Fp	Fireproof
	Fire Protection
Fpm	Feet Per Minute
Frc	Glass Fiber Reinforced Concrete
Frcn	Fractional
Fresnl	Fresnel
FrkIft	Forklift
Frm	Frame
Frng	Furring
Frp	Fiberglass Reinforced Plastic
Frt	Freight
Fs	Fusible Switch
	Forged Steel
Ft Cdls	Foot Candles
Ftg	Footing
	Fitting
Furn	Furnace
Fvnr	Full Voltage Non-Reversing
G	Gallon
Ga	Gauge
Gal	Gallon
Galv	Galvanized
Gd	Good
	Guard
Gfi	Ground Fault Interrupter
Gi	Galvanized Iron

Symbols & Abbreviations

Gip	Galvanized Iron Pipe
Gl	Glass
Glb	Globe
Glv	Galvanized
Glz	Glaze
Glzd	Glazed
Gph	Gallons Per Hour
Gpm	Gallons Per Minute
Gr	Group
Gr-B	Grade B
Gran	Granite
Grav	Gravity
	Gravel
Grc	Galvanized Rigid Conduit
Grd	Ground
Grt	Grate
GSF	Gross Square Footage
Gsm	Galvanized Sheet Metal
Gsp	Galvanized Steel Pipe
Gwb	Gypsum Wall Board
Gyp	Gypsum
H	High
Hc	Hollow Core
Hd	Head
	Heavy Duty
Hdbd	Hardboard
Hdwr	Hardware
Hi	High
Hi-Cap	High Capacity
Hngd	Hinged
Hngr	Hanger
Ho	High Output
Horiz	Horizontal
Hp	Horsepower
	High Pressure
Hps	High Pressure Sodium
Hrdrk	Hardrock
Hs	High Strength
Htd	Heated
Htr	Heater
Hv	High Voltage
Hw	Hot Water
Hwy	Highway
Hyd	Hydraulic
	Hydrant
Hzrd	Hazard
Ib	Iron Body
Ic	Intercom
ICU	Intensive Care Unit
Illum	Illuminator
Imc	Intermediate Metal Conduit
In/Dp	Inch Of Depth
Inc	Incandescent
Incand	Incandescent
Incl	Include
Ins	Insulation
	Insulating
Instr	Instrument
Int	Interior
Intrcm	Intercom
Intrlck	Interlock
Iod	Iodine
Irrig	Irrigation
Iso	Isolation
J-Box	Junction Box
Jkt	Jacket
JSF	Job Square Footage
Jt	Joint
K	Thousand
Kd	Kiln Dried
Kv	Kilovolt
Kva	Kilovolt Amperes
Kvar	Kilovolt Amperes Resistance
Kw	Kilowatt
Lam	Laminated
Laun	Laundry
Lck	Lock
Ld	Load
Ldg	Loading
Ldgrs	Ledgers
Lf	Lineal Foot
Lf/Tr	Lineal Foot Of Tread
Lft	Loft
Lg	Large
	Long
Liq	Liquid

Symbols & Abbreviations

Ll	Live Load		Mon	Month
Lmt	Limit		Mono	Monorail
Lp	Low Pressure			Monolithic
	Loop		MSF	Thousand Square Feet
Lph	Liters Per Hour		Mt	Mount
Lpm	Liters Per Minute		Mtd	Mounted
Ls	Lump Sum		Mtl	Metal
Lt	Light		Mtr	Meter
Lts	Lights		Mu	Masonry Unit
Ltwt	Lightweight		Mull	Mullion
Lvl	Level		Mun	Muntin
Lyr	Layer		Mva	Millivolt Amperes
M	Thousand		N-R	Non-Rated
M/Cfm	Thousand Cubic Foot Per Minute		N-S	Non-Spooled
Mach	Machine		Nema	Nat'l Electrical Mfr's Assoc
Mag	Magnetic		Neop	Neoprene
Mahog	Mahogany		Nlrs	Nailers
Maint	Maintenance		Norm	Normal
Mast	Mastic		Nr	Near
Matl	Material		Nyl	Nylon
Max	Maximum		Nz	Nozzle
Mbh	Thousand Btus Per Hour		O/R	Operating Room
Mbtu	Thousand Btus		Oc	On Center
Mcb	Main Circuit Breaker		Ocbw	On Center Both Ways
Mcc	Motor Control Center		Oh	Overhead
Mcmb	Molded Case Main Breaker		Oper	Operator
Mech	Mechanical			Operated
Med	Medium		Opn	Opening
	Median		OS & Y	Outside Stem & Yoke
Memb	Membrane		Osm	Osmosis
Merc	Mercury		Otlt	Outlet
Mh	Manhole		Ovr	Over
Mi	Mile		Oz	Ounce
Min	Minimum		P	Pole
	Mineral		P-T	Pass-Thru
Mirr	Mirror		Pa	Public Address
Misc	Miscellaneous		Parg	Pargetting
Ml	Millimeter		Part	Partition
Mlo	Main Lug Only		Pb	Paperback
Mm	Million			Panelboard
	Millimeter		Pc	Prime Coat
Mod	Modular		Pe	Polyethylene
	Moderate		Perf	Perforated

Symbols & Abbreviations

Perm	Permanent	Qual	Quality
Pg	Paint Grade	Rad	Radius
Pgbd	Pegboard	Rck	Rack
Ph	Phase	Rdwy	Roadway
Pharm	Pharmacy	Rec	Recreation
Photoelec	Photoelectric		Recessed
Pi	Pressure Injected	Recept	Receptacle
	Pressure Indicator	Recip	Reciprocating
Piv	Post Indicator Valve	Recpt	Receptacle
Pkg	Package	Rect	Rectangular
Pl	Place		Rectifier
Pln	Plan	Refer	Refrigerator
Plstc	Plastic	Refl	Reflective
Plstr	Plaster	Reg	Register
Plts	Plants	Reinf	Reinforced
Pneum	Pneumatic	Rel	Relief
Pnl	Panel	Reloc	Relocatable
Poc	Point Of Connection	Rem	Remove
Polypro	Polypropylene	Res	Residential
Porc	Porcelain	Resusc	Resuscitator
Port	Portable	Ret	Retaining
Pos	Positive		Retainer
Prcst	Precast	Rev	Reverse
Pre-Fab	Pre-Fabricated	Rf	Roof
Pre-Ins	Pre-Insulated	Rig	Rigid
Prefin	Prefinished	Rip	Rip-Rap
Preh	Prehung	Rlld	Rolled
Prep	Prepare	Rsc	Rigid Steel Conduit
	Preparation	Rsr	Riser
Prev	Preventor	Rubb	Rubber
Proj	Projector	Rw	Redwood
Prv	Pressure Reducing Valve	S/By	Standby
Ps	Power Supply	S/O	Shutoff
Psf	Pounds Per Square Foot	S&P	Shielded & Pulled
Psi	Pounds Per Square Inch	Sc	Solid Core
Psig	Pounds Per Square Inch Gauge	Sch	Schedule
Pt	Paper Towel	Scr	Screen
Ptd	Painted		Screw
Ptrn	Pattern	Scrw	Screw
Purif	Purification	Sd	Side
Pvc	Polyvinyl chloride	Sect	Section
Qt	Quart	Sel	Select
Qty	Quantity	Semirec	Semirecessed

Symbols & Abbreviations

Sep	Separator			Square
Sew	Sewage		Srvg	Serving
Sf	Square Foot		Ss	Stainless Steel
Sf/Tr	Square Foot Of Trench		Sta	Station
Sfca	Square Foot Of Contact Area		Stc	Sound Transmission Coefficient
Sffa	Square Foot Form Area		Std	Standard
Sfsa	Square Foot Surface Area		Steril	Sterilizer
Sfwa	Square Foot Wall Area		Stl	Steel
Sg	Select Grade		Stor	Storage
	Scotch Guard		Str	Straight
Sh	Sheath		Str'l	Structural
Shk	Shack		Strnr	Strainer
Shlvs	Shelves		Strt	Straight
Sht	Sheet		Struct	Structure
Shtrprf	Shatterproof		Sty	Story
Sim	Simulated		Subfl	Subfloor
Sk	Sack		Subm	Submerge
Skylt	Skylight		Supt	Support
Sldg	Sliding		Surf	Surface
Slpr	Sleeper		Surg	Surgery
Sm	Sheet Metal		Susp	Suspended
	Small		Svc	Service
Smls	Seamless		Sw	Switch
Smpl	Simplex			Service Weight
Snd	Sand		Swbd	Switchboard
Sndblst	Sandblast		Swgr	Switchgear
Snk	Sink		Swpk	Switchpack
Soff	Soffit		Swr	Sewer
Sog	Slab On Grade		Synth	Synthetic
Sol	Solid		Sys	System
	Solder		T	Ton
Solv	Solvent		T&B	Trenched & Buried
Sp	Steel Pipe		T&C	Threaded And Cut
	Static Pressure		T&G	Tongue & Groove
	Spec		Tbl	Table
	Space		Tc	Thin Coat
Spag	Spaghetti			Termocouple
Spec	Specification		Temp	Temporary
Spl	Split			Tempered
Splsh	Splash			Temperature
Sprd	Spread		Terr	Terrazzo
Sprnklr	Sprinkler		Th	Thick
Sq	Square (100 Sf)		Thrmplstc	Thermoplastic

Symbols & Abbreviations

Thrsh	Threshold	Wknd	Weekend
Tnk	Tank	Wkwy	Walkway
Tp	Toilet Paper	Wl	Wall
Tpl Hng	Triple Hung	Wld	Weld
Trd	Tread	Wp	Waterproof
Trlr	Trailer	Wt	Weight
Trm	Trim		Water Tight
Trtd	Treated	Wthrstrp	Weatherstrip
Tstat	Thermostat	Wtrprf	Waterproof
Tv	Television	Xfmr	Transformer
Twr	Tower	Yd	Yard
Ug	Underground	Yr	Year
Ul	United Labs	Zono	Zonolite
Uncomp	Uncompacted		
Undrcntr	Undercounter		
Undrpn	Underpin		
Unfin	Unfinished		
Unmtd	Unmounted		
Untrtd	Untreated		
Uphol	Upholstered		
Ups	Uninterruptable Power Supply		
Ure	Urethane		
V	Volt		
Vac	Vacuum		
Vap	Vapor		
Vermic	Vermiculite		
Vert	Vertical		
Vf	Vertical Foot		
Vib	Vibration		
Vibr	Vibrator		
Vict	Victaulic		
Vlv	Valve		
Vnt	Vent		
Vnyl	Vinyl		
Vol	Volume		
Vwc	Vinyl Wall Covering		
W	Wide		
	Watt		
W/Tr	With Trench		
Wc	Water Closet		
Wd	Wood		
Wfr	Wafer		
Wht	White		

For Over 50 Years, Saylor Has Been Helping Build Communities.

Leland Saylor Associates
A Certified SDVOSB

Services
- Design Phase Cost Estimating
- Value Engineering
- Construction Phase Cost/Change Management
- Project Controls
- Construction Management
- Construction Inspection
- Construction Market Research

GENERAL REQUIREMENTS - 01

01.0.1000 General Requirements:
General Conditions:

Note: General Conditions will range from 5.5% of project costs for large projects to 15% of project costs for small projects. If manual labor distributables are executed, allow for 0.9% of project cost.

Amount/Size	Repair of Fire Damage %	Alterations & Additions %	Unique Structure %	Institutional Structure %	Commercial Structure %	Public Works Heavy %
Under $200 Thousand	20.0	20.0	18.0	16.5	15.0	15.0
$200 to $500 Thousand	18.0	16.0	14.0	12.0	10.0	10.0
$500 Thousand to $1 Million	16.0	14.0	12.0	10.0	8.0	9.0
$1 to $2 Million	14.0	12.0	10.0	8.0	6.5	7.0
$2 to $5 Million	12.0	10.0	8.0	6.5	5.5	6.5
$5 to $10 Million	10.0	8.0	7.0	6.0	5.0	6.0
$10 to $20 Million	8.0	7.0	6.0	5.5	5.2	5.5
$20 to $50 Million	7.0	6.5	6.0	5.0	5.0	5.0
$50 to $100 Million	6.5	6.0	5.5	4.7	4.5	4.7
$100 Million and Up	6.0	5.5	5.0	4.5	4.5	4.5

Amount/Size	High Rise Housing %	Low Rise Housing %	Single Family Tract %	Single Family Custom %	Single Family Architectural %
Under $200 Thousand	15.0	15.0	15.0	15.0	20.0
$200 to $500 Thousand	10.0	10.0	10.0	12.0	15.0
$500 Thousand to $1 Million	8.0	8.0	8.0	10.0	11.0
$1 to $2 Million	6.5	6.5	6.5	8.0	10.0
$2 to $5 Million	5.5	5.5	5.5	6.5	7.5
$5 to $10 Million	5.0	5.0	5.0	6.0	7.0
$10 to $20 Million	5.0	5.0	5.0	0.0	0.0
$20 to $50 Million	4.5	4.5	4.5	0.0	0.0
$50 to $100 Million	4.5	4.5	4.2	0.0	0.0
$100 Million and Up	4.5	4.5	0.0	0.0	0.0

Broad Categories of General Requirements	Small Jobs *	Large Jobs ** High	Large Jobs ** Medium	Large Jobs ** Low
Mobilization	2.0	1.0	0.3	0.2
Non-distributable labor and supervision	7.0	5.0	3.0	2.2
Permits, licenses and fees	0.4	0.3	0.2	0.1
Temporary utilities, structures, fences	1.5	1.0	0.5	0.2
Material handling equipment	2.0	1.0	0.5	0.2
Other general requirements, including trucks, safety, fuel, scaffolding	2.0	1.0	0.5	0.2
Non-manual labor, distributables, benefits, payroll tax, workers' compensation insurance	2.8	2.0	1.0	1.1
Insurances, comprehensive, builders risk	2.0	2.0	0.8	0.3
TOTAL	**19.7**	**13.3**	**6.8**	**4.5**

* SMALL JOB is less than $100,000.
** Reflects typical average ranges for large jobs.

I. NEGOTIATED PROJECT: Deduct home office overhead & profit and add 2 to 3% to general conditions.
II. ADD TO GENERAL CONDITIONS:
    A. Liquidated damages to specified amount, for duration of project or tight schedules.
    B. Extra expenditures or supervision labor for calendar days short of normal.

01 - GENERAL REQUIREMENTS

Division	Description	Unit	Labor Hours	Labor Cost	Material Cost	Total Cost
01.1000 000	**GENERAL CONDITIONS:**					
01.1000 001	Note: General conditions will range from 5.5% of project costs for large projects to 15% of project costs for small projects. If manual labor distributables are excluded, allow for 0.9% of project cost.					
01.1010 000	**MOBILIZATION, ON & OFF:**					
01.1010 001	Note: Mobilization ranges from 5% of project cost for large projects to 10% of project cost for small projects. Cost varies according to the type of work, site location and site conditions. This cost includes permits, fees, temporary structures, equipment rental and various miscellaneous items. This is a term usually confined to roadworks, dams, bridges or foreign works.					
01.1020 000	**NON-DISTRIBUTABLE LABOR:**					
01.1020 001	Note: The following percentages may be used in early stage estimating for supervision: JOB SIZE .... PERCENTAGE OF JOB 50,000 TO 2,000,000 SF .... 84% 2,000,000 TO 5,000,000 SF .... 1.15% 5,000,000 AND UP .... 1.73% For engineering and layout costs, Use .26% Of the overall job.					
01.1021 000	**NON-DISTRIBUTABLE LABOR:**					
01.1021 001	Note: For fringes, payroll tax and insurance see section 01.1900.					
01.1021 011	Project Manager ($5 million & larger)	MONTH	173.0000	13,468.05		13,468.05
01.1021 021	Superintendent, large projects ($5 million & larger)	MONTH	173.0000	12,630.73		12,630.73
01.1021 031	Superintendent, medium projects ($2 to $5 million)	MONTH	173.0000	11,177.53		11,177.53
01.1021 041	Superintendent, small projects ($5,000 to $2 million)	MONTH	173.0000	9,316.05		9,316.05
01.1021 051	Assistant Superintendent, large projects	MONTH	173.0000	9,072.12		9,072.12
01.1021 061	Civil Engineer	MONTH	173.0000	7,857.66		7,857.66
01.1021 071	Time Keeper	MONTH	173.0000	5,158.86		5,158.86
01.1021 081	Payroll Clerk	MONTH	173.0000	4,541.25		4,541.25
01.1021 091	Secretary	MONTH	173.0000	3,918.45		3,918.45
01.1021 101	Expediter	MONTH	173.0000	5,575.79		5,575.79
01.1021 111	Quality Control Engineer	MONTH	173.0000	10,551.27		10,551.27
01.1021 121	Safety Officer	MONTH	173.0000	10,551.27		10,551.27
01.1030 000	**PERMITS, LICENSES & FEES:**					
01.1030 001	Note: The charges for permits, licenses and meter fees vary. We recommend that you check with local authorities for more exact information. The charges in this section are representative only. Note that building permit and plan check is not required for schools or public projects.					
01.1031 000	**PERMITS, LICENSES & FEES:**					
01.1031 001	Note: The permit fees listed below may vary from one area to another. Please check with your local authorities. Plan check fees are usually based on a percentage of the permit fees. The recommended percentage is 65%, but this also may vary.					
01.1031 011	Permit, first $100,000	EA			1,257.69	1,257.69
01.1031 021	Permit, each $1,000 over $100,000 to $500,000	M			6.92	6.92
01.1031 031	Permit, first $500,000	EA			4,011.01	4,011.01
01.1031 041	Permit, each $1,000 over $500,000 to $1 million	M			5.93	5.93
01.1031 051	Permit, first $1million	EA			6,959.97	6,959.97
01.1031 061	Permit, each $1,000 over $1 million	M			3.95	3.95
01.1031 101	Water meter fee, 3/4" connection	EA			1,464.65	1,464.65
01.1031 111	Water meter fee, 1" connection	EA			1,876.92	1,876.92
01.1031 121	Water meter fee, 1-1/2" connection	EA			3,016.09	3,016.09
01.1031 131	Water meter fee, 2" connection	EA			4,209.54	4,209.54
01.1031 141	Water meter fee, 3" connection	EA			9,113.50	9,113.50
01.1031 151	Water meter fee, 4" connection	EA			12,151.30	12,151.30
01.1031 161	Water meter fee, 6" connection	EA			16,274.13	16,274.13
01.1031 171	Sewer fee, 7 fixture house with laundry outlet	EA			976.44	976.44
01.1031 181	Sewer fee, 5 fixture house with laundry outlet	EA			867.94	867.94
01.1031 191	Sewer fee, major structure, average	/FIX			140.99	140.99

GENERAL REQUIREMENTS - 01

Division	Description	Unit	Labor Hours	Labor Cost	Material Cost	Total Cost
01.1032 000	**PERMITS, MISCELLANEOUS (USE LOCAL CHARGES):**					
01.1032 011	Encroachment					
01.1032 021	Sidewalks					
01.1032 031	Power					
01.1032 041	Blasting					
01.1032 051	Demolition					
01.1032 061	Street & Alley					
01.1032 071	Land use					
01.1040 000	**TEMPORARY UTILITIES, STRUCTURES & FENCES:**					
01.1040 001	Note: For compressors see section 01.1104. For early stage estimating, use 0.20% of job cost to allow for all temporary utilities.					
01.1041 000	**TEMPORARY UTILITIES:**					
01.1041 011	Temporary power pole, in & out only	EA			397.79	397.79
01.1041 021	Temporary power pole, 50 amps	MONTH			625.10	625.10
01.1041 031	Temporary power pole, 100 amps	MONTH			767.17	767.17
01.1041 041	Temporary underground in & out charge	EA			568.27	568.27
01.1041 051	Temporary underground, 100 amps	MONTH			94.37	94.37
01.1041 061	Chemical toilets, serviced, fiberglass	MONTH			160.88	160.88
01.1041 071	Chemical toilets, with sink, fiberglass	MONTH			194.77	194.77
01.1041 081	Water, buy	M/GAL			1.71	1.71
01.1041 091	Telephone	MONTH			488.23	488.23
01.1042 000	**TEMPORARY STRUCTURES:**					
01.1042 011	Temporary shack, 8' x 12', rental	MONTH			152.60	152.60
01.1042 021	Temporary shack, 8' x 12', on & off only	LS			790.84	790.84
01.1042 031	Temporary shack, 8' x 16', rental	MONTH			177.71	177.71
01.1042 041	Temporary shack, 8' x 16', on & off only	LS			894.12	894.12
01.1043 000	**OFFICE TRAILER, RENTALS:**					
01.1043 011	Office trailer, 8' x 20', rental	MONTH			304.82	304.82
01.1043 021	Office trailer, 8' x 20', on & off	LS			677.41	677.41
01.1043 031	Office trailer, 10' x 55', rental	MONTH			721.46	721.46
01.1043 041	Office trailer, 10' x 55', on & off	LS			1,164.95	1,164.95
01.1044 000	**TEMPORARY STRUCTURES, BUY OUT:**					
01.1044 011	Temporary shack, 8' x 12', buy	EA			2,096.08	2,096.08
01.1044 021	Temporary shack, 8' x 16', buy	EA			2,483.37	2,483.37
01.1044 031	Temporary shack, 12' x 24', buy	EA			6,475.15	6,475.15
01.1044 041	Temporary storage, 8' x 40', steel, buy	EA			16,088.47	16,088.47
01.1044 051	Temporary tool bin, 4' x 8', buy	EA			616.23	616.23
01.1045 000	**FENCING & RAILINGS:**					
01.1045 011	Temp wood fence, post in ground	LF	0.3658	39.47	14.39	53.85
01.1045 021	Temp barrier bolted to paving	LF	0.6567	70.85	30.60	101.46
01.1045 031	Add for sidewalk cover	LF	0.1730	18.66	6.33	25.00
01.1045 041	Temporary fence, architectural	LF	2.8491	307.39	91.04	398.43
01.1045 051	Temporary railing, wood	LF	0.2067	22.30	6.83	29.13
01.1045 061	Chain link 6ft high more than 801 ft, minimum 6 months	LF/MON			0.63	0.63
01.1045 071	Chain link 6ft high 501 to 800 ft, minimum 6 months	LF/MON			0.71	0.71
01.1045 081	Chain link 6ft high 201 to 500 ft, minimum 6 months	LF/MON			1.00	1.00
01.1045 091	Chain link 6ft high 100 to 200 ft, minimum 6 months	LF/MON			1.43	1.43
01.1100 000	**EQUIPMENT RENTAL:**					
01.1101 000	**EQUIPMENT RENTAL, MATERIAL HANDLING:**					
01.1101 001	Note: For rental of crawler cranes over 30 days deduct 15% from the total material costs. Prices for truck mounted cranes include on and off charges.					
01.1101 011	Elevator Tower to 200', material, operated	MONTH	173.0000	15,817.39	7,790.19	23,607.58
01.1101 021	Elevator Tower to 200', personnel, operated	MONTH	173.0000	15,817.39	3,387.03	19,204.42
01.1101 031	Erection (2 Hours/LF)	LF	14.0000	1,294.44		1,294.44
01.1101 041	Tower Crane to 138' Radius, medium load, operated	MONTH	173.0000	15,817.39	25,402.83	41,220.22
01.1101 051	Tower Crane to 138' Radius, heavy load, operated	MONTH	173.0000	15,817.39	29,636.63	45,454.02
01.1101 061	Tower crane, on & off only, easy access	LS			63,507.10	63,507.10

01 - GENERAL REQUIREMENTS

Division	Description	Unit	Labor Hours	Labor Cost	Material Cost	Total Cost
01.1101 063	Tower crane, on & off only, average access	LS			71,974.71	71,974.71
01.1101 065	Tower crane, on & off only, difficult access	LS			80,440.84	80,440.84
01.1101 071	Stiff leg derrick, operated	MONTH	173.0000	15,817.39	8,462.52	24,279.91
01.1101 081	Guy derrick, 15 ton, 100' with hoist	MONTH	173.0000	15,817.39	5,641.67	21,459.06
01.1101 091	Gin poles, 2 at 250 ton	MONTH	346.0000	31,634.78	9,872.97	41,507.75
01.1101 101	Derrick, on & off charges	LS			12,411.69	12,411.69
01.1101 111	Truck crane, 35 ton, operator/oiler	MONTH	346.0000	30,468.76	13,185.09	43,653.85
01.1101 121	Truck crane, 60 ton, operator/oiler	MONTH	346.0000	30,468.76	22,850.15	53,318.91
01.1101 131	Truck crane, 90 ton, operator/oiler	MONTH	346.0000	30,468.76	30,609.08	61,077.84
01.1101 141	Truck crane, 140 ton, operator/oiler	MONTH	346.0000	30,468.76	43,066.14	73,534.90
01.1101 151	Handy crane, 10 ton, operated	MONTH	173.0000	15,234.38	3,952.48	19,186.86
01.1101 161	Handy crane, 20 ton, operated	MONTH	86.5000	7,617.19	11,331.93	18,949.12
01.1101 171	Operator for phone on high rise	MONTH	173.0000	15,817.39		15,817.39
01.1101 181	Crawler crane, 70 ton, operator/oiler	MONTH	346.0000	30,468.76	14,013.25	44,482.01
01.1101 191	Crawler crane, 150 ton, operator/oiler	MONTH	346.0000	30,468.76	25,578.71	56,047.47
01.1101 201	Crawler crane, 200 ton, operator/oiler	MONTH	346.0000	30,468.76	30,586.68	61,055.44
01.1101 211	Crawler crane, 70 ton, on/off only	LS			20,211.47	20,211.47
01.1101 221	Crawler crane, 150 ton, on/off only	LS			49,405.78	49,405.78
01.1101 231	Crawler crane, 200 ton, on/off only	LS			59,273.30	59,273.30
01.1101 251	Concrete conveyor, 20" x 42.5', gas, portable	MONTH			1,435.04	1,435.04
01.1101 261	Concrete conveyor, 16" x 48', gas, portable	MONTH			1,984.98	1,984.98
01.1101 271	Concrete pumping, first 50 CY	CY			12.84	12.84
01.1101 281	Concrete pumping, 50-150 CY, add	CY			10.35	10.35
01.1101 291	Concrete pumping, over 150 CY add	CY			8.43	8.43
01.1101 301	Hydraulic Truck Crane, including operator, from yard 5 ton	HR			160.88	160.88
01.1101 311	Hydraulic Truck Crane, including operator, from yard 10 ton	HR			211.73	211.73
01.1101 321	Hydraulic Truck Crane, including operator, from yard 15 ton	HR			237.08	237.08
01.1101 331	Hydraulic Truck Crane, including operator, from yard 30 ton	HR			330.25	330.25
01.1101 341	Hydraulic Truck Crane, including operator, from yard 50 ton	HR			414.92	414.92
01.1101 351	Hydraulic Truck Crane, including operator, from yard 65 ton	HR			482.65	482.65
01.1101 361	Hydraulic Truck Crane, including operator, from yard 75 ton	HR			541.92	541.92
01.1101 371	Lattice Boom Truck Crane, including operator, 65 ton	DAY			2,397.19	2,397.19
01.1101 381	Lattice Boom Truck Crane, including operator, 75 ton	DAY			2,526.77	2,526.77
01.1101 391	Lattice Boom Truck Crane, including operator, 90 ton	DAY			2,656.36	2,656.36
01.1101 401	Lattice Boom Truck Crane, including operator, 150 ton	DAY			3,563.43	3,563.43
01.1101 411	Lattice Boom Truck Crane, including operator, 200 ton	DAY			4,405.68	4,405.68
01.1102 000	**TRUCK & FORKLIFT RENTAL:**					
01.1102 011	1/2 ton pickup rental	DAY			59.27	59.27
01.1102 021	1/2 ton pickup rental	WEEK			237.08	237.08
01.1102 031	1/2 ton pickup rental	MONTH			948.39	948.39
01.1102 041	1 ton flatbed, stakeside, rental	DAY			78.07	78.07
01.1102 051	1 ton flatbed, stakeside, rental	WEEK			390.24	390.24
01.1102 061	1 ton flatbed, stakeside, rental	MONTH			1,170.77	1,170.77
01.1102 071	1.5 ton flatbed, stakeside, 12' bed, rent	DAY			121.91	121.91
01.1102 081	1.5 ton flatbed, stakeside, 12' bed, rent	WEEK			487.75	487.75
01.1102 091	1.5 ton flatbed, stakeside, 12' bed, rent	MONTH			1,463.24	1,463.24
01.1102 101	2 ton flatbed dump truck, 14' bed, rent	DAY			142.26	142.26
01.1102 111	2 ton flatbed dump truck, 14' bed, rent	WEEK			569.05	569.05
01.1102 121	2 ton flatbed dump truck, 14' bed, rent	MONTH			1,673.22	1,673.22
01.1102 131	5 yard dump truck, rental	DAY			321.78	321.78
01.1102 141	5 yard dump truck, rental	WEEK			1,287.07	1,287.07
01.1102 142	Water truck 1,800 gal.	DAY			282.83	282.83
01.1102 143	Water truck 1,800 gal.	WEEK			1,131.26	1,131.26
01.1102 144	Water truck 1,800 gal.	MONTH			3,370.11	3,370.11
01.1102 145	Water truck 2,500 gal.	DAY			392.88	392.88
01.1102 146	Water truck 2,500 gal.	WEEK			1,571.60	1,571.60
01.1102 147	Water truck 2,500 gal.	MONTH			4,714.78	4,714.78

GENERAL REQUIREMENTS - 01

Division	Description	Unit	Labor Hours	Labor Cost	Material Cost	Total Cost
01.1102 148	Water truck 3,500 gal.	DAY			558.87	558.87
01.1102 149	Water truck 3,500 gal.	WEEK			2,235.46	2,235.46
01.1102 150	Water truck 3,500 gal.	MONTH			6,249.11	6,249.11
01.1102 151	Vehicle trailer, double axle	DAY			88.90	88.90
01.1102 152	Vehicle trailer, double axle	WEEK			355.64	355.64
01.1102 153	Vehicle trailer, double axle	MONTH			1,066.92	1,066.92
01.1102 154	Utility trailer, double axle	DAY			88.90	88.90
01.1102 155	Utility trailer, double axle	WEEK			355.64	355.64
01.1102 156	Utility trailer, double axle	MONTH			1,066.92	1,066.92
01.1102 161	Mileage, 1/2-3/4 ton truck	MILE			0.52	0.52
01.1102 171	Mileage, 1 ton truck	MILE			0.64	0.64
01.1102 181	Mileage, 1-1/2 ton truck	MILE			1.08	1.08
01.1102 191	Mileage, 2 ton dump truck	MILE			1.25	1.25
01.1102 201	Forklift, 8' lift, 4000#, gas flotation	DAY			210.86	210.86
01.1102 211	Forklift, 8' lift, 4000#, gas flotation	WEEK			843.37	843.37
01.1102 221	Forklift, 8' lift, 4000#, gas flotation	MONTH			2,108.41	2,108.41
01.1102 231	Forklift, 12' lift, 5000#, gas flotation	DAY			245.54	245.54
01.1102 241	Forklift, 12' lift, 5000#, gas flotation	WEEK			982.25	982.25
01.1102 251	Forklift, 12' lift, 5000#, gas flotation	MONTH			2,455.62	2,455.62
01.1102 261	Forklift, 14' lift, 4000#, gas flotation	DAY			248.95	248.95
01.1102 271	Forklift, 14' lift, 4000#, gas flotation	WEEK			995.78	995.78
01.1102 281	Forklift, 14' lift, 4000#, gas flotation	MONTH			2,489.49	2,489.49
01.1102 291	Forklift, 15' lift, 5000#, gas flotation	DAY			248.95	248.95
01.1102 301	Forklift, 15' lift, 5000#, gas flotation	WEEK			995.78	995.78
01.1102 311	Forklift, 15' lift, 5000#, gas flotation	MONTH			2,489.49	2,489.49
01.1102 321	Forklift, 20' lift, 7000#, gas flotation	DAY			254.03	254.03
01.1102 331	Forklift, 20' lift, 7000#, gas flotation	WEEK			1,007.64	1,007.64
01.1102 341	Forklift, 20' lift, 7000#, gas flotation	MONTH			2,523.34	2,523.34
01.1102 351	Forklift, 30' lift, 10,000#, gas flotation	DAY			372.59	372.59
01.1102 361	Forklift, 30' lift, 10,000#, gas flotation	WEEK			1,490.30	1,490.30
01.1102 371	Forklift, 30' lift, 10,000#, gas flotation	MONTH			3,717.32	3,717.32
01.1102 381	Forklift, 16' lift, 6000#, reach	DAY			248.95	248.95
01.1102 391	Forklift, 16' lift, 6000#, reach	WEEK			995.78	995.78
01.1102 401	Forklift, 16' lift, 6000#, reach	MONTH			2,489.49	2,489.49
01.1102 411	Forklift, 36' lift, 8000#, reach	DAY			407.33	407.33
01.1102 421	Forklift, 36' lift, 8000#, reach	WEEK			1,629.18	1,629.18
01.1102 431	Forklift, 36' lift, 8000#, reach	MONTH			4,072.96	4,072.96
01.1102 441	Forklift, solid, gas, 4000#	DAY			216.78	216.78
01.1102 451	Forklift, solid, gas, 4000#	WEEK			838.28	838.28
01.1102 461	Forklift, solid, gas, 4000#	MONTH			2,384.49	2,384.49
01.1102 471	Forklift, solid, gas, 5000#	DAY			254.03	254.03
01.1102 481	Forklift, solid, gas, 5000#	WEEK			973.76	973.76
01.1102 491	Forklift, solid, gas, 5000#	MONTH			2,794.30	2,794.30
01.1102 501	Forklift, solid, propane, 8000#	DAY			274.36	274.36
01.1102 511	Forklift, solid, propane, 8000#	WEEK			1,049.99	1,049.99
01.1102 521	Forklift, solid, propane, 8000#	MONTH			3,082.21	3,082.21
01.1102 531	Forklift, pneumatic, gas, 5000#	DAY			254.03	254.03
01.1102 541	Forklift, pneumatic, gas, 5000#	WEEK			973.76	973.76
01.1102 551	Forklift, pneumatic, gas, 5000#	MONTH			2,794.30	2,794.30
01.1102 561	Forklift, pneumatic, gas, 6000#	DAY			274.36	274.36
01.1102 571	Forklift, pneumatic, gas, 6000#	WEEK			1,049.99	1,049.99
01.1102 581	Forklift, pneumatic, gas, 6000#	MONTH			3,082.21	3,082.21
01.1102 591	Forklift, 12' lift, 15000#, pneumatic, gas	DAY			381.05	381.05
01.1102 601	Forklift, 12' lift, 15000#, pneumatic, gas	WEEK			1,524.14	1,524.14
01.1102 611	Forklift, 12' lift, 15000#, pneumatic, gas	MONTH			3,810.41	3,810.41
01.1102 621	Forklift, 17.5' lift, 15m#, pneumatic, gas	DAY			397.99	397.99
01.1102 631	Forklift, 17.5' lift, 15m#, pneumatic, gas	WEEK			1,608.85	1,608.85

01 - GENERAL REQUIREMENTS

Division	Description	Unit	Labor Hours	Labor Cost	Material Cost	Total Cost
01.1102 641	Forklift, 17.5' lift, 15m#, pneumatic, gas	MONTH			3,979.77	3,979.77
01.1102 651	Forklift, solid, electric, 4000#	DAY			216.78	216.78
01.1102 661	Forklift, solid, electric, 4000#	WEEK			838.28	838.28
01.1102 671	Forklift, solid, electric, 4000#	MONTH			2,384.49	2,384.49
01.1102 681	Forklift platforms, 4' x 12'	DAY			97.56	97.56
01.1102 691	Forklift platforms, 4' x 12'	WEEK			195.14	195.14
01.1102 701	Forklift platforms, 4' x 12'	MONTH			331.76	331.76
01.1102 711	Forklift boom jib	DAY			64.36	64.36
01.1102 721	Forklift boom jib	WEEK			193.06	193.06
01.1102 731	Forklift boom jib	MONTH			579.19	579.19
01.1102 741	Forklift extension, 6'	DAY			67.73	67.73
01.1102 751	Forklift extension, 6'	WEEK			121.91	121.91
01.1102 761	Forklift extension, 6'	MONTH			203.21	203.21
01.1102 771	Forklift self-dumping hopper	DAY			54.19	54.19
01.1102 781	Forklift self-dumping hopper	WEEK			194.77	194.77
01.1102 791	Forklift self-dumping hopper	MONTH			431.86	431.86
01.1102 801	Rough terrain forward reach high lift loaders, 24 ft lift	DAY			525.03	525.03
01.1102 811	Rough terrain forward reach high lift loaders, 24 ft lift	WEEK			1,998.33	1,998.33
01.1102 821	Rough terrain forward reach high lift loaders, 24 ft lift	MONTH			5,249.92	5,249.92
01.1102 831	Rough terrain forward reach high lift loaders, 36 ft lift	DAY			635.07	635.07
01.1102 841	Rough terrain forward reach high lift loaders, 36 ft lift	WEEK			2,492.85	2,492.85
01.1102 851	Rough terrain forward reach high lift loaders, 36 ft lift	MONTH			6,855.39	6,855.39
01.1102 861	Rough terrain forward reach high lift loaders, 40 ft lift	DAY			787.47	787.47
01.1102 871	Rough terrain forward reach high lift loaders, 40 ft lift	WEEK			3,082.21	3,082.21
01.1102 881	Rough terrain forward reach high lift loaders, 40 ft lift	MONTH			7,874.89	7,874.89
01.1103 000	**SILENCED COMPRESSORS, RENTAL:**					
01.1103 011	Air compressor, 85-100 CFM, gas	DAY			161.68	161.68
01.1103 021	Air compressor, 85-100 CFM, gas	WEEK			483.07	483.07
01.1103 031	Air compressor, 85-100 CFM, gas	MONTH			1,183.02	1,183.02
01.1103 041	Air compressor, 100 CFM, diesel	DAY			143.91	143.91
01.1103 051	Air compressor, 100 CFM, diesel	WEEK			503.86	503.86
01.1103 061	Air compressor, 100 CFM, diesel	MONTH			1,414.09	1,414.09
01.1103 071	Air compressor, 125 CFM, diesel	DAY			216.88	216.88
01.1103 081	Air compressor, 125 CFM, diesel	WEEK			739.39	739.39
01.1103 091	Air compressor, 125 CFM, diesel	MONTH			1,528.02	1,528.02
01.1103 101	Air compressor, 150/160 CFM, diesel	DAY			236.60	236.60
01.1103 111	Air compressor, 150/160 CFM, diesel	WEEK			768.99	768.99
01.1103 121	Air compressor, 150/160 CFM, diesel	MONTH			1,626.62	1,626.62
01.1103 131	Air compressor, 150 CFM, gas	DAY			220.85	220.85
01.1103 141	Air compressor, 150 CFM, gas	WEEK			729.50	729.50
01.1103 151	Air compressor, 150 CFM, gas	MONTH			1,518.22	1,518.22
01.1103 161	Air compressor, 160 CFM, gas	DAY			220.85	220.85
01.1103 171	Air compressor, 160 CFM, gas	WEEK			729.50	729.50
01.1103 181	Air compressor, 160 CFM, gas	MONTH			1,478.74	1,478.74
01.1103 191	Air compressor, 190 CFM, gas	DAY			236.60	236.60
01.1103 201	Air compressor, 190 CFM, gas	WEEK			768.99	768.99
01.1103 211	Air compressor, 190 CFM, gas	MONTH			1,557.61	1,557.61
01.1103 221	Air compressor, 190 CFM, diesel	DAY			203.21	203.21
01.1103 231	Air compressor, 190 CFM, diesel	WEEK			702.79	702.79
01.1103 241	Air compressor, 190 CFM, diesel	MONTH			1,862.89	1,862.89
01.1103 251	Air compressor, 250 CFM, diesel	DAY			220.15	220.15
01.1103 261	Air compressor, 250 CFM, diesel	WEEK			770.55	770.55
01.1103 271	Air compressor, 250 CFM, diesel	MONTH			2,167.71	2,167.71
01.1103 281	Air compressor, 365 CFM, diesel	DAY			308.25	308.25
01.1103 291	Air compressor, 365 CFM, diesel	WEEK			1,100.80	1,100.80
01.1103 301	Air compressor, 365 CFM, diesel	MONTH			2,896.75	2,896.75
01.1103 311	Air compressor, 600 CFM, diesel	DAY			508.07	508.07

GENERAL REQUIREMENTS - 01

Division	Description	Unit	Labor Hours	Labor Cost	Material Cost	Total Cost
01.1103 321	Air compressor, 600 CFM, diesel	WEEK			1,778.20	1,778.20
01.1103 331	Air compressor, 600 CFM, diesel	MONTH			4,657.18	4,657.18
01.1103 341	Air compressor, 750 CFM, diesel	DAY			580.87	580.87
01.1103 351	Air compressor, 750 CFM, diesel	WEEK			1,989.89	1,989.89
01.1103 361	Air compressor, 750 CFM, diesel	MONTH			5,072.10	5,072.10
01.1103 371	Air compressor, 800 CFM, diesel	DAY			729.50	729.50
01.1103 381	Air compressor, 800 CFM, diesel	WEEK			2,119.56	2,119.56
01.1103 391	Air compressor, 800 CFM, diesel	MONTH			5,225.00	5,225.00
01.1103 401	Air compressor, 1000 CFM, diesel	DAY			946.43	946.43
01.1103 411	Air compressor, 1000 CFM, diesel	WEEK			2,858.91	2,858.91
01.1103 421	Air compressor, 1000 CFM, diesel	MONTH			7,492.41	7,492.41
01.1103 431	Air compressor, 1200 CFM, diesel	DAY			1,007.64	1,007.64
01.1103 441	Air compressor, 1200 CFM, diesel	WEEK			3,048.35	3,048.35
01.1103 451	Air compressor, 1200 CFM, diesel	MONTH			6,985.77	6,985.77
01.1103 461	Air compressor, 1400 CFM, diesel	DAY			1,041.53	1,041.53
01.1103 471	Air compressor, 1400 CFM, diesel	WEEK			3,133.02	3,133.02
01.1103 481	Air compressor, 1400 CFM, diesel	MONTH			7,861.33	7,861.33
01.1103 491	Air compressor, 1600 CFM, diesel	DAY			1,083.84	1,083.84
01.1103 501	Air compressor, 1600 CFM, diesel	WEEK			3,454.81	3,454.81
01.1103 511	Air compressor, 1600 CFM, diesel	MONTH			8,501.48	8,501.48
01.1104 000	**PORTABLE AIR COMPRESSORS, RENTAL:**					
01.1104 011	Portable air compressor, 8 CFM, gas	DAY			59.27	59.27
01.1104 021	Portable air compressor, 8 CFM, gas	WEEK			237.08	237.08
01.1104 031	Portable air compressor, 8 CFM, gas	MONTH			592.74	592.74
01.1104 041	Portable air compressor, 375 CFM, diesel	DAY			308.25	308.25
01.1104 051	Portable air compressor, 375 CFM, diesel	WEEK			1,100.80	1,100.80
01.1104 061	Portable air compressor, 375 CFM, diesel	MONTH			2,896.75	2,896.75
01.1104 071	Portable air compressor, 800 CFM, diesel	DAY			580.87	580.87
01.1104 081	Portable air compressor, 800 CFM, diesel	WEEK			1,989.89	1,989.89
01.1104 091	Portable air compressor, 800 CFM, diesel	MONTH			5,072.10	5,072.10
01.1104 101	Portable air compressor, 900 CFM, diesel	DAY			767.00	767.00
01.1104 111	Portable air compressor, 900 CFM, diesel	WEEK			2,607.79	2,607.79
01.1104 121	Portable air compressor, 900 CFM, diesel	MONTH			6,340.52	6,340.52
01.1104 131	Portable air compressor, 1200 CFM, diesel	DAY			1,007.64	1,007.64
01.1104 141	Portable air compressor, 1200 CFM, diesel	WEEK			3,048.35	3,048.35
01.1104 151	Portable air compressor, 1200 CFM, diesel	MONTH			6,985.77	6,985.77
01.1104 161	Portable air compressor, 1600 CFM, diesel	DAY			1,083.84	1,083.84
01.1104 171	Portable air compressor, 1600 CFM, diesel	WEEK			3,454.81	3,454.81
01.1104 181	Portable air compressor, 1600 CFM, diesel	MONTH			8,501.48	8,501.48
01.1104 191	Portable air compressor, 1750 CFM, diesel	DAY			1,789.64	1,789.64
01.1104 201	Portable air compressor, 1750 CFM, diesel	WEEK			4,601.97	4,601.97
01.1104 211	Portable air compressor, 1750 CFM, diesel	MONTH			12,067.38	12,067.38
01.1104 221	Portable air compressor, 2000 CFM, diesel	DAY			2,045.33	2,045.33
01.1104 231	Portable air compressor, 2000 CFM, diesel	WEEK			5,420.10	5,420.10
01.1104 241	Portable air compressor, 2000 CFM, diesel	MONTH			14,317.20	14,317.20
01.1105 000	**ELECTRIC COMPRESSORS, RENTAL:**					
01.1105 011	Electric compressor, 6 CFM, 1.5HP, 110v	DAY			49.28	49.28
01.1105 021	Electric compressor, 6 CFM, 1.5HP, 110v	WEEK			147.86	147.86
01.1105 031	Electric compressor, 6 CFM, 1.5HP, 110v	MONTH			443.67	443.67
01.1105 041	Electric compressor, 35-42 CFM, 10HP, skid	DAY			101.61	101.61
01.1105 051	Electric compressor, 35-42 CFM, 10HP, skid	WEEK			433.74	433.74
01.1105 061	Electric compressor, 35-42 CFM, 10HP, skid	MONTH			808.41	808.41
01.1105 071	Compressor, 60 CFM, 15HP, skid, oil free	DAY			110.09	110.09
01.1105 081	Compressor, 60 CFM, 15HP, skid, oil free	WEEK			788.66	788.66
01.1105 091	Compressor, 60 CFM, 15HP, skid, oil free	MONTH			1,675.95	1,675.95
01.1105 101	Electric compressor, 60 CFM, 15HP, skid	DAY			127.00	127.00
01.1105 111	Electric compressor, 60 CFM, 15HP, skid	WEEK			499.59	499.59

01 - GENERAL REQUIREMENTS

Division	Description	Unit	Labor Hours	Labor Cost	Material Cost	Total Cost
01.1105 121	Electric compressor, 60 CFM, 15HP, skid	MONTH			1,270.15	1,270.15
01.1105 131	Electric compressor, 150 CFM, 40HP, trailer	DAY			186.28	186.28
01.1105 141	Electric compressor, 150 CFM, 40HP, trailer	WEEK			745.16	745.16
01.1105 151	Electric compressor, 150 CFM, 40HP, trailer	MONTH			1,778.20	1,778.20
01.1105 161	Electric compressor, 250-260CFM, 60HP, skid	DAY			254.03	254.03
01.1105 171	Electric compressor, 250-260CFM, 60HP, skid	WEEK			1,163.30	1,163.30
01.1105 181	Electric compressor, 250-260CFM, 60HP, skid	MONTH			2,218.16	2,218.16
01.1105 191	Electric compressor, 300 CFM, 75HP, skid	DAY			296.35	296.35
01.1105 201	Electric compressor, 300 CFM, 75HP, skid	WEEK			1,242.14	1,242.14
01.1105 211	Electric compressor, 300 CFM, 75HP, skid	MONTH			2,415.31	2,415.31
01.1105 221	Electric compressor, 450-460 CFM, 100HP, skid	DAY			313.30	313.30
01.1105 231	Electric compressor, 450-460 CFM, 100HP, skid	WEEK			1,449.18	1,449.18
01.1105 241	Electric compressor, 450-460 CFM, 100HP, skid	MONTH			2,898.34	2,898.34
01.1105 251	Electric compressor, 550-600 CFM, 125HP, skid	DAY			487.75	487.75
01.1105 261	Electric compressor, 550-600 CFM, 125HP, skid	WEEK			2,119.56	2,119.56
01.1105 271	Electric compressor, 550-600 CFM, 125HP, skid	MONTH			3,783.84	3,783.84
01.1106 000	**AIR HOSE & ACCESSORIES, RENTAL:**					
01.1106 011	Air hose, 1/4" ID, 50', coupled	DAY			13.56	13.56
01.1106 021	Air hose, 1/4" ID, 50', coupled	WEEK			33.88	33.88
01.1106 031	Air hose, 1/4" ID, 50', coupled	MONTH			67.73	67.73
01.1106 041	Air hose, 1/2" ID, 50', coupled	DAY			16.08	16.08
01.1106 051	Air hose, 1/2" ID, 50', coupled	WEEK			48.26	48.26
01.1106 061	Air hose, 1/2" ID, 50', coupled	MONTH			96.53	96.53
01.1106 071	Air hose, 3/4" ID, 50', coupled	DAY			23.63	23.63
01.1106 081	Air hose, 3/4" ID, 50', coupled	WEEK			54.19	54.19
01.1106 091	Air hose, 3/4" ID, 50', coupled	MONTH			105.03	105.03
01.1106 101	Air hose, 1" ID, 50', coupled	DAY			17.79	17.79
01.1106 111	Air hose, 1" ID, 50', coupled	WEEK			53.34	53.34
01.1106 121	Air hose, 1" ID, 50', coupled	MONTH			106.68	106.68
01.1106 131	Air hose, 1-1/2" ID, 50', coupled	DAY			59.18	59.18
01.1106 141	Air hose, 1-1/2" ID, 50', coupled	WEEK			134.06	134.06
01.1106 151	Air hose, 1-1/2" ID, 50', coupled	MONTH			236.60	236.60
01.1106 161	Air hose, 2" ID, 50', coupled	DAY			62.66	62.66
01.1106 171	Air hose, 2" ID, 50', coupled	WEEK			93.14	93.14
01.1106 181	Air hose, 2" ID, 50', coupled	MONTH			279.41	279.41
01.1106 191	Air hose, 3" ID, 50', coupled	DAY			147.86	147.86
01.1106 201	Air hose, 3" ID, 50', coupled	WEEK			315.51	315.51
01.1106 211	Air hose, 3" ID, 50', coupled	MONTH			739.39	739.39
01.1106 221	Air manifold, 6 outlet	DAY			65.05	65.05
01.1106 231	Air manifold, 6 outlet	WEEK			104.50	104.50
01.1106 241	Air manifold, 6 outlet	MONTH			173.86	173.86
01.1106 251	Aftercooler, AC2	DAY			98.59	98.59
01.1106 261	Aftercooler, AC2	WEEK			175.47	175.47
01.1106 271	Aftercooler, AC2	MONTH			315.51	315.51
01.1107 000	**AIR TOOLS & ACCESSORIES, RENTAL:**					
01.1107 011	Paving breaker, 30#	DAY			74.53	74.53
01.1107 021	Paving breaker, 30#	WEEK			260.82	260.82
01.1107 031	Paving breaker, 30#	MONTH			782.42	782.42
01.1107 041	Paving breaker, 60/70 #	DAY			77.91	77.91
01.1107 051	Paving breaker, 60/70 #	WEEK			272.64	272.64
01.1107 061	Paving breaker, 60/70 #	MONTH			817.95	817.95
01.1107 071	Paving breaker, 80/90 #	DAY			81.30	81.30
01.1107 081	Paving breaker, 80/90 #	WEEK			284.49	284.49
01.1107 091	Paving breaker, 80/90 #	MONTH			853.53	853.53
01.1107 101	Moil points/chisels, 5/8" to 1-1/4" shank	DAY			14.39	14.39
01.1107 111	Moil points/chisels, 5/8" to 1-1/4" shank	WEEK			43.19	43.19
01.1107 121	Moil points/chisels, 5/8" to 1-1/4" shank	MONTH			129.56	129.56

GENERAL REQUIREMENTS - 01

Division	Description	Unit	Labor Hours	Labor Cost	Material Cost	Total Cost
01.1107 131	Wide chisels, 5/8" to 1-1/4" shank	DAY			18.63	18.63
01.1107 141	Wide chisels, 5/8" to 1-1/4" shank	WEEK			55.89	55.89
01.1107 151	Wide chisels, 5/8" to 1-1/4" shank	MONTH			167.69	167.69
01.1107 161	Spades/asphalt cutter, 1" to 1-1/4" shank	DAY			19.45	19.45
01.1107 171	Spades/asphalt cutter, 1" to 1-1/4" shank	WEEK			58.43	58.43
01.1107 181	Spades/asphalt cutter, 1" to 1-1/4" shank	MONTH			168.51	168.51
01.1107 191	Rock Drill - Jackhammer, 7-25 #	DAY			74.53	74.53
01.1107 201	Rock Drill - Jackhammer, 7-25 #	WEEK			260.82	260.82
01.1107 211	Rock Drill - Jackhammer, 7-25 #	MONTH			670.64	670.64
01.1107 221	Rock Drill - Jackhammer, 30-40 #	DAY			81.30	81.30
01.1107 231	Rock Drill - Jackhammer, 30-40 #	WEEK			284.49	284.49
01.1107 241	Rock Drill - Jackhammer, 30-40 #	MONTH			731.59	731.59
01.1107 251	Rock Drill - Jackhammer, 45 # & over	DAY			88.06	88.06
01.1107 261	Rock Drill - Jackhammer, 45 # & over	WEEK			308.25	308.25
01.1107 271	Rock Drill - Jackhammer, 45 # & over	MONTH			792.57	792.57
01.1107 281	Drill steel, 3' long, 7/8" to 1" shank	DAY			15.73	15.73
01.1107 291	Drill steel, 3' long, 7/8" to 1" shank	WEEK			27.64	27.64
01.1107 301	Drill steel, 3' long, 7/8" to 1" shank	MONTH			45.35	45.35
01.1107 311	Drill steel, 3'-5', 7/8"-1" shank	DAY			31.55	31.55
01.1107 321	Drill steel, 3'-5', 7/8"-1" shank	WEEK			47.35	47.35
01.1107 331	Drill steel, 3'-5', 7/8"-1" shank	MONTH			72.97	72.97
01.1107 341	Drill steel, 5'-8', 7/8"-1" shank	DAY			65.05	65.05
01.1107 351	Drill steel, 5'-8', 7/8"-1" shank	WEEK			116.34	116.34
01.1107 361	Drill steel, 5'-8', 7/8"-1" shank	MONTH			147.86	147.86
01.1107 371	Drill steel, over 8', 7/8-1" shank	DAY			147.86	147.86
01.1107 381	Drill steel, over 8', 7/8-1" shank	WEEK			191.28	191.28
01.1107 391	Drill steel, over 8', 7/8-1" shank	MONTH			306.79	306.79
01.1107 401	Carbide bits, all sizes	DAY			39.41	39.41
01.1107 411	Carbide bits, all sizes	WEEK			78.88	78.88
01.1107 421	Carbide bits, all sizes	MONTH			138.05	138.05
01.1107 431	Chipping hammer	DAY			72.82	72.82
01.1107 441	Chipping hammer	WEEK			251.47	251.47
01.1107 451	Chipping hammer	MONTH			702.79	702.79
01.1107 461	Air impact wrench, 3/4" drive	DAY			52.51	52.51
01.1107 471	Air impact wrench, 3/4" drive	WEEK			157.54	157.54
01.1107 481	Air impact wrench, 3/4" drive	MONTH			472.49	472.49
01.1107 491	Air concrete vibrator	DAY			147.86	147.86
01.1107 501	Air concrete vibrator	WEEK			374.64	374.64
01.1107 511	Air concrete vibrator	MONTH			719.69	719.69
01.1107 521	Air grinders/drills	DAY			53.34	53.34
01.1107 531	Air grinders/drills	WEEK			186.28	186.28
01.1107 541	Air grinders/drills	MONTH			372.59	372.59
01.1107 551	Clay Digger	DAY			76.21	76.21
01.1107 561	Clay Digger	WEEK			228.63	228.63
01.1107 571	Clay Digger	MONTH			685.90	685.90
01.1107 581	Clay Digger Moil Points	DAY			13.56	13.56
01.1107 591	Clay Digger Moil Points	WEEK			27.10	27.10
01.1107 601	Clay Digger Moil Points	MONTH			54.19	54.19
01.1107 611	Clay Digger Clay Spade	DAY			16.49	16.49
01.1107 621	Clay Digger Clay Spade	WEEK			33.02	33.02
01.1107 631	Clay Digger Clay Spade	MONTH			62.66	62.66
01.1107 641	Clay Digger Chisels	DAY			14.82	14.82
01.1107 651	Clay Digger Chisels	WEEK			29.65	29.65
01.1107 661	Clay Digger Chisels	MONTH			52.06	52.06
01.1107 671	Paving Crack Chaser, Bits Extra	DAY			139.71	139.71
01.1107 681	Paving Crack Chaser, Bits Extra	WEEK			419.16	419.16
01.1107 691	Paving Crack Chaser, Bits Extra	MONTH			1,329.40	1,329.40

01 - GENERAL REQUIREMENTS

Division	Description	Unit	Labor Hours	Labor Cost	Material Cost	Total Cost
01.1108 000	**COMPACTION EQUIPMENT RENTAL:**					
01.1108 011	Air backfill tamper	DAY			80.84	80.84
01.1108 021	Air backfill tamper	WEEK			195.17	195.17
01.1108 031	Air backfill tamper	MONTH			561.93	561.93
01.1108 041	Rammer - wacker 100-125 lb	DAY			115.17	115.17
01.1108 051	Rammer - wacker 100-125 lb	WEEK			460.66	460.66
01.1108 061	Rammer - wacker 100-125 lb	MONTH			1,266.76	1,266.76
01.1108 071	Rammer - wacker 150 lb	DAY			153.28	153.28
01.1108 081	Rammer - wacker 150 lb	WEEK			613.06	613.06
01.1108 091	Rammer - wacker 150 lb	MONTH			1,591.93	1,591.93
01.1108 101	Rammer - wacker 250 lb	DAY			165.95	165.95
01.1108 111	Rammer - wacker 250 lb	WEEK			663.87	663.87
01.1108 121	Rammer - wacker 250 lb	MONTH			1,741.77	1,741.77
01.1108 131	Compactor, vibratory, plate type, 150-200 lb	DAY			110.09	110.09
01.1108 141	Compactor, vibratory, plate type, 150-200 lb	WEEK			440.30	440.30
01.1108 151	Compactor, vibratory, plate type, 150-200 lb	MONTH			1,100.80	1,100.80
01.1108 161	Compactor, vibratory, plate type, 200-300 lb	DAY			139.71	139.71
01.1108 163	Compactor, vibratory, plate type, 200-300 lb	WEEK			558.87	558.87
01.1108 165	Compactor, vibratory, plate type, 200-300 lb	MONTH			1,397.18	1,397.18
01.1108 167	Compactor, vibratory, plate type, 300-500 lb	DAY			182.03	182.03
01.1108 169	Compactor, vibratory, plate type, 300-500 lb	WEEK			697.74	697.74
01.1108 171	Compactor, vibratory, plate type, 300-500 lb	MONTH			1,820.56	1,820.56
01.1108 173	Compactor, vibratory, plate type, 750 lb	DAY			189.68	189.68
01.1108 175	Compactor, vibratory, plate type, 750 lb	WEEK			724.80	724.80
01.1108 177	Compactor, vibratory, plate type, 750 lb	MONTH			1,896.74	1,896.74
01.1108 179	Compactor, vibratory, plate type, 1000 lb	DAY			259.09	259.09
01.1108 181	Compactor, vibratory, plate type, 1000 lb	WEEK			1,002.57	1,002.57
01.1108 183	Compactor, vibratory, plate type, 1000 lb	MONTH			2,591.09	2,591.09
01.1108 221	Roller, vibratory, single drum, walk behind, 30"	DAY			142.26	142.26
01.1108 222	Roller, vibratory, single drum, walk behind, 30"	WEEK			569.05	569.05
01.1108 223	Roller, vibratory, single drum, walk behind, 30"	MONTH			1,422.58	1,422.58
01.1108 231	Roller, vibratory, double drum, walk behind, 21"	DAY			142.71	142.71
01.1108 232	Roller, vibratory, double drum, walk behind, 21"	WEEK			570.72	570.72
01.1108 233	Roller, vibratory, double drum, walk behind, 21"	MONTH			1,426.77	1,426.77
01.1108 234	Roller, vibratory, double drum, walk behind, 22 to 27"	DAY			165.95	165.95
01.1108 235	Roller, vibratory, double drum, walk behind, 22 to 27"	WEEK			663.87	663.87
01.1108 236	Roller, vibratory, double drum, walk behind, 22 to 27"	MONTH			1,659.65	1,659.65
01.1108 237	Roller, vibratory, double drum, walk behind, 28 to 29"	DAY			228.63	228.63
01.1108 238	Roller, vibratory, double drum, walk behind, 28 to 29"	WEEK			863.70	863.70
01.1108 239	Roller, vibratory, double drum, walk behind, 28 to 29"	MONTH			2,176.16	2,176.16
01.1108 240	Roller, vibratory, double drum, walk behind, 30"	DAY			248.95	248.95
01.1108 241	Roller, vibratory, double drum, walk behind, 30"	WEEK			938.19	938.19
01.1108 242	Roller, vibratory, double drum, walk behind, 30"	MONTH			2,345.49	2,345.49
01.1108 243	Roller, vibratory, double drum, walk behind, 31 to 36"	DAY			265.89	265.89
01.1108 244	Roller, vibratory, double drum, walk behind, 31 to 36"	WEEK			1,007.64	1,007.64
01.1108 245	Roller, vibratory, double drum, walk behind, 31 to 36"	MONTH			2,582.59	2,582.59
01.1108 246	Roller, vibratory, double drum, walk behind, 39"	DAY			276.90	276.90
01.1108 247	Roller, vibratory, double drum, walk behind, 39"	WEEK			1,049.99	1,049.99
01.1108 248	Roller, vibratory, double drum, walk behind, 39"	MONTH			2,684.22	2,684.22
01.1108 251	Roller, vibratory, double drum, riding, 36"	DAY			270.98	270.98
01.1108 253	Roller, vibratory, double drum, riding, 36"	WEEK			812.89	812.89
01.1108 255	Roller, vibratory, double drum, riding, 36"	MONTH			2,557.21	2,557.21
01.1108 257	Roller, vibratory, double drum, riding, 48"	DAY			293.80	293.80
01.1108 259	Roller, vibratory, double drum, riding, 48"	WEEK			881.49	881.49
01.1108 261	Roller, vibratory, double drum, riding, 48"	MONTH			2,845.14	2,845.14
01.1108 263	Roller, vibratory, double drum, riding, 66"	DAY			457.23	457.23
01.1108 265	Roller, vibratory, double drum, riding, 66"	WEEK			1,600.40	1,600.40

GENERAL REQUIREMENTS - 01

Division	Description	Unit	Labor Hours	Labor Cost	Material Cost	Total Cost
01.1108 267	Roller, vibratory, double drum, riding, 66"	MONTH			4,487.84	4,487.84
01.1108 271	Roller, vibratory, double drum, riding, 84"	DAY			533.44	533.44
01.1108 273	Roller, vibratory, double drum, riding, 84"	WEEK			1,752.79	1,752.79
01.1108 275	Roller, vibratory, double drum, riding, 84"	MONTH			4,826.51	4,826.51
01.1108 281	Roller, vibratory, double drum, riding, 96"	DAY			702.79	702.79
01.1108 283	Roller, vibratory, double drum, riding, 96"	WEEK			241.31	241.31
01.1108 285	Roller, vibratory, double drum, riding, 96"	MONTH			7,028.11	7,028.11
01.1108 371	Roller, vibratory, sheepsfoot, 16 to 20 inch	DAY			375.95	375.95
01.1108 373	Roller, vibratory, sheepsfoot, 16 to 20 inch	WEEK			1,127.87	1,127.87
01.1108 375	Roller, vibratory, sheepsfoot, 16 to 20 inch	MONTH			3,759.62	3,759.62
01.1108 381	Roller, vibratory, sheepsfoot, 24 to 33 inch	DAY			537.68	537.68
01.1108 383	Roller, vibratory, sheepsfoot, 24 to 33 inch	WEEK			1,613.07	1,613.07
01.1108 385	Roller, vibratory, sheepsfoot, 24 to 33 inch	MONTH			5,376.94	5,376.94
01.1108 391	Roller, static, double drum, towable, sheepsfoot	DAY			156.66	156.66
01.1108 393	Roller, static, double drum, towable, sheepsfoot	WEEK			550.40	550.40
01.1108 395	Roller, static, double drum, towable, sheepsfoot	MONTH			1,409.01	1,409.01
01.1108 401	Roller, vibratory, towable, 54", sheepsfoot	DAY			397.99	397.99
01.1108 403	Roller, vibratory, towable, 54", sheepsfoot	WEEK			1,392.90	1,392.90
01.1108 405	Roller, vibratory, towable, 54", sheepsfoot	MONTH			3,979.77	3,979.77
01.1108 411	Roller, vibratory, towable, 54", smooth drum	DAY			372.59	372.59
01.1108 413	Roller, vibratory, towable, 54", smooth drum	WEEK			1,304.01	1,304.01
01.1108 415	Roller, vibratory, towable, 54", smooth drum	MONTH			372.59	372.59
01.1108 421	Roller, vibratory, self propelled, 10 ton	DAY			538.55	538.55
01.1108 423	Roller, vibratory, self propelled, 10 ton	WEEK			2,154.17	2,154.17
01.1108 425	Roller, vibratory, self propelled, 10 ton	MONTH			5,910.41	5,910.41
01.1108 431	Roller, vibratory, 35,000 lb, Force	DAY			697.74	697.74
01.1108 433	Roller, vibratory, 35,000 lb, Force	WEEK			2,154.17	2,154.17
01.1108 435	Roller, vibratory, 35,000 lb, Force	MONTH			5,910.41	5,910.41
01.1108 441	Roller, asphalt, double drum, riding 1-2 ton	DAY			186.28	186.28
01.1108 443	Roller, asphalt, double drum, riding 1-2 ton	WEEK			745.16	745.16
01.1108 445	Roller, asphalt, double drum, riding 1-2 ton	MONTH			1,659.65	1,659.65
01.1108 447	Roller, asphalt, double drum, riding 2-3 ton	DAY			223.58	223.58
01.1108 449	Roller, asphalt, double drum, riding 2-3 ton	WEEK			880.62	880.62
01.1108 451	Roller, asphalt, double drum, riding 2-3 ton	MONTH			2,116.90	2,116.90
01.1108 453	Roller, asphalt, double drum, riding 6-8 ton	DAY			301.45	301.45
01.1108 455	Roller, asphalt, double drum, riding 6-8 ton	WEEK			1,205.77	1,205.77
01.1108 457	Roller, asphalt, double drum, riding 6-8 ton	MONTH			3,014.46	3,014.46
01.1108 459	Roller, asphalt, double drum, riding 8-10 ton	DAY			313.30	313.30
01.1108 461	Roller, asphalt, double drum, riding 8-10 ton	WEEK			1,253.19	1,253.19
01.1108 463	Roller, asphalt, double drum, riding 8-10 ton	MONTH			3,133.02	3,133.02
01.1108 465	Roller, asphalt, double drum, riding 10-12 ton	DAY			359.02	359.02
01.1108 467	Roller, asphalt, double drum, riding 10-12 ton	WEEK			1,436.10	1,436.10
01.1108 469	Roller, asphalt, double drum, riding 10-12 ton	MONTH			3,590.25	3,590.25
01.1109 000	**CONCRETE EQUIPMENT & ACCESSORIES, RENTAL:**					
01.1109 011	Demolition hammer, electric 20 lb	DAY			98.23	98.23
01.1109 013	Demolition hammer, electric 20 lb	WEEK			381.05	381.05
01.1109 015	Demolition hammer, electric 20 lb	MONTH			1,100.80	1,100.80
01.1109 021	Demolition hammer, electric 35 lb	DAY			110.09	110.09
01.1109 023	Demolition hammer, electric 35 lb	WEEK			423.39	423.39
01.1109 025	Demolition hammer, electric 35 lb	MONTH			1,253.19	1,253.19
01.1109 031	Demolition hammer, electric 60 lb	DAY			145.64	145.64
01.1109 033	Demolition hammer, electric 60 lb	WEEK			582.55	582.55
01.1109 035	Demolition hammer, electric 60 lb	MONTH			1,747.74	1,747.74
01.1109 041	Demolition hammer, electric 75 lb	DAY			161.70	161.70
01.1109 043	Demolition hammer, electric 75 lb	WEEK			646.93	646.93
01.1109 045	Demolition hammer, electric 75 lb	MONTH			1,940.78	1,940.78
01.1109 051	Demolition hammer, hydraulic 750 foot/pounds per blow	DAY			762.09	762.09

01 - GENERAL REQUIREMENTS

Division	Description	Unit	Labor Hours	Labor Cost	Material Cost	Total Cost
01.1109 053	Demolition hammer, hydraulic 750 foot/pounds per blow	WEEK			2,709.62	2,709.62
01.1109 055	Demolition hammer, hydraulic 750 foot/pounds per blow	MONTH			9,145.02	9,145.02
01.1109 061	Demolition hammer, hydraulic 1000 foot/pounds per blow	DAY			1,100.80	1,100.80
01.1109 063	Demolition hammer, hydraulic 1000 foot/pounds per blow	WEEK			4,064.48	4,064.48
01.1109 065	Demolition hammer, hydraulic 1000 foot/pounds per blow	MONTH			10,288.16	10,288.16
01.1109 071	Demolition hammer, hydraulic 1400 foot/pounds per blow	DAY			1,270.15	1,270.15
01.1109 073	Demolition hammer, hydraulic 1400 foot/pounds per blow	WEEK			4,233.80	4,233.80
01.1109 075	Demolition hammer, hydraulic 1400 foot/pounds per blow	MONTH			11,007.91	11,007.91
01.1109 081	Rotary hammer drills, 3/4 to 1"	DAY			67.73	67.73
01.1109 083	Rotary hammer drills, 3/4 to 1"	WEEK			203.21	203.21
01.1109 085	Rotary hammer drills, 3/4 to 1"	MONTH			406.47	406.47
01.1109 091	Rotary hammer drills, 1 1/2"	DAY			93.14	93.14
01.1109 093	Rotary hammer drills, 1 1/2"	WEEK			279.41	279.41
01.1109 095	Rotary hammer drills, 1 1/2"	MONTH			558.87	558.87
01.1109 101	Rotary hammer drills, 2"	DAY			118.53	118.53
01.1109 103	Rotary hammer drills, 2"	WEEK			355.64	355.64
01.1109 105	Rotary hammer drills, 2"	MONTH			711.27	711.27
01.1109 111	Rotary drill core boring machine with vacuum rig, bits extra	DAY			155.76	155.76
01.1109 113	Rotary drill core boring machine with vacuum rig, bits extra	WEEK			545.32	545.32
01.1109 115	Rotary drill core boring machine with vacuum rig, bits extra	MONTH			1,402.24	1,402.24
01.1109 121	Diamond core bit, 2"	DAY			93.14	93.14
01.1109 123	Diamond core bit, 2"	WEEK			326.01	326.01
01.1109 125	Diamond core bit, 2"	MONTH			779.03	779.03
01.1109 131	Diamond core bit, 3"	DAY			127.00	127.00
01.1109 133	Diamond core bit, 3"	WEEK			381.05	381.05
01.1109 135	Diamond core bit, 3"	MONTH			931.42	931.42
01.1109 141	Diamond core bit, 4"	DAY			160.88	160.88
01.1109 143	Diamond core bit, 4"	WEEK			563.10	563.10
01.1109 145	Diamond core bit, 4"	MONTH			1,287.07	1,287.07
01.1109 151	Diamond core bit, 5"	DAY			198.15	198.15
01.1109 153	Diamond core bit, 5"	WEEK			693.48	693.48
01.1109 155	Diamond core bit, 5"	MONTH			1,649.48	1,649.48
01.1109 161	Diamond core bit, 6"	DAY			237.08	237.08
01.1109 163	Diamond core bit, 6"	WEEK			829.79	829.79
01.1109 165	Diamond core bit, 6"	MONTH			1,930.63	1,930.63
01.1109 171	Vibrators, air	DAY			123.64	123.64
01.1109 173	Vibrators, air	WEEK			494.49	494.49
01.1109 175	Vibrators, air	MONTH			1,168.51	1,168.51
01.1109 181	Vibrators, electric, 1 to 2-1/2 HP	DAY			98.23	98.23
01.1109 183	Vibrators, electric, 1 to 2-1/2 HP	WEEK			294.67	294.67
01.1109 185	Vibrators, electric, 1 to 2-1/2 HP	MONTH			863.70	863.70
01.1109 191	Mixer, concrete, electric knockdown, 2-1/2 CF	DAY			59.27	59.27
01.1109 193	Mixer, concrete, electric knockdown, 2-1/2 CF	WEEK			209.99	209.99
01.1109 195	Mixer, concrete, electric knockdown, 2-1/2 CF	MONTH			584.29	584.29
01.1109 201	Mixer, concrete, electric knockdown, 3-1/2 CF	DAY			71.14	71.14
01.1109 203	Mixer, concrete, electric knockdown, 3-1/2 CF	WEEK			250.61	250.61
01.1109 205	Mixer, concrete, electric knockdown, 3-1/2 CF	MONTH			651.99	651.99
01.1109 211	Mixer, gas, portable, 2-1/2 CF	DAY			59.27	59.27
01.1109 213	Mixer, gas, portable, 2-1/2 CF	WEEK			209.99	209.99
01.1109 215	Mixer, gas, portable, 2-1/2 CF	MONTH			584.29	584.29
01.1109 221	Mixer, gas or electric portable, 3-1/2 CF	DAY			71.14	71.14
01.1109 223	Mixer, gas or electric portable, 3-1/2 CF	WEEK			250.61	250.61
01.1109 225	Mixer, gas or electric portable, 3-1/2 CF	MONTH			651.99	651.99
01.1109 231	Mixer, gas or electric portable, 6 CF	DAY			81.30	81.30
01.1109 233	Mixer, gas or electric portable, 6 CF	WEEK			291.30	291.30
01.1109 235	Mixer, gas or electric portable, 6 CF	MONTH			801.02	801.02
01.1109 241	Mixer, gas or electric portable 9 CF	DAY			93.14	93.14

GENERAL REQUIREMENTS - 01

Division	Description	Unit	Labor Hours	Labor Cost	Material Cost	Total Cost
01.1109 243	Mixer, gas or electric portable 9 CF	WEEK			335.35	335.35
01.1109 245	Mixer, gas or electric portable 9 CF	MONTH			922.15	922.15
01.1109 251	Plaster Mixer, towable 6 CF	DAY			110.09	110.09
01.1109 253	Plaster Mixer, towable 6 CF	WEEK			389.53	389.53
01.1109 255	Plaster Mixer, towable 6 CF	MONTH			973.76	973.76
01.1109 261	Plaster Mixer, towable 8 CF	DAY			118.53	118.53
01.1109 263	Plaster Mixer, towable 8 CF	WEEK			406.47	406.47
01.1109 265	Plaster Mixer, towable 8 CF	MONTH			1,016.12	1,016.12
01.1109 271	Plaster Mixer 10 CF	DAY			127.00	127.00
01.1109 273	Plaster Mixer 10 CF	WEEK			416.61	416.61
01.1109 275	Plaster Mixer 10 CF	MONTH			1,041.53	1,041.53
01.1109 281	Plaster Mixer 12 CF	DAY			132.09	132.09
01.1109 283	Plaster Mixer 12 CF	WEEK			460.66	460.66
01.1109 285	Plaster Mixer 12 CF	MONTH			1,151.57	1,151.57
01.1109 291	Concrete bucket, 1/3 yard	DAY			59.27	59.27
01.1109 293	Concrete bucket, 1/3 yard	WEEK			152.39	152.39
01.1109 295	Concrete bucket, 1/3 yard	MONTH			287.90	287.90
01.1109 301	Concrete bucket, 1/2 yard	DAY			47.42	47.42
01.1109 303	Concrete bucket, 1/2 yard	WEEK			142.26	142.26
01.1109 305	Concrete bucket, 1/2 yard	MONTH			284.49	284.49
01.1109 311	Concrete bucket, 3/4 yard	DAY			49.09	49.09
01.1109 313	Concrete bucket, 3/4 yard	WEEK			149.02	149.02
01.1109 315	Concrete bucket, 3/4 yard	MONTH			296.35	296.35
01.1109 321	Concrete bucket, 1 to 1-1/2 yard	DAY			50.81	50.81
01.1109 323	Concrete bucket, 1 to 1-1/2 yard	WEEK			152.39	152.39
01.1109 325	Concrete bucket, 1 to 1-1/2 yard	MONTH			304.82	304.82
01.1109 331	Concrete bucket, 2 yards	DAY			64.36	64.36
01.1109 333	Concrete bucket, 2 yards	WEEK			203.21	203.21
01.1109 335	Concrete bucket, 2 yards	MONTH			372.59	372.59
01.1109 341	Concrete troweling machine 20 to 24" electric	DAY			88.06	88.06
01.1109 343	Concrete troweling machine 20 to 24" electric	WEEK			331.92	331.92
01.1109 345	Concrete troweling machine 20 to 24" electric	MONTH			829.79	829.79
01.1109 351	Concrete troweling machine, 36"	DAY			115.17	115.17
01.1109 353	Concrete troweling machine, 36"	WEEK			460.66	460.66
01.1109 355	Concrete troweling machine, 36"	MONTH			1,117.74	1,117.74
01.1109 361	Concrete troweling machine, 44 to 48"	DAY			142.26	142.26
01.1109 363	Concrete troweling machine, 44 to 48"	WEEK			504.68	504.68
01.1109 365	Concrete troweling machine, 44 to 48"	MONTH			1,337.86	1,337.86
01.1109 371	Concrete troweling machine, 48", two wheel riding	DAY			304.82	304.82
01.1109 373	Concrete troweling machine, 48", two wheel riding	WEEK			1,160.06	1,160.06
01.1109 375	Concrete troweling machine, 48", two wheel riding	MONTH			2,921.33	2,921.33
01.1109 381	Concrete saw, gas, 10 HP	DAY			135.48	135.48
01.1109 383	Concrete saw, gas, 10 HP	WEEK			474.20	474.20
01.1109 385	Concrete saw, gas, 10 HP	MONTH			1,320.94	1,320.94
01.1109 391	Concrete saw, gas, self propelled, 18 HP	DAY			194.77	194.77
01.1109 393	Concrete saw, gas, self propelled, 18 HP	WEEK			668.96	668.96
01.1109 395	Concrete saw, gas, self propelled, 18 HP	MONTH			1,896.74	1,896.74
01.1109 401	Concrete saw, gas, self propelled, 36 HP	DAY			254.03	254.03
01.1109 403	Concrete saw, gas, self propelled, 36 HP	WEEK			1,007.64	1,007.64
01.1109 405	Concrete saw, gas, self propelled, 36 HP	MONTH			2,455.62	2,455.62
01.1109 411	Concrete saw, gas, self propelled, 65 HP	DAY			397.99	397.99
01.1109 413	Concrete saw, gas, self propelled, 65 HP	WEEK			1,392.90	1,392.90
01.1109 415	Concrete saw, gas, self propelled, 65 HP	MONTH			3,895.11	3,895.11
01.1109 421	Concrete cart, walk behind 6 CF	DAY			42.34	42.34
01.1109 423	Concrete cart, walk behind 6 CF	WEEK			127.00	127.00
01.1109 425	Concrete cart, walk behind 6 CF	MONTH			381.05	381.05
01.1109 431	Concrete cart, riding, 13 CF	DAY			150.73	150.73

01 - GENERAL REQUIREMENTS

Division	Description	Unit	Labor Hours	Labor Cost	Material Cost	Total Cost
01.1109 433	Concrete cart, riding, 13 CF	WEEK			565.64	565.64
01.1109 435	Concrete cart, riding, 13 CF	MONTH			1,668.12	1,668.12
01.1109 441	Wheelbarrow	DAY			23.70	23.70
01.1109 443	Wheelbarrow	WEEK			71.14	71.14
01.1109 445	Wheelbarrow	MONTH			142.26	142.26
01.1109 451	Jitterbug (tamper)	DAY			18.63	18.63
01.1109 453	Jitterbug (tamper)	WEEK			55.89	55.89
01.1109 455	Jitterbug (tamper)	MONTH			111.78	111.78
01.1109 461	Concrete floor grinder, 5" Cut	DAY			169.36	169.36
01.1109 463	Concrete floor grinder, 5" Cut	WEEK			609.66	609.66
01.1109 465	Concrete floor grinder, 5" Cut	MONTH			1,608.85	1,608.85
01.1109 471	Concrete floor planer, 10" cut	DAY			186.28	186.28
01.1109 473	Concrete floor planer, 10", gas, 9 HP	WEEK			668.96	668.96
01.1109 475	Concrete floor planer, 10", gas, 9 HP	MONTH			1,769.74	1,769.74
01.1109 481	Strip deck membrane remover	DAY			315.51	315.51
01.1109 483	Strip deck membrane remover	WEEK			1,084.44	1,084.44
01.1109 485	Strip deck membrane remover	MONTH			2,366.00	2,366.00
01.1109 491	Concrete floor grinder, 3/4 HP, 100 lb	DAY			433.74	433.74
01.1109 493	Concrete floor grinder, 3/4 HP, 100 lb	WEEK			1,419.59	1,419.59
01.1109 495	Concrete floor grinder, 3/4 HP, 100 lb	MONTH			3,923.63	3,923.63
01.1109 501	Concrete floor grinder, twin head 6 stone, gas	DAY			156.66	156.66
01.1109 503	Concrete floor grinder, twin head 6 stone, gas	WEEK			626.59	626.59
01.1109 505	Concrete floor grinder, twin head 6 stone, gas	MONTH			1,566.52	1,566.52
01.1109 511	Concrete floor grinder, twin head 6 stone, electric	DAY			156.66	156.66
01.1109 513	Concrete floor grinder, twin head 6 stone, electric	WEEK			626.59	626.59
01.1109 515	Concrete floor grinder, twin head 6 stone, electric	MONTH			1,566.52	1,566.52
01.1109 521	Concrete floor grinder, twin head 12 stone, gas	DAY			211.73	211.73
01.1109 523	Concrete floor grinder, twin head 12 stone, gas	WEEK			829.79	829.79
01.1109 525	Concrete floor grinder, twin head 12 stone, gas	MONTH			2,074.55	2,074.55
01.1109 611	Air floor scrabbler, 7 piston	DAY			374.64	374.64
01.1109 621	Air floor scrabbler, 7 piston	WEEK			1,222.45	1,222.45
01.1109 631	Air floor scrabbler, 7 piston	MONTH			1,922.38	1,922.38
01.1109 641	Air floor scrabbler, 3 piston	DAY			128.17	128.17
01.1109 651	Air floor scrabbler, 3 piston	WEEK			414.04	414.04
01.1109 661	Air floor scrabbler, 3 piston	MONTH			920.38	920.38
01.1110 000	**EARTH MOVING & EXCAVATING EQUIPMENT RENTAL:**					
01.1110 011	Posthole digger, gas, 2 man	DAY			84.68	84.68
01.1110 021	Posthole digger, gas, 2 man	WEEK			330.25	330.25
01.1110 031	Posthole digger, gas, 2 man	MONTH			1,151.57	1,151.57
01.1110 041	Posthole digger, hand, clamshell	DAY			15.23	15.23
01.1110 051	Posthole digger, hand, clamshell	WEEK			45.70	45.70
01.1110 061	Posthole digger, hand, clamshell	MONTH			91.45	91.45
01.1110 071	Conveyor belt, electric or gas, 17'	DAY			142.26	142.26
01.1110 081	Conveyor belt, electric or gas, 17'	WEEK			355.64	355.64
01.1110 091	Conveyor belt, electric or gas, 17'	MONTH			1,007.64	1,007.64
01.1110 101	Conveyor belt, electric or gas, 26'	DAY			194.77	194.77
01.1110 111	Conveyor belt, electric or gas, 26'	WEEK			482.65	482.65
01.1110 121	Conveyor belt, electric or gas, 26'	MONTH			1,312.50	1,312.50
01.1110 131	Trencher, riding hydrostatic, 20 hp	DAY			372.59	372.59
01.1110 133	Trencher, riding hydrostatic, 20 hp	WEEK			1,490.30	1,490.30
01.1110 135	Trencher, riding hydrostatic, 20 hp	MONTH			3,725.72	3,725.72
01.1110 141	Trencher, riding hydrostatic, 30 hp	DAY			440.30	440.30
01.1110 143	Trencher, riding hydrostatic, 30 hp	WEEK			1,608.85	1,608.85
01.1110 145	Trencher, riding hydrostatic, 30 hp	MONTH			3,835.83	3,835.83
01.1110 151	Trencher, riding hydrostatic, 40 hp	DAY			618.12	618.12
01.1110 153	Trencher, riding hydrostatic, 40 hp	WEEK			2,472.55	2,472.55
01.1110 155	Trencher, riding hydrostatic, 40 hp	MONTH			6,096.70	6,096.70

GENERAL REQUIREMENTS - 01

Division	Description	Unit	Labor Hours	Labor Cost	Material Cost	Total Cost
01.1110 161	Trencher, riding hydrostatic, 50 hp	DAY			677.41	677.41
01.1110 163	Trencher, riding hydrostatic, 50 hp	WEEK			2,709.62	2,709.62
01.1110 165	Trencher, riding hydrostatic, 50 hp	MONTH			6,748.70	6,748.70
01.1110 171	Trencher, riding hydrostatic, 60 hp	DAY			694.34	694.34
01.1110 173	Trencher, riding hydrostatic, 60 hp	WEEK			2,760.43	2,760.43
01.1110 175	Trencher, riding hydrostatic, 60 hp	MONTH			6,943.44	6,943.44
01.1110 181	Crawler tractor, angle dozer, 42 HP	DAY			541.92	541.92
01.1110 183	Crawler tractor, angle dozer, 42 HP	WEEK			1,625.80	1,625.80
01.1110 185	Crawler tractor, angle dozer, 42 HP	MONTH			5,419.28	5,419.28
01.1110 191	Crawler tractor, 3/4 cy loader bucket, 42 HP	DAY			541.92	541.92
01.1110 193	Crawler tractor, 3/4 cy loader bucket, 42 HP	WEEK			1,625.80	1,625.80
01.1110 195	Crawler tractor, 3/4 cy loader bucket, 42 HP	MONTH			5,419.28	5,419.28
01.1110 201	Crawler tractor, angle dozer, 65 HP	DAY			711.27	711.27
01.1110 203	Crawler tractor, angle dozer, 65 HP	WEEK			2,133.84	2,133.84
01.1110 205	Crawler tractor, angle dozer, 65 HP	MONTH			7,112.79	7,112.79
01.1110 211	Crawler tractor, 1 1/4 cy loader bucket, 65 HP	DAY			711.27	711.27
01.1110 213	Crawler tractor, 1 1/4 cy loader bucket, 65 HP	WEEK			2,133.84	2,133.84
01.1110 215	Crawler tractor, 1 1/4 cy loader bucket, 65 HP	MONTH			7,112.79	7,112.79
01.1110 221	Crawler tractor, dozer or loader, 72 HP	DAY			829.79	829.79
01.1110 223	Crawler tractor, dozer or loader, 72 HP	WEEK			2,489.49	2,489.49
01.1110 225	Crawler tractor, dozer or loader, 72 HP	MONTH			8,044.25	8,044.25
01.1110 231	Crawler tractor, 30,000 lb	DAY			1,185.47	1,185.47
01.1110 233	Crawler tractor, 30,000 lb	WEEK			4,403.15	4,403.15
01.1110 235	Crawler tractor, 30,000 lb	MONTH			11,854.66	11,854.66
01.1110 241	Crawler tractor, 80,000 lb	DAY			1,947.55	1,947.55
01.1110 243	Crawler tractor, 80,000 lb	WEEK			7,790.19	7,790.19
01.1110 245	Crawler tractor, 80,000 lb	MONTH			19,475.52	19,475.52
01.1110 251	Wheel tractors, bare 40 HP	DAY			296.35	296.35
01.1110 253	Wheel tractors, bare 40 HP	WEEK			889.09	889.09
01.1110 255	Wheel tractors, bare 40 HP	MONTH			2,667.32	2,667.32
01.1110 260	**TRACTOR ATTACHMENTS FOR ABOVE:**					
01.1110 261	Box scraper, rake or earth scraper	DAY			64.36	64.36
01.1110 263	Box scraper, rake or earth scraper	WEEK			193.06	193.06
01.1110 265	Box scraper, rake or earth scraper	MONTH			643.53	643.53
01.1110 267	Disc harrow	DAY			113.48	113.48
01.1110 269	Disc harrow	WEEK			397.13	397.13
01.1110 271	Disc harrow	MONTH			1,134.65	1,134.65
01.1110 273	Post hole digger	DAY			164.28	164.28
01.1110 275	Post hole digger	WEEK			658.78	658.78
01.1110 277	Post hole digger	MONTH			1,642.70	1,642.70
01.1110 279	Roto tiller	DAY			220.15	220.15
01.1110 281	Roto tiller	WEEK			880.62	880.62
01.1110 283	Roto tiller	MONTH			2,201.58	2,201.58
01.1110 285	Rotary mower	DAY			143.91	143.91
01.1110 287	Rotary mower	WEEK			525.03	525.03
01.1110 289	Rotary mower	MONTH			1,185.47	1,185.47
01.1110 291	Mini Excavator, up to 10' digging depth	DAY			440.30	440.30
01.1110 293	Mini Excavator, up to 10' digging depth	WEEK			1,685.06	1,685.06
01.1110 295	Mini Excavator, up to 10' digging depth	MONTH			4,259.22	4,259.22
01.1110 301	Excavator, 10,000 lb	DAY			414.92	414.92
01.1110 303	Excavator, 10,000 lb	WEEK			1,659.65	1,659.65
01.1110 305	Excavator, 10,000 lb	MONTH			4,978.95	4,978.95
01.1110 311	Excavator, 40,000 lb	DAY			914.49	914.49
01.1110 313	Excavator, 40,000 lb	WEEK			3,658.01	3,658.01
01.1110 315	Excavator, 40,000 lb	MONTH			10,974.02	10,974.02
01.1110 321	Excavator, 50,000 lb	DAY			1,100.80	1,100.80
01.1110 323	Excavator, 50,000 lb	WEEK			4,403.15	4,403.15

01 - GENERAL REQUIREMENTS

Division	Description	Unit	Labor Hours	Labor Cost	Material Cost	Total Cost
01.1110 325	Excavator, 50,000 lb	MONTH			13,209.47	13,209.47
01.1110 331	Excavator, 80,000 lb	DAY			1,456.44	1,456.44
01.1110 333	Excavator, 80,000 lb	WEEK			5,825.71	5,825.71
01.1110 335	Excavator, 80,000 lb	MONTH			17,477.15	17,477.15
01.1110 341	Excavator, 120,000 lb	DAY			2,794.30	2,794.30
01.1110 343	Excavator, 120,000 lb	WEEK			11,177.25	11,177.25
01.1110 345	Excavator, 120,000 lb	MONTH			30,144.69	30,144.69
01.1110 351	Backhoe loader, wheel type 20 HP	DAY			296.35	296.35
01.1110 353	Backhoe loader, wheel type 20 HP	WEEK			889.09	889.09
01.1110 355	Backhoe loader, wheel type 20 HP	MONTH			2,624.95	2,624.95
01.1110 361	Backhoe loader, wheel type 40 HP	DAY			364.12	364.12
01.1110 363	Backhoe loader, wheel type 40 HP	WEEK			1,456.44	1,456.44
01.1110 365	Backhoe loader, wheel type 40 HP	MONTH			3,641.05	3,641.05
01.1110 371	Backhoe loader, wheel type 60-70 HP	DAY			423.39	423.39
01.1110 373	Backhoe loader, wheel type 60-70 HP	WEEK			1,439.48	1,439.48
01.1110 375	Backhoe loader, wheel type 60-70 HP	MONTH			4,233.80	4,233.80
01.1110 381	Backhoe loader, wheel type 60-70 HP, 4wd	DAY			465.75	465.75
01.1110 383	Backhoe loader, wheel type 60-70 HP, 4wd	WEEK			1,862.89	1,862.89
01.1110 385	Backhoe loader, wheel type 60-70 HP, 4wd	MONTH			4,741.88	4,741.88
01.1110 391	Backhoe loader, wheel type 60-70 HP, extended reach hoe	DAY			508.07	508.07
01.1110 393	Backhoe loader, wheel type 60-70 HP, extended reach hoe	WEEK			2,032.21	2,032.21
01.1110 395	Backhoe loader, wheel type 60-70 HP, extended reach hoe	MONTH			5,080.56	5,080.56
01.1110 401	Backhoe loader, wheel type 100HP,	DAY			626.59	626.59
01.1110 403	Backhoe loader, wheel type 100HP,	WEEK			2,497.96	2,497.96
01.1110 405	Backhoe loader, wheel type 100HP,	MONTH			6,181.34	6,181.34
01.1110 411	Skid steer loader, 700 lb capacity	DAY			330.25	330.25
01.1110 413	Skid steer loader, 700 lb capacity	WEEK			1,151.57	1,151.57
01.1110 415	Skid steer loader, 700 lb capacity	MONTH			3,607.20	3,607.20
01.1110 421	Skid steer loader, 701 to 975 lb capacity	DAY			347.17	347.17
01.1110 423	Skid steer loader, 701 to 975 lb capacity	WEEK			1,219.35	1,219.35
01.1110 425	Skid steer loader, 701 to 975 lb capacity	MONTH			3,810.41	3,810.41
01.1110 431	Skid steer loader, 976 to 1350 lb capacity	DAY			364.12	364.12
01.1110 433	Skid steer loader, 976 to 1350 lb capacity	WEEK			1,287.07	1,287.07
01.1110 435	Skid steer loader, 976 to 1350 lb capacity	MONTH			4,098.30	4,098.30
01.1110 441	Skid steer loader, 1351 to1749 lb capacity	DAY			381.05	381.05
01.1110 443	Skid steer loader, 1351 to1749 lb capacity	WEEK			1,380.22	1,380.22
01.1110 445	Skid steer loader, 1351 to1749 lb capacity	MONTH			4,301.56	4,301.56
01.1110 451	Skid steer loader, 1750 and over capacity	DAY			423.39	423.39
01.1110 453	Skid steer loader, 1750 and over capacity	WEEK			1,515.69	1,515.69
01.1110 455	Skid steer loader, 1750 and over capacity	MONTH			4,708.00	4,708.00
01.1110 461	Skid steer loader trailer	DAY			59.27	59.27
01.1110 463	Skid steer loader trailer	WEEK			177.81	177.81
01.1110 465	Skid steer loader trailer	MONTH			372.59	372.59
01.1110 470	**SKID STEER LOADER ACCESSORIES:**					
01.1110 471	Auger	DAY			101.61	101.61
01.1110 473	Auger	WEEK			355.64	355.64
01.1110 475	Auger	MONTH			973.76	973.76
01.1110 477	Backhoe	DAY			101.61	101.61
01.1110 479	Backhoe	WEEK			355.64	355.64
01.1110 481	Backhoe	MONTH			973.76	973.76
01.1110 483	Rotary broom	DAY			101.61	101.61
01.1110 485	Rotary broom	WEEK			355.64	355.64
01.1110 487	Rotary broom	MONTH			973.76	973.76
01.1110 489	Forklift forks	DAY			132.09	132.09
01.1110 491	Forklift forks	WEEK			465.75	465.75
01.1110 493	Forklift forks	MONTH			1,346.34	1,346.34
01.1110 501	Hydraulic hammer	DAY			270.98	270.98

GENERAL REQUIREMENTS - 01

Division	Description	Unit	Labor Hours	Labor Cost	Material Cost	Total Cost
01.1110 503	Hydraulic hammer	WEEK			948.39	948.39
01.1110 505	Hydraulic hammer	MONTH			2,845.14	2,845.14
01.1110 511	Tree spade	DAY			228.63	228.63
01.1110 513	Tree spade	WEEK			804.43	804.43
01.1110 515	Tree spade	MONTH			2,286.24	2,286.24
01.1110 521	Cold planer attachment	DAY			702.79	702.79
01.1110 523	Cold planer attachment	WEEK			2,628.33	2,628.33
01.1110 525	Cold planer attachment	MONTH			6,054.35	6,054.35
01.1110 621	Hydraulic jaw breaker	DAY			76.21	76.21
01.1110 623	Hydraulic jaw breaker	WEEK			266.74	266.74
01.1110 625	Hydraulic jaw breaker	MONTH			762.09	762.09
01.1110 631	Wheel loader, 13,000 lb, 1.5 cy	DAY			406.47	406.47
01.1110 633	Wheel loader, 13,000 lb, 1.5 cy	WEEK			1,625.80	1,625.80
01.1110 635	Wheel loader, 13,000 lb, 1.5 cy	MONTH			4,755.40	4,755.40
01.1110 641	Wheel loader, 20,000 lb, 2 cy	DAY			516.50	516.50
01.1110 643	Wheel loader, 20,000 lb, 2 cy	WEEK			2,066.07	2,066.07
01.1110 645	Wheel loader, 20,000 lb, 2 cy	MONTH			6,130.56	6,130.56
01.1110 651	Wheel loader, 25,000 lb, 2.75 cy	DAY			609.66	609.66
01.1110 653	Wheel loader, 25,000 lb, 2.75 cy	WEEK			2,438.67	2,438.67
01.1110 655	Wheel loader, 25,000 lb, 2.75 cy	MONTH			7,078.90	7,078.90
01.1110 661	Wheel loader, 30,000 lb, 3.5 cy	DAY			753.62	753.62
01.1110 663	Wheel loader, 30,000 lb, 3.5 cy	WEEK			3,014.46	3,014.46
01.1110 665	Wheel loader, 30,000 lb, 3.5 cy	MONTH			8,772.44	8,772.44
01.1110 671	Motorgrader, blade, 30,000 lb	DAY			1,066.92	1,066.92
01.1110 673	Motorgrader, blade, 30,000 lb	WEEK			4,267.72	4,267.72
01.1110 675	Motorgrader, blade, 30,000 lb	MONTH			10,398.26	10,398.26
01.1111 000	**PUMPS AND ACCESSORIES, RENTAL:**					
01.1111 011	Centrifugal pump, 2", gas	DAY			71.14	71.14
01.1111 021	Centrifugal pump, 2", gas	WEEK			284.49	284.49
01.1111 031	Centrifugal pump, 2", gas	MONTH			728.21	728.21
01.1111 041	Centrifugal pump, 3", gas	DAY			74.53	74.53
01.1111 051	Centrifugal pump, 3", gas	WEEK			298.06	298.06
01.1111 061	Centrifugal pump, 3", gas	MONTH			894.16	894.16
01.1111 071	Centrifugal pump, 6", gas	DAY			177.81	177.81
01.1111 081	Centrifugal pump, 6", gas	WEEK			711.27	711.27
01.1111 091	Centrifugal pump, 6", gas	MONTH			1,778.20	1,778.20
01.1111 101	Trash pump, 2", gas	DAY			71.14	71.14
01.1111 111	Trash pump, 2", gas	WEEK			284.49	284.49
01.1111 121	Trash pump, 2", gas	MONTH			711.27	711.27
01.1111 131	Trash pump, 3", gas	DAY			94.85	94.85
01.1111 141	Trash pump, 3", gas	WEEK			379.34	379.34
01.1111 151	Trash pump, 3", gas	MONTH			948.39	948.39
01.1111 161	Trash pump, 4", gas	DAY			145.64	145.64
01.1111 171	Trash pump, 4", gas	WEEK			582.55	582.55
01.1111 181	Trash pump, 4", gas	MONTH			1,456.44	1,456.44
01.1111 191	Trash pump, 4", diesel	DAY			145.64	145.64
01.1111 201	Trash pump, 4", diesel	WEEK			582.55	582.55
01.1111 211	Trash pump, 4", diesel	MONTH			1,456.44	1,456.44
01.1111 221	Trash pump, low volume, 6", diesel	DAY			220.15	220.15
01.1111 231	Trash pump, low volume, 6", diesel	WEEK			660.46	660.46
01.1111 241	Trash pump, low volume, 6", diesel	MONTH			2,167.71	2,167.71
01.1111 251	Trash pump, high volume, 6", diesel	DAY			220.15	220.15
01.1111 261	Trash pump, high volume, 6", diesel	WEEK			660.46	660.46
01.1111 271	Trash pump, high volume, 6", diesel	MONTH			2,167.71	2,167.71
01.1111 281	Submersible, 2" pump, 110v, float	DAY			54.19	54.19
01.1111 291	Submersible, 2" pump, 110v, float	WEEK			189.68	189.68
01.1111 301	Submersible, 2" pump, 110v, float	MONTH			355.64	355.64

01 - GENERAL REQUIREMENTS

Division	Description	Unit	Labor Hours	Labor Cost	Material Cost	Total Cost
01.1111 311	Diaphragm pump, 2", gas	DAY			81.30	81.30
01.1111 321	Diaphragm pump, 2", gas	WEEK			325.17	325.17
01.1111 331	Diaphragm pump, 2", gas	MONTH			812.89	812.89
01.1111 341	Diaphragm pump, 3", gas	DAY			103.31	103.31
01.1111 351	Diaphragm pump, 3", gas	WEEK			413.21	413.21
01.1111 361	Diaphragm pump, 3", gas	MONTH			1,033.06	1,033.06
01.1111 371	Sump pump, air, 115 GPM	DAY			84.79	84.79
01.1111 381	Sump pump, air, 115 GPM	WEEK			250.43	250.43
01.1111 391	Sump pump, air, 115 GPM	MONTH			561.93	561.93
01.1111 401	Sump pump, air, 200 GPM	DAY			136.05	136.05
01.1111 411	Sump pump, air, 200 GPM	WEEK			333.22	333.22
01.1111 421	Sump pump, air, 200 GPM	MONTH			739.39	739.39
01.1111 431	Hose suction, 2" x 20', coupled	DAY			22.01	22.01
01.1111 441	Hose suction, 2" x 20', coupled	WEEK			66.05	66.05
01.1111 451	Hose suction, 2" x 20', coupled	MONTH			132.09	132.09
01.1111 461	Hose suction, 3" x 20', coupled	DAY			22.01	22.01
01.1111 471	Hose suction, 3" x 20', coupled	WEEK			66.05	66.05
01.1111 481	Hose suction, 3" x 20', coupled	MONTH			132.09	132.09
01.1111 491	Hose suction, 4" x 20', coupled	DAY			24.55	24.55
01.1111 501	Hose suction, 4" x 20', coupled	WEEK			74.53	74.53
01.1111 511	Hose suction, 4" x 20', coupled	MONTH			149.02	149.02
01.1111 521	Hose suction, 6" x 20', coupled	DAY			25.39	25.39
01.1111 531	Hose suction, 6" x 20', coupled	WEEK			76.21	76.21
01.1111 541	Hose suction, 6" x 20', coupled	MONTH			152.39	152.39
01.1111 551	Suction hose, 10" x 25'	DAY			37.23	37.23
01.1111 561	Suction hose, 10" x 25'	WEEK			105.03	105.03
01.1111 571	Suction hose, 10" x 25'	MONTH			209.99	209.99
01.1111 581	Hose, discharge, 3/4" x 50', coupled	DAY			25.35	25.35
01.1111 591	Hose, discharge, 3/4" x 50', coupled	WEEK			52.71	52.71
01.1111 601	Hose, discharge, 3/4" x 50', coupled	MONTH			58.55	58.55
01.1111 611	Hose, discharge, 2" x 50', coupled	DAY			22.01	22.01
01.1111 621	Hose, discharge, 2" x 50', coupled	WEEK			66.05	66.05
01.1111 631	Hose, discharge, 2" x 50', coupled	MONTH			132.09	132.09
01.1111 641	Hose, discharge, 3" x 50', coupled	DAY			25.39	25.39
01.1111 651	Hose, discharge, 3" x 50', coupled	WEEK			76.21	76.21
01.1111 661	Hose, discharge, 3" x 50', coupled	MONTH			152.39	152.39
01.1111 671	Hose, discharge, 4" x 50', coupled	DAY			26.24	26.24
01.1111 681	Hose, discharge, 4" x 50', coupled	WEEK			78.74	78.74
01.1111 691	Hose, discharge, 4" x 50', coupled	MONTH			157.54	157.54
01.1111 701	Hose, discharge, 6" x 50', coupled	DAY			27.93	27.93
01.1111 711	Hose, discharge, 6" x 50', coupled	WEEK			83.81	83.81
01.1111 721	Hose, discharge, 6" x 50', coupled	MONTH			167.69	167.69
01.1111 731	Hose, discharge, 10" x 50'	DAY			38.10	38.10
01.1111 741	Hose, discharge, 10" x 50'	WEEK			105.86	105.86
01.1111 751	Hose, discharge, 10" x 50'	MONTH			211.73	211.73
01.1112 000	**GENERATORS, RENTAL:**					
01.1112 011	Generator, 2750 watt, gas	DAY			66.05	66.05
01.1112 021	Generator, 2750 watt, gas	WEEK			254.03	254.03
01.1112 031	Generator, 2750 watt, gas	MONTH			762.09	762.09
01.1112 041	Generator, 3000 watt, gas	DAY			71.14	71.14
01.1112 051	Generator, 3000 watt, gas	WEEK			284.49	284.49
01.1112 061	Generator, 3000 watt, gas	MONTH			853.53	853.53
01.1112 071	Generator, 3500 watt, gas	DAY			104.17	104.17
01.1112 081	Generator, 3500 watt, gas	WEEK			416.61	416.61
01.1112 091	Generator, 3500 watt, gas	MONTH			1,249.84	1,249.84
01.1112 101	Generator, 5000 watt, gas	DAY			110.09	110.09
01.1112 111	Generator, 5000 watt, gas	WEEK			440.30	440.30

GENERAL REQUIREMENTS - 01

Division	Description	Unit	Labor Hours	Labor Cost	Material Cost	Total Cost
01.1112 121	Generator, 5000 watt, gas	MONTH			1,320.94	1,320.94
01.1112 131	Generator, 10,000 watt, diesel, trailer	DAY			161.70	161.70
01.1112 141	Generator, 10,000 watt, diesel, trailer	WEEK			646.93	646.93
01.1112 151	Generator, 10,000 watt, diesel, trailer	MONTH			1,940.78	1,940.78
01.1112 191	Generator, 12,500 to 15,000 watt, diesel, trailer	DAY			179.52	179.52
01.1112 201	Generator, 12,500 to 15,000 watt, diesel, trailer	WEEK			718.08	718.08
01.1112 211	Generator, 12,500 to 15,000 watt, diesel, trailer	MONTH			2,154.17	2,154.17
01.1112 221	Generator, 25,000 watt, diesel, trailer	DAY			216.78	216.78
01.1112 231	Generator, 25,000 watt, diesel, trailer	WEEK			870.49	870.49
01.1112 241	Generator, 25,000 watt, diesel, trailer	MONTH			2,611.41	2,611.41
01.1112 251	Generator, 30,000 watt, diesel, trailer	DAY			223.58	223.58
01.1112 261	Generator, 30,000 watt, diesel, trailer	WEEK			897.56	897.56
01.1112 271	Generator, 30,000 watt, diesel, trailer	MONTH			2,667.32	2,667.32
01.1112 281	Generator, 40,000 watt, diesel, trailer	DAY			226.92	226.92
01.1112 291	Generator, 40,000 watt, diesel, trailer	WEEK			907.75	907.75
01.1112 301	Generator, 40,000 watt, diesel, trailer	MONTH			2,723.16	2,723.16
01.1112 311	Generator, 50,000 watt, diesel, trailer	DAY			313.30	313.30
01.1112 321	Generator, 50,000 watt, diesel, trailer	WEEK			1,253.19	1,253.19
01.1112 331	Generator, 50,000 watt, diesel, trailer	MONTH			3,759.62	3,759.62
01.1112 341	Generator, 100KW, diesel, trailer/skid	DAY			711.27	711.27
01.1112 351	Generator, 100KW, diesel, trailer/skid	WEEK			2,845.14	2,845.14
01.1112 361	Generator, 100KW, diesel, trailer/skid	MONTH			8,535.36	8,535.36
01.1112 371	Generator, 135KW, diesel, trailer/skid	DAY			880.62	880.62
01.1112 381	Generator, 135KW, diesel, trailer/skid	WEEK			3,014.46	3,014.46
01.1112 391	Generator, 135KW, diesel, trailer/skid	MONTH			9,991.79	9,991.79
01.1112 401	Generator, 150KW, diesel, skid	DAY			931.42	931.42
01.1112 411	Generator, 150KW, diesel, skid	WEEK			3,133.02	3,133.02
01.1112 421	Generator, 150KW, diesel, skid	MONTH			10,161.12	10,161.12
01.1112 431	Generator, 200KW, diesel, skid	DAY			1,049.99	1,049.99
01.1112 441	Generator, 200KW, diesel, skid	WEEK			3,674.95	3,674.95
01.1112 451	Generator, 200KW, diesel, skid	MONTH			10,499.86	10,499.86
01.1112 461	Generator, 300KW, diesel, skid	DAY			1,100.80	1,100.80
01.1112 471	Generator, 300KW, diesel, skid	WEEK			3,852.78	3,852.78
01.1112 481	Generator, 300KW, diesel, skid	MONTH			11,007.91	11,007.91
01.1112 482	Generator, 600KW, diesel, skid	DAY			1,532.62	1,532.62
01.1112 483	Generator, 600KW, diesel, skid	WEEK			5,334.58	5,334.58
01.1112 484	Generator, 600KW, diesel, skid	MONTH			15,326.39	15,326.39
01.1112 485	Generator, 1000KW, diesel, skid	DAY			1,828.99	1,828.99
01.1112 486	Generator, 1000KW, diesel, skid	WEEK			6,401.50	6,401.50
01.1112 487	Generator, 1000KW, diesel, skid	MONTH			18,290.05	18,290.05
01.1112 488	Generator, 1250KW, diesel, skid	DAY			2,116.90	2,116.90
01.1112 489	Generator, 1250KW, diesel, skid	WEEK			7,409.16	7,409.16
01.1112 490	Generator, 1250KW, diesel, skid	MONTH			21,169.03	21,169.03
01.1112 491	Electric extension cord, 50'	DAY			12.70	12.70
01.1112 501	Electric extension cord, 50'	WEEK			38.10	38.10
01.1112 511	Electric extension cord, 50'	MONTH			76.21	76.21
01.1112 521	Electric extension cord, 100'	DAY			25.39	25.39
01.1112 531	Electric extension cord, 100'	WEEK			76.21	76.21
01.1112 541	Electric extension cord, 100'	MONTH			152.39	152.39
01.1112 551	Electric Y's	DAY			7.63	7.63
01.1112 561	Electric Y's	WEEK			22.89	22.89
01.1112 571	Electric Y's	MONTH			45.70	45.70
01.1112 581	Electric pigtail adapters	DAY			7.77	7.77
01.1112 591	Electric pigtail adapters	WEEK			23.35	23.35
01.1112 601	Electric pigtail adapters	MONTH			58.43	58.43
01.1112 611	String lights with guards	DAY			69.00	69.00
01.1112 621	String lights with guards	WEEK			108.42	108.42

01 - GENERAL REQUIREMENTS

Division	Description	Unit	Labor Hours	Labor Cost	Material Cost	Total Cost
01.1112 631	String lights with guards	MONTH			167.61	167.61
01.1112 641	Temporary power box, 4 & 7 receptacle	DAY			45.70	45.70
01.1112 651	Temporary power box, 4 & 7 receptacle	WEEK			91.45	91.45
01.1112 661	Temporary power box, 4 & 7 receptacle	MONTH			274.36	274.36
01.1112 671	Temporary power cord, 6/4, 100'	DAY			49.28	49.28
01.1112 681	Temporary power cord, 6/4, 100'	WEEK			132.11	132.11
01.1112 691	Temporary power cord, 6/4, 100'	MONTH			177.48	177.48
01.1112 701	Temporary power booster	DAY			69.00	69.00
01.1112 711	Temporary power booster	WEEK			153.79	153.79
01.1112 721	Temporary power booster	MONTH			285.91	285.91
01.1113 000	**WELDING EQUIPMENT & ACCESSORIES, RENTAL:**					
01.1113 001	Acetylene outfit, does not include oxygen & acetylene	DAY			91.45	91.45
01.1113 002	Acetylene outfit, does not include oxygen & acetylene	WEEK			364.12	364.12
01.1113 003	Acetylene outfit, does not include oxygen & acetylene	MONTH			1,092.33	1,092.33
01.1113 011	Welder, 175a, gas, with leads	DAY			110.09	110.09
01.1113 021	Welder, 175a, gas, with leads	WEEK			440.30	440.30
01.1113 031	Welder, 175a, gas, with leads	MONTH			1,320.94	1,320.94
01.1113 041	Welder, 135a, gas	DAY			76.21	76.21
01.1113 051	Welder, 135a, gas	WEEK			304.82	304.82
01.1113 061	Welder, 135a, gas	MONTH			914.49	914.49
01.1113 071	Welder, 200a, gas	DAY			110.09	110.09
01.1113 081	Welder, 200a, gas	WEEK			440.30	440.30
01.1113 091	Welder, 200a, gas	MONTH			1,320.94	1,320.94
01.1113 101	Welder, 250a, diesel	DAY			131.25	131.25
01.1113 111	Welder, 250a, diesel	WEEK			525.03	525.03
01.1113 121	Welder, 250a, diesel	MONTH			1,312.50	1,312.50
01.1113 131	Welder, 270a, diesel, 4KW generator	DAY			160.88	160.88
01.1113 141	Welder, 270a, diesel, 4KW generator	WEEK			643.53	643.53
01.1113 151	Welder, 270a, diesel, 4KW generator	MONTH			1,591.93	1,591.93
01.1113 161	Welder, 400a, gas	DAY			174.42	174.42
01.1113 171	Welder, 400a, gas	WEEK			697.74	697.74
01.1113 181	Welder, 400a, gas	MONTH			1,690.16	1,690.16
01.1113 191	Welder, 400a, diesel	DAY			186.28	186.28
01.1113 201	Welder, 400a, diesel	WEEK			719.75	719.75
01.1113 211	Welder, 400a, diesel	MONTH			1,727.39	1,727.39
01.1113 221	Welder, 600a, gas	DAY			228.63	228.63
01.1113 231	Welder, 600a, gas	WEEK			931.42	931.42
01.1113 241	Welder, 600a, gas	MONTH			2,116.90	2,116.90
01.1113 251	Welder, 500a, diesel, dual operator	DAY			254.03	254.03
01.1113 261	Welder, 500a, diesel, dual operator	WEEK			1,049.99	1,049.99
01.1113 271	Welder, 500a, diesel, dual operator	MONTH			2,370.92	2,370.92
01.1113 273	Wire fed welders, 90 amp includes wire	DAY			93.14	93.14
01.1113 275	Wire fed welders, 90 amp includes wire	WEEK			372.59	372.59
01.1113 277	Wire fed welders, 90 amp includes wire	MONTH			1,016.12	1,016.12
01.1113 281	Welding cable, copper, '00', 50'	DAY			21.18	21.18
01.1113 291	Welding cable, copper, '00', 50'	WEEK			84.68	84.68
01.1113 301	Welding cable, copper, '00', 50'	MONTH			254.03	254.03
01.1113 311	Welder's hood	DAY			18.63	18.63
01.1113 321	Welder's hood	WEEK			37.23	37.23
01.1113 331	Welder's hood	MONTH			74.53	74.53
01.1113 341	Arc air cutting torch	DAY			91.45	91.45
01.1113 351	Arc air cutting torch	WEEK			364.12	364.12
01.1113 361	Arc air cutting torch	MONTH			1,092.33	1,092.33
01.1114 000	**SANDBLASTING EQUIPMENT & ACCESSORIES, RENTAL:**					
01.1114 021	Sandblaster, 50# cap, hose, nozzle	WEEK			304.82	304.82
01.1114 031	Sandblaster, 50# cap, hose, nozzle	MONTH			873.83	873.83
01.1114 041	Sandblaster, 50 to 100# cap, remote control	DAY			84.68	84.68

GENERAL REQUIREMENTS - 01

Division	Description	Unit	Labor Hours	Labor Cost	Material Cost	Total Cost
01.1114 051	Sandblaster, 50 to 100# cap, remote control	WEEK			338.69	338.69
01.1114 061	Sandblaster, 50 to 100# cap, remote control	MONTH			1,341.24	1,341.24
01.1114 071	Sandblaster, 150 to 200# cap, remote control	DAY			118.53	118.53
01.1114 081	Sandblaster, 150 to 200# cap, remote control	WEEK			474.20	474.20
01.1114 091	Sandblaster, 150 to 200# cap, remote control	MONTH			1,490.30	1,490.30
01.1114 101	Sandblaster, 300# cap, remote control	DAY			118.53	118.53
01.1114 111	Sandblaster, 300# cap, remote control	WEEK			474.20	474.20
01.1114 121	Sandblaster, 300# cap, remote control	MONTH			1,574.97	1,574.97
01.1114 131	Sandblaster, 600# cap, remote control	DAY			177.81	177.81
01.1114 141	Sandblaster, 600# cap, remote control	WEEK			711.27	711.27
01.1114 151	Sandblaster, 600# cap, remote control	MONTH			1,778.20	1,778.20
01.1114 191	Sandblaster, 6 ton, 2 & 4 outlets	DAY			296.35	296.35
01.1114 201	Sandblaster, 6 ton, 2 & 4 outlets	WEEK			889.09	889.09
01.1114 211	Sandblaster, 6 ton, 2 & 4 outlets	MONTH			2,328.62	2,328.62
01.1114 221	Sandblaster, 8 ton, 2 & 4 outlets	DAY			347.17	347.17
01.1114 231	Sandblaster, 8 ton, 2 & 4 outlets	WEEK			1,083.84	1,083.84
01.1114 241	Sandblaster, 8 ton, 2 & 4 outlets	MONTH			2,540.25	2,540.25
01.1114 251	Twin line hose, 50', extra	DAY			55.89	55.89
01.1114 261	Twin line hose, 50', extra	WEEK			167.69	167.69
01.1114 271	Twin line hose, 50', extra	MONTH			372.59	372.59
01.1114 281	Suction sandblaster	DAY			72.97	72.97
01.1114 291	Suction sandblaster	WEEK			138.05	138.05
01.1114 301	Suction sandblaster	MONTH			266.18	266.18
01.1114 311	Pac unit	DAY			112.38	112.38
01.1114 321	Pac unit	WEEK			226.75	226.75
01.1114 331	Pac unit	MONTH			611.21	611.21
01.1114 341	Pac discharge hose, 1-1/2" x 50'	DAY			207.02	207.02
01.1114 351	Pac discharge hose, 1-1/2" x 50'	WEEK			522.48	522.48
01.1114 361	Pac discharge hose, 1-1/2" x 50'	MONTH			690.07	690.07
01.1114 371	Sandblast hose, 3/4" ID, 50', coupled	DAY			20.33	20.33
01.1114 381	Sandblast hose, 3/4" ID, 50', coupled	WEEK			61.00	61.00
01.1114 391	Sandblast hose, 3/4" ID, 50', coupled	MONTH			142.26	142.26
01.1114 401	Sandblast hose, 3/4" whip hose	DAY			33.49	33.49
01.1114 411	Sandblast hose, 3/4" whip hose	WEEK			57.15	57.15
01.1114 421	Sandblast hose, 3/4" whip hose	MONTH			92.68	92.68
01.1114 431	Sandblast hose, 1" ID, 50', coupled	DAY			63.08	63.08
01.1114 441	Sandblast hose, 1" ID, 50', coupled	WEEK			112.38	112.38
01.1114 451	Sandblast hose, 1" ID, 50', coupled	MONTH			216.88	216.88
01.1114 461	Sandblast hose, 1-1/4" ID, 50', coupled	DAY			74.94	74.94
01.1114 471	Sandblast hose, 1-1/4" ID, 50', coupled	WEEK			124.24	124.24
01.1114 481	Sandblast hose, 1-1/4" ID, 50', coupled	MONTH			266.18	266.18
01.1114 491	Sandblast hose, 1-1/2" ID, 50', coupled	DAY			207.02	207.02
01.1114 501	Sandblast hose, 1-1/2" ID, 50', coupled	WEEK			502.78	502.78
01.1114 511	Sandblast hose, 1-1/2" ID, 50', coupled	MONTH			690.07	690.07
01.1114 521	Sandblast nozzle	DAY			42.34	42.34
01.1114 531	Sandblast nozzle	WEEK			127.00	127.00
01.1114 541	Sandblast nozzle	MONTH			381.05	381.05
01.1114 551	Helmet with air filter	DAY			25.39	25.39
01.1114 561	Helmet with air filter	WEEK			101.61	101.61
01.1114 571	Helmet with air filter	MONTH			304.82	304.82
01.1114 581	Sandblast helmet, air fed, with purifier	DAY			42.34	42.34
01.1114 591	Sandblast helmet, air fed, with purifier	WEEK			127.00	127.00
01.1114 601	Sandblast helmet, air fed, with purifier	MONTH			381.05	381.05
01.1114 641	Wet blast head assembly	DAY			18.43	18.43
01.1114 651	Wet blast head assembly	WEEK			34.74	34.74
01.1114 661	Wet blast head assembly	MONTH			53.24	53.24
01.1114 671	Moisture separator	DAY			69.00	69.00

01 - GENERAL REQUIREMENTS

Division	Description	Unit	Labor Hours	Labor Cost	Material Cost	Total Cost
01.1114 681	Moisture separator	WEEK			114.35	114.35
01.1114 691	Moisture separator	MONTH			335.20	335.20
01.1115 000	**MISCELLANEOUS EQUIPMENT RENTAL:**					
01.1115 011	Pedestal floodlight, one 500w lamp	DAY			38.96	38.96
01.1115 021	Pedestal floodlight, one 500w lamp	WEEK			155.76	155.76
01.1115 031	Pedestal floodlight, one 500w lamp	MONTH			465.75	465.75
01.1115 041	Pedestal floodlight, two 500w lamps	DAY			47.42	47.42
01.1115 051	Pedestal floodlight, two 500w lamps	WEEK			167.69	167.69
01.1115 061	Pedestal floodlight, two 500w lamps	MONTH			491.13	491.13
01.1115 071	Mercury vapor, two 1000w, trailer with tower	DAY			194.77	194.77
01.1115 081	Mercury vapor, two 1000w, trailer with tower	WEEK			584.29	584.29
01.1115 091	Mercury vapor, two 1000w, trailer with tower	MONTH			1,455.58	1,455.58
01.1115 101	4 1000w metal arc lamps	DAY			345.06	345.06
01.1115 111	4 1000w metal arc lamps	WEEK			887.27	887.27
01.1115 121	4 1000w metal arc lamps	MONTH			1,951.95	1,951.95
01.1115 131	Port-a-lite, 1000w metal arc, 110v	DAY			112.38	112.38
01.1115 141	Port-a-lite, 1000w metal arc, 110v	WEEK			374.64	374.64
01.1115 151	Port-a-lite, 1000w metal arc, 110v	MONTH			690.07	690.07
01.1115 161	Port-a-lite, 110v, with 10' tower	DAY			141.98	141.98
01.1115 171	Port-a-lite, 110v, with 10' tower	WEEK			414.04	414.04
01.1115 181	Port-a-lite, 110v, with 10' tower	MONTH			739.39	739.39
01.1115 191	Space heater (lpg), 150,000 BTU	DAY			59.27	59.27
01.1115 201	Space heater (lpg), 150,000 BTU	WEEK			237.08	237.08
01.1115 211	Space heater (lpg), 150,000 BTU	MONTH			651.99	651.99
01.1115 212	Space heater (lpg), 300,000 BTU	DAY			110.09	110.09
01.1115 213	Space heater (lpg), 300,000 BTU	WEEK			440.30	440.30
01.1115 214	Space heater (lpg), 300,000 BTU	MONTH			1,210.87	1,210.87
01.1115 221	Space heater (lpg), 400,000 BTU	DAY			127.00	127.00
01.1115 231	Space heater (lpg), 400,000 BTU	WEEK			508.07	508.07
01.1115 241	Space heater (lpg), 400,000 BTU	MONTH			1,312.50	1,312.50
01.1115 251	Cut off saw, gas	DAY			59.27	59.27
01.1115 261	Cut off saw, gas	WEEK			207.44	207.44
01.1115 271	Cut off saw, gas	MONTH			414.92	414.92
01.1115 281	Cut off saw, electric	DAY			63.49	63.49
01.1115 291	Cut off saw, electric	WEEK			220.15	220.15
01.1115 301	Cut off saw, electric	MONTH			499.59	499.59
01.1115 311	Concrete saw, 10 HP	DAY			135.48	135.48
01.1115 321	Concrete saw, 10 HP	WEEK			474.20	474.20
01.1115 331	Concrete saw, 10 HP	MONTH			1,320.94	1,320.94
01.1115 341	Electromagnet, towable	DAY			443.67	443.67
01.1115 351	Electromagnet, towable	WEEK			857.67	857.67
01.1115 361	Electromagnet, towable	MONTH			2,760.36	2,760.36
01.1115 371	Barrel pump	DAY			27.64	27.64
01.1115 381	Barrel pump	WEEK			53.24	53.24
01.1115 391	Barrel pump	MONTH			98.59	98.59
01.1115 401	Hydrostatic tester, gas	DAY			181.40	181.40
01.1115 411	Hydrostatic tester, gas	WEEK			512.63	512.63
01.1115 421	Hydrostatic tester, gas	MONTH			788.66	788.66
01.1115 431	Industrial truck, golf cart, electric	DAY			216.88	216.88
01.1115 441	Industrial truck, golf cart, electric	WEEK			512.63	512.63
01.1115 451	Industrial truck, golf cart, electric	MONTH			887.27	887.27
01.1115 461	Machinery rollers (1 set)	DAY			59.18	59.18
01.1115 471	Machinery rollers (1 set)	WEEK			128.17	128.17
01.1115 481	Machinery rollers (1 set)	MONTH			414.04	414.04
01.1115 491	Fuel storage tank, 200 & 350 gallon	DAY			128.17	128.17
01.1115 501	Fuel storage tank, 200 & 350 gallon	WEEK			197.19	197.19
01.1115 511	Fuel storage tank, 200 & 350 gallon	MONTH			374.64	374.64

GENERAL REQUIREMENTS - 01

Division	Description	Unit	Labor Hours	Labor Cost	Material Cost	Total Cost
01.1115 521	Fuel storage tank, 350 & 420 gallon	DAY			153.79	153.79
01.1115 531	Fuel storage tank, 350 & 420 gallon	WEEK			276.04	276.04
01.1115 541	Fuel storage tank, 350 & 420 gallon	MONTH			532.37	532.37
01.1115 551	Fuel storage tank, 500 gallon	DAY			187.30	187.30
01.1115 561	Fuel storage tank, 500 gallon	WEEK			345.06	345.06
01.1115 571	Fuel storage tank, 500 gallon	MONTH			591.53	591.53
01.1115 581	Hotsy washer, 4GPM @ 1000 psi	DAY			135.48	135.48
01.1115 591	Hotsy washer, 4GPM @ 1000 psi	WEEK			533.44	533.44
01.1115 601	Hotsy washer, 4GPM @ 1000 psi	MONTH			1,354.81	1,354.81
01.1115 602	Hotsy washer, 6GPM @ 3000 psi	DAY			245.54	245.54
01.1115 603	Hotsy washer, 6GPM @ 3000 psi	WEEK			982.25	982.25
01.1115 604	Hotsy washer, 6GPM @ 3000 psi	MONTH			2,523.34	2,523.34
01.1115 605	Pressure washer, 4 GPM @ 1500 psi	DAY			127.00	127.00
01.1115 606	Pressure washer, 4 GPM @ 1500 psi	WEEK			508.07	508.07
01.1115 607	Pressure washer, 4 GPM @ 1500 psi	MONTH			1,012.72	1,012.72
01.1115 611	Pressure washer, 7 GPM @ 3000 psi	DAY			335.35	335.35
01.1115 621	Pressure washer, 7 GPM @ 3000 psi	WEEK			1,341.24	1,341.24
01.1115 631	Pressure washer, 7 GPM @ 3000 psi	MONTH			3,353.17	3,353.17
01.1115 711	Steam Cleaner, 150 to 240 GPH	DAY			160.88	160.88
01.1115 721	Steam Cleaner, 150 to 240 GPH	WEEK			643.53	643.53
01.1115 731	Steam Cleaner, 150 to 240 GPH	MONTH			1,778.20	1,778.20
01.1115 741	High pressure water blaster, 6,000 psi	DAY			487.75	487.75
01.1115 751	High pressure water blaster, 6,000 psi	WEEK			1,463.24	1,463.24
01.1115 761	High pressure water blaster, 6,000 psi	MONTH			3,658.01	3,658.01
01.1115 771	High pressure water blasters, 10,000 psi	DAY			651.99	651.99
01.1115 781	High pressure water blasters, 10,000 psi	WEEK			1,956.01	1,956.01
01.1115 791	High pressure water blasters, 10,000 psi	MONTH			4,890.06	4,890.06
01.1115 801	High pressure water blasters, 20,000 psi	DAY			821.36	821.36
01.1115 811	High pressure water blasters, 20,000 psi	WEEK			2,464.06	2,464.06
01.1115 821	High pressure water blasters, 20,000 psi	MONTH			7,392.20	7,392.20
01.1800 000	**OTHER GENERAL CONDITIONS:**					
01.1801 000	**MISCELLANEOUS RENTAL & CLEAN-UP:**					
01.1801 001	*Note: Planking for scaffolding is usually furnished by the trade using it.*					
01.1801 011	Safety nets, nylon, 4" mesh, rectangular	SFSA			13.60	13.60
01.1801 021	Safety nets, nylon, 8" mesh, rectangular	SFSA			6.23	6.23
01.1801 031	Scaffold, 1 month, with on/off	SFSA			1.24	1.24
01.1801 041	Scaffold, each additional month	SFSA			0.19	0.19
01.1801 051	Scaffold over 1 story, add	FLOOR	0.0025	0.20		0.20
01.1801 061	Scaffold rolling towers, 5'-8'	MONTH			286.20	286.20
01.1801 071	Scaffold rolling towers, 13'-15'	MONTH			413.21	413.21
01.1801 081	Scaffold rolling towers, 21'-25'	MONTH			609.66	609.66
01.1801 091	Scaffold rolling towers, 25'-30'	MONTH			711.27	711.27
01.1801 101	Scaffold planks, 7'-10'	MONTH			17.61	17.61
01.1801 111	Scaffold planks, aluminum, 10'	MONTH			20.56	20.56
01.1801 121	Scaffold swing stage, 32', on/off, rental	MONTH			1,577.35	1,577.35
01.1801 131	Scaffold swing stage, 32', on/off	LS			3,351.85	3,351.85
01.1801 141	Clean up, progressive	JSF	0.0044	0.36		0.36
01.1801 151	Clean up, final	JSF	0.0030	0.27		0.27
01.1801 161	Clean up, glass, SF of glass	SF	0.0037	0.43		0.43
01.1801 171	Clean & wax resilient floors	JSF	0.0035	0.32		0.32
01.1801 181	Debris removal	SF	0.0037	0.30		0.30
01.1801 182	Debris boxes, 15 CY	LOAD			414.84	414.84
01.1801 183	Debris boxes, 20 CY	LOAD			440.97	440.97
01.1801 184	Debris boxes, 36 CY	LOAD			494.39	494.39
01.1801 185	Debris boxes, 40 CY	LOAD			528.50	528.50
01.1801 191	Sign, "construction", 4' x 8'	EA			1,318.50	1,318.50
01.1801 201	Office expense, general	JSF			0.10	0.10

01 - GENERAL REQUIREMENTS

Division	Description	Unit	Labor Hours	Labor Cost	Material Cost	Total Cost
01.1801 211	Small tools	JSF			0.09	0.09
01.1801 221	Consumable supplies	JSF			0.09	0.09
01.1803 000	**PLATFORMS, TELESCOPING AND SCISSOR LIFTS:**					
01.1803 221	Scissor lift, self propelled, 10 ft	DAY			142.26	142.26
01.1803 231	Scissor lift, self propelled, 10 ft	WEEK			426.77	426.77
01.1803 241	Scissor lift, self propelled, 10 ft	MONTH			1,278.62	1,278.62
01.1803 251	Scissor lift, self propelled, 15 ft	DAY			194.77	194.77
01.1803 261	Scissor lift, self propelled, 15 ft	WEEK			584.29	584.29
01.1803 271	Scissor lift, self propelled, 15 ft	MONTH			1,549.58	1,549.58
01.1803 281	Scissor lift, self propelled, 20 ft	DAY			211.73	211.73
01.1803 291	Scissor lift, self propelled, 20 ft	WEEK			677.41	677.41
01.1803 301	Scissor lift, self propelled, 20 ft	MONTH			2,032.21	2,032.21
01.1803 311	Scissor lift, self propelled, 25 ft	DAY			270.98	270.98
01.1803 321	Scissor lift, self propelled, 25 ft	WEEK			846.74	846.74
01.1803 331	Scissor lift, self propelled, 25 ft	MONTH			2,455.62	2,455.62
01.1803 341	Scissor lift, self propelled, 30 ft	DAY			338.69	338.69
01.1803 351	Scissor lift, self propelled, 30 ft	WEEK			1,033.06	1,033.06
01.1803 361	Scissor lift, self propelled, 30 ft	MONTH			3,099.15	3,099.15
01.1803 371	Scissor lift, self propelled, 35 ft	DAY			372.59	372.59
01.1803 381	Scissor lift, self propelled, 35 ft	WEEK			1,117.74	1,117.74
01.1803 391	Scissor lift, self propelled, 35 ft	MONTH			3,353.17	3,353.17
01.1803 401	Scissor lift, self propelled, 40 ft	DAY			406.47	406.47
01.1803 411	Scissor lift, self propelled, 40 ft	WEEK			1,422.58	1,422.58
01.1803 421	Scissor lift, self propelled, 40 ft	MONTH			4,267.72	4,267.72
01.1803 431	Scissor lift, self propelled, 50 ft	DAY			431.86	431.86
01.1803 441	Scissor lift, self propelled, 50 ft	WEEK			1,608.85	1,608.85
01.1803 451	Scissor lift, self propelled, 50 ft	MONTH			4,826.51	4,826.51
01.1803 461	Scissor lift, self propelled, rough terrain, 25 ft	DAY			270.98	270.98
01.1803 471	Scissor lift, self propelled, rough terrain, 25 ft	WEEK			1,016.12	1,016.12
01.1803 481	Scissor lift, self propelled, rough terrain, 25 ft	MONTH			2,540.25	2,540.25
01.1803 491	Scissor lift, self propelled, rough terrain, 30 ft	DAY			338.69	338.69
01.1803 501	Scissor lift, self propelled, rough terrain, 30 ft	WEEK			1,227.81	1,227.81
01.1803 511	Scissor lift, self propelled, rough terrain, 30 ft	MONTH			3,556.41	3,556.41
01.1803 521	Scissor lift, self propelled, rough terrain, 40 ft	DAY			397.99	397.99
01.1803 531	Scissor lift, self propelled, rough terrain, 40 ft	WEEK			1,591.93	1,591.93
01.1803 541	Scissor lift, self propelled, rough terrain, 40 ft	MONTH			3,979.77	3,979.77
01.1803 551	Scissor lift, self propelled, rough terrain, 50 ft	DAY			436.94	436.94
01.1803 561	Scissor lift, self propelled, rough terrain, 50 ft	WEEK			1,747.74	1,747.74
01.1803 571	Scissor lift, self propelled, rough terrain, 50 ft	MONTH			4,369.29	4,369.29
01.1803 671	Tower platform, not propelled telescoping, cable lift, 15 ft	DAY			84.68	84.68
01.1803 681	Tower platform, not propelled telescoping, cable lift, 15 ft	WEEK			338.69	338.69
01.1803 691	Tower platform, not propelled telescoping, cable lift, 15 ft	MONTH			846.74	846.74
01.1803 701	Tower platform, not propelled telescoping, cable lift, 20 ft	DAY			50.81	50.81
01.1803 711	Tower platform, not propelled telescoping, cable lift, 20 ft	WEEK			406.47	406.47
01.1803 721	Tower platform, not propelled telescoping, cable lift, 20 ft	MONTH			1,016.12	1,016.12
01.1803 731	Tower platform, not propelled telescoping, cable lift, 25 ft	DAY			118.53	118.53
01.1803 741	Tower platform, not propelled telescoping, cable lift, 25 ft	WEEK			474.20	474.20
01.1803 751	Tower platform, not propelled telescoping, cable lift, 25 ft	MONTH			1,185.47	1,185.47
01.1803 761	Tower platform, not propelled telescoping, battery lift, 20 ft	DAY			110.09	110.09
01.1803 771	Tower platform, not propelled telescoping, battery lift, 20 ft	WEEK			440.30	440.30
01.1803 781	Tower platform, not propelled telescoping, battery lift, 20 ft	MONTH			1,100.80	1,100.80
01.1803 791	Tower platform, not propelled telescoping, battery lift, 30 ft	DAY			143.91	143.91
01.1803 801	Tower platform, not propelled telescoping, battery lift, 30 ft	WEEK			575.80	575.80
01.1803 811	Tower platform, not propelled telescoping, battery lift, 30 ft	MONTH			1,439.48	1,439.48
01.1803 821	Tower platform, not propelled telescoping, battery lift, 40 ft	DAY			211.73	211.73
01.1803 831	Tower platform, not propelled telescoping, battery lift, 40 ft	WEEK			719.75	719.75
01.1803 841	Tower platform, not propelled telescoping, battery lift, 40 ft	MONTH			1,862.89	1,862.89

GENERAL REQUIREMENTS - 01

Division	Description	Unit	Labor Hours	Labor Cost	Material Cost	Total Cost
01.1805 000	**PHOTOGRAPHY:**					
01.1805 001	*Note: The following prices are based on 8" x 10" photos.*					
01.1805 011	4 photos/month, 2 prints each	JSF			0.09	0.09
01.1805 021	4 photos, 2 prints each	SET			493.60	493.60
01.1805 031	5 photos, aerial, 1 print, black & white	SET			889.83	889.83
01.1805 041	5 photos, aerial, 1 print, color	SET			1,017.71	1,017.71
01.1900 000	**NON-MANUAL LABOR DISTRIBUTABLES:**					
01.2000 000	**OVERHEAD & PROFIT, BONDS:**					
01.3000 000	**ESCALATION:**					
01.4000 000	**CONTINGENCIES:**					
01.5000 000	**GEOGRAPHICAL DIFFERENCES:**					

02 - SITE WORK

Division	Description	Unit	Labor Hours	Labor Cost	Material Cost	Total Cost
02.1000 000	**DEMOLITION:**					
02.1000 001	*Note: Unless stated otherwise, demolition includes disposal. Other uncommon conditions, such as excessive haul distances, unusual work hours, high voltage lines, limited access or volatile materials must be accounted for by either increased costs or separate allowances.*					
02.1100 000	**SITE DEMOLITION:**					
02.1101 000	**PAVEMENT & MISCELLANEOUS CONCRETE REMOVAL:**					
02.1101 001	*Note: The following prices include disposal for short haul dumps. Adjustments are necessary for long hauls.*					
02.1101 011	Remove pavement, asphaltic concrete, 5,000 to 25,000	SF	0.0038	0.32	0.63	0.95
02.1101 021	Remove pavement, asphaltic concrete, 25,000 to 50,000	SF	0.0038	0.32	0.40	0.72
02.1101 030	Remove pavement, asphaltic concrete, over 50,000	SF	0.0024	0.20	0.28	0.48
02.1101 031	Saw cut concrete, 1" depth	LF	0.0160	1.14	0.15	1.29
02.1101 032	Saw cut concrete, 1-1/2" depth	LF	0.0338	2.41	0.23	2.64
02.1101 033	Saw cut concrete, 2" depth	LF	0.1010	7.20	0.34	7.53
02.1101 034	Saw cut concrete, 3" depth	LF	0.1315	9.37	0.56	9.93
02.1101 035	Pulverize, asphaltic concrete, for recompaction, 4"	SF	0.0015	0.12	0.10	0.23
02.1101 041	Remove concrete slab, 5" max, no rebar	SF	0.0128	0.85	0.59	1.44
02.1101 051	Remove concrete slab, 5" max, with rebar	SF	0.0165	1.10	0.66	1.76
02.1101 061	Remove concrete slab, 9"-12", with rebar	SF	0.0716	4.78	3.17	7.95
02.1101 071	Remove concrete slab, 13"-18", with rebar	SF	0.1289	8.60	4.59	13.19
02.1101 081	Remove concrete curb/gutter, no sawing	LF	0.0282	1.88	1.09	2.98
02.1101 091	Remove concrete curb-planter & batter board	SF	0.0397	2.65	1.36	4.01
02.1101 101	Remove concrete drive, with curb retainer & walk	LF	0.0166	1.11	0.63	1.74
02.1101 111	Remove concrete sidewalk, outside building	SF	0.0147	0.98	0.48	1.46
02.1101 121	Remove concrete catch basin, sump, dry well	EA	3.0190	201.40	124.76	326.16
02.1101 131	Dump charges, low	TON			21.25	21.25
02.1101 141	Dump charges, mid	TON			30.46	30.46
02.1101 151	Dump charges, high	TON			42.25	42.25
02.1102 000	**FENCE & GUARDRAIL REMOVAL:**					
02.1102 011	Remove fence, chain link, 6', dispose	LF	0.0295	1.97	0.51	2.48
02.1102 021	Remove fence, chain link, 6', salvage	LF	0.0406	2.71	0.67	3.38
02.1102 031	Remove fence, wood, 6', dispose	LF	0.0157	1.05	0.20	1.24
02.1102 041	Remove guardrail, dispose	LF	0.0541	4.28	0.65	4.93
02.1102 051	Remove guardrail, salvage	LF	0.0498	3.32	0.89	4.21
02.1103 000	**MANHOLE REMOVAL:**					
02.1103 011	Remove manhole, 6'	EA	4.4640	353.41	153.73	507.15
02.1103 021	Remove manhole, 7'-12'	EA	6.3000	498.77	292.15	790.92
02.1103 031	Remove manhole, break bellow collar, sand fill & plug	EA	1.8600	147.26	121.49	268.74
02.1103 041	Remove manhole, reset to grade, average	EA	1.1690	92.55	143.56	236.11
02.1104 000	**TREE & POLE REMOVAL:**					
02.1104 001	*Note: Tree removal includes removal of stump.*					
02.1104 011	Remove tree, 6"-8", off site disposal	EA	2.3369	185.01	147.51	332.52
02.1104 021	Remove tree, 10"-14", off site disposal	EA	2.3369	185.01	192.44	377.45
02.1104 031	Remove tree, 20"-30", off site disposal	EA	5.8424	462.54	481.29	943.84
02.1104 041	Remove tree, orchard clear	EA	0.5834	46.19	56.00	102.18
02.1104 051	Remove tree, walnut/tap root	EA	1.7210	136.25	96.11	232.36
02.1104 061	Remove tree, burning, large areas	EA	0.4061	27.09	31.98	59.07
02.1104 071	Remove pole, salvage condition	EA	5.4874	366.06	33.89	399.96
02.1105 000	**MISC REMOVAL, RELOCATE TO FACILITATE ABOVE ITEMS:**					
02.1105 011	Remove hydrant, with reset, thrust blocks	EA	6.7590	450.89	136.37	587.26
02.1105 021	Storm drain line to 15", dispose	LF	0.1077	7.18	3.24	10.42
02.1105 031	Sanitary line, 8"-12", complete	LF	0.1060	7.07	3.76	10.83
02.1200 000	**GENERAL BUILDING DEMOLITION:**					
02.1201 000	**BUILDING DEMOLITION:**					
02.1201 001	*Note: For fireproofing on steel, add 30% to the total cost. For salvage of steel, add 50% to the total cost.*					

SITE WORK - 02

Division	Description	Unit	Labor Hours	Labor Cost	Material Cost	Total Cost
02.1201 011	Frame building, single story	SF	0.0635	4.24	0.18	4.42
02.1201 021	Frame building, two story	SF	0.0615	4.10	0.18	4.29
02.1201 031	Frame building, three story	SF	0.0602	4.02	0.18	4.20
02.1201 041	Concrete block building, 2 story, no concrete columns	SF	0.1251	8.35	0.88	9.22
02.1201 051	Concrete block building, 3 story	SF	0.1209	8.07	0.81	8.87
02.1201 061	Concrete building, monolithic, large	SF	0.1411	9.41	3.96	13.38
02.1201 071	Concrete building, precast panel, frame roof	SF	0.1294	8.63	3.04	11.67
02.1201 081	Steel frame building, no fireproof	SF	0.1502	10.02	1.73	11.75
02.1202 000	**CEILING REMOVAL:**					
02.1202 011	Remove ceiling, plaster/lath/frame	SF	0.0320	2.13	0.03	2.17
02.1202 021	Remove ceiling, plaster suspended grid	SF	0.0240	1.60	0.03	1.64
02.1202 031	Remove ceiling, acoustic suspended grid	SF	0.0074	0.49	0.03	0.53
02.1202 041	Remove ceiling, acoustic, salvage	SF	0.0139	0.93	0.03	0.96
02.1202 051	Remove ceiling, area light, salvage	SF	0.0258	1.72	0.03	1.76
02.1202 061	Remove ceiling, tile, 12" x 12", Glue-on, T+G	SF	0.0180	1.20	0.03	1.24
02.1203 000	**ROOF REMOVAL:**					
02.1203 001	*Note: For removal of rigid insulation, add 100% to the costs listed below.*					
02.1203 011	Remove roof, built-up on plywood	SQ	0.9364	62.47	2.16	64.62
02.1203 021	Remove roof, built-up on metal deck	SQ	0.6970	46.50	3.17	49.67
02.1203 031	Remove roof, built-up on gypsum plank	SQ	0.7127	47.54	2.16	49.70
02.1203 041	Remove roof, built-up on concrete	SQ	0.8185	54.60	2.16	56.76
02.1203 061	Remove roof, asphalt shingle	SQ	0.6068	40.48	2.16	42.63
02.1203 071	Remove roof, wood shingle	SQ	0.5746	38.33	2.16	40.49
02.1203 081	Remove roof, skylight	SF	0.1250	8.34	2.88	11.22
02.1204 000	**CUT OPENINGS:**					
02.1204 001	*Note: All openings 3' x 7'.*					
02.1204 011	Cut opening, concrete wall, to 8"	EA	18.1876	1,213.29	57.07	1,270.36
02.1204 021	Cut opening, concrete masonry unit wall, to 8"	EA	12.1247	1,136.69	41.51	1,178.20
02.1204 031	Cut opening, plaster/metal stud	EA	6.7157	629.60	16.86	646.46
02.1204 041	Cut opening, drywall/wood stud	EA	4.4102	413.46	3.23	416.68
02.1204 051	Cut opening, suspended slab, to 8"	EA	7.9411	744.48	45.52	790.00
02.1205 000	**CONCRETE SAWING:**					
02.1205 001	*Note: The following prices may be reduced as much as 50% if a large amount of sawing is to be done at one time. Add 50% if water control is needed. For angle cuts, add 5.5% to the total costs. For overhead cuts, add 103% to the total costs.*					
02.1205 011	Sawing concrete slab, 1"	LF	0.0160	1.14	0.15	1.29
02.1205 021	Sawing concrete slab, 1-1/2"	LF	0.0338	2.41	0.23	2.64
02.1205 031	Sawing concrete slab, 2"	LF	0.1010	7.20	0.33	7.53
02.1205 041	Sawing concrete slab, 3"	LF	0.1315	9.37	0.55	9.93
02.1205 051	Sawing concrete slab, 12"	LF	0.7620	54.31	2.05	56.36
02.1205 061	Sawing concrete wall, 6"	LF	0.3186	22.71	5.58	28.28
02.1205 071	Sawing concrete wall, 8"	LF	0.3867	27.56	6.13	33.69
02.1205 081	Sawing concrete wall, 10"	LF	0.4834	34.45	7.66	42.12
02.1205 091	Sawing concrete wall, 12"	LF	0.5799	41.33	9.21	50.54
02.1206 000	**DRILLING, DIAMOND BIT:**					
02.1206 001	*Note: For angle holes, add 50% to the total costs. For overhead holes, add 100% to the total costs.*					
02.1206 011	Core drilling, floor, 1" diameter	IN/DP	0.0524	3.73	1.13	4.86
02.1206 021	Core drilling, floor, 3" diameter	IN/DP	0.0724	5.16	1.13	6.29
02.1206 031	Core drilling, floor, 6" diameter	IN/DP	0.1148	8.18	1.82	10.00
02.1206 041	Core drilling, floor, 10" diameter	IN/DP	0.2416	17.22	3.91	21.13
02.1206 051	Core drilling, wall, 1" diameter	IN/DP	0.1621	11.55	0.44	11.99
02.1206 061	Core drilling, wall, 3" diameter	IN/DP	0.5741	40.92	1.52	42.44
02.1206 071	Core drilling, wall, 6" diameter	IN/DP	0.9088	64.77	2.39	67.16
02.1206 081	Core drilling, wall, 10" diameter	IN/DP	1.9135	136.38	5.07	141.45
02.1207 000	**FOOTING REMOVAL & MISCELLANEOUS SALVAGE:**					
02.1207 011	Remove concrete foundations, no rebar	CY	0.9337	73.92	50.37	124.29

02 - SITE WORK

Division	Description	Unit	Labor Hours	Labor Cost	Material Cost	Total Cost
02.1207 021	Remove concrete foundation, with rebar	CY	1.3792	109.19	58.89	168.08
02.1207 031	Remove concrete foundation, residence	CY	0.5773	45.70	43.81	89.52
02.1207 041	Remove door & hardware, reuse, storage	EA	1.9934	215.07	3.26	218.33
02.1207 051	Remove wood surrounds, storage	EA	0.9907	106.89	1.59	108.48
02.1207 061	Remove lockset, butts, closer	SET	0.9226	99.54	1.59	101.13
02.1207 071	Remove locker, metal, surface mounted, save	TIER	1.0929	117.91	1.59	119.50
02.1207 081	Remove wire mesh enclosure, save	SF	0.0829	8.94	0.32	9.27
02.1207 091	Remove sash, save glass	SF	0.0412	4.60	0.24	4.84
02.1208 000	**DEMOLITION, WALLS ONLY, BUILDING TO REMAIN:**					
02.1208 011	Remove wall, concrete to 10", reinforced, machine	SF	0.0525	4.16	2.92	7.07
02.1208 021	Remove wall, concrete block, reinforced, machine	SF	0.0378	2.99	1.56	4.55
02.1208 031	Remove wall, stucco/plaster, metal stud	SF	0.0139	1.10	0.02	1.12
02.1208 041	Remove wall, gypsum wall board, metal studs	SF	0.0102	0.81	0.02	0.83
02.1208 051	Remove wall, brick veneer overlaid	SF	0.0949	7.51	0.27	7.78
02.1208 061	Remove wall, brick, 8" solid or 10" cavity	SF	0.0903	7.15	0.45	7.60
02.1208 071	Remove wall, brick, 10-12", reinforced, grout	SF	0.1244	9.85	0.90	10.75
02.1208 081	Remove wall, metal siding, no save	SF	0.0616	4.88	0.33	5.21
02.1208 091	Remove wall, curtain wall, save glass	SF	0.0800	6.33	0.61	6.94
02.1209 000	**DEMOLITION, FLOORS ONLY, BUILDING TO REMAIN:**					
02.1209 011	Remove slab on grade, 4"-5", with mesh	SF	0.0203	1.61	0.27	1.87
02.1209 021	Remove slab on grade, 4"-5", with #4 bars	SF	0.0213	1.69	0.39	2.08
02.1209 031	Remove suspended slab, 6"-8", free fall	SF	0.0378	2.99	0.47	3.47
02.1209 041	Remove slab fill, lightweight concrete, metal deck	SF	0.0258	2.04	0.14	2.18
02.1209 051	Remove insulated topping, no sand blast	SF	0.0194	1.54	0.10	1.64
02.1209 061	Remove ceramic tile, brick plate	SF	0.0249	1.97	0.22	2.19
02.1209 071	Remove resilient flooring	SF	0.0208	1.43	0.10	1.53
02.1209 081	Remove wood floor, subfloor, wear layer	SF	0.0252	1.73	0.10	1.83
02.1209 091	Remove carpet and pad	SF	0.0112	0.77	0.10	0.87
02.1210 000	**DEMOLITION ACCESSORIES:**					
02.1210 011	Chute	LF	0.3186	21.15	14.64	35.78
02.1210 021	Dust partition, 2" x 4" frame, 6 ml plastic	SF	0.0237	1.57	2.95	4.52
02.1210 031	Debris bin, 20 CY, 2 days or 3 day weekend	2 DYS			443.68	443.68
02.1210 041	Debris bin, 30 CY, 2 days or 3 day weekend	2 DYS			501.30	501.30
02.1210 051	Concrete & asphalt debris bin, 7 CY	2 DYS			455.20	455.20
02.1210 061	Dump charges, general bulk fill	TON			43.45	43.45
02.1210 071	Dump charges, concrete, low	CY			30.46	30.46
02.1210 073	Dump charges, concrete, mid	CY			45.65	45.65
02.1210 075	Dump charges, concrete, high	CY			61.79	61.79
02.1210 081	Remove waste oils, 200 gallon, including test	LS			99.55	99.55
02.1211 000	**REMOVE ASBESTOS, FULL PROCEDURE AT PREMIUM TIME:**					
02.1211 011	Remove asbestos, pipe 6" diameter	LF	0.0950	11.02	0.37	11.38
02.1211 021	Remove asbestos, pipe 16" diameter	LF	0.1360	15.77	0.48	16.25
02.1211 031	Remove asbestos, column & beams only	SF	0.3011	34.91	4.89	39.80
02.1211 041	Remove Vinyl Asbestos Tile from floor	SF	0.0295	3.42	0.14	3.56
02.1211 051	Remove asbestos fireproofing from ceiling	SF	0.0220	2.55	0.12	2.67
02.1211 061	Debris bin for asbestos, 20 CY	24 DYS			1,804.47	1,804.47
02.1213 000	**LEAD-BASE PAINT REMOVAL:**					
02.1213 011	Lead base paint removal, door surface	SF	0.0689	6.24	1.87	8.11
02.1213 021	Lead base paint removal, frames, jambs	SF	0.1046	9.47	2.48	11.95
02.1213 031	Lead base paint removal, door trim	SF	0.1181	10.69	3.25	13.94
02.1213 041	Lead base paint removal, window frames & trim	SF	0.1582	14.33	5.99	20.32
02.1213 051	Lead base paint removal, metal/wood siding	SF	0.0608	5.51	1.86	7.36
02.1213 061	Lead base paint removal, cabinet surface	SF	0.1503	13.61	4.79	18.40
02.1213 071	Lead base paint removal, cabinet trim	SF	0.1879	17.01	6.06	23.08
02.1213 081	Lead base paint removal, molding, simple	LF	0.1242	11.25	4.00	15.25
02.1213 091	Lead base paint removal, molding, ornate	LF	0.1774	16.06	5.72	21.78
02.1213 101	Lead base paint removal, base board, 4"	LF	0.1177	10.66	3.48	14.14

SITE WORK - 02

Division	Description	Unit	Labor Hours	Labor Cost	Material Cost	Total Cost
02.1213 111	Lead base paint removal, base board, 6"	LF	0.1471	13.32	4.32	17.64
02.1213 121	Lead base paint removal, pipes, railings & bars, to 2" dia	LF	0.0914	8.28	2.65	10.93
02.1213 131	Lead base paint removal, pipes, railings & bars, to 4" dia	LF	0.1150	10.41	3.47	13.88
02.1213 141	Lead base paint removal, pipes, railings & bars, to 6" dia	LF	0.1392	12.60	4.26	16.87
02.1213 151	Lead base paint removal, pipes, railings & bars, to 8" dia	LF	0.1641	14.86	5.00	19.86
02.1213 161	Lead base paint removal, pipes, railings & bars, to 10" dia	LF	0.1855	16.80	5.77	22.57
02.1213 171	Lead base paint removal, pipes, railings & bars, to 12" dia	LF	0.2085	18.88	6.67	25.55
02.1216 000	**MOVING STRUCTURES:**					
02.1216 011	Move concrete building, 1,000 to 2,000	SF	0.3382	28.91	3.76	32.67
02.1216 021	Move concrete building, 2,000 to 4,000	SF	0.3098	26.48	3.49	29.97
02.1216 031	Move frame building, 1 story, 1,000 to 2,000	SF	0.2908	24.86	3.49	28.35
02.1216 041	Move frame building, 1 story, 2,000 to 3,000	SF	0.2813	24.05	3.38	27.42
02.1216 051	Move frame building, 2 story, 1,000 to 4,000	SF	0.1787	15.28	2.07	17.35
02.1216 061	Move masonry building, 1,000 to 2,000	SF	0.3744	32.00	4.15	36.15
02.1216 071	Move steel frame building, 2,000 to 4,000	SF	0.2210	14.67	5.16	19.83
02.1301 000	**DEMOLITION, MISCELLANEOUS:**					
02.1301 011	Remove door, interior	EA	0.6120	66.03		66.03
02.1301 021	Remove door, exterior	EA	0.8044	86.79		86.79
02.1301 031	Remove door, glass	EA	1.9674	212.26		212.26
02.1301 041	Remove door, overhead	SF	0.0457	4.93		4.93
02.1301 051	Remove window, aluminum	SF	0.0541	5.84		5.84
02.1301 061	Remove window, wood	SF	0.0406	4.38		4.38
02.1301 071	Remove cabinet, base	LF	0.3041	32.81		32.81
02.1301 081	Remove cabinet, wall	LF	0.3419	36.89		36.89
02.1301 091	Remove cabinet, full height	LF	0.5010	54.05		54.05
02.1301 101	Remove counter top	LF	0.3944	42.55		42.55
02.1301 111	Remove shelving, single layer	LF	0.0524	5.65		5.65
02.1401 000	**DEMOLITION, PLUMBING:**					
02.1401 011	Remove bath tub	EA	2.6742	235.33		235.33
02.1401 021	Remove water closet	EA	1.8765	165.13		165.13
02.1401 031	Remove sink	EA	1.5011	132.10		132.10
02.1401 041	Remove lavatory	EA	1.4102	124.10		124.10
02.1401 051	Remove urinal	EA	2.0412	179.63		179.63
02.1401 061	Remove water fountain	EA	2.0044	176.39		176.39
02.1401 071	Remove shower	EA	1.6774	147.61		147.61
02.1401 081	Remove bidet	EA	1.7769	156.37		156.37
02.1401 091	Remove copper, steel or CI piping, to 2"	LF	0.0577	5.08		5.08
02.1401 101	Remove copper, steel or CI piping, to 4"	LF	0.0694	6.11		6.11
02.1401 111	Remove copper, steel or CI piping, to 6"	LF	0.0914	8.04		8.04
02.1401 121	Remove copper, steel or CI piping, to 8"	LF	0.1992	17.53		17.53
02.1401 131	Remove copper, steel or CI piping, to 10"	LF	0.2243	19.74		19.74
02.1501 000	**DEMOLITION, HVAC:**					
02.1501 011	Remove ductwork	#	0.0178	1.27		1.27
02.1501 021	Remove diffuser/register	EA	0.4014	28.61		28.61
02.1501 031	Remove grill	EA	0.3336	23.78		23.78
02.1501 041	Remove damper	EA	0.2667	19.01		19.01
02.1501 051	Remove flexible duct to 10"	LF	0.0441	3.14		3.14
02.1501 061	Remove flexible duct to 14"	LF	0.0579	4.13		4.13
02.1501 071	Remove flexible duct to 18"	LF	0.0967	6.89		6.89
02.1501 081	Remove piping, including insulation, to 1/2"	LF	0.0458	3.26		3.26
02.1501 091	Remove piping, including insulation, to 3"	LF	0.0896	6.39		6.39
02.1501 101	Remove piping, including insulation, to 6"	LF	0.1946	13.87		13.87
02.1501 111	Remove piping, including insulation, to 10"	LF	0.3014	21.48		21.48
02.1601 000	**DEMOLITION, ELECTRICAL:**					
02.1601 011	Remove main switchboard	AMP	0.0311	3.47		3.47
02.1601 021	Remove distribution board/panel	AMP	0.0259	2.89		2.89
02.1601 031	Remove transformer	KVA	0.0774	8.65		8.65

02 - SITE WORK

Division	Description	Unit	Labor Hours	Labor Cost	Material Cost	Total Cost
02.1601 041	Remove conduit, EMT, to 1"	LF	0.0174	1.94		1.94
02.1601 051	Remove conduit, EMT, to 2"	LF	0.0297	3.32		3.32
02.1601 061	Remove conduit, EMT, to 4"	LF	0.0757	8.46		8.46
02.1601 071	Remove conduit, RSC, to 1"	LF	0.0349	3.90		3.90
02.1601 081	Remove conduit, RSC, to 2"	LF	0.0641	7.16		7.16
02.1601 091	Remove conduit, RSC, to 4"	LF	0.1379	15.40		15.40
02.1601 101	Remove wire, 600v, #14	LF	0.0015	0.17		0.17
02.1601 111	Remove wire, 600v, #12	LF	0.0017	0.19		0.19
02.1601 121	Remove wire, 600v, #10	LF	0.0021	0.23		0.23
02.1601 131	Remove wire, 600v, #8	LF	0.0023	0.26		0.26
02.1601 141	Remove wire, 600v, #6	LF	0.0024	0.27		0.27
02.1601 151	Remove wire, 600v, #4	LF	0.0030	0.34		0.34
02.1601 161	Remove wire, 600v, #3	LF	0.0032	0.36		0.36
02.1601 171	Remove wire, 600v, #2	LF	0.0036	0.40		0.40
02.1601 181	Remove wire, 600v, 1/0	LF	0.0045	0.50		0.50
02.1601 191	Remove wire, 600v, 2/0	LF	0.0056	0.63		0.63
02.1601 201	Remove wire, 600v, 3/0	LF	0.0063	0.70		0.70
02.1601 211	Remove wire, 600v, 4/0	LF	0.0069	0.77		0.77
02.1601 221	Remove wire, 600v, 250 MCM	LF	0.0075	0.84		0.84
02.1601 231	Remove wire, 600v, 300 MCM	LF	0.0082	0.92		0.92
02.1601 241	Remove wire, 600v, 350 MCM	LF	0.0089	0.99		0.99
02.1601 251	Remove wire, 600v, 400 MCM	LF	0.0098	1.09		1.09
02.1601 261	Remove wire, 600v, 500 MCM	LF	0.0101	1.13		1.13
02.1601 271	Remove wire, 600v, 600 MCM	LF	0.0111	1.24		1.24
02.1601 281	Remove wire, 600v, 750 MCM	LF	0.0134	1.50		1.50
02.2000 000	**EXCAVATION, FILL & GRADING:**					
02.2001 000	**GRADING:**					
02.2001 001	Note: Material costs include machinery. Installation costs include labor and supervision. Costs do not include auxiliary expenses such as staking, flagmen, lights, etc.					
02.2001 011	Clear, grub, brush, turf, roots, disposal	SF	0.0010	0.09	0.06	0.15
02.2001 021	Clear & grub large area, no disposal	SF	0.0010	0.09	0.03	0.13
02.2001 031	Clear & grub, large products	CY	0.0147	1.34	1.37	2.72
02.2001 041	Strip & stock pile, 6"	CY	0.0175	1.60	1.81	3.41
02.2001 051	Scarify & compact top 6"	SF	0.0024	0.22	0.07	0.29
02.2001 061	Rough grade, machine	SF	0.0010	0.09	0.03	0.13
02.2001 071	Fine grade, machine	SF	0.0019	0.17	0.05	0.22
02.2001 081	Fine grade, hand	SF	0.0093	0.85		0.85
02.2002 000	**ROADWAY EXCAVATION & FILL:**					
02.2002 001	Note: CAT & CAN, 1500' travel, fill in 6" lifts, compact to 95% AASHO standards.					
02.2002 011	Rdwy cut & fill, earth, 500,000 & up	CY	0.0026	0.22	1.24	1.47
02.2002 021	Rdwy cut & fill, earth, 100,000 to 500,000	CY	0.0037	0.31	1.62	1.94
02.2002 031	Rdwy cut & fill, earth, 50,000 to 100,000	CY	0.0045	0.38	1.87	2.25
02.2002 041	Rock & earth conglomerates, 50,000	CY	0.0045	0.38	2.27	2.65
02.2002 051	Ripable rock, large quantities	CY	0.0516	4.54	4.37	8.91
02.2003 000	**SITE CUT & FILL:**					
02.2003 011	Site cut & fill, earth, 250,000 & up	CY	0.0139	1.22	1.36	2.58
02.2003 021	Site cut & fill, earth, 50,000 to 250,000	CY	0.0166	1.46	1.56	3.02
02.2003 031	Site cut & fill, earth, 20,000 to 50,000	CY	0.0213	1.88	1.91	3.79
02.2003 041	Site cut & fill, earth, 5,000 to 20,000	CY	0.0305	2.69	2.72	5.41
02.2003 051	Site cut & fill, rock/earth mix	CY	0.0673	5.93	6.63	12.55
02.2004 000	**CUT FOR BUILDINGS, BACKHOE, TRUCKED:**					
02.2004 001	Note: For trenches 5 feet or deeper, shoring is required. Add 20% to 80% for shoring. Consult the engineer for shoring requirements.					
02.2004 011	Building cut, earth, 500-1,000, site disposal	CY	0.0372	3.28	3.27	6.55
02.2004 021	Building cut, earth, 500-1,000, haul 1 mile	CY	0.0631	5.24	8.99	14.23
02.2004 031	Building cut, earth, 1,000 to 10,000, site disposal	CY	0.0279	2.46	1.98	4.44

SITE WORK - 02

Division	Description	Unit	Labor Hours	Labor Cost	Material Cost	Total Cost
02.2004 041	Building cut, earth, 1,000 to 10,000, haul 1 mile	CY	0.0263	2.19	3.64	5.83
02.2004 051	Building cuts, foundations, 100 to 500	CY	0.1302	11.90	9.03	20.94
02.2004 061	Building cuts, foundations, 500 to 2,000	CY	0.0465	4.25	4.25	8.50
02.2004 071	Building cuts, trench, grade beams	CY	0.1115	10.19	4.81	15.00
02.2004 081	Building cuts, hand excavate & trim	CY	2.0000	142.54	1.96	144.50
02.2004 091	Building cuts, rip rock, dispose	CY	0.0465	4.09	3.37	7.46
02.2004 101	Building cuts, rip rock, haul 1 mile	CY	0.0326	2.71	5.43	8.14
02.2004 111	Building cuts, rock, drill & blast	CY	0.5270	41.72	24.77	66.49
02.2004 121	Building cuts, rock, jackhammered	CY	0.7440	58.02	54.12	112.14
02.2004 131	Building cuts, rip trench, haul 1 mile	TON	0.0593	5.03	22.53	27.56
02.2004 141	Building cuts, hardrock trench, haul 1 mile	CY	0.1395	11.83	20.63	32.46
02.2004 151	Cut for buildings, levee	CY	0.0093	0.82	0.27	1.08
02.2005 000	**ENGINEERED FILL FOR STRUCTURES:**					
02.2005 011	Fill, 1,000 to 2,500, imported one mile	CY	0.0167	1.42	11.71	13.12
02.2005 021	Fill, 2,500 to 10,000, imported one mile	CY	0.0100	0.85	11.37	12.22
02.2005 031	Fill, 10,000 to 25,000, imported one mile	CY	0.0130	1.10	7.02	8.12
02.2005 041	Fill, 25,000 to 50,000, imported one mile	CY	0.0115	0.97	6.51	7.49
02.2005 051	Fill, 100 to 2,500, on site	CY	0.0065	0.55	8.08	8.63
02.2005 061	Fill, 2,500 to 25,000, on site	CY	0.0043	0.36	5.22	5.58
02.2005 071	Fill, 25,000 to 50,000, on site	CY	0.0022	0.19	4.46	4.65
02.2005 081	Fill, compaction by roller	CY	0.0130	1.19	0.83	2.02
02.2005 091	Fill, compaction by sheepsfoot	CY	0.0166	1.52	1.08	2.60
02.2005 101	Fill, dozer spread, no material	CY	0.0093	0.82	0.65	1.46
02.2006 000	**BACKFILL, FOUNDATIONS, WALLS, ETC:**					
02.2006 011	Backfill, machine, no compaction	CY	0.1404	12.84	3.81	16.65
02.2006 021	Backfill, hand, no compaction	CY	1.0000	71.27	6.51	77.78
02.2006 031	Backfill, select import, compacted	CY	0.3803	34.77	15.71	50.48
02.2006 041	Backfill, site material, compacted	CY	0.2449	22.39	7.87	30.26
02.2006 051	Backfill site material, compacted, bridges	CY	0.4337	39.65	12.24	51.89
02.2006 061	Add for lime treatment, 12" native soil, large areas	SF	0.0038	0.30	0.40	0.70
02.2006 071	Lime bulk, material only	TON			132.77	132.77
02.2006 100	**SWPD & STABILIZATION - EROSION CONTROL**					
02.2006 111	Straw bale - Inlet barrier	LF	0.0360	2.57	2.02	4.58
02.2006 121	Filter barrier 12" high filter fabric	LF	0.0430	3.06	1.09	4.16
02.2006 131	Sediment fence- 36" wide mesh fabric	LF	0.1150	8.20	4.21	12.40
02.2006 141	Straw bale barrier	LF	0.0360	2.57	4.21	6.77
02.2006 151	Straw wattles	LF	0.0459	3.27	2.13	5.40
02.2007 000	**SOIL POISONING:**					
02.2007 001	*Note: The following prices assume large areas. For house with slab on grade, price assumes drilling through footing to penetrate under slab area and treating soil.*					
02.2007 011	Soil poison, permanent, inorganic base	SF	0.0019	0.17	0.09	0.27
02.2007 021	Soil poison, temporary, organic base	SF	0.0019	0.17	0.09	0.27
02.2007 031	Tent & fumigate	CF			0.07	0.07
02.2007 041	Soil treatment around house	SF			0.37	0.37
02.2007 051	Soil treatment at house on slab on grade	LF			7.08	7.08
02.2008 000	**POROUS FILL & MISCELLANEOUS ROCK:**					
02.2008 011	Drain rock, 1/2"-3/4"	CY	0.2349	20.69	12.85	33.53
02.2008 021	Bankrun gravel	CY	0.2173	19.14	8.17	27.31
02.2008 031	Pea gravel	CY	0.2349	20.69	18.95	39.63
02.2008 041	Ornamental boulders, etc	CY	1.1573	101.91	75.99	177.90
02.2008 051	Rip rap, machine placed	TON	0.5064	46.30	15.75	62.05
02.2008 061	Rip rap, hand placed	CY	2.0284	144.56	15.75	160.32
02.2008 071	Rip rap, sacked, hand placed	CY	2.0284	178.62	98.21	276.83
02.3000 000	**PILING:**					
02.3001 000	**PILES, CONCRETE:**					
02.3001 011	Piles, precast, 10" square	LF	0.1953	17.91	36.80	54.71

02 - SITE WORK

Division	Description	Unit	Labor Hours	Labor Cost	Material Cost	Total Cost
02.3001 021	Piles, precast, 12" square	LF	0.2229	20.45	42.80	63.25
02.3001 031	Piles, precast, 14" square	LF	0.2381	21.84	43.91	65.75
02.3001 041	Piles, precast, 16" square	LF	0.2505	22.98	47.41	70.39
02.3001 051	Piles, precast, 18" square	LF	0.3310	30.36	48.54	78.90
02.3002 000	**PILES, STEEL H SECTION:**					
02.3002 011	Piles, steel H section, 8"x8"x36#	LF	0.2441	22.39	27.66	50.05
02.3002 021	Piles, steel H section, 10"x10"x57#	LF	0.2744	25.17	43.45	68.62
02.3002 031	Piles, steel H section, 12"x12"x74#	LF	0.2892	26.53	53.32	79.85
02.3002 041	Piles, steel H section, 14"x14"x89#	LF	0.3048	27.96	62.17	90.13
02.3002 051	Piles, steel H section, splice, average	EA	0.3048	27.96	17.48	45.44
02.3003 000	**PILES, PIPE:**					
02.3003 011	Piles, pipe, 12", concrete filled	LF	0.2560	23.48	18.16	41.64
02.3003 021	Piles, pipe, 12", unfilled	LF	0.2284	20.95	15.63	36.58
02.3003 031	Piles, pipe, 16", concrete filled	LF	0.2744	25.17	26.24	51.41
02.3003 041	Piles, pipe, 16", unfilled	LF	0.2442	22.40	22.43	44.83
02.3003 051	Piles, pipe, splicing, average	EA	1.8600	170.62	16.04	186.66
02.3003 061	Piles, pipe, points, average	EA	1.6275	149.29	13.70	162.99
02.3004 000	**PILES, STEEL STEP TAPERED:**					
02.3004 011	Piles, steel tip, 12" butt, concrete fill	LF	0.1796	16.47	15.78	32.25
02.3004 021	Piles, steel tip, 16" butt, concrete fill	LF	0.2035	18.67	25.49	44.16
02.3005 000	**PILES, WOOD:**					
02.3005 011	Piles, wood, 12" butt, untreated to 39'	LF	0.1796	16.47	5.99	22.47
02.3005 021	Piles, wood, 13" butt, untreated to 70'	LF	0.1953	17.91	6.44	24.36
02.3005 031	Piles, wood, 12" butt, treated to 39'	LF	0.1796	16.47	8.47	24.94
02.3005 041	Piles, wood, 13" butt, treated to 70'	LF	0.1953	17.91	9.44	27.35
02.3006 000	**PILES, MISCELLANEOUS ITEMS (INCLUDED ON ALL JOBS):**					
02.3006 001	*Note: Make separate allowance for standby and idle time, access roads, rig mats and special engineering.*					
02.3006 011	Test piles	LS			12,829.63	12,829.63
02.3006 021	Pile driver, truck crane, on/off	LS			11,636.78	11,636.78
02.3006 031	Crane, crawler, 35 tons, on/off	LS			16,243.03	16,243.03
02.3006 041	Crane, crawler, 65 tons, on/off	LS			22,788.69	22,788.69
02.3006 051	Crane, crawler, 100 tons, on/off	LS			32,728.44	32,728.44
02.3006 061	Pile float rig, standby crew, on/off	LS			17,697.56	17,697.56
02.3007 000	**PILES, SOLDIER, STEEL, RECOVERED, NO LAGGING:**					
02.3007 011	Piles, recovered, 15' max, pulled	LF	0.1467	13.46	23.21	36.67
02.3007 021	Piles, recovered, 20' max, pulled	LF	0.1178	10.81	14.92	25.73
02.3007 031	Piles, recovered, 30' max, pulled	LF	0.1095	10.04	14.31	24.36
02.3007 041	Piles, recovered, 50' max, pulled	LF	0.1198	10.99	15.41	26.40
02.3500 000	**CAISSONS & DRILLING:**					
02.3501 000	**BORING & PIERS, DRILLING ONLY, LARGE QUANTITY:**					
02.3501 011	Pier, drill only, 12", 4' deep	LF	0.0130	1.19	1.39	2.58
02.3501 021	Pier, drill, add for each 1' to 6'	EA	0.0028	0.26	0.48	0.74
02.3501 031	Pier, drill only, 16", 2' deep	LF	0.0295	2.70	2.06	4.76
02.3501 041	Pier, drill only, 16", 5' deep	LF	0.0194	1.77	1.75	3.53
02.3501 051	Test borings, 6"	LF	0.2284	20.88	6.34	27.22
02.3502 000	**AUGER HOLES, DRILLING ONLY, LARGE QUANTITY:**					
02.3502 011	Auger hole, drill only, 16"	LF	0.0572	5.23	5.70	10.93
02.3502 021	Auger hole, drill only, 24"	LF	0.0581	5.31	6.38	11.70
02.3502 031	Auger hole, drill only, 36"	LF	0.0650	5.72	9.76	15.48
02.3503 000	**ROCK DRILLING, ACCESSIBLE, NO DISPOSAL:**					
02.3503 011	Rock drilling, 24", no hard rock	LF	0.2606	23.83	24.51	48.34
02.3503 021	Rock drilling, 48", no hard rock	LF	0.5211	47.64	58.88	106.52
02.3503 031	Air tool mining, with rock drilling	CY	5.8704	536.73	245.37	782.10
02.3504 000	**CAISSONS, TO 45', 150 lbs per CY REBAR, NO DISPOSAL:**					
02.3504 001	*Note: For Dewatering add $5.10/LF. Includes drill, rebar, conc, labor.*					
02.3504 011	Caisson, no casings, 18"	LF	0.3180	29.07	16.15	45.22

SITE WORK - 02

Division	Description	Unit	Labor Hours	Labor Cost	Material Cost	Total Cost
02.3504 021	Caisson, no casings, 24"	LF	0.3350	30.63	27.93	58.56
02.3504 031	Caisson, no casings, 30"	LF	0.4280	39.13	41.63	80.76
02.3504 041	Caisson, no casings, 36"	LF	0.5140	47.00	61.52	108.51
02.3504 051	Caisson, no casings, 48"	LF	0.6490	59.34	108.32	167.65
02.3504 061	Caisson, no casings, 60"	LF	0.7210	65.92	166.04	231.96
02.3504 071	Caisson, no casings, 72"	LF	1.1250	102.86	246.28	349.14
02.3504 081	Caisson, casing temporary, 18"	LF	0.5078	46.43	29.38	75.80
02.3504 091	Caisson, casing temporary, 24"	LF	0.5871	53.68	45.75	99.43
02.3504 101	Caisson, casing temporary, 30"	LF	0.7444	68.06	63.66	131.72
02.3504 111	Caisson, casing temporary, 36"	LF	0.8938	81.72	87.98	169.70
02.3504 121	Caisson, casing temporary, 48"	LF	1.1006	100.63	140.00	240.62
02.3504 131	Caisson, casing temporary, 60"	LF	1.2857	117.55	205.36	322.91
02.3504 141	Caisson, casing temporary, 72"	LF	1.7774	162.51	291.72	454.23
02.3504 151	Caisson, cased abandoned, 18"	LF	0.9721	88.88	52.98	141.86
02.3504 161	Caisson, cased abandoned, 24"	LF	1.2105	110.68	77.19	187.86
02.3504 171	Caisson, cased abandoned, 30"	LF	1.4448	132.10	110.11	242.21
02.3504 181	Caisson, cased abandoned, 36"	LF	1.6922	154.72	143.57	298.29
02.3504 191	Caisson, cased abandoned, 48"	LF	2.0195	184.64	210.05	394.69
02.3504 201	Caisson, cased abandoned, 60"	LF	2.6513	242.41	274.64	517.05
02.3504 211	Caisson, cased abandoned, 72"	LF	3.1892	291.59	390.06	681.65
02.3505 000	**BELL FOOTING, ACCESSIBLE, NO DISPOSAL:**					
02.3505 001	Note: Adjust prices where unusual hazards exist.					
02.3505 011	Bell footing, 24", 5' shaft, 5' bell	EA	1.4630	133.76	112.56	246.32
02.3505 021	Add for concrete filled, above item	EA	0.4064	31.70	367.19	398.89
02.3505 031	Bell footing, 30", 6' shaft, 6' bell	EA	2.0051	183.33	156.66	339.99
02.3505 041	Add for concrete filled, above item	EA	0.4472	34.88	411.33	446.21
02.3505 051	Bell footing, 36", 10' shaft, 8' bell	EA	2.5605	234.11	171.34	405.45
02.3505 061	Add for concrete fill, above item	EA	3.2916	300.95	1,410.44	1,711.39
02.4000 000	**SHORING & BULKHEADING:**					
02.4001 000	**SHEET PILING:**					
02.4001 001	Note: Sheet piling is manufactured in many shapes and sizes. Gauges 7 through 12 are normally considered lightweight and are used primarily for trenching and stabilizing conditions. Material costs per square foot can be estimated based on weight and installation, with adjustments for unusual conditions. With lightweight sheet piling, allow for square set bracing as required.					
02.4001 011	Steel sheet piles, left in place, average	SF	0.0670	6.07	29.21	35.29
02.4001 021	Steel sheet piles, pull & save, average	SF	0.0445	4.03	7.62	11.65
02.4001 031	Steel sheet piles, 27#(Z-27), in place	SF	0.0849	7.69	18.76	26.46
02.4001 041	Steel sheet piles, 27#, recycled	SF	0.1410	12.78	8.83	21.61
02.4001 051	Lightweight steel sheet piles, 9 ga, 86#	SF	0.1251	11.34	11.42	22.76
02.4001 061	Lightweight steel sheet piles, 12 ga, 654#	SF	0.1072	9.72	12.56	22.28
02.4001 071	Wood safety sheet pile to 8', braces	SF	0.1015	9.20	3.16	12.36
02.4001 081	Wood safety sheet pile to 12', braces	SF	0.1264	11.46	5.40	16.86
02.4001 091	Wood safety sheet pile to 16', braces	SF	0.1331	12.06	6.28	18.34
02.4002 000	**BULKHEADING & TIEBACK WALLS:**					
02.4002 001	Note: The following items include soldier piles 8' average space, 3" lagging, 2 rows of raker bracing. Hydrostatic heads are not included.					
02.4002 011	Tie back wall, 10'-15', complete	SFWA	0.1881	17.05	30.14	47.18
02.4002 021	Tie back wall, 16'-20', complete	SFWA	0.2025	18.27	32.45	50.72
02.4002 031	Tie back wall, 21'-25', complete	SFWA	0.2170	19.57	34.75	54.32
02.4002 041	Tie back wall, 26'-35', complete	SFWA	0.2459	22.18	39.45	61.63
02.4002 051	Tie back wall, 36'-45', complete	SFWA	0.3327	30.01	53.26	83.27
02.4002 061	Tie backs, drilled, plug, with tie	LF	0.2894	26.10	48.15	74.25
02.4002 071	Rakers, steel	#	0.0116	1.05	1.86	2.90
02.4003 000	**SLURRY TRENCHING:**					
02.4003 001	Note: Prices do not include mats for rigs, traffic controls, bridging, resurfacing or unusual access problems.					
02.4003 011	3,000 psi concrete, 36" trench	SF/TR	0.5572	50.26	114.87	165.13

02 - SITE WORK

Division	Description	Unit	Labor Hours	Labor Cost	Material Cost	Total Cost
02.4003 021	Disp waste excavation, short haul	CY	0.1116	10.07	22.23	32.30
02.4004 000	**PRESSURE INJECTED FOUNDATIONS:**					
02.4004 011	Pressure injected foundation, max 25'x18'x75 T, uncased	VF	0.3038	27.40	48.67	76.07
02.4004 021	Pressure injected foundation, max 25'x24'x150 T, uncased	VF	0.3761	33.92	60.26	94.18
02.4004 031	Pressure injected foundation, max 30'x14'x75 T, cased	VF	0.2749	24.80	44.05	68.84
02.4005 000	**DEWATERING:**					
02.4005 011	Deep well system, 12", cased & graded	VF	0.4774	48.41	76.52	124.93
02.4006 000	**WELL POINTS:**					
02.4006 001	Note: Well point spacing, header and pump size are determined by the expected flow. The following prices assume a 2" well point system at 5' on center, and must be used based on the lineal feet of perimeter of excavation. Prices include in and out costs and an operator 12 hrs per day.					
02.4006 011	Header system to 1000', 1 month	LF	0.4162	42.21	15.23	57.44
02.4006 021	Header system, to 1000', 2 month	LF	0.7347	74.51	27.57	102.07
02.4006 031	Header system, to 1000', 3 month	LF	1.0528	106.76	40.02	146.79
02.4006 041	Header system, to 1000', 4 month	LF	1.3709	139.02	52.34	191.37
02.4006 051	Header system, to 1000', 5 month	LF	1.6880	171.18	64.72	235.90
02.4006 061	Header system, to 1000', 6 month	LF	2.0061	203.44	77.09	280.52
02.4006 071	Header system, over 1000', 1 month	LF	0.2544	25.80	13.32	39.12
02.4006 081	Header system, over 1000', 2 month	LF	0.4265	43.25	19.16	62.42
02.4006 091	Header system, over 1000', 3 month	LF	0.5987	60.71	25.08	85.79
02.4006 101	Header system, over 1000', 4 month	LF	0.7707	78.16	30.99	109.15
02.4006 111	Header system, over 1000', 5 month	LF	0.9431	95.64	36.95	132.59
02.4006 121	Header system, over 1000', 6 month	LF	1.1151	113.08	42.79	155.87
02.4007 000	**UNDERPINNING:**					
02.4007 011	Underpinning, hand mining	CF	0.1034	10.49		10.49
02.4007 021	Underpinning, form, one side	SFFA	0.0422	4.28	6.66	10.94
02.4007 031	Underpinning, 35m# low slump concrete in place	CF	0.4319	43.80	25.71	69.51
02.4007 041	Underpinning, dry pack	CF	0.3472	35.21	55.62	90.83
02.4007 051	Underpinning, temp shoring, 4' oc	SFFA	0.1198	12.15	19.06	31.21
02.4007 061	Underpinning, perm shoring, 4' oc	SFFA	0.2170	22.01	34.75	56.75
02.4007 071	Water line blow off assem, 2"	EA	3.1700	331.99	769.23	1,101.22
02.4007 081	Valve, air release comb assem, 1"	EA	4.5000	456.35	941.10	1,397.44
02.4008 000	**PERMANENT DEWATERING SYSTEM:**					
02.4008 001	Note: A Permanent dewatering system consists of a network of perforated pipes, wrapped in filter fabric and installed in a granular base under the slab. The pipes are in a grid in both directions under the slab. They are connected to laterals that drain into the storm drainage system.					
02.4008 011	Permanent dewatering system, under slab	SF	0.0669	6.78	6.70	13.48
02.5000 000	**SITE UTILITIES:**					
02.5000 001	Note: Excavation and backfill are included with site utilities when applicable, with the exception of conduit. No dewatering, shieldwork, clearing or demolition is included. All trenching is figured to cover the top of pipe by 48". Spreaders are allowed for trenches over 5' deep. No off-site disposal of excess is included. Only conduit is included for electrical runs. See section 16.0000 For wire. Fitting allowance is included with pipe and conduit pricing. Equipment on and off charges are included for average jobs. These prices should be increased for small jobs or where trenching is over 6' deep.					
02.5001 000	**PIPE JACKING (ADD FOR CASING AND/OR PIPE):**					
02.5001 011	Pipe jacking, 2"	LF	0.2936	30.75	19.58	50.33
02.5001 021	Pipe jacking, 3"	LF	0.3375	35.35	23.54	58.89
02.5001 031	Pipe jacking, 4"	LF	0.4585	48.02	30.64	78.66
02.5001 041	Pipe jacking, 6"	LF	0.5744	60.16	38.32	98.47
02.5001 051	Pipe jacking, 8"	LF	0.7759	81.26	51.90	133.16
02.5001 061	Pipe jacking, 10"	LF	1.1289	118.23	75.52	193.75
02.5001 071	Pipe jacking, 12"	LF	1.4399	150.80	93.36	244.16
02.5001 081	Pipe jacking, 16"	LF	1.6949	177.51	110.01	287.52
02.5001 091	Pipe jacking, 18"	LF	1.9185	200.92	124.65	325.57
02.5001 101	Pipe jacking, 24"	LF	2.2091	231.36	143.52	374.88

SITE WORK - 02

Division	Description	Unit	Labor Hours	Labor Cost	Material Cost	Total Cost
02.5001 111	Pipe jacking, 30"	LF	2.8764	301.25	187.18	488.42
02.5001 121	Pipe jacking, 36"	LF	3.4048	356.58	221.44	578.02
02.5001 131	Pipe jacking, 42"	LF	3.7671	394.53	244.66	639.19
02.5001 141	Pipe jacking, 48"	LF	4.1643	436.13	270.37	706.50
02.5001 151	Pipe jacking, 54"	LF	4.5183	473.20	293.36	766.56
02.5001 161	Pipe jacking, 60"	LF	5.3179	556.94	345.84	902.78
02.5001 171	Pipe jacking, 66"	LF	5.5621	582.52	361.63	944.15
02.5001 181	Pipe jacking, 72"	LF	6.1944	648.74	402.90	1,051.64
02.5001 191	Pipe jacking, 84"	LF	7.4994	785.41	485.73	1,271.14
02.5100 000	**STORM DRAINAGE & SANITARY SEWER PIPE:**					
02.5102 000	PIPE, PERFORATED & SOLID, CORRUGATED POLYETHYLENE UNDER DRAIN:					
02.5102 001	Note: The following prices do not include trenching and backfill. For filter cloth on sizes 3" x 8" add 10% to material costs. For filter cloth on sizes 10" to 12" add 15% to the material costs.					
02.5102 011	Corrugated polyetheye pipe, perforated & solid, 3", under drain	LF	0.0880	8.92	0.47	9.40
02.5102 021	Corrugated polyetheye pipe, perforated & solid, 4", under drain	LF	0.0902	9.15	0.63	9.78
02.5102 031	Corrugated polyetheye pipe, perforated & solid, 6", under drain	LF	0.0938	9.51	1.44	10.95
02.5102 041	Corrugated polyetheye pipe, perforated & solid, 8", under drain	LF	0.0964	9.78	2.58	12.36
02.5102 051	Corrugated polyetheye pipe, perforated & solid, 10", under drain	LF	0.1005	10.19	4.98	15.17
02.5102 061	Corrugated polyetheye pipe, perforated & solid, 12", under drain	LF	0.1035	10.50	7.10	17.59
02.5103 000	CAST IRON PIPE, SOIL, SERVICE WEIGHT, 1 HUB, WITH TRENCHING:					
02.5103 011	Cast iron pipe, soil, 1 hub, 1-1/2", service weight	LF	0.1636	16.59	9.95	26.54
02.5103 021	Cast iron pipe, soil, 1 hub, 2", service weight	LF	0.1688	17.12	10.97	28.09
02.5103 031	Cast iron pipe, soil, 1 hub, 3", service weight	LF	0.1800	18.25	13.36	31.61
02.5103 041	Cast iron pipe, soil, 1 hub, 4", service weight	LF	0.1883	19.10	17.72	36.82
02.5103 051	Cast iron pipe, soil, 1 hub, 5", service weight	LF	0.2209	22.40	25.05	47.45
02.5103 061	Cast iron pipe, soil, 1 hub, 6", service weight	LF	0.2537	25.73	30.67	56.39
02.5103 071	Cast iron pipe, soil, 1 hub, 8", service weight	LF	0.2918	29.59	59.04	88.63
02.5103 081	Cast iron pipe, soil, 1 hub, 10", service weight	LF	0.3169	32.14	77.29	109.43
02.5104 000	CAST IRON SOIL PIPE, NO HUB, SERVICE WEIGHT, WITH TRENCHING:					
02.5104 001	Note: For extra heavy cast iron pipe, add 35% to the material costs.					
02.5104 011	Cast iron pipe, no hub, service weight, 2"	LF	0.1686	17.10	11.39	28.48
02.5104 021	Cast iron pipe, no hub, service weight, 3"	LF	0.1854	18.80	12.83	31.63
02.5104 031	Cast iron pipe, no hub, service weight, 4"	LF	0.1939	20.77	17.15	37.92
02.5104 041	Cast iron pipe, no hub, service weight, 5"	LF	0.2275	24.37	26.71	51.09
02.5104 051	Cast iron pipe, no hub, service weight, 6"	LF	0.2613	28.00	31.48	59.48
02.5104 061	Cast iron pipe, no hub, service weight, 8"	LF	0.3005	32.20	56.50	88.70
02.5104 071	Cast iron pipe, no hub, service weight, 10"	LF	0.3262	34.95	81.30	116.25
02.5105 000	CONCRETE PIPE, NON-REINFORCED, WITH TRENCHING:					
02.5105 011	Concrete pipe, non-reinforced, 6"	LF	0.2112	21.42	12.78	34.20
02.5105 021	Concrete pipe, non-reinforced, 8"	LF	0.2207	22.38	13.98	36.36
02.5105 031	Concrete pipe, non-reinforced, 10"	LF	0.2303	23.35	16.06	39.42
02.5105 041	Concrete pipe, non-reinforced, 12"	LF	0.2399	24.33	17.95	42.28
02.5105 051	Concrete pipe, non-reinforced, 15"	LF	0.2784	28.23	22.33	50.57
02.5105 061	Concrete pipe, non-reinforced, 18"	LF	0.3551	36.01	26.71	62.72
02.5105 071	Concrete pipe, non-reinforced, 21"	LF	0.3935	39.90	31.07	70.97
02.5105 081	Concrete pipe, non-reinforced, 24"	LF	0.4511	45.75	35.39	81.14
02.5106 000	CONCRETE PIPE, REINFORCED, CLASS 3, WITH GASKETS, TRENCHING:					
02.5106 001	Note: For class 2 pipe, deduct 5% from the material costs. For class 4, add 10%.					
02.5106 011	Concrete pipe, reinf, class 3, gaskets, 12"	LF	0.2617	26.54	22.98	49.52
02.5106 021	Concrete pipe, reinf, class 3, gaskets, 15"	LF	0.3219	32.64	32.56	65.20
02.5106 031	Concrete pipe, reinf, class 3, gaskets, 18"	LF	0.4000	40.56	42.04	82.60
02.5106 041	Concrete pipe, reinf, class 3, gaskets, 21"	LF	0.4398	44.60	51.52	96.12
02.5106 051	Concrete pipe, reinf, class 3, gaskets, 24"	LF	0.5058	51.29	61.05	112.35
02.5106 061	Concrete pipe, reinf, class 3, gaskets, 27"	LF	0.5864	59.47	76.35	135.81
02.5106 071	Concrete pipe, reinf, class 3, gaskets, 30"	LF	0.5967	60.51	91.63	152.14

02 - SITE WORK

Division	Description	Unit	Labor Hours	Labor Cost	Material Cost	Total Cost
02.5106 081	Concrete pipe, reinf, class 3, gaskets, 33"	LF	0.6282	63.71	106.90	170.60
02.5106 091	Concrete pipe, reinf, class 3, gaskets, 36"	LF	0.6699	67.93	122.17	190.10
02.5106 101	Concrete pipe, reinf, class 3, gaskets, 42"	LF	0.7224	73.26	137.48	210.74
02.5106 111	Concrete pipe, reinf, class 3, gaskets, 48"	LF	0.7852	79.63	152.73	232.36
02.5106 121	Concrete pipe, reinf, class 3, gaskets, 54"	LF	0.8312	84.29	172.22	256.51
02.5106 131	Concrete pipe, reinf, class 3, gaskets, 60"	LF	0.9112	92.40	191.39	283.80
02.5106 141	Concrete pipe, reinf, class 3, gaskets, 66"	LF	1.0215	103.59	210.52	314.11
02.5106 151	Concrete pipe, reinf, class 3, gaskets, 72"	LF	1.1017	118.04	229.62	347.65
02.5106 161	Concrete pipe, reinf, class 3, gaskets, 84"	LF	1.2618	135.19	267.92	403.11
02.5107 000	**CONCRETE PIPE, CAST-IN-PLACE:**					
02.5107 001	*Note: The following prices include excavation, installation and backfill. Paving and grading is not included.*					
02.5107 011	Concrete pipe, cast-in-place, 18"-24"	LF	0.2299	23.31	22.16	45.48
02.5107 021	Concrete pipe, cast-in-place, 30"	LF	0.2757	29.54	26.60	56.14
02.5107 031	Concrete pipe, cast-in-place, 36"	LF	0.3216	34.46	31.05	65.50
02.5107 041	Concrete pipe, cast-in-place, 42"	LF	0.3828	41.01	36.58	77.59
02.5107 051	Concrete pipe, cast-in-place, 48"	LF	0.5217	55.89	46.57	102.46
02.5107 061	Concrete pipe, cast-in-place, 60"	LF	0.6832	73.20	61.01	134.21
02.5107 071	Concrete pipe, cast-in-place, 72"	LF	0.8695	88.18	77.66	165.84
02.5107 081	Concrete pipe, cast-in-place, 84"	LF	1.1179	113.37	99.86	213.22
02.5108 000	**CORRUGATED METAL PIPE, BITUMINOUS COAT, 24" COVER, WITH TRENCH:**					
02.5108 011	Corrugated metal pipe, 10", 16 ga, bitum coat, 24" cover	LF	0.1345	14.41	14.42	28.83
02.5108 021	Corrugated metal pipe, 12", 16 ga, bitum coat, 24" cover	LF	0.1823	19.53	17.62	37.15
02.5108 031	Corrugated metal pipe, 15", 16 ga, bitum coat, 24" cover	LF	0.2112	22.63	21.38	44.01
02.5108 041	Corrugated metal pipe, 18", 16 ga, bitum coat, 24" cover	LF	0.2495	25.30	25.04	50.34
02.5108 051	Corrugated metal pipe, 24", 16 ga, bitum coat, 24" cover	LF	0.3168	32.13	32.34	64.46
02.5108 061	Corrugated metal pipe, 30", 16 ga, bitum coat, 24" cover	LF	0.3742	37.95	32.34	70.28
02.5108 071	Corrugated metal pipe, 15", 14 ga, bitum coat, 24" cover	LF	0.2207	22.38	24.68	47.07
02.5108 081	Corrugated metal pipe, 18", 14 ga, bitum coat, 24" cover	LF	0.2495	25.30	29.01	54.31
02.5108 091	Corrugated metal pipe, 24", 14 ga, bitum coat, 24" cover	LF	0.3168	32.13	37.60	69.73
02.5108 101	Corrugated metal pipe, 30", 14 ga, bitum coat, 24" cover	LF	0.3840	36.63	46.50	83.13
02.5108 111	Corrugated metal pipe, 30", 12 ga, bitum coat, 24" cover	LF	0.3935	37.53	58.45	95.98
02.5108 121	Corrugated metal pipe, 36", 12 ga, bitum coat, 24" cover	LF	0.4799	45.77	69.95	115.72
02.5108 131	Corrugated metal pipe, 48", 12 ga, bitum coat, 24" cover	LF	0.5567	53.10	93.80	146.89
02.5108 141	Corrugated metal pipe, 60", 12 ga, bitum coat, 24" cover	LF	0.8446	80.56	116.12	196.68
02.5108 151	Corrugated metal pipe, 72", 10 ga, bitum coat, 24" cover	LF	1.0269	97.95	168.61	266.56
02.5109 000	**CORRUGATED METAL PIPE, GALVANIZED, 24" COVER, WITH TRENCHING:**					
02.5109 001	*Note: For aluminum pipe, deduct 15% from the labor costs.*					
02.5109 011	Corrugated metal pipe, 8", 16 ga, galv, 24" cover	LF	0.1823	18.49	9.30	27.79
02.5109 021	Corrugated metal pipe, 10", 16 ga, galv, 24" cover	LF	0.1919	19.46	10.30	29.76
02.5109 031	Corrugated metal pipe, 12", 16 ga, galv, 24" cover	LF	0.2016	21.60	11.41	33.01
02.5109 041	Corrugated metal pipe, 15", 16 ga, galv, 24" cover	LF	0.2313	24.78	12.69	37.47
02.5109 051	Corrugated metal pipe, 18", 16 ga, galv, 24" cover	LF	0.2613	28.00	15.79	43.78
02.5109 061	Corrugated metal pipe, 24", 14 ga, galv, 24" cover	LF	0.3287	35.22	19.13	54.35
02.5109 071	Corrugated metal pipe, 30", 14 ga, galv, 24" cover	LF	0.3793	40.64	26.61	67.25
02.5109 081	Corrugated metal pipe, 36", 14 ga, galv, 24" cover	LF	0.4871	52.19	37.79	89.98
02.5109 091	Corrugated metal pipe, 42", 14 ga, galv, 24" cover	LF	0.5057	54.18	41.59	95.77
02.5109 101	Corrugated metal pipe, 48", 14 ga, galv, 24" cover	LF	0.5883	63.03	48.10	111.13
02.5109 111	Corrugated metal pipe, 54", 12 ga, galv, 24" cover	LF	0.7291	78.12	74.31	152.42
02.5109 121	Corrugated metal pipe, 60", 12 ga, galv, 24" cover	LF	0.9273	99.35	77.66	177.01
02.5109 131	Corrugated metal pipe, 66", 12 ga, galv, 24" cover	LF	0.9694	103.86	84.62	188.48
02.5109 141	Corrugated metal pipe, 72", 10 ga, galv, 24" cover	LF	1.1126	112.83	116.24	229.07
02.5109 151	Corrugated metal pipe, 78", 8 ga, galv, 24" cover	LF	1.1568	123.94	172.47	296.41
02.5109 161	Corrugated metal pipe, 84", 8 ga, galv, 24" cover	LF	1.2086	129.49	194.24	323.73
02.5109 171	Corrugated metal pipe, 96", 8 ga, galv, 24" cover	LF	1.2612	135.12	221.97	357.09

SITE WORK - 02

Division	Description	Unit	Labor Hours	Labor Cost	Material Cost	Total Cost
02.5109 181	Corrugated metal pipe, 120", 8 ga, galv, 24" cover	LF	1.4532	155.70	281.24	436.93
02.5110 000	**PIPE, ABS, PLASTIC, WITH TRENCHING:**					
02.5110 011	Pipe, ABS, plastic, 2"	LF	0.1702	17.26	10.51	27.77
02.5110 021	Pipe, ABS, plastic, 3"	LF	0.1784	18.09	15.30	33.40
02.5110 031	Pipe, ABS, plastic, 4"	LF	0.1980	20.08	26.29	46.37
02.5110 041	Pipe, ABS, plastic, 6"	LF	0.2247	24.07	43.33	67.41
02.5110 051	Pipe Tap, ABS, 4"	EA	0.9200	80.96	217.62	298.58
02.5110 061	Pipe Tap, ABS, 6"	EA	1.3500	118.80	290.19	408.99
02.5110 071	Pipe Tap, PVC gravity sewer, 4", ASTM D3034	EA	0.9255	81.44	218.06	299.50
02.5110 081	Pipe Tap, PVC gravity sewer, 6", ASTM D3034	EA	1.3574	119.45	286.91	406.36
02.5110 091	Pipe Tap, PVC gravity sewer, 8", ASTM D3034	EA	1.8510	162.89	608.30	771.19
02.5110 101	Pipe Tap, PVC gravity sewer, 10", ASTM D3034	EA	2.4679	217.18	803.43	1,020.60
02.5110 111	Pipe Tap, PVC gravity sewer, 12", ASTM D3034	EA	3.0849	271.47	1,113.31	1,384.78
02.5110 121	Pipe Tap, PVC gravity sewer, 15", ASTM D3034	EA	4.3189	380.06	1,492.08	1,872.14
02.5111 000	**PIPE, PVC GRAVITY SEWER, ASTM D-3034, WITH TRENCH:**					
02.5111 011	Pipe, PVC gravity sewer, 4", ASTM D3034	LF	0.2026	20.55	10.37	30.92
02.5111 021	Pipe, PVC gravity sewer, 6", ASTM D3034	LF	0.2261	22.93	12.71	35.64
02.5111 031	Pipe, PVC gravity sewer, 8", ASTM D3034	LF	0.2512	25.47	16.18	41.65
02.5111 041	Pipe, PVC gravity sewer 10", ASTM D3034	LF	0.3527	35.77	20.87	56.64
02.5111 051	Pipe, PVC gravity sewer, 12", ASTM D3034	LF	0.4550	46.14	30.01	76.15
02.5111 061	Pipe, PVC gravity sewer, 15", ASTM D3034	LF	0.5487	55.64	53.73	109.37
02.5112 000	**VITRIFIED CLAY PIPE, RING SEAL, WITH TRENCHING:**					
02.5112 011	Clay pipe, 6", ring seal	LF	0.2402	24.36	10.27	34.63
02.5112 021	Clay pipe, 8", ring seal	LF	0.2486	25.21	15.33	40.54
02.5112 031	Clay pipe, 10", ring seal	LF	0.3161	32.06	24.58	56.64
02.5112 041	Clay pipe, 12", ring seal	LF	0.2740	27.79	31.16	58.95
02.5112 051	Clay pipe, 15", ring seal	LF	0.2908	29.49	58.73	88.22
02.5112 061	Clay pipe, 18", ring seal	LF	0.3667	37.19	84.86	122.05
02.5112 071	Clay pipe, 21", ring seal	LF	0.4130	41.88	96.46	138.34
02.5112 081	Clay pipe, 24", ring seal	LF	0.4847	49.15	135.43	184.59
02.5112 091	Clay pipe, 27", ring seal	LF	0.5647	57.27	159.06	216.32
02.5113 000	**PIPE, STORM DRAINAGE, PVC SDR-35, WITH TRENCHING:**					
02.5113 011	Pipe, PVC, SDR-35, 4"	LF	0.2210	22.41	5.12	27.53
02.5113 021	Pipe, PVC, SDR-35, 6"	LF	0.2467	25.02	9.05	34.06
02.5113 031	Pipe, PVC, SDR-35, 8"	LF	0.2716	27.54	15.78	43.32
02.5113 041	Pipe, PVC, SDR-35, 10"	LF	0.3848	39.02	28.22	67.25
02.5113 051	Pipe, PVC, SDR-35, 12"	LF	0.4964	50.34	38.82	89.16
02.5113 061	Pipe, PVC, SDR-35, 15"	LF	0.5986	60.70	60.25	120.95
02.5113 071	Pipe, PVC, SDR-35, 18"	LF	0.7208	73.10	99.86	172.95
02.5113 081	Pipe, PVC, SDR-35, 21"	LF	0.8905	90.31	155.69	246.00
02.5113 091	Pipe, PVC, SDR-35, 24"	LF	1.1858	120.25	216.99	337.24
02.5113 101	Pipe Tap, PVC, SDR-35, 4"	EA	0.9255	81.44	218.06	299.50
02.5113 111	Pipe Tap, PVC, SDR-35, 6"	EA	1.3574	119.45	286.91	406.36
02.5113 121	Pipe Tap, PVC, SDR-35, 8"	EA	1.8510	162.89	608.30	771.19
02.5113 131	Pipe Tap, PVC, SDR-35, 10"	EA	2.4679	217.18	803.43	1,020.60
02.5113 141	Pipe Tap, PVC, SDR-35, 12"	EA	3.0849	271.47	1,113.31	1,384.78
02.5113 151	Pipe Tap, PVC, SDR-35, 15"	EA	4.3189	380.06	1,492.08	1,872.14
02.5300 000	**WATER, STEAM & GAS DISTRIBUTION PIPING:**					
02.5302 000	**ACID WASTE PIPE, POLYPROPYLENE, WITH TRENCH:**					
02.5302 011	Polypropylene acid waste pipe, 2", with trench	LF	0.1702	17.26	17.67	34.93
02.5302 021	Polypropylene acid waste pipe, 3", with trench	LF	0.1784	18.09	22.08	40.17
02.5302 031	Polypropylene acid waste pipe, 4", with trench	LF	0.1980	20.08	26.51	46.58
02.5302 041	Polypropylene acid waste pipe, 6", with trench	LF	0.2247	22.79	45.41	68.19
02.5303 000	**COPPER TUBING, TYPE K HARD, WITH TRENCH:**					
02.5303 001	*Note: For type 'I', deduct 20% from the material costs.*					
02.5303 011	Copper pipe, 'K' hard, 1/2", with trench	LF	0.0695	7.05	12.57	19.62

02 - SITE WORK

Division	Description	Unit	Labor Hours	Labor Cost	Material Cost	Total Cost
02.5303 021	Copper pipe, 'K' hard, 3/4", with trench	LF	0.0800	8.11	17.91	26.02
02.5303 031	Copper pipe, 'K' hard, 1", with trench	LF	0.1432	14.52	21.95	36.48
02.5303 041	Copper pipe, 'K' hard, 1-1/4", with trench	LF	0.1476	14.97	26.63	41.60
02.5303 051	Copper pipe, 'K' hard, 1-1/2", with trench	LF	0.1769	17.94	32.37	50.31
02.5303 061	Copper pipe, 'K' hard, 2", with trench	LF	0.1854	18.80	49.20	68.00
02.5303 071	Copper pipe, 'K' hard, 3", with trench	LF	0.2116	21.46	88.99	110.45
02.5303 081	Copper pipe, 'K' hard, 4", with trench	LF	0.2018	20.46	137.06	157.52
02.5303 091	Copper pipe, 'K' hard, 5", with trench	LF	0.2284	23.16	227.87	251.03
02.5303 101	Copper pipe, 'K' hard, 6", with trench	LF	0.2908	29.49	334.88	364.37
02.5303 111	Pipe Tap, Copper pipe, K, L, M Types hard, 4"	EA	1.2500	110.00	381.75	491.75
02.5304 000	**PVC, SCHEDULE 40, SOLVENT WELD, WITH TRENCH:**					
02.5304 001	*Note: For schedule 80, add 26% to the material costs.*					
02.5304 011	PVC, schedule 40, 1/2", weld, with trench	LF	0.0517	5.54	1.05	6.59
02.5304 021	PVC, schedule 40, 3/4", weld, with trench	LF	0.0679	7.27	1.44	8.72
02.5304 031	PVC, schedule 40, 1", weld, with trench	LF	0.0808	8.66	2.14	10.80
02.5304 041	PVC, schedule 40, 1-1/2", weld, with trench	LF	0.0932	9.99	3.46	13.44
02.5304 051	PVC, schedule 40, 2", weld, with trench	LF	0.1031	11.05	4.53	15.58
02.5304 061	PVC, schedule 40, 2-1/2", weld, with trench	LF	0.1166	12.49	7.23	19.72
02.5304 071	PVC, schedule 40, 3", weld, with trench	LF	0.1312	14.06	8.55	22.61
02.5304 081	PVC, schedule 40, 4", weld, with trench	LF	0.1420	15.21	11.26	26.47
02.5305 000	**PVC, MUNI WATER, AWWA C900, CLASS 150, WITH TRENCHING:**					
02.5305 011	PVC municipal water pipe, 4", '150', C900	LF	0.1468	15.73	27.42	43.14
02.5305 021	PVC municipal water pipe, 6", '150', C900	LF	0.1652	17.70	33.28	50.98
02.5305 031	PVC municipal water pipe, 8", '150', C900	LF	0.1823	19.53	43.96	63.50
02.5305 041	PVC municipal water pipe, 10", '150', C900	LF	0.1919	20.56	57.76	78.32
02.5305 051	PVC municipal water pipe, 12", '150', C900	LF	0.2016	21.60	71.13	92.73
02.5305 061	Pipe Tap, municipal water pipe, 4"',150',C900	EA	0.9255	81.44	218.06	299.50
02.5305 071	Pipe Tap, municipal water pipe, 6"',150',C900	EA	1.3574	119.45	286.91	406.36
02.5305 081	Pipe Tap, municipal water pipe, 8"',150',C900	EA	1.8510	162.89	608.30	771.19
02.5305 091	Pipe Tap, municipal water pipe, 10"',150',C900	EA	2.4679	217.18	803.43	1,020.60
02.5305 101	Pipe Tap, municipal water pipe, 12"',150',C900	EA	3.0849	271.47	1,113.31	1,384.78
02.5305 111	Hot tap for 4" water pipe	EA	15.2100	1,338.48	709.05	2,047.53
02.5305 161	Thrust blocks for 4" water pipe freezing	EA	1.2000	105.60	81.63	187.23
02.5305 171	Thrust blocks for 6" water pipe freezing	EA	2.5000	220.00	163.24	383.24
02.5305 181	Thrust blocks for 8" water pipe freezing	EA	4.0000	352.00	187.06	539.06
02.5305 191	Thrust blocks for 10" water pipe freezing	EA	5.0000	440.00	238.04	678.04
02.5305 201	Thrust blocks for 12" water pipe freezing	EA	7.5000	660.00	340.07	1,000.07
02.5305 121	Hot tap for 6" water pipe	EA	20.8700	1,836.56	962.22	2,798.78
02.5305 131	Hot tap for 8" water pipe	EA	24.1700	2,126.96	1,015.12	3,142.08
02.5305 141	Hot tap for 10" water pipe	EA	29.4200	2,588.96	1,880.76	4,469.72
02.5305 151	Hot tap for 12" water pipe	EA	38.6800	3,403.84	2,776.51	6,180.35
02.5306 000	**DUCTILE IRON PIPE, WITH EXCAVATION & FILL:**					
02.5306 001	*Note: For class 52 pipe, sizes 3" to 4" add 7% to the material costs. For sizes 6" to 24" add 15%*					
02.5306 011	Ductile iron pipe, 350 psi, class 50, 3", excavate & fill	LF	0.2192	23.49	19.79	43.27
02.5306 021	Ductile iron pipe, 350 psi, class 50, 4", excavate & fill	LF	0.2192	23.49	20.80	44.29
02.5306 031	Ductile iron pipe, 350 psi, class 50, 6", excavate & fill	LF	0.2760	29.57	22.47	52.04
02.5306 041	Ductile iron pipe, 350 psi, class 50, 8", excavate & fill	LF	0.2760	29.57	24.86	54.43
02.5306 051	Ductile iron pipe, 350 psi, class 50, 10", excavate & fill	LF	0.3128	33.51	30.57	64.09
02.5306 061	Ductile iron pipe, 350 psi, class 50, 12", excavate & fill	LF	0.3956	42.38	37.21	79.60
02.5306 071	Ductile iron pipe, 350 psi, class 50, 14", excavate & fill	LF	0.4882	52.31	44.62	96.93
02.5306 081	Ductile iron pipe, 350 psi, class 50, 16", excavate & fill	LF	0.5251	56.26	51.11	107.37
02.5306 091	Ductile iron pipe, 350 psi, class 50, 18", excavate & fill	LF	0.6176	66.17	57.97	124.14
02.5306 101	Ductile iron pipe, 350 psi, class 50, 20", excavate & fill	LF	0.6843	73.32	65.15	138.46
02.5306 111	Ductile iron pipe, 350 psi, class 50, 24", excavate & fill	LF	0.6843	73.32	80.28	153.59
02.5306 121	Ductile iron pipe, 350 psi, class 50, 30", excavate & fill	LF	0.6843	73.32	106.02	179.34
02.5306 131	Ductile iron pipe, 350 psi, class 50, 36", excavate & fill	LF	0.7258	77.76	134.50	212.26

SITE WORK - 02

Division	Description	Unit	Labor Hours	Labor Cost	Material Cost	Total Cost
02.5306 141	Ductile iron pipe, 350 psi, class 50, 42", excavate & fill	LF	0.7826	83.85	166.04	249.89
02.5306 151	Ductile iron pipe, 350 psi, class 50, 48", excavate & fill	LF	0.8507	91.14	200.40	291.55
02.5306 161	Ductile iron pipe, 350 psi, class 50, 54", excavate & fill	LF	0.9005	96.48	235.37	331.85
02.5306 171	Pipe Tap, Ductile iron pipe, 350psi, class 50, 4"	EA	3.2083	282.33	303.33	585.66
02.5306 181	Pipe Tap, Ductile iron pipe, 350psi, class 50, 6"	EA	3.2083	282.33	752.27	1,034.60
02.5306 191	Pipe Tap, Ductile iron pipe, 350psi, class 50, 8"	EA	5.6146	494.08	837.18	1,331.27
02.5306 201	Pipe Tap, Ductile iron pipe, 350psi, class 50, 10"	EA	5.6146	494.08	958.52	1,452.61
02.5306 211	Pipe Tap, Ductile iron pipe, 350psi, class 50, 12"	EA	8.8228	776.41	1,176.90	1,953.31
02.5307 000	**STEEL PIPE, GALVANIZED, SCH 40, TREADED & CUT, WITH TRENCHING:**					
02.5307 011	Steel pipe, galvanized, 1/2", schedule 40, threaded & cut	LF	0.0767	8.03	3.08	11.11
02.5307 021	Steel pipe, galvanized, 3/4", schedule 40, threaded & cut	LF	0.0863	9.04	3.45	12.48
02.5307 031	Steel pipe, galvanized, 1", schedule 40, threaded & cut	LF	0.0961	10.06	4.34	14.41
02.5307 041	Steel pipe, galvanized, 1-1/4", schedule 40, threaded & cut	LF	0.1056	11.06	5.53	16.59
02.5307 051	Steel pipe, galvanized, 1-1/2", schedule 40, threaded & cut	LF	0.1151	12.05	6.59	18.65
02.5307 061	Steel pipe, galvanized, 2", schedule 40, threaded & cut	LF	0.1345	14.09	8.68	22.76
02.5307 071	Steel pipe, galvanized, 2-1/2", schedule 40, threaded & cut	LF	0.1535	16.08	12.20	28.28
02.5307 081	Steel pipe, galvanized, 3", schedule 40, threaded & cut	LF	0.1728	18.10	15.79	33.89
02.5307 091	Steel pipe, galvanized, 4", schedule 40, threaded & cut	LF	0.2112	22.12	22.97	45.09
02.5307 101	Steel pipe, galvanized, 5", schedule 40, threaded & cut	LF	0.2303	24.12	33.20	57.32
02.5307 111	Steel pipe, galvanized, 6", schedule 40, threaded & cut	LF	0.2399	25.12	40.09	65.22
02.5307 121	Pipe Tap, Steel pipe galv., 4",sched 40, threaded & cut	EA	2.0595	215.69	177.31	393.00
02.5307 131	Pipe Tap, Steel pipe galv., 5",sched 40, threaded & cut	EA	2.5743	269.61	249.90	519.51
02.5307 141	Pipe Tap, Steel pipe galv., 6",sched 40, threaded & cut	EA	2.5743	269.61	382.65	652.25
02.5308 000	**STEEL PIPE, BLACK, SCH 40, WELDED, TREADED & CUT, WITH TRENCHING:**					
02.5308 001	*Note: For schedule 80, add 39% to the material costs.*					
02.5308 011	Steel pipe, black, A-120, schedule 40, weld, 1/2", threaded & cut	LF	0.0767	8.03	3.17	11.20
02.5308 021	Steel pipe, black, A-120, schedule 40, weld, 3/4", threaded & cut	LF	0.0863	9.04	3.98	13.01
02.5308 031	Steel pipe, black, A-120, schedule 40, weld, 1", threaded & cut	LF	0.0961	10.06	4.94	15.01
02.5308 041	Steel pipe, black, A-120, schedule 40, weld, 1-1/4", threaded & cut	LF	0.1056	11.06	6.18	17.24
02.5308 051	Steel pipe, black, A-120, schedule 40, weld, 1-1/2", threaded & cut	LF	0.1151	12.05	6.74	18.80
02.5308 061	Steel pipe, black, A-120, schedule 40, weld, 2", threaded & cut	LF	0.1345	14.09	8.16	22.25
02.5308 071	Steel pipe, black, A-120, schedule 40, weld, 2-1/2", threaded & cut	LF	0.1535	16.08	10.90	26.98
02.5308 081	Steel pipe, black, A-120, schedule 40, weld, 3", threaded & cut	LF	0.1728	18.10	12.88	30.98
02.5308 091	Steel pipe, black, A-120, schedule 40, weld, 4", threaded & cut	LF	0.2112	20.58	18.30	38.88
02.5308 101	Steel pipe, black, A-120, schedule 40, weld, 5", threaded & cut	LF	0.2303	22.44	25.69	48.12
02.5308 111	Steel pipe, black, A-120, schedule 40, weld, 6", threaded & cut	LF	0.2399	23.37	32.20	55.57
02.5309 000	**STEEL PIPE, BLACK, SCH 40, WITH XTRUCOAT, TREADED & CUT, WITH TRENCH:**					
02.5309 011	Steel pipe, black, wrapped, 1/2", schedule 40, threaded & cut	LF	0.1345	13.73	3.42	17.16
02.5309 021	Steel pipe, black, wrapped, 3/4", schedule 40, threaded & cut	LF	0.1391	14.20	4.18	18.39
02.5309 031	Steel pipe, black, wrapped, 1", schedule 40, threaded & cut	LF	0.1441	14.71	5.20	19.91
02.5309 041	Steel pipe, black, wrapped, 1-1/4", schedule 40, threaded & cut	LF	0.1479	15.10	6.58	21.68
02.5309 051	Steel pipe, black, wrapped, 1-1/2", schedule 40, threaded & cut	LF	0.1593	16.26	7.14	23.41
02.5309 061	Steel pipe, black, wrapped, 2", schedule 40, threaded & cut	LF	0.1631	16.65	8.67	25.32
02.5309 071	Steel pipe, black, wrapped, 2-1/2", schedule 40, threaded & cut	LF	0.1680	17.15	11.55	28.70
02.5309 081	Steel pipe, black, wrapped, 3", schedule 40, threaded & cut	LF	0.1897	19.37	13.78	33.15
02.5309 091	Steel pipe, black, wrapped, 4", schedule 40, threaded & cut	LF	0.2255	23.02	19.41	42.43
02.5310 000	**STEEL PIPE, MORTAR LINED, CEMENT COATED, WITH TRENCH:**					
02.5310 011	Steel pipe, mortar lined, 20", cement coated	LF	0.7284	70.96	234.90	305.86
02.5310 021	Steel pipe, mortar lined, 24", cement coated	LF	0.8376	81.60	278.33	359.93
02.5310 031	Steel pipe, mortar lined, 30", cement coated	LF	1.0120	98.59	343.44	442.03
02.5310 041	Steel pipe, mortar lined, 36", cement coated	LF	1.1167	108.79	408.59	517.38
02.5311 000	**STEEL PIPE, PRE-INSULATED, WITH STEEL JACKET, TRENCHING:**					
02.5311 011	Steel pipe, pre-insulated, steel jacket, 4"	LF	0.7281	74.34	66.64	140.98
02.5311 021	Steel pipe, pre-insulated, steel jacket, 6"	LF	0.9579	97.80	87.00	184.80

02 - SITE WORK

Division	Description	Unit	Labor Hours	Labor Cost	Material Cost	Total Cost
02.5311 031	Steel pipe, pre-insulated, steel jacket, 8"	LF	1.3028	133.02	100.93	233.94
02.5311 041	Steel pipe, pre-insulated, steel jacket, 10"	LF	1.5557	158.84	118.72	277.56
02.5311 051	Steel pipe, pre-insulated, steel jacket, 12"	LF	1.9005	194.04	136.54	330.58
02.5312 000	**STEEL PIPE, PRE-INSULATED, WITH PVC JACKET, TRENCHING:**					
02.5312 011	Steel pipe, pre-insulated, PVC jacket, 2"	LF	0.2299	22.40	26.76	49.16
02.5312 021	Steel pipe, pre-insulated, PVC jacket, 4"	LF	0.4828	47.03	43.98	91.01
02.5312 031	Steel pipe, pre-insulated, PVC jacket, 6"	LF	0.6361	61.97	58.00	119.97
02.5312 041	Steel pipe, pre-insulated, PVC jacket, 8"	LF	0.8660	84.37	67.70	152.07
02.5312 051	Steel pipe, pre-insulated, PVC jacket, 10"	LF	1.0346	100.79	80.02	180.81
02.5312 061	Steel pipe, pre-insulated, PVC jacket, 12"	LF	1.2645	123.19	92.34	215.53
02.5400 000	**VALVES & SPECIALTIES:**					
02.5401 000	CHECK VALVE, BRASS, 125#, SCREWED:					
02.5401 011	Check valve, brass, 125#, 1/2", screw	EA	0.2387	25.00	37.59	62.59
02.5401 021	Check valve, brass, 125#, 3/4", screw	EA	0.2895	30.32	45.95	76.27
02.5401 031	Check valve, brass, 125#, 1", screw	EA	0.3533	37.00	62.84	99.84
02.5401 041	Check valve, brass, 125#, 1-1/4", screw	EA	0.4135	43.31	86.96	130.27
02.5401 051	Check valve, brass, 125#, 1-1/2", screw	EA	0.4801	50.28	103.41	153.69
02.5401 061	Check valve, brass, 125#, 2", screw	EA	0.7767	81.34	151.04	232.38
02.5402 000	**CHECK VALVE, IRON BODY, FLANGED, BOLT & GASKET SET:**					
02.5402 001	*Note: For the following items, add for companion flanges.*					
02.5402 011	Check valve, 125#, iron body, 2", flanged, bolt & gasket	EA	1.2100	122.04	215.47	337.51
02.5402 021	Check valve, 125#, iron body, 2-1/2", flanged, bolt & gasket	EA	1.3500	136.16	271.60	407.76
02.5402 031	Check valve, 125#, iron body, 3", flanged, bolt & gasket	EA	1.5000	151.29	296.70	447.99
02.5402 041	Check valve, 125#, iron body, 4", flanged, bolt & gasket	EA	2.4000	242.06	550.60	792.66
02.5402 051	Check valve, 125#, iron body, 5"-6", flanged, bolt & gasket	EA	3.5000	366.56	796.41	1,162.97
02.5402 061	Check valve, iron body, 8", weld flanged, bolt & gasket	EA	4.6000	481.76	1,441.87	1,923.62
02.5402 071	Check valve, iron body, 10", weld flanged, bolt & gasket	EA	6.8000	712.16	2,462.57	3,174.74
02.5402 081	Check valve, iron body, 12", weld flanged, bolt & gasket	EA	7.5000	785.48	3,825.72	4,611.19
02.5402 091	Check valve, iron body, 14", weld flanged, bolt & gasket	EA	12.3000	1,288.18	4,436.21	5,724.39
02.5403 000	**GATE OR GLOBE VALVE, BRASS, 125#, SCREWED:**					
02.5403 011	Gate/globe valve, brass, 125#, 1", screwed	EA	0.3533	34.42	61.86	96.28
02.5403 021	Gate/globe valve, brass, 125#, 1-1/2", screwed	EA	0.4801	46.77	105.57	152.34
02.5403 031	Gate/globe valve, brass, 125#, 2", screwed	EA	0.7767	75.67	149.34	225.01
02.5404 000	**GATE VALVE, IRON BODY, FLANGED, BOLT & GASKET SET:**					
02.5404 001	*Note: For the following items, add for companion flanges.*					
02.5404 011	Gate valve, 125#, iron body, flanged, 2", bolt & gasket	EA	1.2100	126.72	355.92	482.65
02.5404 021	Gate valve, 125#, iron body, flanged, 2-1/2", bolt & gasket	EA	1.3500	141.39	389.91	531.29
02.5404 031	Gate valve, 125#, iron body, flanged, 3", bolt & gasket	EA	1.5000	157.10	444.88	601.97
02.5404 041	Gate valve, 125#, iron body, flanged, 4", bolt & gasket	EA	2.4000	251.35	651.55	902.90
02.5404 051	Gate valve, 125#, iron body, flanged, 5"-6", bolt & gasket	EA	3.5000	366.56	1,191.65	1,558.21
02.5404 061	Gate valve, 125#, iron body, flanged, 8", bolt & gasket	EA	4.6000	481.76	2,237.19	2,718.95
02.5404 071	Gate valve, 125#, iron body, flanged, 10", bolt & gasket	EA	6.6000	691.22	3,689.44	4,380.66
02.5404 081	Gate valve, 125#, iron body, flanged, 12", bolt & gasket	EA	7.5000	785.48	4,954.58	5,740.05
02.5404 091	Gate valve, 125#, iron body, flanged, 14", bolt & gasket	EA	12.3000	1,288.18	5,779.84	7,068.02
02.5405 000	**GATE VALVE, IRON BODY, MECHANICAL JOINT, WITH ACCESSORIES:**					
02.5405 011	Gate valve, iron body, mech joint, 2", complete	EA	0.9196	96.31	298.39	394.70
02.5405 021	Gate valve, iron body, mech joint, 3", complete	EA	1.1112	116.38	423.09	539.46
02.5405 031	Gate valve, iron body, mech joint, 4", complete	EA	1.4944	156.51	495.22	651.72
02.5405 041	Gate valve, iron body, mech joint, 6", complete	EA	2.2990	240.77	646.07	886.84
02.5405 051	Gate valve, iron body, mech joint, 8", complete	EA	2.6822	280.91	1,000.45	1,281.35
02.5405 061	Gate valve, iron body, mech joint, 10", complete	EA	3.6401	381.23	1,590.80	1,972.03
02.5405 071	Gate valve, iron body, mech joint, 12", complete	EA	4.2914	449.44	1,968.04	2,417.47
02.5405 081	Gate valve, iron body, mech joint, 14", complete	EA	5.2877	553.78	4,460.89	5,014.67
02.5405 091	Gate valve, iron body, mech joint, 16", complete	EA	6.0540	634.04	5,969.74	6,603.77
02.5406 000	**GAS COCK, IRON BODY, SCREWED, 125#:**					
02.5406 011	Gas cock, iron body, 3/4", 125#, screwed	EA	0.3793	39.72	30.04	69.77
02.5406 021	Gas cock, iron body, 1", 125#, screwed	EA	0.5311	55.62	43.18	98.80

SITE WORK - 02

Division	Description	Unit	Labor Hours	Labor Cost	Material Cost	Total Cost
02.5406 031	Gas cock, iron body, 1-1/2", 125#, screwed	EA	0.6828	71.51	84.30	155.81
02.5406 041	Gas cock, iron body, 2", 125#, screwed	EA	0.9862	103.28	132.41	235.70
02.5406 051	Gas cock, iron body, 2-1/2", 125#, screwed	EA	1.2139	127.13	225.03	352.16
02.5406 061	Gas cock, iron body, 3", 125#, screwed	EA	1.5173	162.56	357.97	520.54
02.5406 071	Gas cock, iron body, 4", 125#, screwed	EA	2.1243	227.60	554.40	782.00
02.5407 000	HOSE BIBB & HOSE GATE, WITH CAP:					
02.5407 011	Hose bibb, 3/4", with cap	EA	0.3046	31.90	19.96	51.86
02.5407 021	Hose bibb, 1", with cap	EA	0.3164	33.14	22.74	55.87
02.5407 051	Hose gate, 1", with cap	EA	0.3164	33.14	135.12	168.26
02.5407 061	Hose gate, 1-1/2", with cap	EA	0.4317	45.21	209.64	254.85
02.5407 071	Hose gate, 2", with cap	EA	0.4533	47.47	300.50	347.98
02.5408 000	SPECIALTIES:					
02.5408 011	Fire hydrant, (2) 2-1/2" outlets	EA	7.4576	781.03	1,209.26	1,990.30
02.5408 021	Fire hydrant, (1) 2-1/2"/(1) 4/4 1/2"	EA	7.4576	781.03	1,393.65	2,174.68
02.5408 031	Fire hydrant, (2) 2-1/2"/(1) 4/4 1/2"	EA	8.6411	904.98	1,808.36	2,713.34
02.5408 041	Fire hydrant, bury	VF	3.3002	345.63	420.00	765.63
02.5408 051	Post indicator, for valve, 4"	EA	3.8464	412.10	751.54	1,163.64
02.5408 061	Post indicator, for valve, 6"	EA	5.2651	564.10	1,149.49	1,713.59
02.5408 071	Valve box, cast iron, 4" deep	EA	1.4111	151.19	135.03	286.21
02.5408 081	Add for each extra foot depth	VF	0.3299	35.35	50.26	85.60
02.5408 091	Thrust blocks for 6" fire line	EA	2.5000	220.00	163.24	383.24
02.5408 101	Thrust blocks for 8" fire line	EA	4.0000	352.00	187.06	539.06
02.5408 111	Thrust blocks for 10" fire line	EA	5.0000	440.00	238.04	678.04
02.5408 121	Thrust blocks for 12" fire line	EA	7.5000	660.00	340.07	1,000.07
02.5500 000	MECHANICAL UTILITIES, ACCESSORIES:					
02.5501 000	CATCH BASINS, WITH GRATE, LIGHT DUTY:					
02.5501 001	Note: For heavy duty grade, add 15% to the material costs.					
02.5501 011	Catch basin, 2'x2'x2', grate, light duty	EA	5.2500	500.75	628.47	1,129.21
02.5501 021	Catch basin, 2'x2'x4', grate, light duty	EA	8.1708	779.33	794.20	1,573.53
02.5501 031	Catch basin, 3'x3'x2', grate, light duty	EA	9.3391	890.76	956.20	1,846.97
02.5501 041	Catch basin, 3'x3'x4', grate, light duty	EA	9.7411	929.11	1,125.83	2,054.93
02.5501 051	Add for 6" grade ring	EA	1.7525	167.15	134.91	302.07
02.5502 000	DROP INLETS, PRECAST CONCRETE, THICK WALL, STANDARD GRATES:					
02.5502 011	Drop inlet, precast, 12"x12"x4", thick wall, grate	EA	3.0653	321.03	396.75	717.78
02.5502 021	Drop inlet, precast, 16"x16"x5", thick wall, grate	EA	3.8316	401.28	521.49	922.77
02.5502 031	Drop inlet, precast, 16"x24"x5", thick wall, grate	EA	4.5980	481.55	612.25	1,093.80
02.5502 041	Drop inlet, precast, 24"x24"x5", thick wall, grate	EA	6.1306	642.06	742.60	1,384.66
02.5502 051	Drop inlet, precast, 24"x30"x5", thick wall, grate	EA	6.8969	722.31	912.68	1,634.99
02.5502 061	Drop inlet, precast, 30"x30"x6", thick wall, grate	EA	7.6632	802.57	975.01	1,777.58
02.5502 071	Drop inlet, precast, 36"x36"x6", thick wall, grate	EA	9.1959	963.09	1,037.46	2,000.55
02.5502 081	Drop inlet, precast, 36"x48"x6", thick wall, grate	EA	10.7285	1,123.60	1,349.15	2,472.75
02.5502 091	Drop inlet, precast, 48"x48"x6", thick wall, grate	EA	12.2612	1,284.12	1,774.41	3,058.52
02.5503 000	CURB VALVE BOX, WITH CAST IRON LID:					
02.5503 011	Curb valve box, 8-1/2" dia x 12", cast iron lid	EA	0.5463	52.11	53.61	105.72
02.5503 021	Curb valve box, 8-1/2" diax30", cast iron lid	EA	0.8649	82.49	80.55	163.05
02.5504 000	MANHOLES, WITH LID & RUNGS:					
02.5504 011	Manhole, 4' dia x 6'-8' deep, lid, rungs	EA	14.1338	1,348.08	1,791.56	3,139.64
02.5504 021	Manhole, 4' dia x 9'-12' deep, lid, rungs	EA	18.0637	1,722.92	2,239.55	3,962.46
02.5504 031	Manhole, 4' dia x 13'-16' deep, lid, rungs	EA	21.9935	2,097.74	2,777.00	4,874.74
02.5505 000	CLEANOUTS:					
02.5505 011	Cleanout, 4" to grade	EA	0.8800	83.93	260.90	344.83
02.5505 021	Cleanout, 6" to grade	EA	1.0000	95.38	426.91	522.29
02.5505 031	Cleanout, 8" to grade	EA	1.1200	106.83	487.26	594.09
02.5506 000	AREA DRAINS, PRECAST, LANDSCAPE LIGHT TRAFFIC AREAS:					
02.5506 011	Precast area drain, 12x12x3, thick wall	EA	3.0653	292.37	267.96	560.33
02.5506 021	Precast area drain, 16x16x3, thick wall	EA	3.8316	365.46	348.35	713.81
02.5506 031	Add per foot over 3' high	LF			51.92	51.92

02 - SITE WORK

Division	Description	Unit	Labor Hours	Labor Cost	Material Cost	Total Cost
02.5600 000	**MISCELLANEOUS SITE EQUIPMENT:**					
02.5601 000	**FUEL STORAGE TANKS, UNDERGROUND:**					
02.5601 011	Underground double wall oil tank, 500 gal, coated steel	EA	7.6632	730.92	9,669.08	10,400.00
02.5601 021	Underground tank, steel/glass, 1,000 gal, with access	EA	9.5790	913.65	12,313.92	13,227.56
02.5601 031	Underground tank, steel/glass, 2,000 gal, with access	EA	10.4680	998.44	15,798.97	16,797.41
02.5601 041	Underground tank, steel/glass, 4,000 gal, with access	EA	12.4948	1,191.75	22,692.38	23,884.14
02.5601 051	Underground tank, steel/glass, 6,000 gal, with access	EA	16.3685	1,561.23	27,640.78	29,202.01
02.5601 061	Underground tank, steel/glass, 10,000 gal, with access	EA	19.1580	1,827.29	37,255.86	39,083.15
02.5601 071	Underground tank, steel/glass, 12,000 gal, with access	EA	20.1620	1,923.05	44,042.79	45,965.84
02.5601 081	Underground tank, steel/glass, 20,000 gal, with access	EA	22.9896	2,192.75	68,240.14	70,432.89
02.5602 000	**FIBERGLASS TANKS, UNDERGROUND, DOUBLEWALL, WITH ACCESS:**					
02.5602 011	Underground tank, double wall fiberglass, 550 gal, with access	EA	3.0653	292.37	14,560.45	14,852.82
02.5602 021	Underground tank, double wall fiberglass, 1,000 gal, with access	EA	4.5980	438.56	18,360.27	18,798.83
02.5602 031	Underground tank, double wall fiberglass, 1,500 gal, with access	EA	6.1306	584.74	21,166.24	21,750.98
02.5602 041	Underground tank, double wall fiberglass, 2,000 gal, with access	EA	7.6632	730.92	23,393.86	24,124.77
02.5602 051	Underground tank, double wall fiberglass, 4,000 gal, with access	EA	9.1959	877.10	35,540.53	36,417.63
02.5602 061	Underground tank, double wall fiberglass, 6,000 gal, with access	EA	12.2612	1,169.47	40,250.11	41,419.58
02.5602 071	Underground tank, double wall fiberglass, 8,000 gal, with access	EA	22.9896	2,192.75	50,202.03	52,394.78
02.5602 081	Underground tank, double wall fiberglass, 10,000 gal, with access	EA	34.4844	3,289.12	58,464.27	61,753.39
02.5602 091	Underground tank, double wall fiberglass, 12,000 gal, with access	EA	42.1476	4,020.04	71,497.76	75,517.79
02.5602 101	Underground tank, double wall fiberglass, 15,000 gal, with access	EA	45.9792	4,385.50	87,878.19	92,263.69
02.5602 111	Underground tank, double wall fiberglass, 20,000 gal, with access	EA	53.6424	5,116.41	110,299.55	115,415.96
02.5602 121	Gage system, remote reading inventory	EA	4.2500	405.37	9,393.94	9,799.31
02.5602 131	Leak detection monitor	EA			2,864.13	2,864.13
02.5700 000	**ELECTRICAL DISTRIBUTION, UNDERGROUND:**					
02.5700 001	*Note: For wire and substation equipment, see section 16.0000.*					
02.5701 000	**TRENCH FOR UNDERGROUND CONDUIT, BACKFILL & TAMPED:**					
02.5701 011	Trench, conduit, 1'x2' deep, fill, tamped	LF	0.0250	2.29	1.21	3.50
02.5701 021	Trench, conduit, 1'x3' deep, fill, tamped	LF	0.0352	3.22	1.79	5.01
02.5701 031	Trench, conduit, 2'x3' deep, fill, tamped	LF	0.0455	4.16	3.61	7.77
02.5701 041	Trench, conduit, 2'x4' deep, fill, tamped	LF	0.0550	5.03	4.77	9.80
02.5701 051	Trench, conduit, per cubic yard	CY	0.3200	29.27	16.11	45.38
02.5702 000	**CONCRETE ENVELOPES, COLORED RED:**					
02.5702 011	Concrete envelope, red, 9" x 9"	LF	0.0326	2.98	5.36	8.34
02.5702 021	Concrete envelope, red, 12" x 12"	LF	0.0465	4.25	7.28	11.54
02.5702 031	Concrete envelope, red, 12" x 24"	LF	0.0651	5.95	10.65	16.60
02.5702 041	Concrete envelope, red	CY	0.2325	21.27	159.78	181.05
02.5703 000	**PVC CONDUIT, SCH 40, CONCRETE ENCASED, TRENCH & COVER:**					
02.5703 011	PVC conduit, schedule 40, trenched & buried in concrete, 1-2"	LF	0.0504	4.61	21.99	26.60
02.5703 021	PVC conduit, schedule 40, trenched & buried in concrete, 2-2"	LF	0.0619	5.66	27.04	32.70
02.5703 031	PVC conduit, schedule 40, trenched & buried in concrete, 3-2"	LF	0.0718	6.57	28.03	34.59
02.5703 041	PVC conduit, schedule 40, trenched & buried in concrete, 4-2"	LF	0.0829	7.58	29.03	36.61
02.5703 051	PVC conduit, schedule 40, trenched & buried in concrete, 1-3"	LF	0.0664	6.07	23.50	29.57
02.5703 061	PVC conduit, schedule 40, trenched & buried in concrete, 2-3"	LF	0.0829	7.58	28.46	36.05
02.5703 071	PVC conduit, schedule 40, trenched & buried in concrete, 3-3"	LF	0.0995	9.10	33.97	43.07
02.5703 081	PVC conduit, schedule 40, trenched & buried in concrete, 4-3"	LF	0.1105	10.11	35.96	46.06
02.5703 091	PVC conduit, schedule 40, trenched & buried in concrete, 1-4"	LF	0.1105	10.11	28.17	38.27
02.5703 101	PVC conduit, schedule 40, trenched & buried in concrete, 2-4"	LF	0.1436	13.14	34.19	47.33
02.5703 111	PVC conduit, schedule 40, trenched & buried in concrete, 3-4"	LF	0.1659	15.17	41.06	56.24
02.5703 121	PVC conduit, schedule 40, trenched & buried in concrete, 4-4"	LF	0.1824	16.68	44.21	60.89
02.5703 131	PVC conduit, schedule 40, trenched & buried in concrete, 1-5"	LF	0.1270	11.62	36.42	48.03
02.5703 141	PVC conduit, schedule 40, trenched & buried in concrete, 2-5"	LF	0.1769	16.18	43.34	59.52
02.5703 151	PVC conduit, schedule 40, trenched & buried in concrete, 3-5"	LF	0.2046	18.71	46.43	65.15
02.5703 161	PVC conduit, schedule 40, trenched & buried in concrete, 4-5"	LF	0.2323	21.25	49.01	70.26
02.5703 171	PVC conduit, schedule 40, trenched & buried in concrete, 1-6"	LF	0.1328	12.15	42.09	54.23
02.5703 181	PVC conduit, schedule 40, trenched & buried in concrete, 2-6"	LF	0.1837	16.80	49.04	65.84

SITE WORK - 02

Division	Description	Unit	Labor Hours	Labor Cost	Material Cost	Total Cost
02.5703 191	PVC conduit, schedule 40, trenched & buried in concrete, 4-6"	LF	0.2818	25.78	75.32	101.10
02.5703 201	PVC conduit, schedule 40, trenched & buried in concrete, 6-6"	LF	0.3765	42.06	100.70	142.76
02.5703 211	PVC conduit, schedule 40, trenched & buried in concrete, 8-6"	LF	0.5186	57.93	138.62	196.56
02.5703 221	PVC conduit, schedule 40, trenched & buried in concrete, 10-6"	LF	0.6053	67.62	161.80	229.42
02.5703 231	PVC conduit, schedule 40, trenched & buried in concrete, 12-6"	LF	0.6947	77.60	185.71	263.32
02.5704 000	**FIBER CONDUIT, CLASS II, DIRECT BURIAL:**					
02.5704 011	Underground fiber conduit, class II, 2"	LF	0.0870	9.72	2.33	12.05
02.5704 021	Underground fiber conduit, class II, 3"	LF	0.0921	10.29	2.93	13.22
02.5704 031	Underground fiber conduit, class II, 3-1/2"	LF	0.1023	11.43	4.03	15.46
02.5705 000	**PVC CONDUIT, HEAVYWALL, WITH OUT TRENCHING:**					
02.5705 011	Heavywall PVC conduit, 1/2", no trench	LF	0.0256	2.86	0.50	3.36
02.5705 021	Heavywall PVC conduit, 3/4", no trench	LF	0.0279	3.12	0.63	3.75
02.5705 031	Heavywall PVC conduit, 1", no trench	LF	0.0303	3.38	0.97	4.35
02.5705 041	Heavywall PVC conduit, 1-1/4", no trench	LF	0.0326	3.64	1.28	4.92
02.5705 051	Heavywall PVC conduit, 1-1/2", no trench	LF	0.0349	3.90	1.64	5.54
02.5705 061	Heavywall PVC conduit, 2", no trench	LF	0.0372	4.16	2.11	6.26
02.5705 071	Heavywall PVC conduit, 2-1/2", no trench	LF	0.0419	4.68	3.48	8.16
02.5705 081	Heavywall PVC conduit, 3", no trench	LF	0.0465	5.19	4.56	9.76
02.5705 091	Heavywall PVC conduit, 3-1/2", no trench	LF	0.0558	6.23	5.44	11.67
02.5705 101	Heavywall PVC conduit, 4", no trench	LF	0.0651	7.27	6.41	13.68
02.5705 111	Heavywall PVC conduit, 5", no trench	LF	0.0744	8.31	9.13	17.44
02.5705 121	Heavywall PVC conduit, 6", no trench	LF	0.0930	10.39	11.79	22.18
02.5706 000	**RIGID STEEL CONDUIT:**					
02.5706 011	Rigid steel conduit, 1/2"	LF	0.0385	4.30	1.15	5.45
02.5706 021	Rigid steel conduit, 3/4"	LF	0.0439	4.90	1.33	6.23
02.5706 031	Rigid steel conduit, 1"	LF	0.0576	6.43	1.86	8.29
02.5706 041	Rigid steel conduit, 1-1/4"	LF	0.0653	7.29	2.75	10.05
02.5706 051	Rigid steel conduit, 1-1/2"	LF	0.0784	8.76	3.28	12.04
02.5706 061	Rigid steel conduit, 2"	LF	0.0960	10.72	4.54	15.26
02.5706 071	Rigid steel conduit, 2-1/2"	LF	0.1228	13.72	6.70	20.41
02.5706 081	Rigid steel conduit, 3"	LF	0.1688	18.86	8.80	27.66
02.5706 091	Rigid steel conduit, 3-1/2"	LF	0.2073	23.16	11.93	35.08
02.5706 101	Rigid steel conduit, 4"	LF	0.2379	26.58	14.15	40.73
02.5706 111	Rigid steel conduit, 5"	LF	0.2668	29.80	16.08	45.88
02.5706 121	Rigid steel conduit, 6"	LF	0.2897	32.36	31.54	63.90
02.5706 131	Rigid steel conduit, 2", 90 degree sweep	EA	0.6200	69.26	93.68	162.94
02.5706 141	Rigid steel conduit, 3", 90 degree sweep	EA	0.6975	77.92	168.70	246.62
02.5706 151	Rigid steel conduit, 4", 90 degree sweep	EA	1.8600	207.78	284.30	492.08
02.5706 161	Rigid steel conduit, 5", 90 degree sweep	EA	2.3250	259.73	417.07	676.80
02.5706 171	Rigid steel conduit, 6", 90 degree sweep	EA	2.7900	311.67	438.07	749.74
02.5706 181	Add for pipe wrapping	LF			3.50	3.50
02.5707 000	**ELECTRICAL BOXES AND MANHOLES, CONCRETE:**					
02.5707 011	Pull/J-box, 12" x 12" x 18"	EA	1.5694	175.32	42.65	217.97
02.5707 021	Pull/J-box, 12" x 24" x 18"	EA	1.9563	218.54	57.87	276.41
02.5707 031	Pull/J-box, 12" x 18" round	EA	1.5810	176.61	46.75	223.37
02.5707 041	Pull/J-box, 18" x 24" round	EA	2.0925	233.75	58.40	292.16
02.5707 051	Concrete handhole, 2' x 2' x 3' deep	EA	2.3250	212.67	794.94	1,007.60
02.5707 061	Concrete handhole, 3' x 3' x 3' deep	EA	3.7200	340.27	1,145.72	1,485.99
02.5707 071	Concrete handhole, 4' x 4' x 4' deep	EA	5.3475	489.14	2,560.37	3,049.50
02.5707 081	Concrete manhole, 5'x5'x4', with cover	EA	5.5800	510.40	2,509.39	3,019.80
02.5707 091	Concrete manhole, 4'x6'x6', with cover	EA	6.5100	595.47	3,128.52	3,723.99
02.5707 101	Concrete manhole, 6'x8'x6', with cover	EA	7.9050	723.07	4,977.16	5,700.23
02.5707 111	Concrete manhole, 6'x10'x6', with cover	EA	9.3000	850.67	5,986.05	6,836.72
02.5707 121	Concrete vault, 8' x 10' x 9'	EA	21.0754	1,927.77	6,441.31	8,369.07
02.5707 131	Concrete vault, 8' x 14' x 9'	EA	26.1997	2,396.49	9,869.34	12,265.82
02.5707 141	Vault lid, 4' x 4' x 3'6", galv	EA	3.8192	349.34	548.92	898.26
02.5708 000	**TRANSFORMER PADS:**					

02 - SITE WORK

Division	Description	Unit	Labor Hours	Labor Cost	Material Cost	Total Cost
02.5708 001	Note: For pad mounted transformers, see section 16.0000.					
02.5708 011	Transformer pad, 3' x 4'	EA	3.2511	311.07	368.91	679.98
02.5708 021	Transformer pad, 4' x 5'	EA	3.6798	352.08	560.14	912.23
02.5708 031	Transformer pad, 6' x 8', slab box	EA	5.5800	533.89	2,067.16	2,601.05
02.5708 041	Transformer pad, 6' x 10'	EA	4.7846	457.79	877.97	1,335.76
02.5708 051	Transformer pad, 8' x 10'	EA	5.2638	503.64	1,052.17	1,555.81
02.5708 061	Transformer pad, large sizes	SF	0.0814	7.79	11.55	19.34
02.5709 000	**GROUNDING:**					
02.5709 011	Bare copper ground wire, soft #4	LF	0.0040	0.45	1.38	1.83
02.5709 021	Bare copper ground wire, soft #3/0	LF	0.0240	2.68	4.26	6.94
02.5709 031	Bare copper ground wire, soft #4/0	LF	0.0280	3.13	5.36	8.49
02.5709 041	Ground clamp, 1/2" - 1" ground rod	EA	0.1500	16.76	12.13	28.89
02.5709 051	Cadweld connection, average	EA	1.0000	111.71	19.56	131.27
02.5709 061	Copper ground rod, 5/8" x 10'	EA	0.7500	83.78	28.84	112.63
02.5709 071	Copper ground rod, 3/4" x 10'	EA	0.7500	83.78	46.63	130.41
02.5709 081	Ground rod well	EA	0.7500	83.78	50.52	134.30
02.5800 000	**ELECTRICAL DISTRIBUTION, OVERHEAD:**					
02.5801 000	**POLES, WOOD, BUTT TREATED, CLASS 5:**					
02.5801 001	Note: The following prices include setting with crane or boom and machine augered holes.					
02.5801 011	Pole, wood, 25', butt treated, class 5	EA	5.4762	500.91	124.41	625.32
02.5801 021	Pole, wood, 30', butt treated, class 5	EA	6.4623	591.11	146.90	738.00
02.5801 031	Pole, wood, 35', butt treated, class 5	EA	7.2286	661.20	179.74	840.94
02.5801 041	Pole, wood, 40', butt treated, class 5	EA	7.8864	721.37	228.16	949.52
02.5801 051	Pole, wood, 50', butt treated, class 5	EA	9.8534	901.29	285.18	1,186.47
02.5801 061	Pole, wood, 60', butt treated, class 5	EA	11.8241	1,081.55	342.26	1,423.81
02.5802 000	**CROSS ARMS, WOOD, WITH TYPICAL POLE LINE HARDWARE:**					
02.5802 011	Cross arm, 1 arm, 6', with hardware	EA	1.9172	175.37	27.63	203.00
02.5802 021	Cross arm, 1 arm, 8', with hardware	EA	1.9714	180.32	34.56	214.88
02.5802 031	Cross arm, 2 arm, 6', with hardware	EA	5.3667	490.89	55.30	546.20
02.5802 041	Cross arm, 2 arm, 8', with hardware	EA	5.5314	505.96	69.14	575.10
02.5803 000	**TRANSFORMERS, DISTRIBUTION, 1PH, 5KV, 120/240, OIL, POLE MOUNT:**					
02.5803 011	Transformer, 10kva, 1ph, 5kv, 120/240, pole	EA	6.5166	727.97	624.75	1,352.72
02.5803 021	Transformer, 15kva, 120/240v, 1ph, 5kv, pole	EA	8.3794	936.06	817.00	1,753.07
02.5803 031	Transformer, 25kva, 120/240v, 1ph, 5kv, pole	EA	10.0224	1,119.60	1,025.33	2,144.93
02.5803 041	Transformer, 37.5kva, 120/240v, 1ph, 5kv	EA	11.8842	1,327.58	1,431.17	2,758.76
02.5803 051	Transformer, 50kva, 1ph, 5kv, 120/240v, pole	EA	14.4029	1,608.95	1,618.08	3,227.03
02.5803 061	Transformer, 75kv, 120/240v, 1ph, 5kv, pole	EA	16.7578	1,872.01	2,419.15	4,291.16
02.5803 071	Transformer, 100kva, 1ph, 5kv, 120/240v	EA	21.4125	2,391.99	2,811.62	5,203.61
02.5803 081	Transformer, 167kva, 1ph, 5kv, 120/240v	EA	25.1372	2,808.08	4,165.46	6,973.54
02.5803 091	Transformer, 250kva, 1ph, 5kv, 120/240v	EA	28.8609	3,224.05	8,178.76	11,402.81
02.5804 000	**FUSED CUTOUTS, POLE MOUNTED:**					
02.5804 011	Fused cutout, 50a, 5kv, pole mount	EA	2.0246	226.17	114.91	341.08
02.5804 021	Fused cutout, 100a, 5kv, pole mount	EA	2.0246	226.17	114.91	341.08
02.5804 031	Fused cutout, 250a, 5kv, pole mount	EA	2.5760	287.76	131.78	419.54
02.5805 000	**SWITCHES, DISCONNECT, POLE MOUNT, POLE ARM THROW & LOCK:**					
02.5805 011	Switches, disconnect, 400a, 5kv, set of 3, pole	EA	34.4097	3,843.91	2,242.96	6,086.86
02.5805 021	Switches, disconnect, 600a, 5kv, set of 3, pole	EA	34.4097	3,843.91	3,364.37	7,208.28
02.5805 031	Switches, disconnect, 1200a, 5kv, set of 3, pole	EA	38.6418	4,316.68	6,728.86	11,045.53
02.5806 000	**LIGHTNING ARRESTORS, POLE MOUNTED:**					
02.5806 011	Lightning arrester, pole mounted	EA	3.2445	362.44	16.33	378.77
02.5807 000	**OUTDOOR LIGHTING:**					
02.5807 001	Note: The following prices include base, pole and setting.					
02.5807 011	Light, mercury vapor, 175w, 30' pole	EA	21.8974	2,446.16	1,376.77	3,822.93
02.5807 021	Light, mercury vapor, 400w, 30' pole	EA	21.8974	2,446.16	1,750.07	4,196.23
02.5807 031	Light, mercury vapor, 1000w, 30' pole	EA	22.6943	2,535.18	2,173.47	4,708.65

SITE WORK - 02

Division	Description	Unit	Labor Hours	Labor Cost	Material Cost	Total Cost
02.5807 041	Light, mercury vapor, 175w, 10' pole	EA	10.4275	1,164.86	926.71	2,091.57
02.5807 051	Light, mercury vapor, 175w, on bldg	EA	5.6429	630.37	496.71	1,127.08
02.5807 061	Light, mercury vapor, 400w, on bldg	EA	5.6429	630.37	706.66	1,337.03
02.5808 000	**WALKWAY LIGHTING BOLLARDS:**					
02.5808 011	Wkwy light, 42", 175w mercury vapor	EA	2.7900	311.67	1,024.83	1,336.50
02.5808 021	Wkwy light, 42", 175w, metal halide	EA	2.7900	311.67	1,171.26	1,482.93
02.5808 031	Walkway light, 42", high pressure sodium to 150w	EA	2.7900	311.67	1,250.08	1,561.75
02.5808 041	Walkway light, 42", incandescent to 150w	EA	2.7900	311.67	912.22	1,223.89
02.5809 000	**SPECIALTY OUTDOOR LIGHTING:**					
02.5809 011	150 watt incandescent flush mounted landscape lighting	EA	1.5694	175.32	413.58	588.90
02.5809 021	200 watt incandescent flush mounted landscape lighting	EA	1.5694	175.32	445.98	621.30
02.5809 031	300 watt incandescent flush mounted landscape lighting	EA	1.5694	175.32	463.43	638.75
02.5809 071	Incandescent step light	EA	1.0230	114.28	283.42	397.70
02.5809 081	Fluorescent step light W1 PL-13 lamp	EA	1.0230	114.28	288.69	402.97
02.5809 111	Fluorescent, damp location, two lamp	EA	0.8000	89.37	152.18	241.54
02.5809 211	Adjust floodlight, 75w, PAR 38	EA	0.2000	22.34	89.23	111.57
02.5810 000	**RACEWAY AND WIRE FOR OUTDOOR LIGHTING:**					
02.5810 011	PVC & copper wire, 20 amps	LF	0.0330	3.69	6.27	9.96
02.5810 021	PVC & copper wire, 30 amps	LF	0.0345	3.85	6.56	10.41
02.5810 031	PVC & copper wire, 40 amps	LF	0.0402	4.49	7.71	12.20
02.5810 041	PVC & copper wire, 60 amps	LF	0.0459	5.13	8.78	13.91
02.5810 051	Copper wire, stranded, #12	LF	0.0066	0.74	1.36	2.10
02.5810 061	Copper wire, stranded, #10	LF	0.0075	0.84	1.51	2.35
02.5810 071	Copper wire, stranded, #8	LF	0.0081	0.90	2.04	2.94
02.5811 000	**PARKING LOT LIGHTING:**					
02.5811 011	Typical parking lot lighting, small	SF	0.0076	0.85	0.78	1.63
02.5811 021	Typical parking lot lighting, large	SF	0.0044	0.49	0.51	1.00
02.5812 000	**TRAFFIC SIGNALS:**					
02.5812 011	Signals, 4 lane - "T" intersection	LS			196,773.89	196,773.89
02.5812 021	Signals, 4 lane - "X" intersection	LS			264,887.90	264,887.90
02.5812 031	Signals, 6 lane - "T" intersection	LS			280,024.37	280,024.37
02.5812 041	Signals, 6 lane - "X" intersection	LS			340,570.18	340,570.18
02.5812 051	Traffic signal, pole mounted	LS			15,524.65	15,524.65
02.5812 061	Traffic Controller	LS			18,920.57	18,920.57
02.5814 000	**ATHLETIC FIELD LIGHTING:**					
02.5814 011	Lights, football field, 40 foot candles	LS			330,474.59	330,474.59
02.5814 021	Lights, football field, 100 foot candles	LS			513,964.11	513,964.11
02.5814 031	Playground, diamond, economy	LS			47,740.57	47,740.57
02.5814 041	Tennis court, economy	LS			13,317.11	13,317.11
02.5815 000	**HIGH MAST LIGHTING:**					
02.5815 011	Luminaire, pole top, 400 watt	EA	3.1240	285.75	660.32	946.07
02.5815 021	Luminaire, pole top, 750 watt	EA	3.4410	314.75	696.84	1,011.59
02.5815 031	Luminaire, pole top, 1,000 watt	EA	3.7520	343.20	776.77	1,119.97
02.5815 041	Luminaire, pole top, 1,500 watt	EA	4.0160	367.34	1,007.89	1,375.23
02.5815 051	Pole, galvanized steel, 70 feet high	EA	12.0420	1,101.48	10,685.27	11,786.75
02.5815 061	Pole, galvanized steel, 80 feet high	EA	13.4500	1,230.27	12,732.30	13,962.58
02.5815 071	Pole, galvanized steel, 90 feet high	EA	14.7200	1,346.44	14,923.59	16,270.02
02.5815 081	Pole, galvanized steel, 100 feet high	EA	16.2410	1,485.56	17,093.97	18,579.54
02.5815 091	Pole, footing system for 70 feet high	EA	11.6760	897.07	389.24	1,286.31
02.5815 101	Pole, footing system for 80 feet high	EA	13.6200	1,046.42	455.30	1,501.72
02.5815 111	Pole, footing system for 90 feet high	EA	16.5440	1,271.08	568.21	1,839.29
02.5815 121	Pole, footing system for 100 feet high	EA	19.4570	1,494.88	649.90	2,144.79
02.5815 131	Mounting hoist, 12 fixtures	EA	8.8490	809.42	10,466.30	11,275.72
02.5815 141	Mounting hoist, 16 fixtures	EA	12.5410	1,147.13	12,466.42	13,613.54
02.5815 151	Mounting hoist, 20 fixtures	EA	16.2230	1,483.92	14,808.87	16,292.79
02.5815 161	Mounting brackets for fixture	EA	0.6500	59.46	118.18	177.64

02 - SITE WORK

Division	Description	Unit	Labor Hours	Labor Cost	Material Cost	Total Cost
02.6000 000	**GENERAL SITE WORK, PAVING & WALKS:**					
02.6001 000	**PAVEMENTS, ASPHALTIC:**					
02.6001 001	*Note: See 02.1101*					
02.6001 011	Asphaltic concrete, 2", on 4" base	SF	0.0221	1.70	2.14	3.84
02.6001 021	Asphaltic concrete, 2", on 6" base (parking lot)	SF	0.0274	2.11	2.43	4.54
02.6001 031	Asphaltic concrete, 3", on 8" base (truck & ramp)	SF	0.0348	2.67	2.98	5.66
02.6001 041	Asphaltic concrete, 3", 8" base, 10" sub (streets)	SF	0.0479	3.68	4.10	7.78
02.6001 051	Asphaltic concrete, armor coat, 2 shot, 4" base	SF	0.0152	1.17	0.96	2.12
02.6001 061	Sawcut pavement, asphaltic concrete	SF	0.0200	1.66	0.86	2.53
02.6002 000	**PAVEMENTS, COMPILATION:**					
02.6002 001	*Note: For headers see section 02.6007. The prices are based on quantities under 5,000 square feet. For quantities under 2,000 square feet, add 25%.*					
02.6002 011	Fine grading for roadbed	SF	0.0021	0.19	0.15	0.34
02.6002 021	6" sub base for roadbed	SF	0.0038	0.30	0.75	1.05
02.6002 031	4" base for roadbed	SF	0.0038	0.30	0.45	0.75
02.6002 041	Tack coat	SF	0.0004	0.03	0.08	0.11
02.6002 051	Pavement, 2" asphaltic concrete	SF	0.0084	0.65	2.04	2.69
02.6002 061	Pavement, sealer	SF	0.0008	0.06	0.08	0.14
02.6002 071	Add or deduct per 1" of sub base	INCH			0.16	0.16
02.6002 081	Add or deduct per 1" base rock	INCH			0.22	0.22
02.6002 091	Add or deduct per 1" of asphaltic concrete	INCH			0.84	0.84
02.6003 000	**PAVEMENTS, MISCELLANEOUS:**					
02.6003 011	Tennis court, bases, Asphaltic concrete, color seal	SF			7.35	7.35
02.6003 021	Tennis court, 7200 SF, fence/stripe	UNIT			77,057.49	77,057.49
02.6003 031	Parking, gravel, 6" of 1-1/2" rock	SF			0.61	0.61
02.6004 000	**PAVING MATERIALS & SUPPORT ITEMS:**					
02.6004 001	*Note: The following prices are based on quantities over 50,000 square feet. For quantities between 5,000 and 50,000 square feet, add 5%.*					
02.6004 011	Fine grading	SF	0.0014	0.13	0.03	0.16
02.6004 021	Aggregate, sub base, 1-1/2", class 2 & 3	TON	0.0772	6.11	10.28	16.39
02.6004 031	Aggregate rock, sub base, class 4	TON	0.0772	6.11	9.25	15.37
02.6004 041	Aggregate rock, sub base (non-spec)	TON	0.0860	6.81	9.17	15.98
02.6004 051	Aggregate rock, sub base, 8", class 2 & 3	SF	0.0038	0.30	0.45	0.75
02.6004 061	Aggregate rock, sub base, 12", class 2&3	SF	0.0038	0.30	0.65	0.95
02.6004 071	Aggregate rock, base, class 2 & 3	TON	0.0772	6.11	13.08	19.19
02.6004 081	Aggregate rock, base (non-spec)	TON	0.0972	7.70	9.98	17.68
02.6004 091	Engineered fill, sub base	TON	0.0772	6.11	7.29	13.41
02.6004 101	Aggregate base, chemical treated, 5% by dry weight	TON	0.1502	11.89	17.22	29.11
02.6004 111	Aggregate rock, base, 6", class 2 & 3	SF	0.0038	0.30	0.31	0.61
02.6004 121	Aggregate base, 6", chemical treated, 5% by weight	SF	0.0056	0.44	0.53	0.97
02.6004 131	Add for lime treatment, 12" native soil	SF	0.0040	0.32	0.35	0.66
02.6004 141	Lime bulk, material only	TON			132.77	132.77
02.6004 151	Cement	BBL			32.15	32.15
02.6004 161	Base treatment cement, 134# CY	#			0.03	0.03
02.6004 171	Asphaltic concrete, 1,000 to 5,000	TON	0.1767	13.58	102.96	116.54
02.6004 181	Asphaltic concrete, more than 5,000	TON	0.1767	13.58	95.03	108.61
02.6004 191	Cut back, 1/2", bulk	TON	0.1767	13.58	47.86	61.44
02.6004 201	Cut back, 1/4", bulk	TON	0.1767	13.58	49.14	62.72
02.6004 211	Sheet asphalt	TON	0.1767	13.58	84.97	98.55
02.6004 221	Asphaltic concrete, open graded	TON	0.1767	13.58	100.03	113.61
02.6004 231	Cut back in sacks, 66# sacks	SACK	0.0356	2.83	2.86	5.69
02.6004 241	Seal coat, SC 70, penetrating	TON	0.2764	22.00	450.91	472.91
02.6004 251	Prime coat, MC 250	TON	0.2764	22.00	450.91	472.91
02.6004 261	Prime coat, MC 70-250, 1,000	SY	0.0063	0.50	0.51	1.01
02.6004 271	Prime coat, MC 70-250, 2,000	SY	0.0051	0.41	0.29	0.69
02.6004 281	Prime coat, MC 70-250, 2,500	SY	0.0044	0.35	0.22	0.57
02.6004 291	Prime coat, MC 70-250, 5,000	SY	0.0044	0.35	0.20	0.55

SITE WORK - 02

Division	Description	Unit	Labor Hours	Labor Cost	Material Cost	Total Cost
02.6005 000	**CURBS AND GUTTERS:**					
02.6005 011	Curb forms, fabricated, edge, 6" x 12"	LF	0.0314	2.81	1.44	4.25
02.6005 021	Curb forms, steel, long run, reuse	LF	0.0177	1.59	1.44	3.03
02.6005 031	Curb/gutter, handwork	LF	0.0636	5.70	10.36	16.06
02.6005 041	Curb/gutter, machine work	LF	0.0496	4.44	9.33	13.78
02.6005 051	Curb, vertical face planter, 6"x18"	LF	0.0784	7.02	11.29	18.32
02.6005 061	Curb, vertical face, radial	LF	0.0640	5.73	6.55	12.28
02.6005 071	Curb, 6", asphaltic concrete	LF	0.0850	7.77	2.78	10.55
02.6005 081	Curb, 6", asphaltic concrete with ASB plain fiber	LF	0.0895	8.18	5.11	13.29
02.6005 091	Curb & gutter & 4' sidewalk, mono	LF	0.0916	8.21	16.74	24.95
02.6005 101	Split face granite curb, 6"x18", straight	LF	0.1508	13.51	39.48	52.99
02.6005 111	Split face granite curb, 6"x24", taper	LF	0.2452	21.97	54.71	76.67
02.6005 121	Split face granite curb, 6x24, radial	LF	0.1934	17.32	69.86	87.18
02.6005 131	Curb, driveway, concrete, apron, 4"	SF	0.0337	3.02	1.77	4.79
02.6005 141	Curb, driveway, concrete, apron, 5"	SF	0.0337	3.02	2.36	5.38
02.6006 000	**SIDEWALKS AND CONCRETE FINISHES:**					
02.6006 011	Brick pavers, sand bed 1" compact	SF	0.0696	6.11	3.66	9.78
02.6006 021	Brick pavers, grouted	SF	0.0939	8.25	4.15	12.39
02.6006 031	Precast paver, sand bed, 1" compact	SF	0.0572	5.02	3.00	8.02
02.6006 041	Precast pavers, grouted	SF	0.0854	7.65	3.62	11.27
02.6006 045	Turf Block 6"-12"	SF	0.0622	5.57	2.95	8.52
02.6006 051	Concrete walk, 4", broom finish	SF	0.0215	1.92	3.54	5.46
02.6006 061	Concrete walk, 4", exposed aggregate washed	SF	0.0324	2.90	3.64	6.54
02.6006 071	Concrete walk, 4"-24x24, score, lamp bl	SF	0.0306	2.74	3.49	6.23
02.6006 081	Walk, 4", sandblast, integral color	SF	0.0429	3.84	3.90	7.74
02.6006 091	Concrete walk 4", 80-100% seeded, agg	SF	0.0536	4.80	4.51	9.30
02.6006 101	Concrete walk 4", 100% seeded, color	SF	0.0550	4.93	4.64	9.57
02.6006 111	Concrete bands, 12-24" wide, 7" deep	SF	0.0256	2.29	3.19	5.49
02.6006 121	Finish, steel trowel	SF	0.0102	0.81		0.81
02.6006 131	Finish, broom	SF	0.0102	0.81		0.81
02.6006 141	Concrete hardener, chemical	SF	0.0038	0.30	0.10	0.41
02.6006 151	Concrete hardener, granolithic	SF	0.0038	0.30	0.01	0.31
02.6006 161	Concrete hardener, iron base	SF	0.0074	0.59	0.38	0.97
02.6007 000	**SPECIALTIES (MEDIUM TO LARGE SITES):**					
02.6007 011	Header, redwood treated, 2" x 4"	LF	0.0406	3.88	1.28	5.16
02.6007 021	Header, redwood treated, 2" x 6"	LF	0.0516	4.94	1.36	6.30
02.6007 031	Headers, redwood, 2" x 6"	LF	0.0516	4.94	1.72	6.65
02.6007 041	Dividers, redwood, 2" x 4"	LF	0.0286	2.74	0.86	3.60
02.6007 051	Benders, redwood, 3/8" x 4"	LF	0.0728	6.97	0.86	7.83
02.6007 061	Parking bumper block, precast 3'	EA	0.2514	17.92	29.01	46.92
02.6007 071	Parking bumper block, precast 6'	EA	0.3076	21.92	46.58	68.50
02.6007 081	Bike lock, precast, lockable	EA	0.2717	19.36	26.23	45.59
02.6007 091	Street signs, with pole	EA	1.9625	139.87	147.52	287.39
02.6007 101	Striping, 2 coat, 4" wide	LF	0.0047	0.33	0.05	0.38
02.6007 111	Striping, 2 ct, 4" wide, small quantity	LF	0.0093	0.66	0.05	0.71
02.6007 121	Striping, 1 coat, single line	STALL	0.1584	11.29	0.05	11.34
02.6007 131	Striping, thermoplastic, white bulk	SF	0.1020	7.27	5.69	12.96
02.6007 141	Striping, thermoplastic, yellow bulk	SF	0.1020	7.27	7.53	14.79
02.6007 151	Striping, thermoplastic, white, 4"	LF	0.0189	1.35	0.84	2.19
02.6007 161	Striping, thermoplastic, yellow, 4"	LF	0.0189	1.35	1.49	2.83
02.6007 171	Striping, thermoplastic, white, 8"	LF	0.0189	1.35	2.30	3.65
02.6007 181	Striping, thermoplastic, yellow, 8"	LF	0.0189	1.35	3.69	5.03
02.6007 191	Striping, yellow base, glass embedded	LF	0.0102	0.73	0.31	1.04
02.6007 201	A round buttons, large quantity	EA	0.0489	3.49	1.88	5.36
02.6007 211	B & D buttons, 2 way reflective, large quantity	EA	0.0688	4.90	4.99	9.89
02.6007 221	C buttons, 2 way reflective, large quantity	EA	0.0688	4.90	6.10	11.00
02.6007 231	G & H buttons, 1 way reflective, large quantity	EA	0.0700	4.99	4.62	9.61

02 - SITE WORK

Division	Description	Unit	Labor Hours	Labor Cost	Material Cost	Total Cost
02.6007 241	F reflector stake, steel to 6'	EA	0.1400	9.98	23.66	33.64
02.6007 251	Reverse amber reflector stake, type 1, 2 face	EA	0.1400	9.98	29.10	39.08
02.6007 261	Steel guard rail, type 9, state spec	LF	0.3969	31.42	17.01	48.43
02.6007 271	Guard rail, posts, wood treated	LF	0.3362	26.62	14.29	40.91
02.6007 281	Guard rail, anchors	LF	0.0299	2.37	3.43	5.80
02.6007 291	Guard rail, 2 side, channels & rails	LF	0.3665	29.02	51.03	80.04
02.6007 301	Add for sight screen	LF	0.0272	2.15	3.75	5.90
02.6007 311	Concrete rail continuous	LF	0.2711	21.46	49.88	71.34
02.6007 321	Deep beam median barrier-2 sides	LF	0.3414	27.03	69.79	96.82
02.6007 331	Deep beam guard rail	LF	0.2390	18.92	49.72	68.64
02.6007 341	Tri-guard guide rail	LF	0.2390	18.92	60.31	79.23
02.6007 351	W-beam guard rail	LF	0.2390	18.92	45.42	64.34
02.6007 361	Wire restrained guard rail	LF	0.2390	18.92	54.22	73.14
02.7000 000	**MISCELLANEOUS SITE IMPROVEMENTS:**					
02.7000 001	Note: All fencing prices are based on quantities between 500 and 2,000 lineal feet.					
02.7001 000	**METAL FENCING:**					
02.7001 011	Chain link, 9 gauge, 4'	LF	0.1409	12.28	14.04	26.32
02.7001 021	Chain link, 9 gauge, 6'	LF	0.1850	16.12	15.30	31.43
02.7001 031	Add for barbed wire outrigger	LF	0.0147	1.28	0.86	2.15
02.7001 041	Add for each foot to 10' high	LF	0.0258	2.25	1.57	3.82
02.7001 051	Add for rustake to 6' high	LF	0.0369	3.22	5.16	8.38
02.7001 061	Barbed wire, wood post	LF	0.1630	14.21	5.16	19.37
02.7001 071	Barbed wire, metal post	LF	0.0599	5.22	5.49	10.71
02.7001 081	Pass gate, 3' to 4'	EA	4.8024	418.53	255.62	674.15
02.7001 091	Single leaf gates to 8'	EA	5.2388	456.56	511.25	967.81
02.7001 101	Double leaf gates to 29'	EA	13.0979	1,197.54	994.13	2,191.67
02.7002 000	**WOOD FENCING:**					
02.7002 011	Wood fence, 4', economy	LF	0.0359	3.43	5.15	8.59
02.7002 021	Wood fence, 5', economy	LF	0.0383	3.66	5.80	9.46
02.7002 031	Wood fence, 6', economy	LF	0.0415	3.97	6.40	10.37
02.7002 041	Wood fence, 6', grape stake	LF	0.0445	4.26	6.80	11.06
02.7002 051	Wood fence, 6', cedar rough bd	LF	0.0417	3.99	7.94	11.93
02.7002 061	Wood fence, 6', resawn plywood	LF	0.0479	4.58	10.79	15.37
02.7002 071	Wood fence, 6', architectural quality, plain	LF	0.0896	8.57	19.73	28.30
02.7002 081	Wood fence, 6', architectural quality, decorative	LF	0.4042	36.96	28.72	65.67
02.7003 000	**MISCELLANEOUS IMPROVEMENTS:**					
02.7003 011	Guard rail, steel with wood posts	LF	0.0227	1.92	10.33	12.24
02.7003 021	Steel guard rail, channel 1 side	LF	0.0227	1.92	15.83	17.75
02.7003 031	Headlight glare shield, exposed metal	LF	0.1354	11.44	3.18	14.62
02.7003 041	Wall, precast, with posts, 8' high	LF	0.3474	29.35	25.63	54.98
02.7003 051	Add for barbed wire, 2 strands	LF	0.0239	2.02	0.83	2.85
02.7004 000	**CONCRETE FENCING:**					
02.7004 011	Fence, precast concrete, 10' high	LF	0.6100	43.47	45.75	89.23
02.7004 021	Fence, precast concrete, 12' high	LF	0.8800	62.72	58.28	121.00
02.7004 031	Fence, precast concrete, 14' high	LF	1.1500	81.96	74.31	156.27
02.7400 000	**IRRIGATION, SPRINKLER HEAD SYSTEMS:**					
02.7400 001	Note: For automatic sprinkler systems, add 15% to the total cost. For hose bibbs in lieu of heads, deduct 5% from the total cost.					
02.7401 000	**IRRIGATE LARGE AREAS:**					
02.7401 011	Irrigate, 5,000	SF	0.0077	0.80	0.60	1.40
02.7401 021	Irrigate, 10,000	SF	0.0077	0.80	0.41	1.22
02.7401 031	Irrigate, 25,000	SF	0.0077	0.80	0.37	1.17
02.7401 041	Irrigate, 50,000	SF	0.0057	0.60	0.37	0.96
02.7401 051	Irrigate, 200,000	SF	0.0048	0.50	0.35	0.85
02.7401 061	Irrigate, 500,000	SF	0.0048	0.50	0.35	0.85
02.7402 000	**IRRIGATE SMALL AREAS:**					

SITE WORK - 02

Division	Description	Unit	Labor Hours	Labor Cost	Material Cost	Total Cost
02.7402 011	Irrigate, strip, automatic	SF	0.0220	2.30	1.75	4.05
02.7402 021	Irrigate, lawn, automatic	SF	0.0106	1.11	0.97	2.08
02.7402 031	Irrigate, commercial, manual	SF	0.0077	0.80	0.86	1.67
02.7402 041	Irrigate, residential, manual	SF	0.0057	0.60	0.41	1.01
02.7403 000	**POINT OF CONNECTIONS:**					
02.7403 001	*Note: For meter fees see section 01.1031.*					
02.7403 011	Meter, main line, point of connection	EA	1.4369	150.10	81.53	231.63
02.7404 000	**BACKFLOW PREVENTION ASSEMBLIES:**					
02.7404 011	Backflow preventer, 2" dia	EA	6.5100	680.03	1,091.57	1,771.60
02.7404 021	Backflow preventer, 3" dia	EA	10.2300	1,068.63	2,226.62	3,295.24
02.7404 031	Backflow preventer, 2" dia, commercial with box	EA	11.1600	1,165.77	1,502.06	2,667.83
02.7404 041	Backflow preventer, 3" dia, commercial with box	EA	30.1116	3,145.46	1,826.27	4,971.72
02.7405 000	**MAINLINE PIPE ONLY, WITH TRENCHING:**					
02.7405 001	*Note: The following prices are based on pvc class 315 mainline pipe. For pvc schedule 40, add 10%.*					
02.7405 011	Irrigation pipe, main, 1/2", trench	LF	0.0215	2.25	0.46	2.71
02.7405 021	Irrigation pipe, main, 3/4", trench	LF	0.0241	2.52	0.62	3.14
02.7405 031	Irrigation pipe, main, 1", trench	LF	0.0293	3.06	1.21	4.27
02.7405 041	Irrigation pipe, main, 1-1/2", trench	LF	0.0314	3.28	1.44	4.72
02.7405 051	Irrigation pipe, main, 2", trench	LF	0.0366	3.82	1.90	5.72
02.7405 061	Irrigation pipe, main, 2-1/2", trench	LF	0.0402	4.20	3.05	7.25
02.7405 071	Irrigation pipe, main, 3", trench	LF	0.0456	4.76	3.63	8.39
02.7405 081	Irrigation pipe, main, 4", trench	LF	0.0509	5.32	5.05	10.36
02.7405 091	Irrigation emitter spaghetti line, complete	LF	0.6537	68.29	50.88	119.16
02.7406 000	**LATERAL LINE PIPE ONLY, WITH TRENCHING:**					
02.7406 001	*Note: The following prices are based on pvc class 200 lateral lines.*					
02.7406 011	Irrigation pipe, lateral, 1/2", trench	LF	0.0215	2.25	0.41	2.66
02.7406 021	Irrigation pipe, lateral, 3/4", trench	LF	0.0241	2.52	0.56	3.08
02.7406 031	Irrigation pipe, lateral, 1", trench	LF	0.0293	3.06	1.12	4.18
02.7406 041	Irrigation pipe, lateral, 1-1/2", trench	LF	0.0314	3.28	1.34	4.62
02.7406 051	Irrigation pipe, lateral, 2", trench	LF	0.0366	3.82	1.72	5.54
02.7406 061	Irrigation pipe, lateral, 2-1/2", trench	LF	0.0402	4.20	2.40	6.60
02.7407 000	**VALVES, REMOTE CONTROL, WITH ATMOSPHERIC VACUUM BREAKER:**					
02.7407 011	Valve, remote, 3/4", atmospheric vacuum breaker	EA	1.2887	134.62	65.01	199.63
02.7407 021	Valve, remote, 1", atmospheric vacuum breaker	EA	1.4550	151.99	83.22	235.21
02.7407 031	Valve, remote, 1-1/4", atmospheric vacuum breaker	EA	1.6616	173.57	99.03	272.60
02.7407 041	Valve, remote, 1-1/2", atmospheric vacuum breaker	EA	1.8820	196.59	113.78	310.37
02.7407 051	Valve, remote, 2", atmospheric vacuum breaker	EA	2.0939	218.73	128.36	347.08
02.7407 061	Valve, remote, 2-1/2", atmospheric vacuum breaker	EA	2.4354	254.40	323.99	578.39
02.7408 000	**VALVES, MANUAL CONTROL, WITH ATMOSPHERIC VACUUM BREAKER:**					
02.7408 011	Valve, manual, 3/4", atmospheric vacuum breaker	EA	0.5015	52.39	37.90	90.29
02.7408 021	Valve, manual, 1", atmospheric vacuum breaker	EA	0.8382	87.56	52.92	140.48
02.7408 031	Valve, manual, 1-1/4", atmospheric vacuum breaker	EA	1.0013	104.60	109.57	214.17
02.7408 041	Valve, manual, 1-1/2", atmospheric vacuum breaker	EA	1.1792	123.18	199.28	322.45
02.7408 051	Valve, manual, 2", atmospheric vacuum breaker	EA	1.5035	157.06	240.79	397.84
02.7408 061	Emitter valve assembly, with pressure reducing valve	EA	2.5330	264.60	545.17	809.77
02.7408 071	Quick coupling valves, 1", atmospheric vacuum breaker	EA	0.6706	70.05	84.89	154.94
02.7408 081	Add for concrete valve protector	EA	0.1677	17.52	12.55	30.07
02.7409 000	**SPRINKLER HEADS:**					
02.7409 011	Sprinkler head, bubble type	EA	0.1677	17.52	5.54	23.06
02.7409 021	Sprinkler head, spray type	EA	0.1675	17.50	4.15	21.65
02.7409 031	Sprinkler head, bird, #25	EA	0.2013	21.03	18.37	39.40
02.7409 041	Sprinkler head, bird, #35	EA	0.2012	21.02	37.73	58.75
02.7409 051	Sprinkler system control, residential, 3 station	EA	1.6123	168.42	87.03	255.45
02.7409 061	Sprinkler system control, residential, 6 station	EA	1.6123	168.42	108.68	277.11
02.7409 071	Sprinkler system control, commercial, 6 station	EA	6.7200	701.97	302.68	1,004.65

02 - SITE WORK

Division	Description	Unit	Labor Hours	Labor Cost	Material Cost	Total Cost
02.7409 081	Sprinkler system control, commercial, 12 station	EA	16.1270	1,684.63	462.44	2,147.07
02.7409 091	Sprinkler system master control, 24 station	EA	32.2540	3,369.25	1,303.42	4,672.67
02.7410 000	**WIRING FOR REMOTE CONTROL VALVES:**					
02.7410 001	*Note: The following prices assume that wire is placed in trench with pipe.*					
02.7410 011	Wire, #14, direct burial	LF	0.0028	0.31	0.14	0.45
02.7410 021	Wire, #12, direct burial	LF	0.0028	0.31	0.20	0.51
02.7410 031	Add for trench & backfill	LF	0.0093	1.04	0.84	1.88
02.7600 000	**LANDSCAPING:**					
02.7601 000	**LANDSCAPING:**					
02.7601 001	*Note: For freeways, use one man per work day plus materials. For commercial work, pre-emergence weed control, use $0.025 Per sf up to one acre.*					
02.7601 011	Plant maintenance, automatic, 40,000 SF/30 day	LS			2,783.68	2,783.68
02.7601 021	Plant maintenance, manual, 40,000 SF/30 day	LS			4,175.52	4,175.52
02.7602 000	**GROUND COVER, SHRUBS & TREES:**					
02.7602 011	Ground cover, ice plant, rooted, large	SF	0.0025	0.17	0.20	0.37
02.7602 021	Ground cover, all others, rooted, large	SF	0.0037	0.26	0.20	0.45
02.7602 031	Ground cover, all others, rooted, small	SF	0.0061	0.42	0.28	0.70
02.7602 041	Ground cover, ice plant, cutting, large	SF	0.0025	0.17	0.05	0.22
02.7602 051	Gravel bed, 4" pea gravel	CY	0.6786	47.25	22.56	69.82
02.7602 061	Shrubs, 1 gallon, 1,000 & up	SF	0.0763	5.31	3.98	9.29
02.7602 071	Shrubs, 1 gallon, 500-1,000	EA	0.0948	6.60	4.17	10.77
02.7602 081	Shrubs, 1 gallon, less than 500	EA	0.2608	18.16	11.14	29.30
02.7602 091	Shrub, 5 gallon	EA	0.4128	28.74	18.24	46.99
02.7602 101	Tree, 5 gallon, single staked	EA	0.5022	34.97	23.59	58.56
02.7602 111	Tree, 5 gallon, double staked	EA	0.6216	43.28	29.65	72.93
02.7602 121	Tree, 15 gallon, single staked	EA	1.9698	137.16	65.75	202.90
02.7602 131	Tree, 15 gallon, double staked	EA	2.0897	145.51	71.70	217.21
02.7602 141	Tree, 20/40 box, guyed	EA	4.7601	331.45	262.39	593.84
02.7602 151	Tree, specimen size, 30" box, guyed	EA	6.5119	453.42	601.09	1,054.51
02.7602 161	Tree, 36" box, guyed	EA	7.8152	544.17	1,005.16	1,549.33
02.7602 171	Tree, 42" box, guyed	EA	8.6495	602.26	1,441.20	2,043.46
02.7602 181	Tree, 48" box, guyed	EA	8.1758	569.28	2,010.95	2,580.23
02.7602 191	Tree, field grown	EA	5.8690	408.66	532.45	941.11
02.7602 201	Queen palm, 10' high	EA	8.8450	615.88	318.40	934.27
02.7602 211	Queen palm, 20' high	EA	9.9500	692.82	530.67	1,223.49
02.7602 221	Queen palm, 24' high	EA	17.6900	1,231.75	875.62	2,107.37
02.7602 231	Queen palm, 30' high	EA	23.8300	1,659.28	1,193.99	2,853.27
02.7603 000	**LOAM OR TOP SOIL, IN PLACE:**					
02.7603 011	Loam, under 1,000	CY	0.3029	21.09	23.83	44.92
02.7603 021	Loam, over 2,000	CY	0.2597	18.08	22.94	41.03
02.7603 031	Loam, cultivating	SY	0.0037	0.26	0.21	0.47
02.7603 041	Loam, fine grading, manual	SY	0.0071	0.49	0.21	0.70
02.7603 051	Loam, fine grading, machine	SY	0.0095	0.66	0.60	1.26
02.7604 000	**EDGING AND MULCH:**					
02.7604 011	Edging, redwood benders, 6"	LF	0.0779	5.42	1.05	6.47
02.7604 021	Edging, redwood headers, 6"	LF	0.0430	2.99	1.46	4.46
02.7604 031	Edging, redwood headers, 4"	LF	0.0359	2.50	1.03	3.53
02.7604 041	Mulch, wood chips, 2", hand spread	SY	0.0300	2.09	1.21	3.30
02.7604 051	Mulch, peat moss, 2", hand spread	SY	0.0300	2.09	3.11	5.20
02.7605 000	**PLANT BED, 18" DEEP:**					
02.7605 011	Plant bed, 18", hand prepared	SF	0.0562	3.91	1.43	5.34
02.7605 021	Plant bed, 18", machine prepared	SF	0.0479	3.34	1.43	4.76
02.7606 000	**SEEDING, SOD & FERTILIZER:**					
02.7606 011	Seeding, highway areas, perennial	#	0.0813	5.66	6.55	12.21
02.7606 021	Seeding, highway areas, annual	#	0.0384	2.67	1.43	4.10
02.7606 031	Seeding, lawn, 5,000	SF	0.0075	0.52	0.03	0.56
02.7606 041	Seeding, lawn, 10,000	SF	0.0062	0.43	0.02	0.45

SITE WORK - 02

Division	Description	Unit	Labor Hours	Labor Cost	Material Cost	Total Cost
02.7606 051	Seeding, lawn, 50,000	SF	0.0054	0.38	0.02	0.40
02.7606 061	Seeding, lawn, 100,000	SF	0.0054	0.38	0.02	0.40
02.7606 071	Hydro-seeding, 50,000 to 250,000	SF	0.0038	0.26	0.03	0.30
02.7606 081	Hydro-seeding, 250,000 to 500,000	SF	0.0033	0.23	0.02	0.25
02.7606 091	Hydro-seeding, 500,000 and up	SF	0.0026	0.18	0.02	0.20
02.7606 101	Fertilizer, commercial type	TON	5.1372	357.70	318.44	676.14
02.7606 111	Fertilizer, iron sulphate	#	0.0061	0.42	0.33	0.76
02.7606 121	Fertilizer, manure	CY	0.1449	10.09	15.45	25.54
02.7606 131	Fertilizer, rototiller mixing	SY	0.0061	0.42	0.27	0.69
02.7606 141	Sodding, instant turf, small area	SF	0.0128	0.89	0.83	1.72
02.7606 151	Sodding, instant turf, large area	SF	0.0106	0.74	0.62	1.36
02.7607 000	**LANDSCAPING ACCESSORIES:**					
02.7607 011	Wood & cast iron bench, 8'	EA			1,298.26	1,298.26
02.7607 021	Precast concrete trash receptacle, 22'6"x3'	EA			600.28	600.28
02.7607 031	Precast concrete planter, 5' x 15"	EA			1,322.12	1,322.12
02.7607 041	Precast concrete planter, 4' x 36"	EA			1,326.83	1,326.83
02.7607 051	Wire mesh arm chairs (set of 4)	SET			595.56	595.56
02.7607 061	4' dia table, central support, base	EA			707.50	707.50
02.7607 071	4' diameter table umbrella	EA			607.41	607.41
02.7607 081	Logo for 4' diameter umbrella	EA			119.11	119.11
02.7607 091	4' sq tree grate (2 piece set)	SET			385.89	385.89
02.9100 000	**RAILROAD WORK:**					
02.9101 000	**RAILROAD WORK:**					
02.9101 011	Spur track (90# rail/linear yd)	LF	0.5267	42.85	66.66	109.50
02.9101 021	Spur track bumper	EA	16.9354	1,377.69	2,316.70	3,694.39
02.9101 031	Derails	EA	5.2692	428.65	741.29	1,169.94
02.9101 041	Turn-out with switch points	EA	82.7958	6,735.44	11,120.32	17,855.76
02.9101 051	Railroad, grade & excavate roadbed	CY	0.0231	1.88	2.32	4.20
02.9101 061	Railroad, ballast	CY	0.1409	11.46	19.91	31.38
02.9101 071	Railroad ties, replacement	EA	1.2043	97.97	28.15	126.12
02.9500 000	**MARINE WORK:**					
02.9501 000	**MARINE WORK:**					
02.9501 001	*Note: The following prices for berths include piling and all utilities. The following prices for wharves include fender systems and all utilities.*					
02.9501 011	Covered berth, 28' to 36'	EA	222.3430	22,354.37	8,540.19	30,894.55
02.9501 021	Covered berth, 37' to 44'	EA	361.3073	36,325.84	13,284.78	49,610.62
02.9501 031	Open berth, 28' to 36'	EA	194.5501	19,560.07	6,958.67	26,518.74
02.9501 041	Open berth, 37' to 44'	EA	312.6698	31,435.82	11,070.53	42,506.36
02.9501 051	Timber wharf, wood deck	SF			155.91	155.91
02.9501 061	Timber wharf, concrete deck	SF			141.91	141.91
02.9501 071	Concrete pile & deck wharf	SF			170.28	170.28
02.9501 081	Dolphin, 9 pile, wood, block & wrap	EA			27,478.18	27,478.18
02.9501 091	Dredging, mobilization, maximum	LS			170,343.52	170,343.52
02.9501 101	Dredging, mobilization, minimum	LS			17,034.34	17,034.34
02.9501 111	Dredging	CY			6.96	6.96
02.9501 121	Dredging, levee embankment	CY			5.55	5.55
02.9501 131	Dredging, earthwork, unclassified	CY			5.20	5.20
02.9501 141	Dredging, riprap, bulk	TON			39.47	39.47
02.9501 151	Dredging, riprap, sacked	CY			203.91	203.91
02.9501 161	Dredging, filter material	CY			34.92	34.92

03 - CONCRETE

Division	Description	Unit	Labor Hours	Labor Cost	Material Cost	Total Cost
03.0500 000	**CONCRETE, IN PLACE:**					
03.0500 001	Note: The following prices include forms, rebar and excavation for concrete. For equipment see 1.1115.					
03.0501 000	**FOUNDATIONS:**					
03.0501 011	Foundations, tract housing	CY	2.2741	204.56	219.57	424.12
03.0501 021	Foundations, custom housing, flat lot	CY	4.1349	371.93	369.94	741.87
03.0501 031	Foundations, multi-residence, 2 story	CY	3.5251	317.08	208.90	525.98
03.0501 041	Foundations, school, 1 story, full form	CY	3.8952	350.37	224.28	574.65
03.0501 051	Foundations, school, 1 story, edge form	CY	2.5414	228.60	260.46	489.06
03.0501 061	Foundations, institutional, thru 3 story	CY	3.4631	311.51	261.30	572.80
03.0501 071	Foundations, multi-story commercial	CY	2.1005	188.94	201.40	390.34
03.0501 081	Foundations, heavy engineered structures	CY	4.7936	431.18	259.14	690.32
03.0502 000	**CONCRETE WALLS, STRUCTURAL:**					
03.0502 011	Concrete retaining walls, to 4'	CY	8.4696	761.84	300.06	1,061.90
03.0502 021	Concrete retaining walls, to 8'	CY	9.5289	857.12	286.36	1,143.49
03.0502 023	Precast, retaining sloped walls, to 35'	SF	0.2000	16.27	14.57	30.84
03.0502 025	Precast, retaining walls with geogrid, to 35'	SF	0.2500	20.34	18.64	38.98
03.0502 031	Concrete wall, 8", 12', reinforced	CY	9.8248	883.74	302.31	1,186.05
03.0502 041	Concrete wall, 10", 12', reinforced	CY	7.9559	715.63	286.59	1,002.23
03.0502 051	Concrete wall, 8", over 12', reinforced	CY	10.2182	919.13	298.18	1,217.31
03.0502 061	Concrete wall, 10", over 12', reinforced	CY	8.2751	744.35	283.55	1,027.90
03.0502 071	Concrete wall, slip form, 8"	CY	5.7803	519.94	475.01	994.95
03.0503 000	**CONCRETE WALLS, TILT-UP:**					
03.0503 011	Tilt-up walls, no pilasters, 6"	SF	0.0862	8.46	8.03	16.48
03.0503 021	Tilt-up walls, no pilasters, 8"	SF	0.0880	8.63	9.33	17.96
03.0503 031	Tilt-up walls, with pilasters, 6"	SF	0.1002	9.83	8.51	18.34
03.0503 041	Tilt-up walls, with pilasters, 8"	SF	0.1093	10.72	9.97	20.70
03.0503 051	Tilt-up walls, with pilasters, 6"	CY	4.9261	483.35	402.56	885.91
03.0503 061	Tilt-up walls, with pilasters, 8"	CY	5.1282	503.18	418.44	921.62
03.0503 071	Tilt-up walls, pilasters only	SFFA	0.1504	14.76	16.71	31.47
03.0503 081	Tilt-up walls, pilasters only	CY	7.6336	749.01	504.73	1,253.74
03.0504 000	**SLABS, BEAMS & COLUMNS, STRUCTURAL:**					
03.0504 011	Slab, 1 way beams, 6", with 200# rebar	CY	5.9737	599.46	404.83	1,004.29
03.0504 021	Slab, 2 way beams, 6", with 225# rebar	CY	7.1429	716.79	483.17	1,199.96
03.0504 031	Slab, flat, 8", with double mat	CY	4.4444	446.00	300.34	746.33
03.0504 041	Slab, flat, 12", with 250# rebar	CY	7.0486	707.33	274.22	981.54
03.0504 051	Slab, 7" post tensioned, 1#-1#, tempered bar	CY	5.2083	511.04	352.60	863.64
03.0504 061	Slab, 6"-8", with concurrent steel beam jacketing	CY	5.7803	567.16	391.69	958.85
03.0504 071	Slab, 6"-8", on permanent metal form	CY	5.4054	530.38	365.58	895.96
03.0504 081	Slab, 6"-8", lift slab construction	SF	0.1351	13.26	18.16	31.42
03.0504 091	Slab, 8", concurrent with slip form construction	CY	4.2373	415.76	287.29	703.05
03.0504 101	Slab, precast, 8"	SF	0.1003	9.84	16.12	25.96
03.0504 111	Pan, 30" square, 3"slab, 6" x 12" ribs	CY	5.7803	567.16	392.18	959.35
03.0504 121	Pan, 30" square, 3"slab, 6" x 16" ribs	CY	5.0000	490.60	339.48	830.08
03.0504 131	Pan, 30" square, 3"slab, 6" x 10" ribs	CY	4.4248	434.16	300.34	734.50
03.0504 141	Pan, 20" square, 4-1/2"slab, 6"x12" ribs	CY	4.6296	454.26	313.54	767.79
03.0504 151	Pan, 20" square, 4-1/2", 6"x16" ribs	CY	4.4248	434.16	300.34	734.50
03.0504 161	Pan, 20" square, 4-1/2"slab, 6"x20" ribs	CY	4.2373	415.76	287.29	703.05
03.0504 171	Dome slab, thin shell	SF	0.2941	26.66	19.74	46.40
03.0504 181	Stairs & stairways, concrete tread	CY	10.6383	962.87	721.23	1,684.10
03.0504 191	Stairs & stairways, concrete tread	LF	0.9634	87.20	24.14	111.33
03.0504 201	Stairs, concrete, 22 risers, 4' wide	FLOOR	89.2857	8,081.25	2,255.90	10,337.15
03.0504 211	Beams & girders, 12" x 24"	CY	10.0000	906.60	692.18	1,598.78
03.0504 221	Beams & girders, 18" x 24"	CY	8.3333	755.50	556.01	1,311.51
03.0504 231	Columns, Sonotube form, 425# rebar, 12-18" diameter	CY	6.2500	565.69	417.85	983.54

CONCRETE - 03

Division	Description	Unit	Labor Hours	Labor Cost	Material Cost	Total Cost
03.0504 241	Columns with chamfer, 425#/CY rebar, 12"square	CY	15.5581	1,408.17	649.53	2,057.69
03.0504 251	Columns with chamfer, 425#/CY rebar, 16"square	CY	11.0494	1,000.08	623.11	1,623.19
03.0504 261	Columns with chamfer, 425#/CY rebar, 18"square	CY	9.7463	882.14	617.92	1,500.06
03.0504 271	Columns with chamfer, 425#/CY rebar, 20"square	CY	8.6600	783.81	602.37	1,386.18
03.0504 281	Columns with chamfer, 425#/CY rebar, 24"square	CY	7.2890	659.72	579.96	1,239.68
03.0505 000	**REINFORCED CONCRETE, POURED IN PLACE:**					
03.0505 011	Poured in place, highway structures	CY	5.0100	450.65	339.48	790.13
03.0505 021	Poured in place, hospital wall/slab, steel frame	CY	5.7803	519.94	391.69	911.62
03.0505 031	Poured in place, hospital wall/slab, concrete frame	CY	5.2083	468.49	352.60	821.09
03.0505 041	Poured in place, office building concrete frame walls & slabs	CY	5.0000	449.75	339.48	789.23
03.0505 051	Poured in place, military structures	CY	5.0000	449.75	331.19	780.94
03.0505 061	Poured in place, parking structures	CY	5.2083	468.49	342.29	810.77
03.0505 071	Poured in place, stadiums & auditoriums	CY	5.5866	502.51	367.65	870.17
03.0505 081	Poured in place, school structures	CY	6.3694	572.93	430.96	1,003.88
03.0506 000	**MISCELLANEOUS ITEMS WITH STRUCTURAL CONCRETE:**					
03.0506 011	Lightweight concrete	CY	0.4154	29.61	28.11	57.72
03.0506 021	Waterproofing admix	CY	0.0594	4.23	3.90	8.13
03.0506 031	Color admix, black, 2-8#/sack	CY	0.1433	12.89	9.37	22.26
03.0506 041	Color admix, red/tan or green, 2-8#/sack	CY	0.2967	26.69	19.74	46.43
03.0506 051	Air entrainment, 6 sack mix	SF	0.0226	2.03	1.40	3.44
03.0506 061	Epoxy coat, squeegee applied	SF	0.0389	3.50	2.55	6.05
03.0506 071	Epoxy patch & grout mix	CY	1.2481	112.27	82.99	195.25
03.0506 081	Dust coating, mopped	SF	0.0031	0.28	0.06	0.34
03.0507 000	**SLAB ON GRADE COMBOS:**					
03.0507 001	*Note: The following prices include 2" of sand.*					
03.0507 011	Slab on grade, 4", 4" rock, memb, 6x6 w1.4/w1.4 ewwm	SF	0.0353	3.20	3.48	6.68
03.0507 021	Slab on grade, 5", 4" rock, memb, 6x6 w1.4/w1.4 ewwm	SF	0.0366	3.32	4.00	7.32
03.0507 031	Slab on grade, 6", 6" rock, memb, 6x6 w2.9/w2.9 ewwm	SF	0.0418	3.79	4.85	8.64
03.0507 041	Slab on grade, 5", 4" rock, memb, #4 @ 18" on center each way	SF	0.0402	3.65	4.47	8.11
03.0507 051	Slab on grade, 6", 6" rock, memb, #4 @ 18" on center each way	SF	0.0412	3.74	5.24	8.98
03.0507 061	Kalman floor, 6"slab, 6" rock, #4@ 18" on center each way	SF	0.0653	5.92	5.44	11.36
03.0600 000	**PRECAST CONCRETE:**					
03.0600 001	*Note: For caulking, see 7.9000. For precast utility items, see 2.5700.*					
03.0601 011	Precast concrete, single tees	SF	0.0525	5.27	12.43	17.70
03.0601 021	Precast concrete, double tees	SF	0.0359	3.60	10.22	13.82
03.0601 031	Precast concrete, beams/girders	CY	1.2992	130.37	990.57	1,120.95
03.0601 041	Precast concrete, inverted tee	CY	1.6233	162.90	1,155.34	1,318.24
03.0601 051	Precast concrete, plank, solid, 6"	SF	0.0286	2.87	6.94	9.81
03.0601 061	Precast concrete, plank, hollow, 4"	SF	0.0231	2.32	6.37	8.69
03.0601 071	Precast concrete, plank, hollow, 6"	SF	0.0258	2.59	6.73	9.32
03.0601 081	Precast concrete, plank, hollow, 8"	SF	0.0286	2.87	7.19	10.06
03.0601 091	Precast concrete, plank, hollow, 10"	SF	0.0323	3.24	7.64	10.88
03.0602 000	**PRECAST CONCRETE, ARCHITECTURAL:**					
03.0602 011	Precast panel, vertical, single form	SF	0.1197	11.74	22.09	33.84
03.0602 021	Precast panel, vertical, double form	SF	0.1197	11.74	29.55	41.30
03.0602 031	Precast panel, vertical, single form, exp aggregate	SF	0.1197	11.74	31.95	43.70
03.0602 041	Precast panel, vertical, single form, sandblast	SF	0.1197	11.74	31.95	43.70
03.0602 051	Precast panel, vertical, double form, exp aggregate	SF	0.1197	11.74	47.89	59.63
03.0602 061	Precast panel, vertical, double form, sandblast	SF	0.1197	11.74	47.89	59.63
03.0602 071	Precast panel, vertical, single form, mo-sai fin	SF	0.1197	11.74	30.36	42.11
03.0602 081	Precast panel, vertical, single form, granite fin	SF	0.1197	11.74	55.95	67.70
03.0602 091	Precast coping, 12"x8" wide, sandblast	SF	0.0870	7.83	15.68	23.51
03.0602 101	Precast sill, 12"x6", sandblasted	SF	0.1667	14.99	29.34	44.34
03.0700 000	**SPECIALTY CONCRETE:**					
03.0701 000	**GUNITE:**					

03 - CONCRETE

Division	Description	Unit	Labor Hours	Labor Cost	Material Cost	Total Cost
03.0701 001	Note: The following prices do not include form or bar.					
03.0701 011	Gunite, flat plane, 1" depth	SF	0.0159	1.44	2.42	3.86
03.0701 021	Gunite, curved arch, 1" depth	SF	0.0211	1.91	2.42	4.33
03.0701 031	Gunite, pools, 1" depth	SF	0.0176	1.25	2.42	3.68
03.0701 041	Gunite, bulk, small quantities	CY	7.0423	637.40	261.92	899.32
03.0701 051	Gunite, bulk, large quantities	CY	3.5088	317.58	261.92	579.51
03.0702 000	PRESSURE GROUTING:					
03.0702 011	Pressure grout, large quantity, 50/50 mix	CY	7.6336	544.05	232.46	776.51
03.0703 000	DRY PACKING:					
03.0703 011	Dry packing, 1" thick	SF	0.0714	5.09	3.84	8.93
03.0703 021	Dry packing, embeco, 1" thick	SF	0.1000	7.13	5.72	12.84
03.0704 000	CONCRETE FINISHES:					
03.0704 011	Concrete finish, exposed aggregate, washed	SF	0.0176	1.40	0.41	1.81
03.0704 021	Concrete finish, exposed aggregate, seeded 50%	SF	0.0211	1.68	0.50	2.18
03.0704 031	Concrete finish, exposed aggregate, seeded 100%	SF	0.0333	2.65	0.87	3.52
03.0704 041	Concrete finish, acid etch, 5% solution	SF	0.0054	0.43	0.06	0.49
03.0705 000	CONCRETE FIREPROOFING:					
03.0705 011	Fireproofing, on steel beams to 8"	CY	9.1743	877.80	257.25	1,135.05
03.0705 021	Fireproofing, steel beams 10" & up	CY	7.9365	759.36	222.57	981.94
03.0800 000	EXCAVATION & BACKFILL:					
03.0800 001	Note: This section also includes grading and bases for concrete. For mass excavation, see section 02.2000. For trench over 5' deep, add for shoring, as in section 02.4000. The presence of groundwater, sand, mud or hard pan must be covered by relative unit increases for footing drains.					
03.0801 000	EXCAVATION, SPREAD FOOTINGS AND TRENCHING:					
03.0801 011	Batter boards, foundation layout	EA	0.7449	66.73	7.78	74.51
03.0801 021	Foundations, structural	CY	0.1282	11.72	10.58	22.30
03.0801 031	Spread footings, structural	CY	0.1600	14.63	13.28	27.91
03.0801 041	Trench, structural	CY	0.1538	14.06	12.79	26.85
03.0801 051	Trenching, machine, 18" x 24" deep	LF	0.0192	1.76	1.53	3.29
03.0801 061	Trenching, machine, 24" x 36" deep	LF	0.0333	3.04	2.81	5.85
03.0801 071	Trenching, machine, 36" x 48" deep	LF	0.0638	5.83	5.24	11.08
03.0801 081	Trenching, machine, 48" x 60" deep	LF	0.0957	8.75	7.97	16.72
03.0801 091	Trench machine, minimum 4 hours with move	EACH			507.02	507.02
03.0801 101	Hand labor, trench trimming	SF	0.0230	1.64		1.64
03.0801 111	Trenching, hand, hardpan, rock	CY	1.5625	121.86	12.97	134.83
03.0801 121	Trenching, hand, earth	CY	0.8772	62.52		62.52
03.0802 000	FOOTING BACKFILL:					
03.0802 011	Backfill, hand, no compaction	CY	0.3746	26.70		26.70
03.0802 021	Backfill, hand, water jetted	CY	0.3831	27.30		27.30
03.0802 031	Backfill, hand, rammer compacted	CY	0.3937	28.06	3.76	31.82
03.0802 041	Backfill, mach, no compaction	CY	0.0877	8.02	8.42	16.44
03.0802 051	Backfill, mach, compact 90% aasho	CY	0.1879	17.18	15.76	32.94
03.0803 000	DISPOSAL, EXCESS EARTH:					
03.0803 011	Earth, spot spread, machine	CY	0.0285	2.61	2.81	5.41
03.0803 021	Earth, on site, area spread, machine	CY	0.0323	2.95	3.29	6.25
03.0805 000	GRADING FOR CONCRETE:					
03.0805 011	Rough grading	SF	0.0020	0.18	0.01	0.19
03.0805 021	Fine grading, hand	SF	0.0057	0.41		0.41
03.0805 031	Fine grading, machine	SF	0.0010	0.09	0.03	0.13
03.0806 000	FILL:					
03.0806 011	Capillary fill	CY	0.4545	35.45	14.73	50.18
03.0806 021	Capillary fill, 4" hand grade & roll	SF	0.0056	0.44	0.26	0.69
03.0806 031	Capillary fill, 4" machine grade & roll	SF	0.0010	0.08	0.36	0.44
03.0806 041	Add for each added 1", hand grade	SF	0.0014	0.11	0.06	0.17
03.0806 051	Add for each added 1", mach grade	SF	0.0003	0.02	0.07	0.09

CONCRETE - 03

Division	Description	Unit	Labor Hours	Labor Cost	Material Cost	Total Cost
03.0806 061	Sand fill for concrete, large quantity	CY	0.2857	20.36	13.20	33.56
03.0806 071	Sand fill for concrete, 2" sand cushion	SF	0.0018	0.13	0.10	0.23
03.0806 081	Add for each added 1" sand	SF	0.0009	0.06	0.03	0.10
03.0807 000	**WATERPROOF MEMBRANE:**					
03.0807 011	Membrane, polyethylene, 4 mil	SF	0.0023	0.16	0.05	0.21
03.0807 021	Membrane, polyethylene, 6 mil	SF	0.0023	0.16	0.07	0.23
03.0807 031	Membrane, 2 ply, hot mop	SF	0.0172	1.50	0.20	1.70
03.1000 000	**CONCRETE FORMS:**					
03.1000 001	*Note: The following prices are for large quantities only. They are for quick budget type estimates. For foundations, add 51% to the material costs.*					
03.1001 011	Foundations	CY			266.20	266.20
03.1001 021	Structural walls gang formed	CY			438.64	438.64
03.1001 031	Structural slabs gang formed	CY			522.21	522.21
03.1001 041	Structural beams gang formed	CY			584.85	584.85
03.1001 051	Structural columns gang formed	CY			731.06	731.06
03.1001 061	Slab on grade	CY			333.15	333.15
03.1100 000	**FOUNDATION FORMS:**					
03.1101 000	**FOUNDATION FORMS, TRACT/MULTI-UNIT:**					
03.1101 011	Foundation forms, edge	LF	0.0479	4.58	0.52	5.10
03.1101 021	Foundation forms, full form, 16" panel	SFCA	0.0777	7.43	0.80	8.23
03.1101 031	Foundation, full form, 2-3 story, 24" panel	SFCA	0.0688	6.58	0.66	7.24
03.1102 000	**FOUNDATION FORMS, CUSTOM RESIDENTIAL:**					
03.1102 011	Foundation forms, custom, edge	LF	0.0750	7.18	0.72	7.90
03.1102 021	Foundation full form, custom, 16-24" panel	SFCA	0.1208	11.56	1.02	12.58
03.1103 000	**FOUNDATION FORMS, SCHOOL AND INSTITUTIONAL:**					
03.1103 011	Foundation forms, institutional, slab reveal	LF	0.1035	9.90	1.50	11.40
03.1103 021	Foundation forms, institutional, full form	SFCA	0.1515	14.50	1.55	16.05
03.1103 041	Foundation forms, institutional, keyway	LF	0.0185	1.77	0.38	2.15
03.1200 000	**FOOTING FORMS:**					
03.1201 011	Footing forms, continuous spread, 24" high, 1 use	SFCA	0.1306	12.50	3.03	15.52
03.1201 021	Footing forms, continuous spread, 24" high, 3 uses	SFCA	0.1028	9.84	1.18	11.02
03.1201 031	Footing forms, continuous spread, 24" high, 5 uses	SFCA	0.0972	9.30	0.75	10.05
03.1201 041	Footing forms, pier type, 1 use	SFCA	0.1326	12.69	2.90	15.59
03.1201 061	Footing forms, pile cap, heavy material	SFCA	0.1083	10.36	1.98	12.35
03.1201 071	Footing forms, column & post	SFCA	0.1667	15.95	2.90	18.85
03.1201 081	Footing forms, pilaster/column	SFCA	0.1667	15.95	2.90	18.85
03.1201 091	Footing forms, curbs on foundations	SFCA	0.2413	23.09	5.38	28.47
03.1300 000	**FORMS, SLAB ON GRADE:**					
03.1301 011	Slab edge forms, 2" x 4"	LF	0.0500	4.78	0.56	5.34
03.1301 021	Slab edge forms, 2" x 6"	LF	0.0591	5.65	0.79	6.44
03.1301 031	Slab surface blockout, 1-1/2"x3/4" depressed	LF	0.0521	4.98	0.16	5.15
03.1301 041	Cold joint forms, const/key, wood	LF	0.0500	4.78	0.49	5.27
03.1301 051	Cold joint, const/key, metal	LF	0.0500	4.78	0.84	5.62
03.1301 061	Cold joint forms, keyway, removable	LF	0.0667	6.38	0.48	6.86
03.1301 071	Slab forms, chamfer strip	LF	0.0207	1.98	0.09	2.07
03.1301 081	Set screed posts, hooks & bars	SF	0.0027	0.26	0.01	0.27
03.1400 000	**CONSTRUCTION FORMS, VERTICAL:**					
03.1400 001	*Note: These prices should be used for small quantities. Discount these prices up to 25% for larger quantities, depending on project set-up costs. For battered inside forms, add 5% to the material costs.*					
03.1401 000	**FORMS, WOOD, RETAINING WALLS:**					
03.1401 011	Retaining wall forms, 4' height, 1 use	SFCA	0.1667	15.95	2.82	18.77
03.1401 021	Retaining wall forms, 4' height, 3 uses	SFCA	0.1389	13.29	1.14	14.43
03.1401 031	Retaining wall forms, 4' height, 5 uses	SFCA	0.1333	12.76	0.79	13.55
03.1401 041	Retaining wall forms, 8' height, 1 use	SFCA	0.1667	15.95	2.45	18.39

03 - CONCRETE

Division	Description	Unit	Labor Hours	Labor Cost	Material Cost	Total Cost
03.1401 051	Retaining wall forms, 8' height, 3 uses	SFCA	0.1389	13.29	0.97	14.26
03.1401 061	Retaining wall forms, 8' height, 5 uses	SFCA	0.1333	12.76	0.68	13.44
03.1401 071	Retaining wall forms, 12' height, 1 use	SFCA	0.1845	17.66	2.62	20.28
03.1401 081	Retaining wall forms, 12' height, 3 uses	SFCA	0.1567	14.10	1.06	15.15
03.1401 091	Retaining wall forms, 12' height, 5 uses	SFCA	0.1512	13.60	0.77	14.37
03.1402 000	**FORMS, WOOD, BUILDING WALLS:**					
03.1402 011	Wall forms, 8' height, 1 use	SFCA	0.1667	15.95	2.45	18.40
03.1402 021	Wall forms, 8' height, 3 uses	SFCA	0.1389	13.29	0.97	14.26
03.1402 031	Wall forms, 8' height, 5 uses	SFCA	0.1333	12.75	0.68	13.44
03.1402 041	Wall forms, 12' height, 1 use	SFCA	0.1780	16.01	2.62	18.63
03.1402 051	Wall forms, 12' height, 3 uses	SFCA	0.1503	13.52	1.08	14.60
03.1402 061	Wall forms, 12' height, 5 uses	SFCA	0.1447	13.02	0.77	13.78
03.1402 071	Wall forms, 16' height, 1 use	SFCA	0.1955	17.59	2.52	20.10
03.1402 081	Wall forms, 16' height, 3 uses	SFCA	0.1677	15.08	1.03	16.12
03.1402 091	Wall forms, 16' height, 5 uses	SFCA	0.1622	14.59	0.73	15.32
03.1402 101	Wall form, blockouts & offsets	SFCA	0.1750	16.74	3.11	19.85
03.1403 000	**WALLS, FLYING FORMS:**					
03.1403 011	Flying form, walls to 9', 1 use	SFCA	0.0173	1.56	3.13	4.69
03.1403 021	Flying form, walls to 9', 3 uses	SFCA	0.0173	1.56	2.30	3.85
03.1403 031	Flying form, walls to 9', 5 uses	SFCA	0.0173	1.56	1.47	3.03
03.1403 041	Flying form, walls to 12', 1 use	SFCA	0.0182	1.64	3.46	5.09
03.1403 051	Flying form, walls to 12', 3 uses	SFCA	0.0182	1.64	2.62	4.26
03.1403 061	Flying form, walls to 12', 5 uses	SFCA	0.0182	1.64	1.73	3.37
03.1403 071	Flying form, walls to 16', 1 use	SFCA	0.0196	1.76	3.73	5.50
03.1403 081	Flying form, walls to 16', 3 uses	SFCA	0.0154	1.39	2.93	4.32
03.1403 091	Flying form, walls to 16', 5 uses	SFCA	0.0094	0.85	2.10	2.94
03.1600 000	**COLUMN FORMS:**					
03.1600 001	*Note: These prices should be used for small quantities. Discount these prices up to 25% for larger quantities, depending on project set-up.*					
03.1601 000	**COLUMN FORMS, WOOD, BUILT-UP:**					
03.1601 011	Column forms, square/rect, 1 use	SFCA	0.1250	11.96	2.60	14.56
03.1601 021	Column forms, square/rect, 3 uses	SFCA	0.0972	8.74	0.77	9.51
03.1601 031	Column forms, square/rect, 5 uses	SFCA	0.0917	8.25	0.48	8.72
03.1601 041	Add for 3/4" chamfer, column form	LF	0.0111	1.06	0.14	1.20
03.1602 000	**STEEL FORMS, MULTI-USE:**					
03.1602 011	Column forms, round/rect, rent	SFCA	0.0625	5.62	1.44	7.06
03.1603 000	**FIBREBOARD FORMS:**					
03.1603 011	Fibreform, circular column, 12"	LF	0.1176	10.58	6.53	17.11
03.1603 021	Fibreform, circular column, 14"	LF	0.1176	10.58	9.09	19.67
03.1603 031	Fibreform, circular column, 16"	LF	0.1176	10.58	11.89	22.47
03.1603 041	Fibreform, circular column, 20"	LF	0.1277	11.49	19.54	31.03
03.1603 051	Fibreform, circular column, 24"	LF	0.1277	11.49	25.89	37.37
03.1603 061	Fibreform, circular column, 30"	LF	0.1370	12.32	39.04	51.36
03.1603 071	Fibreform, circular column, 36"	LF	0.1464	13.17	48.62	61.79
03.1604 000	**COLUMN FORMS, METAL (LOST FORM):**					
03.1604 011	Form, circular, 12", lost	SFCA	0.0621	5.59	9.75	15.34
03.1604 021	Form, circular, 16", lost	SFCA	0.0621	5.59	13.12	18.70
03.1604 031	Form, circular, 20", lost	SFCA	0.0630	5.67	16.59	22.25
03.1604 041	Form, circular, 24", lost	SFCA	0.0630	5.67	21.96	27.62
03.1604 051	Form, circular, 30", lost	SFCA	0.0639	5.75	30.78	36.53
03.1604 061	Form, circular, 36", lost	SFCA	0.0658	5.92	36.46	42.38
03.1604 071	Form, circular, 42", lost	SFCA	0.0677	6.09	47.47	53.56
03.1605 000	**COLUMNS FORMS, FLYING, METAL OR FIBERGLASS:**					
03.1605 001	*Note: The following prices assume 4 uses per month.*					
03.1605 011	Column form, flying, 12"	SFCA	0.0621	5.59	3.11	8.69

CONCRETE - 03

Division	Description	Unit	Labor Hours	Labor Cost	Material Cost	Total Cost
03.1605 021	Column form, flying, 16"	SFCA	0.0621	5.59	4.20	9.78
03.1605 031	Column form, flying, 20"	SFCA	0.0621	5.59	5.29	10.87
03.1605 041	Column form, flying, 24"	SFCA	0.0621	5.59	7.10	12.68
03.1605 051	Column form, flying, 30"	SFCA	0.0639	5.75	9.90	15.65
03.1605 061	Column form, flying, 36"	SFCA	0.0658	5.92	11.81	17.73
03.1605 071	Column form, flying, 42"	SFCA	0.0677	6.09	15.25	21.34
03.1605 081	Capital column form, cone	SFCA	0.2413	21.70	9.93	31.63
03.1606 000	**PILASTERS, TILT-UP:**					
03.1606 011	Pilasters, wood formed	LF	0.3333	31.89	2.90	34.79
03.1700 000	**CONSTRUCTION FORMS, HORIZONTAL:**					
03.1700 001	*Note: Discount these prices up to 25% for larger quantities, depending on project set-up costs.*					
03.1701 000	**SLAB FORMS:**					
03.1701 011	Slab form, edge header	LF	0.0593	5.67	1.29	6.96
03.1701 021	Slab form, soffit, ply, 1 use	SFCA	0.2131	19.09	3.02	22.11
03.1701 031	Slab form, soffit, ply, 3 use	SFCA	0.1384	11.83	1.72	13.55
03.1701 041	Slab form, soffit, ply, 5 use	SFCA	0.1234	10.55	1.29	11.83
03.1701 051	Slab form, shoring, rent	SF			0.64	0.64
03.1701 061	Slab form, re-shoring	SF	0.0111	0.93	0.16	1.09
03.1701 071	Slab form, drop head, soffit, 1 use	SFCA	0.3333	31.89	2.25	34.14
03.1701 081	Slab form, screeds	SF	0.0041	0.39	0.08	0.47
03.1702 000	**BEAM AND GIRDER FORMS:**					
03.1702 011	Beam form, sides, 1 use	SFCA	0.3284	31.42	2.48	33.90
03.1702 021	Beam form, sides, 3 uses	SFCA	0.2463	22.15	1.19	23.35
03.1702 031	Beam form, sides, 5 uses	SFCA	0.2299	20.68	0.89	21.57
03.1702 041	Beam form, soffits, 1 use	SFCA	0.3284	31.42	2.63	34.05
03.1702 051	Beam form, soffits, 3 uses	SFCA	0.2463	22.15	2.04	24.20
03.1702 061	Beam form, soffits, 5 uses	SFCA	0.2299	20.68	1.72	22.40
03.1702 071	Beam form, cap head, side	SFCA	0.3333	29.98	2.63	32.61
03.1702 081	Beam form, cap head, soffit	SFCA	0.3333	29.98	2.63	32.61
03.1703 000	**PANS, WAFFLE AND JOIST, SUBCONTRACTOR ERECTED:**					
03.1703 001	*Note: These prices include forms, rental, shoring at 12' height (including beams and girders to nearest wall. General contractor provides ledgers at walls).*					
03.1703 011	Metal pans, 1 use, 10-20,000 square feet	SFCA	0.0682	6.84	5.49	12.33
03.1703 021	Metal pans, 3 use, 10-20,000 square feet	SFCA	0.0626	6.28	4.20	10.48
03.1703 031	Metal pans, 5 use, 10-20,000 square feet	SFCA	0.0617	6.19	3.18	9.37
03.1703 041	Fiberglass pans, 1 use	SFCA	0.0682	6.45	5.97	12.43
03.1703 051	Fiberglass pans, 3 uses	SFCA	0.0626	5.92	5.71	11.63
03.1703 061	Fiberglass pans, 5 uses	SFCA	0.0617	5.84	4.72	10.56
03.1704 000	**STRUCTURAL SLABS WITH BEAMS, COMPOSITE, FORMS:**					
03.1704 011	Slab forms, 1 way beams, 1 use	SFCA	0.0562	5.06	4.53	9.59
03.1704 021	Slab forms, 1 way beams, 3 uses	SFCA	0.0562	5.06	4.37	9.43
03.1704 031	Slab forms, 2 way beams, 1 use	SFCA	0.0673	6.05	5.39	11.45
03.1704 041	Slab forms, 2 way beams, 3 uses	SFCA	0.0673	6.05	5.00	11.05
03.1705 000	**STRUCTURAL SLAB AND BEAM FLYING FORMS:**					
03.1705 011	Slab, flying forms, soffits, 1 use	SFCA	0.0173	1.56	5.07	6.62
03.1705 021	Slab, flying forms, soffits, 3 uses	SFCA	0.0169	1.52	3.68	5.20
03.1705 031	Slab, flying forms, soffits, 5 uses	SFCA	0.0168	1.51	2.55	4.06
03.1705 041	Beam, flying forms, side, 1 use	SFCA	0.0447	4.02	5.63	9.65
03.1705 051	Beam, flying forms, side, 3 uses	SFCA	0.0438	3.94	4.35	8.29
03.1705 061	Beam, flying forms, side, 5 uses	SFCA	0.0428	3.85	3.22	7.07
03.1800 000	**CONSTRUCTION FORMS, HORIZONTAL, HEAVY DUTY:**					
03.1800 001	*Note: Prices include stripping, cleaning, and oiling forms, and shoring erection. Add shoring material. These prices should be used for small quantities. Discount these prices up to 25% for larger quantities depending on project set-up costs.*					

03 - CONCRETE

Division	Description	Unit	Labor Hours	Labor Cost	Material Cost	Total Cost
03.1801 000	**SLAB FORMS, HEAVY DUTY:**					
03.1801 011	Slab form, edge	LF	0.0816	7.81	3.09	10.89
03.1801 021	Slab form, soffit, ply, 1 use	SFCA	0.2367	21.20	7.14	28.35
03.1801 031	Slab form, soffit, ply, 3 uses	SFCA	0.1619	13.84	4.12	17.95
03.1801 041	Slab form, soffit, ply, 5 uses	SFCA	0.1469	12.56	3.09	15.64
03.1801 051	Slab form, shoring, rent	SF			1.53	1.53
03.1801 061	Slab form, re-shoring	SF	0.0363	3.03	0.30	3.33
03.1801 071	Slab form, drop head, soffit, 1 use	SFCA	0.3556	34.02	5.25	39.28
03.1801 081	Slab form, screeds	SF	0.0264	2.53	0.08	2.61
03.1802 000	**BEAM AND GIRDER FORMS, HEAVY DUTY:**					
03.1802 011	Beam form, sides, 1 use	SFCA	0.3507	33.55	5.85	39.40
03.1802 021	Beam form, sides, 3 uses	SFCA	0.2690	24.20	2.80	26.99
03.1802 031	Beam form, sides, 5 uses	SFCA	0.2526	22.72	2.17	24.89
03.1802 041	Beam form, soffits, 1 use	SFCA	0.3507	33.55	6.17	39.73
03.1802 051	Beam form, soffits, 3 uses	SFCA	0.2690	24.20	4.76	28.95
03.1802 061	Beam form, soffits, 5 uses	SFCA	0.2526	22.72	4.00	26.72
03.1802 071	Beam form, cap head, side	SFCA	0.3560	32.02	6.21	38.23
03.1802 081	Beam form, cap head, soffit	SFCA	0.3560	32.02	6.17	38.19
03.1803 000	**PANS, WAFFLE & JOIST, SUBCONTRACTOR ERECTED, HEAVY DUTY:**					
03.1803 001	*Note: These prices include forms, rental, shoring at 12' height (including beams and girders to nearest wall. General contractor provides ledgers at walls).*					
03.1803 011	Metal pans, 1 use, 10-20,000 square feet	SFCA	0.0888	8.91	8.25	17.16
03.1803 021	Metal pans, 3 use, 10-20,000 square feet	SFCA	0.0832	8.35	6.29	14.64
03.1803 031	Metal pans, 5 use, 10-20,000 square feet	SFCA	0.0823	8.26	4.76	13.01
03.1803 041	Fiberglass pans, 1 use	SFCA	0.0898	8.50	8.98	17.47
03.1803 051	Fiberglass pans, 3 uses	SFCA	0.0842	7.97	8.52	16.49
03.1803 061	Fiberglass pans, 5 uses	SFCA	0.0833	7.88	7.09	14.97
03.1804 000	**STRUCTURAL SLABS WITH BEAMS, COMPOSITE, FORMS, HEAVY DUTY:**					
03.1804 011	Slab forms, 1 way beams, 1 use	SFCA	0.0789	7.10	7.69	14.79
03.1804 021	Slab forms, 1 way beams, 3 uses	SFCA	0.0789	7.10	7.31	14.40
03.1804 031	Slab forms, 2 way beams, 1 use	SFCA	0.0900	8.10	9.20	17.29
03.1804 041	Slab forms, 2 way beams, 3 uses	SFCA	0.0900	8.10	8.52	16.62
03.1805 000	**STRUCTURAL SLAB & BEAM FLYING FORMS, HEAVY DUTY:**					
03.1805 011	Slab, flying forms, soffits, 1 use	SFCA	0.0401	3.61	8.68	12.28
03.1805 021	Slab, flying forms, soffits, 3 uses	SFCA	0.0396	3.56	6.29	9.85
03.1805 031	Slab, flying forms, soffits, 5 uses	SFCA	0.0396	3.56	4.35	7.91
03.1805 041	Beam, flying forms, side, 1 use	SFCA	0.0674	6.06	9.51	15.57
03.1805 051	Beam, flying forms, side, 3 uses	SFCA	0.0665	5.98	7.28	13.27
03.1805 061	Beam, flying forms, side, 5 uses	SFCA	0.0656	5.90	5.49	11.39
03.1850 000	**STAIR AND RAMP FORMS:**					
03.1850 011	Stair forms, landing soffits	SFCA	0.2000	20.11	4.37	24.48
03.1850 021	Stair forms, sloping soffits	SFCA	0.2941	29.57	4.76	34.33
03.1850 031	Stair forms, risers	LF	0.2941	29.57	4.74	34.32
03.1900 000	**MISCELLANEOUS CONCRETE FORMS & FORM SPECIALTIES:**					
03.1900 001	*Note: For architectural detail, multi-story batter walls, radial walls or architectural board forming, add 150% to the total form costs.*					
03.1901 000	**CONSTRUCTION JOINTS:**					
03.1901 011	Control joint	LF	0.0258	2.47	1.39	3.86
03.1901 021	Expansion joint	LF	0.0833	7.97	1.50	9.47
03.1901 031	Keyed 4" patent with stakes	LF	0.0415	3.97	2.24	6.21
03.1901 041	Key joints, recess form	LF	0.0295	2.97	1.39	4.36
03.1902 000	**SPECIAL FORMS:**					
03.1902 011	Fiberglass architectural forms	SFCA	0.1111	9.99	20.64	30.64
03.1903 000	**FORMS, SPECIALTIES, GENERAL:**					
03.1903 011	Forms, curbs, equipment bases	LF	0.1481	14.17	2.28	16.46

CONCRETE - 03

Division	Description	Unit	Labor Hours	Labor Cost	Material Cost	Total Cost
03.1903 021	Forms, slab depressions	LF	0.0673	6.44	1.25	7.69
03.1903 031	Add for rough board forms	SFCA	0.0105	1.00	6.75	7.75
03.1903 041	Form detailing	SFCA			0.35	0.35
03.1905 000	**LINERS FOR FORMS:**					
03.1905 011	Form liner, foam rubber	SFCA	0.1750	16.74	7.70	24.45
03.1905 021	Form liner, plastic	SFCA	0.1750	16.74	8.33	25.07
03.1905 031	Form liner, metal	SFCA	0.1750	16.74	19.15	35.89
03.1905 041	Form liner, architectural board	SFCA	0.1750	16.74	12.13	28.88
03.1905 051	Form liner, fiberglass, reusable	SFCA	0.1750	16.74	11.90	28.64
03.2000 000	**REINFORCING STEEL:**					
03.2000 001	Note: Average weight of post tensioned reinforcing slabs is as follows: Building Slabs 1#/SFSA (50# Bridge Slabs 2.5#/SFSA; AASHO Standards). Balance as detailed.					
03.2100 000	**REINFORCING STEEL, IN PLACE:**					
03.2101 000	**CAISSON, REINFORCING, WITH SPIRAL:**					
03.2101 011	Caisson, 16" diameter, with 4 #6 bars	LF	0.0185	1.74	7.91	9.65
03.2101 021	Caisson, 24" diameter, with 6 #6 bars	LF	0.0350	3.30	10.17	13.47
03.2101 031	Caisson, 36" diameter, with 8 #8 bars	LF	0.0709	6.68	18.51	25.19
03.2102 000	**PRE-STRESS AND POST TENSIONING:**					
03.2102 011	Steel bars	#	0.0157	1.48	1.05	2.53
03.2102 021	Tubing & wire rope	#	0.0194	1.83	1.45	3.28
03.2102 031	Wire rope without tubing	#	0.0185	1.74	1.19	2.93
03.2200 000	**REINFORCING STEEL, BUILT-UP COST:**					
03.2201 000	**BARS, LIGHTWEIGHT, #3-#5:**					
03.2201 011	Rebar, lightweight, #3-#5, 0-5,000	#	0.0079	0.74	0.59	1.34
03.2201 021	Rebar, lightweight, #3-#5, 5,000 to 20,000	#	0.0067	0.63	0.55	1.18
03.2201 031	Rebar, lightweight, #3-#5, 20,000-50,000	#	0.0061	0.57	0.55	1.13
03.2201 041	Rebar, lightweight, #3-#5, 50,000-100,000	#	0.0055	0.52	0.55	1.07
03.2201 051	Rebar, lightweight, #3-#5, 100,000 & up	#	0.0051	0.48	0.55	1.03
03.2202 000	**BARS, MEDIUM WT, #6-#8:**					
03.2202 011	Rebar, medium wt, #6-#8, 5,000-50,000	#	0.0061	0.57	0.59	1.17
03.2202 021	Rebar, medium wt, #6-#8, 50,000-100,000	#	0.0055	0.52	0.55	1.07
03.2202 031	Rebar, medium wt, #8-#8, 100,000-250,000 & up	#	0.0051	0.48	0.55	1.03
03.2203 000	**BARS, HEAVY, #9 AND UP:**					
03.2203 011	Rebar, heavy, #9 and up, less than 100,000	#	0.0051	0.48	0.55	1.03
03.2203 021	Rebar, heavy, #9 and up, 100,000-250,000	#	0.0044	0.41	0.55	0.97
03.2203 031	Rebar, heavy, #9 and up, 250,000 and up	#	0.0040	0.38	0.55	0.93
03.2204 000	**REINFORCING STEEL, ACCESSORIES AND SPECIALTIES:**					
03.2204 011	Reinforcing steel, chairs, average	EA	0.0102	0.96	0.55	1.51
03.2204 021	Reinforcing steel, spirals	#	0.0067	0.63	0.76	1.39
03.2204 031	Reinforcing steel, splices	EA	0.1667	15.71	15.71	31.42
03.2204 041	Reinforcing steel, stirrups	#	0.0250	2.36	0.69	3.05
03.2205 000	**REINFORCING BAR, WELDING:**					
03.2205 011	Rebar, weld, #4 bar, 24 welds/day	WELD	0.5000	47.11	4.51	51.62
03.2205 021	Rebar, weld, #6 bar, 12 welds/day	WELD	0.6667	62.82	9.20	72.02
03.2205 031	Rebar, weld, #8 bar, 10 welds/day	WELD	0.8000	75.38	11.03	86.41
03.2205 041	Rebar, weld, #9 bar, 8 welds/day	WELD	1.0000	94.22	13.86	108.08
03.2205 051	Rebar, weld, #10 bar, 6 welds/day	WELD	1.3333	125.62	18.53	144.16
03.2205 061	Rebar, weld, #14 bar, 4 welds/day	WELD	2.0000	188.44	27.85	216.29
03.2205 071	Rebar, weld, #18 bar, 1 welds/day	WELD	8.0000	753.76	37.11	790.87
03.2206 000	**REINFORCING STEEL, WIRE MESH:**					
03.2206 011	Mesh, 4/4, W2/W2, for slab	SF	0.0040	0.38	0.19	0.56
03.2206 021	Mesh, 4/4, W2.9/W2.9, for slab	SF	0.0047	0.44	0.32	0.76
03.2206 031	Mesh, 6/6, W1.4/W1.4, for slab	SF	0.0020	0.19	0.32	0.51
03.2206 041	Mesh, 6/6, W2/W2, for slab	SF	0.0028	0.26	0.14	0.41

03 - CONCRETE

Division	Description	Unit	Labor Hours	Labor Cost	Material Cost	Total Cost
03.2206 051	Mesh, 6/6, W2.9/W2.9, for slab	SF	0.0038	0.36	0.15	0.51
03.2206 061	Mesh, cut lengthwise	SF	0.0019	0.18		0.18
03.2206 071	Mesh, 2/2, 14/14, beams & columns, galvanized	SF	0.0312	2.94	0.19	3.13
03.2206 081	Mesh, 2/2, 12/12, beams & columns, galvanized	SF	0.0312	2.94	0.33	3.27
03.2208 000	**REBAR, NON-CORROSIVE, RESIN & FIBERGLASS:**					
03.2208 011	Rebar, non-corrosive, resin & fiberglass, #2, 1/4" diameter	SF	0.0067	0.63	0.45	1.08
03.2208 021	Rebar, non-corrosive, resin & fiberglass, #3, 3/8" diameter	SF	0.0067	0.63	0.66	1.29
03.2208 031	Rebar, non-corrosive, resin & fiberglass, #4, 1/2" diameter	SF	0.0067	0.63	0.97	1.61
03.2208 041	Rebar, non-corrosive, resin & fiberglass, #5, 5/8" diameter	SF	0.0067	0.63	1.28	1.91
03.2208 051	Rebar, non-corrosive, resin & fiberglass, #6, 3/4" diameter	SF	0.0055	0.52	1.70	2.22
03.2208 061	Rebar, non-corrosive, resin & fiberglass, #7, 7/8" diameter	SF	0.0055	0.52	2.08	2.59
03.2208 071	Rebar, non-corrosive, resin & fiberglass, #8, 1" diameter	SF	0.0055	0.52	2.63	3.15
03.2208 081	Rebar, non-corrosive, resin & fiberglass, #9, 1-1/8" diameter	SF	0.0055	0.52	2.97	3.49
03.3000 000	**READYMIX CONCRETE:**					
03.3000 001	Note: The following prices do not include forms, finishing or rebar. Prices are subject to quoted discounts of 5 to 25%.					
03.3001 000	**DESIGN MIX POSTED PRICE & SALES TAX:**					
03.3001 011	Ready mix, 2,000 PSI, 1-1/2" aggregate, 4.4 sack mix	CY			118.42	118.42
03.3001 021	Ready mix, 2,500 PSI, 1-1/2" aggregate, 4.8 sack mix	CY			121.53	121.53
03.3001 031	Ready mix, 3,000 PSI, 1-1/2" aggregate, 5.2 sack mix	CY			124.65	124.65
03.3001 041	Ready mix, 3,500 PSI, 1-1/2" aggregate, 5.8 sack mix	CY			129.33	129.33
03.3001 051	Ready mix, 4,000 PSI, 1-1/2" aggregate, 6.5 sack mix	CY			134.81	134.81
03.3001 061	Ready mix, 2,000 PSI, 3/4" aggregate, 4.7 sack mix	CY			120.75	120.75
03.3001 071	Ready mix, 2,500 PSI, 3/4" aggregate, 5.1 sack mix	CY			123.87	123.87
03.3001 081	Ready mix, 3,000 PSI, 3/4" aggregate, 5.5 sack mix	CY			127.01	127.01
03.3001 091	Ready mix, 3,500 PSI, 3/4" aggregate, 6.3 sack mix	CY			133.25	133.25
03.3001 101	Ready mix, 4,000 PSI, 3/4" aggregate, 6.8 sack mix	CY			137.16	137.16
03.3002 000	**ADDERS FOR DESIGN MIX:**					
03.3002 011	Hi early strength	CY			12.61	12.61
03.3002 021	Lightweight aggregate	CY			54.86	54.86
03.3002 031	Granite aggregate	CY			14.10	14.10
03.3002 041	White cement	CY			201.32	201.32
03.3002 051	Calcium chloride, 1%	CY			0.71	0.71
03.3002 061	Short loads (add each cubic yard under 9)	EA			70.95	70.95
03.3002 071	Stand-by charge (5 minutes per cubic yard-no charge)	MIN			1.72	1.72
03.3002 081	Stand-by charge (over time)	HR			106.39	106.39
03.3002 091	Pumping quality, mix	CY			9.71	9.71
03.3002 101	Fiber reinforcement	CY			13.69	13.69
03.3100 000	**CONCRETE PLACEMENT:**					
03.3101 000	**POUR FOUNDATIONS:**					
03.3101 001	Foundation, tract housing	CY	0.2222	16.58	126.70	143.28
03.3101 011	Foundation, residential, truck access	CY	0.2564	21.40	126.70	148.10
03.3101 021	Foundation, residential, hillside	CY	0.4211	35.99	126.70	162.69
03.3101 023	Foundations, multi-residence, 2 story	CY	0.2564	19.14	130.05	149.19
03.3101 025	Foundations, school, 1 story, full form	CY	0.7142	53.30	135.79	189.09
03.3101 027	Foundations, school, 1 story, edge form	CY	0.6522	48.67	138.38	187.05
03.3101 031	Foundation, institutional, thru 3 story	CY	1.0527	78.56	135.79	214.36
03.3101 041	Foundations, multi-story commercial	CY	0.7500	55.97	135.79	191.77
03.3101 051	Foundations, heavy engineered structures	CY	0.9231	68.89	135.79	204.68
03.3101 061	Foundations, retaining wall footing	CY	0.8750	65.30	140.03	205.33
03.3102 000	**POUR WALLS:**					
03.3102 011	Concrete retaining walls, to 4'	CY	0.7500	55.97	141.29	197.26
03.3102 021	Concrete retaining walls, to 8'	CY	0.8243	61.52	141.29	202.81
03.3102 031	Concrete wall, 8", to 12'	CY	0.6580	49.11	137.22	186.33
03.3102 041	Concrete wall, 10", to 12'	CY	0.5635	42.05	137.22	179.27

CONCRETE - 03

Division	Description	Unit	Labor Hours	Labor Cost	Material Cost	Total Cost
03.3102 051	Concrete wall, 12", to 12'	CY	0.4928	36.78	137.22	174.00
03.3102 061	Concrete wall, 8", over 12'	CY	0.6580	49.11	137.22	186.33
03.3102 071	Concrete wall, 10", over 12'	CY	0.5635	42.05	137.22	179.27
03.3103 000	**POUR STRUCTURAL SLAB:**					
03.3103 011	Slab, suspended, 1 story, pumped	CY	0.8750	65.30	148.43	213.74
03.3103 021	Slab, beam & flat slab, pumped	CY	0.8583	64.05	148.43	212.49
03.3103 031	Slab, pan joist & slab, pumped	CY	0.8250	61.57	148.43	210.00
03.3103 041	Slab, fill on metal deck, 3" to 5", pumped	CY	1.2500	93.29	148.43	241.72
03.3104 000	**OTHER CONCRETE POURS:**					
03.3104 011	Concrete basement walls, pumped	CY	1.3158	98.20	143.18	241.38
03.3104 021	Beams & slabs, suspended, pumped	CY	0.7353	54.88	148.43	203.31
03.3104 031	Columns, 20" & under(cross-section), pumped	CY	1.5000	111.95	148.43	260.38
03.3104 035	Columns, 20" square feet & over(cross-section), pumped	CY	0.9000	67.17	148.43	215.60
03.3104 041	Mass concrete, on grade	CY	0.2000	14.93	143.18	158.11
03.3104 051	Pile caps & bases	CY	0.6000	44.78	143.18	187.96
03.3104 061	Slabs, on grade 4" to 5" thick, direct chute	CY	0.3333	24.87	128.39	153.27
03.3104 071	Increase for each added inch thick	SF	0.0015	0.11	0.35	0.46
03.3104 081	Add for checkerboard pours	SF	0.0020	0.18	0.01	0.19
03.3104 091	Stairs on grade, pumped	CY	1.6130	144.17	148.43	292.60
03.3104 095	Stairs & landings, suspended, pumped	CY	2.4999	223.44	148.43	371.88
03.3104 101	Steps, concrete	LF/TR	0.1668	14.91	0.58	15.49
03.3104 111	Concrete test	EA			184.59	184.59
03.3105 000	**LIGHTWEIGHT FILLS, METAL DECK:**					
03.3105 011	Lightweight fill, 2-1/4", 1-1/2" deck	SF	0.0263	1.96	2.41	4.38
03.3105 021	Lightweight fill, 2-1/4", 3" deck	SF	0.0286	2.13	3.07	5.21
03.3105 031	Lightweight fill, 3-1/4", 1-1/2" deck	SF	0.0294	2.19	3.07	5.27
03.3105 041	Lightweight concrete fill, 3-1/4", 3" deck	SF	0.0323	2.41	3.53	5.94
03.3105 051	Lightweight fill, 3-1/4", 4-1/2" deck	SF	0.0345	2.57	4.01	6.59
03.3105 061	Lightweight concrete fills, pour	CY	0.5000	37.32	220.90	258.22
03.3600 000	**SLAB FINISHES:**					
03.3601 011	Float only	SF	0.0062	0.49		0.49
03.3601 021	Trowel, steel, by machine	SF	0.0093	0.74		0.74
03.3601 031	Trowel, steel, by hand	SF	0.0130	1.03		1.03
03.3601 033	Trowel small area, pour strips, etc.	SF	0.0250	1.99		1.99
03.3601 035	Trowel stair landings	SF	0.0400	3.18		3.18
03.3601 037	Trowel stair treads & nosing	SF	0.2685	21.38		21.38
03.3601 039	Trowel steel pan treads	SF	0.0800	6.37		6.37
03.3601 041	Broom finish	SF	0.0111	0.88		0.88
03.3601 051	Scoring	LF	0.0043	0.34		0.34
03.3601 061	Exposed aggregate, washed	SF	0.0160	1.27	0.05	1.32
03.3601 071	Seeded aggregate, native material	SF	0.0178	1.42	0.15	1.57
03.3601 081	Non-slip finish	SF	0.0013	0.10	0.51	0.61
03.3601 091	Hardener, sprayed	SF	0.0006	0.05	0.30	0.35
03.3601 101	Color, tans and browns	SF	0.0013	0.10	1.13	1.23
03.3601 111	Color, reds & greens	SF	0.0013	0.10	2.23	2.33
03.3601 121	Curing compound	SF			0.03	0.03
03.3601 131	Treads and risers	LF/TR	0.0500	3.98		3.98
03.3601 141	Treads & risers with abrasives	LF/TR	0.0667	5.31	2.30	7.61
03.3601 151	Wax	SF	0.0019	0.15	0.07	0.22
03.3700 000	**VERTICAL SURFACE FINISHES:**					
03.3701 011	Cut back ties & patch	SF	0.0093	0.66	0.30	0.96
03.3701 021	Patch & remove fins	LF	0.0062	0.44	0.02	0.47
03.3701 031	Patch & grind smooth	SF	0.0185	1.32	0.02	1.34
03.3701 041	Patch & sack, simple	SF	0.0099	0.71	0.07	0.78
03.3701 051	Sandblast, light	SF	0.0269	1.92	0.19	2.10

03 - CONCRETE

Division	Description	Unit	Labor Hours	Labor Cost	Material Cost	Total Cost
03.3701 061	Sandblast, medium	SF	0.0383	2.73	0.41	3.14
03.3701 071	Sandblast, heavy	SF	0.0464	3.31	0.73	4.04
03.3701 081	Bush hammer, light	SF	0.0271	1.93	0.73	2.66
03.3701 091	Bush hammer, medium	SF	0.0465	3.31	1.52	4.83
03.3701 101	Bush hammer, heavy	SF	0.0659	4.70	2.85	7.55
03.3701 111	Needle gun treatment	SF	0.0465	3.31	1.89	5.20
03.3701 121	Wire brush, green concrete	SF	0.0154	1.10	0.02	1.12
03.3701 131	Wash with acid & rinse	SF	0.0090	0.64	0.06	0.70
03.3800 000	**MISCELLANEOUS CONCRETE FINISHES:**					
03.3801 000	**ADDERS TO FINISHES:**					
03.3801 011	Monolithic topping 1/16"	SF	0.0190	1.51	0.08	1.59
03.3801 021	Monolithic topping 3/16"	SF	0.0200	1.59	0.32	1.92
03.3801 031	Monolithic topping 1/2"	SF	0.0211	1.68	0.64	2.32
03.3801 041	White cement	SACK			21.62	21.62
03.3801 051	Felton sand	CY			72.48	72.48
03.3801 061	Integral colors	#			2.12	2.12
03.3802 000	**ADDERS FOR SPECIAL WEAR SURFACES:**					
03.3802 011	Mono rock, 3/8", wear course	SF	0.0184	1.46	0.43	1.89
03.3802 021	Kalman, 3/4", wear course	SF	0.0333	2.65	0.90	3.56
03.3802 031	Commercial surface hardeners	SF	0.0016	0.13	0.01	0.14
03.3802 035	Sealer, 1 coat	SF	0.0004	0.03	0.03	0.06
03.3802 041	Traffic surface, waterproof	SF	0.0452	3.22	1.26	4.49
03.3802 051	Pressure grouting	CF	0.3055	21.77	7.24	29.01
03.3900 000	**MISCELLANEOUS CONCRETE ITEMS & ACCESSORIES:**					
03.3901 011	Sandblasting, whip clean up	SF	0.0054	0.38	0.27	0.65
03.3901 021	Sandblasting, dust controlled	SF	0.0150	1.07	0.86	1.93
03.3901 031	Inserts, unistrut, average	LF	0.0262	2.83	4.86	7.69
03.3901 041	Water stop, rubber, 9"	LF	0.0483	5.21	7.02	12.23
03.3901 051	Water stop, rubber, 12"	LF	0.0698	7.53	10.58	18.11
03.3901 061	Water stop, PVC, 9"	LF	0.0691	7.46	7.59	15.04
03.3901 071	Snap ties, average	EA	0.0049	0.53	0.71	1.24
03.3901 081	Hair pins, rental, month average	EA			0.14	0.14
03.3901 091	Saw cut concrete, 1" depth	LF	0.0160	1.14	0.15	1.29
03.3901 101	Saw cut concrete, 1-1/2" depth	LF	0.0338	2.41	0.23	2.64
03.3901 111	Saw cut concrete, 2" depth	LF	0.1010	7.20	0.34	7.53
03.3901 121	Saw cut concrete, 3" depth	LF	0.1315	9.37	0.56	9.93
03.3901 131	Visqueen membrane, 6 mil	SF	0.0023	0.16	0.03	0.20
03.3901 141	Embedded iron, install	#	0.0625	6.74		6.74
03.3901 151	Foundations, accessories	CY	0.0133	1.43	13.88	15.32
03.3901 153	Bollards, pipe, 6", 8' long	LF	1.2500	134.86	444.16	579.02
03.3901 155	Bollards, pipe, 8", 8' long	LF	1.2500	134.86	583.52	718.38
03.3901 161	Wall, accessories	CY	0.0526	5.68	16.16	21.83
03.3901 171	Structural slab, accessories	CY	0.1739	18.76	18.52	37.28
03.3901 181	Column & beam, accessories	CY	0.1950	21.04	20.80	41.83
03.4000 000	**TILT UP CONSTRUCTION:**					
03.4001 000	**TILT UP CASTING:**					
03.4001 011	Tilt-up, bed preparation	SF	0.0021	0.15		0.15
03.4001 021	Tilt-up, edge forms, flat	LF	0.0625	6.17	1.09	7.26
03.4001 031	Tilt-up, edge forms, with key way	LF	0.0833	8.22	1.35	9.57
03.4001 041	Tilt-up, blockout forms, plain	LF	0.2000	19.75	1.93	21.67
03.4001 051	Tilt-up, blockout forms, with key way	LF	0.2500	24.68	2.03	26.71
03.4001 061	Tilt-up, chamfers	LF	0.0078	0.77	0.37	1.14
03.4001 071	Tilt-up, bond breaker	SF	0.0067	0.48	0.15	0.63
03.4001 081	Tilt-up, inserts, bracing	EA	0.5000	50.28	15.36	65.64
03.4001 091	Tilt-up, inserts, lifting	EA	0.5000	50.28	18.52	68.80

CONCRETE - 03

Division	Description	Unit	Labor Hours	Labor Cost	Material Cost	Total Cost
03.4001 101	Tilt-up, embedded metal, install	#	0.0500	4.78		4.78
03.4001 111	Tilt-up, set, door frames, install	EA	2.0000	191.36		191.36
03.4001 121	Tilt-up, pour concrete	CY	0.3333	29.79		29.79
03.4001 131	Tilt-up, finish concrete	SF	0.0160	1.27	0.02	1.30
03.4002 000	**TILT-UP, SPECIAL FINISHES:**					
03.4002 011	Washed aggregate	SF	0.0233	1.81	0.38	2.19
03.4002 021	Seeded aggregate, Mexican pebbles	SF	0.0074	0.57	0.80	1.37
03.4002 031	Seeded aggregate, palos verdes pebbles	SF	0.0074	0.57	0.43	1.00
03.4002 041	Seeded aggregate, quartz pebbles	SF	0.0047	0.36	0.38	0.75
03.4002 051	Aggregate, adhesive, 3/8"-3/4"	SF	0.0200	1.55	0.51	2.06
03.4002 061	Aggregate, adhesive, 3/4"-1"	SF	0.0400	3.10	1.72	4.82
03.4002 071	Exposed aggregate	SF	0.0120	0.93	0.37	1.30
03.4003 000	**TILT UP ERECTION, CREW OF 7 MEN, TRUCK CRANE:**					
03.4003 011	Tilt-up, panel lift, average	SF	0.0102	1.02	0.37	1.39
03.4003 021	Tilt-up, panel lift, each lift	EA	2.8905	290.06	150.62	440.68
03.4003 031	Tilt-up shoring & bracing	SF	0.0019	0.19	0.08	0.27
03.4003 041	Tilt-up, leveling pads	EA	0.3204	27.38	7.59	34.97
03.4003 051	Tilt-up, grout foundation to panel joint with embeco grout	LF	0.1483	11.81	7.47	19.28
03.4500 000	**COMPILATION OF IN PLACE COST:**					
03.4501 000	**FOUNDATIONS, INSTITUTIONAL:**					
03.4501 011	Foundations, institutional, forms, 1 use	SFCA	0.0909	9.14	1.79	10.93
03.4501 021	Foundations, institutional, forms, 3 uses	SFCA	0.0728	7.32	1.26	8.58
03.4501 031	Foundations, institutional, forms, 5 uses	SFCA	0.0756	7.60	1.13	8.73
03.4501 041	Foundations, institutional, reinforced, 60#-90# per cubic yard	#	0.0052	0.49	0.57	1.06
03.4501 051	Foundations, institutional, trench, machine, by subcontractor	CY	0.1905	15.50	19.01	34.51
03.4501 061	Foundations, institutional, trench/backfill, hand	CY	1.1628	94.59	15.05	109.65
03.4501 071	Foundations, institutional, concrete	CY	0.5882	43.90	166.28	210.18
03.4501 081	Foundations, institutional, embedded steel, install	#	0.0584	5.50		5.50
03.4501 091	Foundations, institutional, struct backfill, machine	CY	0.2196	17.86	23.24	41.11
03.4502 000	**CONCRETE STRUCTURAL WALL, INCLUDING ACCESSORIES:**					
03.4502 011	Wall, forms, 1 use	SFCA	0.1143	11.49	2.97	14.46
03.4502 021	Wall, forms, 3 uses	SFCA	0.1081	10.87	1.66	12.53
03.4502 031	Wall, forms, 5 uses	SFCA	0.1081	10.87	1.13	12.00
03.4502 041	Wall, reinforced steel, 75#-100# per cubic yard	#	0.0062	0.58	0.58	1.16
03.4502 051	Wall, concrete, placed	CY	0.6670	49.78	169.27	219.04
03.4502 061	Strip & stock pile forms	SF	0.0130	1.24		1.24
03.4502 071	Wall, patch & sack, simple	SF	0.0213	1.52	0.07	1.59
03.4502 081	Wall, embedded steel, install	#	0.0581	5.47		5.47
03.4502 091	Hoist or pump, average	CY	0.0528	3.94	10.57	14.51
03.4503 000	**STRUCTURAL BEAMS & SLAB COMBINED, 2 WAY BEAM:**					
03.4503 011	Beam & slab, forms, 1 use	SFCA	0.1280	12.87	2.30	15.17
03.4503 021	Beam & slab, forms, 3 uses	SFCA	0.1152	11.58	1.62	13.21
03.4503 031	Beam & slab, forms, 5 uses	SFCA	0.1115	11.21	1.36	12.57
03.4503 041	Beam & slab, forms, shoring, rent	SF	0.0077	0.69	0.32	1.02
03.4503 061	Beam & slab, concrete, placed	CY	0.7145	53.32	138.51	191.83
03.4503 071	Beam & slab, finish	SF	0.0157	1.25	0.02	1.27
03.4503 081	Strip & stockpile forms	SF	0.0102	0.98		0.98
03.4503 091	Beam & slab, patch & sack, simple	SF	0.0074	0.53	0.08	0.61
03.4503 101	Beam & slab, embedded steel, install	#	0.0525	4.95		4.95
03.4503 111	Hoist or pump, concrete	CY	0.0479	3.57	10.79	14.36
03.5000 000	**CEMENTITIOUS DECKS, LIGHTWEIGHT & INSULATING CONCRETE:**					
03.5100 000	**INSULATING CONCRETE, INTERIOR:**					
03.5101 000	**INSULATING CONCRETE, RESIDENTIAL, 1-1/2":**					
03.5101 011	Insulating concrete, residential, under 3,000 square feet	SF	0.0120	0.99	0.63	1.61
03.5101 021	Insulating concrete, residential, 3,000-10,000 square feet	SF	0.0111	0.91	0.63	1.54

03 - CONCRETE

Division	Description	Unit	Labor Hours	Labor Cost	Material Cost	Total Cost
03.5101 031	Insulating concrete, residential, over 10,000 square feet	SF	0.0074	0.61	0.63	1.23
03.5102 000	**INSULATING CONCRETE, INCLUDING PUMPING:**					
03.5102 011	Insulating concrete, interior, 100# per cubic foot	CY	1.6667	136.99	161.00	297.98
03.5102 021	Insulating concrete, roof, 32# per cubic foot	CY	1.4286	117.42	115.69	233.11
03.5103 000	**INSULATING CONCRETE, 3-1/4":**					
03.5103 011	Lightweight concrete fill, trowel finish	SF	0.0167	1.33	3.18	4.51
03.5103 021	Lightweight vermiculite, screed & float	SF	0.0143	1.14	2.06	3.20
03.5103 031	Lightweight gypsum deck, screed & float	SF	0.0143	1.14	1.94	3.08
03.5200 000	**CONCRETE DECK, EXTERIOR:**					
03.5201 000	**CONCRETE WALKING SURFACE, TROWELED:**					
03.5201 011	Walk, lightweight concrete, 2-5/8"	SF	0.0310	2.47	1.61	4.08
03.5201 021	Walk, lightweight concrete, 4"	SF	0.0319	2.54	2.27	4.81
03.5201 031	Add for 4,000 psi mix	SF			0.31	0.31
03.5201 041	Add for 5,000 psi mix	SF			0.51	0.51
03.5201 051	Cool-deck, pool side	SF	0.0400	3.18	3.34	6.52
03.5300 000	**INSULATING DECKS:**					
03.5301 000	**TWO HOUR SYSTEM, WITH 24 GA METAL DECK, REINFORCED & LIGHTWEIGHT:**					
03.5301 011	Zonolite, 2-1/2" thick	SF	0.0465	4.98	3.02	7.99
03.5301 021	Zonolite, 2-1/2", 1-1/2" styrene	SF	0.0439	4.70	3.96	8.66
03.5302 000	**TEE SYSTEM, 2-1/2" GYPSUM WALL BOARD WITH REINFORCING & CEILING BOARDS:**					
03.5302 011	Tee system, with 1/2" sheet rock	SF	0.0474	5.08	3.32	8.39
03.5302 021	Tee system, with 1" wood fiberboard	SF	0.0474	5.08	3.70	8.77
03.5302 031	Tee system, with 1" acoustical board	SF	0.0474	5.08	4.20	9.27
03.5302 041	Add for 1" urethane board	SF	0.0042	0.45	0.86	1.31
03.5400 000	**FIBER DECK:**					
03.5401 000	**FIBER, T & G AND CEMENTITIOUS PLANKS:**					
03.5401 011	Fiber deck, 2" thickness	SF	0.0157	1.68	3.09	4.77
03.5401 021	Fiber deck, 2-1/2" thickness	SF	0.0157	1.68	3.56	5.24
03.5401 031	Fiber deck, 3" thickness	SF	0.0157	1.68	5.18	6.87
03.6000 000	**EPOXY INJECTION, REPAIR:**					
03.6001 011	Epoxy injection, with ports & cap seal, 4" slab	LF	0.3250	28.03	14.13	42.16
03.6001 021	Epoxy injection, with ports & cap seal, 6" slab	LF	0.3670	31.66	14.92	46.57
03.6001 031	Epoxy injection, with ports & cap seal, 8" slab	LF	0.4420	38.13	15.32	53.45
03.6001 041	Epoxy injection, parking structure, 12", overhead	LF	0.5850	50.46	22.98	73.44
03.6001 071	Epoxy injection, tilt-up, to 8' high	LF	0.5500	47.44	14.92	62.36
03.6001 081	Epoxy injection, tilt-up, above 8' high	LF	0.7650	65.99	14.92	80.90

MASONRY - 04

Division	Description	Unit	Labor Hours	Labor Cost	Material Cost	Total Cost
04.0000 001	Note: Any condition other than ideal must be accounted for by increasing costs. Use the adders and deductors when applicable. The following costs include mortar and standard reinforcing. Cuts, lintels, coping and sills, jambs and heads must be included from the adders at the end of the section.					
04.1000 000	**BRICK MASONRY:**					
04.1001 000	**BRICK VENEER:**					
04.1001 011	Veneer, 4", standard brick, commercial	SF	0.0871	7.65	6.81	14.46
04.1001 021	Veneer, 4", standard brick, residential	SF	0.0783	6.88	4.19	11.07
04.1001 031	Veneer, 4", modular	SF	0.1228	10.78	7.22	18.00
04.1001 041	Veneer, jumbo brick, 4"x4"x12"	SF	0.0864	7.59	4.84	12.42
04.1001 051	Veneer, jumbo brick, 6"x4"x12"	SF	0.0961	8.44	6.89	15.33
04.1001 061	Veneer, jumbo brick, 8"x4"x12"	SF	0.1063	9.33	8.99	18.32
04.1001 071	Veneer, face brick, select modular	SF	0.1321	11.60	7.92	19.52
04.1001 081	Veneer, Norman brick	SF	0.1305	11.46	6.57	18.03
04.1001 091	Veneer, Roman brick	SF	0.1856	16.30	7.62	23.92
04.1001 101	Veneer, glazed brick	SF	0.2100	18.44	12.18	30.62
04.1002 000	**BRICK WALLS:**					
04.1002 011	Brick wall, 10" wall, cavity	SF	0.2241	19.68	13.47	33.15
04.1002 021	Brick wall, 10", reinforced #4/24", OCBW	SF	0.2425	21.29	15.16	36.45
04.1002 031	Brick wall, 13", reinforced #4/24", OCBW	SF	0.2651	23.28	18.27	41.55
04.1002 041	Brick wall, 16", reinforced #4/24", OCBW	SF	0.2818	24.74	22.63	47.37
04.1002 051	Brick wall, 20", reinforced #4/24", OCBW	SF	0.3553	31.20	25.35	56.55
04.1003 000	**ADDERS FOR BRICK MASONRY:**					
04.1003 001	Note: For extra reinforcing, see section 03.2000. COMMON BOND, FULL HEADER EACH 7TH COURSE .... 15% SHORT RUNS AND/OR CUT-UP WORK .... 20% PILASTERS, PER SQUARE FOOT OF PILASTER .... 10% DEDUCT FOR RESIDENTIAL TRACTS .... 18% BASKETWEAVE PATTERN .... 20% FLEMISH BOND .... 33% HERRINGBONE PATTERN .... 30% SOLDIER COURSE .... 20% STACKED BOND .... 15% INSTITUTIONAL INSPECTION .... 20% WALLS OVER ONE STORY .... 10% MODULAR BRICK .... 15% ARCHES .... 30% CAPS & COPING .... 15% CUTS .... 50% HEADER COURSE .... 33% JAMBS .... 15% LINTELS OVER OPENINGS .... 30% SILLS, CUT .... 50% DOVETAIL ANCHORS 5%					
04.1003 031	Add for zonolite filled walls	SF	0.0058	0.51	0.18	0.69
04.2000 000	**CONCRETE MASONRY UNIT:**					
04.2001 000	**CONCRETE MASONRY UNIT:**					
04.2001 001	Note: SPECIAL COLORS .... 10% STACK BOND .... 16% SCORED BLOCK .... 9% SCULPTURED BLOCK .... 14% CUT UP PROJECTS .... 20% INSTITUTIONAL INSPECTION .... 10% SPECIAL GLAZING OF ENDS & SHAPING 100%					
04.2001 011	Concrete masonry unit, 4x8x16, #4 bar, 32" OCBW	SF	0.0766	6.73	8.46	15.18
04.2001 021	Concrete masonry unit, 6x8x16, #4 bar, 32" OCBW	SF	0.0823	7.23	9.81	17.04
04.2001 031	Concrete masonry unit, 8x8x16, #4 bar, 32" OCBW	SF	0.0870	7.64	10.33	17.97
04.2001 041	Concrete masonry unit, 12x8x16, #4 bar, 32" OCBW	SF	0.0970	8.52	16.07	24.58
04.2001 051	Concrete masonry unit, 4x8x16, #4 bar, 32" OCBW, filled	SF	0.0824	7.24	8.96	16.20

04 - MASONRY

Division	Description	Unit	Labor Hours	Labor Cost	Material Cost	Total Cost
04.2001 061	Concrete masonry unit, 6x8x16, #4 bar, 32" OCBW, filled	SF	0.0842	7.39	11.31	18.70
04.2001 071	Concrete masonry unit, 8x8x16, #4 bar, 32" OCBW, filled	SF	0.0889	7.81	12.15	19.95
04.2001 081	Concrete masonry unit, 12x8x16, #4 bar, 32" OCBW, filled	SF	0.1077	9.46	15.33	24.79
04.2001 091	Concrete masonry unit, 8x4x16, #4 bar, 32" OCBW, filled	SF	0.1171	10.28	17.42	27.70
04.2001 101	Concrete masonry unit, 12x4x16, #4 bar, 32" OCBW, filled	SF	0.1356	11.91	19.08	30.98
04.2001 111	Concrete masonry unit, grout lock, 6x8x24, #4 bar, filled	SF	0.0571	5.01	12.11	17.13
04.2001 121	Concrete masonry unit, grout lock, 8x8x16, #4 bar, filled	SF	0.0693	6.09	12.81	18.90
04.2001 131	Concrete masonry unit, grout lock, 12x8x16, #4 bar, filled	SF	0.0823	7.23	16.46	23.68
04.2001 141	Concrete masonry unit, slumpstone, 8x4x16, #4 bar, filled	SF	0.1262	11.08	21.06	32.15
04.2001 151	Concrete masonry unit slumpstone, 8x8x16, #4 bar, filled	SF	0.1061	9.32	16.39	25.70
04.2001 161	Concrete masonry unit splitface, 8x4x16, #4 bar, filled	SF	0.1261	11.07	19.03	30.10
04.2001 171	Concrete masonry unit glazed 1 side, 4x8x16, reinforced, filled	SF	0.1170	10.27	14.35	24.63
04.2001 181	Concrete masonry unit glazed 1 side, 6x8x16, reinforced, filled	SF	0.1251	10.99	16.51	27.50
04.2001 191	Concrete masonry unit glazed 1 side, 8x8x16, reinforced, filled	SF	0.1357	11.92	19.32	31.23
04.2001 201	Concrete masonry unit glazed 1 side, 4x4x16, reinforced, filled	SF	0.1338	11.75	21.26	33.01
04.2001 211	Concrete masonry unit glazed 1 side, 6x4x16, reinforced, filled	SF	0.1554	13.65	24.82	38.47
04.2001 221	Concrete masonry unit glazed 1 side, 8x4x16, reinforced, filled	SF	0.1653	14.51	28.37	42.89
04.2001 231	Concrete masonry unit glazed 1 side, 12x8x16, reinforced, filled	SF	0.1926	16.91	23.34	40.25
04.2001 241	Add for glazing both sides	SF			7.85	7.85
04.2001 251	Screen block 4" x 12" x 12"	SF	0.0991	8.70	8.24	16.94
04.2003 000	**ADDERS FOR CONCRETE MASONRY UNIT:**					
04.2003 021	Add for pilasters per SF of pilaster	SF			34.39	34.39
04.2003 061	Add for sill blocks	LF	0.0756	6.64	3.71	10.35
04.2003 071	Add for cutting blocks	LF	0.1244	10.92	4.29	15.21
04.2003 081	Add for bond beams @ door and sash openings	LF	0.1682	14.77	8.15	22.92
04.2003 091	Add for lintels, over openings	LF	0.2844	24.97	15.89	40.87
04.2003 101	Add for concrete masonry unit, #4 bar, 32" OCBW	SF	0.0047	0.41	0.43	0.84
04.2003 111	Add for concrete masonry unit, #5 bar, 24" OCBW	SF	0.0069	0.61	0.76	1.36
04.2003 121	Add for concrete masonry unit, #5 bar, 16" OCBW	SF	0.0115	1.01	1.29	2.30
04.2004 000	**CLAY BACKING TILE:**					
04.2004 001	Note: Not used in critical seismic zones, may be used as blocking at columns as a plaster base.					
04.2004 011	Clay tile, load bearing, 4", 12" x 12"	SF	0.0852	7.48	3.42	10.91
04.2004 021	Clay tile, load bearing, 6", 12" x 12"	SF	0.0935	8.21	3.98	12.19
04.2004 031	Clay tile, load bearing, 8", 12" x 12"	SF	0.1047	9.19	4.57	13.77
04.2004 041	Clay tile, non-load bearing, 4", 12" x 12"	SF	0.0768	6.74	3.36	10.10
04.2004 051	Clay tile, non-load bearing, 6", 12" x 12"	SF	0.0853	7.49	3.56	11.05
04.2004 061	Clay tile, non-load bearing, 8", 12" x 12"	SF	0.0934	8.20	5.63	13.83
04.2005 000	**CLAY FACING TILE (GLAZED STRUCTURAL):**					
04.2005 001	Note: For large areas, deduct 15% from the total costs.					
04.2005 011	Tile, glazed 1 side, 2" x 6" x 12"	SF	0.1442	12.66	7.17	19.83
04.2005 021	Tile, glazed 1 side, 4" x 6" x 12"	SF	0.1647	14.46	10.55	25.01
04.2005 031	Tile, glazed 2 sides, 4" x 6" x 12"	SF	0.1898	16.67	12.25	28.92
04.2005 041	Tile, glazed 1 side, 6" x 6" x 12"	SF	0.1738	15.26	10.14	25.40
04.2005 051	Tile, glazed 1 side, 3" x 6" x 12"	SF	0.1673	14.69	7.49	22.18
04.2005 061	Tile, glazed 1 side, base	LF	0.1673	14.69	10.18	24.87
04.2005 071	Tile, glazed 2 sides, cap	SF	0.1701	14.94	9.91	24.84
04.2006 000	**CERAMIC VENEER:**					
04.2006 011	Ceramic veneer facing, vertical	SF	0.2328	20.44	13.70	34.14
04.2006 021	Ceramic veneer facing, horizontal on precast panels	SF	0.1542	13.54	13.70	27.24
04.2006 031	Brick plate, applied on walls	SF	0.1475	12.95	7.95	20.90
04.2007 000	**PAVERS AND FLOOR TILE:**					
04.2007 001	Note: For special brick patterns, add 30% to the labor costs.					
04.2007 011	Quarry tile, unglazed, floor, 16"	SF	0.1636	14.37	5.48	19.85
04.2007 021	Quarry tile, unglazed, floor, 12"	SF	0.1412	12.40	5.73	18.13

MASONRY - 04

Division	Description	Unit	Labor Hours	Labor Cost	Material Cost	Total Cost
04.2007 031	Quarry tile, unglazed, base, 6"	LF	0.1552	13.63	5.91	19.53
04.2007 041	Quarry tile, glazed, floor, 6"	SF	0.1637	14.37	7.21	21.58
04.2007 051	Quarry tile, glazed, base, 6"	LF	0.1552	13.63	7.02	20.65
04.2007 061	Brick plate, glazed	SF	0.1319	11.58	7.78	19.36
04.2007 071	Brick, grouted on concrete substratum	SF	0.1124	9.87	6.93	16.80
04.2007 081	Brick, sand laid, no grout	SF	0.1047	9.19	6.22	15.41
04.2007 091	Brick pavers, grouted on concrete substratum	SF	0.1087	9.54	7.80	17.35
04.2007 101	Slate	SF	0.2019	17.73	11.34	29.07
04.2007 111	Terrazzo tiles, standard	SF	0.1355	11.21	10.57	21.78
04.2007 121	Terrazzo tiles, granite chips	SF	0.1614	13.35	32.35	45.70
04.2007 131	Brick steps	SF	0.2452	20.29	9.92	30.20
04.2008 000	**GLASS MASONRY UNIT:**					
04.2008 001	Note: For quantities between 1000 and 5000 sf, deduct 10% from the total costs. Quantities over 5000 sf, deduct 20% from the total costs. For special colors, add 10% to the total costs. For special decorations, add 400% to total cost.					
04.2008 011	Glass masonry unit, clear, 4" x 6" x 6"	SF	0.3637	30.09	31.87	61.96
04.2008 021	Glass masonry unit, clear, 4" x 8" x 8"	SF	0.2496	20.65	30.53	51.18
04.2008 031	Glass masonry unit, clear, 4" x 12" x 12"	SF	0.1823	15.08	25.96	41.04
04.3000 000	**ARCHITECTURAL STONEWORK:**					
04.3001 000	**STONE, ROUGH:**					
04.3001 011	Veneer, lava stone, average 4" thick	SF	0.2205	18.24	8.07	26.31
04.3001 021	Veneer, Arizona stone, average 4" thick	SF	0.2481	20.53	14.15	34.67
04.3001 031	Veneer, rubble stone, average 4" thick	SF	0.2626	21.72	7.27	29.00
04.3001 041	Veneer, Palos Verdes & driftwood	SF	0.2020	16.71	7.06	23.77
04.3001 051	Veneer, other common varieties	SF	0.2300	19.03	10.54	29.57
04.3001 061	Veneer, imported East Coast type	SF	0.2346	19.41	15.15	34.55
04.3002 000	**STONE, CUT:**					
04.3002 001	Note: Prices for cut stone are representative only. You can expect wide variations in price. For honed finish, add 10% to the total costs. For polished finish, add 30% to the total costs.					
04.3002 011	Granite, 3/4" thick	SF	0.3386	28.01	27.69	55.71
04.3002 021	Granite, 1-1/4" thick	SF	0.3793	31.38	28.38	59.76
04.3002 031	Limestone, 2" thick	SF	0.3402	28.14	19.52	47.67
04.3002 041	Limestone, 3" thick	SF	0.4255	35.20	20.75	55.96
04.3002 051	Marble, 7/8" thick	SF	0.3352	27.73	26.47	54.20
04.3002 061	Marble, 1-1/4" thick	SF	0.3738	30.92	30.26	61.18
04.3002 071	Travertine	SF	0.3100	27.22	23.05	50.27
04.3002 081	Base travertine 9" x 3/4"	SF	0.4400	38.64	32.85	71.49
04.3002 091	Columns, travertine (not curved)	SF	0.5100	44.78	28.80	73.58
04.3002 101	Base curved radius 4'6"	SF	0.6500	57.08	48.23	105.31
04.3002 111	Sandstone, 2" thick	SF	0.2910	25.55	17.19	42.74
04.3002 121	Sandstone, 3" thick	SF	0.3238	28.43	20.39	48.82
04.3003 000	**MASONRY AND STONE SPECIALTIES:**					
04.3003 011	Floor, marble, 7/8"	SF	0.1750	15.37	26.58	41.95
04.3003 021	Thresholds, marble, 1-1/4"	LF	0.2137	18.76	23.10	41.86
04.3003 031	Base, marble, 7/8" x 6" high	LF	0.2137	18.76	23.87	42.63
04.3003 041	Columns, marble, plain	CF	0.9162	80.45	223.66	304.11
04.3003 051	Columns, marble, fluted	CF	0.9889	86.84	347.99	434.82
04.3003 061	Window stools, marble	LF	0.1455	12.78	22.32	35.09
04.3003 071	Toilet partitions, marble	EA	9.5192	835.88	1,045.31	1,881.19
04.3003 081	Stair treads, 1-1/4" x 11", marble	LF	0.3057	26.84	50.90	77.74
04.3003 091	Limestone, roughcut, large block	CF	0.3471	30.48	55.37	85.85
04.3004 000	**ARTIFICIAL STONE WORK:**					
04.3004 011	Facing panel, terrazzo	SF	0.2671	23.45	9.08	32.53
04.3004 021	Palos Verdes, cast plaster	SF	0.1455	12.78	4.80	17.58
04.3004 031	Brick, all types	SF	0.1225	10.76	3.82	14.57

04 - MASONRY

Division	Description	Unit	Labor Hours	Labor Cost	Material Cost	Total Cost
04.3004 041	Cast stone, facing	SF	0.2671	23.45	9.79	33.24
04.3004 051	Cast plaster, simulated stone, rough	SF	0.1455	12.78	4.95	17.73
04.4000 000	**MASONRY ACCESSORIES & MISCELLANEOUS WORK:**					
04.4001 000	**MASONRY WALL TIES & REINFORCING:**					
04.4001 011	Masonry wall ties, galvanized	EA	0.0046	0.40	0.15	0.55
04.4001 021	Masonry wall ties, copper coated	EA	0.0046	0.40	0.18	0.59
04.4001 031	Masonry wall ties, z, galvanized	EA	0.0046	0.40	0.18	0.59
04.4001 041	Masonry wall ties, z, copper coated	EA	0.0046	0.40	0.22	0.62
04.4001 051	Masonry reinforced, truss, 6" wide	LF	0.0024	0.21	0.66	0.87
04.4001 061	Masonry reinforced, ladder, 6" wide	LF	0.0024	0.21	0.45	0.66
04.4001 071	Masonry reinforced, truss, 10" wide	LF	0.0031	0.27	0.74	1.01
04.4001 081	Masonry reinforced, ladder, 10" wide	LF	0.0031	0.27	0.45	0.72
04.4002 000	**MASONRY WALL FINISHES:**					
04.4002 011	Masonry, acid etch	SF	0.0083	0.73	0.05	0.77
04.4002 021	Masonry, wash	SF	0.0083	0.73	0.05	0.77
04.4002 031	Masonry, steam clean	SF	0.0157	1.38	0.32	1.70
04.4002 041	Masonry, existing, sandblasting	SF	0.0117	1.03	0.83	1.85
04.4002 051	Masonry, new, light sandblasting	SF	0.0095	0.83	0.40	1.24
04.4002 061	Masonry, medium sandblasting	SF	0.0109	0.96	0.44	1.39
04.4002 071	Masonry, heavy sandblasting	SF	0.0164	1.44	0.60	2.04
04.4003 000	**MASONRY POINTING AND WATERPROOFING:**					
04.4003 011	Masonry pointing, brick	SF	0.0130	1.14	0.17	1.31
04.4003 021	Masonry pointing, concrete block	SF	0.0083	0.73	0.17	0.90
04.4003 031	Masonry repointing, brick	SF	0.0158	1.39	0.17	1.56
04.4003 041	Masonry, waterproofing, flood coat	SF	0.0047	0.41	0.47	0.88
04.5000 000	**FIREPLACES:**					
04.5000 001	Note: The following prices are based on fireplaces with 15' stacks. For prefabricated metal fireplaces, see section 07.6023.					
04.5001 000	**COMMON BRICK FIREPLACES:**					
04.5001 001	Note: The following item uses 900 bricks.					
04.5001 011	Fireplace, 30" box, to mantle	EA	15.2671	1,340.60	936.45	2,277.05
04.5001 021	Fireplace, 36" box, to mantle	EA	21.5104	1,888.83	1,072.66	2,961.49
04.5001 031	Fireplace, 42" box, to 9'	EA	31.8940	2,800.61	1,719.68	4,520.29
04.5001 041	Fireplace, 48" box, to 9'	EA	37.4569	3,289.09	2,009.15	5,298.24
04.5002 000	**COMMON BRICK FIREPLACE, ADDERS:**					
04.5002 011	Add for full width stack to 15'	EA	6.9228	607.89	359.13	967.03
04.5002 021	Add for stone face to 9'	EA	7.7186	677.77	400.21	1,077.98
04.5002 031	Add for raised hearth	EA	2.9278	257.09	151.77	408.86
04.5002 041	Add for face brick mantle to ceiling	EA	2.9278	257.09	151.77	408.86
04.5002 051	Add for extra 10' stack (2 story)	EA	5.3232	467.43	276.00	743.43
04.5003 000	**PRE-FABRICATED FIREPLACES (QUANTITY):**					
04.5003 011	Fireplace, 30" box, simulated brick	EA	10.9126	958.24	565.95	1,524.18
04.5003 021	Add for face & hearth, pre-fabricated	EA	4.5246	397.31	234.68	631.99
04.5003 031	Add for carpentry, pre-fabricated	EA	1.7299	151.90	89.68	241.58
04.5004 000	**PRE-FABRICATED FIREPLACE, FLUES:**					
04.5004 011	Flues for pre-fab fireplace, average	EA	5.3232	467.43	276.00	743.43
04.5004 021	Patent flues, 12', average	EA	3.5188	308.99	76.37	385.36
04.5004 041	Pre-fabricated flue, 8"x10"x24" long	LF	0.1279	11.23	5.63	16.86
04.5004 051	Pre-fabricated flue, 12"x12"x24" long	LF	0.1463	12.85	6.67	19.51
04.5005 000	**FIREBRICK:**					
04.5005 011	Firebrick, industrial	SF	0.4024	35.33	11.81	47.15
04.6000 000	**PARGETING:**					
04.6001 000	**PARGET:**					
04.6001 011	Parget, cement, 2 coats, 1/2"	SF	0.0223	1.96	0.57	2.53
04.6001 021	Parget, cement, waterproof, 2 coats, 1/2"	SF	0.0223	1.96	0.75	2.71

METALS - 05

Division	Description	Unit	Labor Hours	Labor Cost	Material Cost	Total Cost
05.1000 000	**STRUCTURAL STEEL:**					
05.1001 000	STRUCTURAL SHAPES (A-36):					
05.1001 011	Base US price, FOB Pittsburgh or Chicago, freight extra	CWT			52.56	52.56
05.1001 021	Base price, foreign, FOB dock	CWT			51.99	51.99
05.1002 000	COLUMN SHAPES (A-36), 10" & LARGER:					
05.1002 011	Column shapes, 10" & up, 5,000#	#	0.0052	0.49	2.28	2.77
05.1002 021	Column shapes, 10" & up, 5,000-10,000#	#	0.0047	0.44	1.91	2.35
05.1002 031	Column shapes, 10" & up, 10,000-20,000#	#	0.0044	0.41	1.70	2.12
05.1002 041	Column shapes, 10" & up, 20,000-50,000#	#	0.0042	0.40	1.46	1.86
05.1002 051	Column shapes, 10" & up, 50,000-300,000#	#	0.0038	0.36	1.46	1.82
05.1002 061	Column shapes, 10" & up, 300,000-1,000,000#	#	0.0033	0.31	1.35	1.66
05.1002 071	Column shapes, 10" & up, 1,000,000# & up	#	0.0029	0.27	1.35	1.63
05.1003 000	BEAMS & GIRDERS, 8" & SMALLER, WELDED:					
05.1003 011	Beams & girders, up to 8", under 5,000#	#	0.0081	0.76	2.28	3.04
05.1003 021	Beams & girders, up to 8", 5,000-10,000#	#	0.0069	0.65	1.74	2.39
05.1003 031	Beams & girders, up to 8", 10,000-20,000#	#	0.0061	0.57	1.47	2.05
05.1003 041	Beams & girders, up to 8", 20,000-50,000#	#	0.0057	0.54	1.43	1.97
05.1003 051	Beams & girders, up to 8", 50,000-300,000#	#	0.0053	0.50	1.38	1.88
05.1003 061	Beams & girders, up to 8", 300,000-1,000,000#	#	0.0051	0.48	1.38	1.87
05.1003 071	Deduct for bolted construction	#	0.0013	0.12		0.12
05.1004 000	SHAPES, 10" & LARGER, WELDED:					
05.1004 011	Shapes, 10" & up, under 5,000#	#	0.0069	0.65	2.04	2.69
05.1004 021	Shapes, 10" & up, 5,000-10,000#	#	0.0057	0.54	1.51	2.04
05.1004 031	Shapes, 10" & up, 10,000-20,000#	#	0.0057	0.54	1.46	2.00
05.1004 041	Shapes, 10" & up, 20,000-50,000#	#	0.0053	0.50	1.42	1.92
05.1004 051	Shapes, 10" & up, 50,000-300,000#	#	0.0051	0.48	1.23	1.72
05.1004 061	Shapes, 10" & up, 300,000-1,000,000#	#	0.0046	0.43	0.96	1.40
05.1004 071	Shapes, 10" & up, 1,000,000-3,000,000#	#	0.0044	0.41	0.96	1.38
05.1004 081	Shapes, 10" & up, 3,000,000# or more	#	0.0044	0.41	0.93	1.34
05.1004 091	Deduct for bolted construction	#	0.0013	0.12		0.12
05.1005 000	ADDERS & DEDUCTORS FOR STRUCTURAL SHAPES:					
05.1005 011	Deduct for jumbo columns	#			0.04	0.04
05.1005 021	Deduct for foreign steel	#			0.04	0.04
05.1005 031	Add for high strength steel, A-572	#	0.0012	0.11	0.04	0.16
05.1005 041	Add for high strength steel, A-188	#	0.0012	0.11	0.04	0.16
05.1005 051	Add for high strength steel, A-441, A-440, A-212	#	0.0012	0.11	0.04	0.16
05.1005 061	Add for high strength steel, A-588 corten	#	0.0012	0.11	0.15	0.27
05.1005 071	Add for light shapes & junior sizes	#	0.0027	0.25	0.43	0.68
05.1005 081	Add for tube shapes	#	0.0007	0.07	0.45	0.51
05.1005 091	Add for warehouse purchase	#			0.14	0.14
05.1005 101	Adder; 16 to 30 stories	#	0.0013	0.12	0.08	0.20
05.1005 111	Adder; 31 to 45 stories	#	0.0017	0.16	0.08	0.24
05.1005 121	Adder; 46 stories & over	#	0.0020	0.19	0.12	0.31
05.1100 000	**STRUCTURAL STEEL SPECIALTIES:**					
05.1101 000	TRUSSES:					
05.1101 011	Trusses, light steel, bolted	#	0.0053	0.50	1.74	2.24
05.1101 021	Trusses, light steel, welded	#	0.0057	0.54	1.82	2.36
05.1101 031	Trusses, heavy steel, bolted	#	0.0035	0.33	1.39	1.72
05.1101 041	Trusses, heavy steel, welded	#	0.0046	0.43	1.49	1.92
05.1102 000	SPACE FRAME SYSTEM, 10,000 SF & UP:					
05.1102 011	Space frame, 4' modular, 20# live load	SF	0.0917	8.64	11.33	19.97
05.1102 021	Space frame, 5' modular, 20# live load	SF	0.0770	7.25	9.82	17.08
05.1102 031	Space frame, 4' modular, 35# live load	SF	0.0985	9.28	13.45	22.73
05.1102 041	Space frame, 5' modular, 35# live load	SF	0.0837	7.89	11.75	19.64
05.1103 000	DETAIL STEEL:					

05 - METALS

Division	Description	Unit	Labor Hours	Labor Cost	Material Cost	Total Cost
05.1103 001	Note: The following prices are based on 7% of shapes bolted, 5% welded.					
05.1103 011	Base plates, milled	#	0.0048	0.45	2.32	2.77
05.1103 021	Tie plates & angles	#	0.0081	0.76	2.21	2.97
05.1103 031	Gussets	#	0.0081	0.76	2.21	2.97
05.1103 041	Tie rod with clevis & turn buckles	#	0.0089	0.84	2.22	3.06
05.1103 051	Clips & angles, attached	#	0.0081	0.76	2.48	3.25
05.1103 061	Shear studs, welded	EA	0.0574	5.41	1.38	6.79
05.1103 071	Moment connections	EA	4.0250	379.24	247.08	626.32
05.1104 000	**MISCELLANEOUS COLUMNS:**					
05.1104 011	Pipe columns, 3"-6"	#	0.0075	0.71	1.59	2.30
05.1104 021	Pipe columns, 8"-12"	#	0.0037	0.35	1.52	1.86
05.1104 031	Tube columns, 3" x 4" x 5/16"	#	0.0055	0.52	2.05	2.57
05.1104 041	Concrete encasement for columns	CF	0.4255	40.09	15.68	55.77
05.1104 051	Fireproof jackets and fill	SF	0.0483	4.55	17.94	22.49
05.1104 061	False work support/ton supported	TON	4.4471	419.01	527.70	946.71
05.1105 000	**MISCELLANEOUS SUPPORT ITEMS:**					
05.1105 011	High strength bolts, 3/4" dia	EA	0.1565	14.75	3.16	17.91
05.1105 021	High strength bolts, 7/8" dia	EA	0.1565	14.75	3.81	18.55
05.1105 031	Welds, 1/8", 1 pass	LF	0.1140	10.74	0.15	10.89
05.1105 041	Welds, 1/4", 3 pass	LF	0.2655	25.02	0.57	25.58
05.1105 051	Welds, 3/8", 6 pass	LF	0.5469	51.53	0.95	52.48
05.1105 061	Welds, 1/2", 10 pass	LF	0.9126	85.99	1.34	87.33
05.1105 071	Galvanizing, 2 oz	#			0.35	0.35
05.1105 091	Field paint, 1 coat	#	0.0022	0.21	0.15	0.36
05.1105 101	Field touchup, painting	#	0.0026	0.24	0.16	0.41
05.1106 000	**JOISTS, OPEN WEB:**					
05.1106 011	Joists, "H" series	#	0.0064	0.60	1.36	1.97
05.1106 021	Joists, "J" series	#	0.0050	0.47	1.36	1.83
05.1106 031	Joists, "H" series, high strength	#	0.0060	0.57	1.53	2.09
05.1106 041	Joists, "LH" or "LJ" series	#	0.0053	0.50	0.99	1.49
05.1106 051	Joists, "LH" or "LJ" series, high strength	#	0.0053	0.50	1.35	1.85
05.1106 061	Built/up heavy girders	#	0.0053	0.50	1.08	1.58
05.1106 071	Built/up heavy box girders	#	0.0074	0.70	1.07	1.77
05.1107 000	**STEEL BUILDING FRAME, 14' EAVE, NO FOUNDATIONS OR SLAB:**					
05.1107 011	Steel buildings, 40,000 SF and up	SF	0.0808	7.61	19.78	27.40
05.1107 021	Steel buildings, 10,000-40,000 SF	SF	0.0841	7.92	21.88	29.80
05.1107 031	Steel buildings, 5,000-10,000 SF	SF	0.0872	8.22	22.75	30.96
05.1107 041	Steel buildings, less than 5,000 SF	SF	0.0896	8.44	23.04	31.49
05.1108 000	**STEEL BUILDINGS, ADDERS:**					
05.1108 001	Note: For sheet metal siding or roofing, see section 07.6000. For sash see section 08.5000.					
05.1108 011	Steel building, frame, 3.8-4.4#/SF	#	0.0044	0.41	1.96	2.38
05.1108 021	Bents only	#	0.0035	0.33	1.86	2.19
05.1108 031	Purlins & eave struts only	#	0.0053	0.50	2.08	2.58
05.1108 041	Galvanized siding, 26 ga	SF	0.0501	4.72	3.78	8.50
05.1108 051	Galvanized roofing, 26 ga	SF	0.0442	4.16	3.78	7.95
05.1108 061	Steel building door & frame, 3' x 7', no hardware or paint	EA	2.2417	211.21	672.32	883.53
05.1108 071	Steel building door & frame, over head, 8' x 10', prime painted	EA	5.8852	554.50	2,604.69	3,159.19
05.3000 000	**DECKING & SIDING:**					
05.3000 001	Note: The installed prices are based on the square foot coverage shown in the 25,000 to 50,000 square foot range. Quantities under 15,000 square feet are subject to special pricing. Quantities under 2,000 sf............ add 50% Between 2,000 and 12,000 sf........ add 25% Between 12,000 and 25,000 sf....... add 10% For quantities over 50,000 sf.......deduct 6% For galvanizing............. add $.18 per sf.					

METALS - 05

Division	Description	Unit	Labor Hours	Labor Cost	Material Cost	Total Cost
05.3001 000	**DECKING, 25M-50M SF:**					
05.3001 011	Roof decking, typical, 22 ga	#	0.0028	0.26	1.56	1.82
05.3001 021	Floor decking, typical, 18 ga	#	0.0032	0.30	1.56	1.86
05.3003 000	**FLOOR DECKING, STANDARD RIB, SIMPLE, 1-1/2":**					
05.3003 001	*Note: The maximum span for the items below is 6'6".*					
05.3003 011	Floor deck, 22 ga, Q-3, 1-1/2", simple	SF	0.0098	0.92	2.46	3.38
05.3003 021	Floor deck, 20 ga, Q-3, 1-1/2", simple	SF	0.0105	0.99	2.68	3.67
05.3003 031	Floor deck, 18 ga, Q-3, 1-1/2", simple	SF	0.0110	1.04	2.90	3.94
05.3003 041	Floor deck, 16 ga, Q-3, 1-1/2", simple	SF	0.0117	1.10	3.34	4.44
05.3004 000	**FLOOR DECKING, STANDARD RIB, SIMPLE 3":**					
05.3004 001	*Note: The maximum span for the items below is 10'.*					
05.3004 011	Floor deck, 22 ga, Q-21, 3", simple	SF	0.0105	0.99	2.76	3.75
05.3004 021	Floor deck, 20 ga, Q-21, 3", simple	SF	0.0111	1.05	3.08	4.12
05.3004 031	Floor deck, 18 ga, Q-21, 3", simple	SF	0.0115	1.08	3.34	4.42
05.3004 041	Floor deck, 16 ga, Q-21, 3", simple	SF	0.0120	1.13	3.78	4.91
05.3004 051	Decking pour stops	LF	0.0420	3.96	1.16	5.12
05.3005 000	**FLOOR DECKING, DOUBLE FLUTED, CELLULAR, 3":**					
05.3005 011	Floor deck, 18-18 ga, 3", RK, cellular	SF	0.0190	1.79	5.46	7.25
05.3005 021	Floor deck, 18-16 ga, 3", RK, cellular	SF	0.0200	1.88	6.29	8.17
05.3006 000	**FLOOR DECKING, FLUTED & FLAT, CELLULAR, 1-1/2":**					
05.3006 011	Floor deck, 20-20 ga, 1-1/2", UKX	SF	0.0164	1.55	4.61	6.16
05.3006 021	Floor deck, 18-18 ga, 1-1/2", UKX	SF	0.0171	1.61	5.02	6.64
05.3006 031	Floor deck, 18-16 ga, 1-1/2", UKX	SF	0.0180	1.70	5.71	7.41
05.3006 041	Floor deck, 16-16 ga, 1-1/2", UKX	SF	0.0193	1.82	6.13	7.95
05.3007 000	**FLOOR DECKING, FLUTED & FLAT, CELLULAR, 3":**					
05.3007 011	Floor deck, 20-20 ga, 3", NKX, cellular	SF	0.0173	1.63	5.26	6.89
05.3007 021	Floor deck, 18-18 ga, 3", NKX, cellular	SF	0.0179	1.69	5.77	7.45
05.3007 031	Floor deck, 18-16 ga, 3", NKX, cellular	SF	0.0190	1.79	6.61	8.40
05.3008 000	**FLOOR DECKING, STANDARD, SIMPLE, 4-1/2":**					
05.3008 001	*Note: The maximum span for the items below is 13'3".*					
05.3008 011	Floor deck, 20 ga, 4 1/2", Q-21, simple	SF	0.0137	1.29	4.93	6.22
05.3008 021	Floor deck, 18 ga, 4 1/2", Q-21, simple	SF	0.0142	1.34	5.32	6.66
05.3008 031	Floor deck, 16 ga, 4 1/2", Q-21, simple	SF	0.0152	1.43	6.05	7.49
05.3009 000	**FLOOR DECKING, FLUTED & FLAT, CELLULAR, 4-1/2":**					
05.3009 011	Floor deck, 18-18 ga, 4-1/2", FKX, cellular	SF	0.0184	1.73	7.48	9.21
05.3009 021	Floor deck, 18-16 ga, 4-1/2", FKX, cellular	SF	0.0209	1.97	7.65	9.62
05.3009 031	Floor deck, 16-16 ga, 4-1/2", FKX, cellular	SF	0.0218	2.05	8.02	10.08
05.3010 000	**ROOF DECKING, STANDARD RIB, SIMPLE, 1-1/2":**					
05.3010 011	Roof deck, 22 ga, Q-3, 1-1/2", simple	SF	0.0105	0.99	2.46	3.45
05.3010 021	Roof deck, 20 ga, Q-3, 1-1/2", simple	SF	0.0105	0.99	2.68	3.67
05.3010 031	Roof deck, 18 ga, Q-3, 1-1/2", simple	SF	0.0116	1.09	2.90	4.00
05.3010 041	Roof deck, 16 ga, Q-3, 1-1/2", simple	SF	0.0121	1.14	3.34	4.48
05.3011 000	**ROOF DECKING, STANDARD RIB, SIMPLE, 3":**					
05.3011 011	Roof deck, 22 ga, Q-21, 3", simple	SF	0.0108	1.02	2.71	3.73
05.3011 021	Roof deck, 20 ga, Q-21, 3", simple	SF	0.0112	1.06	3.03	4.08
05.3011 031	Roof deck, 18 ga, Q-21, 3", simple	SF	0.0117	1.10	3.28	4.38
05.3011 041	Roof deck, 16 ga, Q-21, 3", simple	SF	0.0125	1.18	3.71	4.89
05.3012 000	**ROOF DECKING, ACOUSTIC, SIMPLE 1-1/2":**					
05.3012 011	Roof deck, 22 ga, 'B', 1-1/2", simple	SF	0.0110	1.04	3.86	4.90
05.3012 021	Roof deck, 20 ga, 'B', 1-1/2", simple	SF	0.0114	1.07	4.18	5.26
05.3012 031	Roof deck, 18 ga, 'B', 1-1/2", simple	SF	0.0121	1.14	4.40	5.54
05.3012 041	Sheet metal toe board, 4"	LF	0.0340	3.20	4.06	7.26
05.3013 000	**ROOF DECKING, ACOUSTIC, SIMPLE 3":**					
05.3013 011	Roof deck, 20 ga, C-3, 3", simple	SF	0.0110	1.04	4.51	5.55
05.3013 021	Roof deck, 18 ga, C-3, 3", simple	SF	0.0116	1.09	4.72	5.81

05 - METALS

Division	Description	Unit	Labor Hours	Labor Cost	Material Cost	Total Cost
05.3013 031	Roof deck, 16 ga, C-3, 3", simple	SF	0.0126	1.19	5.13	6.32
05.3013 041	Roof deck, 14 ga, C-3, 3", simple	SF	0.0133	1.25	5.87	7.13
05.3014 000	**ROOF DECKING, CURRUFORM & TUFCOR, WITH REINFORCING:**					
05.3014 011	Roof deck, 26 ga, 5' span	SF	0.0105	0.99	1.95	2.94
05.3014 021	Roof deck, 20 ga, 8' span	SF	0.0110	1.04	2.15	3.19
05.3015 000	**SIDING, METAL WALLS:**					
05.3015 011	Siding, baked enamel	SF	0.0465	4.38	39.40	43.78
05.3015 031	Siding, porcelain	SF	0.0490	4.62	43.78	48.39
05.3016 000	**SIDING, INSULATED METAL WALLS:**					
05.3016 011	Siding, insulated metal, baked enamel	SF	0.0700	6.60	40.76	47.36
05.3016 021	Siding, insulated metal, porcelain	SF	0.0790	7.44	56.43	63.88
05.5000 000	**MISCELLANEOUS IRON:**					
05.5001 000	**IRON, MISCELLANEOUS:**					
05.5001 011	Anchor bolts, hook, 1/2" x 8"	EA	0.1650	15.55	4.12	19.66
05.5001 021	Anchor bolts, hook, 5/8" x 10"	EA	0.1740	16.39	4.66	21.06
05.5001 031	Anchor bolts, hook, 3/4" x 12"	EA	0.1870	17.62	7.15	24.77
05.5001 041	Anchor bolts, hook, 1" x 15"	EA	0.1920	18.09	11.99	30.08
05.5001 051	Anchor bolts, hook, 1-1/4" x 18"	EA	0.2000	18.84	27.75	46.60
05.5001 061	Anchor bolts, hook, 1-1/2" x 24"	EA	0.2110	19.88	51.38	71.26
05.5001 071	Expansion bolt, 1/2"	EA	0.1835	17.29	3.97	21.26
05.5001 081	Expansion bolt, 5/8"	EA	0.1920	18.09	4.81	22.90
05.5001 091	Expansion bolt, 3/4"	EA	0.2024	19.07	6.09	25.16
05.5001 101	Expansion bolt, 7/8"	EA	0.2232	21.03	7.64	28.67
05.5001 111	Expansion bolt, 1"	EA	0.2344	22.09	10.34	32.43
05.5001 121	High strength bolt, ASTM 373, 3/4" x 2"	EA	0.1157	10.90	4.81	15.71
05.5001 131	High strength bolt, ASTM 373, 7/8" x 3"	EA	0.1157	10.90	5.51	16.41
05.5002 000	**STAIRS, WITHOUT RAILINGS:**					
05.5002 011	Steel stair, concrete tread, 44"	RISER	0.3162	29.79	354.24	384.03
05.5002 021	Steel stair, concrete tread, 44", switch	RISER	0.3798	35.78	405.53	441.31
05.5002 031	Steel pan, concrete fill, 44"	RISER	1.0766	101.44	707.78	809.22
05.5002 041	Steel pan, concrete fill, 48"	RISER	1.1392	107.34	776.15	883.49
05.5002 051	Steel stringer, ladder type	RISER	0.6333	59.67	413.01	472.68
05.5002 061	Cast iron stair, circular, 32"	RISER	0.4434	41.78	488.51	530.29
05.5002 071	Steel stair, concrete tread, 14 riser	FLT	6.7609	637.01	4,959.46	5,596.47
05.5002 081	Steel stair pan, concrete fill, 18 riser	FLT	15.5000	1,460.41	12,740.11	14,200.52
05.5002 091	Steel stair, steel tread, 18 riser	FLT	20.4000	1,922.09	10,829.08	12,751.17
05.5003 000	**LADDERS, GALVANIZED STEEL:**					
05.5003 011	Ladder, 2-1/2"x3/8" bar, 3/4" rung	LF	0.2829	26.65	105.06	131.71
05.5003 021	Ladder, hook bends only	EA	1.7528	165.15	634.78	799.93
05.5003 031	Ladder, protective cage only	LF	0.4640	43.72	169.53	213.25
05.5003 041	Ships ladder, plaster & tread, 4"	LF	0.2923	27.54	124.90	152.44
05.5004 000	**PIPE RAILING, WELDED, WITH KICK PLATE:**					
05.5004 011	Pipe rail, 2 high, 1-1/2", with kick	LF	0.1820	17.15	68.89	86.04
05.5004 021	Pipe rail, 3 high, 1-1/2", with kick	LF	0.1820	17.15	88.60	105.75
05.5004 031	Pipe rail, 4 high, 1-1/2", with kick	LF	0.1820	17.15	108.30	125.45
05.5004 041	Pipe rail, 1-1/2", wall type	LF	0.1820	17.15	32.23	49.38
05.5004 051	Pipe rail, 2 high, 1-1/2", kick, galvanized	LF	0.1820	17.15	78.44	95.59
05.5004 061	Pipe rail, 3 high, 1-1/2", kick, galvanized	LF	0.1820	17.15	97.03	114.18
05.5004 071	Pipe rail, 4 high, 1-1/2", kick, galvanized	LF	0.1820	17.15	118.84	135.99
05.5004 081	Pipe rail, 1-1/2", wall, galvanized	LF	0.1820	17.15	31.87	49.02
05.5005 000	**FLATBAR RAILING, WELDED:**					
05.5005 011	Angle & flatbar railing	LF	0.1700	16.02	40.21	56.22
05.5005 021	Flat & square bar, ornamental	LF	0.1840	17.34	48.32	65.66
05.5006 000	**WROUGHT IRON RAILINGS:**					
05.5006 011	Wrought iron rail, stock pattern, 6' high	LF	0.1931	18.19	170.08	188.27

METALS - 05

Division	Description	Unit	Labor Hours	Labor Cost	Material Cost	Total Cost
05.5006 021	Wrought iron rail, custom pattern, 8' high	LF	0.4097	38.60	376.48	415.08
05.5006 031	Wrought iron rail, stair, 42" high	LF	0.2002	18.86	117.13	135.99
05.5007 000	**ALUMINUM & BRASS RAILING:**					
05.5007 001	*Note: For anodizing, add 10% to the total costs. For bronze anodizing, add 20%. For black anodizing, add 30%.*					
05.5007 011	Aluminum railing, stock patterns	LF	0.2480	23.37	111.78	135.14
05.5007 021	Aluminum railing, architectural designs	LF	0.2940	27.70	161.99	189.70
05.5007 031	Brass railing, architectural designs	LF	0.3251	30.63	178.12	208.75
05.5008 000	**DECORATIVE HANDRAILS, EASY DESIGNS:**					
05.5008 011	Rail, brass/bronze, floor mounted	LF	0.3870	36.46	379.02	415.48
05.5008 021	Rail, aluminum, floor mounted	LF	0.3220	30.34	138.39	168.73
05.5008 031	Rail, stainless steel, floor mounted	LF	0.4255	40.09	231.77	271.87
05.5008 041	Rail, brass/bronze, wall mounted	LF	0.2576	24.27	132.94	157.21
05.5008 051	Rail, aluminum, wall mounted	LF	0.2150	20.26	49.87	70.13
05.5008 061	Rail, stainless steel, wall mounted	LF	0.3415	32.18	83.54	115.72
05.5009 000	**WIRE MESH:**					
05.5009 011	Mesh & frame grill, heavy protection	SF	0.0723	6.81	34.28	41.09
05.5009 021	Mesh, partitions, heavy protection	SF	0.0723	6.81	10.39	17.20
05.5009 031	Mesh, standard partitions	SF	0.0792	7.46	7.67	15.14
05.5009 041	Mesh window guards, galvanized	SF	0.0723	6.81	18.92	25.73
05.5009 051	Grate, welded steel, 1"x1/8", galvanized, 50#/SF	SF	0.1182	11.14	45.75	56.88
05.5009 061	Grate, steel, 1-1/4"x3/16", 98#/SF	SF	0.1488	14.02	54.56	68.58
05.5009 071	Grate, steel, 1-1/2"x3/16", 105#/SF	SF	0.1860	17.52	66.99	84.52
05.5009 081	Grate, aluminum, 1" x 1/8", 19#/SF	SF	0.1023	9.64	42.01	51.65
05.5009 091	Grate, aluminum, 1-1/4" x 1/8", 33#/SF	SF	0.1395	13.14	63.09	76.24
05.5009 101	Grate, aluminum, 1-1/4" x 3/16", 38#/SF	SF	0.1767	16.65	67.87	84.52
05.5009 111	Add for edge banding, steel	LF	0.1033	9.73	6.55	16.28
05.5009 121	Add for edge banding, aluminum	LF	0.1219	11.49	8.61	20.09
05.5010 000	**FRAMING:**					
05.5010 011	Canopy framing	#	0.0498	4.69	4.62	9.31
05.5010 021	Frame supports, steel	#	0.0184	1.73	3.84	5.57
05.5010 031	Frame supports, aluminum	#	0.0592	5.58	22.39	27.97
05.5011 000	**FIRE ESCAPES:**					
05.5011 011	Fire escapes, ladder & balcony	FLOOR	7.2521	683.29	4,584.51	5,267.80
05.5012 000	**EMBEDDED STEEL:**					
05.5012 011	Light steel, embedded	#	0.0288	2.71	3.88	6.60
05.5012 021	Heavy steel, embedded	#	0.0209	1.97	2.73	4.69
05.5012 031	Trench covers with embedded frames	#	0.0201	1.89	2.60	4.49
05.5012 041	Carpenter's iron, general	#	0.0314	2.96	3.25	6.21
05.5012 051	Light door frames, U channel, anchor	#	0.0288	2.71	4.37	7.08
05.5012 061	Heavy door frames, U channel, anchor	#	0.0209	1.97	3.01	4.98
05.5012 071	Thresholds with anchors	#	0.0288	2.71	4.37	7.08
05.5012 081	Clips & L's, tiedowns, etc	#	0.0314	2.96	3.65	6.61
05.5013 000	**COLUMNS, TUBE AND PIPE:**					
05.5013 011	Column, pipe, 3", with plates	#	0.0100	0.94	1.74	2.68
05.5013 021	Column, pipe, 4", with plates	#	0.0094	0.89	1.56	2.45
05.5013 031	Column, pipe, 6", with plates	#	0.0090	0.85	1.45	2.30
05.5013 041	Column, tube, 4" x 4" x 3/8"	#	0.0073	0.69	2.40	3.08
05.5014 000	**ROOF HATCHES:**					
05.5014 011	Roof hatch, 30" x 48"	EA	2.5603	241.23	1,588.59	1,829.82
05.5014 021	Roof hatch, 30" x 72"	EA	3.2678	307.89	2,515.29	2,823.19
05.5014 031	Ceiling hatch, 30" x 30"	EA	2.1762	205.04	899.09	1,104.13
05.5015 000	**CATCH BASIN AND MANHOLE ACCESSORIES:**					
05.5015 011	Catch basin, grate & frame, 18" square, 500#	EA	0.8331	78.49	523.21	601.70
05.5015 021	Catch basin, grate & frame, 24" square, 500#	EA	1.3950	131.44	634.63	766.07

05 - METALS

Division	Description	Unit	Labor Hours	Labor Cost	Material Cost	Total Cost
05.5015 031	Catch basin, grate & frame, 30" square, 500#	EA	1.9540	184.11	1,288.14	1,472.25
05.5015 041	Manhole, cover & frame, 24" dia	EA	1.9540	184.11	1,249.07	1,433.18
05.5015 051	Manhole, cover & frame, 30" dia	EA	1.3950	131.44	704.67	836.11
05.5015 071	Manhole, steps, wrought iron/aluminum	EA	0.4218	39.74	28.63	68.37
05.5015 081	Manhole ladder rungs, galvanized	EA	0.3562	33.56	24.33	57.89
05.5015 091	Trench drain, 8"	LF	0.2139	20.15	98.21	118.36
05.5016 000	**MISCELLANEOUS SPECIALTIES:**					
05.5016 011	Gray iron foundry items	#	0.0122	1.15	2.60	3.75
05.5016 021	Stair nosings	LF	0.0723	6.81	13.35	20.16
05.5016 031	Stainless steel corridor wall rail, 1/2"	LF	0.0930	8.76	80.21	88.97
05.5016 041	Add for galvanizing	#			1.37	1.37
05.5017 000	**SUPPORTS, SPECIAL:**					
05.5017 011	Toilet partition supports	EA	0.9765	92.01	134.35	226.36
05.5017 021	Surgical light supports, single	EA	1.1625	109.53	341.99	451.52
05.5017 031	Surgical light supports, double	EA	2.3250	219.06	683.94	903.00
05.5017 041	X-ray track supports, double	EA	5.6916	536.26	1,289.00	1,825.26
05.5017 051	Elevator beams	#	0.0107	1.01	3.08	4.08
05.5018 000	**COLUMN BASES:**					
05.5018 011	Column base, 16" x 16" x 2-1/2", cast iron	EA	0.6149	57.94	171.55	229.48
05.5018 021	Column base, 32" x 32" x 3-3/4", cast iron	EA	1.2904	121.58	688.50	810.08
05.5018 031	Wood column base, 8" x 8", cast aluminum	EA	0.1844	17.37	186.14	203.52
05.5018 041	Wood column base, 12" x 12", cast aluminum	EA	0.1844	17.37	296.46	313.84
05.5019 000	**CORNER GUARDS:**					
05.5019 011	Corner guard, wheel, 2' 6" high	EA	0.2499	23.55	71.17	94.72
05.5019 021	Corner guard, wheel, 5' high	EA	0.4998	47.09	142.28	189.37
05.5020 000	**CASTINGS, MISCELLANEOUS:**					
05.5020 011	Iron casting, lightweight section	#	0.0040	0.38	2.29	2.67
05.5020 021	Iron casting, heavy section	#	0.0040	0.38	2.02	2.40
05.5021 000	**MISCELLANEOUS ORNAMENTAL METAL:**					
05.5021 011	Sight screen, aluminum	SF	0.1314	12.38	58.26	70.64
05.5021 021	Sight screen, extruded metal	SF	0.1314	12.38	44.80	57.19
05.5021 031	Sun screen, aluminum, manual	SF	0.0653	6.15	58.62	64.77
05.5021 041	Sun screen, aluminum, motorized	SF	0.0862	8.12	105.75	113.87
05.5021 051	Wrought iron gate, 6' x 7'	EA	3.7547	353.77	2,696.01	3,049.78
05.5021 061	Brass/bronze work	#	0.0618	5.82	34.15	39.97
05.5021 071	Aluminum work	#	0.1227	11.56	27.75	39.31
05.5021 081	Stainless steel, 304 series	#	0.0618	5.82	15.46	21.28
05.5022 000	**EXPANSION CONTROL & JOINTS:**					
05.5022 001	*Note: For bronze or stainless, add 2% to the material costs.*					
05.5022 011	Expansion joint, 1-1/2", floor	LF	0.1153	10.86	63.29	74.16
05.5022 021	Expansion joint, 1-1/2", wall	LF	0.1404	13.23	44.65	57.88
05.5022 031	Expansion joint, gymnasium base	LF	0.0948	8.93	28.16	37.09
05.5022 041	Expansion joint, roof, 1-1/2"	LF	0.1360	12.81	88.77	101.59
05.5022 051	Expansion joints, 4", floor	LF	0.2314	21.80	122.88	144.69
05.5022 061	Expansion joints, 4", wall	LF	0.2808	26.46	86.14	112.60
05.5022 071	Expansion joints, 4", roof	LF	0.2605	24.54	130.34	154.88

CARPENTRY - 06

Division	Description	Unit	Labor Hours	Labor Cost	Material Cost	Total Cost
06.0000 000	**CARPENTRY:**					
06.0000 001	Note: In the following section, material prices are based on commercial structures using standard or better grade lumber. Labor prices are based on institutional grade framing.					
06.1000 000	**ROUGH CARPENTRY:**					
06.1000 001	Note: The following percentages may be used to adjust the pricing in this section to fit job requirements: Construction grade lumber . . . . add 19% to material Select structural lumber . . . . add 40% to material Forest Stewardship Council Approved (LEED) . . . . . add 15% to material Kiln dried lumber . . . . add 15% to material Hand selected lumber . . . . add 19% to material Institutional structures . . . . add 25% to material Residential by builder . . . . deduct 15% from labor Residential by framer . . . . deduct 30% from labor Note that the following sections (.1001, .1002, and .1003) are given in square foot area measure, and are pre-factored for board foot measure for quick estimates to eliminate board foot measure take-off by the estimator.					
06.1001 000	**STUD WALLS, 16" OC, INSTALLED:**					
06.1001 001	Note: The following prices include single plates, blocking, headers and diagonal bracing. Prices are by the square foot, already factored for board feet.					
06.1001 011	Wall, 2" x 4" x 8', 16" OC, 1.1 BF/SF	SF	0.0218	2.22	0.75	2.97
06.1001 021	Wall, 2" x 4" x 10', 16" OC, 1.0 BF/SF	SF	0.0215	2.19	0.69	2.87
06.1001 031	Wall, 3" x 4" x 8', 16" OC, 1.5 BF/SF	SF	0.0296	3.01	1.03	4.04
06.1001 041	Wall, 2" x 6" x 8', 16" OC, 1.7 BF/SF	SF	0.0315	3.21	1.30	4.51
06.1001 051	Wall, 2" x 6" x 10', 16" OC, 1.5 BF/SF	SF	0.0336	3.42	1.15	4.57
06.1001 061	Wall, 2" x 6" x 8', 12" OC, 2.0 BF/SF	SF	0.0394	4.01	1.53	5.54
06.1001 071	Wall, 2" x 4", 2" x 6" plate, staggered	SF	0.0357	3.63	2.03	5.67
06.1002 000	**JOIST, FLOOR, 16" OC, (MINIMUM 3000 SF):**					
06.1002 001	Note: The following prices include rim, joists and blocking. Prices are by square foot, already factored for board feet.					
06.1002 011	Floor joist, 2" x 6", 16" OC, 1.1 BF/SF	SF	0.0183	1.86	0.84	2.70
06.1002 021	Floor joist, 2" x 8", 16" OC, 1.55 BF/SF	SF	0.0289	2.94	1.22	4.16
06.1002 031	Floor joist, 2" x 10", 16"oc, 1.85 BF/SF	SF	0.0301	3.06	1.83	4.89
06.1002 041	Floor joist, 2" x 12", 16" OC, 2.1 BF/SF	SF	0.0331	3.37	2.12	5.49
06.1002 051	Floor joist, 2" x 14", 16" OC, 2.42 BF/SF	SF	0.0394	4.01	2.49	6.50
06.1003 000	**RAFTERS, 24" OC:**					
06.1003 001	Note: The following prices include freeze block, ribbon & strong backs. Prices are by the square foot, already factored for board feet.					
06.1003 011	Rafters, 2" x 4", 24" OC, 0.46 BF/SF	SF	0.0364	3.70	0.32	4.02
06.1003 021	Rafters, 2" x 6", 24" OC, 0.68 BF/SF	SF	0.0451	4.59	0.52	5.11
06.1003 031	Rafters, 2" x 8", 24" OC, 0.90 BF/SF	SF	0.0621	6.32	0.71	7.03
06.1003 041	Rafters, 2" x 10", 24" OC, 1.16 BF/SF	SF	0.0773	7.87	1.15	9.01
06.1003 051	Rafters, 2" x 12", 24" OC, 1.37 BF/SF	SF	0.0924	9.40	1.38	10.79
06.1100 000	**VERTICAL FRAMING, WALLS, PER 1,000 BOARD FEET:**					
06.1101 000	**STUD WALLS, COMBINED, 16" OC:**					
06.1101 001	Note: The following prices include plates, blocks, bracing and mud sills.					
06.1101 011	Stud wall, 2" x 4", 2'-6', 16" OC	MBF	19.7003	2,005.10	685.51	2,690.61
06.1101 021	Stud wall, 2" x 4", 8', 16" OC	MBF	19.7003	2,005.10	685.51	2,690.61
06.1101 031	Stud wall, 3" x 4", 8', 16" OC	MBF	19.7003	2,005.10	685.51	2,690.61
06.1101 041	Stud wall, 2" x 6", 8', 16" OC	MBF	19.7003	2,005.10	765.46	2,770.56
06.1101 051	Stud wall, 2" x 4", 10'-16', 16" OC	MBF	21.5000	2,188.27	685.51	2,873.78
06.1101 061	Stud wall, 3" x 4", 10'-16', 16" OC	MBF	22.4266	2,282.58	685.51	2,968.09
06.1101 071	Stud wall, 2" x 6", 10'-16', 16" OC	MBF	22.4266	2,282.58	765.46	3,048.04
06.1102 000	**SLOPING STUD WALLS, COMBINED, 16" OC:**					
06.1102 001	Note: The following prices include blocks, bracing and mid blocks, with flat bottom plate and sloping top plate.					
06.1102 011	Stud wall, sloping, 2" x 4", 8'-16', 16" OC	MBF	30.7887	3,133.67	685.51	3,819.18

06 - CARPENTRY

Division	Description	Unit	Labor Hours	Labor Cost	Material Cost	Total Cost
06.1102 021	Stud wall, sloping, 3" x 4", 8'-16', 16" OC	MBF	30.3481	3,088.83	685.51	3,774.34
06.1102 031	Stud wall, sloping, 2" x 6", 8'-16', 16" OC	MBF	30.3481	3,088.83	765.46	3,854.29
06.1103 000	UNDERPINNING:					
06.1103 011	Pier caps, field grade redwood, 2" x 6"	MBF	67.3129	6,851.11	765.46	7,616.57
06.1103 021	Jack studs, 2" x 4"	MBF	42.0818	4,283.09	685.51	4,968.60
06.1103 031	Jack post, 4" x 4"	MBF	38.8170	3,950.79	797.15	4,747.94
06.1104 000	MUD SILLS:					
06.1104 011	Mud sill, bolt, 2" x 4" or 2" x 6", field grade redwood	MBF	19.6974	2,004.80	1,028.26	3,033.07
06.1104 021	Mud sill, bolt, 3" x 4", field grade redwood	MBF	39.4402	4,014.22	1,028.26	5,042.49
06.1104 031	Mud sill, shot, 2" x 4" or 2" x 6", field grade redwood	MBF	16.7400	1,703.80	1,028.26	2,732.06
06.1104 041	Mud sill, shot, 3" x 4", field grade redwood	MBF	29.0611	2,957.84	1,028.26	3,986.10
06.1104 051	Mud sill, grout, 2" x 4" or 2" x 6"	LF	0.0679	6.91	2.84	9.75
06.1104 061	Mud sill, grout, 2" x 8"	LF	0.0727	7.40	3.17	10.57
06.1104 071	Anchor bolt, embedded, 8" x 5/8"	EA	0.0698	7.10	3.05	10.16
06.1104 081	Shot fasteners	EA	0.0364	3.70	1.91	5.61
06.1105 000	STUD WALL COMPONENTS:					
06.1105 011	Plate, 2" x 4" or 2" x 6"	MBF	18.6000	1,893.11	685.51	2,578.62
06.1105 021	Studs, 2" x 4" or 2" x 6", 8'	MBF	20.4600	2,082.42	685.51	2,767.93
06.1105 031	Studs, 3" x 4", 8'	MBF	20.4600	2,082.42	685.51	2,767.93
06.1105 041	Studs, 2" x 4" or 2" x 6", 10'-16'	MBF	22.3200	2,271.73	685.51	2,957.24
06.1105 051	Studs, 3" x 4", 10'-16'	MBF	22.3200	2,271.73	685.51	2,957.24
06.1105 061	Blocking/brace, 2" x 4"	MBF	46.2068	4,702.93	685.51	5,388.44
06.1105 071	Blocking, 2" x 6"	MBF	44.6079	4,540.19	765.46	5,305.65
06.1105 081	Bracing, diagonal, let in, 1" x 4'	MBF	42.4305	4,318.58	685.51	5,004.09
06.1105 091	Headers, 4" x 12"	MBF	33.8568	3,445.95	1,008.29	4,454.24
06.1106 000	POSTS AND COLUMNS:					
06.1106 001	Note: The following prices include handling, unloading, cutting, notching, dapping, erecting and bolting.					
06.1106 011	Post, 4" x 4" to 4" x 8", construction grade	MBF	60.1710	6,124.20	797.15	6,921.35
06.1106 021	Post, 6" x 6" to 6" x 12", construction grade	MBF	45.1283	4,593.16	1,221.98	5,815.14
06.1106 031	Post, 8" x 8" to 8" x 12", construction grade	MBF	43.1463	4,391.43	1,221.98	5,613.41
06.1106 041	Post, 12" x 12", construction grade	MBF	39.9900	4,070.18	1,421.98	5,492.16
06.1106 051	Post, 4" x 4" to 4" x 8", select grade	MBF	60.1710	6,124.20	797.15	6,921.35
06.1106 061	Post, 6" x 6" to 6" x 12", select grade	MBF	45.1283	4,593.16	1,221.98	5,815.14
06.1106 071	Post, 8" x 8" to 8" x 12", select grade	MBF	43.1463	4,391.43	1,221.98	5,613.41
06.1106 081	Post, 12" x 12", select grade	MBF	39.9900	4,070.18	1,421.98	5,492.16
06.1200 000	HORIZONTAL FRAMING, PER 1,000 BOARD FEET:					
06.1201 000	BEAMS:					
06.1201 011	Beams, 4" x 6" to 4" x 12", to 20', construction grade	MBF	30.2665	3,080.52	797.15	3,877.67
06.1201 021	Beams, 6" x 6" to 6" x 12", to 20', construction grade	MBF	29.0113	2,952.77	1,221.98	4,174.75
06.1201 031	Beams, 8" x 8" to 8" x 12", to 20', construction grade	MBF	27.7561	2,825.02	1,221.98	4,047.00
06.1201 041	Beams, 4" x 6" to 4" x 12", to 20', select grade	MBF	30.2665	3,080.52	897.52	3,978.04
06.1201 051	Beams, 6" x 6" to 6" x 12", to 20', select grade	MBF	29.0113	2,952.77	1,221.98	4,174.75
06.1201 061	Beams, 8" x 8" to 8" x 12", to 20', select grade	MBF	27.7561	2,825.02	1,221.98	4,047.00
06.1202 000	FLOOR JOISTS & BLOCKING INCLUDING RIM JOISTS:					
06.1202 011	Floor joist, block, 2" x 6", rim joist	MBF	19.2000	1,954.18	765.46	2,719.64
06.1202 021	Floor joist, block, 2" x 8", rim joist	MBF	18.2000	1,852.40	786.46	2,638.86
06.1202 031	Floor joist, block, 2" x 10", rim joist	MBF	17.1882	1,749.41	988.21	2,737.62
06.1202 041	Floor joist, block, 2" x 12", rim joist	MBF	16.5483	1,684.29	1,008.29	2,692.58
06.1202 051	Floor joist, block, 2" x 14", rim joist	MBF	16.5483	1,684.29	1,028.46	2,712.75
06.1202 061	Floor joist, block, 4" x 14", rim joist	MBF	18.1000	1,842.22	1,687.77	3,529.99
06.1202 071	Floor joist, block, 3" x 14", rim joist	MBF	18.1000	1,842.22	1,687.77	3,529.99
06.1203 000	PURLINS:					
06.1203 011	Purlin, 4" x 14"-16" x 20'	MBF	19.6751	2,002.53	1,687.77	3,690.30
06.1203 021	Purlin, 4" x 14"-16" x 24'	MBF	19.3439	1,968.82	1,687.77	3,656.59

CARPENTRY - 06

Division	Description	Unit	Labor Hours	Labor Cost	Material Cost	Total Cost
06.1203 031	Purlin, 6" x 10"	MBF	29.2019	2,972.17	1,454.88	4,427.04
06.1203 041	Purlin, 6" x 12"	MBF	28.4175	2,892.33	1,687.77	4,580.10
06.1204 000	**RAFTERS & OUTRIGGERS:**					
06.1204 001	*Note: For 5, 6, 8 or 12/12 pitch, add 50% to the labor costs. For rafter framing with dormers and cut up sections, add 75% to the labor costs.*					
06.1204 011	Rafter, 2" x 4", 4/12 pitch	MBF	40.9091	4,163.73	685.51	4,849.24
06.1204 021	Rafter, 2" x 6", 4/12 pitch	MBF	34.1538	3,476.17	765.46	4,241.63
06.1204 031	Rafter, 2" x 8", 4/12 pitch	MBF	34.8837	3,550.46	786.46	4,336.92
06.1204 041	Rafter, 2" x 10"-12", 4/12 pitch	MBF	34.4675	3,508.10	988.21	4,496.31
06.1205 000	**ROOF JOISTS & BLOCKING, INCLUDING RIM JOISTS:**					
06.1205 011	Roof joist, block, 2" x 4", rim joist	MBF	21.6181	2,200.29	685.51	2,885.80
06.1205 021	Roof joist, block, 2" x 6", rim joist	MBF	21.5306	2,191.38	765.46	2,956.84
06.1205 031	Roof joist, block, 2" x 8", rim joist	MBF	21.1816	2,155.86	786.46	2,942.32
06.1205 041	Roof joist, block, 2" x 10", rim joist	MBF	18.9162	1,925.29	988.21	2,913.50
06.1205 051	Roof joist, block, 2" x 12", rim joist	MBF	18.3058	1,863.16	1,008.29	2,871.45
06.1300 000	**MISCELLANEOUS FRAMING & MATERIALS:**					
06.1301 000	**FURRING:**					
06.1301 011	Furring, 2" x 4", nailed to masonry	MBF	43.0000	4,376.54	685.51	5,062.05
06.1301 021	Furring, 1" x 3", machine nailed, ceiling	MBF	38.9703	3,966.40	1,063.80	5,030.20
06.1301 031	Furring, 1" x 3", nailed to concrete	MBF	56.3000	5,730.21	1,063.80	6,794.02
06.1302 000	**INSULATION BOARD:**					
06.1302 001	*Note: The following items are installed by carpenters. See also section 07.2001.*					
06.1302 011	Insulation, fiberboard, 1/2", 4' x 8', wall	SF	0.0099	1.01	0.54	1.55
06.1302 021	Insulation, fiberboard, 3/4", 4' x 8', wall	SF	0.0107	1.09	0.71	1.80
06.1302 031	Insulation, fiberboard, 1", 2' x 8', roof	SF	0.0156	1.59	1.18	2.77
06.1302 041	Insulation, fiberboard, 2", 2' x 8', roof	SF	0.0188	1.91	1.49	3.41
06.1303 000	**STRIPPING:**					
06.1303 011	Strip, 1" x 4"-6", machine nailed	MBF	44.3105	4,509.92	1,550.23	6,060.15
06.1303 021	Strip, 1" x 4"-6", soffit, machine nailed	MBF	55.2949	5,627.91	1,550.23	7,178.14
06.1304 000	**ADDERS & MISCELLANEOUS MATERIAL:**					
06.1304 011	Asphalt felt, 15#, walls	SF	0.0040	0.41	0.24	0.64
06.1304 021	Backing for other trades	MBF	47.5747	4,842.15	1,216.27	6,058.43
06.1304 031	Beams, solid timbers, 6" and up	MBF	35.6810	3,631.61	1,895.69	5,527.31
06.1304 041	Sisalkraft flashing	SF	0.0084	0.85	0.25	1.10
06.1304 051	Gypsum sheathing, 5/8"	SF	0.0159	1.62	0.55	2.17
06.1304 061	Headers, solid timbers, 4" & up	MBF	35.6810	3,631.61	1,477.90	5,109.51
06.1304 071	Ledgers/nailers, bolted, 2" x 6-8-10"	MBF	52.2873	5,321.80	1,000.01	6,321.81
06.1304 081	Post/mullion, solid timbers, to 16'	MBF	47.5747	4,842.15	1,947.37	6,789.52
06.1304 091	Ribbons and strong backs	MBF	40.1698	4,088.48	1,003.05	5,091.54
06.1304 101	Furring	MBF	39.9900	4,070.18	1,503.61	5,573.79
06.1304 111	Stripping	MBF	39.9900	4,070.18	1,550.23	5,620.41
06.1304 121	Grounds	MBF	51.1500	5,206.05	1,003.05	6,209.10
06.1304 131	Building paper	SF	0.0028	0.28	0.24	0.52
06.1305 000	**ADDERS TO ROUGH FRAMING:**					
06.1305 001	*Note: For backing and bracing add approximately 5% to total rough carpentry quantities and labor. Attaching to metal frame, add 20% to the labor costs.*					
06.1305 031	Add for bracing, miscellaneous	MBF	33.0049	3,359.24	943.82	4,303.06
06.1305 041	Add for framing, miscellaneous	MBF	33.0049	3,359.24	943.82	4,303.06
06.1400 000	**SHEATHING:**					
06.1401 000	**FLOOR SHEATHING:**					
06.1401 011	Floor sheathing, diagonal, 1" x 6"-8"	MBF	14.5164	1,477.48	765.46	2,242.94
06.1401 021	Floor sheathing, tongue & groove, 2" x 6"-8"	MBF	15.0002	1,526.72	765.46	2,292.18
06.1401 031	Floor sheathing, tongue & groove, 2" select Douglas fir	MBF	13.7108	1,395.49	2,810.11	4,205.60
06.1401 041	Deck, timber, 3"-4", laminated, tongue & groove	MBF	12.8234	1,305.17	797.15	2,102.32

06 - CARPENTRY

Division	Description	Unit	Labor Hours	Labor Cost	Material Cost	Total Cost
06.1401 051	Deck, timber, 2" x 6", laminated, tongue & groove	MBF	16.5000	1,679.37	765.46	2,444.83
06.1401 061	Deck, composition board, (TREX), 2"x6", natural	LF	0.0165	1.68	4.08	5.76
06.1402 000	**WALL SHEATHING:**					
06.1402 011	Wall sheathing, diagonal, 1"	MBF	23.0651	2,347.57	1,947.94	4,295.51
06.1403 000	**ROOF SHEATHING:**					
06.1403 011	Roof sheathing, 1-1/8" plywood	MBF	15.0002	1,526.72	777.56	2,304.28
06.1404 000	**SHEATHING, FLOOR, PLYWOOD:**					
06.1404 011	Floor, 1/2", C-D exterior, hand nailed	MSF	10.6735	1,086.35	894.20	1,980.55
06.1404 021	Floor, 1/2", C-D exterior, machine nailed	MSF	9.2225	938.67	894.20	1,832.86
06.1404 031	Floor, 5/8", C-D exterior, hand nailed	MSF	11.2245	1,142.43	1,065.92	2,208.35
06.1404 041	Floor, 5/8", C-D exterior, machine nailed	MSF	9.6908	986.33	1,065.92	2,052.25
06.1404 051	Floor, 3/4", C-D exterior, hand nailed	MSF	12.1682	1,238.48	1,249.98	2,488.46
06.1404 061	Floor, 3/4", C-D exterior, machine nailed	MSF	10.4930	1,067.98	1,249.98	2,317.96
06.1404 071	Floor, 5/8", C-D exterior, tongue & groove, hand nailed	MSF	11.3504	1,155.24	1,199.16	2,354.40
06.1404 081	Floor, 5/8", C-D exterior, tongue & groove, machine nailed	MSF	9.7978	997.22	1,199.16	2,196.38
06.1404 091	Floor, 3/4", C-D exterior, tongue & groove, hand nailed	MSF	12.8694	1,309.85	1,406.23	2,716.07
06.1404 101	Floor, 3/4", C-D exterior, tongue & groove, machine nailed	MSF	11.0890	1,128.64	1,406.23	2,534.86
06.1404 111	Floor, 1-1/8", C-D exterior, tongue & groove, hand nailed	MSF	18.0748	1,839.65	2,293.19	4,132.84
06.1404 121	Floor, 1-1/8", C-D exterior, tongue & groove, machine nailed	MSF	17.0689	1,737.27	2,293.19	4,030.46
06.1404 131	Floor, 3/8", particle board underlay, hand nailed	MSF	10.6735	1,086.35	615.50	1,701.85
06.1404 141	Floor, 3/8", particle board underlay, machine nailed	MSF	9.2225	938.67	615.50	1,554.17
06.1404 151	Floor, 1/2", particle board underlay, hand nailed	MSF	10.6735	1,086.35	745.16	1,831.51
06.1404 161	Floor, 1/2", particle board underlay, machine nailed	MSF	9.2225	938.67	745.16	1,683.83
06.1404 171	Floor, 3/4", particle board underlay, hand nailed	MSF	12.1682	1,238.48	1,041.65	2,280.13
06.1404 181	Floor, 3/4", particle board underlay, machine nailed	MSF	10.4930	1,067.98	1,041.65	2,109.63
06.1405 000	**SHEATHING, WALL, PLYWOOD:**					
06.1405 001	Note: Do not deduct for opening less than 6'8". For machine nailing, deduct 15% from the labor costs.					
06.1405 011	Wall, 5/16", C-D exterior, hand nailed	MSF	21.4779	2,186.02	727.55	2,913.57
06.1405 021	Wall, 3/8", C-D exterior, hand nailed	MSF	21.4779	2,186.02	765.84	2,951.86
06.1405 031	Wall, 1/2", C-D exterior, hand nailed	MSF	22.4543	2,285.40	927.17	3,212.57
06.1405 041	Wall, 5/8", C-D exterior, hand nailed	MSF	25.3839	2,583.57	1,105.23	3,688.80
06.1405 051	Wall, 3/4", C-D exterior, hand nailed	MSF	25.3839	2,583.57	1,296.07	3,879.64
06.1406 000	**SHEATHING, ROOF:**					
06.1406 001	Note: The following prices do not include stocking to roof. For structural grade plywood . . . add 20% to material costs For shear wall nailing . . . . add 20% to labor costs For 9' sheets. . . . . add 15% to material costs Allow 5% of material costs for waste.					
06.1406 011	Roof, 3/8", C-D exterior, hand nailed	MSF	13.4734	1,371.32	927.17	2,298.49
06.1406 021	Roof, 3/8", C-D exterior, machine nailed	MSF	10.7787	1,097.06	927.17	2,024.23
06.1406 031	Roof, 1/2", C-D exterior, hand nailed	MSF	13.6684	1,391.17	1,105.23	2,496.40
06.1406 041	Roof, 1/2", C-D exterior, machine nailed	MSF	10.9347	1,112.93	1,105.23	2,218.16
06.1406 051	Roof, 5/8", C-D exterior, hand nailed	MSF	15.1325	1,540.19	1,105.23	2,645.41
06.1406 061	Roof, 5/8", C-D exterior, machine nailed	MSF	12.6919	1,291.78	1,105.23	2,397.01
06.1406 071	Roof, 3/4", C-D exterior, hand nailed	MSF	15.8165	1,609.80	1,296.07	2,905.87
06.1406 081	Roof, 3/4", C-D exterior, machine nailed	MSF	13.3746	1,361.27	1,296.07	2,657.34
06.1406 091	Roof, 1-1/8", C-D exterior, machine nailed	MSF	17.0740	1,737.79	2,377.75	4,115.54
06.1500 000	**CARPENTRY SPECIALTIES:**					
06.1500 001	Note: The treatment for house on slab on grade includes drilling through footing to penetrate to under-slab area and treating soil.					
06.1501 000	**WOOD TREATING:**					
06.1501 011	Fire proofing treatment	MBF			479.72	479.72
06.1501 021	Fungus and termite treatment	MBF			403.86	403.86
06.1501 031	Tent and fumigate existing house	MCF			139.42	139.42
06.1501 041	Soil treatment around house	SF			0.57	0.57

CARPENTRY - 06

Division	Description	Unit	Labor Hours	Labor Cost	Material Cost	Total Cost
06.1501 051	House on slab on grade (per running foot)	FT			13.95	13.95
06.2000 000	**FINISH CARPENTRY:**					
06.2000 001	Note: The following percentages may be used to adjust pricing on this section to fit job requirements: Residential by builder.......... deduct 15% from material Residential by finish contractor ... deduct 30% from material					
06.2001 000	**FACIA:**					
06.2001 011	Redwood facia, 2" x 3"-4", KD, clear heart	MBF	42.2685	4,302.09	3,025.52	7,327.61
06.2001 021	Redwood facia, 2" x 6"-8", KD, clear heart	MBF	42.2685	4,302.09	3,125.94	7,428.03
06.2001 031	Redwood facia, 2" x 10", KD, clear heart	MBF	42.2685	4,302.09	3,352.07	7,654.15
06.2001 041	Redwood facia, 2" x 12", KD, clear heart	MBF	42.2685	4,302.09	3,830.86	8,132.94
06.2001 051	Cedar facia, 2" x 4", KD, select	MBF	42.2685	4,302.09	2,786.16	7,088.25
06.2001 061	Cedar facia, 2" x 6", KD, select	MBF	42.2685	4,302.09	2,865.94	7,168.03
06.2001 071	Cedar facia, 2" x 8", KD, select	MBF	42.2685	4,302.09	2,916.80	7,218.88
06.2001 081	Cedar facia, 2" x 10", KD, select	MBF	42.2685	4,302.09	2,938.48	7,240.57
06.2001 091	Cedar facia, 2" x 12", KD, select	MBF	42.2685	4,302.09	3,032.82	7,334.91
06.2001 101	Douglas fir facia, 2", KD	MBF	42.2685	4,302.09	2,880.45	7,182.54
06.2002 000	**TRIM, VERTICAL & HORIZONTAL AVERAGE:**					
06.2002 011	Redwood trim, 1" x 4", clear heart	MBF	97.6500	9,938.82	2,556.98	12,495.79
06.2002 021	Redwood trim, 1" x 6", clear heart	MBF	68.3550	6,957.17	2,663.19	9,620.36
06.2002 031	Redwood trim, 1" x 8", clear heart	MBF	51.2663	5,217.88	2,663.19	7,881.07
06.2002 041	Redwood trim, 1" x 10", clear heart	MBF	51.2663	5,217.88	2,945.27	8,163.15
06.2002 051	Redwood trim, 1" x 12", clear heart	MBF	51.2663	5,217.88	3,431.28	8,649.16
06.2002 061	Redwood trim, 2", average	MBF	24.4125	2,484.70	3,206.40	5,691.10
06.2002 071	Redwood deck, 2 x 4, 2 x 6, construction heart	MSF	32.0000	3,256.96	1,232.97	4,489.93
06.2003 000	**SIDING, WOOD:**					
06.2003 001	Note: For green lumber, deduct 20% from the total costs.					
06.2003 011	Siding, redwood, bevel, rustic, 1", KD clear	MBF	33.0000	3,358.74	4,157.70	7,516.44
06.2003 021	Siding, cedar, bevel, rustic, 1", KD clear	MBF	32.0000	3,256.96	2,238.82	5,495.78
06.2003 031	Siding, redwood board & batten, rustic, KD clear	MBF	26.9700	2,745.01	6,657.93	9,402.93
06.2003 041	Siding, redwood tongue & groove, 1" x 6", KD clear	MBF	25.0000	2,544.50	6,114.73	8,659.23
06.2003 051	Add for mitred/pattern nailed	MBF	22.2261	2,262.17		2,262.17
06.2003 061	Add for resawn finish	MBF			144.60	144.60
06.2004 000	**SIDING, PLYWOOD & MISCELLANEOUS:**					
06.2004 001	Note: The following prices are for premium grade siding.					
06.2004 011	Siding, redwood plywood, 3/8", plain, rough	MSF	30.0536	3,058.86	2,775.13	5,833.98
06.2004 021	Siding, redwood plywood, 3/8", planktex	MSF	30.0536	3,058.86	2,775.13	5,833.98
06.2004 031	Siding, redwood plywood, 3/8", prestain	MSF	30.0536	3,058.86	2,872.28	5,931.13
06.2004 041	Siding, redwood plywood, 5/8", T-1-11	MSF	30.8717	3,142.12	3,073.29	6,215.41
06.2004 051	Siding, cedar plywood, 5/8", T-1-11	MSF	30.8717	3,142.12	2,645.49	5,787.61
06.2004 061	Siding, redwood plywood, 5/8", board & batten shiplap	MSF	30.8717	3,142.12	3,243.11	6,385.24
06.2004 071	Siding, fir plywood, 3/8", rough sawn	MSF	30.0536	3,058.86	1,671.95	4,730.81
06.2004 081	Siding, fir plywood, 3/8" planktex	MSF	30.0536	3,058.86	1,902.47	4,961.32
06.2004 091	Siding, fir plywood, 5/8", board & batten, rough	MSF	30.0536	3,058.86	1,999.14	5,058.00
06.2004 101	Siding, fir plywood, 5/8", T-1-11, rough	MSF	30.0536	3,058.86	1,999.14	5,058.00
06.2004 111	Add for prefinish of plywood siding	MSF			349.45	349.45
06.2004 121	Siding, hardboard, tempered & primed	MSF	35.6534	3,628.80	1,623.32	5,252.12
06.2004 131	Siding, cedar shingles, 7" exposed	SQ	6.2919	640.39	309.37	949.76
06.2004 141	Siding, cedar shingles, 3 ply-back	SQ	3.0051	305.86	344.16	650.02
06.2005 000	**SIDING, HARDBOARD:**					
06.2005 011	Siding, hardboard, 4' x 8', 1/8" tempered	SF	0.0360	3.66	0.82	4.48
06.2005 021	Siding, hardboard, 4' x 8', 1/4" tempered	SF	0.0360	3.66	1.02	4.69
06.2005 031	Siding, hardboard, 4' x 8', 3/8" tempered & prime coated	SF	0.0360	3.66	1.55	5.21
06.2006 000	**SIDING, GlasWeld PANELS, WOOD OR METAL STOPS:**					
06.2006 011	Siding, GlasWeld, 1/8", 1 side finished	SF	0.0492	5.01	6.42	11.43
06.2006 021	Siding, GlasWeld, 1/4", 2 side finished	SF	0.0553	5.63	8.14	13.77

06 - CARPENTRY

Division	Description	Unit	Labor Hours	Labor Cost	Material Cost	Total Cost
06.2007 000	**MOLDING:**					
06.2007 001	*Note: For horizontal molding, small forms, add 20% to the labor costs.*					
06.2007 011	Base, 1-5/8", prefinished, hardwood	LF	0.0246	2.50	1.95	4.46
06.2007 021	Base, 2-1/4", softwood	LF	0.0246	2.50	1.58	4.09
06.2007 031	Base, 3-1/4", softwood	LF	0.0246	2.50	1.84	4.35
06.2007 041	Molding, chair rail	LF	0.0383	3.90	1.70	5.59
06.2007 051	Molding, cove, 3/4"	LF	0.0219	2.23	1.02	3.25
06.2007 061	Molding, cove, 1-3/4"	LF	0.0219	2.23	1.21	3.44
06.2007 071	Molding, picture	LF	0.0383	3.90	1.25	5.14
06.2007 081	Molding, shoe	LF	0.0274	2.79	0.73	3.52
06.2007 091	Molding, stucco	LF	0.0302	3.07	1.04	4.12
06.2007 101	Molding, quarter round, 3/4"	LF	0.0233	2.37	0.90	3.27
06.2008 000	**TRIM:**					
06.2008 011	Trim, apron, 1-3/8"	LF	0.0284	2.89	1.17	4.06
06.2008 021	Trim, jamb, 2" pine with stop	LF	0.0233	2.37	2.79	5.16
06.2008 031	Trim, jamb, 4-5/8" pine	LF	0.0997	10.15	4.13	14.28
06.2008 041	Trim, oval casing, 2-1/4"	LF	0.0399	4.06	1.58	5.64
06.2008 051	Oval casing or streamline, 2-5/8"	LF	0.0399	4.06	1.73	5.79
06.2008 061	Trim, stool, 6" milled redwood	LF	0.0532	5.41	11.82	17.23
06.2008 071	Trim, stool, 4-1/2", flat pine	LF	0.0532	5.41	1.88	7.29
06.2008 081	Trim, stop, sash, 3/8" x 3/4"	LF	0.0284	2.89	0.76	3.65
06.2008 091	Trim, stop, 1-3/8"	LF	0.0284	2.89	1.11	4.00
06.2009 000	**MILLING CHARGES:**					
06.2009 011	Knife grinding, custom trim	SETUP			357.54	357.54
06.2010 000	**FLOORING & SHELVING:**					
06.2010 011	Flooring, pine, 1" x 4"	MSF	20.3177	2,067.94	2,299.34	4,367.27
06.2010 021	Shelving, pine, 1" x 12", clear	LF	0.0459	4.67	4.09	8.76
06.2010 031	Hook strip, pine, 4-1/2", clear	LF	0.0459	4.67	2.75	7.42
06.2010 041	Shelving, particle board, 3/4 x 11-1/4	LF	0.0459	4.67	1.77	6.45
06.2011 000	**PANELING:**					
06.2011 001	*Note: For book matched paneling ...... add 40% to material costs* *For top grade workmanship........ add 100% to labor costs* *For 10' sheets .... add 30% to material costs*					
06.2011 011	Panel, birch or ash, select	SF	0.0460	4.68	2.20	6.88
06.2011 021	Panel, birch or ash, economy	SF	0.0422	4.30	1.43	5.72
06.2011 031	Panel, mahogany, rotary cut, economy	SF	0.0384	3.91	1.42	5.32
06.2011 041	Panel, mahogany, ribbon cut, good	89	0.0384	3.91	1.64	5.55
06.2011 051	Panel, mahogany, Philippine, select	SF	0.0384	3.91	1.58	5.49
06.2011 061	Panel, mahogany, African, unfinished, good	SF	0.0566	5.76	3.53	9.29
06.2011 071	Panel, teak, unfinished, select	SF	0.0566	5.76	3.00	8.76
06.2011 081	Panel, teak, finished v plank, good	SF	0.0575	5.85	5.02	10.87
06.2011 091	Panel, walnut, finished, domestic	SF	0.0537	5.47	4.77	10.24
06.2011 101	Panel, walnut, finished, select	SF	0.0575	5.85	7.78	13.64
06.2011 111	Panel, hardboard, embossed/printed		0.0419	4.26	1.34	5.60
06.2011 121	Panel, hardboard, printed finish	SF	0.0419	4.26	1.03	5.30
06.2011 131	Panel, Douglas fir, clear, vertical grain	SF	0.0358	3.64	3.41	7.06
06.2011 141	Panel, redwood, clear all heart	SF	0.0035	0.36	3.50	3.86
06.2100 000	**CARPENTRY, INSTALLATION ONLY:**					
06.2101 000	**CARPENTRY, INSTALLATION ONLY:**					
06.2101 011	Cabinets, base, modular	LF	0.3906	39.76		39.76
06.2101 021	Cabinets, wall	LF	0.7440	75.72		75.72
06.2101 031	Cabinets, full height	LF	0.3033	30.87		30.87
06.2101 041	Doors, wood, pre-hung	EA	0.5500	55.98		55.98
06.2101 051	Doors, job hung, residential & commercial	EA	1.0100	102.80		102.80
06.2101 061	Doors, job hung, institutional	EA	1.0629	108.18		108.18

CARPENTRY - 06

Division	Description	Unit	Labor Hours	Labor Cost	Material Cost	Total Cost
06.2101 071	Frames, wood, to 7' h & 6' w	EA	1.6700	169.97		169.97
06.2101 081	Hardware	DOOR	1.1625	118.32		118.32
06.2101 091	Sash	SF	0.0372	3.79		3.79
06.2101 101	Toilet accessories, average	EA	0.3720	37.86		37.86
06.2101 111	Frames, hollow metal, single	EA	0.8138	82.83		82.83
06.2101 121	Frames, hollow metal, pair	EA	1.0776	109.68		109.68
06.3000 000	**GLU-LAM BEAMS, TRUSSES & HEAVY TIMBER:**					
06.3001 000	**GLU-LAMS, 7,000 BF OR MORE:**					
06.3001 001	*Note: For quantities under 7,000 bf, add 33% to total.*					
06.3001 011	Glu-lams, industrial finish	MBF	6.2430	635.41	2,568.48	3,203.89
06.3001 021	Glu-lams, arch finish, wrapped	MBF	6.2430	635.41	2,846.96	3,482.37
06.3001 031	Glu-lam, arch, wrap, curved, 32' rad	MBF	6.7518	687.20	5,274.54	5,961.73
06.3001 041	Glu-lam, arch, Tudor, 10', 31' rad	MBF	9.0024	916.26	6,958.79	7,875.06
06.3001 051	Add for minor camber	EA			100.54	100.54
06.3001 061	Add for delivery (min 40m#)	MILE			3.71	3.71
06.3001 071	Add for delivery, 60' (min 40m#)	MILE			11.32	11.32
06.3001 081	Add for steel connect, shoes, bolts	#			2.75	2.75
06.3002 000	**BOWSTRING TRUSSES:**					
06.3002 011	Bowstring trusses, under 3,000 BF	MBF	11.2530	1,145.33	2,500.56	3,645.89
06.3002 021	Bowstring trusses, over 3,000 BF	MBF	9.9297	1,010.64	2,500.56	3,511.21
06.3003 000	**TRUSSES, RESIDENTIAL, & HEAVY TIMBER:**					
06.3003 011	Residential trusses, quantity	LF	0.0216	2.20	5.11	7.31
06.3003 021	Residential trusses, single house	LF	0.0217	2.21	5.69	7.90
06.3003 031	Heavy timber trusses	MBF	22.5060	2,290.66	2,548.17	4,838.83
06.3004 000	**TRUSS JOISTS, IN PLACE:**					
06.3004 001	*Note: Quantities under 2,000 LF .... add 50% to total cost* *Between 2,000 and 3,000 LF .... add 20% to total cost*					
06.3004 011	Truss joist, simple, over 3,000 lf	LF	0.0533	5.42	8.31	13.74
06.3004 021	Truss joist, compound, over 3000 lf	LF	0.0760	7.74	11.99	19.73
06.3004 031	Truss hardware, shoes, etc	#	1.0000	101.78	6.40	108.18
06.3005 000	**HEAVY TIMBERS, MILLED & LAMS:**					
06.3005 011	6" lumber & larger	MBF	10.0739	1,025.32	2,925.85	3,951.17
06.3005 021	2" & larger, mill constructed	MBF	13.1277	1,336.14	1,731.18	3,067.32
06.3005 031	2" laminates, mill constructed	MBF	23.2668	2,368.09	1,822.48	4,190.57
06.5000 000	**STAIRS, WOOD:**					
06.5000 001	*Note: For metal rails see section 05.5004.*					
06.5001 000	**STAIRS, CIRCULAR, 36" WIDE RISER:**					
06.5001 001	*Note: Add for rail.*					
06.5001 011	Stair, circular, hardwood, closed riser, mill	RISER	3.5936	365.76	515.64	881.40
06.5001 021	Stairs, circular, hardwood, open riser, mill	RISER	3.5936	365.76	459.56	825.32
06.5001 031	Stairs, semi-circular, 36" riser, job	RISER	1.1972	121.85	171.65	293.51
06.5002 000	**STAIRS, STRAIGHT & SWITCHBACK:**					
06.5002 011	Stair, straight, 36" riser, hardwood	RISER	1.1972	121.85	134.10	255.95
06.5002 021	Stair, switchback, 36" riser, hardwood	RISER	1.4386	146.42	145.10	291.53
06.5002 031	Stair, fourway, 36" riser, hardwood	RISER	1.7976	182.96	191.31	374.27
06.5003 000	**STAIR LANDINGS:**					
06.5003 011	Stair landings, finished hardwood	SF	0.3003	30.56	24.71	55.27
06.5004 000	**STAIR RAILINGS, HARDWARE NOT INCLUDED:**					
06.5004 011	Hand rail, fir, 1-5/8" x 1-3/4", wall	LF	0.1843	18.76	2.21	20.97
06.5004 021	Hand rail, fir, 1-5/8" x 2-5/8", wall	LF	0.1843	18.76	2.86	21.62
06.5004 031	Hand rail, custom, hardwood, straight	LF	0.3003	30.56	31.97	62.54
06.5004 041	Hand rail, custom, hardwood, curve	LF	0.6006	61.13	142.63	203.76
06.6000 000	**ROUGH HARDWARE:**					
06.6000 001	*Note: Allow $60.00 Per 1,000 board feet of lumber for general construction.*					
06.6001 000	**FRAMING HARDWARE:**					

06 - CARPENTRY

Division	Description	Unit	Labor Hours	Labor Cost	Material Cost	Total Cost
06.6001 011	Joist hanger, 2" x 4", nails	EA	0.0355	3.61	1.24	4.85
06.6001 021	Joist hanger, 2" x 6"-8", nails	EA	0.0440	4.48	1.54	6.02
06.6001 031	Joist hanger, 2" x 12", nails	EA	0.0502	5.11	1.90	7.01
06.6001 041	Joist hanger, 4" x 10"-12"	EA	0.0539	5.49	4.45	9.93
06.6001 051	Joist hanger, 4" x 14"-16"	EA	0.0662	6.74	4.77	11.51
06.6001 061	Joist hanger, 6" x 10"-14", heavy duty	EA	0.0662	6.74	15.23	21.97
06.6001 071	Post anchor, 4" x 4"-6"	EA	0.0722	7.35	7.70	15.05
06.6001 081	Post cap, 16 ga, galvanized, 4" x 4"	EA	0.0722	7.35	7.36	14.70
06.6001 091	Post base	EA	0.0869	8.84	7.91	16.75
06.6001 101	Post base & cap combination	EA	0.0539	5.49	14.98	20.47
06.6001 111	Framing clips	EA	0.0404	4.11	2.59	6.71
06.6001 121	Ply clips, 1/2", 5/8", 3/4", aluminum	EA	0.0098	1.00	0.24	1.23
06.6001 131	Tie straps, 2-3/8" x 12", 20 ga	EA	0.2033	20.69	2.44	23.13
06.6001 141	Tie straps, 2-3/8" x 23", 16 ga	EA	0.2033	20.69	4.16	24.85
06.6001 151	Strap anchor, 3" x 33", 1/4" plate	EA	0.2657	27.04	23.85	50.90
06.6001 161	Strap anchor, 3" x 45", 1/4" plate	EA	0.2719	27.67	35.33	63.01
06.6001 171	Brace, 12 ga galvanized, 2" x 8'	EA	0.0797	8.11	31.47	39.58
06.6001 181	Brace, 12 ga galvanized, 1-1/4' x 12'	EA	0.0797	8.11	36.71	44.83
06.6001 191	Bridge, metal, 2" x 8", 12" OC	EA	0.0196	1.99	0.95	2.95
06.6001 201	Bridge, metal, 2" x 8", 16" OC	EA	0.0196	1.99	1.06	3.05
06.6001 211	Bridge, metal, 2" x 8", 24" OC	EA	0.0196	1.99	1.29	3.29
06.6001 221	Bridge, metal, 2" x 16", 12" OC	EA	0.0196	1.99	1.06	3.05
06.6001 231	Bridge, metal, 2" x 16", 16" OC	EA	0.0196	1.99	1.06	3.05
06.6001 241	Bridge, metal, 2" x 16", 24" OC	EA	0.0196	1.99	1.29	3.29
06.6001 251	Hanger, stainless steel, 2" x 4"	EA	0.0527	5.36	6.66	12.02
06.6001 261	Hanger, stainless steel, 2" x 8"	EA	0.0539	5.49	9.59	15.08
06.6001 271	Hanger, stainless steel, 2"x14"	EA	0.0600	6.11	19.69	25.79
06.6001 281	Hanger, stainless steel, 4" x 4"	EA	0.0539	5.49	8.08	13.56
06.6001 291	Hanger, stainless steel, 4" x 8"	EA	0.0600	6.11	13.62	19.73
06.6001 301	Hanger, stainless steel, 4" x 16"	EA	0.0698	7.10	28.21	35.32
06.6001 311	Metal bridging, 16" OC	SET			1.61	1.61
06.6001 321	Joist hanger, 4" x 4"	EA			3.17	3.17
06.6001 331	Joist hanger, 4" x 16"	EA			8.41	8.41
06.6002 000	**NAILS:**					
06.6002 011	Nails, common, average	100#			142.04	142.04
06.6002 021	Nails, galvanized, average	100#			242.95	242.95
06.6002 031	Nails, aluminum, average	100#			1,040.63	1,040.63
06.6002 041	Nails, copper, average	100#			946.51	946.51
06.6002 051	Nails, stainless steel, average	100#			1,193.21	1,193.21
06.6002 061	Nails, add for cement coating	100#			14.56	14.56
06.6002 071	Nails, add for zinc plating	100#			26.71	26.71
06.6003 000	**TIMBER CONNECTORS:**					
06.6003 011	Split rings, 2-5/8"	EA	0.0599	6.10	2.72	8.81
06.6003 021	Split rings, 4"	EA	0.0654	6.66	3.90	10.55
06.6003 031	Toothed rings, 2-5/8"	EA	0.0498	5.07	1.18	6.25
06.6003 041	Toothed rings, 4"	EA	0.0544	5.54	1.29	6.83
06.6003 051	Shear plates, 2-5/8"	EA	0.0792	8.06	3.26	11.32
06.6003 061	Shear plates, 4"	EA	0.1004	10.22	7.11	17.33
06.6003 071	Glu-lam beam seats, 5-1/4" x 1-1/2"	EA	0.4362	44.40	32.03	76.43
06.6003 081	Glu-lam beam seats, 6-3/4" x 1-3/4"	EA	0.4950	50.38	36.34	86.72
06.6003 091	Glu-lam beam seats, 8-3/4" x 1-3/4"	EA	0.5538	56.37	50.11	106.48
06.6003 101	Laminated hangers with nails, 3-1/8 x 13-1/2	EA	0.6028	61.35	44.23	105.58
06.6003 111	Laminated hangers with nails, 5-1/8 x 13-1/2	EA	0.6616	67.34	47.54	114.88
06.6003 121	Laminated hangers with nails, 6-3/4 x 13-1/2	EA	0.7204	73.32	50.09	123.41
06.6003 131	Laminated hangers with nails, 8-3/4 x 13-1/2	EA	0.7792	79.31	54.39	133.70

CARPENTRY - 06

Division	Description	Unit	Labor Hours	Labor Cost	Material Cost	Total Cost
06.6003 141	Laminated saddle hangers, 3-1/8 x 13-1/2	EA	1.0511	106.98	77.13	184.12
06.6003 151	Laminated saddle hangers, 5-1/8 x 13-1/2	EA	1.1099	112.97	103.02	215.99
06.6003 161	Laminated saddle hangers, 6-3/4 x 13-1/2	EA	1.1687	118.95	147.56	266.51
06.6003 171	Laminated saddle hangers, 8-3/4 x 13-1/2	EA	1.2275	124.93	162.91	287.84
06.6003 181	Hinge connectors, 5-1/8" x 13"	EA	1.3969	142.18	102.54	244.72
06.6003 191	Hinge connectors, 6-3/4" x 13"	EA	1.4557	148.16	205.47	353.63
06.6003 201	Hinge connectors, 8-3/4" x 13"	EA	1.5145	154.15	244.44	398.59
06.6004 000	**BOLTS:**					
06.6004 011	Foundation bolts, 8" x 1/2" dia	EA	0.0698	7.10	1.45	8.55
06.6004 021	Foundation bolts, 8" x 5/8" dia	EA	0.0698	7.10	2.90	10.00
06.6004 031	Foundation bolts, 10" x 3/4" dia	EA	0.0698	7.10	4.38	11.48
06.6004 041	Foundation bolts, 10" x 1/2" dia	EA	0.0698	7.10	1.50	8.61
06.6004 051	Stud bolts, 4" x 1/2" dia	EA	0.0940	9.57	2.19	11.76
06.6004 061	Stud bolts, 8" x 3/4" dia	EA	0.1188	12.09	4.68	16.77
06.6004 071	Ledger bolts, 8" x 5/8" dia	EA	0.0746	7.59	2.56	10.15

07 - THERMAL & MOISTURE PROTECTION

Division	Description	Unit	Labor Hours	Labor Cost	Material Cost	Total Cost
07.1000 000	**WATERPROOFING:**					
07.1000 001	*Note: For the following section, add 50% for small areas, such as small decks.*					
07.1001 000	**WATERPROOFING, HOT COATINGS, VERTICAL SURFACES:**					
07.1001 011	Bitumals, walls, per coat	SF	0.0130	1.14	0.05	1.19
07.1001 021	Hot mop, 15# felt, 1 ply	SF	0.0130	1.14	0.12	1.25
07.1001 031	Hot mop, 15# felt, 2 ply	SF	0.0183	1.60	0.27	1.87
07.1001 041	Hot mop, 15# felt, 3 ply	SF	0.0248	2.17	0.41	2.58
07.1001 051	Hot mop, 30# felt, 1 ply	SF	0.0130	1.14	0.86	2.00
07.1001 061	Hot mop, 30# felt, 2 ply	SF	0.0183	1.60	1.00	2.60
07.1001 071	Hot mop, 30# felt, 3 ply	SF	0.0248	2.17	1.11	3.28
07.1001 081	Hot mop, glass fiber, 1 ply	SF	0.0130	1.14	0.77	1.90
07.1001 091	Hot mop, glass fiber, 2 ply	SF	0.0183	1.60	0.94	2.54
07.1001 101	Hot mop, glass fiber, 3 ply	SF	0.0248	2.17	1.09	3.26
07.1001 111	Add for 1/2" asphalt fiberboard	SF	0.0072	0.63	0.38	1.01
07.1002 000	**WATERPROOFING, COLD APPLICATION, VERTICALS:**					
07.1002 011	Waterproof, 1/32" butyl	SF	0.0226	1.97	0.80	2.77
07.1002 021	Waterproof, 1/16" butyl	SF	0.0226	1.97	1.17	3.15
07.1002 031	Elastomeric membrane, 60 mil butyl, wall	SF	0.0054	0.47	1.02	1.50
07.1002 051	Add for butyl with nylon	SF	0.0050	0.44	0.25	0.68
07.1002 061	Waterproof, 1/32" neoprene	SF	0.0226	1.97	0.80	2.77
07.1002 071	Waterproof, 1/16" neoprene	SF	0.0226	1.97	1.21	3.18
07.1002 081	Add for neoprene, nylon, extra coat	SF	0.0050	0.44	0.25	0.68
07.1002 101	Waterproof, Kraft paper	SF	0.0033	0.29	0.01	0.30
07.1002 111	Elastomeric polyurethane, 60 mil, horizontal	SF	0.0226	1.97	1.42	3.40
07.1002 121	Elastomeric polyurethane, 60 mil, vertical	SF	0.0300	2.62	1.50	4.12
07.1002 141	Waterproof, 60 mil Bituthene membrane	SF	0.0226	1.97	1.21	3.18
07.1003 000	**WATERPROOFING, EMULSION, VERTICAL SURFACES:**					
07.1003 011	Waterproof, asphalt mastic, 1/16"	SF	0.0119	1.04	0.12	1.16
07.1003 021	Waterproof, asphalt mastic, 1/8"	SF	0.0140	1.22	0.33	1.56
07.1003 031	Waterproof, asphalt paint, brush/coat	SF	0.0054	0.47	0.12	0.59
07.1003 041	Waterproof, asphalt paint, spray/coat	SF	0.0022	0.19	0.12	0.31
07.1003 051	Waterproof, silicone, spray/coat	SF	0.0022	0.19	0.12	0.31
07.1003 061	Waterproof, silicone, spray, 2 coat	SF	0.0054	0.47	0.28	0.75
07.1003 071	Add for each extra coat, silicone	SF	0.0027	0.24	0.06	0.30
07.1003 081	Add for 1/2" fiberboard protection	SF	0.0065	0.57	0.33	0.90
07.1004 000	**WATERPROOFING, SPECIALTIES, VERTICAL SURFACES:**					
07.1004 011	Waterproof, bentonite panels, 3/16"	SF	0.0144	1.26	1.63	2.88
07.1004 021	Waterproof, bentonite panels, 3/16", coated	SF	0.0144	1.26	1.94	3.20
07.1004 031	Waterproof, bentonite panel, 9/16"	SF	0.0144	1.26	2.08	3.34
07.1004 051	Waterproof, bentonite, 3/8"	SF	0.0159	1.39	2.80	4.19
07.1004 061	Waterproof, interior, cement pargeting, 1/2", 2 coat	SF	0.0223	1.95	0.38	2.32
07.1004 071	Waterproof, exterior, cement pargeting, 1/2", 2 coat	SF	0.0223	1.95	0.53	2.48
07.1004 081	Waterproof, ironite, bushhammer preparation	SF	0.0215	1.88	0.12	2.00
07.1004 091	Waterproof, ironite, 2 coats	SF	0.0369	3.22	0.82	4.04
07.1004 101	Waterproof, ironite, 4 coats	SF	0.0563	4.92	1.16	6.08
07.1004 111	Waterproof, ironite, 6 coats	SF	0.0706	6.16	2.35	8.51
07.1004 121	Polymeric waterproof membrane, 1/16", vertical	SF	0.0226	1.97	1.72	3.70
07.1005 000	**WATERPROOFING, HOT COATINGS, HORIZONTAL SURFACE:**					
07.1005 011	Hot mop, deck, 15# felt, 2 ply	SF	0.0120	1.05	0.22	1.27
07.1005 021	Hot mop, deck, 15# felt, 3 ply	SF	0.0167	1.46	0.28	1.74
07.1005 031	Hot mop, deck, 30# felt, 2 ply	SF	0.0120	1.05	0.28	1.33
07.1005 041	Hot mop, deck, 30# felt, 3 ply	SF	0.0167	1.46	0.34	1.80
07.1005 051	Add for fiberglass fabric	SF			0.27	0.27
07.1006 000	**WATERPROOFING, COLD APPLICATION, HORIZONTAL SURFACES:**					
07.1006 011	Waterproof, deck, 1/32" butyl	SF	0.0054	0.47	0.78	1.25
07.1006 021	Waterproof, deck, 1/16" butyl	SF	0.0054	0.47	1.06	1.53
07.1006 031	Add for butyl with nylon	SF			0.27	0.27

THERMAL & MOISTURE PROTECTION - 07

Division	Description	Unit	Labor Hours	Labor Cost	Material Cost	Total Cost
07.1006 041	Elastomeric membrane, 60 mil butyl, floor	SF	0.0054	0.47	1.05	1.52
07.1006 051	Waterproof, deck, 1/32" neoprene	SF	0.0076	0.66	1.52	2.18
07.1006 061	Waterproof, deck, 1/16" neoprene	SF	0.0660	5.76	0.94	6.70
07.1006 071	Add for neoprene with nylon	SF			0.28	0.28
07.1006 081	Waterproof, shower pans	SF	0.0226	1.97	1.15	3.13
07.1008 000	**WATERPROOFING, EMULSION, HORIZONTAL SURFACES:**					
07.1008 011	Waterproof, deck, asphalt mastic, 1/16"	SF	0.0119	1.04	0.05	1.09
07.1008 021	Waterproof, deck, asphalt mastic, 1/8"	SF	0.0119	1.04	0.11	1.15
07.1008 031	Polyethylene asphalt, 60 mil, sheet, waterproof	SF	0.0226	1.97	1.33	3.30
07.1009 000	**WATERPROOFING, SPECIALTY, HORIZONTAL SURFACES:**					
07.1009 011	Waterproof, deck, Dex-o-tex, non-wear	SF	0.1100	9.61	1.12	10.73
07.1009 021	Waterproof, deck, Dex-o-tex, walk on	SF	0.0900	7.86	2.79	10.65
07.1009 041	Polyurethane traffic deck coat	SF	0.0220	1.92	1.86	3.79
07.1009 051	Siloxane sealer ultra violet protection	SF	0.0029	0.25	0.36	0.61
07.1009 061	Siloxane sealer, economy, floor/wall	SF	0.0017	0.15	0.20	0.35
07.1009 071	Siloxane sealer, best, floor/wall	SF	0.0031	0.27	0.41	0.68
07.1009 081	Siloxane, anti graffiti/fungus	SF	0.0027	0.24	0.34	0.58
07.1009 091	Polymeric waterproof membrane, 1/16", horizontal	SF	0.0226	1.97	1.44	3.42
07.1009 101	Waterproof, 60 mil bituthene membrane	SF	0.0230	2.01	1.35	3.36
07.2000 000	**THERMAL & SOUND INSULATION:**					
07.2000 001	Note: For prefab construction see section 13.0000. For insulating concrete, see section 03.5100. For the following items: Double prices for areas under 100 square feet Double prices for existing ceiling (attic) Triple prices for existing wall or for flat ceiling (blown) For overhead space . . . . add 10% to the total cost For tight enclosed space . . . . add 15% to the total cost For one hour fire rating . . . . .add 15% to the total cost					
07.2001 000	**INSULATION BOARD, DECKS:**					
07.2001 011	Insulation board deck, 1", rigid mineral fiber	SF	0.0033	0.29	0.77	1.06
07.2001 021	Insulation board deck, 1-1/2", rigid mineral fiber	SF	0.0043	0.38	1.11	1.48
07.2001 031	Insulation board deck, 2", rigid mineral fiber	SF	0.0043	0.38	1.49	1.87
07.2001 041	Insulation board deck, 6", rigid mineral fiber	SF	0.0078	0.68	4.59	5.27
07.2001 051	Insulation board deck, 1-1/2", rigid fiberglass	SF	0.0046	0.40	1.94	2.34
07.2001 061	Insulation board deck, 2", rigid fiberglass	SF	0.0043	0.38	2.57	2.94
07.2001 071	Insulation board deck, 3", rigid fiberglass	SF	0.0046	0.40	3.39	3.79
07.2001 081	Insulation board deck, 2", rigid firtex, 32" OC	SF	0.0054	0.47	1.89	2.36
07.2001 091	Insulation board deck, 2", rigid tectum, painted	SF	0.0086	0.75	3.39	4.14
07.2001 101	Insulation board deck, 2-1/2", rigid tectum, painted	SF	0.0097	0.85	3.94	4.79
07.2001 111	Insulation board deck, 1-1/4", rigid urethane, R9	SF	0.0038	0.33	1.94	2.27
07.2001 121	Insulation board deck, 1-1/2", rigid urethane, R11	SF	0.0046	0.40	2.49	2.89
07.2001 131	Insulation board deck, 2", rigid urethane, R14	SF	0.0055	0.48	3.20	3.68
07.2001 141	Insulation board deck, 1", rigid urethane, R7	SF	0.0030	0.26	1.42	1.68
07.2001 151	Insulation board deck, 1-1/2", rigid urethane, R11	SF	0.0042	0.37	2.22	2.59
07.2001 161	Insulation board deck, 2", rigid urethane, R14	SF	0.0055	0.48	2.49	2.97
07.2001 171	Insulation board deck, 2-1/2", rigid urethane, R19	SF	0.0072	0.63	3.19	3.82
07.2001 181	Insulation board deck, 3", rigid urethane, R25	SF	0.0080	0.70	4.19	4.89
07.2001 191	Insulation board deck, 1-1/2" rigid Styrofoam	SF	0.0046	0.40	1.89	2.29
07.2001 201	Insulation board deck, 2" rigid Styrofoam	SF	0.0051	0.45	2.41	2.86
07.2001 211	Insulation board deck, 3" rigid Styrofoam	SF	0.0055	0.48	3.65	4.13
07.2002 000	**INSULATION, FIBERGLASS & MINERAL, BLOWN:**					
07.2002 011	Insulation, blown mineral, 4", R11	SF	0.0022	0.21	0.14	0.35
07.2002 021	Insulation, blown mineral, 6", R17	SF	0.0033	0.31	0.21	0.52
07.2002 031	Insulation, blown fiberglass, 4", R13	SF	0.0022	0.21	0.14	0.35
07.2002 041	Insulation, blown fiberglass, 6", R19	SF	0.0033	0.31	0.21	0.52
07.2002 051	Insulation, blown fiberglass, 9", R24	SF	0.0043	0.41	0.32	0.73
07.2003 000	**INSULATION, PELLETIZED, LOOSE:**					

07 - THERMAL & MOISTURE PROTECTION

Division	Description	Unit	Labor Hours	Labor Cost	Material Cost	Total Cost
07.2003 011	Insulation, expanded glass beads, 4"	SF	0.0033	0.31	0.15	0.47
07.2003 021	Insulation, expanded glass beads, 6"	SF	0.0043	0.41	0.21	0.62
07.2003 031	Insulation, expanded shale beads, 4"	SF	0.0033	0.31	0.17	0.48
07.2003 041	Insulation, expanded shale beads, 6"	SF	0.0043	0.41	0.21	0.62
07.2003 051	Insulation, vermiculite beads, 4"	SF	0.0033	0.36	0.48	0.83
07.2003 061	Insulation, vermiculite beads, 6"	SF	0.0043	0.46	0.70	1.16
07.2003 071	Insulation, polystyrene beads, 4"	SF	0.0033	0.36	0.51	0.87
07.2003 081	Insulation, polystyrene beads, 6"	SF	0.0043	0.46	0.81	1.27
07.2004 000	**INSULATION, BATT, WALL & CEILING:**					
07.2004 011	Batts, 2-1/2", mineral fiber, R7	SF	0.0041	0.44	0.27	0.71
07.2004 021	Batts, 3", mineral fiber, R11	SF	0.0041	0.44	0.35	0.80
07.2004 031	Batts, 3-1/2", mineral fiber, R13	SF	0.0041	0.44	0.38	0.82
07.2004 041	Batts, 3-1/2", fiberglass, R13	SF	0.0041	0.44	0.45	0.90
07.2004 051	Batts, 6", fiberglass, R19	SF	0.0041	0.44	0.56	1.01
07.2004 061	Add for supporting batts on wire	SF	0.0021	0.23	0.01	0.24
07.2004 071	Batts, 8-9" fiberglass, R30	SF	0.0084	0.91	0.49	1.39
07.2004 081	Insulation blown fiber 11"	SF	0.0057	0.61	0.37	0.98
07.2005 000	**INSULATION, FOAM OR SPRAYED:**					
07.2005 011	Insulation, urethane foam, 1", R7	SF	0.0108	1.17	0.79	1.95
07.2005 021	Insulation, urethane foam, 1-1/3", R9	SF	0.0119	1.28	0.87	2.16
07.2005 031	Insulation, urethane foam, 1-1/2", R11	SF	0.0119	1.28	1.02	2.30
07.2005 041	Insulation, urethane foam, 1 3/4", R13	SF	0.0140	1.51	1.14	2.65
07.2005 051	Insulation, urethane foam, 2", R14	SF	0.0161	1.74	1.26	3.00
07.2005 061	Insulation, urethane foam, 2-1/4", R17	SF	0.0161	1.74	1.39	3.13
07.2005 071	Insulation, urethane foam, 2 3/4", R20	SF	0.0183	1.97	1.65	3.62
07.2005 081	Insulation, urethane foam, 3", R21	SF	0.0204	2.20	1.76	3.96
07.2005 091	Insulation, urethane foam, 3-1/4", R24	SF	0.0204	2.20	2.07	4.27
07.2005 101	Insulation, urethane foam, 4", R29	SF	0.0265	2.86	2.39	5.25
07.2006 000	**INSULATION, RIGID FIBERGLASS:**					
07.2006 011	Insulation, wall, rigid fiberglass, 1-1/2"	SF	0.0046	0.50	1.01	1.50
07.2006 021	Insulation, wall, rigid fiberglass, 2"	SF	0.0046	0.50	1.29	1.79
07.2006 031	Insulation, wall, rigid fiberglass, 2-1/2"	SF	0.0046	0.50	1.64	2.13
07.2006 041	Insulation, wall, rigid fiberglass, 3"	SF	0.0046	0.50	2.03	2.52
07.2006 051	Insulation, wall, rigid fiberglass, 4"	SF	0.0046	0.50	2.39	2.89
07.2007 000	**INSULATION, RIGID STYROFOAM:**					
07.2007 011	Insulation, rigid Styrofoam, 1"	SF	0.0046	0.50	0.82	1.32
07.2007 021	Insulation, rigid Styrofoam, 1-1/2"	SF	0.0046	0.50	1.22	1.71
07.2007 031	Insulation, rigid Styrofoam, 2"	SF	0.0046	0.50	1.57	2.07
07.2008 000	**INSULATION CORES:**					
07.2008 011	Vermiculite or Perlite, 4"	SF	0.0022	0.24	0.23	0.47
07.2008 021	Vermiculite or Perlite, 6"	SF	0.0029	0.31	0.40	0.71
07.2008 031	Vermiculite or Perlite, 8"	SF	0.0037	0.40	0.58	0.97
07.2008 041	Vermiculite or Perlite, 10"	SF	0.0044	0.47	0.79	1.26
07.2009 000	**ACOUSTIC INSULATION:**					
07.2009 011	Insulation, sound board, 1/2", vertical	SF	0.0076	0.82	0.24	1.06
07.2009 021	Insulation, sound board, 1/2", horizontal	SF	0.0065	0.70	0.24	0.94
07.2009 031	Insulation, acoustic board, 2", horizontal	SF	0.0183	1.97	1.77	3.75
07.2009 041	Insulation, cork, 1/2", horizontal	SF	0.0119	1.28	0.77	2.06
07.2009 051	Insulation, noise barrier, batt, 2-1/2"	SF	0.0043	0.46	0.50	0.96
07.2009 061	Insulation, noise barrier, batt, 3-1/2"	SF	0.0043	0.46	0.53	1.00
07.2010 000	**INSULATION, REFRIGERATION:**					
07.2010 011	Insulation, refrigeration, polystyrene, 2"	SF	0.0097	0.85	0.82	1.67
07.2010 021	Insulation, refrigeration, polystyrene, 3 5/8"	SF	0.0161	1.41	1.45	2.86
07.2010 031	Insulation, refrigeration, polyurethane, 2"	SF	0.0161	1.41	1.76	3.17
07.2010 041	Insulation, refrigeration, polyurethane, 3 5/8"	SF	0.0183	1.60	2.90	4.50
07.2010 051	Insulation, refrigeration, cork, 1"	SF	0.0119	1.04	1.32	2.36
07.2010 061	Insulation, refrigeration, cork, 2"	SF	0.0140	1.22	2.15	3.37

THERMAL & MOISTURE PROTECTION - 07

Division	Description	Unit	Labor Hours	Labor Cost	Material Cost	Total Cost
07.2010 071	Insulation, refrigeration, cork, 3"	SF	0.0183	1.60	3.00	4.60
07.2500 000	**FIREPROOFING, SPRAYED:**					
07.2500 011	Fireproofing, columns, 1-3/8", no finish	SYCA	0.2793	26.08	13.26	39.34
07.2500 021	Fireproofing metal deck, 1/2"	SYCA	0.1790	16.72	9.77	26.48
07.2500 031	Fireproofing beams & girders, 3/4"-1"	SYCA	0.2192	20.47	12.06	32.53
07.2500 041	Fireproofing, Monokote, per ton steel	TON	2.4049	224.57	150.06	374.62
07.2500 051	Fireproofing, Monokote, per SF building	SF	0.0182	1.70	0.95	2.65
07.3000 000	**ROOFING:**					
07.3001 000	**SHINGLES:**					
07.3001 011	Shingles, composition asphalt, 240#	SQ	1.1131	97.20	78.81	176.00
07.3001 021	Shingles, composition asphalt, 240#, 'A'	SQ	1.1691	102.09	101.51	203.60
07.3001 031	Shingles, composition asphalt, 300#	SQ	1.1691	102.09	112.15	214.24
07.3001 041	Shingles, composition asphalt, 325#, 'A'	SQ	1.1691	102.09	128.20	230.29
07.3001 051	Shingles, composition asphalt, president	SQ	1.0700	93.43	174.44	267.87
07.3001 061	Shingles, valley roll	LF	0.0173	1.51	0.54	2.05
07.3001 071	Shingles, aluminum tabs, 020"	SF	0.0347	3.03	3.42	6.45
07.3001 081	Shingles, aluminum tabs, 030"	SF	0.0347	3.03	3.98	7.01
07.3001 091	Shingles, porcelain enamel, 18 ga	SQ	3.7806	330.12	440.20	770.32
07.3001 101	Shingles, fiberglass tabs, 300#	SQ	1.2028	129.77	129.09	258.86
07.3002 000	**TILE, CLAY:**					
07.3002 011	Tile, clay, Spanish, 2 piece	SQ	7.7143	673.61	422.68	1,096.30
07.3002 021	Tile, clay, flat red	SQ	4.2968	375.20	179.73	554.93
07.3002 031	Tile, clay, glazed, interlock	SQ	4.6651	407.36	293.07	700.43
07.3002 041	Tile, clay, Spanish	SQ	4.2356	369.85	254.04	623.89
07.3002 051	Add for 'riness system', galvanized iron	SQ	0.9420	82.26	36.79	119.04
07.3002 061	Add for 'tile-tie system', copper	SQ	0.9420	82.26	55.24	137.50
07.3002 091	Fire vent, skylight, 4'8" x 9'6"	EA	6.5217	568.37	1,501.78	2,070.15
07.3003 000	**TILE, CONCRETE:**					
07.3003 011	Tile, concrete, premium	SQ	4.5802	399.94	367.95	767.89
07.3003 021	Tile, concrete, flat	SQ	3.9330	343.43	246.84	590.27
07.3003 031	Tile, concrete, interlock	SQ	3.6177	315.90	226.23	542.13
07.3004 000	**STONE:**					
07.3004 011	Slate	SQ	2.9545	257.99	701.66	959.65
07.3005 000	**SHINGLES & SHAKES, WOOD, 30# FELT:**					
07.3005 011	Shingles, wood, 5" exposed, roof, premium	SQ	1.5113	131.97	236.97	368.94
07.3005 021	Shingles, wood, 5" exposed, roof, fireproof, premium	SQ	1.5113	131.97	297.30	429.26
07.3005 031	Shingles, roof, shakertown, panel	SQ	0.6984	60.98	365.55	426.53
07.3005 041	Shakes, wood, 1/2" butt, roof	SQ	1.2954	113.11	185.83	298.95
07.3005 051	Shakes, wood, 3/4" butt, roof	SQ	1.2954	113.11	207.93	321.04
07.3005 061	Shakes, wood, medium, B grade	SQ	1.7600	153.68	211.24	364.92
07.3005 071	Shakes, wood, heavy, C grade	SQ	1.9300	168.53	258.15	426.68
07.3005 081	Felt, 30#	SQ	0.0417	3.64	22.56	26.21
07.3006 000	**BUILT/UP ROOFING:**					
07.3006 001	*Note: For chopped up work or multiple levels, add 100% to total costs.*					
07.3006 011	Built-up, 3 ply, low rise	SQ	1.1345	99.06	142.35	241.41
07.3006 021	Built-up, 3 ply, high rise	SQ	2.5525	222.88	142.35	365.23
07.3006 031	Built-up, 4 ply, low rise	SQ	1.2375	108.06	165.31	273.37
07.3006 041	Built-up, 4 ply, high rise	SQ	2.7844	243.13	165.31	408.44
07.3006 051	Built-up, 5 ply, 20 year, low rise	SQ	1.4089	123.03	194.43	317.45
07.3006 061	Built-up, 5 ply, 20 year, high rise	SQ	3.1700	276.80	194.43	471.23
07.3006 071	Add for light rock dress off	SQ	0.0708	6.18	9.76	15.95
07.3006 081	Add for heavy rock dress off	SQ	0.0708	6.18	11.00	17.19
07.3006 091	Add for coal tar pitch roof	SQ	1.2555	109.63	211.47	321.10
07.3006 101	Add for aluminizing	SQ	0.0763	6.66	33.52	40.18
07.3006 111	Cap sheet, 90#, parapet walls	SQ	1.7191	150.11	82.10	232.22
07.3007 000	**PLASTIC ROOFING:**					

07 - THERMAL & MOISTURE PROTECTION

Division	Description	Unit	Labor Hours	Labor Cost	Material Cost	Total Cost
07.3007 001	Note: For the acrylic roof items listed below, prices are based on installation over existing roof, assuming acrylic polyester flashing. Allow 70% for elongation. Average tensile strength of fabric is 100 pounds per inch, 1914 psi minimum (Metacrylics).					
07.3007 011	Roof, elastomeric membrane, 1/16"	SQ	3.6440	318.19	235.73	553.92
07.3007 021	Roof, elastomeric membrane, 1/32"	SQ	3.6440	318.19	209.61	527.81
07.3007 031	Roof, elastomeric, loose, trocal	SQ	1.2945	113.04	223.95	336.98
07.3007 041	Roof, neoprene (gaco-auto guard)	SF	0.0623	5.44	3.37	8.81
07.3007 051	Roof, bituthene, 1/16", selfseal	SQ	3.6440	318.19	220.07	538.26
07.3007 061	Roof, silicone, 3 ply, rolled	SQ	4.4682	390.16	298.74	688.91
07.3007 071	Urethane foam, silicone cover, 1" thick	SQ	1.9000	165.91	564.49	730.39
07.3007 081	Roof, acrylic, existing cap sheet	SQ	0.8399	73.34	266.07	339.41
07.3007 091	Roof, acrylic, existing gravel	SQ	1.6800	146.70	266.07	412.76
07.3007 101	Roof, acrylic, existing deck	SQ	0.8329	72.73	272.35	345.08
07.3007 111	Roof, acrylic, existing corrugated metal	SQ	0.7636	66.68	266.07	332.74
07.3007 121	Roof, acrylic, new	SQ	1.6800	146.70	266.07	412.76
07.3008 000	**CORRUGATED ROOFING:**					
07.3008 011	Roof, corrugated aluminum, 020"	SF	0.0275	2.40	1.43	3.83
07.3008 021	Roof, corrugated aluminum, 032"	SF	0.0319	2.79	2.39	5.17
07.3008 031	Roof, corrugated composition, 3/16", non-walk	SF	0.0373	3.26	2.09	5.34
07.3008 041	Roof, corrugated composition, 3/8"	SF	0.0413	3.61	2.52	6.13
07.3008 051	Roof, corrugated fiberglass, 8 oz	SF	0.0373	3.26	3.46	6.72
07.3008 061	Roof, corrugated galvanized iron, 26 ga	SF	0.0413	3.61	2.47	6.07
07.3009 000	**CORRUGATED SIDING:**					
07.3009 011	Siding, corrugated aluminum, 032"	SF	0.0413	3.61	3.87	7.48
07.3009 021	Siding, corrugated aluminum, 032", painted	SF	0.0413	3.61	4.64	8.25
07.3009 031	Siding, aluminum, simulated wood, insulated	SF	0.0413	3.61	4.32	7.93
07.3009 041	Siding, corrugated composition, 3/8"	SF	0.0552	4.82	3.00	7.82
07.3009 051	Siding, corrugated fiberglass, 8 oz	SF	0.0413	3.61	3.45	7.06
07.3009 061	Siding, zip-rib, 032", over 100,000 SF	SF	0.0275	2.40	3.87	6.27
07.3009 071	Siding, zip-rib, 032", under 100,000 SF	SF	0.0319	2.79	5.20	7.99
07.3009 081	Siding, zip-rib, 032", under 15,000, painted	SF	0.0510	4.45	5.70	10.15
07.3009 091	Siding, corrugated galvanized iron, 26 ga	SF	0.0455	3.97	2.71	6.68
07.3010 000	**ROOF, COPPER CLAD STAINLESS STEEL:**					
07.3010 011	Roof, copper clad stainless steel, 16 oz panels	SF	0.0753	6.58	9.39	15.96
07.3010 021	Roof, copper clad stainless steel, 24 oz panels	SF	0.0933	8.15	13.66	21.80
07.3011 000	**ROOF, COPPER, STANDING SEAM:**					
07.3011 011	Roof, 16 oz copper, 16" panels	SF	0.0753	6.58	8.75	15.32
07.3012 000	**ROOFING, SHEET METAL, CUSTOM FABRICATED:**					
07.3012 011	Roof, fabricated, aluminum, .032", .456#/SF	SF	0.0432	3.77	3.30	7.07
07.3012 021	Roof, fabricated, 16 oz copper, .020", 1.0#/SF	SF	0.0951	8.30	12.22	20.52
07.3012 031	Roof, fabricated, .015" copper clad stainless steel, .787#/SF	SF	0.0475	4.15	5.72	9.86
07.3012 041	Roof, .020" lead coated copper, 1.25#/SF	SF	0.0642	5.61	8.54	14.15
07.3012 051	Roof, galvanized sheet metal, .026", .906#/SF	SF	0.0441	3.85	2.46	6.31
07.3012 061	Roof, fabricated, stainless steel, .018", .787#/SF	SF	0.0612	5.34	4.67	10.02
07.3012 071	Roof, fabricated, terne, .018", .787#/SF	SF	0.0621	5.42	3.19	8.62
07.3012 081	Roof, titanaloy, .027", 1.025#/SF	SF	0.0477	4.17	3.53	7.69
07.3012 091	Roof, fabricated, stainless steel, copper coated galvanized sheet metal	SF	0.0599	5.23	4.78	10.01
07.3013 000	**ROOFING SPECIALTIES:**					
07.3013 011	Cant strip, 3" fiber	LF	0.0182	1.59	0.47	2.06
07.3013 021	Cant strip, 4" fiber, wood	LF	0.0182	1.59	0.59	2.18
07.3013 031	Bonds, roof, 15 year	SQ			6.54	6.54
07.3013 041	Bonds, roof, 20 year	SQ			7.93	7.93
07.3100 000	**COMPOSITE BUILDING PANELS:**					
07.3100 001	Note: For storefront systems, see section 08.9000.					
07.3101 000	**ALUMINUM E-Z WALL SYSTEM:**					
07.3101 011	Aluminum E-Z wall, 1,000-4,000 SF	SF	0.1191	10.40	13.47	23.87

THERMAL & MOISTURE PROTECTION - 07

Division	Description	Unit	Labor Hours	Labor Cost	Material Cost	Total Cost
07.3101 021	Aluminum E-Z wall, 4,000-10,000 SF	SF	0.1145	10.00	12.78	22.78
07.3101 031	Aluminum E-Z wall, over 10,000 SF	SF	0.1032	9.01	11.36	20.37
07.3102 000	**COPPER VENEER PANELS:**					
07.3102 001	*Note: For the following three sections, the standard panel size is 2' x 8'. Prices are based on 10,000 square feet or more and include all joint members.*					
07.3102 011	Copper veneer panel, laminated, 3/8" ply	SF	0.0669	5.84	28.83	34.67
07.3102 021	Copper veneer panel, laminated, 3/4" ply	SF	0.0669	5.84	41.83	47.68
07.3103 000	**STRUCTURAL PANELS, HONEYCOMB CORE:**					
07.3103 011	Structural panel, 1", copper, paper core	SF	0.0748	6.53	38.60	45.14
07.3103 021	Structural panel, 2", copper, paper core	SF	0.0748	6.53	40.71	47.24
07.3103 031	Structural panel, 3", copper, paper core	SF	0.0805	7.03	42.72	49.75
07.3104 000	**INSULATED PANELS, FOAM CORE:**					
07.3104 011	Insulated panel, 1", copper, foam core	SF	0.0748	6.53	40.71	47.24
07.3104 021	Insulated panel, 2", copper, foam core	SF	0.0748	6.53	42.72	49.25
07.3104 031	Insulated panel, 3", copper, foam core	SF	0.0805	7.03	44.92	51.95
07.3105 000	**PANELS, COMPOSITION, EXPOSED AGGREGATE FACE:**					
07.3105 001	*Note: The following panels are prefab units, made in fullwall, window and door components, 60' wide to 25' high.*					
07.3105 011	Veneer panel, granowall, 1/4"	SF	0.0532	4.65	5.04	9.69
07.3105 021	Insulated panel, granowall, 1", color 1 side	SF	0.0584	5.10	7.27	12.37
07.3105 031	Structural wall unit, granostrut	SF	0.0676	5.90	11.73	17.63
07.3105 041	Projected facia, granostrut	SF	0.0604	5.27	9.10	14.37
07.3105 051	Facia/soffit assembly, granostrut	SF	0.0676	5.90	8.94	14.84
07.3200 000	**STONE PANELS, MANUFACTURED:**					
07.3200 001	*Note: The following prices are based on 10,000 square feet or more. Panels are 24' wide to 20' high, with a sandblasted finish.* *Quantities under 10,000 SF . . . . . . add 15% to the total costs* *For Smooth finish . . . . add 30% to the total costs*					
07.3201 000	**FACESPAN PANELS, TO 15':**					
07.3201 011	Facespan panel, flat, 3/4"	SF	0.0870	7.60	12.52	20.12
07.3201 021	Facespan panel, rib one, 1"	SF	0.0972	8.49	13.93	22.42
07.3201 031	Facespan panel, rib 2-4, 1-1/2"	SF	0.0972	8.49	17.86	26.35
07.3201 041	Facespan panel, rib 5, 2 3/4"	SF	0.1218	10.64	19.35	29.98
07.3202 000	**CORSPAN WALL PANELS, TO 20':**					
07.3202 011	Corspan panel, striated, 1-1/2"	SF	0.0850	7.42	17.35	24.77
07.3202 021	Corspan panel, flat/recessed, 3"	SF	0.0921	8.04	20.32	28.36
07.3202 031	Corspan panel, rib 1, 3-1/4"	SF	0.0921	8.04	24.20	32.24
07.3202 041	Corspan panel, rib 2, 3-3/4"	SF	0.0972	8.49	27.16	35.65
07.3202 051	Corspan panel, rib 6, 2-1/2"	SF	0.0870	7.60	25.22	32.82
07.3203 000	**COMPOSITE BUILDING PANELS:**					
07.3203 011	Colorlith panel, 1/4"	SF	0.0532	4.65	10.20	14.85
07.3203 021	Colorlith panel, 5/8"	SF	0.0870	7.60	15.47	23.06
07.3203 031	Stonehenge panel	SF	0.0870	7.60	15.62	23.22
07.3203 041	Splitwood panel, natural	SF	0.0492	4.30	9.10	13.40
07.3203 051	Splitwood panel, factory coated	SF	0.0492	4.30	10.60	14.90
07.3203 061	Santone panel, 1/4"	SF	0.0492	4.30	8.75	13.04
07.3203 071	Kleftstone panel, 5/8"	SF	0.0870	7.60	14.45	22.05
07.3300 000	**MINERAL FIBER PANELS, CURTAIN WALLS:**					
07.3300 001	*Note: The following items are laminated and pressed panels, sizes are 4' wide and from 4' to 10' high.*					
07.3301 000	**GlasWeld PANELS:**					
07.3301 011	GlasWeld panel, one face, 1/4"	SF	0.1240	10.83	13.11	23.94
07.3301 021	GlasWeld panel, two face, 1/4"	SF	0.1240	10.83	17.38	28.21
07.3301 031	GlasWeld panel, one face, 1", insulated	SF	0.1488	12.99	18.13	31.12
07.3301 041	GlasWeld panel, two face, 1", insulated	SF	0.1488	12.99	20.39	33.38
07.3302 000	**PERMASTONE PANELS:**					
07.3302 011	Permastone panel, 1/8", color 1 side	SF	0.0532	4.65	5.47	10.12

07 - THERMAL & MOISTURE PROTECTION

Division	Description	Unit	Labor Hours	Labor Cost	Material Cost	Total Cost
07.3302 021	Permastone panel, 1/4", color 1 side	SF	0.0532	4.65	8.73	13.37
07.3302 031	Permastone panel, 1", insulated, color 1 side	SF	0.0584	5.10	9.67	14.77
07.3302 041	Permastone panel, 1/4", color 2 side	SF	0.0532	4.65	10.75	15.39
07.3302 051	Permastone panel, trim, aluminum, 1/8"/1/4"	SF	0.0195	1.70	2.38	4.08
07.3303 000	**SCULPTURED PANELS:**					
07.3303 011	Relief panel, facade, 2-1/2"	SF	0.1023	8.93	15.04	23.97
07.3303 021	Panel, qasal, 1/4"	SF	0.0604	5.27	9.93	15.21
07.3303 031	Panel, qasal, 1/2"	SF	0.0604	5.27	11.92	17.19
07.3303 041	Panel, qasal, 19/32"	SF	0.0676	5.90	14.84	20.74
07.3304 000	**PRE-CAST AGGREGATE CO-POLYMER PANELS:**					
07.3304 011	Panel, co-polymer, aggregate face, 1"	SF	0.1023	8.93	18.93	27.86
07.3305 000	**PORCELAIN METAL PANELS:**					
07.3305 011	Porcelain on steel, 1-1/2" insulated panel	SF	0.0676	5.90	26.64	32.54
07.3305 021	Porcelain on steel, 1/2" panel	SF	0.0604	5.27	17.89	23.16
07.3305 031	Porcelain on aluminum, 1-1/2" insulated panel	SF	0.0676	5.90	28.30	34.20
07.3305 041	Porcelain on aluminum, 1" panel	SF	0.0604	5.27	25.09	30.37
07.4000 000	**EXTERIOR INSULATION FINISH SYSTEM (EIFS):**					
07.4100 011	EIFS, Expanded polystyrene insulation, 1"	SF	0.1300	12.14	3.09	15.23
07.4100 021	EIFS, Expanded polystyrene insulation, 2"	SF	0.1400	13.07	3.44	16.51
07.4100 031	EIFS, Expanded polystyrene insulation, 3"	SF	0.1450	13.54	3.78	17.32
07.4100 041	EIFS, Expanded polystyrene insulation, 4"	SF	0.1500	14.01	4.14	18.15
07.4100 071	EIFS, Expanded polystyrene insulation, 1", on 3.4# metal lath	SF	0.2370	22.13	3.50	25.63
07.4100 081	EIFS, Expanded polystyrene insulation, 2", on 3.4# metal lath	SF	0.2470	23.06	3.88	26.94
07.4100 091	EIFS, Expanded polystyrene insulation, 3", on 3.4# metal lath	SF	0.2520	23.53	4.22	27.76
07.4100 101	EIFS, Expanded polystyrene insulation, 4", on 3.4# metal lath	SF	0.2570	24.00	4.59	28.59
07.6000 000	**SHEET METAL & FABRICATED SKYLIGHTS:**					
07.6000 001	*Note: For the following section, you may deduct 25% from the total cost for residential tracts.*					
07.6001 000	**COPING & WALL CAPS, ALL 16" GIRTH:**					
07.6001 011	Coping, galvanized sheet metal, 26 ga	LF	0.0514	4.48	3.54	8.02
07.6001 021	Coping, aluminum, .032 ga	LF	0.0514	4.48	2.65	7.13
07.6001 031	Coping, copper, 16 oz	LF	0.0532	4.64	8.82	13.46
07.6001 041	Coping, copper clad stainless steel, .015 ga	LF	0.0549	4.78	8.82	13.61
07.6002 000	**GUTTERS:**					
07.6002 011	Gutter, galvanized iron facia, 5"	LF	0.0514	4.48	1.86	6.34
07.6002 021	Gutter, aluminum facia, 5"	LF	0.0514	4.48	1.26	5.74
07.6002 031	Gutter, copper facia, 5"	LF	0.0532	4.64	9.07	13.70
07.6002 041	Gutter, galvanized iron, 4" offset gutter, standard	LF	0.0514	4.48	1.70	6.18
07.6002 051	Gutter, aluminum, 4" offset gutter, standard	LF	0.0514	4.48	1.33	5.81
07.6002 061	Gutter, aluminum, 5" offset gutter, standard	LF	0.0514	4.48	2.13	6.61
07.6002 071	Gutter, copper, 4" offset gutter, standard	LF	0.0532	4.64	9.19	13.83
07.6002 081	Gutter, galvanized iron, 6" box, shop fabricated	LF	0.0696	6.07	2.75	8.82
07.6002 091	Gutter, copper, 6" box, shop fabricated	LF	0.0696	6.07	11.93	18.00
07.6003 000	**DOWNSPOUTS:**					
07.6003 011	Downspout, galvanized iron, 2" x 3", standard	LF	0.0323	2.81	1.88	4.70
07.6003 021	Downspout, aluminum, 2" x 3", standard	LF	0.0323	2.81	1.42	4.24
07.6003 031	Downspout, copper, 2"x3", standard	LF	0.0323	2.81	7.18	10.00
07.6003 041	Downspout, copper clad stainless steel, 2" x 3", standard	LF	0.0392	3.42	7.18	10.60
07.6003 051	Downspout, galvanized iron, 3" x 4", fabricated	LF	0.0444	3.87	2.64	6.51
07.6003 061	Downspout, aluminum, 3" x 4", fabricated	LF	0.0444	3.87	1.95	5.82
07.6003 071	Downspout, copper, 3" x 4", fabricated	LF	0.0470	4.10	10.07	14.17
07.6003 081	Downspout, copper clad stainless steel, 3" x 4", fabricated	LF	0.0522	4.55	10.07	14.62
07.6003 091	Downspout, galvanized iron, 2", round	LF	0.0323	2.81	1.04	3.86
07.6003 101	Downspout, galvanized iron, 3" round	LF	0.0323	2.81	1.31	4.12
07.6003 111	Downspout, galvanized iron, 6" round	LF	0.0696	6.07	6.36	12.43
07.6003 121	Add for over 2 stories	LF	0.0184	1.60		1.60

THERMAL & MOISTURE PROTECTION - 07

Division	Description	Unit	Labor Hours	Labor Cost	Material Cost	Total Cost
07.6004 000	**GRAVEL STOP:**					
07.6004 011	Gravel stop, galvanized iron, 4"	LF	0.0140	1.22	0.72	1.94
07.6004 021	Gravel stop, aluminum, 4"	LF	0.0140	1.22	0.97	2.19
07.6004 031	Gravel stop, copper, 4"	LF	0.0149	1.30	2.50	3.80
07.6004 041	Gravel stop, copper clad stainless steel, 4"	LF	0.0157	1.37	2.50	3.87
07.6004 051	Gravel stop, facia, galvanized iron, 10"	LF	0.0444	3.87	2.19	6.05
07.6004 061	Gravel stop, facia, aluminum, 10"	LF	0.0444	3.87	3.06	6.93
07.6004 071	Gravel stop, facia, copper, 10"	LF	0.0479	4.17	5.76	9.93
07.6004 081	Gravel stop, facia, copper clad stainless steel, 10"	LF	0.0532	4.64	5.76	10.39
07.6005 000	**FLASHINGS:**					
07.6005 011	Flashing, galvanized iron, 6", 26 ga	LF	0.0270	2.35	0.97	3.32
07.6005 021	Flashing, galvanized iron, 12", 26 ga	LF	0.0296	2.58	1.46	4.04
07.6005 031	Flashing, galvanized iron, 18", 26 ga	LF	0.0323	2.81	1.97	4.78
07.6005 041	Flashing, aluminum, 6", .024"	LF	0.0270	2.35	1.20	3.56
07.6005 051	Flashing, aluminum, 12", .024"	LF	0.0296	2.58	1.70	4.28
07.6005 061	Flashing, aluminum, 18", .024"	LF	0.0323	2.81	2.19	5.00
07.6005 071	Flashing, copper, 6", 16 oz	LF	0.0270	2.35	3.40	5.75
07.6005 081	Flashing, copper, 12", 16 oz	LF	0.0296	2.58	5.18	7.76
07.6005 091	Flashing, copper, 18", 16 oz	LF	0.0323	2.81	6.89	9.71
07.6005 101	Flashing, copper clad stainless steel, 6", .015"	LF	0.0270	2.35	3.40	5.75
07.6005 111	Flashing, copper clad stainless steel, 12", .015"	LF	0.0296	2.58	5.18	7.76
07.6005 121	Flashing, 18" wide	LF	0.0323	2.81	6.89	9.71
07.6005 131	Flashing, lead, 12", 12#/SF	LF	0.0358	3.12	3.40	6.52
07.6005 141	Flashing, lead, 12", 25#/SF	LF	0.0444	3.87	6.59	10.46
07.6006 000	**ROOF FLASHING MISCELLANEOUS:**					
07.6006 011	Roof flashing, galvanized iron, 26 ga	SF	0.0436	3.80	1.97	5.77
07.6006 021	Roof flashing, aluminum, .032"	SF	0.0436	3.80	2.19	5.99
07.6006 031	Roof flashing, copper, 16 oz	SF	0.0436	3.80	6.91	10.71
07.6006 041	Roof flashing, copper clad stainless steel, .015"	SF	0.0436	3.80	5.24	9.04
07.6006 051	Roof flashing, terne, 26 ga	SF	0.0436	3.80	2.06	5.86
07.6006 061	Roof flashing, lead, 25#/SF	SF	0.0974	8.49	3.80	12.29
07.6006 071	Roof flashing, fabric/copper, 5 oz	SF	0.0192	1.67	1.56	3.23
07.6006 081	Roof flashing, fabric/aluminum, .005"	SF	0.0192	1.67	0.89	2.56
07.6006 091	Roof flashing, mastic/copper, 5 oz	SF	0.0157	1.37	1.50	2.87
07.6006 101	Roof flashing, mastic/aluminum, .005"	SF	0.0157	1.37	0.72	2.09
07.6006 111	Flash, 7 oz lead coated copper, fabric backing	SF	0.0174	1.52	1.78	3.30
07.6006 121	Sheet metal, fabricated, 26 ga galvanized iron	SF	0.0704	6.14	1.42	7.56
07.6006 131	Sheet metal, fabricated, .032 aluminum	SF	0.0704	6.14	2.43	8.56
07.6006 141	Sheet metal, fabricated, 16 oz copper	SF	0.0741	6.46	4.95	11.41
07.6006 151	Sheet metal, fabricated, 12# lead	SF	0.0704	6.14	3.36	9.50
07.6006 161	Sheet metal, fabricated, .015 stainless steel	SF	0.0890	7.76	6.80	14.56
07.6007 000	**REGLETS:**					
07.6007 011	Reglets, galvanized iron, 26 ga	LF	0.0270	2.35	1.05	3.41
07.6007 021	Reglets, aluminum, .032"	LF	0.0270	2.35	1.31	3.66
07.6007 031	Reglets, copper, 16 oz	LF	0.0270	2.35	3.30	5.65
07.6007 041	Reglets, copper clad stainless steel, .015"	LF	0.0270	2.35	3.71	6.06
07.6007 051	Reglets, PVC	LF	0.0270	2.35	1.20	3.56
07.6007 061	Add for neoprene gasket	LF	0.0088	0.77	0.65	1.41
07.6008 000	**COUNTER FLASH FOR REGLETS, ETC:**					
07.6008 011	Counter flashing, galvanized iron, 8", 26 ga	LF	0.0549	4.78	1.71	6.50
07.6008 021	Counter flashing, aluminum, 8", .032"	LF	0.0540	4.71	1.82	6.52
07.6008 031	Counter flashing, copper, 8", 16 oz	LF	0.0584	5.09	5.01	10.10
07.6008 041	Counter flashing, copper clad stainless steel, 8", .015"	LF	0.0644	5.61	5.17	10.78
07.6009 000	**SHEET METAL SPECIALTIES:**					
07.6009 011	Roof safe & cap, 4", 28 ga galvanized sheet metal	EA	0.1479	12.89	11.59	24.48
07.6009 021	Roof safe & cap, 4", aluminum, 032"	EA	0.1479	12.89	13.55	26.44
07.6009 031	Plumber's flashing cone, galvanized sheet metal, 2"	EA	0.1827	15.92	8.02	23.95

07 - THERMAL & MOISTURE PROTECTION

Division	Description	Unit	Labor Hours	Labor Cost	Material Cost	Total Cost
07.6009 041	Plumber's flashing cone, galvanized sheet metal, 3"	EA	0.1879	16.38	10.79	27.16
07.6009 051	Plumber's flashing cone, galvanized sheet metal, 4"	EA	0.1931	16.83	15.22	32.05
07.6009 061	Scupper, galvanized sheet metal, 28 ga	EA	0.2870	25.01	17.69	42.70
07.6009 071	Scupper, aluminum, .032"	EA	0.2870	25.01	19.87	44.88
07.6010 000	**SHEET METAL VENTS:**					
07.6010 011	Vent, foundation, galvanized iron, 6" x 16"	EA	0.0418	3.64	2.28	5.92
07.6010 021	Vent, foundation, galvanized iron, 8" x 16"	EA	0.0418	3.64	2.59	6.23
07.6010 031	Vent, foundation, galvanized iron, 10" x 16"	EA	0.0418	3.64	3.16	6.80
07.6010 041	Vent, foundation, aluminum, 6" x 16"	EA	0.0418	3.64	2.14	5.78
07.6010 051	Vent, foundation, aluminum, 8" x 16"	EA	0.0418	3.64	2.39	6.04
07.6010 061	Vent, foundation, aluminum, 10" x 16"	EA	0.0418	3.64	2.96	6.60
07.6010 071	Vent, frieze, galvanized sheet metal, 4" x 24"	EA	0.0505	4.40	2.28	6.68
07.6010 081	Vent, frieze, galvanized sheet metal, 6" x 24"	EA	0.0505	4.40	2.59	6.99
07.6010 091	Vent, frieze, aluminum, 4" x 24"	EA	0.0505	4.40	2.14	6.54
07.6010 101	Vent, frieze, aluminum, 6" x 24"	EA	0.0505	4.40	2.39	6.79
07.6010 111	Vent, attic, galvanized sheet metal, 14" x 24"	EA	0.2957	25.77	19.15	44.92
07.6010 121	Vent, block/brick, galvanized sheet metal, 8" x 16"	EA	0.3827	33.35	19.03	52.39
07.6010 131	Vent, block/brick, galvanized sheet metal, 12" x 16"	EA	0.3827	33.35	24.63	57.98
07.6010 141	Vent, block/brick, aluminum, 8" x 16"	EA	0.3827	33.35	21.84	55.20
07.6010 151	Vent, block/brick, aluminum, 12" x 16"	EA	0.3827	33.35	27.39	60.75
07.6011 000	**LOUVERS & SCREENS:**					
07.6011 011	Louvers, door	EA	0.8088	70.49	36.53	107.02
07.6011 021	Louvers, fixed, galvanized sheet metal	SF	0.1827	15.92	9.16	25.08
07.6011 031	Louvers, manual, galvanized sheet metal	SF	0.2261	19.70	11.94	31.65
07.6011 041	Screens, cooling tower, galvanized sheet metal	SF	0.2261	19.70	9.50	29.21
07.6011 051	Screens, bird	SF	0.0262	2.28	1.91	4.19
07.6011 061	Screens, insect	SF	0.0218	1.90	1.26	3.16
07.6012 000	**GRAVITY VENTILATORS:**					
07.6012 011	Gravity ventilator, 8", galvanized sheet metal	EA	0.5392	46.99	181.33	228.33
07.6012 021	Gravity ventilator, 18", galvanized sheet metal	EA	0.5392	46.99	294.37	341.36
07.6012 031	Gravity ventilator, 24", galvanized sheet metal	EA	0.7044	61.39	410.30	471.69
07.6012 041	Gravity ventilator, 36", galvanized sheet metal	EA	1.0957	95.49	770.13	865.62
07.6012 051	Gravity ventilator, 48", galvanized sheet metal	EA	1.8261	159.14	1,228.13	1,387.27
07.6012 061	Gravity ventilator, 60", galvanized sheet metal	EA	3.9130	341.02	2,060.76	2,401.78
07.6012 071	Add for hand damper	EA			61.08	61.08
07.6012 081	Add for motor damper	EA			91.66	91.66
07.6013 000	**MUSHROOM VENTILATORS, MOTORIZED:**					
07.6013 001	Note: For 2 speeds, add 25% to the material costs. For back draft damper, add 40% to the material costs.					
07.6013 011	Mushroom ventilator, 8", to 180 CFM	EA	2.8000	244.02	451.67	695.69
07.6013 021	Mushroom ventilator, 12", to 360 CFM	EA	2.8000	244.02	732.04	976.06
07.6013 031	Mushroom ventilator, 18", to 1000 CFM	EA	3.6696	319.81	1,152.67	1,472.47
07.6013 041	Mushroom ventilator, 24", to 1200 CFM	EA	4.5826	399.37	1,619.94	2,019.31
07.6014 000	**EXPANSION JOINTS:**					
07.6014 011	Expansion joint, dry wall, aluminum cover, 2"	LF	0.0522	4.55	19.15	23.70
07.6014 021	Expansion joint, plaster wall, aluminum cover, 2"	LF	0.0522	4.55	16.96	21.51
07.6014 031	Expansion joint, floor, aluminum cover, 2"	LF	0.0522	4.55	19.15	23.70
07.6014 041	Expansion joint, concrete wall, aluminum cover, 4"-6"	LF	0.3131	27.29	40.34	67.63
07.6014 051	Expansion joint, concrete floor, aluminum cover, 4"-6"	LF	0.3131	27.29	38.90	66.19
07.6014 061	Expansion joint, roof, neoprene & aluminum, 2"	LF	0.0696	6.07	27.27	33.33
07.6015 000	**ROOF HATCHES:**					
07.6015 001	Note: Add for curbs on hatches & skylights.					
07.6015 011	Hatch frame/cover, 3' x 3'6", galvanized	EA	4.8861	425.82	552.12	977.94
07.6015 021	Hatch frame/cover, 3' x 5'4", galvanized	EA	5.5905	487.21	861.69	1,348.90
07.6015 031	Hatch frame/cover, 3' x 9'6", galvanized	EA	6.2938	548.50	1,391.61	1,940.12
07.6016 000	**FIRE/SMOKE VENT, AUTOMATIC, 160 DEGREE:**					
07.6016 011	Smoke vent, 420#, 4'8" x 4'8", galvanized	EA	5.4243	472.73	1,464.47	1,937.20

THERMAL & MOISTURE PROTECTION - 07

Division	Description	Unit	Labor Hours	Labor Cost	Material Cost	Total Cost
07.6016 021	Smoke vent, 540#, 4'8" x 7'2", galvanized	EA	6.2938	548.50	1,870.07	2,418.58
07.6016 031	Smoke vent, 650#, 4'8" x 9', galvanized	EA	7.1634	624.29	2,309.48	2,933.77
07.6016 041	Smoke vent, 450#, 6' x 6', galvanized	EA	6.2938	548.50	1,585.64	2,134.15
07.6016 051	Smoke vent, 590#, 6' x 9'6", galvanized	EA	8.0329	700.07	2,041.92	2,741.98
07.6016 061	Smoke vent, 1075#, 6'6"x1'4", galvanized	EA	12.1737	1,060.94	4,264.06	5,324.99
07.6017 000	FIRE VENT, AUTOMATIC, 160 DEGREE:					
07.6017 011	Fire vent, 260#, 4'8" x 4'8", aluminum	EA	4.5652	397.86	1,560.26	1,958.12
07.6017 021	Fire vent, 320#, 4'8" x 7'2", aluminum	EA	5.4348	473.64	1,771.49	2,245.13
07.6017 031	Fire vent, 380#, 4'8" x 9'6", aluminum	EA	6.5217	568.37	2,005.27	2,573.64
07.6017 041	Fire vent, 330#, 6' x 6', aluminum	EA	5.6521	492.58	1,802.42	2,295.00
07.6017 051	Fire vent, 425#, 6' x 9'6", aluminum	EA	7.1739	625.21	2,320.68	2,945.89
07.6017 061	Fire vent, 800#, 6'6" x 1'4", aluminum	EA	11.3042	985.16	4,373.92	5,359.08
07.6018 000	FIRE VENT, SKYLIGHT, MELT-OUT, 205 DEGREE:					
07.6018 011	Fire vent, skylight, 3' x 3'	EA	3.4782	303.13	405.51	708.64
07.6018 021	Fire vent, skylight, 3' x 4'8"	EA	3.7391	325.86	557.58	883.44
07.6018 031	Fire vent, skylight, 3'6" x 6'	EA	4.0000	348.60	664.62	1,013.22
07.6018 041	Fire vent, skylight, 3'6" x 7'2"	EA	4.2608	371.33	794.14	1,165.47
07.6018 051	Fire vent, skylight, 4'8" x 4'8"	EA	4.5652	397.86	692.76	1,090.62
07.6018 061	Fire vent, skylight, 4'8" x 6'	EA	5.0000	435.75	833.63	1,269.38
07.6018 071	Fire vent, skylight, 4'8" x 7'2"	EA	5.4348	473.64	985.73	1,459.37
07.6019 000	SKYLIGHTS, ALUMINUM FRAME, PLASTIC DOME:					
07.6019 011	Skylight, 2'x2', aluminum frame, plastic dome	EA	1.1322	98.67	155.53	254.20
07.6019 021	Skylight, 4'x4', aluminum frame, plastic dome	EA	1.1322	98.67	316.50	415.17
07.6019 031	Skylight, 5'x5', aluminum frame, plastic dome	EA	1.1322	98.67	598.88	697.55
07.6019 041	Skylight, 3'x6', aluminum frame, plastic dome	EA	1.1322	98.67	396.63	495.30
07.6019 051	Skylight, 7'x7', aluminum frame, plastic dome	EA	1.5279	133.16	1,547.02	1,680.18
07.6019 061	Skylight, 8'x10', aluminum frame, plastic dome	EA	1.5279	133.16	2,798.81	2,931.97
07.6020 000	SKYLIGHTS, ALUMINUM FRAME, PYRAMID DOME:					
07.6020 011	Skylight, 2'x2', aluminum frame, pyramid dome	EA	1.1322	98.67	225.78	324.45
07.6020 021	Skylight, 3'x3', aluminum frame, pyramid dome	EA	1.1322	98.67	254.30	352.97
07.6020 031	Skylight, 4'x4', aluminum frame, pyramid dome	EA	1.1322	98.67	459.33	558.00
07.6021 000	SKYLIGHTS, DOUBLE GLAZED ALUMINUM FRAME:					
07.6021 011	Skylight, 2'x2', double glazed aluminum frame	EA	1.2870	112.16	272.76	384.93
07.6021 021	Skylight, 3'x3', double glazed aluminum frame	EA	1.6174	140.96	441.22	582.17
07.6021 031	Skylight, 4'x4', double glazed aluminum frame	EA	1.9305	168.24	680.88	849.12
07.6021 041	Skylight, 6'x6', double glazed aluminum frame	EA	2.2957	200.07	1,703.48	1,903.55
07.6021 051	Skylight, 4'x8', double glazed aluminum frame	EA	2.2957	200.07	1,623.34	1,823.41
07.6022 000	FABRICATED SKYLIGHTS:					
07.6022 011	Fabricated skylight, steel frame, laminated glass, 20' span	SF			100.06	100.06
07.6022 021	Fabricated skylight, steel frame, laminated glass, 30' span	SF			123.37	123.37
07.6022 031	Fabricated skylight, steel frame, laminated glass, 40' span	SF			144.44	144.44
07.6023 000	PREFAB METAL FIREPLACES:					
07.6023 011	Prefabricated fireplace, 36"x24" opening	EA	3.7373	325.71	3,277.94	3,603.65
07.6023 021	Prefabricated fireplace, 42"x27" opening	EA	3.7373	325.71	5,151.06	5,476.76
07.6023 031	Prefabricated, free standing fireplace	EA	4.0583	353.68	3,512.08	3,865.76
07.6023 041	Add for log lighter	EA	3.2034	279.18	60.37	339.55
07.6023 051	Add for patent flue, from 6' up	EA	3.2034	279.18	66.35	345.53
07.6023 061	Add for exterior stack, simulated brick	LF	0.2853	24.86	28.83	53.69
07.6023 071	Add for extension, high rise	LF	0.1419	12.37	19.35	31.71
07.7000 000	ARCHITECTURAL SHEET METAL:					
07.7001 000	PRE-FINISHED METAL FACIA & MANSARDS:					
07.7001 011	Beam & batten, straight & simple	SF	0.0693	6.04	7.19	13.23
07.7001 021	Beam & batten, curved & complex	SF	0.1386	12.08	7.76	19.84
07.7001 031	Add for coping/gravel stop/flashing	LF	0.0501	4.37	3.03	7.40
07.7002 000	WAINSCOT, GALVANIZED SHEET METAL:					
07.7002 011	Wainscot, galvanized sheet metal	SF	0.0266	2.32	1.65	3.97

07 - THERMAL & MOISTURE PROTECTION

Division	Description	Unit	Labor Hours	Labor Cost	Material Cost	Total Cost
07.7003 000	**SCREENS & METAL LOUVERS:**					
07.7003 011	Louvers, aluminum, fixed blade	SF	0.0601	5.24	18.37	23.61
07.7003 021	Louvers, aluminum, manual	SF	0.0747	6.51	21.95	28.46
07.7003 031	Screens, insect, with frame	SF	0.0251	2.19	1.51	3.70
07.7003 041	Screens, bird, with frame	SF	0.0251	2.19	2.37	4.56
07.7004 000	**ARCHITECTURAL FACADE SCREENS:**					
07.7004 011	Screen, arch facade, aluminum	SF	0.0866	7.55	24.88	32.43
07.7004 021	Add for enamel or light anodizing	SF			7.85	7.85
07.7004 031	Add for porcelain/heavy anodizing	SF			7.08	7.08
07.7005 000	**DOOR LOUVERS & CORNER GUARDS:**					
07.7005 011	Louvers, door	EA	0.1906	16.61	100.02	116.63
07.7005 021	Corner guard, stainless steel, 40" length	LF	0.2325	20.26	57.15	77.41
07.7005 031	Corner guard, stainless steel, 48" length	LF	0.2325	20.26	57.15	77.41
07.7005 041	Ceiling access hatch, 30" x 30"	EA	1.0896	94.96	287.89	382.85
07.9000 000	**CAULKING & SEALANTS:**					
07.9001 000	**CAULKING, GUN GRADE:**					
07.9001 001	Note: For caulking above 14 stories, add 20% to the labor costs. For caulking between 6 and 14 stories, add 10% to the material costs. For 2 part caulk, add 15% to the labor costs.					
07.9001 011	Caulk, linseed oil base, per gallon	EA			16.06	16.06
07.9001 021	Caulk, linseed oil base, per tube	EA			1.64	1.64
07.9001 031	Caulk, linseed base, 1/8" x 1/8"	LF	0.0103	1.11	0.02	1.13
07.9001 041	Caulk, linseed base, 1/4" x 1/4"	LF	0.0154	1.66	0.10	1.76
07.9001 051	Caulk, linseed base, 1/2" x 1/2"	LF	0.0205	2.21	0.23	2.44
07.9001 061	Caulk, linseed base, 3/4" x 3/4"	LF	0.0307	3.31	0.38	3.69
07.9001 071	Caulk, linseed base, 1" x 1"	LF	0.0410	4.42	1.01	5.44
07.9001 081	Caulk, butyl base, per gallon	EA			27.02	27.02
07.9001 091	Caulk, butyl base, per tube	EA			3.39	3.39
07.9001 101	Caulk, butyl base, 1/8" x 1/8"	LF	0.0103	1.11	0.02	1.13
07.9001 111	Caulk, butyl base, 1/4" x 1/4"	LF	0.0154	1.66	0.10	1.76
07.9001 121	Caulk, butyl base, 1/2" x 1/2"	LF	0.0205	2.21	0.40	2.61
07.9001 131	Caulk, butyl base, 3/4" x 3/4"	LF	0.0307	3.31	1.31	4.63
07.9001 141	Caulk, butyl base, 1" x 1"	LF	0.0410	4.42	1.69	6.12
07.9001 151	Caulk, solvent acrylic, per gallon	EA			58.10	58.10
07.9001 161	Caulk, solvent acrylic, per tube	EA			7.02	7.02
07.9001 171	Caulk, acrylic, 1/8" x 1/8"	LF	0.0103	1.11	0.06	1.18
07.9001 181	Caulk, acrylic, 1/4" x 1/4"	LF	0.0154	1.66	0.29	1.95
07.9001 191	Caulk, acrylic, 1/2" x 1/2"	LF	0.0205	2.21	0.96	3.17
07.9001 201	Caulk, acrylic, 3/4" x 3/4"	LF	0.0307	3.31	2.57	5.88
07.9001 211	Caulk, acrylic, 1" x 1"	LF	0.0410	4.42	3.48	7.90
07.9001 221	Caulk, polysulfide/urethane, per gallon	EA			69.63	69.63
07.9001 231	Caulk, polysulfide/urethane, per tube	EA			8.61	8.61
07.9001 241	Caulk, polysulfide, 1/8" x 1/8"	LF	0.0103	1.11	0.34	1.46
07.9001 251	Caulk, polysulfide, 1/4" x 1/4"	LF	0.0154	1.66	0.31	1.97
07.9001 261	Caulk, polysulfide, 1/2" x 1/2"	LF	0.0205	2.21	1.11	3.32
07.9001 271	Caulk, polysulfide, 3/4" x 3/4"	LF	0.0307	3.31	3.16	6.47
07.9001 281	Caulk, polysulfide, 1" x 1"	LF	0.0410	4.42	4.56	8.98
07.9001 291	Caulk, silicone, per gallon	EA			91.43	91.43
07.9001 301	Caulk, silicone, per tube	EA			11.34	11.34
07.9001 311	Caulk, silicone, 1/8" x 1/8"	LF	0.0103	1.11	0.10	1.21
07.9001 321	Caulk, silicone, 1/4" x 1/4"	LF	0.0154	1.66	0.34	2.01
07.9001 331	Caulk, silicone, 1/2" x 1/2"	LF	0.0205	2.21	1.49	3.70
07.9001 341	Caulk, silicone, 3/4" x 3/4"	LF	0.0307	3.31	4.02	7.33
07.9001 351	Caulk, silicone, 1" x 1"	LF	0.0410	4.42	5.90	10.32
07.9001 361	Caulk, mildew resistant, per gallon	EA			117.64	117.64
07.9001 371	Caulk, mildew resistant, per tube	EA			14.53	14.53
07.9001 381	Caulk, mildew resistant, 1/8" x 1/8"	LF	0.0103	1.11	0.13	1.24

THERMAL & MOISTURE PROTECTION - 07

Division	Description	Unit	Labor Hours	Labor Cost	Material Cost	Total Cost
07.9001 391	Caulk, mildew resistant, 1/4" x 1/4"	LF	0.0154	1.66	0.38	2.04
07.9001 401	Caulk, mildew resistant, 1/2" x 1/2"	LF	0.0205	2.21	2.31	4.52
07.9001 411	Caulk, mildew resistant, 3/4" x 3/4"	LF	0.0307	3.31	5.16	8.47
07.9001 421	Caulk, mildew resistant, 1" x 1"	LF	0.0410	4.42	7.72	12.14
07.9001 431	Caulk, elastomeric, for concrete	LF	0.0171	1.84	2.71	4.55
07.9002 000	**SEALANTS, SELF-LEVELING:**					
07.9002 011	Polysulfide polymer, 1/4" x 3/8"	LF	0.0286	3.09	1.65	4.73
07.9002 021	Acrylic latex polymer, 1/4"x3/8"	LF	0.0286	3.09	1.45	4.54
07.9002 031	Polyurethane, 1/4" x 3/8"	LF	0.0286	3.09	1.83	4.92
07.9003 000	**POLYISOBUTYLENE TAPES:**					
07.9003 011	Polybutene tape	LF	0.0231	2.49	0.31	2.80
07.9003 021	Polisobutyl/butyl tape, preformed	LF	0.0175	1.89	0.18	2.07
07.9004 000	**THERMOFIBER:**					
07.9004 011	Thermofiber, 1/8", 6"x8", glue back	SF	0.0161	1.74	2.35	4.09
07.9004 021	Thermofiber, 1" thick, 16" x 48"	SF	0.0417	4.50	0.31	4.81
07.9004 031	Thermofiber, 1-1/2" thick, 16" x 48"	SF	0.0500	5.39	0.34	5.74
07.9004 041	Thermofiber, 2" thick, 16" x 48"	SF	0.0583	6.29	0.53	6.82
07.9004 051	Thermofiber, 3" thick, 16" x 48"	SF	0.0667	7.20	0.65	7.84
07.9005 000	**CAULKING, ACOUSTICAL:**					
07.9005 011	Caulk, butyl rubber, 1/4" x 1/2"	LF	0.0231	2.49	0.53	3.02
07.9006 000	**NEOPRENE GASKETS, CLOSED CELL:**					
07.9006 011	Neoprene gasket, 1/8" x 2"	LF	0.0175	1.89	1.34	3.22
07.9006 021	Neoprene gasket, 1/8" x 6"	LF	0.0222	2.40	3.02	5.41
07.9006 031	Neoprene gasket, 1/4" x 2"	LF	0.0185	2.00	1.45	3.45
07.9006 041	Neoprene gasket, 1/4" x 6"	LF	0.0222	2.40	3.23	5.63
07.9006 051	Neoprene gasket, 1/2" x 6"	LF	0.0194	2.09	4.75	6.85
07.9006 061	Neoprene gasket, 1/2" x 12"	LF	0.0240	2.59	9.12	11.71
07.9007 000	**BACKING RODS:**					
07.9007 011	Backing rod, butyl, 3/8"	LF	0.0222	2.40	0.23	2.62
07.9007 021	Backing rod, butyl, 1/2"	LF	0.0231	2.49	0.38	2.87
07.9007 031	Backing rod, polyethylene, 3/8"	LF	0.0175	1.89	0.22	2.10
07.9007 041	Backing rod, polyethylene, 1/2"	LF	0.0185	2.00	0.25	2.24
07.9008 000	**WEATHER-STRIPPING, METAL:**					
07.9008 011	Weatherstrip, exterior anodized aluminum, neoprene	LF	0.0231	2.49	5.51	8.00
07.9008 021	Weatherstrip, exterior anodized aluminum, neoprene, adjust	LF	0.0341	3.68	12.28	15.95
07.9008 031	Weatherstrip, bronze Y angle, neoprene	LF	0.0534	5.76	10.12	15.88
07.9008 041	Weatherstrip, bronze Z bar & angle	LF	0.0534	5.76	8.02	13.78
07.9008 051	Weatherstrip, astragal, adjust mortise	LF	0.0848	9.15	18.97	28.12

08 - DOORS, WINDOWS & GLASS

Division	Description	Unit	Labor Hours	Labor Cost	Material Cost	Total Cost
08.0000 000	**DOORS, WINDOWS & GLASS:**					
08.0000 001	Note: Materials are generally supplied FOB job, with labor supplied by the general contractor. The following prices are based on purchase of 25 or more per order. For larger amounts the cost may be reduced as much as 15%. For smaller amounts, add 5%. For installing metal doors in concrete, add 100% to unit man hours. Remember to include the costs for special work as shown in section 08.1000.					
08.1000 000	**HOLLOW METAL DOORS & FRAMES:**					
08.1000 001	Note: Door sizes are expressed in width then height, where the feet & inches are not expressed. For example, 2'8" by 6'8" is expressed "2868".					
08.1001 000	**FRAMES, 16 GA, 4-5/8", PRIME COAT, NON-RATED:**					
08.1001 001	Note: The following items are standard quality and are prefabricated.					
08.1001 011	Frame, 16 ga, to 2868, prime coat, non-rated	EA	0.8138	87.80	103.00	190.80
08.1001 021	Frame, 16 ga, 3068, prime coat, non-rated	EA	0.8138	87.80	103.00	190.80
08.1001 031	Frame, 16 ga, 4068, prime coat, non-rated	EA	0.8952	96.58	129.40	225.98
08.1001 041	Frame, 16 ga, to 2870, prime coat, non-rated	EA	0.8138	87.80	124.17	211.97
08.1001 051	Frame, 16 ga, 3070, prime coat, non-rated	EA	0.8138	87.80	126.39	214.19
08.1001 061	Frame, 16 ga, 4070, prime coat, non-rated	EA	0.8952	96.58	129.54	226.12
08.1001 071	Frame, 16 ga, 5070, prime coat, non-rated	EA	1.0776	116.26	134.94	251.20
08.1001 081	Frame, 16 ga, 6070, prime coat, non-rated	EA	1.0776	116.26	146.90	263.16
08.1001 091	Frame, 16 ga, 8080, prime coat, non-rated	EA	1.1253	121.41	190.50	311.91
08.1002 000	**FRAMES, 16 GA, 4-7/8", PRIME COAT, NON-RATED:**					
08.1002 001	Note: The following items are standard quality and are prefabricated.					
08.1002 011	Frame, custom, 16 ga, 2868, prime coat, non-rated	EA	0.8138	87.80	159.43	247.23
08.1002 021	Frame, custom, 16 ga, 3068, prime coat, non-rated	EA	0.8138	87.80	123.69	211.49
08.1002 031	Frame, custom, 16 ga, 4068, prime coat, non-rated	EA	0.8952	96.58	130.97	227.56
08.1002 041	Frame, custom, 16 ga, to 2870, prime coat, non-rated	EA	0.8138	87.80	165.78	253.58
08.1002 051	Frame, custom, 16 ga, 3070, prime coat, non-rated	EA	0.8138	87.80	127.92	215.73
08.1002 061	Frame, custom, 16 ga, 4070, prime coat, non-rated	EA	0.8952	96.58	131.09	227.67
08.1002 071	Frame, custom, 16 ga, 5070, prime coat, non-rated	EA	1.0776	116.26	135.58	251.84
08.1002 081	Frame, custom, 16 ga, 6070, prime coat, non-rated	EA	1.0776	116.26	146.90	263.16
08.1002 091	Frame, custom, 16 ga, 8080, prime coat, non-rated	PAIR	1.1718	126.43	200.79	327.21
08.1003 000	**FRAMES, 14 GA, 4-7/8", PRIME COAT, NON-RATED:**					
08.1003 011	Frame, to 3070, 14 ga, prime coat, non-rated	EA	0.8138	87.80	190.89	278.69
08.1003 021	Frame, 3670, 14 ga, prime coat, non-rated	EA	0.8138	87.80	195.93	283.73
08.1003 031	Frame, 3070, 14 ga, prime coat, non-rated	EA	0.8138	87.80	192.87	280.67
08.1003 041	Frame, 4070, 14 ga, prime coat, non-rated	EA	0.8952	96.58	198.01	294.59
08.1003 051	Frame, 5070, 14 ga, prime coat, non-rated	EA	1.0776	116.26	219.28	335.54
08.1003 061	Frame, 6070, 14 ga, prime coat, non-rated	EA	1.2520	135.08	237.78	372.85
08.1003 071	Frame, 8080, 14 ga, prime coat, non-rated	EA	1.4471	156.13	249.35	405.48
08.1003 090	**FRAMES, ROLL FORMED 20GA, PRE-FINISHED, NON-RATED, WITH EMBOSSED HARDBOARD DOORS:**					
08.1003 091	Note: The doors in this section are 1-3/4" embossed hardboard, oak/walnut legacy, prefinished and prepared for cylinder lock. See 08.2012, "ADDERS FOR WOOD DOORS" for door upgrades and hardware. Frames are prefinished. The following prices are based on purchase of 25 or more per order. For larger amounts the cost may be reduced as much as 10%. For smaller amounts add 10%.					
08.1003 101	Hardboard door, with 20 ga frame 3068	EA	0.7467	80.56	190.98	271.54
08.1003 111	Hardboard door, with 20 ga frame 4068	EA	0.8300	89.55	230.04	319.59
08.1003 121	Hardboard door, with 20 ga frame 6068	EA	1.1000	118.68	384.20	502.88
08.1003 131	Hardboard door, with 20 ga frame 3070	EA	0.7467	80.56	205.08	285.64
08.1003 141	Hardboard door, with 20 ga frame 4070	EA	0.8300	89.55	211.16	300.71
08.1003 151	Hardboard door, with 20 ga frame 6070	EA	1.1000	118.68	406.29	524.97
08.1003 161	Hardboard door, with 20 ga frame 8080	EA	1.4200	153.20	538.21	691.41
08.1003 171	Hardboard door, with 18 ga frame 3068	EA	0.7467	80.56	201.62	282.19
08.1003 181	Hardboard door, with 18 ga frame 4068	EA	0.8300	89.55	240.71	330.26

DOORS, WINDOWS & GLASS - 08

Division	Description	Unit	Labor Hours	Labor Cost	Material Cost	Total Cost
08.1003 191	Hardboard door, with 18 ga frame 6068	EA	1.1000	118.68	395.07	513.75
08.1003 201	Hardboard door, with 18 ga frame 3070	EA	0.7467	80.56	233.21	313.77
08.1003 211	Hardboard door, with 18 ga frame 4070	EA	0.8300	89.55	239.32	328.87
08.1003 221	Hardboard door, with 18 ga frame 6070	EA	1.1000	118.68	417.14	535.82
08.1003 231	Hardboard door, with 18 ga frame 8080	EA	1.4200	153.20	549.06	702.27
08.1004 000	**DOORS, 18 GA, 1-3/4", PRIME COAT, NON-RATED:**					
08.1004 001	*Note: The following items are custom fabricated.*					
08.1004 011	Door, 18 ga, to 2868, prime coat, non-rated	EA	0.6975	75.25	289.06	364.31
08.1004 021	Door, 18 ga, 3068, prime coat, non-rated	EA	0.6975	75.25	304.43	379.69
08.1004 031	Door, 18 ga, 4068, prime coat, non-rated	EA	0.8952	96.58	344.90	441.48
08.1004 041	Door, 18 ga, to 2870, prime coat, non-rated	EA	0.6975	75.25	300.25	375.51
08.1004 051	Door, 18 ga, 3070, prime coat, non-rated	EA	0.6975	75.25	300.25	375.51
08.1004 061	Door, 18 ga, 4070, prime coat, non-rated	EA	0.8370	90.30	357.01	447.31
08.1004 071	Door, 18 ga, 4080, prime coat, non-rated	EA	1.1346	122.41	367.79	490.20
08.1005 000	**DOORS, 18 GA, CUSTOM, 1-3/4", PRIME COAT, NON-RATED:**					
08.1005 011	Door, custom, to 2868, prime coat, non-rated	EA	0.6975	75.25	378.93	454.18
08.1005 021	Door, custom, 3068, prime coat, non-rated	EA	0.6975	75.25	397.39	472.64
08.1005 031	Door, custom, 4068, prime coat, non-rated	EA	0.8952	96.58	563.79	660.37
08.1005 041	Door, custom, to 2870, prime coat, non-rated	EA	0.6975	75.25	391.26	466.51
08.1005 051	Door, custom, 3070, prime coat, non-rated	EA	0.6975	75.25	459.04	534.30
08.1005 061	Door, custom, 4070, prime coat, non-rated	EA	0.8370	90.30	520.65	610.96
08.1005 071	Door, custom, 4080, prime coat, non-rated	EA	1.1346	122.41	589.96	712.37
08.1006 000	**DOORS, 16 GA, 1-3/4" PRIME COATED, NON-RATED:**					
08.1006 011	Door, to 3070, 16 ga, prime coat, non-rated	EA	0.6975	75.25	385.12	460.37
08.1006 021	Door, 3670, 16 ga, prime coat, non-rated	EA	0.6975	75.25	477.52	552.77
08.1006 031	Door, 4070, 16 ga, prime coat, non-rated	EA	0.8370	90.30	539.17	629.48
08.1006 041	Door, 4080, 16 ga, prime coat, non-rated	EA	1.1160	120.41	613.11	733.51
08.1007 000	**MISCELLANEOUS ADDERS FOR FRAMES:**					
08.1007 011	Add for galvanizing	EA			80.58	80.58
08.1007 021	Add for A label, 3 hours	EA			34.77	34.77
08.1007 031	Add for B label, 1-1/2 hours	EA			30.36	30.36
08.1007 041	Add for C label, 1 hour	EA			26.86	26.86
08.1007 051	Add for stainless steel	EA			288.98	288.98
08.1007 061	Add for baked enamel	EA			22.81	22.81
08.1007 071	Add for porcelain enamel	EA	0.2976	32.11	68.46	100.57
08.1007 081	Add for special dapping	EA	0.6975	75.25	31.56	106.82
08.1007 091	Add for lengthening sections	SF			6.92	6.92
08.1007 101	Add for bronze	EA			941.44	941.44
08.1008 000	**MISCELLANEOUS ADDERS FOR DOORS:**					
08.1008 011	Add for galvanizing	EA			67.64	67.64
08.1008 021	Add for A label, 3 hours	EA			164.46	164.46
08.1008 031	Add for A label, 1-1/2 hours	EA			65.77	65.77
08.1008 041	Add for C label, 1 hour	EA			32.95	32.95
08.1008 051	Add for cutouts to 4 SF	EA			95.66	95.66
08.1008 061	Add for cutouts over 4 SF	SF			21.23	21.23
08.1008 071	Add for stainless steel	EA			561.06	561.06
08.1008 081	Add for baked enamel	EA			22.81	22.81
08.1008 091	Add for porcelain enamel	EA			187.49	187.49
08.1008 101	Add for special dapping	EA			34.76	34.76
08.1008 111	Add for bronze	EA			2,824.41	2,824.41
08.1008 121	Add for 10" x 10" vision light	EA			99.26	99.26
08.1008 131	Add for half glass opening	EA			123.96	123.96

08 - DOORS, WINDOWS & GLASS

Division	Description	Unit	Labor Hours	Labor Cost	Material Cost	Total Cost
08.2000 000	**WOOD DOORS & FRAMES:**					
08.2000 002	Note: For the following sections, it is assumed that materials are supplied fob job site, with labor being supplied by the general contractor. Prices are for quantities of 25 or more. For quantities in excess of 250, cost may be reduced as much as 15%. For quantities less than 25, add 5%. The following prices for residential doors & frames are based on frames at 68, doors at 1-3/8" fj jambs and oval casing. Add 5% for 70 in lieu of 68, add 35% for 80 in lieu of 68. Dapping is included for all standard hardware, with hinges applied. No other hardware is included. The following prices for commercial doors and frames are based on 4-7/8" thick walls and 1-3/4" doors.					
08.2001 000	**DOORS & FRAMES, PREHUNG, HOLLOW, INTERIOR:**					
08.2001 001	Note: The following items are paint grade 1-3/8" thick.					
08.2001 011	Door, hollow core, to 2468, prehung, paint grade	EA	0.4650	50.17	109.64	159.81
08.2001 021	Door, hollow core, 2668, prehung, paint grade	EA	0.4650	50.17	110.19	160.36
08.2001 031	Door, hollow core, 2868, prehung, paint grade	EA	0.4650	50.17	115.51	165.68
08.2001 041	Door, hollow core, 3068, prehung, paint grade	EA	0.4650	50.17	120.80	170.97
08.2001 051	Door, hollow core, 3668, prehung, paint grade	EA	0.4650	50.17	131.40	181.57
08.2002 000	**DOORS & FRAMES, PREHUNG, 1-3/8", PAINT GRADE, INTERIOR:**					
08.2002 001	Note: The following are hollow core masonite.					
08.2002 011	Door, hollow core, interior, 2468x1-3/8", prehung, paint grade	EA	0.6975	75.25	93.02	168.27
08.2002 021	Door, hollow core, interior, 2668x1-3/8", prehung, paint grade	EA	0.6975	75.25	96.20	171.45
08.2002 031	Door, hollow core, interior, 2868x1-3/8", prehung, paint grade	EA	0.6975	75.25	99.34	174.59
08.2002 041	Door, hollow core, interior, 3068x1-3/8", prehung, paint grade	EA	0.6975	75.25	102.48	177.74
08.2002 051	Door, hollow core, interior, 3668x1-3/8", prehung, paint grade	EA	0.6975	75.25	105.63	180.88
08.2003 000	**DOORS & FRAMES, PREHUNG, SOLID CORE, 1-3/8", PAINT GRADE, INTERIOR:**					
08.2003 001	Note: The following prices include solid jamb & casing.					
08.2003 011	Door, solid core, interior, 2468x1-3/8", prehung, paint grade	EA	0.6975	75.25	148.16	223.41
08.2003 021	Door, solid core, interior, 2668x1-3/8", prehung, paint grade	EA	0.6975	75.25	151.74	226.99
08.2003 031	Door, solid core, interior, 2868x1-3/8", prehung, paint grade	EA	0.6975	75.25	165.04	240.30
08.2003 041	Door, solid core, interior, 3068x1-3/8", prehung, paint grade	EA	0.6975	75.25	160.62	235.87
08.2003 051	Door, solid core, interior, 3668x1-3/8", prehung, paint grade	EA	0.6975	75.25	169.57	244.82
08.2004 000	**DOORS & FRAMES, PREHUNG, EXTERIOR:**					
08.2004 001	Note: The following prices include rabbet jamb, exterior molding and fir sill.					
08.2004 011	Door, solid core, exterior, 3068x1-3/8", prehung	EA	0.9300	100.34	172.28	272.61
08.2004 021	Door, solid core, exterior, 3068x1-3/4", prehung	EA	0.9300	100.34	182.42	282.76
08.2004 031	Door, exterior, 9 lite, x buck, 3'x1-3/4"	EA	0.9300	100.34	235.12	335.46
08.2004 041	Door, exterior, 12 lite, x buck, 3'x1-3/4"	EA	0.9300	100.34	267.54	367.88
08.2004 051	Door, exterior, Dutch, 3068x1-3/4", prehung	EA	1.3950	150.51	389.80	540.31
08.2004 061	Door, exterior, 12 lite, x buck, 3690x1-3/4"	EA	0.9300	100.34	332.41	432.75
08.2005 000	**DOORS & FRAMES, PREHUNG, PAINT GRADE, MISC:**					
08.2005 011	Door, pocket, 2870 x1-3/8", prehung, paint grade	EA	1.1625	125.42	149.34	274.76
08.2005 021	Door, by pass slide, to 50, prehung, paint grade	EA	1.3950	150.51	179.72	330.23
08.2005 031	Door, by pass slide, 60, prehung, paint grade	EA	1.3950	150.51	202.15	352.66
08.2005 041	Door, by pass slide, 80, prehung, paint grade	EA	1.5810	170.57	213.38	383.95
08.2005 051	Door, bifold, wood, 50, prehung, paint grade	EA	1.3950	150.51	151.39	301.90
08.2005 061	Door, bifold, wood, to 60, prehung, paint grade	EA	1.3950	150.51	163.30	313.80
08.2005 071	Door, bifold, wood, 80, prehung, paint grade	EA	1.5810	170.57	212.07	382.64
08.2005 081	Door, bifold, metal, 60, prehung, paint grade	EA	1.3950	150.51	120.12	270.62
08.2005 091	Door, bifold, metal, 80, prehung, paint grade	EA	1.3950	150.51	144.97	295.48
08.2005 101	Door, full louver, 2068, prehung, paint grade	EA	0.6975	75.25	149.82	225.08
08.2005 111	Door, full louver, 2668, prehung, paint grade	EA	0.6975	75.25	160.14	235.39
08.2005 121	Door, full louver, 3068, prehung, paint grade	EA	0.6975	75.25	168.75	244.00
08.2006 000	**DOORS & FRAMES, JOB HUNG, HOLLOW CORE, 1-3/8", INTERIOR:**					
08.2006 011	Door, hollow core, to 2470 x 1-3/8", paint grade	EA	1.1625	125.42	84.40	209.82
08.2006 021	Door, hollow core, 2670 x 1-3/8", paint grade	EA	1.1625	125.42	89.18	214.60
08.2006 031	Door, hollow core, 2870 x 1/3/8", paint grade	EA	1.1625	125.42	89.48	214.90

DOORS, WINDOWS & GLASS - 08

Division	Description	Unit	Labor Hours	Labor Cost	Material Cost	Total Cost
08.2006 041	Door, hollow core, 3070 x 1-3/8", paint grade	EA	1.1625	125.42	92.25	217.67
08.2006 051	Door, hollow core, 3670 x 1-3/8", paint grade	EA	1.1625	125.42	96.03	221.45
08.2007 000	**DOORS & FRAMES, JOB HUNG, SOLID CORE, 1-3/8", INTERIOR:**					
08.2007 011	Door, solid core, interior, to 2470 x 1-3/8", paint grade	EA	1.3950	150.51	132.07	282.58
08.2007 021	Door, solid core, interior, 2670 x 1-3/8", paint grade	EA	1.3950	150.51	133.97	284.48
08.2007 031	Door, solid core, interior, 2870 x 1-3/8", paint grade	EA	1.3950	150.51	133.97	284.48
08.2007 041	Door, solid core, interior, 3070 x 1-3/8", paint grade	EA	1.3950	150.51	142.50	293.01
08.2007 051	Door, solid core, interior, 3670 x 1-3/8", paint grade	EA	1.3950	150.51	162.40	312.90
08.2008 000	**ADDERS & DEDUCTORS FOR DOORS & FRAMES:**					
08.2008 011	Deduct for Philippine mahogany	EA			5.45	5.45
08.2008 021	Add for stain grade birch	EA			11.85	11.85
08.2008 031	Add for red oak	EA			13.42	13.42
08.2008 041	Add for walnut	EA			51.44	51.44
08.2008 051	Add for Formica clad	EA			108.38	108.38
08.2008 061	Add for stain grade ash	EA			15.65	15.65
08.2008 071	Add for plant-ons	EA			13.95	13.95
08.2008 081	Add for prefinishing	EA	0.1270	13.70	17.87	31.57
08.2008 091	Deduct for hardboard	EA			15.08	15.08
08.2008 101	Refinishing, average	EA	1.3950	150.51	17.87	168.37
08.2009 000	**DOORS & FRAMES, PREHUNG, SOLID CORE, PAINT GRADE, INTERIOR:**					
08.2009 011	Door, solid core, interior, to 2868, prehung, paint grade	EA	0.9300	100.34	143.12	243.46
08.2009 021	Door, solid core, interior, 3068, prehung, paint grade	EA	0.9300	100.34	145.73	246.07
08.2009 031	Door, solid core, interior, 3668, prehung, paint grade	EA	0.9300	100.34	163.63	263.97
08.2009 041	Door, solid core, interior, to 2870, prehung, paint grade	EA	1.0230	110.37	158.33	268.70
08.2009 051	Door, solid core, interior, 3070, prehung, paint grade	EA	1.0230	110.37	163.60	273.97
08.2009 061	Door, solid core, interior, 3670, prehung, paint grade	EA	1.0230	110.37	170.30	280.68
08.2009 071	Door, solid core, interior, 3090, prehung, paint grade	EA	1.3950	150.51	207.21	357.72
08.2009 081	Door, solid core, interior, 3690, prehung, paint grade	EA	1.3950	150.51	212.35	362.86
08.2010 000	**DOORS, JOB HUNG, SOLID, 1-3/4", PAINT GRADE:**					
08.2010 001	*Note: Prices include labor (not frame cost) for job fitting to hollow metal frames.*					
08.2010 011	Door, solid core, to 2868, paint grade	EA	1.8600	200.68	114.39	315.06
08.2010 021	Door, solid core, 3068, paint grade	EA	1.8600	200.68	117.02	317.69
08.2010 031	Door, solid core, 3668, paint grade	EA	1.8600	200.68	161.87	362.54
08.2010 041	Door, solid core, to 2870, paint grade	EA	1.9530	210.71	117.02	327.73
08.2010 051	Door, solid core, 3070, paint grade	EA	1.9530	210.71	125.02	335.73
08.2010 061	Door, solid core, 3670, paint grade	EA	1.9530	210.71	169.69	380.40
08.2010 071	Door, solid core, 3090, paint grade	EA	2.3250	250.84	207.12	457.96
08.2010 081	Door, solid core, 3690, paint grade	EA	2.3250	250.84	216.72	467.57
08.2011 000	**DOORS, JOB HUNG, INSTITUTIONAL., SOLID CORE, 1-3/4" STAIN GRADE:**					
08.2011 011	Door, institutional, solid core, 2870, stain grade	EA	1.8600	200.68	226.85	427.53
08.2011 021	Door, institutional, solid core, 3070, stain grade	EA	1.8600	200.68	237.01	437.68
08.2011 031	Door, institutional, solid core, 3670, stain grade	EA	1.8600	200.68	247.19	447.86
08.2011 041	Door, institutional, solid core, 2880, stain grade	EA	1.9530	210.71	338.61	549.32
08.2011 051	Door, institutional, solid core, 3080, stain grade	EA	1.9530	210.71	352.12	562.83
08.2011 061	Door, institutional, solid core, 3680, stain grade	EA	1.9530	210.71	365.70	576.41
08.2011 071	Door, institutional, solid core, 3090, stain grade	EA	2.3250	250.84	507.88	758.73
08.2011 081	Door, institutional, solid core, 3690, stain grade	EA	2.3250	250.84	524.81	775.66
08.2012 000	**ADDERS FOR WOOD DOORS:**					
08.2012 011	Add for stain grade birch	EA	0.2311	24.93	18.41	43.34
08.2012 021	Add for stain grade ash	EA	0.3023	32.62	21.52	54.14
08.2012 031	Add for walnut	EA	0.6840	73.80	83.07	156.87
08.2012 041	Add for Formica clad	EA			108.90	108.90
08.2012 051	Add for prefinishing	EA			27.70	27.70
08.2012 061	Add for A label, 3 hour	EA			201.71	201.71

08 - DOORS, WINDOWS & GLASS

Division	Description	Unit	Labor Hours	Labor Cost	Material Cost	Total Cost
08.2012 071	Add for B label, 1-1/2 hour	EA			154.92	154.92
08.2012 081	Add for C label, 1 hour	EA			108.08	108.08
08.2012 091	Add for X label, 20 minute	EA			11.28	11.28
08.2012 101	Add for sound proof, STC 40	EA	0.0847	9.14	212.99	222.13
08.2012 111	Add for sound proof, STC 45	EA	0.1135	12.25	258.18	270.43
08.2012 121	Add for sound proof, STC 51	EA	0.3451	37.23	516.38	553.61
08.2013 000	**JAMB & TRIM SETS (FRAMES):**					
08.2013 001	*Note: The following prices assume solid stock, integral stop, with matching wood casing.*					
08.2013 011	Jamb & trim, paint grade pine/fir	EA	1.1625	125.42	32.85	158.27
08.2013 021	Jamb & trim, stain grade pine/fir	EA	1.1625	125.42	41.05	166.47
08.2013 031	Jamb & trim, birch or ash	EA	1.3020	140.47	58.26	198.73
08.2013 041	Jamb & trim, walnut	EA	1.3950	150.51	91.23	241.74
08.2100 000	**WOOD GARAGE DOORS:**					
08.2101 000	**GARAGE DOORS, WOOD, SPRING BALANCED:**					
08.2101 011	Garage door, 7' x 8', economy, spring	EA	4.1306	445.65	234.95	680.60
08.2101 021	Garage door, 7' x 16', economy, spring	EA	5.8398	630.06	319.54	949.59
08.2101 031	Garage door, 7' x 16', custom, spring	EA	6.1958	668.46	397.82	1,066.28
08.2101 041	Garage door, 7' x 16', premium, spring	EA	7.0496	760.58	463.63	1,224.21
08.2102 000	**GARAGE DOORS, WOOD, TRACK OPERATED:**					
08.2102 011	Garage door, 7' x 8', economy, track	EA	4.2730	461.01	244.39	705.40
08.2102 021	Garage door, 7' x 16', economy, track	EA	5.9822	645.42	328.91	974.33
08.2102 031	Garage door, 7' x 16', custom, track	EA	6.4086	691.42	410.44	1,101.86
08.2102 041	Garage door, 7' x 16', premium, track	EA	7.4769	806.68	498.13	1,304.81
08.2102 051	Garage door, 7' x 16', segment, track	EA	8.1891	883.52	541.98	1,425.50
08.2103 000	**ADDERS FOR WOOD GARAGE DOORS:**					
08.2103 011	Add for motor operation, premium grade	EA	2.7900	301.01	431.91	732.92
08.2103 021	Add for motor operation, standard grade	EA	2.7900	301.01	259.83	560.84
08.2200 000	**WOOD DOOR SPECIALTIES:**					
08.2201 000	**DOORS, DECORATOR, TO 3680, 1-3/4":**					
08.2201 011	Door, carved & relief fir	EA	2.6350	284.29	644.60	928.89
08.2201 021	Door, carved & relief hardwood	EA	2.6350	284.29	1,160.32	1,444.61
08.2201 031	Door, redwood slab	EA	2.6350	284.29	219.11	503.40
08.2201 041	French 1 lite or store door	EA	2.6350	284.29	283.01	567.30
08.2201 051	Door, French multi-lite	EA	2.6350	284.29	241.68	525.97
08.2201 061	Door, wardrobe, economy, per panel	EA	0.6975	75.25	59.23	134.49
08.2201 071	Door, wardrobe, custom, per panel	EA	0.6975	75.25	77.30	152.55
08.2202 000	**DOORS, FIR, TO 3070, 1-3/4":**					
08.2202 011	Door, fir, 1 panel, 1 lite	EA	1.3950	150.51	248.04	398.55
08.2202 021	Door, fir, 1 panel, 9 lite	EA	1.3950	150.51	281.39	431.89
08.2202 031	Door, fir, 1 panel, 12 diamond lite	EA	1.3950	150.51	298.89	449.40
08.2202 041	Door, fir, x buck, 1 lite	EA	1.3950	150.51	271.85	422.36
08.2202 051	Door, fir, x buck, 9 lite	EA	1.3950	150.51	303.67	454.17
08.2202 061	Door, fir, x buck, 12 diamond lite	EA	1.3950	150.51	319.54	470.05
08.2202 071	Add for Dutch door	EA	0.9300	100.34	66.76	167.10
08.2202 081	Add for Dutch shelf	EA	0.3069	33.11	189.16	222.27
08.3000 000	**SPECIAL DOORS:**					
08.3001 000	**ACCESS DOORS:**					
08.3001 001	*Note: For roof hatches, see section 07.6015. The following items are installed in ceilings, walls or acoustic tile.*					
08.3001 011	Access panel, aluminum, 12" x 12"	EA	0.1700	16.02	34.47	50.48
08.3001 021	Access panel, aluminum, 24" x 24"	EA	0.2329	21.94	68.93	90.88
08.3001 031	Access panel, galvanized sheet metal, 12" x 12"	EA	0.1863	17.55	22.63	40.18
08.3001 041	Access panel, galvanized sheet metal, 24" x 24"	EA	0.2521	23.75	45.28	69.04
08.3001 051	Access panel, fire rated, 12"x12"	EA	0.1863	17.55	93.15	110.70

DOORS, WINDOWS & GLASS - 08

Division	Description	Unit	Labor Hours	Labor Cost	Material Cost	Total Cost
08.3001 061	Access panel, fire rated, 24"x24"	EA	0.2521	23.75	138.48	162.23
08.3002 000	**ROLL-UP DOOR, CHAIN OPERATED, STEEL, 20 GA, GALVANIZED:**					
08.3002 011	Door, steel roll-up, 8'x10', 20 ga, galvanized	EA	22.7874	2,147.03	1,243.10	3,390.13
08.3002 021	Door, steel roll-up, 10'x10', 20 ga, galvanized	EA	23.1435	2,180.58	985.73	3,166.31
08.3002 031	Door, steel roll-up, 12'x10', 20 ga, galvanized	EA	23.8557	2,247.68	1,099.18	3,346.86
08.3002 041	Door, steel roll-up, 12'x12', 20 ga, galvanized	EA	24.2117	2,281.23	1,349.28	3,630.51
08.3002 051	Door, steel roll-up, 14'x14', 20 ga, galvanized	EA	24.2117	2,281.23	1,686.55	3,967.78
08.3002 061	Door, steel roll-up, 18'x14', 20 ga, galvanized	EA	30.9772	2,918.67	2,393.21	5,311.88
08.3002 071	Door, steel roll-up, large, 20 ga, galvanized	SF	0.1359	12.80	12.42	25.22
08.3002 081	Add for motor operation	EA	5.6973	536.80	1,087.06	1,623.86
08.3002 091	Add for fusible link	EA	1.4956	140.92	276.21	417.12
08.3003 000	**GRILL, ROLL-UP, CRANK OPERATED, ALUMINUM:**					
08.3003 011	Grill, aluminum roll-up, 8' x 8', crank	EA	12.1060	1,140.63	1,384.96	2,525.59
08.3003 021	Grill, aluminum roll-up, 10' x 10', crank	EA	16.0228	1,509.67	1,607.36	3,117.03
08.3003 031	Grill, aluminum roll-up, 10' x 12', crank	EA	24.2117	2,281.23	1,983.40	4,264.62
08.3003 041	Grill, aluminum roll-up, 18' x 8', crank	EA	25.9922	2,448.99	2,256.15	4,705.14
08.3003 051	Add for motor operation	EA	2.6350	248.27	860.56	1,108.83
08.3004 000	**DOOR, COUNTER, PUSH-UP, ALUMINUM:**					
08.3004 011	Door, push-up counter, aluminum, 4' x 4'	EA	7.4400	701.00	556.02	1,257.02
08.3004 021	Door, push-up counter, aluminum, 6' x 4'	EA	7.4400	701.00	749.38	1,450.38
08.3004 031	Door, push-up counter, aluminum, 10' x 4'	EA	8.3700	788.62	820.87	1,609.49
08.3004 041	Add for motor operation	EA	1.8873	177.82	463.35	641.17
08.3005 000	**OVERHEAD DOORS, SECTIONAL, PRIME COATED, MANUAL:**					
08.3005 001	Note: For bronze anodized aluminum, add 100% to the material costs.					
08.3005 011	Door, steel over head, 8' x 8', prime coat, manual	EA	4.9140	463.00	537.94	1,000.93
08.3005 021	Door, steel over head, 10'x10', prime coat, manual	EA	7.1921	677.64	670.33	1,347.97
08.3005 031	Door, steel over head, 12'x12', prime coat, manual	EA	8.1891	771.58	740.94	1,512.52
08.3005 041	Door, steel over head, 12'x14', prime coat, manual	EA	10.3256	972.88	800.64	1,773.52
08.3005 051	Add for chain operation	EA	0.7479	70.47	81.39	151.86
08.3005 061	Add for motor operation	EA	1.4956	140.92	1,017.76	1,158.68
08.3005 071	Sliding door with track, 12' x 14'	EA	3.8130	359.26	1,023.99	1,383.25
08.3005 081	Sliding door with track, 14' x 16'	EA	4.3710	411.84	1,315.73	1,727.57
08.3005 091	Sliding door with track, 16' x 20'	EA	5.3010	499.46	1,672.97	2,172.43
08.3005 101	Industrial door to 50' x 30', complete	SF	0.5673	53.45	27.53	80.98
08.3005 111	Industrial door, 90' x 30', complete	SF	0.6045	56.96	31.94	88.90
08.3006 000	**SLIDING FIRE DOORS WITH HARDWARE, FUSIBLE LINK:**					
08.3006 011	Fire door, sliding, 4' x 7', with hardware	EA	21.3631	2,012.83	1,096.34	3,109.17
08.3006 021	Fire door, sliding, 6' x 7', with hardware	EA	23.8557	2,247.68	1,425.26	3,672.95
08.3006 031	Fire door, sliding, 10' x 10', with hardware	EA	50.9160	4,797.31	2,946.49	7,743.79
08.3006 041	Fire door, sliding, with hardware	SF	0.5253	49.49	32.27	81.77
08.3006 051	Fire shutter door, stainless steel center 4'x4'	EA	12.0000	1,130.64	5,824.49	6,955.13
08.3006 061	Fire shutter door, stainless steel center 4'x6'	EA	14.4000	1,356.77	6,107.36	7,464.13
08.3006 071	Fire shutter door, stainless steel center 4'x10'	EA	18.7500	1,766.63	8,026.14	9,792.76
08.3007 000	**FIRE DOORS, ROLL-UP:**					
08.3007 011	Fire door, roll-up, 6070, 4 hour	EA	23.1435	2,180.58	1,136.35	3,316.93
08.3007 021	Fire door, roll-up, 5080, 3 hour	EA	23.1435	2,180.58	1,181.96	3,362.54
08.3007 031	Fire door, roll-up	SF	0.5567	52.45	33.15	85.60
08.3008 000	**VAULT DOORS, MINIMUM SECURITY:**					
08.3008 001	Note: For bank vault doors, see section 11.1101.					
08.3008 011	Vault door, 3070, 2 hour	EA	16.7400	1,577.24	2,208.79	3,786.03
08.3008 021	Vault door, 4070, 2 hour	EA	16.7400	1,577.24	3,158.55	4,735.79
08.3008 031	Vault door, 3070, 4 hour	EA	18.6000	1,752.49	2,475.54	4,228.03
08.3008 041	Vault door, 4070, 4 hour	EA	18.6000	1,752.49	3,350.55	5,103.04
08.3009 000	**REVOLVING DOORS:**					
08.3009 011	Revolving door, 7', aluminum	EA	49.6341	4,676.52	19,633.41	24,309.94

08 - DOORS, WINDOWS & GLASS

Division	Description	Unit	Labor Hours	Labor Cost	Material Cost	Total Cost
08.3009 021	Revolving door, 7', stainless steel, satin finish	EA	49.6341	4,676.52	39,473.57	44,150.09
08.3009 031	Revolving door, 7', stainless steel, mirror finish	EA	73.9031	6,963.15	47,740.28	54,703.43
08.3009 041	Revolving door, 7', bronze satin finish	EA	72.4926	6,830.25	38,964.59	45,794.84
08.3009 051	Revolving door, 7', bronze mirror	EA	94.0788	8,864.10	47,326.91	56,191.01
08.3010 000	**REFRIGERATOR DOORS WITH HARDWARE & FRAME, 30 DEGREE:**					
08.3010 011	Refrigerator door, 3066, economy, plywood	EA	13.6015	1,281.53	401.37	1,682.90
08.3010 021	Refrigerator door, 3066, custom, galvanized face	EA	16.2365	1,529.80	688.08	2,217.88
08.3010 031	Refrigerator door, 3066, stainless steel, chrome	EA	17.3759	1,637.16	1,225.70	2,862.86
08.3010 041	Refrigerator door, 4066, economy, plywood	EA	15.8804	1,496.25	498.08	1,994.33
08.3010 051	Refrigerator door, 4066, custom, galvanized face	EA	17.3759	1,637.16	842.17	2,479.32
08.3010 061	Refrigerator door, 4066, stainless steel, chrome	EA	18.5153	1,744.51	1,648.55	3,393.07
08.4000 000	**WINDOWS & DOORS, GLAZED:**					
08.4001 000	**WINDOWS, VINYL, INSULATED GLASS WITH 1/2" AIR:**					
08.4001 001	Note: The following items are residential/light commercial grade Intermediate quality with white or almond vinyl; clear insulated glass with 1/2" air, no grid Apply all adder percentages to base price only. Bronze or grey glass add: 15% For low-E glazing add: 15%, Tempered add: 40% For economy quality materials deduct 15% X means active panel, O means stationary panel Window sizes are expressed in width then height, where the feet & inches are not shown. For example, 2'8" by '6'8" is expressed "2868".					
08.4001 011	Window, vinyl, sliding, XO, 2016, clear insulated glass	EA	0.3667	39.56	193.17	232.73
08.4001 021	Window, vinyl, sliding, XO, 2630, clear insulated glass	EA	0.3667	39.56	263.09	302.66
08.4001 031	Window, vinyl, sliding, XO, 3040, clear insulated glass	EA	0.4417	47.66	305.16	352.82
08.4001 041	Window, vinyl, sliding, XO, 4040, clear insulated glass	EA	0.5167	55.75	332.97	388.72
08.4001 051	Window, vinyl, sliding, XO, 5050, clear insulated glass	EA	0.5833	62.93	406.44	469.37
08.4001 061	Window, vinyl, sliding, XO, 6040, clear insulated glass	EA	0.5833	62.93	393.93	456.86
08.4001 071	Window, vinyl, sliding, XO, 6050, clear insulated glass	EA	0.7333	79.12	458.47	537.59
08.4001 081	Window, vinyl, sliding, clear insulated glass, average cost	SF	0.0257	2.77	27.66	30.43
08.4002 000	**WINDOWS, VINYL, CASEMENT, INSULATED GLASS WITH 1/2" AIR, WITH SCREEN:**					
08.4002 011	Window, vinyl, casement, X, 1620, clear insulated glass	EA	0.3667	39.56	315.23	354.79
08.4002 021	Window, vinyl, casement, X, 2026, clear insulated glass	EA	0.3667	39.56	407.90	447.46
08.4002 031	Window, vinyl, casement, X, 2050, clear insulated glass	EA	0.3667	39.56	445.93	485.49
08.4002 041	Window, vinyl, casement, X, 2630, clear insulated glass	EA	0.3667	39.56	385.24	424.80
08.4002 051	Window, vinyl, casement, X, 2646, clear insulated glass	EA	0.4417	47.66	444.22	491.88
08.4002 061	Window, vinyl, casement, X, 3040, clear insulated glass	EA	0.4417	47.66	436.25	483.91
08.4002 071	Window, vinyl, casement, clear insulated glass, average cost	SF	0.0257	2.77	57.55	60.33
08.4003 000	**WINDOWS, VINYL, SINGLE HUNG, INSULATED GLASS WITH 1/2" AIR, WITH SCREEN:**					
08.4003 011	Window, vinyl, single hung, 2030, clear insulated glass	EA	0.3667	39.56	234.45	274.01
08.4003 021	Window, vinyl, single hung, 2650, clear insulated glass	EA	0.4417	47.66	301.94	349.60
08.4003 031	Window, vinyl, single hung, 3060, clear insulated glass	EA	0.5167	55.75	352.68	408.43
08.4003 041	Window, vinyl, single hung, 4070, clear insulated glass	EA	0.6500	70.13	427.69	497.82
08.4003 051	Window, vinyl, single hung, clear insulated glass, average cost	SF	0.0257	2.77	24.54	27.31
08.4004 000	**WINDOWS, VINYL, FIXED, INSULATED GLASS WITH 1/2" AIR:**					
08.4004 011	Window, vinyl, fixed, 2020, clear insulated glass	EA	0.2933	31.64	113.58	145.22
08.4004 021	Window, vinyl, fixed, 2650, clear insulated glass	EA	0.3533	38.12	212.95	251.06
08.4004 031	Window, vinyl, fixed, 3030, clear insulated glass	EA	0.2933	31.64	171.69	203.33
08.4004 041	Window, vinyl, fixed, 4040, clear insulated glass	EA	0.4133	44.59	236.46	281.05
08.4004 051	Window, vinyl, fixed, 5050, clear insulated glass	EA	0.4667	50.35	327.09	377.45
08.4004 061	Window, vinyl, fixed, 6060, clear insulated glass	EA	0.7067	76.25	532.15	608.39
08.4004 071	Window, vinyl, fixed, 7050, clear insulated glass	EA	0.7067	76.25	526.37	602.62
08.4004 081	Window, vinyl, fixed, 8050, clear insulated glass	EA	0.7067	76.25	582.65	658.90
08.4004 091	Window, vinyl, fixed, 10040, clear insulated glass	EA	0.7067	76.25	601.57	677.82

DOORS, WINDOWS & GLASS - 08

Division	Description	Unit	Labor Hours	Labor Cost	Material Cost	Total Cost
08.4004 101	Window, vinyl, fixed, clear insulated glass, average cost	SF	0.0206	2.22	16.87	19.10
08.4005 000	**DOORS, VINYL, SLIDING, INSULATED GLASS WITH 1/2" AIR, WITH SCREEN:**					
08.4005 001	*Note: The following items are glazed with tempered glass.*					
08.4005 011	Door, vinyl, sliding, 5068, XO, clear insulated glass	EA	2.0935	225.87	752.50	978.37
08.4005 021	Door, vinyl, sliding, 6068, XO, clear insulated glass	EA	2.3502	253.56	815.87	1,069.44
08.4005 031	Door, vinyl, sliding, 8068, XO, clear insulated glass	EA	2.6305	283.80	1,016.11	1,299.92
08.4005 041	Door, vinyl, sliding, 10068, OXXO, clear insulated glass	EA	2.8487	307.35	1,707.54	2,014.89
08.4005 051	Door, vinyl, sliding, 12068,OXO, clear insulated glass	EA	4.0594	437.97	1,763.54	2,201.51
08.4005 061	Door, vinyl, sliding, 12068, OXXO, clear insulated glass	EA	4.0594	437.97	1,879.70	2,317.67
08.4005 071	Door, vinyl, sliding, 6080, XO, clear insulated glass	EA	2.3502	253.56	1,189.58	1,443.15
08.4005 081	Door, vinyl, sliding, 8080, XO, clear insulated glass	EA	2.8487	307.35	1,259.91	1,567.26
08.4005 091	Door, vinyl, sliding, clear insulated glass, average cost	SF	0.0526	5.68	22.13	27.80
08.5000 000	**WINDOWS, ALUMINUM, INSULATED GLASS WITH 1/2" AIR SPACE:**					
08.5000 001	*Note: The following items are residential/light commercial grade Intermediate quality with clear, white, or bronze aluminum; clear insulated glass with 1/2" air, no grid* *Apply all adder percentages to base price only.* *For clear single glazing deduct: 20%, Bronze or grey glass add: 20%* *For low-E glazing add: 15%, Tempered add: 60%* *For thermally broken frame add 20%* *For economy quality materials deduct 20%, Premium add 50%*					
08.5001 000	**WINDOWS, ALUMINUM, SLIDING, INSULATED GLASS WITH 1/2" AIR, WITH SCREEN:**					
08.5001 011	Window, aluminum, sliding, XO, 2016, clear insulated glass	EA	0.3671	39.61	75.69	115.30
08.5001 021	Window, aluminum, sliding, XO, 2030, clear insulated glass	EA	0.3671	39.61	99.71	139.31
08.5001 031	Window, aluminum, sliding, XO, 3040, clear insulated glass	EA	0.4401	47.48	141.00	188.49
08.5001 041	Window, aluminum, sliding, XO, 4040, clear insulated glass	EA	0.5142	55.48	157.22	212.70
08.5001 051	Window, aluminum, sliding, XO, 5040, clear insulated glass	EA	0.5142	55.48	176.72	232.20
08.5001 061	Window, aluminum, sliding, XO, 6040, clear insulated glass	EA	0.5873	63.36	192.99	256.35
08.5001 071	Window, aluminum, sliding, XOX, 8040, clear insulated glass	EA	0.7341	79.20	289.43	368.63
08.5001 081	Window, aluminum, sliding, XOX, 10040, clear insulated glass	EA	0.8802	94.96	331.00	425.96
08.5001 091	Window, aluminum, sliding, clear insulated glass, average cost	SF	0.0256	2.76	12.19	14.96
08.5002 000	**WINDOWS, ALUMINUM, CASEMENT, INSULATED GLASS WITH 1/2" AIR, WITH SCREEN:**					
08.5002 011	Window, aluminum, casement, X, 2026, clear insulated glass	EA	0.3671	39.61	193.62	233.22
08.5002 021	Window, aluminum, casement, X, 2640, clear insulated glass	EA	0.3671	39.61	258.76	298.37
08.5002 031	Window, aluminum, casement, XO, 4050, clear insulated glass	EA	0.5142	55.48	338.34	393.82
08.5002 041	Window, aluminum, casement, XX, 4050, clear insulated glass	EA	0.5142	55.48	424.61	480.09
08.5002 051	Window, aluminum, casement, XO, 6030, clear insulated glass	EA	0.5142	55.48	312.35	367.83
08.5002 061	Window, aluminum, casement, XX, 6030, clear insulated glass	EA	0.5142	55.48	395.40	450.87
08.5002 071	Window, aluminum, casement, clear insulated glass, average cost	SF	0.0256	2.76	26.93	29.69
08.5003 000	**WINDOWS, ALUMINUM, SINGLE HUNG, INSULATED GLASS WITH 1/2" AIR, WITH SCREEN:**					
08.5003 011	Window, aluminum, single hung, 2030, clear insulated glass	EA	0.3671	39.61	122.59	162.20
08.5003 021	Window, aluminum, single hung, 2640, clear insulated glass	EA	0.3671	39.61	156.31	195.91
08.5003 031	Window, aluminum, single hung, 3060, clear insulated glass	EA	0.5142	55.48	203.40	258.88
08.5003 041	Window, aluminum, single hung, 4080, clear insulated glass	EA	0.7341	79.20	287.67	366.87
08.5003 051	Window, aluminum, single hung, clear insulated glass, average cost	SF	0.0256	1.71	14.08	15.78
08.5004 000	**WINDOWS, ALUMINUM, FIXED, INSULATED GLASS WITH 1/2" AIR:**					
08.5004 011	Window, aluminum, fixed, 2020, clear insulated glass	EA	0.2933	31.64	59.89	91.53
08.5004 021	Window, aluminum, fixed, 2650, clear insulated glass	EA	0.3533	38.12	111.77	149.89
08.5004 031	Window, aluminum, fixed, 3030, clear insulated glass	EA	0.2933	31.64	90.06	121.71
08.5004 041	Window, aluminum, fixed, 4040, clear insulated glass	EA	0.4133	44.59	124.79	169.38
08.5004 051	Window, aluminum, fixed, 5050, clear insulated glass	EA	0.4667	50.35	183.69	234.04
08.5004 061	Window, aluminum, fixed, 6060, clear insulated glass	EA	0.7067	76.25	309.55	385.80

08 - DOORS, WINDOWS & GLASS

Division	Description	Unit	Labor Hours	Labor Cost	Material Cost	Total Cost
08.5004 071	Window, aluminum, fixed, 7050, clear insulated glass	EA	0.5867	63.30	309.76	373.06
08.5004 081	Window, aluminum, fixed, 8050, clear insulated glass	EA	0.7067	76.25	343.89	420.14
08.5004 091	Window, aluminum, fixed, 10040, clear insulated glass	EA	0.7067	76.25	353.06	429.30
08.5004 101	Window, aluminum, fixed, clear insulated glass, average cost	SF	0.0206	2.22	9.33	11.55
08.5005 000	**DOORS, ALUMINUM, SLIDING, INSULATED GLASS WITH 1/2" AIR, WITH SCREEN:**					
08.5005 001	Note: The following items are glazed with tempered glass.					
08.5005 002	Note: X means active panel, O means stationary panel					
08.5005 011	Door, aluminum, sliding, 5068, XO, clear insulated glass	EA	2.0935	225.87	469.45	695.32
08.5005 021	Door, aluminum, sliding, 6068, XO, clear insulated glass	EA	2.3502	253.56	508.99	762.56
08.5005 031	Door, aluminum, sliding, 8068, XO, clear insulated glass	EA	2.6350	284.29	633.90	918.19
08.5005 041	Door, aluminum, sliding, 10068, OXXO, clear insulated glass	EA	2.8487	307.35	1,065.25	1,372.59
08.5005 051	Door, aluminum, sliding, 12068, OXO, clear insulated glass	EA	4.0594	437.97	1,100.20	1,538.17
08.5005 061	Door, aluminum, sliding, 12068, OXXO, clear insulated glass	EA	4.0594	437.97	1,172.63	1,610.60
08.5005 071	Door, aluminum, sliding, 6080, XO, clear insulated glass	EA	2.8000	302.09	631.18	933.28
08.5005 081	Door, aluminum, sliding, 8080, XO, clear insulated glass	EA	2.6350	284.29	785.99	1,070.28
08.5005 091	Door, aluminum, sliding, clear insulated glass, average cost	SF	0.0526	5.68	13.52	19.19
08.5006 000	**METAL SASH, UNGLAZED:**					
08.5006 001	Note: For commercial sash glazing, see section 08.8000 Glass & Glazing.					
08.5006 011	Sash, steel, industrial, vented 50%	SF	0.0612	6.60	17.05	23.65
08.5006 021	Sash, steel, projected, vented 100%	SF	0.0612	6.60	20.50	27.10
08.5006 031	Sash, steel, industrial, fixed 100%	SF	0.0612	6.60	13.01	19.62
08.5006 041	Sash, aluminum, industrial, vented 50%	SF	0.0677	7.30	20.15	27.45
08.5006 051	Sash, aluminum, projected, vented 100%	SF	0.0677	7.30	25.97	33.27
08.5006 061	Sash, aluminum, industrial, fixed 100%	SF	0.0612	6.60	13.61	20.21
08.5006 071	Add for screens	SF			4.82	4.82
08.6000 000	**WINDOWS, WOOD, INSULATED GLASS:**					
08.6001 000	WINDOWS, WOOD, INSULATED GLASS WITH 1/4" AIR, WITH SCREEN:					
08.6001 001	Note: The following windows are unfinished softwood, built to order; prices include standard frame with exterior mould, 1-3/8" sash, glazed clear, weatherstrip, hardware and screens					
08.6001 010	**WINDOWS, WOOD, DOUBLE HUNG, 2 LITE, INSULATED GLASS WITH 1/2" AIR, WITH SCREEN:**					
08.6001 021	Window, wood, double hung, 1626, 2 lite, clear insulated glass	EA	0.9542	102.95	280.38	383.33
08.6001 031	Window, wood, double hung, 2026, 2 lite, clear insulated glass	EA	0.9542	102.95	284.11	387.05
08.6001 041	Window, wood, double hung, 2030, 2 lite, clear insulated glass	EA	0.9542	102.95	298.96	401.91
08.6001 051	Window, wood, double hung, 2036, 2 lite, clear insulated glass	EA	0.9542	102.95	343.50	446.45
08.6001 061	Window, wood, double hung, 2640, 2 lite, clear insulated glass	EA	0.9542	102.95	376.93	479.88
08.6001 071	Window, wood, double hung, 2646, 2 lite, clear insulated glass	EA	1.0272	110.82	402.93	513.76
08.6001 081	Window, wood, double hung, 3040, 2 lite, clear insulated glass	EA	1.0272	110.82	419.63	530.46
08.6001 091	Window, wood, double hung, 3046, 2 lite, clear insulated glass	EA	1.0272	110.82	449.35	560.17
08.6001 101	Window, wood, double hung, 3050, 2 lite, clear insulated glass	EA	1.0272	110.82	477.20	588.02
08.6001 111	Window, wood, double hung, 3060, 2 lite, clear insulated glass	EA	1.1742	126.68	592.31	719.00
08.6001 121	Window, wood, double hung, 4046, 2 lite, clear insulated glass	EA	1.3933	150.32	542.21	692.53
08.6001 131	Window, wood, double hung, 4060, 2 lite, clear insulated glass	EA	1.3933	150.32	724.18	874.50
08.6001 141	Window, wood, double hung, 2 lite, clear insulated glass, average cost	SF			41.45	41.45
08.6002 000	**WINDOWS, WOOD, CASEMENT, 2 LITE, INSULATED GLASS WITH 1/4" AIR, WITH SCREEN:**					
08.6002 011	Window, wood, casement, 1620, 1 lite, clear insulated glass	EA	1.0272	110.82	226.53	337.36
08.6002 021	Window, wood, casement, 2030, 1 lite, clear insulated glass	EA	1.1002	118.70	291.52	410.22
08.6002 031	Window, wood, casement, 5040, XX, 2 lite, clear insulated glass	EA	1.9065	205.69	683.33	889.02
08.6002 041	Window, wood, casement, 8050, XOX, 3 lite, clear insulated glass	EA	2.0535	221.55	1,245.91	1,467.47
08.6002 051	Window, wood, casement, 10050, XOX, 3 lite, clear insulated glass	EA	2.6405	284.88	1,448.31	1,733.19
08.6002 061	Window, wood, casement, 2 lite, clear insulated glass, average cost	SF			43.66	43.66
08.6003 000	**WINDOWS, WOOD, FIXED, 1 LITE, INSULATED GLASS WITH 1/4" AIR, WITH SCREEN:**					

DOORS, WINDOWS & GLASS - 08

Division	Description	Unit	Labor Hours	Labor Cost	Material Cost	Total Cost
08.6003 011	Window, wood, fixed, 2020, 1 lite, clear insulated glass	EA	0.8218	88.66	172.69	261.35
08.6003 021	Window, wood, fixed, 3036, 1 lite, clear insulated glass	EA	0.8218	88.66	282.25	370.91
08.6003 031	Window, wood, fixed, 5060, 1 lite, clear insulated glass	EA	1.1002	118.70	664.75	783.45
08.6003 041	Window, wood, fixed, average cost/SF	SF			30.71	30.71
08.6005 000	**WINDOWS, WOOD, VINYL CLAD, INSULATED GLASS WITH 1/2" AIR SPACE:**					
08.6005 001	Note: The following windows are white or tan vinyl clad exterior, unfinished softwood interior; prices include standard frame depth, HP insulated glass, with strip, hardware and screen.					
08.6005 010	**WINDOWS, WOOD, VINYL CLAD, DOUBLE HUNG, 2 LITE, INSULATED GLASS WITH 1/2" AIR, WITH SCREEN:**					
08.6005 011	Window, wood, vinyl clad, double hung, 18210, 2 lite, clear insulated glass	EA	0.3667	39.56	290.63	330.19
08.6005 021	Window, wood, vinyl clad, double hung, 2032, 2 lite, clear insulated glass	EA	0.3667	39.56	309.79	349.35
08.6005 031	Window, wood, vinyl clad, double hung, 24310, 2 lite, clear insulated glass	EA	0.3667	39.56	350.34	389.90
08.6005 041	Window, wood, vinyl clad, double hung, 2842, 2 lite, clear insulated glass	EA	0.4417	47.66	397.67	445.32
08.6005 051	Window, wood, vinyl clad, double hung, 3046, 2 lite, clear insulated glass	EA	0.4417	47.66	439.35	487.00
08.6005 061	Window, wood, vinyl clad, double hung, 3452, 2 lite, clear insulated glass	EA	0.5167	55.75	503.54	559.28
08.6005 071	Window, wood, vinyl clad, double hung, 3856, 2 lite, clear insulated glass	EA	0.5833	62.93	554.21	617.14
08.6005 081	Window, wood, vinyl clad, double hung, 3862, 2 lite, clear insulated glass	EA	0.5833	62.93	629.69	692.63
08.6005 091	Window, wood, vinyl clad, double hung, 2 lite, clear insulated glass, average	SF			37.63	37.63
08.6006 000	**WINDOWS, WOOD, VINYL CLAD, CASEMENT, 2 LITE, INSULATED GLASS WITH 1/2" AIR, WITH SCREEN:**					
08.6006 011	Window, wood, vinyl clad, casement, 1818, 1 lite, clear insulated glass	EA	0.3667	39.56	223.05	262.61
08.6006 021	Window, wood, vinyl clad, casement, 18210, 1 lite, clear insulated glass	EA	0.3667	39.56	247.82	287.38
08.6006 031	Window, wood, vinyl clad, casement, 2333, 1 lite, clear insulated glass	EA	0.3667	39.56	291.75	331.31
08.6006 041	Window, wood, vinyl clad, casement, 23310, 1 lite, clear insulated glass	EA	0.3667	39.56	333.44	373.00
08.6006 051	Window, wood, vinyl clad, casement, 23410, 1 lite, clear insulated glass	EA	0.4417	47.66	397.67	445.32
08.6006 061	Window, wood, vinyl clad, casement, 23510, 1 lite, clear insulated glass	EA	0.4417	47.66	460.73	508.38
08.6006 071	Window, wood, vinyl clad, casement, 210510, 1 lite, clear insulated glass	EA	0.5167	55.75	558.70	614.45
08.6006 081	Window, wood, vinyl clad, casement, 1 lite, clear insulated glass, average	SF			44.37	44.37
08.6007 000	**WINDOWS, WOOD, VINYL CLAD, FIXED, 1 LITE, INSULATED GLASS WITH 1/4" AIR, WITH SCREEN:**					
08.6007 011	Window, wood, vinyl clad, fixed, 3320, 1 lite, clear insulated glass	EA	0.2933	31.64	289.52	321.17
08.6007 021	Window, wood, vinyl clad, fixed,31043, 1 lite, clear insulated glass	EA	0.4133	44.59	465.23	509.82
08.6007 031	Window, wood, vinyl clad, fixed, 310510, 1 lite, clear insulated glass	EA	0.4667	50.35	619.56	669.91
08.6007 041	Window, wood, vinyl clad, fixed, 510410, 1 lite, clear insulated glass	EA	0.5200	56.10	827.96	884.07
08.6007 051	Window, wood, vinyl clad, fixed, 1 lite, clear insulated glass, average	SF			32.57	32.57
08.7000 000	**FINISH HARDWARE:**					
08.7001 000	**HARDWARE, IN PLACE COSTS, AVERAGE**					
08.7001 011	Hardware, tract, 3 bedroom, 2 bath	UNIT	8.1840	882.97	3,055.48	3,938.45
08.7001 021	Hardware, custom, 4 bedroom, 3 bath	UNIT	19.6416	2,119.13	2,251.64	4,370.77
08.7001 031	Hardware, commercial, economy	SF	0.0028	0.30	0.24	0.54
08.7001 041	Hardware, commercial, standard	SF	0.0046	0.50	0.60	1.09
08.7001 051	Hardware, institutional, with out panic hardware	DOOR	2.2077	238.19	467.48	705.67
08.7001 061	Hardware, hospital, with out panic hardware	DOOR	2.8131	303.51	580.82	884.33
08.7001 071	Hardware, office, with out panic hardware	DOOR	1.7448	188.25	350.69	538.93
08.7002 000	**GENERAL HARDWARE ITEMS:**					
08.7002 001	Note: Install prices are for labor to install in pre-machined doors. Commercial hardware is priced as dull chrome (26D). Residential locksets priced as polished brass (US3), hinges as dull brass (US4). Many other finishes are available. Letters in the lockset descriptions below refer to Schlage product codes for standard of quality.					
08.7002 011	Lockset, decorative entry, 'E/B'	EA	0.7392	79.75	406.87	486.62
08.7002 021	Lockset, commercial, mortise, 'L'	EA	0.4200	45.31	295.93	341.24
08.7002 031	Lockset, commercial , 'D'	EA	0.4200	45.31	207.52	252.83
08.7002 041	Lockset, commercial, 'S'	EA	0.3381	36.48	122.28	158.76
08.7002 051	Lockset, residential, good, 'A'	EA	0.2713	29.27	88.37	117.64

08 - DOORS, WINDOWS & GLASS

Division	Description	Unit	Labor Hours	Labor Cost	Material Cost	Total Cost
08.7002 061	Lockset, residential, economy, 'F'	EA	0.2710	29.24	31.81	61.04
08.7002 071	Latchset, commercial, mortise, 'L'	EA	0.4000	43.16	225.02	268.18
08.7002 081	Latchset, commercial, 'D'	EA	0.3781	40.79	154.13	194.93
08.7002 091	Latchset, commercial, 'S'	EA	0.3381	36.48	74.11	110.59
08.7002 101	Latchset, residential, 'A'	EA	0.2713	29.27	42.11	71.38
08.7002 111	Latchset, residential, privacy, 'A'	EA	0.2713	29.27	49.35	78.62
08.7002 112	Latchset, residential, 'F'	EA	0.2713	29.27	16.97	46.24
08.7002 121	Dummy trim, pair, 'L'	EA	0.3781	40.79	213.72	254.51
08.7002 131	Dummy trim, single, 'D'	EA	0.2849	30.74	69.86	100.59
08.7002 141	Add for electrified mortise lock	EA			159.26	159.26
08.7002 151	Add for lead lined mortise lock	EA			123.29	123.29
08.7002 161	Single cylinder, residential, economy, deadbolt, 'B160'	EA	0.2713	29.27	29.81	59.08
08.7002 171	Double cylinder residential, economy, deadbolt, 'B162'	EA	0.2713	29.27	42.11	71.38
08.7002 181	Single cylinder, residential, standard, deadbolt, 'B460'	EA	0.2713	29.27	57.54	86.81
08.7002 191	Double cylinder residential, standard, deadbolt, 'B462'	EA	0.2713	29.27	72.98	102.25
08.7002 201	Single cylinder, residential, standard, deadbolt, 'B560'	EA	0.3100	33.45	107.86	141.31
08.7002 202	Double cylinder residential, standard, deadbolt, 'B562'	EA	0.3100	33.45	128.45	161.90
08.7002 211	Add lever handle, residential, economy, 'F'	EA			6.92	6.92
08.7002 221	Add lever handle, residential, standard, 'A'	EA			21.55	21.55
08.7002 231	Hinge, butt, 4.5x4.5, standard weight	PR	0.3452	37.24	16.94	54.18
08.7002 241	Hinge, butt, 4.5x4.5, standard weight, ball bearing	PR	0.3452	37.24	30.74	67.98
08.7002 251	Hinge, butt, 4.5x4.5, heavy weight, ball bearing	PR	0.3452	37.24	66.60	103.84
08.7002 261	Hinge, butt, 4.5x4.5, standard weight, ball bearing, brass	PR	0.3452	37.24	68.97	106.21
08.7002 271	Add elec thru wire or monitor	EA			169.05	169.05
08.7002 281	Hinge, butt, 4" x 4", residential	PR	0.3452	37.24	5.00	42.24
08.7002 291	Hinge, butt, 3.5x3.5, residential	PR	0.2713	29.27	3.53	32.80
08.7002 301	Hinge, swing clear, 4.5, standard weight, ball bearing	PR	0.5625	60.69	121.21	181.89
08.7002 311	Hinge, butt, spring, 4.5x4.5 standard weight	EA	0.2713	29.27	39.71	68.98
08.7002 321	Hinge, butt, spring, 4x4, residential	EA	0.2713	29.27	30.71	59.98
08.7002 331	Hinge, double acting, 7"	PR	0.6419	69.25	137.85	207.10
08.7002 341	Hinge, pivot set, t & b, interior to 3'	EA	0.6903	74.48	202.56	277.04
08.7002 351	Hinge, pivot set, t & b, exterior to 3'	EA	0.6903	74.48	290.76	365.24
08.7002 361	Hinge, pivot set, t & b, exterior over 3'	EA	0.8547	92.21	409.24	501.45
08.7002 371	Intermediate pivot, standard	EA	0.2713	29.27	145.86	175.13
08.7002 381	Hinge, floor residential	ST	0.6903	74.48	110.15	184.63
08.7002 391	Closer, economy	EA	0.6410	69.16	58.84	128.00
08.7002 401	Closer, standard	EA	0.7122	76.84	103.93	180.77
08.7002 411	Closer, heavy duty	EA	0.9259	99.90	189.49	289.39
08.7002 421	Closer, floor mounted, hold open, w/cover	ST	1.2463	134.46	840.68	975.14
08.7002 422	Closer, floor mounted, non-hold open, w/cover	ST	1.2463	134.46	778.80	913.27
08.7002 423	Closer, floor mounted, install cement case	EA	2.0000	215.78		215.78
08.7002 431	Closer, smoke activated	EA	2.5538	275.53	837.93	1,113.46
08.7002 441	Panic devices, good, single door	EA	2.3250	250.84	467.99	718.83
08.7002 451	Panic devices, good, double door	PR	3.4875	376.27	983.46	1,359.72
08.7002 461	Panic devices, best, single door	EA	2.3250	250.84	1,075.11	1,325.96
08.7002 471	Panic devices, best, double door	PR	3.4875	376.27	2,235.20	2,611.46
08.7002 481	Panic device, best, fire labeled, single door	EA	2.3250	250.84	1,180.71	1,431.55
08.7002 491	Panic device, best, fire labeled, double door	PR	3.4875	376.27	2,539.85	2,916.12
08.7002 501	Kick plate, 10"x34", 16 ga bronze	EA	0.4704	50.75	70.80	121.55
08.7002 511	Kick plate, 10"x34", 18 ga stainless steel	EA	0.4704	50.75	54.28	105.03
08.7002 521	Push plate, 4"x16", wrought bronze	EA	0.2000	21.58	18.80	40.38
08.7002 531	Pull, 4" x 16', wrought bronze	EA	0.2000	21.58	49.64	71.22
08.7002 541	Pull bar, 30"	EA	0.1352	14.59	132.41	147.00
08.7002 551	Flush bolt, automatic, with label	EA	0.7834	84.52	212.78	297.30
08.7002 561	Flush bolt, extension	EA	0.3698	39.90	36.03	75.93

DOORS, WINDOWS & GLASS - 08

Division	Description	Unit	Labor Hours	Labor Cost	Material Cost	Total Cost
08.7002 571	Surface bolt, 4"	EA	0.3342	36.06	8.37	44.42
08.7002 581	Surface bolt, 6"	EA	0.3342	36.06	11.31	47.37
08.7002 591	Surface bolt, 12" brass, decorative	PR	0.4000	43.16	235.64	278.79
08.7002 601	Cremone bolt with egg handle, brass	EA	2.0000	215.78	685.51	901.29
08.7002 611	Dust proof strike	EA	0.2923	31.54	20.81	52.35
08.7002 621	Auto door bottom, aluminum, 36", rabbet	EA	0.5233	56.46	47.22	103.67
08.7002 631	Auto door bottom, residential, surface mounted	EA	0.2923	31.54	24.97	56.50
08.7002 641	Letter drop plate	EA	0.3781	40.79	47.22	88.01
08.7002 651	Door stop, rubber, tip, screw base	EA	0.0631	6.81	8.00	14.80
08.7002 661	Door stop & holder, floor mounted	EA	0.2923	31.54	28.94	60.48
08.7002 671	Door stop & holder, wall mounted	EA	0.2923	31.54	28.94	60.48
08.7002 681	Door stop/holder, overhead mounted	EA	0.6055	65.33	183.27	248.60
08.7002 691	Door privacy device, viewer	EA	0.2000	21.58	20.95	42.52
08.7002 701	Door interview panel, grill	EA	0.3279	35.38	44.87	80.25
08.7002 711	Threshold, anodized aluminum, 6"x3'	EA	0.2000	21.58	30.74	52.32
08.7002 721	Threshold, bronze, polished, 6"x3'	EA	0.2000	21.58	114.12	135.70
08.7002 731	Smoke seal, adhesive, 3x7 door	DOOR	0.0074	0.80	17.94	18.73
08.7002 741	Economy weatherstrip, threshold, 3x7 door	DOOR	1.6025	172.89	22.62	195.51
08.7002 742	Weatherstrip, with interlock threshold, 3x7 door	DOOR	1.6025	172.89	50.46	223.35
08.7002 751	Cabinet hardware, residential	LF	0.1142	12.32	10.10	22.42
08.7002 761	Cabinet hardware, commercial	LF	0.1498	16.16	17.67	33.83
08.7002 771	Cabinet hardware, institutional	LF	0.1891	20.40	21.72	42.12
08.7003 000	**ELECTRICALLY OPERATED HARDWARE DEVICES:**					
08.7003 001	*Note: Electrical connection extra.*					
08.7003 011	Electromagnetic door lock	EA	3.8000	409.98	764.83	1,174.81
08.7003 021	Magnetic card reader	EA	3.5000	377.62	695.28	1,072.89
08.7003 031	Magnetic door holder	EA	1.7000	183.41	135.08	318.49
08.7003 041	Electric strike, economy	EA	0.3781	40.79	96.70	137.49
08.7003 051	Electric strike, good	EA	0.3781	40.79	430.80	471.59
08.8000 000	**GLASS & GLAZING:**					
08.8000 001	*Note: The following prices are based on small quantities. For quantities over 300 square feet, see section 08.9000.*					
08.8001 000	**JOB GLAZING, SINGLE GLAZED:**					
08.8001 001	*Note: For lites under 2 SF, add 30% to the total costs.*					
08.8001 011	Job, 3/32" (2.5mm) single strength, clear, float	SF	0.0284	2.77	3.97	6.74
08.8001 021	Job, 1/8" (3mm) double strength, clear, float	SF	0.0284	2.77	4.24	7.01
08.8001 031	Job, 1/8" (3mm) tempered float	SF	0.0353	3.44	7.89	11.34
08.8001 041	Job, 1/8" (3mm) obscure	SF	0.0353	3.44	6.91	10.35
08.8001 051	Job, 3/16" (5mm), clear, float	SF	0.0484	4.72	6.37	11.09
08.8001 061	Job, 3/16" (5mm), clear, tempered	SF	0.0353	3.44	8.30	11.74
08.8001 071	Job, 1/4" (6mm), clear, float	SF	0.0484	4.72	6.63	11.35
08.8001 081	Job, 1/4" (6mm) clear, wire	SF	0.0484	4.72	31.57	36.29
08.8001 091	Job, 1/4" (6mm), bronze/gray, float	SF	0.0484	4.72	8.99	13.71
08.8001 101	Job, 1/4" obscure	SF	0.0484	4.72	13.76	18.48
08.8001 111	Job, 1/4" obscure, wire	SF	0.0484	4.72	15.67	20.39
08.8001 121	Job, 1/4" (6mm) float, Solex	SF	0.0484	4.72	11.53	16.25
08.8001 131	Job, 1/4" clear laminated (.030)	SF	0.0484	4.72	23.51	28.23
08.8001 141	Job, 1/4" (6mm) clear, float, tempered	SF	0.0484	4.72	8.70	13.42
08.8001 161	Job, 1/4" spandrel, custom color	SF	0.0559	5.45	28.52	33.97
08.8001 171	Job, 1/4" spandrel, plain	SF	0.0559	5.45	20.36	25.81
08.8001 181	Job, 1/4" (6mm) float, temp, reflective, bronze/grey	SF	0.0484	4.72	22.44	27.16
08.8001 191	Job, 1/4" (6mm) float, reflective, solar bronze/grey	SF	0.0484	4.72	20.53	25.25
08.8001 201	Job, Greylite 14	SF	0.0484	4.72	15.90	20.62
08.8001 211	Job, Greylite 31	SF	0.0484	4.72	15.21	19.93
08.8001 221	Job, 3/8" clear, float, clear	SF	0.0591	5.76	15.90	21.66

08 - DOORS, WINDOWS & GLASS

Division	Description	Unit	Labor Hours	Labor Cost	Material Cost	Total Cost
08.8001 231	Job, 3/8" float, bronze/gray	SF	0.0591	5.76	26.05	31.81
08.8001 251	Job, 1/2" clear, float, tempered	SF	0.0763	7.44	38.32	45.76
08.8001 261	Job, 1/2" float, tempered bronze/gray	SF	0.0763	7.44	44.13	51.57
08.8002 000	**GLAZING SPECIALTIES:**					
08.8002 011	Glazing, acrylic sheet, 1/8" clear	SF	0.0355	3.46	9.65	13.11
08.8002 021	Glazing, acrylic sheet, 1/4" clear	SF	0.0484	4.72	13.24	17.96
08.8002 031	Glazing, acrylic sheet, 1/4" color	SF	0.0484	4.72	15.38	20.10
08.8002 041	Glazing, acrylic sheet, 1/4" shatterproof	SF	0.0484	4.72	40.24	44.96
08.8002 051	Glazing, 3/4" bullet resistant	SF	0.7133	69.56	81.87	151.43
08.8002 061	Glazing, 2" bullet resistant	SF	0.8481	82.71	97.37	180.07
08.8002 063	Glazing, service window, secure, 2'x3'	EA	1.6000	156.03	4,011.94	4,167.97
08.8002 065	Glazing, service window, secure, draft free	EA	1.6000	156.03	5,377.90	5,533.93
08.8002 071	Glazing, faceted glass, leaded	SF	0.5565	54.27	171.41	225.68
08.8002 081	Glazing, artistic murals, leaded	SF	0.5565	54.27	187.66	241.93
08.8002 091	Glazing, decor glass, small lites	SF	0.1676	16.34	19.52	35.87
08.8002 101	Glazing, decorative glass, panels	SF	0.0559	5.45	9.65	15.10
08.8002 111	Glazing, insulated glass, 1/8", B, 2 layer	SF	0.1096	10.69	17.31	28.00
08.8002 121	Glazing, insulated, 1/8" float, 2 layer	SF	0.1096	10.69	24.21	34.90
08.8002 131	Glazing, insulated, 1/4" float, 2 layer	SF	0.1274	12.42	27.30	39.73
08.8002 141	Glazing, insulated, 1/4" float, tempered, vented	SF	0.1266	12.35	53.41	65.76
08.8002 151	Glazing, insulated, 1/4" float, tinted	SF	0.1268	12.37	44.79	57.15
08.8002 161	Glazing, mirrors, sheet	SF	0.0919	8.96	11.76	20.72
08.8002 171	Glazing, mirrors, float	SF	0.0997	9.72	17.74	27.46
08.8002 181	Glazing, mirrors, wall	SF	0.0919	8.96	16.75	25.71
08.8002 191	Glazing, mirror, panel, 1-way, in wood	SF	0.1093	10.66	23.61	34.27
08.8002 201	Glazing, shower door, 24"x72", wire/tempered	EA			279.38	279.38
08.8002 211	Glazing, shower door, panel & lite	EA			502.79	502.79
08.8002 221	Glazing, shower door, average cost/SF	SF			25.01	25.01
08.8002 231	Glazing, shower door, 24"x72", plastic	EA			230.40	230.40
08.8002 241	Glazing, shower door, plastic, 1 pnl/lite	EA			356.06	356.06
08.8002 251	Glazing, shower door, average cost/SF	SF			18.81	18.81
08.9000 000	**CURTAIN WALL & STOREFRONT SYSTEMS:**					
08.9001 000	**STOREFRONT, CLEAR GLASS & CLEAR ANODIZED ALUMINUM:**					
08.9001 011	Storefront system, stub wall to 8'	SF	0.0971	9.47	28.44	37.90
08.9001 021	Storefront system, floor 8' to 10'	SF	0.1135	11.07	31.43	42.50
08.9002 000	**ADDERS:**					
08.9002 011	Add for tint	SF			2.33	2.33
08.9002 021	Add for bronze anodized aluminum	SF			2.55	2.55
08.9002 031	Add for black anodized aluminum	SF			6.19	6.19
08.9002 041	Add for extra aluminum sections	LF			14.93	14.93
08.9003 000	**ENTRANCES, CONCEALED CLOSER, CENTER PIVOT:**					
08.9003 001	Note: The following items should be added to those in the sections immediately preceding, without credit for square foot of contact area.					
08.9003 011	Entry, to 3' x 7', narrow stile	EA	5.9087	576.22	959.73	1,535.95
08.9003 021	Entrance, to 3' x 7', heavy section	EA	6.2164	606.22	1,126.75	1,732.98
08.9003 031	Entrance, to 3' x 7', 1/2" tempered	EA	7.6927	750.19	2,380.42	3,130.61
08.9003 041	Add for center stop	EA	0.2341	22.83	31.59	54.42
08.9003 051	Add for floor check	EA	1.2309	120.04	256.45	376.49
08.9003 061	Add for bronze anodized aluminum	EA			188.85	188.85
08.9003 071	Add for black anodized aluminum	EA			274.34	274.34
08.9003 081	Add for large sizes	SF	0.1170	11.41	21.07	32.48
08.9003 091	Add for automatic opener	EA	18.7561	2,023.60	3,618.15	5,641.74
08.9004 000	**STOREFRONT, CLEAR POLISHED PLATE, CLEAR ANODIZED ALUMINUM:**					
08.9004 011	Storefront system, stub wall to 9'	SF	0.1152	11.23	24.72	35.95

DOORS, WINDOWS & GLASS - 08

Division	Description	Unit	Labor Hours	Labor Cost	Material Cost	Total Cost
08.9004 021	Storefront system, stub wall to 13'	SF	0.1255	12.24	25.60	37.84
08.9004 031	Storefront system, floor to 9'	SF	0.1186	11.57	26.60	38.16
08.9004 041	Storefront system, floor to 13'	SF	0.1332	12.99	28.60	41.59
08.9005 000	**ADDERS:**					
08.9005 011	Add for tint	SF			1.89	1.89
08.9005 021	Add for bronze anodized aluminum	SF			2.27	2.27
08.9005 031	Add for black anodized aluminum	SF			4.81	4.81
08.9005 041	Add for extra aluminum sections	LF	0.0761	8.21	12.93	21.14
08.9006 000	**ENTRANCES, CONCEALED CLOSER:**					
08.9006 001	*Note: The following items should be added to those in the sections immediately preceding, without 'credit for square feet of contact area.*					
08.9006 011	Entrance, 3070, aluminum & glass, hidden closer	EA	7.2003	702.17	1,237.95	1,940.12
08.9006 021	Entrance, 3070, 5/8" tempered, herculite	EA	8.4317	822.26	2,639.21	3,461.47
08.9006 031	Add for floor check	EA	1.3480	131.46	305.61	437.07
08.9006 041	Add for black anodized aluminum	EA			285.12	285.12
08.9006 051	Add for bronze anodized aluminum	EA			213.81	213.81
08.9006 061	Add for larger sizes	SF	0.1760	17.16	24.59	41.76
08.9006 071	Add for automatic opener	EA	18.7561	1,829.09	3,618.15	5,447.24
08.9007 000	**STOREFRONT, CLEAR PLATE GLASS ANODIZED ALUMINUM:**					
08.9007 011	Storefront system, floor to 9'	SF	0.1358	13.24	36.46	49.70
08.9007 021	Storefront system, floor to 11'	SF	0.1452	14.16	37.34	51.49
08.9007 031	Storefront system, floor to 13'	SF	0.1547	15.09	38.41	53.50
08.9007 041	Anod bronze frame, 1/4" tint plate	SF	0.1547	15.09	27.17	42.26
08.9007 051	Decorative storefront	SF	0.1672	16.31	29.40	45.70
08.9007 061	Decorative curved storefront	SF	0.3183	31.04	55.86	86.90
08.9008 000	**ADDERS:**					
08.9008 011	Add for tint	SF			2.00	2.00
08.9008 021	Add for bronze anodized aluminum	SF			2.44	2.44
08.9008 031	Add for black anodized aluminum	SF			5.44	5.44
08.9008 041	Add for extra aluminum sections	LF			18.26	18.26
08.9008 051	Add for lacquered bronze	SF	0.0705	6.88	12.65	19.52
08.9008 061	Add for oil-rubbed bronze	SF	0.0819	7.99	13.44	21.42
08.9008 071	Add for seismic bracing	SF	0.0222	2.40	3.55	5.95
08.9008 081	Add for automatic opener	EA	18.7561	2,023.60	4,830.07	6,853.66
08.9009 000	**FRAMING, STEEL STOREFRONT:**					
08.9009 001	*Note: For steel glazing, use the adders given for aluminum.*					
08.9009 011	Storefront, primed steel framing	SF	0.0620	6.05	19.81	25.86
08.9009 021	Storefront, stainless steel framing	SF	0.0929	9.06	53.49	62.55
08.9010 000	**STEEL STOREFRONT FRAMING, DOOR & SASH INSERTION:**					
08.9010 011	Storefront, stainless steel door, 1 lite, frame	EA	7.3850	720.19	3,826.13	4,546.32
08.9010 021	Storefront, stainless steel door, 1 lite, frame, pair	EA	10.2578	1,000.34	7,420.42	8,420.76
08.9010 031	Automatic opener	EA	18.7561	1,829.09	3,569.89	5,398.98
08.9100 000	**CURTAIN WALLS & EXTERIOR PANEL SYSTEMS:**					
08.9100 001	*Note: The following items are for mid-rise and high-rise construction.*					
08.9101 000	**CURTAIN WALLS, IN PLACE WITH OUT GLAZING (FRAME ONLY):**					
08.9101 001	*Note: The following prices include installation and fasteners, but do not include structural steel, see section 05.0000. Prices are based on the gross square footage of exterior skin. Glazing must be added to determine total system cost. For sloped or butt glazing, add 30% to the total costs.*					
08.9101 011	Curtain wall frame, bronze anodized aluminum	SF			25.59	25.59
08.9101 021	Curtain wall frame, black anodized aluminum	SF			29.47	29.47
08.9101 031	Curtain wall frame, aluminum, sloping section	SF			40.80	40.80
08.9101 041	Curtain wall frame, steel, painted	SF			24.84	24.84
08.9101 051	Curtain wall frame, steel, porcelain enamel	SF			26.82	26.82

08 - DOORS, WINDOWS & GLASS

Division	Description	Unit	Labor Hours	Labor Cost	Material Cost	Total Cost
08.9102 000	**CURTAIN WALL GLAZING:**					
08.9102 001	*Note: The following items are for glazed installation in the frames listed above. Most systems include a combination of two or more of the following items. Determine the gross square footage of surface area for each item used and add the resultant cost to the cost of frames. The total square footage of these items should equal the total square footage of the frame.*					
08.9102 011	Plate, 1/4", clear	SF			9.04	9.04
08.9102 021	Plate, 1/4", tinted	SF			11.11	11.11
08.9102 031	Plate, 1/4" solar cool, reflective	SF			20.83	20.83
08.9102 041	Plate, 1/4", veri-tran, reflective	SF			27.82	27.82
08.9102 051	Tempered plate, 1/4", clear	SF			16.64	16.64
08.9102 061	Tempered plate, 1/4", tinted	SF			19.49	19.49
08.9102 071	Tempered plate, 1/4", solar cool reflective	SF			31.32	31.32
08.9102 081	Double glazed, 5/8", clear	SF			16.64	16.64
08.9102 091	Double glazed, 5/8", tinted	SF			18.08	18.08
08.9102 101	Double glazed, 5/8", solar cool reflective	SF			22.20	22.20
08.9102 111	Double glazed, tempered, 5/8", clear	SF			28.53	28.53
08.9102 121	Double glazed, tempered, 5/8", tinted	SF			30.61	30.61
08.9102 131	Double glazed, tempered, 5/8", solar cool	SF			43.18	43.18
08.9102 141	Double glazed, 1", clear	SF			18.08	18.08
08.9102 151	Double glazed, 1", tinted	SF			19.49	19.49
08.9102 161	Double glazed, 1", solar cool reflective	SF			25.75	25.75
08.9102 171	Double glazed, tempered, 1", clear	SF			32.69	32.69
08.9102 181	Double glazed, tempered, 1", tinted	SF			35.49	35.49
08.9102 191	Double glazed, tempered, 1", solar cool	SF			18.08	18.08
08.9102 201	Spandrel	SF			19.49	19.49
08.9102 211	GlasWeld, 1/4", finished 1 side	SF			34.14	34.14
08.9102 221	GlasWeld, 1", insulated, finished 1 side, dual	SF			44.38	44.38
08.9102 231	Mirawal, 5/16", finished 1 side, dual	SF			34.79	34.79
08.9102 241	Mirawal, 1", insulated, finished 1 side, dual	SF			45.60	45.60
08.9102 251	Aluca bond, glazed installation	SF			50.08	50.08
08.9102 261	Add for sloped glazing 30%					
08.9102 271	Add for butt glazing 30%					
08.9103 000	**PREFINISHED & INSULATED BUILDING PANELS WITH SUPPORT FRAME:**					
08.9103 011	Aluca bond	SF			103.40	103.40
08.9104 000	**PREFINISHED & INSULATED BUILDING PANELS, STRUCTURAL STUDS:**					
08.9104 001	*Note: The following prices do not include structural studs unless otherwise noted. For studs see section 09.1100. For sash see section 08.9106. The GlasWeld item below is installed over gwb or ply sheathing. The price does not include sheathing but does include adhesive and matching trim.*					
08.9104 011	Aluca bond	SF			54.88	54.88
08.9104 021	Dryvit, with studs & 3" insulation	SF			33.30	33.30
08.9104 031	GlasWeld, 1/4" finished 1 side	SF			23.92	23.92
08.9105 000	**PRECAST CONCRETE & FIBERGLASS REINFORCED CONCRETE PANELS:**					
08.9105 001	*Note: The following prices are for furnished and installed panels including fasteners. Structural steel supports are not included. For structural steel see section 05.1000. For sash see section 08.9106.*					
08.9105 011	Precast panel, 1 form, standard finish	SF			40.29	40.29
08.9105 021	Precast panel, 2 form, standard finish	SF			53.88	53.88
08.9105 031	Precast panel, 1 form, sandblast	SF			64.95	64.95
08.9105 041	Precast panel, 2 form, sandblast	SF			71.62	71.62
08.9105 051	Precast panel, 1 form, mo-sai finish	SF			56.09	56.09
08.9105 061	Precast panel, 1 form, granite overlay	SF			82.71	82.71
08.9105 071	GFRC panel, 1 form, standard finish	SF			46.00	46.00
08.9105 081	GFRC panel, 2 form, standard finish	SF			53.83	53.83
08.9105 091	GFRC panel, 1 form, sandblast	SF			66.04	66.04

DOORS, WINDOWS & GLASS - 08

Division	Description	Unit	Labor Hours	Labor Cost	Material Cost	Total Cost
08.9105 101	GFRC panel, 2 form, sandblast	SF			70.20	70.20
08.9105 111	GFRC panel, 1 form, mo-sai finish	SF			53.88	53.88
08.9105 121	GFRC panel, 1 form, granite overlay	SF			82.84	82.84
08.9106 000	**SASH, METAL, GLAZED, IN EXTERIOR PANEL SYSTEMS:**					
08.9106 001	*Note: The following items are to be used with all systems except curtain walls.*					
08.9106 011	Sash, steel, fixed, 1/4" clear	SF	0.1097	10.70	19.10	29.80
08.9106 021	Sash, steel, fixed, 1/4" tint	SF	0.1097	10.70	20.39	31.09
08.9106 031	Sash, steel, fixed, 1/4" reflective	SF	0.1097	10.70	23.56	34.25
08.9106 041	Sash, steel, 50% vent, 1/4" clear	SF	0.1097	10.70	22.79	33.49
08.9106 051	Sash, steel, 50% vent, 1/4" tint	SF	0.1097	10.70	24.06	34.76
08.9106 061	Sash, steel, 50% vent, 1/4" reflective	SF	0.1097	10.70	27.25	37.94
08.9106 071	Sash, steel, 100% vent, 1/4" clear	SF	0.1097	10.70	25.94	36.64
08.9106 081	Sash, steel, 100% vent, 1/4" tint	SF	0.1097	10.70	27.21	37.91
08.9106 091	Sash, steel, 100% vent, 1/4" reflective	SF	0.1097	10.70	30.38	41.07
08.9106 101	Sash, aluminum, fixed, 1/4" clear	SF	0.1097	10.70	19.63	30.32
08.9106 111	Sash, aluminum, fixed, 1/4" tint	SF	0.1097	10.70	20.89	31.59
08.9106 121	Sash, aluminum, fixed, 1/4" reflective	SF	0.1097	10.70	24.13	34.83
08.9106 131	Sash, aluminum, 50% vent, 1/4" clear	SF	0.1161	11.32	25.57	36.89
08.9106 141	Sash, aluminum, 50% vent, 1/4" tint	SF	0.1161	11.32	26.87	38.19
08.9106 151	Sash, aluminum, 50% vent, 1/4" reflective	SF	0.1161	11.32	30.08	41.40
08.9106 161	Sash, aluminum, 100% vent, 1/4" clear	SF	0.1161	11.32	30.86	42.18
08.9106 171	Sash, aluminum, 100% vent, 1/4" tint	SF	0.1161	11.32	32.14	43.46
08.9106 181	Sash, aluminum, 100% vent, 1/4" reflective	SF	0.1161	11.32	35.29	46.61

09 - FINISHES

Division	Description	Unit	Labor Hours	Labor Cost	Material Cost	Total Cost
09.1000 000	**LATH, PLASTER, STUDDING & FURRING:**					
09.1000 001	Note: Remember to add 5% to all material for laps and waste. The studs listed below are spaced 16" on center unless otherwise noted. Prices include top and bottom track and bridging. For drywall studs see section 09.2001. In plaster and stucco, scaffolding is assumed to be provided by the subcontractor up to 12', by the general contractor above 12'. Planks are assumed to be provided by the subcontractor. Note that stucco prices may vary greatly if configurations are irregular. For areas under 1,000 sf . . . . . . add 20% to the material costs For galvanized studs . . . . add 5% to the material costs For painted sfs studs . . . . . deduct 15% from the material costs For 24" spacing . . . . deduct 20% from the material costs					
09.1100 000	**STUDS:**					
09.1101 000	**STUDS, STANDARD, 'C' STRUCTURAL, 1-5/8", 16"OC:**					
09.1101 011	Studs, 'C' structural, 1-5/8", 16"OC, 14 ga, 2-1/2"	SF	0.0475	4.26	2.00	6.26
09.1101 021	Studs, 'C' structural, 1-5/8", 16"OC, 14 ga, 3-5/8"	SF	0.0500	4.48	2.62	7.10
09.1101 031	Studs, 'C' structural, 1-5/8", 16"OC, 14 ga, 4"	SF	0.0534	4.78	2.74	7.52
09.1101 041	Studs, 'C' structural, 1-5/8", 16"OC, 14 ga, 6"	SF	0.0561	5.03	3.42	8.45
09.1101 051	Studs, 'C' structural, 1-5/8", 16"OC, 14 ga, 8"	SF	0.0615	5.51	4.17	9.68
09.1101 111	Studs, 'C' structural, 1-5/8", 16"OC, 16 ga, 2-1/2"	SF	0.0475	4.26	1.75	6.00
09.1101 121	Studs, 'C' structural, 1-5/8", 16"OC, 16 ga, 3-5/8"	SF	0.0500	4.48	2.00	6.48
09.1101 131	Studs, 'C' structural, 1-5/8", 16"OC, 16 ga, 4"	SF	0.0534	4.78	2.10	6.89
09.1101 141	Studs, 'C' structural, 1-5/8", 16"OC, 16 ga, 6"	SF	0.0561	5.03	2.64	7.66
09.1101 151	Studs, 'C' structural, 1-5/8", 16"OC, 16 ga, 8"	SF	0.0615	5.51	3.12	8.63
09.1101 211	Studs, 'C' structural, 1-5/8", 16"OC, 18 ga, 2-1/2"	SF	0.0475	4.26	1.46	5.71
09.1101 221	Studs, 'C' structural, 1-5/8", 16"OC, 18 ga, 3-5/8"	SF	0.0500	4.48	1.72	6.20
09.1101 231	Studs, 'C' structural, 1-5/8", 16"OC, 18 ga, 4"	SF	0.0534	4.78	1.82	6.60
09.1101 241	Studs, 'C' structural, 1-5/8", 16"OC, 18 ga, 6"	SF	0.0561	5.03	2.22	7.25
09.1101 251	Studs, 'C' structural, 1-5/8", 16"OC, 18 ga, 8"	SF	0.0615	5.51	2.78	8.28
09.1101 311	Studs, 'C' structural, 1-5/8", 16"OC, 20 ga, 2-1/2"	SF	0.0475	4.26	1.49	5.75
09.1101 321	Studs, 'C' structural, 1-5/8", 16"OC, 20 ga, 3-5/8"	SF	0.0500	4.48	1.63	6.11
09.1101 331	Studs, 'C' structural, 1-5/8", 16"OC, 20 ga, 4"	SF	0.0534	4.78	1.65	6.44
09.1101 341	Studs, 'C' structural, 1-5/8", 16"OC, 20 ga, 6"	SF	0.0561	5.03	1.94	6.97
09.1101 351	Studs, 'C' structural, 1-5/8", 16"OC, 20 ga, 8"	SF	0.0615	5.51	2.15	7.66
09.1102 000	**STUDS, 'C' STRUCTURAL, 1-3/8", 16"OC:**					
09.1102 011	Studs, 'C' structural, 1-3/8", 16"OC, 14 ga, 2-1/2"	SF	0.0475	4.26	2.02	6.28
09.1102 021	Studs, 'C' structural, 1-3/8", 16"OC, 14 ga, 3-5/8"	SF	0.0500	4.48	2.43	6.91
09.1102 031	Studs, 'C' structural, 1-3/8", 16"OC, 14 ga, 4"	SF	0.0534	4.78	2.59	7.37
09.1102 041	Studs, 'C' structural, 1-3/8", 16"OC, 14 ga, 6"	SF	0.0561	5.03	3.24	8.26
09.1102 051	Studs, 'C' structural, 1-3/8", 16"OC, 14 ga, 8"	SF	0.0615	5.51	3.89	9.39
09.1102 111	Studs, 'C' structural, 1-3/8", 16"OC, 16 ga, 2-1/2"	SF	0.0475	4.26	1.60	5.85
09.1102 121	Studs, 'C' structural, 1-3/8", 16"OC, 16 ga, 3-5/8"	SF	0.0500	4.48	1.92	6.40
09.1102 131	Studs, 'C' structural, 1-3/8", 16"OC, 16 ga, 4"	SF	0.0534	4.78	1.98	6.76
09.1102 141	Studs, 'C' structural, 1-3/8", 16"OC, 16 ga, 6"	SF	0.0561	5.03	2.45	7.48
09.1102 151	Studs, 'C' structural, 1-3/8", 16"OC, 16 ga, 8"	SF	0.0615	5.51	3.02	8.53
09.1102 211	Studs, 'C' structural, 1-3/8", 16"OC, 18 ga, 2-1/2"	SF	0.0475	4.26	1.34	5.60
09.1102 221	Studs, 'C' structural, 1-3/8", 16"OC, 18 ga, 3-5/8"	SF	0.0500	4.48	1.62	6.10
09.1102 231	Studs, 'C' structural, 1-3/8", 16"OC, 18 ga, 4"	SF	0.0534	4.78	1.72	6.51
09.1102 241	Studs, 'C' structural, 1-3/8", 16"OC, 18 ga, 6"	SF	0.0561	5.03	2.13	7.15
09.1102 251	Studs, 'C' structural, 1-3/8", 16"OC, 18 ga, 8"	SF	0.0615	5.51	2.54	8.05
09.1103 000	**STUDS OR TRACK, STANDARD FLANGE STRUCTURAL, 1-1/4", 16"OC:**					
09.1103 011	Studs, channel, 1", 16"OC, 14 ga, 2-1/2"	SF	0.0475	4.26	1.67	5.92
09.1103 021	Studs, channel, 1", 16"OC, 14 ga, 3-5/8"	SF	0.0500	4.48	1.91	6.39
09.1103 031	Studs, channel, 1", 16"OC, 14 ga, 4"	SF	0.0534	4.78	2.02	6.81
09.1103 041	Studs, channel, 1", 16"OC, 14 ga, 6"	SF	0.0561	5.03	2.74	7.77
09.1103 051	Studs, channel, 1", 16"OC, 14 ga, 8"	SF	0.0615	5.51	3.41	8.92
09.1103 061	Studs, channel, 1", 16"OC, 16 ga, 2-1/2"	SF	0.0475	4.26	1.18	5.43
09.1103 071	Studs, channel, 1", 16"OC, 16 ga, 3-5/8"	SF	0.0500	4.48	1.49	5.97

FINISHES - 09

Division	Description	Unit	Labor Hours	Labor Cost	Material Cost	Total Cost
09.1103 081	Studs, channel, 1", 16"OC, 16 ga, 4"	SF	0.0534	4.78	1.60	6.38
09.1103 091	Studs, channel, 1", 16"OC, 16 ga, 6"	SF	0.0561	5.03	2.10	7.13
09.1103 101	Studs, channel, 1", 16"OC, 16 ga, 8"	SF	0.0615	5.51	2.64	8.15
09.1103 111	Studs, channel, 1", 16"OC, 18 ga, 2-1/2"	SF	0.0475	4.26	1.03	5.28
09.1103 121	Studs, channel, 1", 16"OC, 18 ga, 3-5/8"	SF	0.0500	4.48	1.25	5.73
09.1103 131	Studs, channel, 1", 16"OC, 18 ga, 4"	SF	0.0534	4.78	1.34	6.12
09.1103 141	Studs, channel, 1", 16"OC, 18 ga, 6"	SF	0.0561	5.03	1.82	6.84
09.1103 151	Studs, channel, 1", 16"OC, 18 ga, 8"	SF	0.0615	5.51	2.22	7.73
09.1103 161	Studs, channel, 1", 16"OC, 20 ga, 2-1/2"	SF	0.0475	4.26	1.02	5.27
09.1103 171	Studs, channel, 1", 16"OC, 20 ga, 3-5/8"	SF	0.0500	4.48	1.20	5.68
09.1103 181	Studs, channel, 1", 16"OC, 20 ga, 4"	SF	0.0534	4.78	1.21	6.00
09.1103 191	Studs, channel, 1", 16"OC, 20 ga, 6"	SF	0.0561	5.03	1.56	6.59
09.1103 201	Studs, channel, 1", 16"OC, 20 ga, 8"	SF	0.0615	5.51	1.83	7.34
09.1104 000	**SHAFTWALL, STUDS, CH, 24" OC:**					
09.1104 001	*Note: Costs include finished gypsum wall board*					
09.1104 011	Shaftwall, studs, CH, 24"OC, 20 ga, 2-1/2"	SF	0.0475	4.26	2.10	6.36
09.1104 021	Shaftwall, studs, CH, 24"OC, 20 ga, 4"	SF	0.0534	4.78	2.15	6.93
09.1104 031	Shaftwall, studs, CH, 24"OC, 20 ga, 6"	SF	0.0561	5.03	2.74	7.77
09.1104 041	Shaftwall, studs, CH, 24"OC, 25 ga, 2-1/2"	SF	0.0475	4.26	1.27	5.53
09.1104 051	Shaftwall, studs, CH, 24"OC, 25 ga, 4"	SF	0.0534	4.78	1.45	6.23
09.1104 061	Shaftwall, studs, CH, 24"OC, 25 ga, 6"	SF	0.0561	5.03	1.83	6.85
09.1200 000	**FURRING:**					
09.1201 000	**FURRING CHANNELS, CEILINGS:**					
09.1201 011	Furring, channels, ceiling, 1-1/2" x 3/4", 16" OC	SF	0.0206	1.85	0.98	2.83
09.1201 021	Hat channel, ceiling, 1-1/2"x7/8", 16" OC	SF	0.0212	1.90	1.09	2.99
09.1201 031	Furring, ceiling, 3-1/4x1-1/2x3/4, triple hung	SF	0.0394	3.53	1.43	4.96
09.1201 041	Furring, ceiling, 1-1/2"x3/4", coffered, double hung	SF	0.0513	4.60	1.33	5.93
09.1201 051	Furring, ceiling, 2 layers gypsum wall board, studs, 25ga	SF	0.0744	6.66	1.82	8.48
09.1201 061	Furring, ceiling, 3 layers gypsum wall board, studs, 25ga	SF	0.1111	9.95	2.46	12.42
09.1201 071	Furring, ceiling, 4 layers gypsum wall board, studs, 25ga	SF	0.1481	13.27	3.17	16.44
09.1202 000	**FURRING CHANNEL, WALL:**					
09.1202 011	Hat channel, 3/4" x 5/8", 16" OC	SF	0.0137	1.23	0.76	1.99
09.1202 021	Furring, channels, wall, 3/4"x 3/4", 16" OC	SF	0.0137	1.23	0.67	1.90
09.1202 031	Furring, hat channel, articulated wall	SF	0.0496	4.44	0.76	5.21
09.1202 041	Furring, wall, 2 layers gypsum wall board, studs, 25ga	SF	0.0744	6.66	1.82	8.48
09.1202 051	Furring, wall, 3 layers gypsum wall board, studs, 25ga	SF	0.1111	9.95	2.46	12.42
09.1202 061	Furring, wall, 4 layers gypsum wall board, studs, 25ga	SF	0.1481	13.27	3.13	16.40
09.1300 000	**LATHING:**					
09.1301 000	**METAL LATH:**					
09.1301 011	Lath, metal, 2.5#, painted, wood frame	SY	0.0833	7.78	3.75	11.53
09.1301 021	Lath, metal, 3.4#, painted, wood frame	SY	0.0833	7.78	4.28	12.06
09.1301 031	Lath, metal, 3.4#, galvanized, wood frame	SY	0.0833	7.78	5.18	12.96
09.1302 000	**WIRE MESH & SPECIAL LATH:**					
09.1302 001	*Note: For wiring to steel frames or screwing to steel studs, add 5% to the total costs. For wiring to steel ceiling frame, add 15% to the total costs.*					
09.1302 011	Mesh, 1" x 18 ga, 15# felt, wire	SY	0.0775	7.24	4.61	11.85
09.1302 021	Mesh, 1-1/2"x17 ga, 15# felt, wire	SY	0.0775	7.24	3.70	10.94
09.1302 031	Lath, stucco rite, 2" x 16 ga	SY	0.0758	7.08	4.17	11.25
09.1302 041	Lath, paperback with felt, mesh, wire	SY	0.0618	5.77	4.11	9.89
09.1302 051	Lath, aqua, 2" x 16 ga	SY	0.0758	7.08	4.46	11.54
09.1302 061	Lath, rock, 3/8", wood frame	SY	0.0653	6.10	3.31	9.41
09.1302 071	Lath, rock, 3/8", steel frame	SY	0.0758	7.08	3.31	10.39
09.1302 081	Lath, gypsum, asphalt coated, 1/2", nailed	SY	0.0549	5.13	4.30	9.43
09.1302 091	Lath, gypsum, lead lined, 4#	SY	0.2901	27.09	71.01	98.10
09.1302 101	Lath, gypsum, lead lined, 2#	SY	0.2901	27.09	56.36	83.45
09.1302 111	Lath, gypsum, lead lined, 1#	SY	0.2901	27.09	56.36	83.45

09 - FINISHES

Division	Description	Unit	Labor Hours	Labor Cost	Material Cost	Total Cost
09.1400 000	**PLASTER & LATH:**					
09.1400 001	Note: Scaffolding is provided by the subcontractor to 12'. Above 12' scaffolding is provided by the general contractor. Planks are supplied by the subcontractor. Remember to add 5% to materials for waste. With irregular configurations, stucco may vary greatly in cost.					
09.1401 000	**PLASTER & LATH, IN PLACE, AVERAGE COSTS:**					
09.1401 011	Plaster & lath, tract, residential	SY	0.2828	26.41	25.72	52.13
09.1401 021	Plaster & lath, condo & apartment 2 story	SY	0.4242	39.61	25.72	65.33
09.1401 031	Plaster & lath, condo & apartment 3 story	SY	0.5000	46.69	25.72	72.41
09.1401 041	Plaster & lath, residential, 3 coat exterior, custom	SY	0.5656	52.82	38.61	91.42
09.1401 051	Plaster & lath, commercial, 3 coat exterior	SY	0.5710	53.32	50.81	104.13
09.1401 061	Plaster & lath, commercial, 3 coat interior	SY	0.6257	58.43	36.23	94.66
09.1401 071	Plaster & lath, commercial, 3 coat exterior soffits	SY	0.6510	60.79	40.53	101.33
09.1401 081	Plaster & lath, partitions	SY	0.8105	75.68	42.57	118.25
09.1402 000	**EXTERIOR STUCCO & PLASTER, WITHOUT LATH OR FURRING:**					
09.1402 001	Note: Because of current propensity for complicated plaster interior and exterior shapes, plaster is given without furring or lath. Both must be added to complete the plaster price. Do not forget shaping of plaster requires extensive plaster accessories and trim.					
09.1402 011	Stucco, residential, textured, soffit, wide & low	SY	0.4414	41.22	23.68	64.90
09.1402 021	Stucco, residential, textured, soffit, high & narrow	SY	0.5736	53.56	24.04	77.60
09.1402 031	Stucco, residential, textured	SY	0.3252	30.37	24.32	54.69
09.1402 041	Stucco, residential, float finish	SY	0.3850	35.95	30.37	66.32
09.1402 051	Stucco, institutional, float finish	SY	0.4187	39.10	46.71	85.81
09.1402 061	Stucco, institutional, dash finish	SY	0.4242	39.61	53.28	92.89
09.1402 071	Stucco, institutional, float finish, soffit, wide & low	SY	0.3500	32.68	30.37	63.05
09.1402 081	Stucco, institutional, float finish, soffit, high & narrow	SY	0.4646	43.38	30.37	73.75
09.1402 091	Stucco, institutional, float finish, articulated, wood	SY	0.8913	83.23	30.37	113.60
09.1402 101	Cement plaster, 2 coat, for paint	SY	0.2588	24.17	15.93	40.09
09.1402 111	Add for dash coat	SY	0.0476	4.44	1.01	5.45
09.1402 121	Stucco, run mold, ogee 6"	LF	0.1575	14.71	21.40	36.11
09.1402 131	Stucco, run mold, ogee 9"	LF	0.2300	21.48	26.37	47.85
09.1402 141	Stucco, run mold, ogee 12"	LF	0.3000	28.01	31.31	59.33
09.1402 151	Stucco, run mold, ogee 15"	LF	0.3300	30.82	38.74	69.55
09.1402 161	Stucco, rustication, 1"	LF	0.0800	7.47	8.77	16.24
09.1402 171	Stucco, rustication, 2"	LF	0.0900	8.40	10.99	19.39
09.1402 181	Stucco, quoins, 12"x12"	LF	0.5436	50.76	30.73	81.49
09.1402 191	Stucco, quoins, 16"x16"	LF	0.4900	45.76	30.73	76.49
09.1403 000	**INTERIOR PLASTER WALLS, WITHOUT LATH OR FURRING:**					
09.1403 001	Note: Because of current propensity for complicated plaster interior and exterior shapes. Plaster is given without furring or lath. Both must be added to complete the plaster price. Do not forget shaping of plaster requires extensive plaster accessories and trim.					
09.1403 011	Plaster, gypsum, interior	SY	0.2718	25.38	8.17	33.55
09.1403 021	Cement plaster, Keenes, interior	SY	0.2718	25.38	9.76	35.14
09.1403 031	Plaster, structo lite, interior	SY	0.2718	25.38	9.35	34.73
09.1403 041	Cement plaster, interior	SY	0.1843	17.21	7.31	24.52
09.1403 051	Stucco, interior, 2 coat, residential	SY	0.1575	14.71	8.37	23.08
09.1403 061	Plaster, acoustic, 1/2" on brown coat	SY	0.3407	31.81	14.41	46.23
09.1403 071	Plaster, scratch coat, for tile	SY	0.1519	14.18	5.25	19.43
09.1403 081	Plaster, brown coat, for tile	SY	0.2171	20.27	7.17	27.44
09.1404 000	**INTERIOR PLASTER CEILINGS, WITHOUT LATH OR FURRING:**					
09.1404 001	Note: Because of current propensity for complicated plaster interior and exterior shapes. Plaster is given without furring or lath. Both must be added to complete the plaster price. Do not forget shaping of plaster requires extensive plaster accessories and trim.					
09.1404 011	Plaster, gypsum, ceilings	SY	0.2718	25.38	7.25	32.63
09.1404 021	Plaster, Keenes cement, ceilings	SY	0.2718	25.38	9.00	34.38
09.1404 031	Plaster, structo lite, ceilings	SY	0.2718	25.38	8.43	33.81

FINISHES - 09

Division	Description	Unit	Labor Hours	Labor Cost	Material Cost	Total Cost
09.1404 041	Plaster, cement, ceilings	SY	0.2718	25.38	6.54	31.92
09.1404 051	Plaster, acoustic, 4 coat, ceilings	SY	0.6126	57.20	17.57	74.78
09.1404 061	Stucco, 2 coat, board lath, ceiling	SY	0.1849	17.27	10.56	27.83
09.1405 000	**PLASTER, THIN COAT, WITHOUT LATH OR FURRING:**					
09.1405 011	Plaster, thin, 1/2" board, wood studs	SY	0.2066	19.29	8.38	27.68
09.1405 021	Plaster, thin, 5/8" board, wood studs	SY	0.2066	19.29	11.10	30.40
09.1405 031	Plaster, thin, 1/2" board, metal stud	SY	0.2346	21.91	9.46	31.37
09.1405 041	Plaster, thin, 5/8" board, metal stud	SY	0.2346	21.91	12.36	34.26
09.1406 000	**INTERIOR THIN COAT PLASTER CEILINGS, WITHOUT LATH, FURRING:**					
09.1406 011	Plaster, thin, ceiling, 1/2" gypsum wall board, nailed	SY	0.2343	25.09	8.38	33.47
09.1406 021	Plaster, thin, ceiling, 5/8" gypsum wall board, nailed	SY	0.2343	25.09	11.10	36.19
09.1406 031	Plaster, thin, ceiling, 1/2" gypsum wall board, screw	SY	0.2604	27.88	9.46	37.34
09.1406 041	Plaster, thin, ceiling, 5/8" gypsum wall board, screw	SY	0.2604	27.88	12.36	40.24
09.1406 051	Add for bonding, concrete or masonry	SF			0.39	0.39
09.1408 000	**PLASTERING ACCESSORIES, MISCELLANEOUS:**					
09.1408 011	Bullnose, 3/4"	LF	0.0459	4.29	0.81	5.10
09.1408 021	Bullnose, 1-1/2", galvanized	LF	0.0456	4.26	2.61	6.87
09.1408 031	Bullnose, 1-1/2", stainless	LF	0.0779	7.27	6.52	13.80
09.1408 041	Casing/stop, square nose, galvanized	LF	0.0320	2.99	0.46	3.45
09.1408 051	Corner bead, short nose	LF	0.0355	3.31	0.45	3.77
09.1408 061	Corner bead, expanded, galvanized, zinc nose	LF	0.0355	3.31	0.80	4.11
09.1408 071	Expansion joint, 3/4", galvanized	LF	0.0355	3.31	1.46	4.77
09.1408 081	Expansion joint, 1-1/2", galvanized	LF	0.0378	3.53	1.97	5.50
09.1408 091	Screed, base	LF	0.0355	3.31	0.36	3.67
09.1408 101	Vent, 1-1/2", galvanized	LF	0.0476	4.44	2.43	6.87
09.1408 111	Vent, 4", galvanized	LF	0.0676	6.31	2.87	9.18
09.1408 121	Drip pacific #2	LF	0.0355	3.31	0.46	3.78
09.1409 000	**ACCESS DOORS:**					
09.1409 001	*ACCESS DOOR, 24"X30":*					
09.1409 011	Access door, 12" x 12"	EA	1.3020	121.58	157.56	279.14
09.1409 021	Access door, 18" x 18"	EA	1.3020	121.58	196.26	317.84
09.1409 031	Access door, 20" x 30"	EA	1.3020	121.58	214.88	336.46
09.1409 041	Access door, 24" x 24"	EA	1.3020	121.58	229.23	350.81
09.1409 051	Access door, 24" x 30"	EA	1.3020	121.58	236.36	357.94
09.1410 000	**PLASTER MOLDINGS & ORNAMENTS, INTERIOR:**					
09.1410 011	Plaster molding, 2"	LF	0.1078	10.07	7.67	17.73
09.1410 021	Plaster molding, 4"	LF	0.1283	11.98	14.27	26.25
09.1410 031	Plaster molding, 6"	LF	0.1502	14.03	17.08	31.10
09.2000 000	**GYPSUM WALL BOARD, STUDDING & FURRING:**					
09.2000 001	*Note: For metal studs see section 09.1000. For wood studs see section 06.0000.*					
09.2001 000	**STUDS, METAL DRYWALL, 16" OC:**					
09.2001 001	*Note: For galvanized studs, add 5% to the material costs. This section, 09.2000, relates to interior partition work not exceeding 12' above floor level.*					
09.2001 011	Drywall stud, 25 ga, 16" OC, 1-5/8"	SF	0.0176	1.88	0.65	2.53
09.2001 021	Drywall stud, 25 ga, 16" OC, 2-1/2"	SF	0.0187	2.00	0.67	2.67
09.2001 031	Drywall stud, 25 ga, 16" OC, 3-5/8"	SF	0.0198	2.12	0.82	2.94
09.2001 041	Drywall stud, 25 ga, 16" OC, 4"	SF	0.0215	2.30	0.90	3.20
09.2001 051	Drywall stud, 25 ga, 16" OC, 6"	SF	0.0226	2.42	1.04	3.46
09.2001 061	Drywall stud, 20 ga, 16" OC, 1-5/8"	SF	0.0197	2.11	1.30	3.40
09.2001 071	Drywall stud, 20 ga, 16" OC, 2-1/2"	SF	0.0197	2.11	1.38	3.49
09.2001 081	Drywall stud, 20 ga, 16" OC, 3-5/8"	SF	0.0218	2.33	1.47	3.80
09.2001 091	Drywall stud, 20 ga, 16" OC, 4"	SF	0.0236	2.53	1.64	4.17
09.2001 101	Drywall stud, 20 ga, 16" OC, 6"	SF	0.0249	2.67	1.92	4.59
09.2004 000	**RESIDENTIAL GYPSUM WALL BOARD, WALLS, HANG, TAPE, TEXTURE:**					
09.2004 001	*Note: All gypsum wall board prices are given for 8' heights. You must add for structural studs and height as shown in section 09.2013.*					
09.2004 011	Gypsum wall board, 1/4", wall, residential, taped & textured	SF	0.0074	0.79	0.48	1.27

09 - FINISHES

Division	Description	Unit	Labor Hours	Labor Cost	Material Cost	Total Cost
09.2004 021	Gypsum wall board, 3/8", wall, residential, taped & textured	SF	0.0074	0.79	0.73	1.52
09.2004 031	Gypsum wall board, 1/2", wall, residential, taped & textured	SF	0.0082	0.88	0.79	1.67
09.2004 041	Gypsum wall board, fire resistant, 1/2", wall, residential, taped & textured	SF	0.0082	0.88	0.91	1.79
09.2004 051	Gypsum wall board, water resistant, 1/2", wall, residential, taped & textured	SF	0.0082	0.88	1.08	1.95
09.2004 061	Gypsum wall board, 5/8", wall, residential, taped & textured	SF	0.0082	0.88	0.87	1.75
09.2004 071	Gypsum wall board, fire resistant, 5/8", wall, residential, taped & textured	SF	0.0082	0.88	1.05	1.93
09.2004 081	Gypsum wall board, water resistant, 5/8", wall, residential, taped & textured	SF	0.0082	0.88	1.13	2.01
09.2004 091	Gypsum wall board, 2 hour, 5/8", wall, residential, taped & textured	SF	0.0140	1.50	1.95	3.44
09.2005 000	**RESIDENTIAL, GYPSUM BOARD, CEILINGS, HANG, TAPE, TEXTURE:**					
09.2005 011	Gypsum wall board, 1/4", ceilings, residential, tape & texture	SF	0.0081	0.87	0.48	1.35
09.2005 021	Gypsum wall board, 3/8", ceilings, residential, tape & texture	SF	0.0081	0.87	0.73	1.59
09.2005 031	Gypsum wall board, 1/2", ceilings, residential, tape & texture	SF	0.0081	0.87	0.79	1.66
09.2005 041	Gypsum wall board, fire resistant, 1/2", ceilings, residential, tape & texture	SF	0.0081	0.87	0.91	1.78
09.2005 051	Gypsum wall board, water resistant, 1/2", ceilings, residential, tape & texture	SF	0.0081	0.87	1.08	1.94
09.2005 061	Gypsum wall board, 5/8", ceilings, residential, tape & texture	SF	0.0083	0.89	0.87	1.76
09.2005 071	Gypsum wall board, fire resistant, 5/8", ceilings, residential, tape & texture	SF	0.0083	0.89	1.05	1.94
09.2005 081	Gypsum wall board, water resistant, 5/8", ceilings, residential, tape & texture	SF	0.0083	0.89	1.13	2.02
09.2005 091	Gypsum wall board, 2 hour, 5/8", ceilings, residential, tape & texture	SF	0.0085	0.91	1.95	2.86
09.2006 000	**RESIDENTIAL GYPSUM BOARD, MISCELLANEOUS:**					
09.2006 011	Gypsum wall board, asphalt core sheathing, 1/2", residential	SF	0.0041	0.44	0.91	1.35
09.2006 021	Gypsum wall board, vinyl clad, good, 5/8", residential	SF	0.0099	1.06	3.27	4.33
09.2006 031	Gypsum wall board, vinyl clad, best, 5/8", residential	SF	0.0099	1.06	3.85	4.91
09.2006 041	Gypsum sound board, 1/4", residential	SF	0.0058	0.62	0.48	1.10
09.2006 051	Fiber sound board, 1/2", residential	SF	0.0058	0.62	0.83	1.45
09.2006 061	Gypsum wall board, partition, 5/8" each side, residential	SF	0.0508	5.44	3.73	9.17
09.2007 000	**COMMERCIAL GYPSUM WALL BOARD, WALLS, HANG, TAPE, TEXTURE:**					
09.2007 001	*Note: All gypsum wall board prices are given in 8' heights. You must add adders under section 09.2013 for both height and structural studs.*					
09.2007 011	Gypsum wall board, 1/4", 8', commercial, wall, tape & texture	SF	0.0115	1.23	0.48	1.71
09.2007 021	Gypsum wall board, 3/8", 8', commercial, wall, tape & texture	SF	0.0123	1.32	0.73	2.04
09.2007 031	Gypsum wall board, 1/2", 8', commercial, wall, tape & texture	SF	0.0140	1.50	0.79	2.29
09.2007 041	Gypsum wall board, fire resistant, 1/2", 8', commercial, wall, tape & texture	SF	0.0140	1.50	0.91	2.41
09.2007 051	Gypsum wall board, water resistant, 1/2", 8', commercial, wall, tape & texture	SF	0.0140	1.50	1.08	2.58
09.2007 061	Gypsum wall board, 5/8", 8', commercial, wall, tape & texture	SF	0.0140	1.50	0.87	2.37
09.2007 071	Gypsum wall board, fire resistant, 5/8", 8', commercial, wall, tape & texture	SF	0.0140	1.50	1.05	2.55
09.2007 081	Gypsum wall board, water resistant, 5/8", 8', commercial, wall, tape & texture	SF	0.0140	1.50	1.13	2.63
09.2007 091	Gypsum wall board, 2 hour, 5/8", 8', commercial, wall, tape & texture	SF	0.0188	2.01	1.95	3.96
09.2008 000	**COMMERCIAL GYPSUM BOARD, CEILINGS, HANG, TAPE, TEXTURE:**					
09.2008 011	Gypsum wall board, 1/4", ceilings, commercial, tape & texture	SF	0.0121	1.30	0.48	1.78
09.2008 021	Gypsum wall board, 3/8", ceilings, commercial, tape & texture	SF	0.0130	1.39	0.73	2.12
09.2008 031	Gypsum wall board, 1/2", ceilings, commercial, tape & texture	SF	0.0149	1.60	0.79	2.39
09.2008 041	Gypsum wall board, fire resistant, 1/2", ceilings, commercial	SF	0.0149	1.60	0.91	2.50
09.2008 051	Gypsum wall board, water resistant, 1/2", ceilings, commercial	SF	0.0149	1.60	1.08	2.67
09.2008 061	Gypsum wall board, 5/8", ceilings, commercial	SF	0.0149	1.60	0.87	2.46
09.2008 071	Gypsum wall board, fire resistant, 5/8", ceilings, commercial	SF	0.0149	1.60	1.05	2.65
09.2008 081	Gypsum wall board, water resistant, 5/8", ceilings, commercial	SF	0.0149	1.60	1.13	2.72
09.2008 091	Gypsum wall board, 2 hour, 5/8", ceilings, commercial	SF	0.0196	2.10	1.95	4.04
09.2009 000	**COMMERCIAL GYPSUM BOARD, MISCELLANEOUS:**					
09.2009 001	*Note: All gypsum wall board prices are given in 8' heights. You must add adders under section 09.2013 for both height and structural studs.*					
09.2009 011	Gypsum wall board, 1/2" asphalt core sheath, commercial	SF	0.0066	0.71	0.91	1.61
09.2009 021	Gypsum wall board, vinyl clad, good, 5/8", commercial	SF	0.0147	1.57	3.27	4.84
09.2009 031	Gypsum wall board, vinyl clad, best, 5/8", commercial	SF	0.0173	1.85	3.85	5.70
09.2009 041	Gypsum sound board, 1/4", commercial	SF	0.0066	0.71	0.48	1.19
09.2009 051	Fiber sound board, 1/2", commercial	SF	0.0066	0.71	0.83	1.54
09.2009 061	Gypsum wall board, partition, 5/8", each side, commercial	SF	0.0557	5.96	3.73	9.70

FINISHES - 09

Division	Description	Unit	Labor Hours	Labor Cost	Material Cost	Total Cost
09.2010 000	**INSTITUTIONAL GYPSUM BOARD, WALL, HANG, TAPE, TEXTURE:**					
09.2010 001	*Note: All gypsum wall board prices are given in 8' heights. You must add adders under section 09.2013 for both height and structural studs.*					
09.2010 011	Gypsum wall board, 1/4", 8', institutional, wall, tape & texture	SF	0.0156	1.67	0.57	2.24
09.2010 021	Gypsum wall board, 3/8", 8', institutional, wall, tape & texture	SF	0.0156	1.67	0.79	2.46
09.2010 031	Gypsum wall board, 1/2", 8', institutional, wall, tape & texture	SF	0.0164	1.76	0.84	2.60
09.2010 041	Gypsum wall board, fire resistant, 1/2", 8', institutional, wall, tape & texture	SF	0.0164	1.76	1.04	2.79
09.2010 051	Gypsum wall board, water resistant, 1/2", 8', institutional, wall, tape & texture	SF	0.0164	1.76	1.12	2.87
09.2010 061	Gypsum wall board, 5/8", 8', institutional, wall, tape & texture	SF	0.0164	1.76	1.04	2.79
09.2010 071	Gypsum wall board, fire resistant, 5/8", 8', institutional, wall, tape & texture	SF	0.0164	1.76	1.12	2.87
09.2010 081	Gypsum wall board, water resistant, 5/8", 8', institutional, wall, tape & texture	SF	0.0164	1.76	1.27	3.03
09.2010 091	Gypsum wall board, 2 hour, 5/8", 8', institutional, wall, tape & texture	SF	0.0197	2.11	2.14	4.25
09.2010 101	Gypsum wall board, 2 hour, shaft liner, 1", institutional	SF	0.0068	0.73	3.20	3.93
09.2011 000	**INSTITUTIONAL GYPSUM BOARD, CEILINGS, HANG, TAPE TEXTURE:**					
09.2011 011	Gypsum wall board, 1/4", ceilings, institutional, tape & texture	SF	0.0168	1.80	0.57	2.37
09.2011 021	Gypsum wall board, 3/8", ceilings, institutional, tape & texture	SF	0.0168	1.80	0.79	2.59
09.2011 031	Gypsum wall board, 1/2", ceilings, institutional, tape & texture	SF	0.0177	1.90	0.84	2.74
09.2011 041	Gypsum wall board, fire X, 1/2", ceilings, institutional, tape & texture	SF	0.0177	1.90	1.04	2.93
09.2011 051	Gypsum wall board, water resistant, 1/2", ceilings, institutional, tape & texture	SF	0.0177	1.90	1.12	3.01
09.2011 061	Gypsum wall board, 5/8", ceilings, institutional, tape & texture	SF	0.0177	1.90	1.04	2.93
09.2011 071	Gypsum wall board, fire X, 5/8", ceilings, institutional, tape & texture	SF	0.0177	1.90	1.12	3.01
09.2011 081	Gypsum wall board, water resistant, 5/8", ceilings, institutional, tape & texture	SF	0.0177	1.90	1.27	3.17
09.2011 091	Gypsum wall board, 2 hour, 5/8", ceilings, institutional, tape & texture	SF	0.0205	2.19	2.14	4.33
09.2012 000	**INSTITUTIONAL GYPSUM BOARD, MISCELLANEOUS:**					
09.2012 011	Gypsum wall board, asphalt core sheath, 1/2", institutional	SF	0.0082	0.88	0.91	1.79
09.2012 021	Gypsum wall board, vinyl clad, good, 5/8", institutional	SF	0.0197	2.11	3.27	5.38
09.2012 031	Gypsum wall board, vinyl clad, best, 5/8", institutional	SF	0.0197	2.11	3.85	5.96
09.2012 041	Gypsum sound board, 1/4", institutional	SF	0.0066	0.71	0.48	1.19
09.2012 051	Fiber sound board, 1/2", institutional	SF	0.0066	0.71	0.83	1.54
09.2012 061	Gypsum wall board, partition, 5/8", each side, institutional	SF	0.0614	6.57	3.73	10.31
09.2013 000	**ADDERS FOR ALL GRADES OF GYPSUM WALL BOARD:**					
09.2013 011	Add for fire rated back taping	SF	0.0033	0.35	0.11	0.47
09.2013 021	Add for 10' wall	SF	0.0017	0.18	0.11	0.29
09.2013 031	Add for 12' wall	SF	0.0033	0.35	0.11	0.47
09.2013 041	Add for 16' wall	SF	0.0041	0.44	0.11	0.55
09.2013 051	Add for application to metal studs, 25 ga	SF	0.0019	0.20	0.11	0.32
09.2013 061	Add for application to metal studs, 20 ga	SF	0.0027	0.29	0.11	0.40
09.2013 071	Add for application to metal studs, 16 ga	SF	0.0041	0.44	0.15	0.59
09.2013 081	Add for resilient clip system	SF	0.0099	1.06	0.57	1.63
09.2100 000	**GYPSUM WALL BOARD SPECIALTIES:**					
09.2101 000	**GYPSUM BOARD, ACOUSTICAL TEXTURES:**					
09.2101 011	Gypsum wall board, simulated acoustic sprayed	SF	0.0024	0.26	0.22	0.48
09.2102 000	**GYPSUM BOARD, TAPING & TEXTURING ONLY:**					
09.2102 011	Gypsum wall board, taping	SF	0.0033	0.35	0.19	0.55
09.2102 021	Gypsum wall board, taping and sanding	SF	0.0058	0.62	0.19	0.82
09.2102 031	Gypsum wall board, tape & texture, residential	SF	0.0033	0.35	0.19	0.55
09.2102 041	Gypsum wall board, tape & sand, institutional	SF	0.0066	0.71	0.19	0.90
09.2102 051	Gypsum wall board, tape & texture, 15' or more	SF	0.0066	0.71	0.19	0.90
09.2102 061	Add for sand finish, no texture	SF	0.0033	0.35		0.35
09.2103 000	**GYPSUM WALL BOARD TRIM:**					
09.2103 011	Gypsum wall board, trim, corner beads	LF	0.0147	1.57	0.41	1.99
09.2103 021	Gypsum wall board, trim, stop/casing	LF	0.0222	2.38	0.73	3.10
09.2103 031	Gypsum wall board, trim, jamb casing	LF	0.0229	2.45	0.84	3.29
09.2104 000	**DRAFTSTOP:**					
09.2104 011	Draftstop, gypsum wall board, 5/8", taped only	SF	0.0110	1.18	1.05	2.23

09 - FINISHES

Division	Description	Unit	Labor Hours	Labor Cost	Material Cost	Total Cost
09.3000 000	**CERAMIC TILE:**					
09.3000 001	*Note: For brown and scratch coat see section 09.1403. For institutional applications, material costs may increase from 60 to 200%, depending on specification of the tile. The following prices reflect standard utility grade materials, standard colors and size from manufacturer's palettes.*					
09.3001 000	**CERAMIC TILE, RESIDENTIAL:**					
09.3001 001	*Note: For single family custom residence, add 60% to the material costs.*					
09.3001 011	Tile, ceramic, residential, shower & tub, mastic	SF	0.0659	5.45	6.98	12.43
09.3001 021	Tile, ceramic, residential, shower & tub, mortar	SF	0.0811	6.71	7.67	14.38
09.3001 031	Tile, ceramic, residential, counter & splash, mastic	SF	0.3259	26.96	11.87	38.83
09.3001 041	Tile, ceramic, residential, counter & splash, mortar	SF	0.4330	35.82	17.83	53.65
09.3001 051	Tile, ceramic, residential, wall hung, counter, splash, mastic	SF	0.8377	69.30	43.18	112.48
09.3001 061	Tile, ceramic, residential, wall hung, counter, splash, mortar	SF	1.0435	86.33	54.61	140.94
09.3002 000	**CERAMIC TILE, FLOORS, COMMERCIAL & INSTITUTIONAL:**					
09.3002 011	Tile, ceramic, commercial & institutional, floor, unglazed, 1"x1", mortar	SF	0.0912	7.54	6.67	14.21
09.3002 021	Tile, ceramic, commercial & institutional, floor, glazed, 1" x 1", mortar	SF	0.0992	8.21	6.72	14.93
09.3002 031	Tile, ceramic, commercial & institutional, floor, glazed, 4" x 4", mortar	SF	0.0899	7.44	6.51	13.95
09.3002 041	Tile, ceramic, commercial & institutional, floor, glazed, 4" x 4", mastic	SF	0.0547	4.53	6.34	10.86
09.3002 051	Tile, ceramic, commercial & institutional, bull nose base, 4", mortar	LF	0.0650	5.38	8.66	14.03
09.3002 061	Tile, ceramic, commercial & institutional, bull nose base, 4", mastic	LF	0.0417	3.45	7.53	10.98
09.3003 000	**CERAMIC TILE, WALL, COMMERCIAL & INSTITUTIONAL:**					
09.3003 011	Tile, ceramic, commercial & institutional, wall, 1" x 1", mortar	SF	0.1271	10.51	6.83	17.34
09.3003 021	Tile, ceramic, commercial & institutional, wall, 1" x 1", mastic	SF	0.0659	5.45	6.72	12.18
09.3003 031	Tile, ceramic, commercial & institutional, wall, 2" x 1", mortar	SF	0.1045	8.65	6.47	15.11
09.3003 041	Tile, ceramic, commercial & institutional, wall, 2" x 1", mastic	SF	0.0585	4.84	6.31	11.15
09.3003 051	Tile, ceramic, commercial & institutional, wall, 4" x 4", mortar	SF	0.1019	8.43	6.02	14.45
09.3003 061	Tile, ceramic, commercial & institutional, wall, 4" x 4", mastic	SF	0.0585	4.84	5.92	10.76
09.3003 071	Tile, ceramic, commercial & institutional, wall, 6" x 3", mortar	SF	0.0992	8.21	6.47	14.67
09.3003 081	Tile, ceramic, commercial & institutional, wall, 6" x 3", mastic	SF	0.0505	4.18	6.31	10.49
09.3003 091	Tile, ceramic, commercial & institutional, wall, 6" x 4", mortar	SF	0.0912	7.54	6.31	13.86
09.3003 101	Tile, ceramic, commercial & institutional, wall, 6" x 4", mastic	SF	0.0487	4.03	6.31	10.34
09.3004 000	**TILE ACCESSORIES:**					
09.3004 011	Soap & grab, mortar	EA	0.2006	19.17	15.64	34.81
09.3004 021	Soap & grab, mastic	EA	0.1463	13.98	15.02	29.00
09.3004 031	Paper holder, mortar	EA	0.2169	17.94	15.64	33.58
09.3004 041	Paper holder, mastic	EA	0.1596	13.20	15.02	28.22
09.3004 051	Epoxy grout, 1/16" joint	SF	0.0100	0.96	1.43	2.38
09.4000 000	**TERRAZZO:**					
09.4001 000	**STANDARD TERRAZZO, MUD:**					
09.4001 001	*Note: The following prices are based on amounts of 3200 square feet or more. For quantities between 200 & 1600 SF .... add 100% to labor cost. For quantities between 1600 & 3200 SF .... add 25% to labor cost*					
09.4001 011	Terrazzo floor, mud, bonded, 2", bed only	SF			6.04	6.04
09.4001 021	Terrazzo floor, mud, bonded, 2", conductive	SF	0.1297	12.40	7.71	20.11
09.4001 031	Terrazzo floor, mud, bonded, 2", decorative	SF	0.1401	13.39	10.51	23.90
09.4001 041	Terrazzo floor, mud, bonded, 3", 15#, with felt	SF	0.1349	12.89	6.59	19.48
09.4001 051	Terrazzo floor, mono, 3-1/2", with mesh	SF	0.1314	12.56	7.13	19.69
09.4001 061	Terrazzo wainscot, mud	SF	0.2697	25.78	10.51	36.29
09.4002 000	**ADDERS FOR TERRAZZO FLOORS:**					
09.4002 011	Add for white cement grout	SF			0.44	0.44
09.4002 021	Add for non-slip abrasive, heavy	SF			0.95	0.95
09.4002 031	Add for non-slip abrasive, light	SF			0.40	0.40
09.4002 041	Add for stairs, mud set	LF			8.62	8.62
09.4002 051	Add for countertops, mud set	LF			30.52	30.52
09.4002 061	Add for cove base, mud set	LF			14.66	14.66
09.4003 000	**THINSET TERRAZZO:**					
09.4003 001	*Note: The following prices are based on amounts of 1000 square feet or more.*					

FINISHES - 09

Division	Description	Unit	Labor Hours	Labor Cost	Material Cost	Total Cost
09.4003 011	Floor, polyester, 3/8", polished	SF	0.1238	11.83	6.81	18.64
09.4003 021	Floor, epoxy, conductive, 3/8", polish	SF	0.1146	10.95	7.34	18.30
09.4003 031	Floor, polyester, 3/8", unpolished	SF	0.0798	7.63	4.33	11.96
09.4003 041	Floor, polyester, 3/8", industrial	SF	0.0717	6.85	3.95	10.80
09.4003 051	Floor, latex, polished, 3/8"	SF	0.1515	14.48	8.34	22.82
09.4003 061	Floor, neoprene, 3/8", polished	SF	0.1463	13.98	8.14	22.12
09.4003 071	Floor, epoxy, 3/8", chemical resistant, polish	SF	0.1208	11.55	6.70	18.25
09.4003 081	Base, neoprene, 6"	LF	0.1433	13.70	7.98	21.68
09.4003 091	Wainscot, neoprene	SF	0.1525	14.58	8.47	23.04
09.4003 101	Add for waterproof membrane	SF	0.0236	2.26	1.16	3.41
09.4004 000	**TERRAZZO ACCESSORIES:**					
09.4004 011	Divider strip, brass, 12 ga	LF	0.0053	0.51	4.80	5.31
09.4004 021	Divider strip, white metal, 12 ga	LF	0.0053	0.51	1.90	2.41
09.4004 031	Divider strip, brass, 4' OC, each way	SF	0.0030	0.29	2.68	2.97
09.4004 041	Divider strip, brass, 2' OC, each way	SF	0.0056	0.54	5.08	5.62
09.5000 000	**ACOUSTIC TREATMENT:**					
09.5000 001	*Note: The following prices are based on amounts of 5000 square feet or more. Prices are based on non-rated ceilings unless otherwise noted. For quantities between 2500 and 5000 square feet .... add 50% For quantities under 2500 square feet .... add 80%*					
09.5001 000	**ACOUSTIC ADDERS:**					
09.5001 011	Add for earthquake sway bracing	SF			0.24	0.24
09.5002 000	**CEILING SYSTEM, SUSPENDED T BAR, 5/8" BD, STC 40:**					
09.5002 011	T-bar, 2x2, 5/8" board, 55 NRC, STC 40	SF	0.0168	1.81	2.21	4.03
09.5002 021	T-bar, 2x4, 5/8" board, 55 NRC, STC 40	SF	0.0121	1.31	1.95	3.26
09.5002 031	T-bar, 4x4, 5/8" board, 55 NRC, STC 40	SF	0.0115	1.24	2.33	3.57
09.5002 041	T-bar, 2x2, each room divided	SF	0.0270	2.91	2.21	5.13
09.5002 051	T-bar, 2x4, each room divided	SF	0.0254	2.74	1.95	4.69
09.5002 061	T-bar, 4x4, each room divided	SF	0.0267	2.88	2.33	5.21
09.5002 071	Deduct for no STC rating	SF			0.11	0.11
09.5002 081	Ceiling, suspended, coffered, 4x5, with fixture	SF	0.0451	4.87	6.42	11.29
09.5003 000	**SUSPENDED CEILING, DIRECT HUNG, CONCEALED SPLINE:**					
09.5003 011	T-bar, 1x1, mineral board, direct hung	SF	0.0270	2.91	4.06	6.97
09.5003 021	T-bar, 2x2, mineral board, direct hung	SF	0.0254	2.74	6.36	9.10
09.5003 031	T-bar, 1x1, 6# fiberboard, direct hung	SF	0.0328	3.54	5.74	9.27
09.5003 041	T-bar, 1x1, metal pan tile, direct hung	SF	0.0352	3.80	7.10	10.90
09.5003 051	T-bar, 1x1, aluminum pan tile direct hung	SF	0.0311	3.36	10.49	13.84
09.5003 061	T-bar, 1x1, vinyl diaphragm, direct hung	SF	0.0304	3.28	5.37	8.65
09.5003 071	T-bar, 2x2, tile, pop-out, direct hung	SF	0.0304	3.28	5.37	8.65
09.5004 000	**TILE, WOOD FIBER, RANDOM PERFORATION, ADHESIVE:**					
09.5004 011	Tile, wood fiber, 1/2"x12"x12"	SF	0.0112	1.21	1.82	3.02
09.5004 021	Tile, wood fiber, 1/2"x12"x24"	SF	0.0112	1.21	1.70	2.91
09.5004 031	Tile, wood fiber, 3/4"x12"x12"	SF	0.0112	1.21	2.35	3.56
09.5004 041	Tile, wood fiber, 1/2"x12"x12", fissured	SF	0.0112	1.21	1.82	3.02
09.5004 051	Tile, wood fiber, 3/4"x12"x12", fissured	SF	0.0112	1.21	2.60	3.81
09.5004 061	Add for fireproofing	SF			0.14	0.14
09.5005 000	**TILE, MINERAL FIBER, FISSURED, ADHESIVE:**					
09.5005 011	Tile, mineral, 1/2"x12"x12", fissured	SF	0.0112	1.21	2.77	3.98
09.5005 021	Tile, mineral, 5/8"x12"x12", fissured	SF	0.0112	1.21	3.22	4.43
09.5005 031	Tile, mineral, 3/4"x12"x12", fissured	SF	0.0112	1.21	3.30	4.51
09.5006 000	**TILE, VINYL DIAPHRAGM, WASHABLE, ADHESIVE:**					
09.5006 011	Tile, vinyl, 1/2x12x12, spray paint	SF	0.0140	1.51	2.60	4.11
09.5006 021	Tile, vinyl, 1/2"x12"x12", bonded	SF	0.0140	1.51	3.30	4.81
09.5006 031	Add for 1 hour fire rating	SF			0.35	0.35
09.5006 041	Add for stapling tile to furring	SF	0.0025	0.27	0.01	0.28
09.5006 051	Add for plastic overlay, acoustic tile	SF			0.15	0.15
09.5006 061	Add for enamel paint, acoustic tile	SF			0.81	0.81

09 - FINISHES

Division	Description	Unit	Labor Hours	Labor Cost	Material Cost	Total Cost
09.5006 071	Add for colored steel grid	SF			0.51	0.51
09.5006 081	Add for colored aluminum grid	SF			1.22	1.22
09.5007 000	**BOARD, ACOUSTIC:**					
09.5007 001	Note: The following prices must be added to grid prices to determine total system cost.					
09.5007 011	Board, fiberglass, 5/8"	SF	0.0041	0.44	1.22	1.66
09.5007 021	Board, fiberglass, 3/4"	SF	0.0050	0.54	1.27	1.81
09.5007 031	Board, mineral fiber, 5/8"	SF	0.0041	0.44	1.22	1.66
09.5007 041	Board, mineral fiber, 5/8, vinyl face	SF	0.0058	0.63	2.77	3.40
09.5007 051	Board, mineral fiber, 5/8", 1-2 hour	SF	0.0050	0.54	1.43	1.97
09.5007 061	Board, mineral fiber, 5/8", 4 hour	SF	0.0058	0.63	2.00	2.63
09.5007 071	Board, wood fiber, 3/4"	SF	0.0058	0.63	1.58	2.21
09.5007 081	Board, mineral fiber, 5/8", air distribution	SF	0.0058	0.63	1.70	2.33
09.5007 091	Board, mineral fiber, 5/8", aluminum face	SF	0.0081	0.87	2.33	3.20
09.5008 000	**CEILING, T BAR SUSPENSION, GRID ONLY:**					
09.5008 011	T-bar ceiling, 2'x2', no divisions	SF	0.0066	0.71	1.03	1.74
09.5008 021	T-bar ceiling, 2'x4', no divisions	SF	0.0058	0.63	0.97	1.60
09.5008 031	T-bar ceiling, 4'x4', no divisions	SF	0.0050	0.54	0.93	1.46
09.5008 041	T-bar ceiling, 2'x2', each room divided	SF	0.0169	1.82	1.03	2.85
09.5008 051	T-bar ceiling, 2'x4', each room divided	SF	0.0158	1.70	0.97	2.68
09.5008 061	T-bar ceiling, 4'x4', each room divided	SF	0.0131	1.41	0.93	2.34
09.5009 000	**CEILING, T BAR SPLINE, CONCEALED, GRID ONLY:**					
09.5009 011	T-bar spline, 1' x 1', grid only	SF	0.0270	2.91	1.21	4.13
09.5009 021	T-bar spline, 1' x 2', grid only	SF	0.0270	2.91	1.26	4.17
09.5009 031	T-bar spline, 2'x2', grid, pop-out	SF	0.0270	2.91	1.86	4.77
09.5009 041	T-bar spline, 2'x2', grid, pop-out	SF	0.0270	2.91	1.86	4.77
09.5009 051	T-bar spline, 1'x1', one room	SF	0.0372	4.01	1.21	5.23
09.5009 061	T-bar spline, 1'x2', one room	SF	0.0375	4.05	1.26	5.31
09.5009 071	T-bar spline, 2'x2', one room	SF	0.0367	3.96	1.19	5.15
09.5009 091	Add for second suspension system	SF	0.0147	1.59	0.45	2.04
09.5009 101	Add for hanging wire, 12 ga, 4'	SQ	0.1553	16.76	0.13	16.88
09.5009 111	Add for hanging wire, 12 ga, 8'	SQ	0.1553	16.76	0.24	17.00
09.5700 000	**WOOD CEILINGS:**					
09.5700 001	Note: The following prices are based on amounts of 5000 square feet or more. Prices do not reflect custom work, but include suspension system					
09.5700 011	Ceiling System, wood, proprietary system, average	SF	0.1713	18.49	16.81	35.31
09.6000 000	**WOOD FLOORING:**					
09.6000 001	Note: The following prices are based on amounts of 5000 square feet or more. Prices do not reflect custom work.					
09.6001 000	**WOOD FLOORING:**					
09.6001 011	Oak floor, commercial, 25/32"x2-1/4", unfinished	SF	0.0419	4.52	5.35	9.87
09.6001 021	Oak floor, select, 25/32"x2-1/4", unfinished	SF	0.0445	4.80	5.44	10.25
09.6001 031	Oak plank floor, 25/32", tract	SF	0.0445	4.80	7.64	12.45
09.6001 041	Oak plank floor, 25/32", institutional	SF	0.0469	5.06	5.63	10.69
09.6001 051	1/2" pre-finish plank floor	SF	0.0559	6.03	4.69	10.73
09.6001 061	Oak parquet floor, 5/16", prefinished	SF	0.0559	6.03	5.42	11.45
09.6001 071	Oak parquet floor, 25/32", prefinished	SF	0.0575	6.21	5.84	12.04
09.6001 081	Maple parquet floor, 5/16", prefinished	SF	0.0600	6.48	9.06	15.53
09.6001 091	Walnut parquet floor, 5/16", prefinished	SF	0.0624	6.74	9.24	15.98
09.6001 101	Teak parquet floor, 5/16", prefinished	SF	0.0664	7.17	9.62	16.79
09.6001 111	Maple gym floor, 25/32", 2x4 sleepers	SF	0.0623	6.73	10.05	16.78
09.6001 121	Gym floor, 25/32" on steel springs	SF	0.0909	9.81	11.59	21.40
09.6001 131	Gym floor, 25/32", on steel channels	SF	0.0811	8.76	10.33	19.09
09.6001 141	Floor, Douglas fir, 'B'	SF	0.0469	5.06	4.03	9.09
09.6001 151	Floor, Douglas fir, 'C'	SF	0.0469	5.06	3.39	8.45
09.6001 161	Floor, mill, 2x4", 20-30d, 5.3 BF/SF	SF	0.0401	4.33	3.55	7.88
09.6001 171	Floor, mill, 2x6, 20-30 d, 8 BF/SF	SF	0.0655	7.07	5.66	12.74

FINISHES - 09

Division	Description	Unit	Labor Hours	Labor Cost	Material Cost	Total Cost
09.6001 181	Mill, knock down, sanding	SF	0.0058	0.63	0.09	0.71
09.6002 000	**FLOORING, WOOD BLOCK:**					
09.6002 011	Floor, wood block, natural, 1-1/2"	SF	0.0386	4.17	4.26	8.43
09.6002 021	Floor, wood block, treated, 2"	SF	0.0173	1.87	4.61	6.48
09.6002 031	Floor, wood block, treated, 2-1/2"	SF	0.0188	2.03	4.85	6.88
09.6002 041	Floor, wood block, treated, 3"	SF	0.0197	2.13	5.15	7.28
09.6003 000	**FLOOR SANDING & FINISHING ONLY:**					
09.6003 011	Floor, sand, fill, 2 coats lacquer	SF	0.0132	1.43	0.47	1.89
09.6003 021	Floor, waxing	SF	0.0058	0.63	0.16	0.79
09.6003 031	Floor, pegging	SF	0.0058	0.63	0.11	0.73
09.6003 041	Floor, institutional, total finishing	SF	0.0345	3.72	0.34	4.06
09.6003 051	Floor, refinish existing, average	SF	0.0332	3.58	1.06	4.65
09.7000 000	**RESILIENT FLOORING:**					
09.7000 001	Note: The labor prices in this section are based on the following productivity figures: Tile . . . . 450 SF PER MAN DAY Linoleum, Coved . . . . . . 30 SY PER MAN DAY Linoleum, Flat . . . . 60 SY PER MAN DAY					
09.7001 000	**TILE, RESILIENT:**					
09.7001 011	Tile, asphalt, 1/8", 'B'	SF	0.0178	1.22	1.53	2.75
09.7001 021	Tile, asphalt, 1/8", 'C'	SF	0.0178	1.22	1.70	2.92
09.7001 031	Tile, vinyl, 1/8", grease resistant	SF	0.0178	1.22	4.88	6.10
09.7001 041	Tile, cork, 5/16"	SF	0.0178	1.22	6.96	8.18
09.7001 051	Tile, cork, 3/16"	SF	0.0178	1.22	4.63	5.85
09.7001 061	Tile, vinyl composition, 1/16", standard	SF	0.0178	1.22	1.30	2.52
09.7001 071	Tile, vinyl composition, 3/32", standard	SF	0.0178	1.22	1.70	2.92
09.7001 081	Tile, vinyl composition, 1/8", standard	SF	0.0178	1.22	1.91	3.13
09.7001 091	Tile, vinyl composition, 1/16", metallic	SF	0.0178	1.22	1.62	2.84
09.7001 101	Tile, vinyl solid, 1/16", standard	SF	0.0178	1.22	4.54	5.76
09.7001 111	Tile, vinyl solid, 1/8", heavy duty	SF	0.0178	1.22	6.71	7.93
09.7001 121	Tile, vinyl, decorative, 125", best	SF	0.0356	2.44	13.76	16.20
09.7002 000	**SHEET GOODS, RESILIENT:**					
09.7002 011	Vinyl, .065", with cove	SY	0.1565	10.72	24.31	35.04
09.7002 021	Vinyl, good, .040", with cove	SY	0.1565	10.72	19.90	30.62
09.7002 031	Vinyl, metallic, .065", with cove	SY	0.1565	10.72	30.81	41.54
09.7002 041	Vinyl, metallic, .090", custom	SY	0.0375	2.57	42.88	45.45
09.7002 051	Vinyl, flat, with out cove, labor only	SY	0.0900	6.17		6.17
09.7002 061	Vinyl, flat, cove, labor only	LF	0.3564	24.42		24.42
09.7002 071	Linoleum, inlaid, standard grade with out cove	SY	0.1564	10.72	26.30	37.02
09.7002 081	Linoleum, inlaid, heavy duty, 1/8", with out cove	SY	0.1907	13.07	33.93	47.00
09.7002 091	Border, linoleum	LF	0.1211	8.30	3.04	11.34
09.7002 101	Floor, PVC, edge sealed, hot weld, small area	SF	0.0604	4.14	2.18	6.32
09.7002 111	Floor, PVC, edge sealed, small area	SF	0.0727	4.98	2.78	7.76
09.7003 000	**FLOORING, RESILIENT, BASE:**					
09.7003 001	Note: For wood in combination with resilient flooring see section 06.2000.					
09.7003 011	Base, top set, vinyl, 6"	LF	0.0147	1.01	1.62	2.62
09.7003 021	Base, top set, vinyl, 4"	LF	0.0147	1.01	1.11	2.12
09.7003 031	Base, top set, vinyl, 2-1/2"	LF	0.0147	1.01	0.70	1.71
09.7003 041	Base, top set, rubber, 6"	LF	0.0147	1.01	2.24	3.25
09.7003 051	Base, cove set, rubber, 4"	LF	0.0147	1.01	1.61	2.61
09.7003 061	Base, top set, rubber, 2-1/2"	LF	0.0147	1.01	1.11	2.12
09.7004 000	**FLOORING, SEAMLESS:**					
09.7004 011	Floor, seamless resilient, large area	SF	0.0502	3.44	1.55	4.99
09.7004 021	Floor, seamless resilient, small area	SF	0.1382	9.47	4.52	13.99
09.7004 031	Floor, seamless, 4" base, chemical resistant	SF	0.0604	4.14	1.90	6.04
09.7004 041	Elastomeric/poly cove base, large area	SF	0.0502	3.44	1.55	4.99
09.7004 051	Elastomeric/poly cove base, small area	SF	0.0584	4.00	1.83	5.83

09 - FINISHES

Division	Description	Unit	Labor Hours	Labor Cost	Material Cost	Total Cost
09.7004 061	Wainscot, seamless, cove, large area	SF	0.0584	4.00	1.55	5.55
09.7004 071	Wainscot, seamless, cove, small area	SF	0.0737	5.05	1.83	6.88
09.7004 081	Tread, seamless, riser, nosing, large	LF	0.0931	6.38	2.44	8.82
09.7004 091	Tread, seamless, riser, nosing, small	LF	0.0931	6.38	3.07	9.45
09.7005 000	**FLOORING, RESILIENT, SPECIALTIES:**					
09.7005 011	Floor, magnesite, medium area	SF	0.0584	4.00	3.55	7.55
09.7005 021	Floor, magnesite, large area	SF	0.0451	3.09	3.10	6.19
09.7005 031	Base, magnesite, medium area	LF	0.0604	4.14	3.57	7.71
09.7005 041	Base, magnesite, large area	LF	0.0502	3.44	3.23	6.67
09.7005 051	Treads, rubber, molded, 12" x 5/16"	LF	0.0829	5.68	10.26	15.94
09.7005 061	Treads, rubber, molded, 12" x 3/16"	LF	0.0727	4.98	9.75	14.74
09.7005 071	Treads, rubber, grip strip, 5/16"	LF	0.1555	10.66	15.09	25.75
09.7005 081	Treads, rubber, grip strip, 3/16"	LF	0.1341	9.19	11.20	20.39
09.7005 091	Treads, vinyl, molded, 12" x 1/8"	LF	0.0348	2.38	4.05	6.44
09.7005 101	Treads, vinyl, molded, 12" x 1/4"	LF	0.0655	4.49	6.97	11.46
09.7005 111	Risers, rubber, 7" x 1/8"	LF	0.0410	2.81	3.03	5.84
09.7005 121	Risers, vinyl, 7" x 1/8"	LF	0.0430	2.95	2.61	5.55
09.7005 131	Underlayment, particle board	SF	0.0058	0.40	0.31	0.71
09.7005 141	Asphalt plank, 1/2" x 12" x 24"	SF	0.0132	0.90	2.42	3.33
09.7005 151	Corner guards, rubber, 2-3/4"x1/4"	LF	0.0372	2.55	4.49	7.04
09.7005 161	Corner guard, vinyl, 2-3/4"x1-1/4"	LF	0.0372	2.55	4.49	7.04
09.8000 000	**PAINTING & WALL COVERING:**					
09.8001 000	**PAINTING & WALL COVERING, IN PLACE:**					
09.8001 001	Note: The following production rates are used: Brush Painting ..... 2,000 SF MAN DAY Roller Painting ..... 3,000 SF MAN DAY Spray Painting ..... 5,000 SF MAN DAY Airless Painting .... 8,000 SF MAN DAY SF/GAL TYPICAL: PRIME COAT 200, FINISH COATS 250					
09.8001 011	Multi residential, economy, 1 flat, 2 enamel	SF	0.0120	1.09	0.33	1.42
09.8001 021	Multi residential, economy, 2 flat, 3 enamel	SF	0.0170	1.54	0.56	2.10
09.8001 031	Custom residential, 2 flat, 3 enamel	SF	0.0250	2.26	0.81	3.07
09.8001 041	Commercial & industrial, 2 flat, 2 enamel	SF	0.0120	1.09	0.33	1.42
09.8001 051	Frame & schools, 2 flat, 3 enamel	SF	0.0180	1.63	0.81	2.44
09.8002 000	**PAINTING, EXTERIOR, NO HAND CUT-IN, LARGE AREAS:**					
09.8002 011	Paint, concrete, prime	SF	0.0042	0.38	0.10	0.48
09.8002 021	Paint, concrete, prime + 1 finish	SF	0.0080	0.72	0.18	0.91
09.8002 031	Paint, concrete, prime + 2 finish	SF	0.0117	1.06	0.25	1.31
09.8002 041	Paint, plaster, prime	SF	0.0042	0.38	0.10	0.48
09.8002 051	Paint, plaster, prime + 1 finish	SF	0.0080	0.72	0.15	0.88
09.8002 061	Paint, plaster, prime + 2 finish	SF	0.0117	1.06	0.25	1.31
09.8002 071	Paint, masonry, prime	SF	0.0047	0.43	0.15	0.58
09.8002 081	Paint, masonry, prime + 1 finish	SF	0.0084	0.76	0.15	0.91
09.8002 091	Paint, masonry, prime + 2 finish	SF	0.0120	1.09	0.25	1.33
09.8002 101	Paint, wood siding, prime	SF	0.0038	0.34	0.09	0.43
09.8002 111	Paint, wood siding, prime + 1 finish	SF	0.0080	0.72	0.15	0.88
09.8002 121	Paint, wood siding, prime + 2 finish	SF	0.0117	1.06	0.25	1.31
09.8002 131	Stain, wood siding	SF	0.0038	0.34	0.10	0.44
09.8002 141	Stain, wood siding, + 2 seal coats	SF	0.0112	1.01	0.10	1.11
09.8002 151	Stain, shingle siding	SF	0.0038	0.34	0.09	0.43
09.8002 161	Stain, shingle siding, + 2 seal coats	SF	0.0112	1.01	0.10	1.11
09.8002 171	Silicone seal spray, 1 coat	SF	0.0038	0.34	0.27	0.61
09.8002 181	Silicone seal spray, 2 coat	SF	0.0070	0.63	0.67	1.30
09.8002 191	Polyurethane coating, 1/8"	SF	0.0372	3.37	1.74	5.10
09.8002 201	Paint, sheet metal, prime	SF	0.0075	0.68	0.09	0.77
09.8002 211	Paint, sheet metal, prime + 1 finish	SF	0.0126	1.14	0.15	1.29
09.8002 221	Paint, sheet metal, prime + 2 finish	SF	0.0186	1.68	0.20	1.89
09.8002 231	Paint, wood trim, prime	LF	0.0047	0.43	0.10	0.52

FINISHES - 09

Division	Description	Unit	Labor Hours	Labor Cost	Material Cost	Total Cost
09.8002 241	Paint, wood trim, prime + 1 finish	LF	0.0082	0.74	0.20	0.95
09.8002 251	Paint, wood trim, prime + 2 finish	LF	0.0131	1.19	0.27	1.46
09.8003 000	**INTERIOR PAINTING, AIRLESS:**					
09.8003 001	*Note: conc-gypsum-plaster = Concrete, gypsum wall board, plaster. Fin = finish*					
09.8003 011	Paint, conc-gypsum-plaster, interior, prime, brush	SF	0.0038	0.34	0.04	0.39
09.8003 021	Paint, conc-gypsum-plaster, interior, prime+1 fin, brush	SF	0.0070	0.63	0.13	0.76
09.8003 031	Paint, conc-gypsum-plaster, interior, prime+2 fin, brush	SF	0.0112	1.01	0.23	1.24
09.8003 041	Paint, conc-gypsum-plaster, interior, prime, roller	SF	0.0025	0.23	0.04	0.27
09.8003 051	Paint, conc-gypsum-plaster, interior, prime+1 fin, roll	SF	0.0047	0.43	0.13	0.55
09.8003 061	Paint, conc-gypsum-plaster, interior, prime+2 fin, roll	SF	0.0075	0.68	0.23	0.91
09.8003 071	Paint, conc-gypsum-plaster, interior, prime, spray	SF	0.0015	0.14	0.04	0.18
09.8003 081	Paint, conc-gypsum-plaster, interior, prime+1 fin, spray	SF	0.0028	0.25	0.13	0.38
09.8003 091	Paint, conc-gypsum-plaster, interior, prime+2 fin, spray	SF	0.0045	0.41	0.23	0.63
09.8003 101	Paint, conc-gypsum-plaster, interior, prime, airless	SF	0.0010	0.09	0.09	0.18
09.8003 111	Paint, conc-gypsum-plaster, interior, prime+1 fin, airless	SF	0.0018	0.16	0.15	0.31
09.8003 121	Paint, conc-gypsum-plaster, interior, prime+2 fin, airless	SF	0.0028	0.25	0.25	0.50
09.8003 123	Paint, cmu, interior, prime, brush	SF	0.0042	0.38	0.04	0.42
09.8003 125	Paint, cmu, interior, prime + 1 fin, brush	SF	0.0080	0.72	0.15	0.88
09.8003 127	Paint, cmu, interior, prime + 2 fin, brush	SF	0.0118	1.07	0.25	1.32
09.8003 131	Paint, cmu, interior, prime, rolled	SF	0.0028	0.25	0.04	0.30
09.8003 141	Paint, cmu, interior, prime + 1 fin, roll	SF	0.0053	0.48	0.15	0.63
09.8003 151	Paint, cmu, interior, prime + 2 fin, roll	SF	0.0079	0.72	0.25	0.96
09.8003 161	Stain, paneling, brush	SF	0.0038	0.34	0.09	0.43
09.8003 171	Stain, paneling, + 2 seal coats, brush	SF	0.0112	1.01	0.10	1.11
09.8003 181	Paint, sheet metal, interior, prime, brush	SF	0.0070	0.63	0.09	0.72
09.8003 191	Paint, sheet metal, interior, prime+1 fin, brush	SF	0.0121	1.10	0.15	1.25
09.8003 201	Paint, sheet metal, interior, prime+2 fin, brush	SF	0.0182	1.65	0.20	1.85
09.8003 211	Paint, wood trim, interior, prime, brush	LF	0.0042	0.38	0.10	0.48
09.8003 221	Paint, wood trim, interior, prime+1 fin, brush	LF	0.0077	0.70	0.20	0.90
09.8003 231	Paint, wood trim, interior, prime+2 fin, brush	LF	0.0117	1.06	0.27	1.33
09.8004 000	**MISCELLANEOUS PAINTING ITEMS:**					
09.8004 001	*Note: For institutional finish on cabinets, add 30% to the total costs. Price does not include painting interior of cabinets.*					
09.8004 011	Cabinets, exterior, stain & varnish	LF	0.2086	18.89	4.83	23.72
09.8004 021	Cabinets, exterior, lacquer, 2 coats	LF	0.1897	17.18	3.65	20.83
09.8004 031	Cabinets, stain & varnish, 3 coats	LF	0.2756	24.96	9.75	34.71
09.8004 041	Door & trim, lacquer, spray 2 coats	EA	0.4137	37.46	4.83	42.29
09.8004 051	Door & trim, 2 coats, residential	EA	0.2241	20.29	6.09	26.38
09.8004 061	Door & trim, 3 coats, institutional	EA	0.4825	43.69	14.56	58.25
09.8004 071	Sash & trim, wood sash, 2 coats	SF	0.0258	2.34	0.63	2.96
09.8004 081	Sash & trim, steel sash, 2 coats	SF	0.0345	3.12	0.81	3.93
09.8004 091	Hand rail, architectural, 3 coats	LF	0.0449	4.07	1.20	5.26
09.8004 101	Hand rail, pipe, 3 coats	LF	0.0449	4.07	1.20	5.26
09.8004 111	Paint wall louver, 24" x 8"	EA	0.2840	25.72	7.34	33.06
09.8004 121	Paint roll-up doors, 2 sides	LF	0.0270	2.44	1.20	3.64
09.8004 131	Ceraglazed e-p coating on gypsum wall board	SF	0.0140	1.27	1.20	2.46
09.8004 141	Paint guard rail	LF	0.0350	3.17	1.20	4.37
09.8004 151	Paint ladder	LF	0.0600	5.43	1.20	6.63
09.8004 161	Graffiti resistant coating	SF	0.0146	1.32	0.68	2.00
09.8005 000	**PAINTING STEEL:**					
09.8005 011	Paint, structural steel, average cost/ton	TON	0.6936	62.81	26.88	89.68
09.8005 021	Paint, structural steel, average cost/SF building	SF	0.0042	0.38	0.10	0.48
09.8005 031	Paint, structural steel, 1 coat, brush	SF	0.0056	0.51	0.03	0.54
09.8005 041	Paint, structural steel, 2 coats, brush	SF	0.0103	0.93	0.10	1.03
09.8005 051	Paint, structural steel, 1 coat, spray	SF	0.0028	0.25	0.11	0.36
09.8005 061	Paint, structural steel, 2 coats, spray, airless	SF	0.0052	0.47	0.25	0.72
09.8005 071	Paint, structural steel, 1 epoxy, touch up	SF	0.0070	0.63	0.30	0.94

09 - FINISHES

Division	Description	Unit	Labor Hours	Labor Cost	Material Cost	Total Cost
09.8005 081	Paint, pipe to 6", prime	LF	0.0070	0.63	0.04	0.68
09.8005 091	Paint, pipe to 6", prime + 1 finish	LF	0.0131	1.19	0.14	1.33
09.8005 101	Paint, pipe to 6", prime + 2 finish	LF	0.0186	1.68	0.23	1.91
09.8005 111	Paint, pipe to 12", prime	LF	0.0140	1.27	0.04	1.31
09.8005 121	Paint, pipe to 12", prime + 1 finish	LF	0.0256	2.32	0.14	2.46
09.8005 131	Paint, pipe to 12", prime + 2 finish	LF	0.0326	2.95	0.25	3.20
09.8005 141	Paint, pipe over 12", prime	SF	0.0066	0.60	0.03	0.63
09.8005 151	Paint, pipe over 12", prime + 1 finish	SF	0.0140	1.27	0.10	1.36
09.8005 161	Paint, pipe over 12", prime + 2 finish	SF	0.0186	1.68	0.14	1.82
09.8005 171	Paint, steel tank, prime	SF	0.0056	0.51	0.03	0.54
09.8005 181	Paint, steel tank, prime + 1 finish	SF	0.0103	0.93	0.10	1.03
09.8005 191	Paint, steel tank, prime + 2 finish	SF	0.0121	1.10	0.14	1.24
09.8006 000	**SANDBLASTING, IN CONJUNCTION WITH PAINTING:**					
09.8006 001	*Note: Prices include compressor time. Material prices are based on sand, a b & c pattern Swedish pictorial. For cost of clean blast abrasive multiply sand cost by 3. Codes sp1 through sp10 refer to steel structures painting council coding.*					
09.8006 011	Whip blast, SP 7, 6000 SF/day	SF	0.0025	0.23	0.17	0.40
09.8006 021	Sandblast, commercial, SP 6, 3200 SF/day	SF	0.0038	0.34	0.37	0.71
09.8006 031	Sandblast, near white, SP 10, 2200 SF/day	SF	0.0074	0.67	0.83	1.50
09.8006 041	Sandblast, white metal, SP 5, 1400 SF/day	SF	0.0099	0.90	1.11	2.01
09.8006 051	Wire brush, incidental, SP 1	SF	0.0034	0.31		0.31
09.8006 061	Wire brush, power, SP 2	SF	0.0052	0.47		0.47
09.8006 071	Acid clean & etch, water rinse	SF	0.0048	0.43		0.43
09.8007 000	**WALL COVERINGS:**					
09.8007 001	*Note: For quantities under 500 sf, add 20% to the total costs, less than 250 sf double time.*					
09.8007 021	Paper hanging, labor, normal conditions	SF	0.0105	0.95		0.95
09.8007 031	Wall paper, 36 SF/roll, average	ROLL	0.3100	28.07	34.13	62.20
09.8007 041	Wall cover, vinyl, 7 oz, light	SF	0.0068	0.62	1.83	2.45
09.8007 051	Wall cover, vinyl, 14 oz, medium	SF	0.0086	0.78	2.00	2.78
09.8007 061	Wall cover, vinyl, 22 oz, heavy	SF	0.0099	0.90	2.27	3.17
09.8007 071	Fiber board backing for vinyl	SF	0.0025	0.23	0.20	0.43
09.8007 081	Wall cover, vinyl, 14 oz, aluminum back	SF	0.0253	2.29	1.69	3.98
09.8007 091	Wall cover, linen, acrylic back	SF	0.0135	1.22	1.42	2.65
09.8007 101	Wall cover, glass cloth	SF	0.0093	0.84	1.55	2.39
09.8007 111	Wall cover, felt	SF	0.0191	1.73	2.71	4.43
09.8007 121	Wall cover, cork sheathing, 1/8"	SF	0.0191	1.73	2.16	3.89
09.8007 131	Wall cover, flexible wood, veneer	SF	0.0345	3.12	5.12	8.24
09.9000 000	**PLASTIC & FACTORY FINISH WALL SURFACES:**					
09.9001 000	**LAM PLASTICS, STD PATTERNS & COLORS, WITHOUT BACKING:**					
09.9001 011	Laminated plastic cover, adhesive, 1/16"	SF	0.0246	2.23	2.36	4.59
09.9001 021	Laminated plastic, adhesive, 1/32", vertical surface	SF	0.0246	2.23	2.50	4.73
09.9001 031	Laminated plastic, adhesive, 1/16", acid resistant	SF	0.0246	2.23	2.36	4.59
09.9001 041	Laminated plastic, adhesive, .020", back sheet	SF	0.0105	0.95	0.63	1.58
09.9001 051	Add for post formed	SF	0.0025	0.23	0.18	0.41
09.9002 000	**WALL COVER, HARDBOARD, PHOTO REPRODUCTION WITHOUT BACKING:**					
09.9002 001	*Note: For patterns and deep colors, add 20% to the material costs.*					
09.9002 011	Wall cover, plastic/hardboard, 1/8", trim	SF	0.0154	1.39	2.75	4.14
09.9002 021	Wall cover, plastic/hardboard, 1/4", trim	SF	0.0154	1.39	3.64	5.04
09.9002 031	Wall cover, plastic/pegboard, 1/8", trim	SF	0.0154	1.39	0.91	2.30
09.9002 041	Wall cover, plastic/pegboard, 1/4", trim	SF	0.0154	1.39	0.83	2.22
09.9003 000	**WALL COVERING, TILE:**					
09.9003 011	Plastic tile, 4-1/4"x4-1/4"x.11"	SF	0.0246	2.23	1.79	4.02
09.9003 021	Plastic tile, 4-1/4"x4-1/4"x.05"	SF	0.0246	2.23	1.15	3.38
09.9003 031	Aluminum tile, 4-1/2" x 4-1/2"	SF	0.0295	2.67	8.02	10.69
09.9003 041	Copper/aluminum tile, 4-1/4"x4-1/4"	SF	0.0295	2.67	8.02	10.69
09.9003 051	Stain steel tile, 4-1/4"x4-1/4"	SF	0.0350	3.17	16.30	19.46

SPECIALTIES - 10

Division	Description	Unit	Labor Hours	Labor Cost	Material Cost	Total Cost
10.1000 000	**CHALK & TACK BOARDS:**					
10.1001 000	**CHALK & TACK BOARD, IN PLACE:**					
10.1001 011	Chalk board, institutional, with trim, map, rail	SF	0.0524	5.65	24.68	30.33
10.1001 021	Chalk board, institutional, without trim, average	SF	0.0263	2.84	10.35	13.19
10.1001 041	Chalk board, vertical sliding	SF	0.2480	26.76	98.47	125.23
10.1001 051	Chalk board, horizontal sliding	SF	0.1621	17.49	63.99	81.47
10.1001 061	Chalk board, swing leaf panels	SF	0.1416	15.28	216.41	231.68
10.1001 071	Chalk board, reversible, roll, 4x8	EA	4.6298	499.51	1,018.05	1,517.56
10.1002 000	**CHALK BOARDS, WITHOUT FRAME:**					
10.1002 011	Chalk board, hardboard, tempered, 1/4"	SF	0.0181	1.95	15.84	17.79
10.1002 021	Chalk board, hardboard, tempered, 1/2"	SF	0.0244	2.63	17.95	20.59
10.1002 031	Chalk board, metal, 24 ga, 1/4"	SF	0.0244	2.63	20.34	22.97
10.1002 041	Chalk board, metal, 24 ga, 1/2"	SF	0.0286	3.09	20.66	23.75
10.1002 051	Chalk board, slate, 3/8"	SF	0.0530	5.72	27.07	32.79
10.1002 061	Chalk board, adhesive	GAL			12.00	12.00
10.1003 000	**TACK BOARDS, WITHOUT FRAME:**					
10.1003 011	Tack board, cork, unbacked, 1/8"	SF	0.0066	0.71	7.11	7.82
10.1003 021	Tack board, cork, unbacked, 1/4"	SF	0.0115	1.24	13.63	14.87
10.1003 031	Tack board, cork, 1/8", burlap back	SF	0.0140	1.51	14.39	15.91
10.1003 041	Tack board, cork, 1/4", burlap back	SF	0.0246	2.65	14.70	17.36
10.1003 051	Tack board, vinyl cork, 1/4", burlap back	SF	0.0254	2.74	15.69	18.43
10.1003 061	Tack board, vinyl/fiberboard, 1/2"	SF	0.0197	2.13	14.57	16.69
10.1003 071	Tack board, 1/4" vinyl cork, 1/4" hardboard	SF	0.0328	3.54	25.82	29.36
10.1003 081	Tack board, 1/8" vinyl cork, 3/8" fiberboard	SF	0.0254	2.74	15.69	18.43
10.1003 091	Tack board, adhesive	GAL			22.20	22.20
10.1004 000	**FRAMES, TRIM, TRAYS & RAILS:**					
10.1004 011	Chalk board, aluminum frame, trim	LF	0.0084	0.91	5.25	6.15
10.1004 021	Aluminum chalk tray	LF	0.0244	2.63	11.96	14.59
10.1004 031	Aluminum map & display rail, deluxe	LF	0.0106	1.14	6.34	7.48
10.1500 000	**TOILET PARTITIONS & COMPARTMENTS:**					
10.1501 000	**TOILET PARTITIONS:**					
10.1501 011	Toilet partition, baked enamel, floor mounted	EA	1.3056	140.86	714.44	855.31
10.1501 021	Toilet partition, baked enamel, ceiling mounted	EA	1.5230	164.32	754.17	918.49
10.1501 031	Toilet partition, porcelain enamel, floor mounted	EA	1.5960	172.19	1,055.53	1,227.72
10.1501 041	Toilet partition, porcelain enamel, ceiling mounted	EA	1.8618	200.87	1,111.33	1,312.20
10.1501 051	Toilet partition, laminated plastic, floor mounted	EA	1.5960	172.19	923.86	1,096.05
10.1501 061	Toilet partition, laminated plastic, ceiling mounted	EA	1.8618	200.87	938.62	1,139.49
10.1501 071	Toilet partition, stainless steel, floor mounted	EA	1.5960	172.19	2,423.32	2,595.51
10.1501 081	Toilet partition, stainless steel, ceiling mounted	EA	1.8618	200.87	2,250.24	2,451.11
10.1501 091	Add for best quality	EA			162.45	162.45
10.1502 000	**URINAL SCREENS:**					
10.1502 011	Urinal screen, baked enamel, wall mounted	EA	0.4490	48.44	257.83	306.28
10.1502 021	Urinal screen, porcelain enamel, wall mounted	EA	0.5487	59.20	485.19	544.39
10.1502 031	Urinal screen, laminated plastic, wall mounted	EA	0.5487	59.20	357.70	416.90
10.1502 041	Urinal screen, stainless steel, wall mounted	EA	0.5487	59.20	656.97	716.17
10.1502 051	Add for best quality	EA			45.17	45.17
10.1502 061	Add for floor mounted	EA	0.2242	24.19	65.72	89.91
10.1503 000	**SIGHT SCREENS, 3' X 7':**					
10.1503 011	Sight screen, baked enamel, floor mounted	EA	0.9421	101.64	406.10	507.74
10.1503 021	Sight screen, porcelain enamel, floor mounted	EA	1.1518	124.27	539.63	663.89
10.1503 031	Sight screen, laminated plastic, floor mounted	EA	1.1518	124.27	527.95	652.22
10.1503 041	Sight screen, stainless steel, floor mounted	EA	1.1518	124.27	828.27	952.54
10.1503 051	Add for best quality	EA			88.13	88.13
10.1504 000	**ACCESSORIES:**					
10.1504 011	Coat hooks & door stop	EA			18.30	18.30
10.1504 021	Purse shelf, chrome, 5" x 14"	EA	0.1310	14.13	38.15	52.28

10 - SPECIALTIES

Division	Description	Unit	Labor Hours	Labor Cost	Material Cost	Total Cost
10.1504 031	Toilet paper dispenser, chrome	EA	0.0788	8.50	22.77	31.28
10.1504 041	Seat cover dispenser, chrome	EA	0.4594	49.56	133.51	183.08
10.1505 000	**DRESSING CUBICLES, 80" HIGH, W/CURTAIN:**					
10.1505 011	Dressing cubicle, baked enamel, floor mounted	EA	1.0486	113.13	1,320.20	1,433.34
10.1505 021	Dressing cubicle, porcelain enamel, floor mounted	EA	1.2820	138.31	1,954.67	2,092.99
10.1505 031	Dressing cubicle, laminated plastic, floor mounted	EA	1.2820	138.31	1,716.30	1,854.62
10.1505 041	Dressing cubicle, stainless steel, floor mounted	EA	1.2820	138.31	2,446.19	2,584.50
10.1506 000	**SHOWER COMPARTMENTS, WITH RECEPTOR, WITHOUT PLUMBING:**					
10.1506 011	Shower compartment, baked enamel, 1 entry	EA	5.5058	668.73	1,308.78	1,977.52
10.1506 021	Shower compartment, porcelain enamel, 1 entry	EA	6.1616	748.39	1,700.29	2,448.68
10.1506 031	Shower compartment, fiberglass, 1 entry	EA	5.5058	668.73	1,558.19	2,226.92
10.1506 041	Shower compartment, stainless steel, 1 entry	EA	10.4878	1,273.85	2,344.99	3,618.84
10.1506 051	Add for soap dish	EA	0.0614	6.62	14.88	21.50
10.1506 061	Add for curtain rod	LF	0.0481	5.19	11.67	16.86
10.1506 071	Add for shower door	EA	0.9177	99.01	222.57	321.58
10.2000 000	**PARTITIONS; FOLDING, RELOCATABLE & DEMOUNTABLE:**					
10.2000 001	Note: The prices listed below for relocatable partitions are for preliminary estimates only. They are based on a minimum of 100 lineal feet of partition, 8-12' ceilings, 3 doors and frames, 5 starters, 4 corners and 4' high base. The prices for folding partitions do not include structural supports, architectural trim, placement of track in concrete floor or electric circuits.					
10.2001 000	**STANDARD RELOCATABLE PARTITIONS:**					
10.2001 011	Relocatable partition, 1/2" gypsum wall board, STC 38	SF	0.0236	2.55	8.63	11.18
10.2001 021	Relocatable partition, 5/8" gypsum wall board, STC 40	SF	0.0266	2.87	9.62	12.49
10.2001 031	Relocatable partition, 1/2"-5/8" gypsum wall board, STC 45	SF	0.0379	4.09	13.99	18.08
10.2001 041	Relocatable partition, 1/2" gypsum wall board, vinyl wall cover, 2 sides	SF	0.0389	4.20	14.23	18.43
10.2001 051	Relocatable partition, 5/8" gypsum wall board, vinyl wall cover, 2 sides	SF	0.0502	5.42	18.29	23.70
10.2001 061	Relocatable partition, metallic gypsum wall board, baked enamel	SF	0.0471	5.08	17.35	22.43
10.2001 071	Relocatable partition, 5/8" gypsum wall board, STC 30	SF	0.0727	7.84	26.61	34.45
10.2001 081	Relocatable partition, 5/8" gypsum wall board, STC 43, 2 hour	SF	0.0788	8.50	29.00	37.50
10.2001 091	Relocatable partition, gypsum wall board, studs, unfinished	SF	0.0246	2.65	9.11	11.77
10.2001 101	Relocatable partition, gypsum wall board, stud, 3", unfinished	SF	0.0318	3.43	11.50	14.93
10.2001 111	Add for factory vinyl, 15-22 oz	SF	0.0123	1.33	4.28	5.61
10.2001 121	Cubicle/welding booth, 5'x5'x5'	SF	0.0614	6.62	6.99	13.62
10.2001 131	Divider, 5'-6', 18" glass top, 2" vinyl	SF	0.0522	5.63	5.97	11.60
10.2002 000	**RELOCATABLE PARTITIONS, ACCESSORIES:**					
10.2002 011	Windows, frame & glass, 3'6" x 2'	EA	0.4983	48.59	488.35	536.94
10.2002 021	Windows, frame & glass, 3'6" x 4'	EA	0.6558	63.95	805.41	869.36
10.2002 031	Door, hollow core, prefinished, 3' x 7'	EA	1.0486	113.13	308.16	421.29
10.2002 041	Door jamb, metal, 3' x 7'	EA	0.7867	84.88	369.84	454.71
10.2002 051	Door jamb, metal, 6' x 7'	EA	1.2450	134.32	451.99	586.32
10.2002 061	Add for passage set	EA	0.3018	32.56	65.72	98.28
10.2002 071	Add for lock set	EA	0.3929	42.39	85.44	127.83
10.2002 081	Corner	EA	0.4850	52.33	105.41	157.74
10.2002 091	Corner, metal	EA	0.9177	99.01	199.60	298.61
10.2002 101	Starter	EA	0.1709	18.44	36.95	55.39
10.2002 111	Starter, metal	EA	0.3540	38.19	76.93	115.13
10.2002 121	Metal base, not supplied with wall	LF	0.0103	1.11	2.27	3.38
10.2003 000	**ACCORDION PARTITIONS:**					
10.2003 011	Accordion partition, vinyl, 8', STC 36, economy	SF	0.0594	6.41	21.70	28.10
10.2003 021	Accordion partition, vinyl, 30x17, STC 41, economy	SF	0.0778	8.39	28.28	36.68
10.2003 031	Accordion partition, vinyl, 30x17, STC 43, good	SF	0.0921	9.94	33.60	43.54
10.2003 041	Accordion partition, vinyl, 30x17, STC 44, better	SF	0.1013	10.93	37.10	48.03
10.2003 051	Accordion partition, vinyl, large opening, STC 45, better	SF	0.1116	12.04	40.91	52.95
10.2003 061	Accordion partition, vinyl, large opening, STC 47, best	SF	0.1146	12.36	48.02	60.38
10.2003 071	Accordion partition, wood slat, birch or ash, prefinished	SF	0.0594	6.41	21.70	28.10

SPECIALTIES - 10

Division	Description	Unit	Labor Hours	Labor Cost	Material Cost	Total Cost
10.2004 000	**FOLDING PARTITIONS, OPERABLE WALLS, LEAF:**					
10.2004 011	Folding partition, vinyl and metal panel, STC 52	SF	0.2497	26.94	67.84	94.78
10.2004 021	Folding partition, vinyl and metal panel, STC 48	SF	0.2517	27.16	61.99	89.14
10.2004 031	Folding partition, vinyl and wood panel, STC 40	SF	0.1832	19.77	49.92	69.69
10.2004 041	Add for laminated plastic	SF	0.0123	1.33	4.17	5.49
10.2004 051	Add for wood veneer	SF	0.0154	1.66	5.45	7.11
10.2004 061	Add for chalk board	SF	0.0123	1.33	4.44	5.77
10.2005 000	**FOLDING PARTITIONS, OPERABLE WALLS, COMMERCIAL:**					
10.2005 001	Note: The following prices are based on 60' x 25' walls, top hung, with no bottom track required.					
10.2005 011	Folding partition, aluminum slat, 13#/SF	SF	0.0839	9.05	22.51	31.56
10.2005 021	Folding partition, steel slat, 16#/SF	SF	0.0573	6.18	15.39	21.58
10.2005 031	Folding partition, wood, side coil, single, crank	SF	0.1177	12.70	31.60	44.30
10.2005 041	Folding partition, wood, side coil, single, motor	SF	0.2497	26.94	67.03	93.97
10.2005 051	Folding partition, wood, side coil, 2 crank	SF	0.2098	22.64	56.39	79.02
10.2005 061	Folding partition, wood, side coil, 2 motor	SF	0.4195	45.26	112.97	158.23
10.2005 071	Folding partition, motor, under 30' x 9'	EA	7.2102	777.91	1,944.06	2,721.97
10.2005 081	Folding partition, motor drive, over 30'	EA	14.4203	1,555.81	3,888.12	5,443.92
10.2006 000	**DEMOUNTABLE PARTITIONS:**					
10.2006 011	Demountable partitions, air wall	SF	0.1054	11.37	29.67	41.05
10.2006 021	Demountable modular panels, spring mount	SF	0.0594	6.41	17.94	24.35
10.2006 031	Add for 1 hour doors & hardware	SF	0.0594	6.41	17.94	24.35
10.4000 000	**TOILET ACCESSORIES:**					
10.4000 001	Note: Labor costs for the following items can be found in section 06.2101.					
10.4001 000	**PAPER TOWEL & WASTE COMBINATION:**					
10.4001 011	Paper towel & waste combination, stainless steel, laminated, 14"x24", recessed	EA			516.66	516.66
10.4001 021	Paper towel & waste combination, stainless steel, 17"x54", semi-recessed	EA			779.00	779.00
10.4001 031	Paper towel & waste combination, stainless, laminated, 12"x72", recessed	EA			900.85	900.85
10.4001 041	Paper towel & waste combination, 12"x72", stainless steel trim, surface	EA			1,298.32	1,298.32
10.4002 000	**PAPER TOWEL DISPENSERS:**					
10.4002 011	Towel dispenser, stainless steel, 12"x17", surface	EA			127.18	127.18
10.4002 021	Towel dispenser, stainless steel, 12" x 15", recessed	EA			246.46	246.46
10.4002 031	Towel dispenser, stainless steel, 14" x 26", recessed	EA			368.28	368.28
10.4002 041	Towel dispenser, laminated, 14"x26", recessed	EA			556.40	556.40
10.4002 051	Towel dispenser, stainless steel trim, 17"x28"	EA			877.06	877.06
10.4002 061	Towel & soap combination, stainless steel	EA			556.44	556.44
10.4002 071	Towel & soap combination, laminated	EA			654.44	654.44
10.4002 081	Towel, soap & mirror, stainless steel	EA			779.00	779.00
10.4002 091	Towel, soap & mirror, laminated	EA			839.95	839.95
10.4003 000	**WASTE RECEPTACLES:**					
10.4003 011	Waste receptacle, stainless steel, 3 gal, recessed	EA			380.87	380.87
10.4003 021	Waste receptacle, stainless steel, 12 gal, semi-recessed	EA			479.17	479.17
10.4003 031	Waste receptacle, stainless steel, 18 gal, semi-recessed	EA			577.51	577.51
10.4003 041	Waste receptacle, stainless steel, 10 gal, recessed	EA			577.51	577.51
10.4003 051	Waste receptacle, laminated, 3 gal, recessed	EA			606.95	606.95
10.4003 061	Waste receptacle, laminated, 10 gal, recessed	EA			759.29	759.29
10.4003 071	Waste receptacle, stainless steel, 13 gal, top, unmounted	EA			380.87	380.87
10.4004 000	**TOILET SEAT COVER DISPENSERS:**					
10.4004 011	Seat cover dispenser, stainless steel, surface	EA			79.49	79.49
10.4004 021	Seat cover dispenser, stainless steel, recessed	EA			182.05	182.05
10.4004 031	Seat cover dispenser, laminated, recess	EA			348.75	348.75
10.4004 041	Seat cover dispenser, stainless steel, partition mounted	EA			371.80	371.80
10.4005 000	**TOILET PAPER DISPENSERS:**					
10.4005 011	Toilet paper dispenser, aluminum, single, surface	EA			28.18	28.18
10.4005 021	Toilet paper dispenser, stainless steel, single, surface	EA			51.29	51.29

10 - SPECIALTIES

Division	Description	Unit	Labor Hours	Labor Cost	Material Cost	Total Cost
10.4005 031	Toilet paper dispenser, stainless steel, single, recessed	EA			38.47	38.47
10.4005 041	Folded tissue cabinet, stainless steel, surface	EA			51.29	51.29
10.4005 051	Toilet paper dispenser, aluminum, double, surface	EA			48.72	48.72
10.4005 061	Toilet paper dispenser, stainless steel, double, surface	EA			84.58	84.58
10.4005 071	Toilet paper dispenser, stainless steel, double, recessed	EA			87.18	87.18
10.4005 081	Toilet paper & seat cover dispenser combination, recessed	EA			751.30	751.30
10.4005 091	Toilet paper, seat cover & napkin dispenser combination, recessed	EA			835.89	835.89
10.4006 000	**FEMININE NAPKIN DISPENSERS:**					
10.4006 011	Napkin dispenser, stainless steel, surface	EA			769.22	769.22
10.4006 021	Napkin dispenser, stainless steel, recessed	EA			841.07	841.07
10.4006 031	Napkin dispenser, laminated, recessed	EA			994.89	994.89
10.4006 041	Napkin, dispenser & disposal, stainless steel	EA			1,161.58	1,161.58
10.4007 000	**FEMININE NAPKIN DISPOSAL:**					
10.4007 011	Napkin disposal, stainless steel, surface	EA			89.76	89.76
10.4007 021	Napkin disposal, stainless steel, recessed, small	EA			210.25	210.25
10.4007 031	Napkin disposal, stainless steel, partition mounted	EA			325.68	325.68
10.4007 041	Napkin disposal, stainless steel, recessed, large	EA			325.68	325.68
10.4007 051	Napkin disposal, laminated, recessed	EA			474.33	474.33
10.4008 000	**SOAP DISPENSERS:**					
10.4008 011	Soap dispenser, plastic, liquid, surface, stainless steel lid	EA			48.72	48.72
10.4008 021	Soap dispenser, stainless steel, liquid & powder, surface	EA			100.05	100.05
10.4008 031	Soap dispenser, counter mounted, tank below	EA			115.41	115.41
10.4008 041	Soap dispenser, stainless steel, liquid, recessed	EA			166.64	166.64
10.4008 051	Soap dispenser, stainless steel, powder, recessed	SS			253.82	253.82
10.4008 061	Soap dispenser, laminated, liquid, recessed	EA			253.82	253.82
10.4008 071	Soap dispenser, stainless steel, liquid, recessed, shelf	EA			325.68	325.68
10.4008 081	Soap dispenser, stainless steel, leaf, surface	EA			328.23	328.23
10.4008 091	Soap dispenser, stainless steel, leaf, recessed	EA			356.43	356.43
10.4009 000	**FACIAL TISSUE DISPENSERS:**					
10.4009 011	Tissue dispenser, stainless steel, recessed	EA			38.47	38.47
10.4009 021	Tissue dispenser, stainless steel, surface	EA			41.04	41.04
10.4009 031	Tissue dispenser, laminated, recessed	EA			125.65	125.65
10.4009 041	Tissue dispenser, stainless steel, recessed, electric outlet	EA			189.71	189.71
10.4009 051	Tissue dispenser, stainless steel, recessed, with shelf	EA			230.78	230.78
10.4009 061	Tissue dispenser, stainless steel, recessed, shelf, outlet	EA			276.94	276.94
10.4010 000	**GRAB BARS:**					
10.4010 001	*Note: For concealed mounting, add 10% to the total costs. For peened grip, add 20% to the total costs.*					
10.4010 011	Grab bar, stainless steel, 1-1/2" x 24", exposed mounted	EA			82.06	82.06
10.4010 021	Grab bar, stainless steel, 1-1/2" x 48", exposed mounted	EA			123.08	123.08
10.4010 031	Grab bar, swing away, exposed, floor mounted	EA			910.27	910.27
10.4010 041	Wheel chair compartment, exposed mounted	EA			158.96	158.96
10.4010 051	Grab bar, horizontal tub, exposed mounted	EA			197.42	197.42
10.4010 061	Grab bar, 2 way tub, exposed mounted	EA			333.39	333.39
10.4011 000	**MISCELLANEOUS TOILET ACCESSORIES:**					
10.4011 011	Towel bar, stainless steel, 18"	EA			71.78	71.78
10.4011 021	Towel bar, stainless steel, 24"	EA			76.92	76.92
10.4011 031	Towel bar, stainless steel, 30"	EA			82.06	82.06
10.4011 041	Shower rod, stainless steel, 1" x 6'	EA			53.83	53.83
10.4011 051	Shower rod flanges, pair	PAIR			38.47	38.47
10.4011 061	Shower curtain, 70"x7'2", with hooks	EA			61.56	61.56
10.4011 071	Robe hook	EA			46.09	46.09
10.4011 081	Mirror, stainless steel frame, tilt	EA			206.16	206.16
10.4011 091	Mirror, stainless steel frame, 16" x 24"	EA			101.55	101.55
10.4011 101	Mirror, stainless steel frame, 16" x 24", shelf	EA			147.76	147.76
10.4011 111	Medicine cabinet & mirror, baked enamel, surface	EA			161.54	161.54
10.4011 121	Medicine cabinet & mirror, baked enamel, recessed	EA			226.05	226.05

SPECIALTIES - 10

Division	Description	Unit	Labor Hours	Labor Cost	Material Cost	Total Cost
10.4011 131	Medicine cabinet & mirror, stainless steel, recessed	EA			656.05	656.05
10.4011 141	Clothes line, retractable	EA			49.60	49.60
10.4011 151	Shower seat, stainless steel	EA			242.56	242.56
10.4011 161	Ash tray, stainless steel, wall mounted	EA			64.51	64.51
10.4011 171	Ash tray, stainless steel, surface, small	EA			190.15	190.15
10.4011 181	Ash tray, stainless steel, surface, large	EA			281.83	281.83
10.4011 191	Ash tray, stainless steel, recessed	EA			305.60	305.60
10.4011 201	Ash tray, laminated, recessed	EA			461.75	461.75
10.4011 211	Towel ring, stainless steel	EA			50.93	50.93
10.4011 221	Towel fin, stainless steel	EA			37.36	37.36
10.4011 231	Shelf, stainless steel, 18"	EA			108.64	108.64
10.4011 241	Shelf, stainless steel, 24"	EA			118.82	118.82
10.4011 251	Shelf, stainless steel, 30"	EA			129.04	129.04
10.4011 261	Towel shelf, stainless steel, 18"	EA			142.58	142.58
10.4011 271	Towel shelf, stainless steel, 24"	EA			156.21	156.21
10.4011 281	Bottle opener, stainless steel	EA			10.22	10.22
10.4011 291	Blade disposal, stainless steel	EA			44.16	44.16
10.4011 301	Electric hand dryer, surface, 40 second cycle	EA			2,061.71	2,061.71
10.5000 000	**MISCELLANEOUS BUILDING SPECIALTIES:**					
10.5001 000	**LINEN & GARBAGE CHUTES:**					
10.5001 001	*Note: The following prices assume prefab units with roof vents, 1 1/2 hour 'B' doors, discharge and sprinkler systems.*					
10.5001 011	Chute, aluminum, 20', light duty	FLOOR	4.7034	507.45	939.43	1,446.87
10.5001 021	Chute, 18 ga galvanized steel, 24"	FLOOR	4.7034	507.45	1,159.13	1,666.58
10.5001 031	Chute, 18 ga galvanized steel, 30"	FLOOR	4.7034	507.45	1,390.24	1,897.69
10.5001 041	Chute, 18 ga stainless steel, 24", heavy duty	FLOOR	4.7034	507.45	1,929.23	2,436.68
10.5001 051	Chute, 18 ga stainless steel, 30", heavy duty	FLOOR	4.7034	507.45	2,124.23	2,631.68
10.5001 061	Chute door, stainless steel rim, manual	EA	1.6794	181.19	473.43	654.62
10.5001 071	Pneumatic vertical	FLOOR	16.1249	1,739.72	10,813.49	12,553.20
10.5001 081	Pneumatic horizontal	LF	1.6794	181.19	1,590.33	1,771.52
10.5002 000	**FLAG POLES WITH FOUNDATIONS:**					
10.5002 011	Flag pole, fiberglass, 30', with foundation	EA	17.3087	1,867.44	2,210.05	4,077.48
10.5002 021	Flag pole, fiberglass, 35', with foundation	EA	23.1354	2,496.08	3,382.55	5,878.63
10.5002 031	Flag pole, fiberglass, 39', with foundation	EA	26.0276	2,808.12	3,836.06	6,644.18
10.5002 041	Flag pole, fiberglass, 60', with foundation	EA	34.7035	3,744.16	9,315.68	13,059.84
10.5002 051	Flag pole, aluminum, 30', with foundation	EA	17.3087	1,867.44	2,367.21	4,234.65
10.5002 061	Flag pole, aluminum, 35', with foundation	EA	23.1354	2,496.08	3,358.18	5,854.26
10.5002 071	Flag pole, aluminum, 40', with foundation	EA	26.0276	2,808.12	4,406.58	7,214.69
10.5002 081	Flag pole, aluminum, 50', with foundation	EA	31.8113	3,432.12	6,897.88	10,330.00
10.5002 091	Flag pole, aluminum, 60', with foundation	EA	34.7035	3,744.16	13,155.91	16,900.07
10.5002 101	Flag pole, aluminum, 70', with foundation	EA	42.4154	4,576.20	15,215.35	19,791.55
10.5003 000	**DIRECTORIES:**					
10.5003 001	*Note: For bronze or stainless steel .... add 100% to total costs* *For illumination .... add 50% to total costs* *For lettered header .... add 15% to total costs* *For recessing .... add 10% to total costs*					
10.5003 011	Directory, aluminum & glass, 24" x 36"	EA	3.1656	341.54	949.76	1,291.29
10.5003 021	Directory, aluminum & glass, 48" x 36"	EA	4.7492	512.39	1,371.90	1,884.30
10.5003 031	Directory, aluminum & glass, 60" x 36"	EA	5.2763	569.26	1,899.50	2,468.76
10.5003 041	Directory, aluminum & glass, 72" x 48"	EA	6.8599	740.11	2,279.48	3,019.59
10.5004 000	**DISPLAY CASES:**					
10.5004 011	Display case, glass, wood pedestal, 48x36 no frame	EA	3.9579	427.02	6,880.29	7,307.31
10.5004 021	Display case, glass, wood pedestal, 72x48 no frame	EA	4.7492	512.39	7,892.13	8,404.52
10.5004 031	Display case, glass, 72x48, recessed front	EA	5.0127	540.82	2,630.69	3,171.51

10 - SPECIALTIES

Division	Description	Unit	Labor Hours	Labor Cost	Material Cost	Total Cost
10.5005 000	**METAL LETTERS & PLAQUES:**					
10.5005 001	Note: For black aluminum or satin bright face … add 10% to total cost For gold and black aluminum …………………… add 15% For duranodic color or hard coat finish …………… add 20% For chrome plate or verdigris …………………… add 15% For nickel silver …………………………… add 20%					
10.5005 011	Letters, aluminum, 2", gothic block	EA	0.4185	45.15	18.22	63.37
10.5005 021	Letters, aluminum, 6", gothic block	EA	0.4185	45.15	30.49	75.64
10.5005 031	Letters, aluminum, 12", gothic block	EA	0.4185	45.15	74.70	119.85
10.5005 041	Letters, bronze, 2", gothic	EA	0.4185	45.15	21.86	67.01
10.5005 051	Letters, bronze, 6", gothic	EA	0.4185	45.15	67.69	112.84
10.5005 061	Letters, bronze, 12", gothic	EA	0.4185	45.15	169.30	214.45
10.5005 071	Letters, stainless steel, 6"	EA			215.10	215.10
10.5005 081	Plaque, aluminum, 24" x 24"	EA	0.6975	75.25	1,720.07	1,795.32
10.5005 091	Plaque, aluminum, 24" x 36"	EA	0.6975	75.25	2,661.56	2,736.81
10.5005 101	Plaque, bronze, 24" x 24"	EA	0.6975	75.25	2,105.43	2,180.68
10.5005 111	Plaque, bronze, 24" x 36"	EA	0.6975	75.25	3,241.63	3,316.88
10.5006 000	**GRAPHICS, SIGNS:**					
10.5006 011	Letters & numbers, plastic, 1"	PLATE	0.1793	19.34	2.40	21.74
10.5006 021	Sign, engraved brass	PLATE	0.1793	19.34	2.96	22.31
10.5006 031	Sign, hand lettered	SQ IN	0.1770	19.10	2.82	21.92
10.5006 041	Sign, hand painted, gold leaf	SQ IN	0.3632	39.19	4.47	43.65
10.5006 051	Sign, porcelain enamel, 12"x2-1/2"x6", neon	EA	0.7063	76.20	270.03	346.23
10.5006 061	Sign, porcelain enamel, 24"x5"x8", neon	EA	1.4135	152.50	568.38	720.88
10.5006 071	Sign, porcelain enamel, 36"x6 1/2"x10", neon	EA	2.8268	304.98	888.64	1,193.62
10.5006 081	Sign, baked enamel, 12"x3" deep, neon	EA	0.7063	76.20	152.21	228.42
10.5006 091	Sign, baked enamel, 24"x6" deep, neon	EA	1.1777	127.06	336.00	463.06
10.5006 101	Sign, baked enamel, 36"x8" deep, neon	EA	2.1197	228.69	612.29	840.98
10.5007 000	**TURNSTILES:**					
10.5007 011	Turnstile, non-register	EA	1.3463	145.25	1,374.98	1,520.23
10.5007 021	Turnstile, register	EA	1.3463	145.25	1,887.59	2,032.84
10.5007 031	Turnstile, register, portable	EA	0.3049	32.90	2,156.04	2,188.93
10.5007 041	Turnstile ticket collection box	EA	0.3049	32.90	410.76	443.66
10.5008 000	**EXTINGUISHERS:**					
10.5008 001	Note: The following prices are based on quantities of 20 or more.					
10.5008 011	Extinguisher, CO2, 5#, swivel horn	EA	0.3479	37.53	207.19	244.72
10.5008 021	Extinguisher, CO2, 10#, hose & 'H' horn	EA	0.3479	37.53	315.70	353.24
10.5008 031	Extinguisher, CO2, 15#, hose & 'H' horn	EA	0.3479	37.53	353.95	391.49
10.5008 041	Extinguisher, CO2, 20#, hose & 'H' horn	EA	0.3479	37.53	440.87	478.41
10.5008 051	Extinguisher, CO2, 50#, wheeled cart	EA	0.3479	37.53	1,754.04	1,791.57
10.5008 061	Extinguisher, monoamm phos, 5#, nozzle	EA	0.3479	37.53	106.76	144.29
10.5008 071	Extinguisher, monoamm, 10#, short hose	EA	0.3479	37.53	170.58	208.12
10.5008 081	Extinguisher, monoamm phos, 20#, nozzle	EA	0.3479	37.53	218.45	255.98
10.5008 091	Extinguisher, halon 1211, 5#, nozzle	EA	0.3479	37.53	189.69	227.23
10.5008 101	Extinguisher, halon 1211, 10#, tall hose	EA	0.3479	37.53	291.73	329.27
10.5008 111	Extinguisher, halon 1211, 20#, hose	EA	0.3479	37.53	424.12	461.66
10.5009 000	**EXTINGUISHER CABINETS:**					
10.5009 011	Extinguisher cabinet, steel, 12" x 27"	EA	0.9502	102.52	138.73	241.24
10.5009 021	Extinguisher cabinet, steel, 20" x 30"	EA	0.9502	102.52	173.77	276.29
10.5009 031	Extinguisher cabinet, aluminum, 12" x 27"	EA	0.9502	102.52	180.33	282.84
10.5009 041	Extinguisher cabinet, aluminum, 20" x 30"	EA	0.9502	102.52	225.85	328.37
10.5009 051	Extinguisher cabinet, stainless steel, 12" x 27"	EA	0.9502	102.52	416.23	518.74
10.5009 061	Extinguisher cabinet, stainless steel, 20" x 30"	EA	0.9502	102.52	521.34	623.86
10.5010 000	**FIRE HOSE CABINETS, WITH GLASS DOOR:**					
10.5010 011	Hose cabinet, steel, 22" x 30" x 5"	EA	1.1605	125.21	248.17	373.38
10.5010 021	Hose cabinet, steel, 22x30x5, 50' hose & nozzle	EA	1.1605	125.21	567.64	692.84
10.5010 031	Hose cabinet, steel, 24" x 30" x 8"	EA	1.1605	125.21	260.30	385.51

SPECIALTIES - 10

Division	Description	Unit	Labor Hours	Labor Cost	Material Cost	Total Cost
10.5010 041	Hose cabinet, steel, 24x30x8, 75' hose & nozzle	EA	1.1605	125.21	704.82	830.03
10.5010 051	Hose cabinet, aluminum, 24" x 30" x 8"	EA	1.1605	125.21	338.45	463.66
10.5010 061	Hose cabinet, aluminum, 24x30x8, 75' hose & nozzle	EA	1.1605	125.21	782.88	908.09
10.5010 071	Hose cabinet, stainless steel, 24" x 30" x 8"	EA	1.1605	125.21	781.15	906.36
10.5010 081	Hose cabinet, stainless steel, 24x30x8, 75' hose & nozzle	EA	1.1605	125.21	1,225.58	1,350.79
10.5011 000	**MAIL CHUTES & COLLECTION BOXES:**					
10.5011 001	*Note: For bronze mail chutes, use prices given for stainless steel.*					
10.5011 011	Mail chute, aluminum & glass, 8-3/4"x3-1/2"	FLOOR	3.5781	470.31	799.36	1,269.66
10.5011 021	Mail chute, aluminum & glass, 14-1/4"x4-5/8"	FLOOR	4.2942	564.43	1,315.09	1,879.52
10.5011 031	Mail chute, aluminum & glass, 14-1/4"x8-5/8"	FLOOR	4.2942	564.43	1,682.70	2,247.13
10.5011 041	Mail chute, stainless steel, 8-3/4" x 3-1/2"	FLOOR	3.5781	470.31	1,097.17	1,567.48
10.5011 051	Mail chute, stainless steel, 14-1/4" x 4-5/8"	FLOOR	4.8564	638.33	1,940.31	2,578.64
10.5011 061	Mail chute, stainless steel, 14-1/4" x 8-5/8"	FLOOR	4.8564	638.33	2,342.98	2,981.30
10.5011 071	Collection boxes, aluminum	EA	3.9873	524.09	1,621.38	2,145.47
10.5011 081	Collection boxes, stainless steel or bronze	EA	0.7243	95.20	2,007.88	2,103.08
10.5012 000	**MAIL BOXES:**					
10.5012 011	Mail box, aluminum, gang, front load	EA	0.1433	15.46	23.17	38.63
10.5012 021	Mail box, aluminum, gang, front load, intercom	EA	0.1985	21.42	31.72	53.13
10.5012 031	Add for bronze or stainless steel	EA	0.0727	7.84	10.58	18.43
10.5013 000	**SEATING:**					
10.5013 011	Seating, theater, economy	EA	0.3090	33.34	134.32	167.66
10.5013 021	Seating, theater, loge, rocking	EA	0.4890	52.76	211.09	263.85
10.5013 031	Seating, auditorium	EA	0.6118	66.01	263.85	329.86
10.5013 041	Bleachers, 18" aluminum seat, steel support	EA	0.3540	38.19	148.59	186.78
10.5013 051	Bleachers, plastic seat, steel support	EA	0.0594	6.41	23.31	29.72
10.5013 061	Bleachers, 18" folding seat, lacquer	EA	0.1228	13.25	49.38	62.63
10.5013 071	Add for under structure	EA	0.0993	10.71	36.61	47.32
10.5013 081	Table, folding, with seats	EA	2.0925	225.76	1,139.39	1,365.15
10.5014 000	**METAL LOCKERS:**					
10.5014 011	Locker, 12" x 12" x 72", 1 tier	EA	0.2259	19.69	235.31	255.00
10.5014 021	Locker, 15" x 18" x 72", 1 tier	EA	0.2766	24.11	256.11	280.22
10.5014 031	Locker, 12" x 12" x 30", 2 tier	TIER	0.2259	19.69	297.90	317.58
10.5014 041	Locker, 12" x 12" x 24", 3 tier	TIER	0.2259	19.69	339.82	359.51
10.5014 051	Locker, 12" x 12" x 12", 5 high	TIER	0.2889	25.18	313.47	338.65
10.5014 061	Locker, 15" x 15" x 12", 5 high	TIER	0.2889	25.18	371.27	396.45
10.5015 000	**SHELVING & BINS:**					
10.5015 011	Shelves, metal, 5, 30"x8"x84", library	EA	0.8479	91.48	550.74	642.22
10.5015 021	Shelves, metal, 7, 36"x8"x84", library	EA	0.9420	101.63	563.97	665.61
10.5015 031	Shelves, metal, 7, 36"x12"x84", industrial	EA	0.8479	91.48	304.83	396.31
10.5015 041	Shelves, metal, 5, 36"x24"x84", industrial	EA	0.9420	101.63	424.59	526.23
10.5016 000	**SCALES, STEEL, PLATFORM:**					
10.5016 011	Scale, 5 ton, 6' x 8', dial/readout	EA	34.9572	3,771.53	18,092.44	21,863.97
10.5016 021	Scale, 25 ton, 10'x24', dial/readout	EA	56.5343	6,099.49	26,707.25	32,806.74
10.5016 031	Scale, 50 ton, 10'x60', dial/readout	EA	89.5125	9,657.50	43,194.39	52,851.89
10.5016 041	Scale, 80 ton, 10'x60', dial/readout	EA	89.5125	9,657.50	52,607.04	62,264.54
10.5017 000	**TELEPHONE ENCLOSURES:**					
10.5017 011	Phone enclosure, shelf style	EA	1.5233	164.35	1,567.60	1,731.95
10.5017 021	Phone enclosure, desk style	EA	0.9913	106.95	1,019.40	1,126.35
10.5017 031	Phone enclosure, full height	EA	3.9499	426.15	4,065.44	4,491.59
10.5018 000	**CHAIN LINK PARTITIONS & GATES:**					
10.5018 011	Chain link partition with pipe frame	SF	0.0322	2.81	1.95	4.76
10.5018 021	Chain link gate, 3'x4'	EA	4.8024	418.53	493.46	911.99
10.5019 000	**RACKING, INDUSTRIAL:**					
10.5019 001	*Note: Codes may require in-rack sprinklers. See section 15.5500.*					
10.5019 011	Racking, 2-tier, single row	LF	0.6189	66.77	85.59	152.36
10.5019 021	Racking, 2-tier, back to back	LF	1.2379	133.56	171.20	304.76
10.5019 031	Racking, 3-tier, single row	LF	0.8253	89.04	114.10	203.15

10 - SPECIALTIES

Division	Description	Unit	Labor Hours	Labor Cost	Material Cost	Total Cost
10.5019 041	Racking, 3-tier, back to back	LF	1.6506	178.08	228.22	406.30
10.5019 051	Add for wall brackets	LF	0.0165	1.78	2.30	4.08
10.5020 000	**AWNINGS & CANOPIES:**					
10.5020 011	Awning, canvas, with frame, average	SF	0.1650	17.80	21.08	38.88
10.5020 021	Awning, canvas, with frame, custom	SF	0.2600	28.05	31.58	59.63
10.5020 031	Canopy, entrance, aluminum	SF	0.1500	16.18	13.18	29.36
10.5020 041	Canopy, wall hung, aluminum	SF	0.0900	9.71	7.89	17.60

EQUIPMENT - 11

Division	Description	Unit	Labor Hours	Labor Cost	Material Cost	Total Cost
11.1100 000	**BANK EQUIPMENT:**					
11.1101 000	**VAULTS & VAULT DOORS:**					
11.1101 001	Note: The vault doors listed below are for safekeeping, such as in safe deposit boxes.					
11.1101 011	Vault door, 78"x44"x3-1/2", steel, class 1	EA	12.2800	1,157.02	37,673.09	38,830.11
11.1101 021	Vault door, 78"x44"x3-1/2", steel, class 2	EA	12.2800	1,157.02	42,534.11	43,691.14
11.1101 031	Vault door, 78"x44"x3-1/2", steel, class 3	EA	12.2800	1,157.02	51,040.96	52,197.99
11.1101 041	Vault door, 78"x44"x3-1/2", 6 hour	EA	12.2800	1,157.02	28,432.50	29,589.52
11.1101 051	Vault, 78" x 33", 1 hour, single	EA	6.3500	598.30	5,004.99	5,603.28
11.1101 061	Vault, 84" x 51-1/8", 2 hour, single	EA	6.3500	598.30	6,099.89	6,698.18
11.1101 071	Vault, 84" x 51-1/8", 4 hour, single	EA	6.2300	586.99	6,375.97	6,962.96
11.1101 081	Vault, 84" x 51-1/8", 6 hour, single	EA	6.3100	594.53	7,400.08	7,994.61
11.1101 091	Modular vaults, class I, 15 minute	EA	103.5200	9,753.65	35,395.97	45,149.62
11.1101 101	Modular vaults, class II, 1 hour	EA	103.5100	9,752.71	58,993.25	68,745.97
11.1101 111	Modular vaults, class III, 2 hour	EA	103.5600	9,757.42	82,590.52	92,347.94
11.1102 000	**DRIVE- & WALK-UP TELLER WINDOWS:**					
11.1102 011	Teller window, drive-up, manual	EA	13.6046	1,281.83	11,628.88	12,910.70
11.1102 021	Teller window, drive-up, motorized	EA	13.6046	1,281.83	13,096.99	14,378.82
11.1102 031	Teller window, walk-up, one teller	EA	10.8835	1,025.44	7,994.69	9,020.14
11.1102 041	Teller window, walk-up, 2 teller	EA	10.8835	1,025.44	8,832.72	9,858.17
11.1103 000	**NIGHT DEPOSIT DOORS:**					
11.1103 011	Night deposit, bag/envelope, illuminated	EA	11.2700	1,061.86	6,583.64	7,645.50
11.1103 021	Night deposit, envelope	EA	3.1099	293.01	2,720.71	3,013.73
11.1103 031	Night deposit, bag, flush mount	EA	5.1110	481.56	3,324.80	3,806.35
11.1104 000	**TELLER COUNTERS & CHECK DESKS:**					
11.1104 011	Teller counter, modular component	LF	1.0300	97.05	397.33	494.38
11.1104 021	Check desk, round, 48", 4 person	EA	6.2370	587.65	2,911.91	3,499.56
11.1104 031	Check desk, square, 48", 4 person	EA	6.2954	602.34	2,454.06	3,056.41
11.1104 041	Check desk, 72" x 24", 4 person	EA	3.8120	364.73	2,916.62	3,281.35
11.1104 051	Check desk, 72" x 36", 8 person	EA	4.9870	477.16	3,624.54	4,101.70
11.1105 000	**SAFE DEPOSIT BOXES, MODULAR UNITS:**					
11.1105 011	Safe deposit, 42 openings, 2" x 5"	MOD	3.9740	380.23	3,884.08	4,264.31
11.1105 021	Safe deposit, 30 openings, 2" x 5"	MOD	2.9681	283.99	2,774.99	3,058.98
11.1105 031	Safe deposit, 18 openings, 5" x 5"	MOD	2.4731	236.63	2,005.74	2,242.37
11.1105 041	Safe deposit, base, 32" x 24" x 3"	EA	0.8998	86.09	259.55	345.64
11.1105 051	Safe deposit, canopy top	EA	0.4950	47.36	66.65	114.01
11.1105 061	Safe deposit, 9 openings, 5"x10 3/8"	EA	2.9021	273.44	1,510.20	1,783.63
11.1105 071	Safe deposit, 15 openings, 3"x10 3/8"	EA	2.8760	270.98	1,746.21	2,017.18
11.1105 081	Safe deposit, 1 section, 3 openings, 5"x10"	EA	5.2300	492.77	1,533.82	2,026.59
11.1105 091	Deposit, 1 section, 3 openings, 10"x10 3/8"	EA	5.3210	501.34	1,533.82	2,035.16
11.1106 000	**SAFES:**					
11.1106 011	Safe, floor, 10"x24", minimum security	EA	1.1777	127.06	633.15	760.21
11.1106 021	Safe, wall, minimum security	EA	8.3897	905.16	3,315.82	4,220.98
11.1106 031	Safe, cabinet, medium security, full door	EA	12.0899	1,304.38	4,778.00	6,082.38
11.1106 041	Book drop, minimum security	EA	3.0005	323.72	1,185.77	1,509.49
11.1106 051	Night depository	EA	4.8941	528.02	1,934.16	2,462.18
11.1200 000	**ECCLESIASTICAL EQUIPMENT:**					
11.1201 000	**LECTERNS:**					
11.1201 011	Lectern, economy, 16" x 24"	EA	3.2981	315.56	816.71	1,132.27
11.1201 021	Lectern, good, 16" x 24"	EA	3.2981	315.56	1,537.32	1,852.88
11.1201 031	Lectern, best, 16" x 24"	EA	3.2981	315.56	2,882.58	3,198.14
11.1202 000	**PULPITS:**					
11.1202 011	Pulpit, economy	EA	3.5336	338.09	912.71	1,250.81
11.1202 021	Pulpit, good	EA	3.5336	338.09	1,601.43	1,939.53
11.1202 031	Pulpit, best	EA	3.5336	338.09	3,523.21	3,861.31
11.1203 000	**ARKS:**					

11 - EQUIPMENT

Division	Description	Unit	Labor Hours	Labor Cost	Material Cost	Total Cost
11.1203 011	Ark, with curtain, economy	EA	4.2399	405.67	944.82	1,350.49
11.1203 021	Ark, with curtain, good	EA	4.2399	405.67	1,249.16	1,654.83
11.1203 031	Ark, with curtain, best	EA	4.2399	405.67	2,081.88	2,487.56
11.1203 041	Ark, with doors, economy	EA	6.5953	631.04	1,072.89	1,703.93
11.1203 051	Ark, with doors, good	EA	6.5953	631.04	1,857.68	2,488.72
11.1203 061	Ark, with doors, best	EA	6.5953	631.04	2,914.61	3,545.65
11.1204 000	**PEWS:**					
11.1204 011	Pew, bench, economy	LF	0.2596	24.84	56.00	80.83
11.1204 021	Pew, bench, good	LF	0.2596	24.84	60.82	85.66
11.1204 031	Pew, bench, best	LF	0.2596	24.84	70.40	95.24
11.1204 041	Pew, seat, economy	LF	0.3060	29.28	70.40	99.68
11.1204 051	Pew, seat, good	LF	0.3060	29.28	86.40	115.68
11.1204 061	Pew, seat, best	LF	0.3060	29.28	102.46	131.74
11.1205 000	**KNEELERS:**					
11.1205 011	Kneeler, good	LF	0.1272	12.17	15.24	27.41
11.1205 021	Kneeler, best	LF	0.1272	12.17	19.30	31.47
11.1205 031	Kneeler, serenity	LF	0.1272	12.17	29.74	41.91
11.1206 000	**CATHEDRAL CHAIRS, SERENITY:**					
11.1206 011	Cathedral chair, shaped wood, economy	EA	0.1651	15.80	192.14	207.94
11.1206 021	Cathedral chair, shaped wood, book rack	EA	0.1651	15.80	208.10	223.89
11.1206 031	Cathedral chair, wood, book rack, kneeler	EA	0.1651	15.80	240.15	255.94
11.1206 041	Cathedral chair, upholstered, economy	EA	0.1882	18.01	201.77	219.78
11.1206 051	Cathedral chair, upholstered, book rack	EA	0.1882	18.01	217.82	235.83
11.1206 061	Cathedral chair, upholstered, book rack, kneeler	EA	0.1882	18.01	249.76	267.77
11.1207 000	**CONFESSIONALS, SINGLE:**					
11.1207 011	Confessional, single, with curtain, economy	EA	6.5953	631.04	3,523.21	4,154.25
11.1207 021	Confessional, single, with curtain, good	EA	6.5953	631.04	4,163.79	4,794.83
11.1207 031	Confessional, single, with curtain, best	EA	6.5953	631.04	4,804.38	5,435.41
11.1207 041	Confessional, single, with door, economy	EA	6.5953	631.04	4,323.92	4,954.96
11.1207 051	Confessional, single, with door, good	EA	6.5953	631.04	4,884.46	5,515.49
11.1207 061	Confessional, single, with door, best	EA	6.5953	631.04	5,445.02	6,076.06
11.1208 000	**CONFESSIONALS, DOUBLE:**					
11.1208 011	Confessional, double, with curtain, economy	EA	9.8934	946.60	5,605.18	6,551.78
11.1208 021	Confessional, double, with curtain, good	EA	9.8934	946.60	6,774.22	7,720.82
11.1208 031	Confessional, double, with curtain, best	EA	9.8934	946.60	8,007.40	8,954.00
11.1208 041	Confessional, double, with doors, economy	EA	9.8934	946.60	7,046.57	7,993.17
11.1208 051	Confessional, double, with doors, good	EA	9.8934	946.60	8,007.40	8,954.00
11.1208 061	Confessional, double, with doors, best	EA	9.8934	946.60	9,288.54	10,235.14
11.1209 000	**COMMUNION RAILS, HARDWOOD, WITH STANDARDS:**					
11.1209 011	Communion rail, hardwood, economy	LF	0.2355	22.53	63.99	86.52
11.1209 021	Communion rail, hardwood, good	LF	0.2355	22.53	79.96	102.50
11.1209 031	Communion rail, hardwood, best	LF	0.2355	22.53	95.99	118.52
11.1210 000	**COMMUNION RAILS, CARVED, WITH STANDARDS:**					
11.1210 011	Communion rail, carved oak, economy	LF	0.2355	22.53	112.01	134.55
11.1210 021	Communion rail, carved oak, good	LF	0.2355	22.53	160.08	182.61
11.1210 031	Communion rail, carved oak, best	LF	0.2355	22.53	224.15	246.69
11.1211 000	**COMMUNION RAILS, METAL, WITH STANDARDS:**					
11.1211 011	Communion rail, bronze or stainless steel	LF	0.2828	26.65	140.88	167.53
11.1212 000	**ALTARS, HARDWOOD:**					
11.1212 011	Altar, hardwood, economy	EA	4.0518	437.15	1,120.96	1,558.11
11.1212 021	Altar, hardwood, good	EA	4.0518	437.15	1,357.97	1,795.12
11.1212 031	Altar, hardwood, best	EA	4.4754	482.85	1,761.58	2,244.43
11.1213 000	**ALTARS, HARDWOOD, CARVED:**					
11.1213 011	Altar, carved hardwood, economy	EA	4.0518	437.15	4,528.98	4,966.13
11.1213 021	Altar, carved hardwood, good	EA	4.0518	437.15	8,052.20	8,489.35
11.1213 031	Altar, carved hardwood, best	EA	4.4754	482.85	9,304.54	9,787.39

EQUIPMENT - 11

Division	Description	Unit	Labor Hours	Labor Cost	Material Cost	Total Cost
11.1214 000	**ALTARS, MARBLE OR GRANITE:**					
11.1214 011	Altar, marble or granite, economy	EA	23.5557	2,270.77	6,405.87	8,676.64
11.1214 021	Altar, marble or granite, good	EA	23.5557	2,270.77	9,608.87	11,879.64
11.1214 031	Altar, marble or granite, best	EA	23.5557	2,270.77	14,413.37	16,684.14
11.1215 000	**ALTARS, HARDWOOD, WITH MARBLE BASE & LEGS:**					
11.1215 011	Altar, hardwood, marble base, good	EA	18.8449	1,816.65	3,523.21	5,339.86
11.1215 021	Altar, hardwood, marble base, best	EA	18.8449	1,816.65	5,124.66	6,941.31
11.1216 000	**STAINED GLASS, INCLUDING ARTWORK:**					
11.1216 011	Stained glass, simple artwork	SF	0.1673	16.32	67.78	84.10
11.1216 021	Stained glass, moderate artwork	SF	0.2786	27.17	91.50	118.67
11.1216 031	Stained glass, elaborate artwork	SF	0.5562	54.24	118.64	172.88
11.1217 000	**FACET GLASS, INCLUDING ART WORK:**					
11.1217 011	Facet glass, with simple artwork	SF	0.1673	16.32	50.83	67.15
11.1217 021	Facet glass, with moderate artwork	SF	0.2786	27.17	77.90	105.07
11.1217 031	Facet glass, with elaborate artwork	SF	0.5562	54.24	108.43	162.67
11.1218 000	**COLORED GLASS:**					
11.1218 011	Colored glass, single pane	SF	0.0558	5.44	6.68	12.12
11.1218 021	Colored glass, patterned	SF	0.1673	16.32	13.52	29.84
11.1218 031	Colored glass, small pieces	SF	0.2228	21.73	37.26	58.99
11.1300 000	**EDUCATIONAL EQUIPMENT:**					
11.1300 001	*Note: For chalk & tack boards see section 10.1000.*					
11.1301 000	**WARDROBES:**					
11.1301 011	Wardrobe, teacher, 40"x78"x26-1/4"	EA	0.7440	70.10	1,279.66	1,349.76
11.1301 021	Wardrobe, student, 40"x78"x26-1/4"	EA	0.5580	52.57	838.58	891.15
11.1302 000	**SEATING:**					
11.1302 011	Seating, pedestal, folding arm	EA	0.2342	25.27	141.45	166.71
11.1302 021	Seating, horizontal 2 section, 5 chair	SEAT	0.2514	27.12	152.03	179.15
11.1303 000	**TABLES:**					
11.1303 011	Table, fixed pedestal, 48" x 16"	EA	0.9753	105.23	589.65	694.88
11.1303 021	Table, fixed pedestal, chair, 48"x16"	EA	1.3438	144.98	812.33	957.31
11.1304 000	**PROJECTION SCREENS:**					
11.1304 011	Slide screen, pull, 70"x70", ceiling	SF	0.0138	1.49	6.88	8.36
11.1304 021	Slide screen, electric, ceiling	SF	0.0567	6.12	59.06	65.18
11.1305 000	**DRAFTING FURNITURE:**					
11.1305 011	Draft table, steel base, 60"x37-1/2"	EA			657.53	657.53
11.1305 021	Draft table, hardwood, 60"x37-1/2"	EA			1,356.90	1,356.90
11.1305 031	Draft table, 2 station, 10 drawer, flexible	EA			918.89	918.89
11.1305 041	Draft table, 1 station, 6 drawer, flexible	EA			734.25	734.25
11.1305 051	Desk, metal frame, mechanical drawing	EA			985.33	985.33
11.1305 061	Desk, wood, mechanical drawing	EA			492.62	492.62
11.1305 071	Tracing table, pedestal, 24" x 22"	EA			879.56	879.56
11.1306 000	**FILES:**					
11.1306 011	File cabinet, steel, 10 drawer	EA			2,389.03	2,389.03
11.1306 021	File cabinet, wood, 10 drawer	EA			1,857.06	1,857.06
11.1306 031	Files, modular, 8 tube, 48"	EA			145.37	145.37
11.1306 041	Files, vertical, 26 binder, 44 3/4"	EA			2,651.80	2,651.80
11.1307 000	**AUDIO-VISUAL EQUIPMENT:**					
11.1307 011	Video tape recorder	EA	4.1558	448.37	5,258.64	5,707.01
11.1307 021	Camera	EA	4.4522	480.35	2,102.70	2,583.04
11.1307 031	Monitor	EA	4.2047	453.65	931.49	1,385.14
11.1308 000	**STUDY CARREL, PLASTIC LAMINATED WOOD:**					
11.1308 011	Study carrel, 48"x30"x54", 1 station	EA	0.9891	106.71	1,003.98	1,110.70
11.1308 021	Study carrel, 73"x30"x47", 2 station	EA	1.4841	160.12	1,757.27	1,917.39
11.1308 031	Study carrel, 66"x66"x47", 4 station	EA	2.2257	240.13	2,079.90	2,320.03
11.1309 000	**AUDIO-VISUAL EQUIPMENT:**					
11.1309 011	Tape recorder	EA			581.10	581.10

11 - EQUIPMENT

Division	Description	Unit	Labor Hours	Labor Cost	Material Cost	Total Cost
11.1309 021	Head set	SET			56.11	56.11
11.1309 031	Projector, movie, 8 mm	EA			686.78	686.78
11.1309 041	Projector, slide, carousel	EA			528.27	528.27
11.1310 000	**AUDIO-VISUAL CENTER, MOBILE, WITH CONTROL PANELS:**					
11.1310 001	*Note: The following item has 10 listening stations with earphones.*					
11.1310 011	Folding table, elect, ear phone, 4x8	EA			1,447.42	1,447.42
11.1310 021	Stack chairs	EA			154.58	154.58
11.1311 000	**ACCESSORIES FOR CARRELS:**					
11.1311 011	Rear projection module with light	EA			351.24	351.24
11.1311 021	Power column, study carrel	EA			70.21	70.21
11.1400 000	**OBSERVATORIES & PLANETARIUMS:**					
11.1401 000	**DOME, OBSERVATION, REVOLVING, SHELL ONLY:**					
11.1401 011	Dome, observation, revolving, 12', 800#, shell	EA	30.9175	3,335.69	10,237.23	13,572.92
11.1401 021	Dome, observation, revolving, 12' base, shell	EA	13.6040	1,467.74	3,802.34	5,270.08
11.1401 031	Dome, observation, revolving, 18', 2500#, shell	EA	49.2201	5,310.36	28,225.55	33,535.91
11.1401 041	Dome, observation, revolving, 18' base, shell	EA	24.7339	2,668.54	11,114.63	13,783.17
11.1401 051	Dome, observation, revolving, 20', 4500#, shell	EA	91.5151	9,873.56	50,601.31	60,474.87
11.1401 061	Dome, observation, revolving, 20' base, shell	EA	33.3907	3,602.52	16,964.58	20,567.10
11.1402 000	**TELESCOPES:**					
11.1402 011	Telescope, reflector, 6", portable	EA			2,166.55	2,166.55
11.1402 021	Telescope, reflector, 8", stationary	EA			7,410.14	7,410.14
11.1402 031	Telescope, refraction, 4", portable	EA			4,187.12	4,187.12
11.1402 041	Telescope, refraction 4", stationary	EA			4,334.20	4,334.20
11.1402 051	Telescope, refraction, 6", stationary	EA			17,272.21	17,272.21
11.1403 000	**PLANETARIUM CLASS ROOM EQUIPMENT:**					
11.1403 011	Instrument control console system 512	EA			102,070.74	102,070.74
11.1403 021	Add for automatic controls	EA			52,758.59	52,758.59
11.1403 031	Hemispherical screen, 30' dia	EA			65,948.32	65,948.32
11.1403 041	Seating, table arm & reclinable	EA			380.63	380.63
11.1403 051	Special effects projector	EA			2,407.96	2,407.96
11.1403 061	Stereo sound system	ROOM			10,274.74	10,274.74
11.1403 071	Cove lighting	ROOM			19,411.04	19,411.04
11.1501 000	**VOCATIONAL SHOP EQUIPMENT:**					
11.1501 011	Welding booth, glassweld panel	SF			15.99	15.99
11.1501 021	Welding booth, fireproof panel	SF			13.12	13.12
11.1600 000	**FOOD SERVICE EQUIPMENT:**					
11.1600 001	*Note: The prices for commercial and institutional equipment do not include final electrical or mechanical connections. Note that for preliminary and schematic estimates, the unit price for food service equipment must be at least $90 per square foot.*					
11.1601 000	**FOOD SERVICE EQUIPMENT, RESIDENTIAL:**					
11.1601 001	*Note: The following prices are based on quantities of 100 to 500 each purchased over one year. For more than 500, deduct 10%.*					
11.1601 011	Oven, single, self clean	EA			1,323.03	1,323.03
11.1601 021	Oven, double, self clean	EA			1,541.35	1,541.35
11.1601 031	Oven, microwave, built-in	EA			2,452.35	2,452.35
11.1601 041	Cook top	EA			890.96	890.96
11.1601 051	Range & oven, drop-in	EA			1,276.24	1,276.24
11.1601 061	Range hood with microwave	EA			906.50	906.50
11.1601 071	Hood	EA			155.91	155.91
11.1601 081	Hood with microwave	EA			906.50	906.50
11.1601 091	Electric grill	EA			913.22	913.22
11.1601 101	Refrigerator, 12 CF	EA			933.25	933.25
11.1601 111	Refrigerator, 18 CF	EA			1,218.35	1,218.35
11.1601 121	Refrigerator with icemaker	EA			1,173.81	1,173.81
11.1601 131	Freezer, 16 CF	EA			861.96	861.96
11.1601 141	Dishwasher, built-in	EA			741.67	741.67

EQUIPMENT - 11

Division	Description	Unit	Labor Hours	Labor Cost	Material Cost	Total Cost
11.1601 151	Garbage disposal	EA			218.29	218.29
11.1601 161	Trash compactor	EA			652.61	652.61
11.1601 171	Washer	EA			815.22	815.22
11.1601 181	Dryer, electric	EA			648.14	648.14
11.1601 191	Dryer, gas	EA			1,176.06	1,176.06
11.1602 000	**IN-PLACE COST, FABRICATED ITEMS:**					
11.1602 001	*Note: Prices for fabricated food service fixtures may vary as much as $200.00 Per foot, depending on style, materials, accessories and built-in items. The following prices are mean average costs for most common configurations. Factory manufactured buy-out items are listed in the latter part of this section. The price for sandwich preparation with open storage does not include ductwork or exhaust fan.*					
11.1602 021	Tables, counter with sink, shelves & racks, economy	LF	0.7530	90.10	401.70	491.81
11.1602 031	Tables, counter with sink, shelves & racks, best	LF	1.1492	137.51	647.23	784.75
11.1602 041	Pot washer table, loading apron & sink, economy	LF	0.6341	75.88	357.07	432.95
11.1602 051	Pot washer table, loading apron & sink, best	LF	0.7926	94.84	446.31	541.15
11.1602 061	Sandwich prep, open storage	LF	1.0304	123.30	580.25	703.55
11.1602 071	Sandwich prep, closed storage, undercounter refrigerator	LF	1.7039	203.89	985.93	1,189.82
11.1602 081	Sandwich prep, closed storage, plastic laminated face, refrigerator	LF	1.7832	213.38	1,004.31	1,217.68
11.1602 091	Service counter, open storage	LF	1.7039	203.89	985.93	1,189.82
11.1602 101	Service center, closed storage	LF	2.1795	260.80	985.93	1,246.73
11.1602 131	Cook's table, with sink, 6'	EA	5.1525	616.55	5,657.99	6,274.54
11.1602 141	Vegetable preparation table, with sink, 12'	EA	6.6203	792.19	5,939.35	6,731.54
11.1602 151	Food service equipment, average cost	SF			505.06	505.06
11.1603 000	**FABRICATED FIXTURES, BUILT-UP, UNIT COST:**					
11.1603 001	*Note: The following basic fixtures are 30" x 32" x 6'-21', free standing or wall mounted with top and frame. Combine these items with the adders listed below for complete cost.*					
11.1603 011	Table & counter, stainless steel, rolled edge, 6" splash	LF	1.0270	122.89	394.43	517.32
11.1603 021	Table & counter, stainless steel, straight edge, 6" splash	LF	1.0270	122.89	309.97	432.86
11.1603 031	Serving fixture, stainless steel top, galvanized frame stiles, economy	LF	1.3689	163.80	338.12	501.92
11.1603 041	Serving fixture, all stainless steel, best	LF	1.3689	163.80	479.05	642.85
11.1604 000	**ADDERS FOR BASIC BUILT-UP FIXTURES:**					
11.1604 011	Shelves, stainless steel, 85/SF, base	LF			78.33	78.33
11.1604 021	Shelves, galvanized iron, 375/SF, base	LF			42.73	42.73
11.1604 031	Angle or pipe stretchers	LF			26.68	26.68
11.1604 041	Tray slide, stainless steel	LF			90.84	90.84
11.1604 051	Display shelf, sneeze guard	LF			110.43	110.43
11.1604 061	Each additional shelf	LF			65.93	65.93
11.1604 071	Plastic laminate on plywood	LF			55.32	55.32
11.1604 081	Stainless steel facing, 18 ga, 3' high	LF			82.03	82.03
11.1604 091	Stainless steel facing, 22 ga, 3' high	LF			52.06	52.06
11.1604 101	Marine front coping	LF			20.58	20.58
11.1604 111	Stainless steel dirty dish table, 12" splash	LF			41.00	41.00
11.1604 121	Slop gutter, 4" square, with sump	LF			78.12	78.12
11.1605 000	**SHOP INSTALLED ACCESSORIES:**					
11.1605 001	*Note: The following items should be added separately to the basic fixture costs above.*					
11.1605 011	Joint & miter, weld & polish	EA			238.30	238.30
11.1605 021	Door, stainless steel, to 24"	EA			344.39	344.39
11.1605 031	Drawer, stainless steel face, removable pan	EA			313.46	313.46
11.1605 041	Drawer, stainless steel face, galvanized iron pan	EA			217.97	217.97
11.1605 051	Punch-out	EA			78.40	78.40
11.1605 061	Ventilating grill, 24" x 12"	EA			115.72	115.72
11.1605 071	Maple cutting board, laminated, 2"	EA			33.76	33.76
11.1605 081	Richlite cutting board, laminated, 2"	EA			46.03	46.03
11.1605 091	Pot washer sink, 24" x 24" x 24"	EA			1,261.90	1,261.90
11.1605 101	Vegetable sink	EA			1,177.84	1,177.84

11 - EQUIPMENT

Division	Description	Unit	Labor Hours	Labor Cost	Material Cost	Total Cost
11.1605 111	Mixer valve	EA			185.86	185.86
11.1605 121	Lever operated drain valve	EA			150.17	150.17
11.1605 131	Disposer, with stainless steel cone, 1 hp	EA	4.6210	552.95	1,913.99	2,466.94
11.1605 141	Disposer, with stainless steel cone, 1-1/2 hp	EA	4.6210	552.95	2,626.39	3,179.34
11.1605 151	Disposer, with stainless steel cone, 3 hp	EA	4.6210	552.95	3,996.65	4,549.60
11.1605 161	Pot washer, sink mounted	EA	5.3915	645.15	2,336.13	2,981.27
11.1605 171	Water heater, electric, sink mounted	EA	4.6210	552.95	1,116.45	1,669.40
11.1605 181	Conveyors, soiled dish, 16'	EA	63.5388	7,603.05	10,570.86	18,173.91
11.1605 191	Electric sub panel, 100a, prewired case	EA			655.18	655.18
11.1605 201	Display, fluorescent fixture, 6"	EA			275.83	275.83
11.1605 211	Shelf heater, infra-red	EA	3.0810	368.67	386.63	755.30
11.1605 221	Pot & pan rack, 6', table mounted	EA			696.51	696.51
11.1605 231	Shelf rack, 12', table mounted	LF			81.86	81.86
11.1605 241	Shelf, stainless steel, wall mounted	SF	0.2311	27.65	40.23	67.89
11.1605 251	Equipment base, 4", galvanized iron channel	LF	0.4003	47.90	14.64	62.54
11.1605 261	Glass rack dispenser, self level	EA	2.4643	294.88	1,015.02	1,309.89
11.1605 271	Cup & plate dispenser, heated	EA	1.5401	184.29	773.32	957.61
11.1605 281	Hot food well, electric, 12" x 20"	EA	2.3105	276.47	628.28	904.76
11.1605 291	Bain marie, gas or steam, 6'	EA	3.4659	414.73	4,011.77	4,426.50
11.1605 301	Bain marie, electric	EA	3.4659	414.73	3,818.40	4,233.13
11.1605 311	Drop-in deep fryer, electric	EA	3.0810	368.67	1,691.60	2,060.27
11.1605 321	Griddle, built-in, 4', electric	EA	3.4659	414.73	3,383.37	3,798.10
11.1605 331	Hotplate, built-in, 12"x20", electric	EA	1.5401	184.29	289.91	474.20
11.1605 341	Drop-in water cooler	EA	1.5401	184.29	1,643.34	1,827.63
11.1606 000	**REFRIGERATOR, UNDER THE COUNTER:**					
11.1606 011	Refrigerator, without compressor	LF			406.95	406.95
11.1606 021	Add for each door	EA			481.54	481.54
11.1606 031	Add for each drawer	EA			425.25	425.25
11.1606 041	Add for each 16" x 16" coil	EA			276.68	276.68
11.1606 051	Add for remote compressor, 1/4 hp	EA	14.6335	1,751.04	1,063.30	2,814.34
11.1606 061	Refrigerator, 5 CF, undercounter, with compressor	EA	3.4659	414.73	2,126.69	2,541.42
11.1606 071	Cold pan, 6', non-refrigerating	EA	1.9258	230.44	773.32	1,003.76
11.1606 081	Cold pan, 6', refrigerating	EA	5.3915	645.15	1,208.36	1,853.51
11.1606 091	Soft drink dispenser, 4 spout, 1 remote	EA	30.8060	3,686.25	10,730.32	14,416.56
11.1607 000	**BUY-OUT EQUIPMENT:**					
11.1607 001	Note: The item canopy hood, ss, 4'3" x 9'6" listed below includes filters, automatic fire extinguisher and detergent washdown.					
11.1607 011	Cart, self level, glass, dish, cup	EA			2,795.41	2,795.41
11.1607 021	Cart, clean dish stacking	EA			1,048.78	1,048.78
11.1607 031	Cart, delivery, hot & cold tray	EA			11,706.74	11,706.74
11.1607 041	Coffee urn, twin, 8 gallon	EA	1.5404	184.32	3,721.69	3,906.01
11.1607 051	Coffee urn, twin, 12 gallon	EA	2.1398	256.05	5,725.73	5,981.78
11.1607 061	Coffee urn, twin, 20 gallon	EA	2.3535	281.62	7,850.06	8,131.67
11.1607 071	Coffee urn, twin, 40 gallon	EA	2.5888	309.78	9,115.44	9,425.22
11.1607 081	Cart, utility	EA			587.93	587.93
11.1607 091	Canopy hood, stainless steel, with filters	SFSA	0.3774	45.16	138.24	183.40
11.1607 101	Canopy hood, stainless steel, 4'3" x 9'6"	EA	18.3340	2,193.85	11,525.22	13,719.07
11.1607 111	Can washer	EA	2.4643	294.88	2,719.67	3,014.55
11.1607 121	Can crusher	EA	1.3864	165.90	4,819.02	4,984.92
11.1607 131	Deep fryer, electric	EA	2.0022	239.58	1,908.54	2,148.12
11.1607 141	Freezer, roll-in, thru, 1 section	EA	1.8600	222.57	7,759.24	7,981.81
11.1607 151	Freezer, roll-in, thru, 2 section	EA	1.8600	222.57	11,242.31	11,464.88
11.1607 161	Freezer, roll-in, thru, 3 section	EA	1.8600	222.57	13,814.95	14,037.51
11.1607 171	Hot food cab, roll-in, thru, 1 section	EA	1.8600	222.57	13,212.11	13,434.68
11.1607 181	Hot food cab, roll-in, thru, 2 section	EA	1.8600	222.57	9,207.61	9,430.18
11.1607 191	Hot food cab, roll-in, thru, 3 section	EA	1.8600	222.57	12,311.35	12,533.92
11.1607 201	Griddle & skillet, 4', electric, tilt	EA	4.6210	552.95	8,788.91	9,341.86

EQUIPMENT - 11

Division	Description	Unit	Labor Hours	Labor Cost	Material Cost	Total Cost
11.1607 211	Glass rack dispenser	EA			1,344.93	1,344.93
11.1607 221	Hose reel & spray	EA	1.8484	221.18	882.64	1,103.82
11.1607 231	Ice machine	EA	2.6189	313.38	6,441.36	6,754.74
11.1607 241	Kettle filler, wall standard	EA	2.3105	276.47	667.91	944.38
11.1607 251	Mixer, 20 quart, bench mounted	EA	1.8484	221.18	2,920.06	3,141.24
11.1607 261	Milk dispenser, refrigerated, self-level cart	EA	3.0810	368.67	4,675.98	5,044.65
11.1607 271	Micro spray cart washer, without drain	EA	3.6969	442.37	906.50	1,348.87
11.1607 281	Oven, microwave, large, with timer	EA	1.9258	230.44	3,793.38	4,023.82
11.1607 291	Oven, microwave, small, with timer	EA	1.9258	230.44	1,550.64	1,781.08
11.1607 301	Oven, gas, convection, 3 deck, bake	EA	8.3179	995.32	15,268.58	16,263.90
11.1607 311	Oven, gas, convection, 2 deck, bake	EA	4.7748	571.35	11,880.80	12,452.15
11.1607 321	Oven, gas, convection, 1 deck, bake	EA	4.7748	571.35	8,349.99	8,921.34
11.1607 331	Oven, convection, 3 deck, roast	EA	9.8579	1,179.60	21,471.53	22,651.13
11.1607 341	Oven, convection, 2 deck, roast	EA	8.1641	976.92	15,745.76	16,722.68
11.1607 351	Oven, convection, 1 deck, roast	EA	6.6231	792.52	10,497.18	11,289.70
11.1607 361	Pot sink, single	EA	1.9258	230.44	715.67	946.11
11.1607 371	Pot sink, double	EA	3.0810	368.67	1,145.05	1,513.72
11.1607 381	Pot filler	EA	3.4659	414.73	214.60	629.33
11.1607 391	Peeler, vegetable	EA	3.8506	460.76	4,699.85	5,160.61
11.1607 401	Pot rack, mobile	EA	0.9242	110.59	2,027.77	2,138.36
11.1607 411	Pan rack, roll-in	EA	0.9242	110.59	1,407.54	1,518.13
11.1607 421	Refrigerator, 1 section reach-in, pass thru	EA	2.7900	333.85	3,986.48	4,320.33
11.1607 431	Refrigerator, 2 section, reach-in, pass thru	EA	2.7900	333.85	5,365.96	5,699.81
11.1607 441	Refrigerator, 3 section, reach-in, pass thru	EA	2.7900	333.85	6,814.38	7,148.23
11.1607 451	Range match, gas, fryer, 36"	EA	4.1590	497.67	3,106.19	3,603.85
11.1607 461	Range, electric, hot top, oven 36" deep	EA	6.1611	737.24	3,817.06	4,554.30
11.1607 471	Range, hot top, oven under, gas	EA	4.1590	497.67	2,910.50	3,408.17
11.1607 481	Range, spreader plate, cabinet, 24"	EA	1.5401	184.29	906.50	1,090.79
11.1607 491	Range, spreader plate, cabinet, 18"	EA	1.5401	184.29	763.37	947.66
11.1607 501	Roll warmer, 2 drawer	EA	1.3864	165.90	1,560.24	1,726.14
11.1607 511	Roll warmer, 3 drawer	EA	1.3864	165.90	1,798.83	1,964.73
11.1607 521	Racks, chrome wire, 2x5, 5 shelves	EA			747.25	747.25
11.1607 531	Racks, stainless steel, per shelf to 5 shelves	LF			30.32	30.32
11.1607 541	Steam kettle, 40 gal, tilt, motor	EA	4.1590	497.67	11,556.36	12,054.03
11.1607 551	Steam kettle, 20 gal, tilt, motor	EA	4.1590	497.67	9,065.69	9,563.36
11.1607 561	Steam kettle, 10 gal, tilt, motor	EA	2.7727	331.78	2,051.62	2,383.41
11.1607 571	Steam kettle, 40 gal, wall mounted	EA	4.1590	497.67	8,808.03	9,305.70
11.1607 581	Steam kettle, 60 gal, wall mounted	EA	4.1590	497.67	9,752.75	10,250.41
11.1607 591	Steam cooker, 1 compartment, small	EA	4.1590	497.67	3,017.43	3,515.09
11.1607 601	Steam cooker, 1 compartment, large	EA	4.1590	497.67	6,203.93	6,701.60
11.1607 611	Steam cooker, 2 compartment, large	EA	4.1590	497.67	11,394.05	11,891.71
11.1607 621	Soak sink, mobile, 24" x 24"	EA	1.5401	184.29	1,321.61	1,505.90
11.1607 631	Slicing machine	EA	1.1554	138.26	2,266.41	2,404.67
11.1607 641	Soft ice cream & shake machine	EA	5.2368	626.64	13,407.72	14,034.36
11.1607 651	Mobile stand, mixer & slicer	EA	2.4643	294.88	1,431.37	1,726.25
11.1607 661	Steam table, 6 well, heated over shelf	EA	4.1590	497.67	6,679.97	7,177.63
11.1607 671	Baker's table, 6', refrigerated	EA	4.7748	571.35	10,497.18	11,068.53
11.1607 681	Toaster, conveyor	EA	2.4643	294.88	3,101.35	3,396.23
11.1607 691	Tray dispenser, self level, mobile	EA	1.3864	165.90	1,932.40	2,098.30
11.1607 701	Tray maker conveyor, 36'	EA	3.6969	442.37	22,425.75	22,868.12
11.1607 711	Table, stainless steel, 6', with shelf under	EA	1.3864	165.90	2,767.39	2,933.28
11.1607 721	Work center & cabinet stainless steel, 6'	EA	4.1590	497.67	6,403.18	6,900.85
11.1607 731	Water station, chilled	EA	4.7748	571.35	4,151.09	4,722.45
11.1607 741	Washer, utensil, pass-thru	EA			20,520.89	20,520.89
11.1607 751	Washer, glassware, pass-thru	EA			20,423.19	20,423.19
11.1607 761	Dishwasher, peg-belt type	EA			24,429.62	24,429.62
11.1607 771	Dishwasher, 30' flight, without access	EA			55,177.17	55,177.17

11 - EQUIPMENT

Division	Description	Unit	Labor Hours	Labor Cost	Material Cost	Total Cost
11.1607 781	Dishwasher, 26' flight, without access	EA			52,418.34	52,418.34
11.1607 791	Dishwasher, 22' flight, without access	EA			49,659.45	49,659.45
11.1607 801	Dishwasher, 18' flight, without access	EA			46,900.58	46,900.58
11.1607 811	Refrigerator, wall, stainless steel, glass door, 5', no compressor	EA			4,031.90	4,031.90
11.1607 821	Walk-in cooler, metal, 8', prefab	SF			285.25	285.25
11.1607 831	Refrigerator shelving	LF			24.43	24.43
11.1608 000	**PREFAB KITCHEN, WITH ELECT RANGE, REFER & TOP:**					
11.1608 011	Kitchen unit, 60", no wall cabinets	EA	2.3250	278.21	4,000.35	4,278.56
11.1608 021	Kitchen unit, 72", no wall cabinets	EA	2.7900	333.85	4,345.16	4,679.01
11.1608 031	Kitchen unit, 84", no wall cabinets	EA	3.2550	389.49	4,910.72	5,300.21
11.1608 041	Add for disposal	EA			344.82	344.82
11.1608 051	Add for microwave	EA			3,172.67	3,172.67
11.1608 061	Add for hot water dispenser	EA			413.80	413.80
11.1608 071	Add for porcelain colors	EA			206.91	206.91
11.1700 000	**GYMNASIUM & PLAYGROUND EQUIPMENT:**					
11.1701 000	**FIELD EQUIPMENT:**					
11.1701 011	Basketball, post & steel backstop, single	EA	4.8450	454.22	1,086.24	1,540.46
11.1701 021	Basketball, post & steel backstop, double	EA	5.1969	487.21	1,525.18	2,012.38
11.1701 031	Add for fiberglass backstop	EA			65.79	65.79
11.1701 041	Baseball backstop, 34x10, with hood	EA	14.4997	1,359.35	3,280.83	4,640.18
11.1701 051	Baseball backstop, 60x15, with hood	EA	25.4482	2,385.77	9,184.19	11,569.96
11.1701 061	Football goal, single post, two	SET	11.8365	1,109.67	3,910.67	5,020.34
11.1701 071	Soccer goal, two	SET	15.3876	1,442.59	3,039.42	4,482.01
11.1701 081	Tennis post, two	SET	2.4856	233.03	384.04	617.06
11.1701 091	Tennis net, nylon	EA			331.32	331.32
11.1701 101	Tennis net, metal	EA			702.22	702.22
11.1701 111	Volley ball post, two	SET	2.6098	244.67	329.13	573.80
11.1701 121	Tether ball post	EA	1.3672	128.18	137.99	266.17
11.1701 131	Add for ground sock, tether ball	EA	1.3672	128.18	105.63	233.81
11.1701 141	Court striping, tennis, basketball	EA			344.61	344.61
11.1701 151	Court striping, volleyball	EA			206.75	206.75
11.1701 161	Swings, 10' high, 4 seats	SET	8.0785	757.36	1,422.03	2,179.39
11.1701 171	Swings, 10' high, 6 seats	SET	9.6322	903.02	1,744.60	2,647.62
11.1701 181	Horizontal ladder, 8' x 16'	EA	4.9712	466.05	917.32	1,383.37
11.1701 191	Horizontal bar, single, 6-1/2'	EA	2.3614	221.38	270.98	492.36
11.1702 000	**BLEACHERS, ON CONCRETE RISER, BENCHES ONLY:**					
11.1702 011	Benches, fiberglass	SEAT			32.69	32.69
11.1702 021	Benches, wood	SEAT			61.95	61.95
11.1702 031	Benches, aluminum	SEAT			68.88	68.88
11.1702 041	Benches, fiberglass, with back	SEAT			82.68	82.68
11.1702 051	Individual seats	SEAT			117.10	117.10
11.1702 061	Benches, port, 16 rows, 500 minimum	SEAT			65.43	65.43
11.1703 000	**OUTDOOR EQUIPMENT, MISCELLANEOUS:**					
11.1703 011	Benches, wood slats, 6'	EA	0.4650	43.59	808.11	851.71
11.1703 021	Benches, precast concrete, 6'	EA	0.6975	65.39	1,818.27	1,883.67
11.1703 031	Picnic table & benches, 6'	EA	0.9300	87.19	1,817.25	1,904.44
11.1703 041	Bicycle rack, galvanized iron, 10', 1 side	EA	0.9300	87.19	1,401.79	1,488.97
11.1703 051	Bicycle rack, galvanized iron, 10', 2 side	EA	0.9300	87.19	1,609.51	1,696.70
11.1703 061	Bicycle rack, precast concrete, single	EA			228.29	228.29
11.1704 000	**ATHLETIC FIELD, SYNTHETIC SURFACE:**					
11.1704 011	Uniturf, embossed, running, 3/8"	SF			10.10	10.10
11.1704 021	Turf, 3 layer	SF			15.21	15.21
11.1704 031	Turf, 2 layer	SF			13.30	13.30
11.1704 041	Running track, volcanic cinder, 7"	SF			2.62	2.62
11.1704 051	Running track, bitumen/cork, 2"	SF			4.17	4.17
11.1704 061	Track, asphaltic concrete, 1/4 mi, 55,000 SF, 1/4" synthetic	EA			638,393.89	638,393.89
11.1704 071	Track, 1/4 mi, 55,000 SF, 2" cinder, 6" curb	EA			179,667.71	179,667.71

EQUIPMENT - 11

Division	Description	Unit	Labor Hours	Labor Cost	Material Cost	Total Cost
11.1704 081	Track, bitumen & cork, 1/4 mi, 55,000 SF	EA			229,363.08	229,363.08
11.1704 091	Tennis court, asphaltic concrete base, all weather	SF			4.74	4.74
11.1704 101	Score board	EA			19,113.53	19,113.53
11.1705 000	**BASKETBALL BACKSTOPS:**					
11.1705 011	Backstop, wall, out-rigger, fixed	EA	8.4072	788.18	818.90	1,607.08
11.1705 021	Backstop, wall, out-rigger, swing	EA	16.0575	1,505.39	2,510.48	4,015.87
11.1705 031	Backstop, ceiling, swing up, manual	EA	21.5083	2,016.40	5,048.90	7,065.30
11.1705 041	Add for glass, fan backstop	EA			1,146.49	1,146.49
11.1705 051	Add for glass, rectangular backs	EA			1,369.24	1,369.24
11.1705 061	Add for power operation	EA			910.59	910.59
11.1706 000	**GYM WALLS:**					
11.1706 011	Padded gym wall	SF			10.78	10.78
11.1707 000	**GYM FLOORS, NO SUB-FLOOR OR BASE INCLUDED:**					
11.1707 011	Synthetic gym floor, 3/16"	SF			8.88	8.88
11.1707 021	Synthetic gym floor, 3/8"	SF			9.75	9.75
11.1707 031	Gym floor, maple, wood, spring	SF			25.66	25.66
11.1707 041	Gym floor, rubber cushion, maple	SF			14.36	14.36
11.1707 051	Gym floor, maple over sleepers	SF			20.62	20.62
11.1707 061	Add for apparatus inserts	EA			68.77	68.77
11.1708 000	**GYM SEATING:**					
11.1708 011	Bleachers, telescoping, manual	SEAT	0.1763	19.02	84.04	103.06
11.1708 021	Bleachers, portable, hydraulic	SEAT			85.43	85.43
11.1709 000	**SCORE BOARDS:**					
11.1709 011	Score board, basketball, economy	EA	7.1765	774.27	3,182.05	3,956.32
11.1709 021	Score board, basketball, good	EA	13.1569	1,419.50	3,196.96	4,616.46
11.1709 031	Score board, basketball, best	EA	20.9321	2,258.36	4,758.33	7,016.70
11.1800 000	**INDUSTRIAL EQUIPMENT:**					
11.1801 000	**SERVICE STATION:**					
11.1801 001	*Note: The prices for the next two items are rough averages. The price for the first item does not include the cost of land.*					
11.1801 011	Service station, 3 island, 4 tanks	EA			739,643.95	739,643.95
11.1801 021	Service station, remold to self serve, conversion	EA			211,326.83	211,326.83
11.1801 031	Air compressor, 2 hp, with receiver	EA	5.1469	366.82	3,719.33	4,086.15
11.1801 041	Air compressor, 3 hp, with receiver	EA	5.1469	366.82	3,909.52	4,276.34
11.1801 051	Gas pump, full size, 1/2 hp	EA	2.5729	183.37	2,747.22	2,930.59
11.1801 061	Gas pump, submerged turbine, 1/3 hp	EA	0.9651	92.05	1,299.60	1,391.65
11.1801 071	Gas pump, submerged turbine, 3/4 hp	EA	0.9651	92.05	1,504.63	1,596.68
11.1801 081	Gas dispenser, computing, single hose	EA	1.0720	94.34	4,617.47	4,711.81
11.1801 091	Add to above for vapor recovery	EA			950.93	950.93
11.1801 101	Gas dispenser, computing, dual hose	EA	1.0720	94.34	9,234.97	9,329.30
11.1801 111	Fill boxes, 12", cast iron	EA	0.9624	84.69	72.88	157.57
11.1801 121	Air & water, bibbs, underground reels	EA	0.8580	75.50	496.58	572.08
11.1801 131	Add for electric thermal unit	EA			200.77	200.77
11.1801 141	Hoist, 1 post, 8000#, semihydraulic	EA	8.5779	820.73	4,194.84	5,015.57
11.1801 151	Hoist, 2 post, 8000#, semihydraulic	EA	18.4037	1,760.87	6,915.81	8,676.67
11.1801 155	Truck lift, 50,000 lb, 4 post	EA	211.6000	20,245.89	117,649.89	137,895.78
11.1801 161	Cash box & pedestal stand	EA	1.2866	91.70	200.77	292.47
11.1801 171	Tire changer, air operated	EA	1.2864	113.20	2,019.68	2,132.88
11.1801 181	Lube, oil & air, reels & remote pump	EA	9.6494	849.15	2,430.21	3,279.36
11.1801 183	Trapeze, air & electrical, drop	EA	3.6000	316.80	1,810.00	2,126.80
11.1801 185	Reel bank, 3 services	EA	11.8000	1,038.40	5,902.18	6,940.58
11.1801 186	Reel bank, 4 services	EA	17.7000	1,557.60	8,853.24	10,410.84
11.1801 187	Reel bank, 5 services	EA	23.6000	2,076.80	11,804.36	13,881.16
11.1801 191	Exhaust fume system, underground	STA.			1,315.50	1,315.50
11.1801 201	Island metal furring for 3 island	ONE S	1.8600	177.96	1,267.90	1,445.87
11.1801 211	Dynamometer	EA	112.9820	10,329.94	79,531.56	89,861.50
11.1801 221	Tank, waste oil with pump, 240 gal	EA	12.4000	1,091.20	6,197.29	7,288.49

11 - EQUIPMENT

Division	Description	Unit	Labor Hours	Labor Cost	Material Cost	Total Cost
11.1802 000	**SHOP EQUIPMENT, MISCELLANEOUS:**					
11.1802 011	Sawdust collector	EA	11.0408	1,056.38	14,624.93	15,681.32
11.1802 021	Sawdust collector, large capacity	EA	11.0408	1,056.38	19,384.61	20,440.99
11.1802 031	Paint spray booths	SF			15.04	15.04
11.1802 041	Paint spray booth car	EA	30.0790	2,877.96	65,897.62	68,775.57
11.1802 051	Paint spray booth, 60' bus	EA	721.8600	65,999.66	284,041.47	350,041.13
11.1803 000	**VEHICLE MAINTENANCE EQUIPMENT, HEAVY DUTY:**					
11.1803 011	Cabinet, storage, shop	EA	1.3500	118.80	914.95	1,033.75
11.1803 021	Workbench, steel top, 6'	EA	0.7870	69.26	522.80	592.05
11.1803 031	Vise, machinist, swivel base, 4"	EA	0.7307	64.30	485.48	549.78
11.1803 041	Pump, air piston	EA	7.1669	630.69	4,761.44	5,392.13
11.1803 051	Pump, air piston, with hoist	EA	10.1180	890.38	6,722.07	7,612.46
11.1803 061	Tank, storage, cube, 500 gal	EA	2.8106	247.33	1,867.23	2,114.57
11.1803 071	Tank, storage, cube, 1,000 gal	EA	3.9348	346.26	2,614.15	2,960.41
11.1803 081	Washer, vehicle, gantry	EA	109.6121	9,645.86	72,822.23	82,468.09
11.1803 091	Water reclamation system	EA	19.6740	1,731.31	13,070.61	14,801.92
11.1803 101	Washer, high pressure, hot water	EA	27.4030	2,411.46	18,205.53	20,616.99
11.1803 111	Lift, ramp, surface mounted, 50,000#	EA	168.0720	14,790.34	111,660.74	126,451.08
11.1803 121	Buffer & grinder, 10", with dust collector	EA	12.9286	1,137.72	8,589.28	9,727.00
11.1803 131	Drill press variable speed, 20"	EA	15.7392	1,385.05	10,456.54	11,841.59
11.1803 141	Lathe, brake drum/rotor	EA	92.7487	8,161.89	61,618.82	69,780.70
11.1803 151	Hoist, chain, electric, 2 ton	EA	10.9612	964.59	7,282.22	8,246.81
11.1803 161	Hoist, chain, electric, 3 ton	EA	13.3502	1,174.82	8,869.36	10,044.18
11.1803 171	Tank, water oil, with pump, 240 gal	EA	8.8533	779.09	5,881.75	6,660.84
11.1803 181	Mounter & demounter, tire, truck	EA	30.4947	2,683.53	20,259.52	22,943.05
11.1901 000	**PARKING LOT EQUIPMENT:**					
11.1901 011	Automatic gate, automatic arm, 8'	EA	1.8600	177.96	4,235.00	4,412.96
11.1901 021	Traffic detector	EA	2.3250	222.46	1,355.19	1,577.65
11.1901 031	Ticket dispenser, control unit	EA	1.5500	148.30	5,048.15	5,196.45
11.1901 041	Gate operator card or coin	EA	5.5800	510.18	1,355.19	1,865.37
11.2000 000	**MATERIAL HANDLING EQUIPMENT:**					
11.2000 001	*Note: Dock levelers don't include forms, concrete reinforcing or embedded metal.*					
11.2001 000	**MATERIAL HANDLING EQUIPMENT:**					
11.2001 011	Conveyor power riser, 12' floors	EA	33.4800	3,612.16	9,051.90	12,664.05
11.2001 021	Conveyor, roller, 12"	LF			42.27	42.27
11.2001 031	Conveyor, roller, 24"	LF			50.37	50.37
11.2001 041	Dock bumper, 10" x 4-1/2" x 2'	EA	0.5869	63.32	104.82	168.14
11.2001 051	Dock bumper, 10" x 4-1/2" x 3'	EA	0.5869	63.32	149.59	212.91
11.2001 061	Dock bumper, 12" x 4-1/2" x 6'	EA	1.5159	163.55	270.81	434.36
11.2001 071	Dock leveler, medium duty, manual	EA	9.6525	1,041.41	6,801.39	7,842.80
11.2001 081	Dock leveler, heavy duty, hydraulic	EA	7.6260	822.77	8,669.87	9,492.64
11.2001 091	Dock leveler, heavy duty, hydraulic	EA	7.6260	822.77	8,669.87	9,492.64
11.2001 101	Scissor lift, 56", 2000#, 4x4 deck	EA			9,991.57	9,991.57
11.2001 111	Scissors lift, 56", 2000#, 4x7 deck	EA			12,993.50	12,993.50
11.2001 121	Scissors lift, 60", 5000#, 6x8 deck	EA			15,233.79	15,233.79
11.2001 131	Load dock, accordion door, 7'6"x8'	EA	7.8841	850.62	1,351.90	2,202.51
11.2100 000	**LABORATORY EQUIPMENT:**					
11.2101 000	**LAB FURNITURE:**					
11.2101 001	*Note: The prices on following items include sink, fixtures & acid resistant tops.*					
11.2101 011	Instructor's table, 12'x36"x36"	EA	2.7900	301.01	3,451.89	3,752.91
11.2101 021	Table, 8 student, 15' x 5' x 36"	EA	2.7900	301.01	7,281.42	7,582.43
11.2102 000	**LABORATORY CABINETS, BASE UNITS:**					
11.2102 011	Steel laboratory cabinet, base	LF	1.6100	173.70	297.97	471.67
11.2102 021	Wood laboratory cabinet, base	LF	1.6100	173.70	245.74	419.44
11.2102 031	Plastic laminated laboratory cabinet, base	LF	1.6100	173.70	222.92	396.62

EQUIPMENT - 11

Division	Description	Unit	Labor Hours	Labor Cost	Material Cost	Total Cost
11.2103 000	**LAB CABINETS, KNEE SPACE DRAWER UNITS:**					
11.2103 011	Steel laboratory cabinet, knee space, drawers	LF	1.0230	110.37	190.89	301.26
11.2103 021	Wood laboratory cabinet, knee space, drawers	LF	1.0220	110.26	159.01	269.27
11.2103 031	Plastic laminated laboratory cab, knee space, drawers	LF	1.0210	110.16	143.12	253.28
11.2104 000	**LAB CABINETS, WALL HUNG:**					
11.2104 011	Steel laboratory cabinet, wall hung	LF	0.7820	84.37	218.97	303.34
11.2104 021	Wood laboratory cabinet, wall hung	LF	0.7820	84.37	228.07	312.44
11.2104 031	Plastic laminated laboratory cabinet, wall hung	LF	0.7820	84.37	144.00	228.37
11.2105 000	**LABORATORY CABINETS, TOPS, REAGENT SHELF, IN PLACE:**					
11.2105 001	*Note: Labor charges for the following items are included in the prices for base units.*					
11.2105 011	Plastic laminate laboratory cabinet top	LF			81.81	81.81
11.2105 021	Epoxy resin, acid resistant cabinet top	LF			190.89	190.89
11.2105 031	Stainless steel laboratory cabinet top	LF			274.95	274.95
11.2105 041	Chemical resist laboratory cabinet top	LF			109.04	109.04
11.2106 000	**CENTRIFUGES:**					
11.2106 011	Centrifuge, high speed, portable	EA			2,794.96	2,794.96
11.2106 021	Centrifuge, ultra hi speed, explosion proof	EA			12,725.04	12,725.04
11.2106 031	Centrifuge, ultra hi speed, refrigerated	EA			20,223.76	20,223.76
11.2107 000	**REAGENT RACKS:**					
11.2107 011	Rack, metal upright, wall, 1 tier	EA	0.4630	49.95	109.04	158.99
11.2107 021	Rack, metal upright center, 1 tier	EA	0.4630	49.95	134.05	184.01
11.2107 031	Laboratory storage, wardrobe	LF	0.6565	70.83	305.55	376.38
11.2108 000	**LABORATORY EQUIPMENT, MISCELLANEOUS:**					
11.2108 011	Fume hood	LF	3.4100	367.90	1,363.39	1,731.29
11.2108 021	Add for stainless steel	LF			681.68	681.68
11.2108 031	Water distiller, 10 GPM	EA	3.4114	414.35	8,606.29	9,020.64
11.2108 041	Water tank, stainless steel, 50 gal	EA	3.4114	414.35	9,392.63	9,806.98
11.2108 051	Washer	EA	6.8225	828.66	11,190.64	12,019.30
11.2108 061	Portable water distiller, 18 liters per hour	EA	1.3950	150.51	1,442.09	1,592.60
11.2108 071	Fume hood laminar flow	LF	3.4100	367.90	2,006.13	2,374.04
11.2200 000	**LIBRARY EQUIPMENT:**					
11.2201 000	**STUDY CARRELS, HARDWOOD:**					
11.2201 011	Carrel, 36"x24"x29", 2 face, hardware	EA	1.0509	113.38	508.15	621.53
11.2201 021	Carrel, individual lights, hardware	EA	0.6310	68.08	90.64	158.72
11.2201 031	Carrel, power post receptacle, hardware	EA	0.6310	68.08	25.30	93.38
11.2202 000	**LIBRARY TABLES & CHAIRS:**					
11.2202 011	Table, 60" x 36" x 39"	EA			1,621.27	1,621.27
11.2202 021	Table, 48", round	EA			1,468.81	1,468.81
11.2202 031	Chairs, wood	EA			274.03	274.03
11.2203 000	**CARD CATALOG CABINET:**					
11.2203 011	Card catalog, wood, 60 tray	EA			1,239.89	1,239.89
11.2203 021	Card catalog, wood, 30 tray	EA			713.31	713.31
11.2203 031	Charging counters, hardwood	LF	0.6310	68.08	180.65	248.73
11.2203 041	Charging counter top	SF	0.1051	11.34	16.05	27.39
11.2204 000	**LIBRARY SHELVING:**					
11.2204 011	Shelving, metal, bracket, 3 tier, 3x7	EA	0.5780	62.36	209.66	272.02
11.2204 021	Shelving, metal, bracket, 5 tier, 3x7	EA	0.6310	68.08	225.83	293.91
11.2204 031	Shelving, metal, bracket, 7 tier, 3x7	EA	0.6980	75.31	244.05	319.36
11.2205 011	Loss prevention device, walk thru, expandable	EA	6.0000	647.34	20,934.99	21,582.33
11.2205 021	Loss prevention device, walk thru, non-expandable	EA	5.4000	582.61	16,740.54	17,323.15
11.2300 000	**HOSPITAL EQUIPMENT:**					
11.2300 001	*Note: The following prices do not include costs for mechanical or electrical hookup.*					
11.2301 000	**NURSES MONITORING EQUIPMENT, CCU/ICU:**					
11.2301 011	Cardioscope, 1 channel, bed side	EA			5,357.98	5,357.98
11.2301 021	Blood pressure monitor, bed side, with readout	EA			8,226.54	8,226.54

11 - EQUIPMENT

Division	Description	Unit	Labor Hours	Labor Cost	Material Cost	Total Cost
11.2301 031	Coronary care unit bed & nurse display, 8 bed	EA			167,394.86	167,394.86
11.2301 041	Telemetry, wireless, 8 bed	EA			457,832.93	457,832.93
11.2302 000	**NURSING ACUTE CARE EQUIPMENT:**					
11.2302 011	Modular wall unit, 1 bed core, prewire	EA	7.2327	780.34	6,087.81	6,868.15
11.2302 021	Modular wall unit, 24" nurse treatment	EA	1.6711	180.29	807.71	988.01
11.2302 031	Modular wall unit, 24" storage center	EA	1.8381	198.31	874.45	1,072.77
11.2302 041	Patient communication & convenience unit	EA			1,887.26	1,887.26
11.2302 051	Cubicle track & curtain, to 8' high	LF	0.0408	4.40	14.77	19.18
11.2302 061	Mirror, medicine cabinet, stainless steel, recessed light	EA	1.2174	131.35	581.77	713.11
11.2303 000	**NURSING STATION/CORE EQUIPMENT:**					
11.2303 011	Nurse chart desk	EA			1,327.47	1,327.47
11.2303 021	Nurse call, 40 station, two way	EA	61.5846	6,644.36	25,020.25	31,664.61
11.2303 031	Nourishment station, 84" x 80", stainless steel	EA	3.7200	401.35	22,692.77	23,094.12
11.2303 041	Medical preparation cabinet, 72"x80", stainless steel, lock	EA	3.7200	401.35	10,473.53	10,874.88
11.2303 051	Intravenous prep center, 60" x 80", stainless steel	EA	3.7200	401.35	9,018.89	9,420.24
11.2303 061	Sani-prep maintenance station, stainless steel	EA	16.1123	1,738.36	6,545.95	8,284.30
11.2303 071	Multi-use tote carts, average cost	EA			1,838.43	1,838.43
11.2303 081	Modular storage systems	LF			418.03	418.03
11.2304 000	**CORRIDORS:**					
11.2304 011	Hospital corner guard, plastic, 8'	EA	0.4530	48.87	163.65	212.52
11.2304 021	Hospital corner guard, plastic, 4'	EA	0.3376	36.42	93.52	129.94
11.2304 031	Hospital corner guard, stainless steel, 8'	EA	1.4520	156.66	394.97	551.62
11.2304 041	Hospital corner guard, stainless steel, 4'	EA	0.7855	84.75	197.48	282.22
11.2305 000	**SURGERY TABLES & ACCESSORIES:**					
11.2305 011	Surgery table, electric, stationary	EA	23.7150	2,558.61	27,461.50	30,020.12
11.2305 021	Surgery service island, complete	EA	19.2975	2,082.01	9,941.34	12,023.35
11.2305 031	Surgical monitor, conduct casters	EA			10,362.57	10,362.57
11.2305 041	Conductivity meter, current leak detector	EA	2.6854	289.73	1,147.00	1,436.73
11.2305 051	Surgical clock, stainless steel, auto reset	EA	2.7571	297.46	1,385.44	1,682.90
11.2306 000	**OBSTETRICAL & NURSERY EQUIPMENT:**					
11.2306 011	Delivery table	EA	18.6186	2,008.76	7,953.01	9,961.77
11.2306 021	Incubator, isolation servo-care	EA			7,354.00	7,354.00
11.2306 031	Incubator, warming	EA			2,406.76	2,406.76
11.2307 000	**SURGICAL LIGHTING:**					
11.2307 011	Surgery light, 3 arm, surface mounted	EA	26.7997	2,993.79	31,056.66	34,050.45
11.2307 021	Surgery light, 2 arm, surface mounted	EA	24.6761	2,756.57	21,568.29	24,324.86
11.2307 031	Surgery light, 1 arm, surface mounted	EA	17.9026	1,999.90	15,045.63	17,045.53
11.2307 041	Add for auxiliary light head	EA	7.4400	831.12	7,827.21	8,658.33
11.2307 051	Surgery light intensity control, 600w	EA	5.5800	623.34	869.70	1,493.04
11.2307 061	Surgery light intensity control, 300w	EA	3.7200	401.35	652.22	1,053.57
11.2308 000	**SCRUB & CLEAN ROOM EQUIPMENT:**					
11.2308 011	Scrub station, stainless steel, 3 bay, base mounted	EA	14.1825	1,722.61	11,335.84	13,058.44
11.2308 021	Scrub station, stainless steel base mounted, 1 bay	EA	9.3000	1,129.58	8,097.02	9,226.60
11.2308 031	Solution warm cab, stainless steel, 24"x30"x74"	EA	2.3250	282.39	8,372.67	8,655.06
11.2308 041	Surgery storage console, stainless steel, 12'	EA	1.8600	200.68	5,576.10	5,776.78
11.2308 051	Suture & drug cabinet, stainless steel, 36"x18"x81"	EA	1.3950	150.51	1,844.27	1,994.78
11.2308 061	Instrument cabinet, stainless steel, 48"x18"x60"	EA	1.3950	150.51	2,501.06	2,651.57
11.2308 071	Sterilizer, 16" x 16" x 26"	EA	4.6500	501.69	25,861.87	26,363.56
11.2309 000	**CARDIAC EMERGENCY EQUIPMENT:**					
11.2309 011	Monitor & resuscitator unit with DC defibrillator	EA			20,262.41	20,262.41
11.2309 021	Mobile emergency utility, crash cart	EA			11,052.14	11,052.14
11.2309 031	DC defibrillator	EA			5,894.27	5,894.27
11.2309 041	External cardiac compressor	EA			6,631.35	6,631.35
11.2310 000	**EMERGENCY ACCESSORIES:**					
11.2310 011	Treatment cabinet, mobile unit	EA			2,026.24	2,026.24
11.2310 021	Treatment cabinet, with suction compressor	EA			3,592.08	3,592.08
11.2310 031	Exam lights, 22", ceiling mounted	EA	6.2775	701.26	3,946.18	4,647.44

EQUIPMENT - 11

Division	Description	Unit	Labor Hours	Labor Cost	Material Cost	Total Cost
11.2310 041	Plaster sink with trap & fittings	EA	4.7025	571.17	1,578.86	2,150.03
11.2310 051	Splint cabinet with plaster bins	EA	2.3868	257.51	1,376.44	1,633.95
11.2311 000	**STERILIZERS, RECESSED, WITH AUTOMATIC DOORS:**					
11.2311 011	Sterilizer, 24"x36"x36", 1 door, steam	EA	231.9520	28,172.89	104,661.25	132,834.14
11.2311 021	Sterilizer, 24"x36"x48", 1 door, steam	EA	231.9520	28,172.89	106,667.59	134,840.48
11.2311 031	Sterilizer, gas, 24"x36"x60", 1 door	EA	187.3533	22,755.93	125,409.44	148,165.37
11.2311 041	Sterilizer, gas, 24"x36"x60", pass-thru	EA	187.3533	22,755.93	131,540.81	154,296.74
11.2311 051	Sterilizer, steam, 24"x36"x48", pass-thru	EA	231.9520	28,172.89	133,932.41	162,105.30
11.2311 061	Sterilizer, steam, 24"x36"x60", pass-thru	EA	231.9520	28,172.89	136,293.63	164,466.52
11.2311 071	Sterilizer, gas/cryotherm, 24"x36"x60", pass-thru	EA	201.0809	24,423.29	112,407.55	136,830.83
11.2311 081	Sterilizer, gas/cryotherm, 24"x36"x48", 1 door	EA	92.4487	11,228.82	95,666.04	106,894.86
11.2311 091	Gas aerator, 24" x 36" x 60"	EA	25.0063	3,037.27	11,958.24	14,995.50
11.2311 101	Gas aerator, 24" x 36" x 48"	EA	13.4054	1,628.22	6,457.44	8,085.66
11.2312 000	**STERILIZER ACCESSORIES:**					
11.2312 011	Loading car & carriage, large	EA			7,892.41	7,892.41
11.2312 021	Loading car & carriage, medium	EA			6,457.44	6,457.44
11.2312 031	Steam generator, 10 BHP/110 PSIG/208v, 3ph	EA	10.1185	1,228.99	5,680.11	6,909.11
11.2313 000	**INSTRUMENT & UTENSIL WASHER/STERILIZER:**					
11.2313 011	Instrument cleaner, sonic, 12x11x24, 1 compartment	EA	47.2054	5,733.57	18,535.26	24,268.83
11.2313 021	Instrument cleaner, sonic, 12x11x24, 2 compartment	EA	47.2054	5,733.57	25,003.56	30,737.13
11.2313 031	Glass washer, steam, 24x20x24, pass-thru	EA	68.1248	8,274.44	39,331.77	47,606.20
11.2313 041	Utensil washer, 27x27x60, free stand	EA	26.8817	3,265.05	16,741.53	20,006.58
11.2313 051	Utensil washer, conveyor, pass-thru	EA	100.1667	12,166.25	56,801.64	68,967.89
11.2313 061	Drying oven, 25"x25"x50", steam	EA	22.8199	2,771.71	13,871.55	16,643.26
11.2313 071	Charging table, 5', stainless steel, 2 sink	EA	5.6394	684.96	2,630.79	3,315.75
11.2314 000	**HOSPITAL CART WASH, PIT MOUNTED:**					
11.2314 011	Cart washer, 92"x72"x98", pass-thru, pit	EA	195.4953	23,744.86	109,011.40	132,756.25
11.2314 021	Cart washer, 92x72x196, automatic, 2 stage	EA	236.3130	28,702.58	273,275.24	301,977.82
11.2315 000	**CENTRAL PHARMACY EQUIPMENT:**					
11.2315 011	Pharmacy unit with basic modules	LF			671.68	671.68
11.2315 021	Medication refrigerator, 115 CF, stainless steel	EA	2.7452	333.43	4,655.77	4,989.20
11.2315 031	Water purifier, reverse osmosis, 6 liters per minute	EA	6.8508	832.10	3,846.03	4,678.13
11.2315 041	Lam flow hood, 36" work area	EA	13.0738	1,587.94	8,744.75	10,332.69
11.2315 051	Hi-density storage, caster/track	EA	2.1483	260.93	3,643.59	3,904.52
11.2316 000	**CENTRAL LABORATORY EQUIPMENT:**					
11.2316 011	Laboratory work counter, with base units	LF			591.70	591.70
11.2316 021	Laboratory sterilizer, 16"x26", 208v, 3ph	EA	6.8746	834.99	7,287.32	8,122.31
11.2316 031	Liquid nitrogen refrigerator, 30 CF, 190 deg	EA	7.1611	869.79	32,388.13	33,257.92
11.2316 041	Water distribution, steam, 10 GPH, wall	EA	9.7868	1,188.70	17,937.31	19,126.01
11.2316 051	Water tank, stainless steel, 50 gal, wall mounted	EA	8.7843	1,066.94	10,762.41	11,829.36
11.2316 061	Specimen pass-thru box, stainless steel	EA	1.1586	140.72	222.66	363.39
11.2317 000	**X-RAY EQUIPMENT:**					
11.2317 011	X-ray, ceiling mounted, telescoping	EA	46.4033	5,183.71	43,724.00	48,907.72
11.2317 021	X-ray, wall mounted, chest	EA	22.9152	2,559.86	12,955.21	15,515.07
11.2317 031	Mobile x-ray unit	EA			54,156.20	54,156.20
11.2317 041	X-ray control unit	EA	25.3500	2,831.85	14,331.67	17,163.52
11.2317 051	Multix table	EA	48.3368	5,399.70	27,327.50	32,727.20
11.2318 000	**X-RAY PROCESSING EQUIPMENT:**					
11.2318 011	Auto film processor, complete	EA			43,027.93	43,027.93
11.2318 021	Development tank, 10 gal, stainless steel, 2 compartment, mix valve	EA			2,325.85	2,325.85
11.2318 031	X-ray pass box, 2 compartment, ro-in frame	EA			1,943.21	1,943.21
11.2318 041	X-ray film loading bin	EA			789.45	789.45
11.2318 051	Revolving door, 36x80, safe hinge	EA			4,007.96	4,007.96
11.2319 000	**X-RAY VIEWING EQUIPMENT:**					
11.2319 011	X-ray film illuminator, wet, drip tray	EA			386.61	386.61
11.2319 021	X-ray film illuminator, 1 panel, 14x17	EA			308.64	308.64
11.2319 031	X-ray film illuminator, 2 panel, 30x18	EA			587.01	587.01

11 - EQUIPMENT

Division	Description	Unit	Labor Hours	Labor Cost	Material Cost	Total Cost
11.2319 041	X-ray film illuminator, multibank, 4/4	EA			2,797.51	2,797.51
11.2319 051	X-ray film illuminator, multibank, 6/6	EA			3,724.59	3,724.59
11.2319 061	X-ray shield, view glass, deluxe	EA			1,275.28	1,275.28
11.2320 000	**NUCLEAR EQUIPMENT:**					
11.2320 011	Gamma camera, 10" view, complete	EA			481,450.31	481,450.31
11.2320 021	Gamma camera, 15" view, complete	EA			542,178.18	542,178.18
11.2321 000	**ULTRA SOUND EQUIPMENT:**					
11.2321 011	Ultra sound unit complete	EA			280,663.82	280,663.82
11.2322 000	**HYDROTHERAPY UNITS:**					
11.2322 011	Hubbard tank, 400 gal, twin eject	EA			32,792.95	32,792.95
11.2322 021	Treatment/wade tank, 1000 gal	EA			40,080.32	40,080.32
11.2322 031	Whirlpool, stainless steel, 85 gal, leg & hip	EA			6,194.21	6,194.21
11.2322 041	Whirlpool, stainless steel, 80 gal, arm, leg & hip	EA			5,344.00	5,344.00
11.2322 051	Whirlpool, stainless steel, 25 gal, arm	EA			3,947.31	3,947.31
11.2322 061	Moisture heat therapy unit, table	EA			5,167.69	5,167.69
11.2322 071	Mobile paraffin bath	EA			1,893.70	1,893.70
11.2322 081	Mobile sitz bath	EA			516.36	516.36
11.2323 000	**INHALATION THERAPY EQUIPMENT:**					
11.2323 011	Ventilator, complete	EA			17,559.95	17,559.95
11.2323 021	Suction unit, stainless steel cabinet	EA			1,618.42	1,618.42
11.2323 031	IPPB inhaler	EA			2,134.68	2,134.68
11.2323 041	Air volume tester, lung	EA			8,607.79	8,607.79
11.2324 000	**PHYSICAL THERAPY EQUIPMENT:**					
11.2324 011	Exercise unit, complete	EA			8,798.38	8,798.38
11.2324 021	Exercise chair	EA			2,582.21	2,582.21
11.2324 031	Treadmill, motorized, 5 speed	EA			5,597.92	5,597.92
11.2324 041	Treadmill, adjustable angle	EA			1,205.10	1,205.10
11.2324 051	Rowing machine	EA			1,119.55	1,119.55
11.2324 061	Rehabilitation loom	EA			1,807.29	1,807.29
11.2325 000	**LAUNDRY EQUIPMENT:**					
11.2325 011	Unloading washer, 60"x44", 400#	EA	158.7340	17,125.81	117,513.40	134,639.21
11.2325 021	Extractor, 200#, with accessories	EA	129.4739	13,968.94	94,636.15	108,605.09
11.2325 031	Washer/extractor, 600#	EA	232.0164	25,032.25	171,591.23	196,623.48
11.2325 041	Dryer, gas or steam, 200/400#	EA	104.8735	11,314.80	77,552.87	88,867.67
11.2325 051	Flat ironer, 6 roller	EA	235.8486	25,445.71	174,511.24	199,956.95
11.2325 061	Ironer/folder, 2 lane	EA	95.8102	10,336.96	70,057.30	80,394.26
11.2325 071	Folder, 3 lane	EA	59.5576	6,425.67	44,095.25	50,520.92
11.2400 000	**DENTAL EQUIPMENT:**					
11.2400 001	Note: The following prices do not include costs for mechanical or electrical hook-up.					
11.2401 000	**DENTAL CHAIRS & INSTRUMENTATION UNITS:**					
11.2401 011	Dental chair, deluxe, with lift	EA			11,428.98	11,428.98
11.2401 021	Dental chair, standard, tilt	EA			6,651.68	6,651.68
11.2401 031	Instrument unit, 4 port, tray, chair mounted	EA			7,465.40	7,465.40
11.2401 041	Assistant instrument unit, chair mounted	EA			5,684.04	5,684.04
11.2401 051	Mobile instrument unit, cabinet	EA			8,546.38	8,546.38
11.2401 061	Mobile assistant's unit, cabinet	EA			7,343.92	7,343.92
11.2401 071	Instrument unit, cabinet, wall mounted	EA			21,372.13	21,372.13
11.2402 000	**DENTAL LIGHTING & X-RAY:**					
11.2402 011	Dental light, chair or unit mounted	EA			1,939.21	1,939.21
11.2402 021	Dental light, ceiling mounted	EA			1,939.21	1,939.21
11.2402 031	Dental x-ray, wall mounted	EA			15,821.55	15,821.55
11.2402 041	Dental x-ray, wall, extra remote heavy duty	EA			25,011.72	25,011.72
11.2403 000	**DENTAL EQUIPMENT, MISCELLANEOUS:**					
11.2403 011	Dental sterilizer, chemiclave	EA			2,228.70	2,228.70
11.2403 021	Dental sterilizer, vibraclean	EA			1,359.43	1,359.43

EQUIPMENT - 11

Division	Description	Unit	Labor Hours	Labor Cost	Material Cost	Total Cost
11.2403 031	Dental compressor with dryer	EA			4,033.04	4,033.04
11.2404 000	**DENTAL EQUIPMENT, LABORATORY, MISCELLANEOUS:**					
11.2404 011	Dust collector, pedestal	EA	4.0818	495.78	2,825.81	3,321.58
11.2404 021	Waxing unit, 3 compartment	EA			111.91	111.91
11.2404 031	Pneumatic curing unit	EA			239.84	239.84
11.2404 041	Double pneumatic press	EA			1,679.27	1,679.27
11.2404 051	Curing tank assembly	EA	1.8480	224.46	1,305.61	1,530.07
11.2404 061	Boilout assembly	EA	1.5181	184.39	1,072.86	1,257.25
11.2404 071	Plaster bin, 4 compartment, 300#	EA	1.0506	127.61	744.86	872.47
11.2405 000	**DENTAL LABORATORY FURNITURE, METAL:**					
11.2405 011	Dental laboratory tech bench, 5 drawer	EA	2.0032	216.13	1,415.49	1,631.62
11.2405 021	Dental laboratory, 2 door cab, 36"x24"x36"	EA	2.5780	278.14	1,603.14	1,881.28
11.2405 031	Dental laboratory, 1 door cab, 24"x24"x36"	EA	2.0055	216.37	1,246.86	1,463.24
11.2405 041	Dental laboratory, 1 door cabinet, corner	EA	3.1508	339.94	1,959.43	2,299.37
11.2501 000	**MORTUARY EQUIPMENT:**					
11.2501 011	Mortuary refrigerator, 4 place, 2 tier	EA	14.5474	1,569.52	31,082.23	32,651.75
11.2501 021	Mortuary refrigerator, 6 place, 2 tier	EA	18.1849	1,961.97	40,883.14	42,845.11
11.2501 031	Mortuary refrigerator, 10 place, 2 tier	EA	31.6411	3,413.76	57,491.21	60,904.97
11.2501 041	Autopsy table	EA			11,699.68	11,699.68
11.2600 000	**PRISON EQUIPMENT:**					
11.2601 000	**CELL & CORRIDOR CONSTRUCTION:**					
11.2601 011	Steel plate wall lining	SF	0.1654	17.85	20.04	37.89
11.2601 021	Steel plate ceiling lining	SF	0.3853	41.57	20.04	61.61
11.2601 031	Bar walls, normal security	SF	0.1654	17.85	32.18	50.02
11.2601 041	Bar door, hinged, key locking device only	SF	0.5507	59.42	120.59	180.00
11.2601 051	Bar door, hinged, key locking device only	SF	0.6055	65.33	132.62	197.95
11.2601 061	Panel door, hinged, key locking device only	SF	0.4959	53.50	100.48	153.98
11.2601 071	Panel door, sliding, key locking device only	SF	0.4959	53.50	104.54	158.04
11.2601 081	Cells, average door	EA	27.6400	2,982.08	7,079.21	10,061.29
11.2601 091	Sallyport, average door	EA	27.6400	2,982.08	5,899.32	8,881.40
11.2601 101	Air locking device	EA	27.6400	2,982.08	7,551.12	10,533.20
11.2602 000	**REMOTE ELECTRIC CONTROLS:**					
11.2602 011	Add to total cost, per door	EA	7.3417	792.10	150.38	942.47
11.2603 000	**CELL ACCESSORIES, BUILT-INS:**					
11.2603 011	Single bunk, stainless steel	EA	3.4139	368.33	1,045.79	1,414.12
11.2603 021	Double bunks, stainless steel	EA	4.5706	493.12	1,790.06	2,283.18
11.2701 000	**CENTRAL VACUUM SYSTEM:**					
11.2701 011	Vacuum, central residential, 8 outlet max	INLET			333.80	333.80
11.2701 021	Vacuum, central commercial, 1 station/1200 SF	STATI			1,289.82	1,289.82
11.2702 000	**VACUUM, CENTRAL COMMERCIAL, COMPONENTS:**					
11.2702 011	Vacuum, motor closed, 1200 CFM/1200 SF	EA			1,385.42	1,385.42
11.2702 021	Vacuum, motor closed, 2400 CFM/2500 SF	EA			2,937.83	2,937.83
11.2702 031	Vacuum, motor closed, 5000 CFM/50,000 SF	EA			6,069.91	6,069.91
11.2702 041	**Filter separator, 10 bag (2:1)**	EA			3,110.79	3,110.79
11.2702 051	Steel tubing, 1-1/2"	LF			9.96	9.96
11.2702 061	Steel tubing, 2-1/2"	LF			11.63	11.63
11.2702 071	Inlets	EA			38.38	38.38
11.2702 081	Portable pickup with wet separator	EA			910.46	910.46
11.2800 000	**STAGE EQUIPMENT:**					
11.2801 000	**STAGE EQUIPMENT, IN PLACE:**					
11.2801 001	*Note: The following prices include lighting, control, rigging and drops.*					
11.2801 011	Large stage, professional/college	GSF			842.80	842.80
11.2801 021	Medium stage, college/high school	GSF			505.65	505.65
11.2801 031	Medium stage, community theater	GSF			421.34	421.34
11.2801 041	Small stage, junior high school	GSF			337.10	337.10
11.2801 051	Small stage, elementary school	GSF			235.96	235.96

11 - EQUIPMENT

Division	Description	Unit	Labor Hours	Labor Cost	Material Cost	Total Cost
11.2801 061	Minimum equip stage, multipurpose	GSF			134.78	134.78
11.2802 000	**CONTROL CENTERS-LIGHTING:**					
11.2802 011	Five scene, 2 sub-scene, 390 pot	LS			202,288.74	202,288.74
11.2802 021	One scene, 2 sub-scene, 24 pot	LS			104,515.82	104,515.82
11.2802 031	Patch board, 3 2500w dim, 15 non-dim	LS			25,960.35	25,960.35
11.2803 000	**LIGHTING INSTRUMENTS:**					
11.2803 011	Footlights, disappearing, reflect	PER 5			1,028.22	1,028.22
11.2803 021	Add for motorized	PER 5			1,769.94	1,769.94
11.2803 031	Border lights, 1 row, reflective, 3 color	PER 8			859.67	859.67
11.2803 041	Border lights, 1 row, roundel, 3 color	PER 8			1,196.81	1,196.81
11.2803 051	Border lights, 2 row, roundel, 4 color	PER 8			3,034.25	3,034.25
11.2803 061	Ball, mirrored, rotating, 30"	EA	14.3754	1,605.88	2,697.33	4,303.21
11.2803 071	Spotlight, follow, carbon arc	EA			7,417.24	7,417.24
11.2803 081	Spotlight, follow, quartz halogen	EA			2,359.98	2,359.98
11.2803 091	Quartz spot, elliptical, iodine lamp, 3000w	EA			2,191.38	2,191.38
11.2803 101	Quartz spot, elliptical, iodine lamp, 1500w	EA			758.54	758.54
11.2803 111	Quartz spot, elliptical, iodine lamp, 500w	EA			471.95	471.95
11.2803 121	Quartz spot, fresnel, iodine lamp, 1000w	EA			539.33	539.33
11.2803 131	Quartz spot, fresnel, iodine lamp, 500w	EA			320.22	320.22
11.2803 141	Color wheel, motorized, 20"	EA			320.22	320.22
11.2803 151	Quartz beam projection, iodine lamp, 1500w	EA			522.45	522.45
11.2803 161	Floodlight, utility, incandescent, 1000w	EA			286.50	286.50
11.2803 171	Flood, scoop, iodine lamp, 18", 750w	EA			320.22	320.22
11.2804 000	**LIGHTING & CONTROL ACCESSORIES:**					
11.2804 011	Light tower, 4' x 6' x 15'	EA			8,428.63	8,428.63
11.2804 021	Light stand, cast iron base, 24"	EA			387.69	387.69
11.2804 031	Plug strip	LF			57.25	57.25
11.2804 041	Floor pocket, plug, 4 outlet, 100a	EA			269.64	269.64
11.2804 051	Floor pocket, plug, 2 outlet, 50a	EA			235.96	235.96
11.2804 061	Wall pocket, surface plug, 3 outlet, 50a	EA			84.25	84.25
11.2804 071	Wall pocket, surface plug, 1 outlet, 50a	EA			185.36	185.36
11.2804 081	Wall pocket, flush, 2 outlet, 100a	EA			326.96	326.96
11.2804 091	Wall pocket, flush, 4 outlet, 50a	EA			286.50	286.50
11.2804 101	Catwalk, wood, metal pipe mount rail	LF	0.9414	101.57	23.45	125.02
11.2805 000	**STAGE RIGGING, CURTAINS & DROPS:**					
11.2805 001	Note: The price for the scissors lift at the end of this section does not include costs for any structural work.					
11.2805 011	Acoustic cloud, adjust, wood frame	SF	0.0777	8.38	4.36	12.75
11.2805 021	Curtain track, heavy duty, straight	LF	0.2000	21.58	17.37	38.95
11.2805 031	Curtain track, medium duty, straight	LF	0.2000	21.58	10.93	32.51
11.2805 041	Curtain track, heavy duty, curved	LF	0.3005	32.42	69.72	102.15
11.2805 051	Curtain track, medium duty, curved	LF	0.3005	32.42	17.98	50.40
11.2805 061	Add for electric driven heavy duty	EA	2.5027	279.58	1,547.59	1,827.17
11.2805 071	Add for electric driven medium duty	EA	2.5027	279.58	1,486.89	1,766.47
11.2805 081	T bar rigg, counter weight, 4 loft block, 55x35	SET			53,943.60	53,943.60
11.2805 091	Rigg, wire guard, 4 loft block, 45x30	SET			47,200.70	47,200.70
11.2805 101	T bar rigg, 4 loft block, caster mounted	SET			57,315.11	57,315.11
11.2805 111	Add for electric control	LS			53,943.60	53,943.60
11.2805 121	Curtain, fireproof, straight, lift, 23x47	LS			75,858.21	75,858.21
11.2805 131	Curtain, trip, 23' x 47'	LS			72,486.75	72,486.75
11.2805 141	Curtain, main, velour, heavy	SY			26.08	26.08
11.2805 151	Curtain, main, velour, medium	SY			18.30	18.30
11.2805 161	Curtain, light weight, cyclorama	SY			13.74	13.74
11.2805 171	Drops, velour, heavy, 6' x 40'	EA			2,160.73	2,160.73
11.2805 181	Drops, velour, medium, 6' x 40'	EA			2,022.86	2,022.86
11.2805 191	Wings, velour, heavy, 10' x 30'	EA			1,517.11	1,517.11
11.2805 201	Wings, velour, medium, 10' x 30'	EA			1,415.94	1,415.94

EQUIPMENT - 11

Division	Description	Unit	Labor Hours	Labor Cost	Material Cost	Total Cost
11.2805 211	Scissors lift, 5'x10', 12' lift	SECTI			20,228.80	20,228.80
11.2805 221	Scissors lift, 15'x40' with 15' lift	EA			212,129.53	212,129.53
11.2806 000	**TELEVISION STUDIO LIGHTING, INCLUDING LAMPS:**					
11.2806 011	Studio, 8000 SF, very well equipped	PKG	143.7568	16,059.07	659,181.91	675,240.98
11.2806 021	Studio, 6, 60'x72', well equipped	PKG	40.2517	4,496.52	156,113.15	160,609.67
11.2806 031	Studio, medium equipment	PKG	28.7518	3,211.86	121,383.79	124,595.66
11.2806 041	Studios, 2, 20'x30', medium equipped	PKG	13.8007	1,541.68	28,019.38	29,561.06
11.2806 051	Studio, portable, minimum equipment, no control	PKG	5.7500	642.33	4,383.20	5,025.54
11.2901 000	**GARBAGE COMPACTORS:**					
11.2901 011	Garbage compactor, 1-1/2 CY	EA	9.3000	1,129.58	15,760.34	16,889.92
11.2901 021	Garbage compactor, 2 CY	EA	9.3000	1,129.58	23,400.03	24,529.61
11.2901 031	Garbage compactor, 2-1/2" CY	EA	11.1600	1,355.49	28,709.01	30,064.50
11.3000 000	**WINDOW WASHING EQUIPMENT, POWERED:**					
11.3000 001	*Note: The following items constitute a spider staging system.*					
11.3000 011	Rolling davit, tilting type	EACH			107,074.92	107,074.92
11.3000 021	Trackage	LF			107.04	107.04
11.3000 031	Rolling unit stepladder, 56' long	EA			107,074.92	107,074.92
11.3000 041	Modulated platform	EA			74,952.44	74,952.44
11.3000 051	Turntable	EA			8,566.00	8,566.00
11.3000 061	Trollies	EA			6,424.48	6,424.48
11.4001 000	**VOCATIONAL EQUIPMENT:**					
11.4001 011	Drill press, floor, 12", 1/2 hp	EA	1.9408	209.39	1,210.97	1,420.36
11.4001 021	Grinders, double wheel, 1 hp	EA	1.9408	209.39	960.44	1,169.84
11.4001 031	Jointer, 4", 1 hp	EA	1.9408	209.39	3,000.11	3,209.50
11.4001 041	Lathe, wood, 10", 1/2 hp	EA	1.9408	209.39	1,156.61	1,366.00
11.4001 051	Planer, 13"x6", 3/4 hp	EA	1.9408	209.39	3,392.10	3,601.49
11.4001 061	Band saw, woodcutting, 14", 3/4 hp	EA	1.9408	209.39	1,909.19	2,118.58
11.4001 071	Band saw, metal cutting, 14", 3/4 hp	EA	1.9408	209.39	3,818.30	4,027.69
11.4001 081	Radial arm saw, 10", 1-1/2 hp	EA	1.9408	209.39	1,778.29	1,987.69
11.4001 091	Scroll saw, 24", 1/2 hp	EA	1.9408	209.39	2,727.37	2,936.76
11.4001 101	Table saw, 10", 2 hp	EA	1.9408	209.39	3,491.04	3,700.43
11.4001 111	Potter's wheel, motorized	EA	1.9408	209.39	1,375.42	1,584.81
11.4001 121	Kiln, 16 CF to 2000 degrees	EA	1.9408	209.39	4,323.39	4,532.78

12 - FURNISHINGS

Division	Description	Unit	Labor Hours	Labor Cost	Material Cost	Total Cost
12.1101 000	**BLINDS & SHADES:**					
12.1101 011	Shades, standard roller	SF			2.84	2.84
12.1101 021	Shades, extra quality	SF			3.11	3.11
12.1101 031	Shades, decorator	SF			11.38	11.38
12.1101 041	Blinds, horizontal	SF			6.49	6.49
12.1101 051	Blinds, horizontal, mini	SF			8.15	8.15
12.1101 061	Blinds, vertical, PVC	SF			7.79	7.79
12.1101 071	Blinds, vertical, fabric	SF			10.39	10.39
12.3000 000	**CABINETS & LAMINATED PLASTIC TOPS:**					
12.3000 001	Note: For plastic laminated cabinets, use the prices listed in section 12.3501.					
12.3001 000	**CABINETS, MULTI-UNIT, ECONOMY:**					
12.3001 011	Cabinets, hardwood, economy, base	LF	0.2112	22.79	46.02	68.81
12.3001 021	Cabinets, hardwood, economy, wall	LF	0.3732	40.26	29.71	69.97
12.3001 031	Cabinets, hardwood, economy, full height	LF	0.4796	51.74	114.71	166.46
12.3002 000	**CABINETS, HARDWOOD, CUSTOM:**					
12.3002 011	Cabinets, hardwood, custom, base	LF	0.2112	22.79	66.22	89.01
12.3002 021	Cabinets, hardwood, custom, wall	LF	0.3732	40.26	47.99	88.26
12.3002 031	Cabinets, hardwood, custom, full height	LF	0.4796	51.74	115.95	167.70
12.3003 000	**CABINETS, PREMIUM, INSTITUTIONAL:**					
12.3003 011	Cabinets, birch, premium, base	LF	0.4138	44.64	132.87	177.52
12.3003 021	Cabinets, birch, premium, wall	LF	0.7258	78.31	111.49	189.79
12.3003 031	Cabinets, birch, premium, full height	LF	0.5224	56.36	188.47	244.83
12.3004 000	**ADDERS FOR MILL MADE CABINETS:**					
12.3004 001	Note: For field finishing see section 09.0000.					
12.3004 011	Add for ash	LF			12.13	12.13
12.3004 021	Add for walnut	LF			66.07	66.07
12.3004 031	Add for overlay	LF			8.12	8.12
12.3004 041	Add for flush overlay	LF			12.13	12.13
12.3004 051	Add for edge banding	LF			8.12	8.12
12.3004 061	Add for pre-finishing, exterior	LF			6.21	6.21
12.3004 071	Add for pre-finishing, interior	LF			13.33	13.33
12.3004 081	Add for standard hardware	LF			5.52	5.52
12.3004 091	Add for institutional hardware	LF			8.12	8.12
12.3004 101	Add for roller guides	SET			6.21	6.21
12.3004 111	Add for roller guides, full suspension	SET			30.88	30.88
12.3005 000	**CABINET UNITS, PREMIUM:**					
12.3005 011	Cabinet, open, with one shelf	LF			83.06	83.06
12.3005 021	Cabinet, with door & one shelf	LF			102.98	102.98
12.3005 031	Cabinet, door, 1 shelf & 1 drawer	LF			142.96	142.96
12.3005 041	Cabinet, sink	LF			91.42	91.42
12.3005 051	Cabinet, 4 drawers	LF			164.59	164.59
12.3005 061	Add for apron (knee space)	LF			23.22	23.22
12.3005 071	Add for backsplash	LF			13.33	13.33
12.3006 000	**CABINETS, METAL, COMMERCIAL:**					
12.3006 001	Note: The following prices include hardware and plastic tops.					
12.3006 011	Cabinet, metal, base, with door, shelf	LF	0.2202	23.76	132.03	155.79
12.3006 021	Cabinet, metal, wall, with door, 2 shelves	LF	0.2644	28.53	97.23	125.76
12.3006 031	Cabinet, metal, full, with door, 5 shelves	LF	0.3455	37.28	229.11	266.39
12.3006 041	Cabinet, metal, library shelving	LF	0.1932	20.84	85.04	105.88
12.3006 051	Cabinet, metal, wardrobe, 4' wide	EA	0.5835	62.95	555.78	618.73
12.3007 000	**CABINETS, WOOD, FORMICA FACED, SCHOOL:**					
12.3007 001	Note: The following prices include hardware and plastic tops.					
12.3007 011	Cabinet, wood, base, with door, 1 shelf	LF	0.3661	39.50	252.92	292.42
12.3007 021	Cabinet, wood, wall, with door, 2 shelves	LF	0.4396	47.43	183.64	231.07
12.3007 031	Cabinet, wood, full, with doors	LF	0.4523	48.80	300.15	348.95
12.3007 041	Add for laboratory cabinets	LF	0.2527	27.26	59.30	86.57

FURNISHINGS - 12

Division	Description	Unit	Labor Hours	Labor Cost	Material Cost	Total Cost
12.3008 000	**CABINETS, WOOD & METAL, HOSPITAL:**					
12.3008 001	*Note: The following prices include hardware and plastic tops.*					
12.3008 011	Cabinet, wood & metal, with door, 1 shelf, 1 drawer	LF	0.7407	79.91	191.77	271.69
12.3008 021	Cabinet, wood & metal, wall, with door, 2 shelves	LF	0.4935	53.24	112.39	165.64
12.3008 031	Cabinet, wood & metal, full, with doors	LF	0.3855	41.59	231.89	273.48
12.3009 000	**CABINETS, FURNITURE GRADE, LABORATORY:**					
12.3009 001	*Note: The following prices include hardware and plastic tops.*					
12.3009 011	Cabinet, furniture, base, laboratory	LF	1.3013	140.40	268.54	408.94
12.3009 021	Cabinet, furniture, wall, laboratory	LF	0.6285	67.81	127.92	195.73
12.3009 031	Cabinet, furniture, wardrobe, laboratory	LF	0.5385	58.10	287.22	345.32
12.3009 041	Cabinet, furniture, laboratory island	LF	1.5705	169.44	383.99	553.43
12.3009 051	Cabinet, furniture, fume hood	LF	3.4995	377.56	530.78	908.34
12.3009 061	Cabinet, furniture, fume hood, stainless steel	LF	3.3653	363.08	708.76	1,071.84
12.3009 071	Add for premium quality	LF	0.2423	26.14	30.57	56.71
12.3500 000	**LAMINATED PLASTIC & SIMULATED MARBLE TOPS:**					
12.3500 001	*Note: For ceramic tile tops see section 09.3001. For solid surface color, add 11% to the total costs. For solid core color, add 25% to the total costs.*					
12.3501 000	**LAMINATED PLASTIC & SIMULATED MARBLE TOPS:**					
12.3501 011	Laminated plastic top, multires	LF			36.32	36.32
12.3501 021	Laminated plastic top, small projects	LF			41.93	41.93
12.3501 031	Laminated plastic top, custom jobs	LF			55.98	55.98
12.3501 041	Vanity top, cultured marble, no bowl	LF			50.13	50.13
12.3501 051	Add to vanity for molded bowl	EA			46.48	46.48
12.3501 061	Add to vanity, molded clam shell	EA			81.34	81.34
12.3501 071	Acid proof tops	LF			112.45	112.45
12.3501 081	Komar simulated marble, molded section	SF			20.98	20.98
12.4000 000	**CARPETS:**					
12.4001 000	**CARPETS, WITH 50 OZ PAD:**					
12.4001 011	Carpet, 30 oz polyester, with pad	SY	0.1147	7.86	27.50	35.36
12.4001 021	Carpet, 35 oz polyester, with pad	SY	0.1147	7.86	30.53	38.40
12.4001 031	Carpet, 40 oz polyester, with pad	SY	0.1147	7.86	32.54	40.40
12.4001 041	Carpet, 50 oz polyester, with pad	SY	0.1147	7.86	38.19	46.05
12.4001 051	Carpet, 20 oz nylon shag, pad	SY	0.1147	7.86	15.99	23.85
12.4001 061	Carpet, 24 oz nylon shag, pad	SY	0.1147	7.86	18.22	26.08
12.4001 071	Carpet, 20 oz nylon filament, pad	SY	0.1147	7.86	15.99	23.85
12.4001 081	Carpet, 24 oz nylon filament, pad	SY	0.1147	7.86	22.10	29.96
12.4001 091	Carpet, 15 oz nylon level loop, rubber back	SY	0.0983	6.74	18.22	24.95
12.4001 101	Carpet, 21 oz nylon level loop, rubber back	SY	0.0983	6.74	22.10	28.83
12.4001 111	Carpet, 28 oz nylon level loop, rubber back	SY	0.0983	6.74	30.14	36.88
12.4001 121	Carpet, 15 oz nylon level loop, pad	SY	0.1147	7.86	19.09	26.95
12.4001 131	Carpet, 21 oz nylon level loop, pad	SY	0.1147	7.86	23.19	31.05
12.4001 141	Carpet, 28 oz nylon level loop, pad	SY	0.1147	7.86	31.76	39.62
12.4001 151	Carpet, 48 oz nylon level loop	SY	0.1147	7.86	38.41	46.27
12.4001 161	Carpet, 21 oz antron nylon, anti-static	SY	0.1147	7.86	30.14	38.00
12.4001 171	Carpet, wool commercial, pad, standard weight	SY	0.1147	7.86	64.51	72.37
12.4001 181	Carpet, exterior, premium, without pad	SY	0.1313	9.00	29.77	38.76
12.4002 000	**CARPET PADS:**					
12.4002 011	Carpet pad, 40 oz jute	SY	0.0244	1.67	3.98	5.66
12.4002 021	Carpet pad, 50 oz jute/hair	SY	0.0244	1.67	5.58	7.25
12.4002 031	Carpet pad, 50 oz hair	SY	0.0244	1.67	7.00	8.67
12.4002 041	Carpet pad, rubber waffle, 72 oz	SY	0.0244	1.67	5.10	6.77
12.4002 051	Carpet pad, rubber waffle, 100 oz	SY	0.0244	1.67	9.64	11.31
12.4002 061	Carpet pad, rubber slab, 72 oz	SY	0.0244	1.67	6.44	8.11
12.4002 071	Carpet pad, rubber slab, 88 oz	SY	0.0244	1.67	9.64	11.31
12.4002 081	Carpet pad, 15# urethane	SY	0.0244	1.67	3.29	4.97
12.4002 091	Carpet pad, 2# urethane	SY	0.0244	1.67	5.29	6.96
12.4002 101	Carpet pad, 25# urethane	SY	0.0244	1.67	6.21	7.88

12 - FURNISHINGS

Division	Description	Unit	Labor Hours	Labor Cost	Material Cost	Total Cost
12.4002 111	Carpet pad, urethane rebound	SY	0.0244	1.67	6.78	8.46
12.4003 000	**CARPET, AVERAGE ALLOWANCES:**					
12.4003 011	Carpet, average, housing	SY			24.58	24.58
12.4003 021	Carpet, average, commercial	SY			34.79	34.79
12.4003 031	Carpet, average, school	SY			37.94	37.94
12.4003 041	Carpet, average, hotel/motel, theater	SY			41.13	41.13
12.4003 051	Carpet, average, custom housing	SY			60.13	60.13
12.4004 000	**FLOOR MATS:**					
12.4004 011	Floor mat, rubber, 1/4", recessed	SF	0.0399	3.30	7.91	11.21
12.4004 021	Floor mat, rubber, 1/2", recessed	SF	0.0502	4.15	12.50	16.65
12.4004 031	Floor mat, vinyl, 1/4", recessed	SF	0.0399	3.30	11.63	14.93
12.4004 041	Floor mat, vinyl, 1/2", recessed	SF	0.0502	4.15	15.39	19.55
12.4004 051	Walk off mat with serrated filler	SF	0.0781	6.46	80.30	86.76
12.5000 000	**DRAPERIES & CURTAINS:**					
12.5001 000	DRAPERIES & CURTAINS:					
12.5001 001	Note: The prices in the following items are based on the measurement of window glass only. Allowances have already been made for overlap and pleating.					
12.5001 011	Curtain, lead mesh (soundproof)	SY			61.36	61.36
12.5001 021	Curtain, lead mesh, 25#/SF, x-ray	SY			70.13	70.13
12.5001 031	Draperies, window, multires	LF			26.21	26.21
12.5001 041	Draperies, sliding door	LF			26.21	26.21
12.5001 051	Draperies, window, custom residential	LF			70.13	70.13
12.5001 061	Draperies, sliding door, custom	LF			56.98	56.98
12.5001 071	Draperies, lining	LF			4.36	4.36
12.5001 081	Draperies, fiberglass	LF			69.24	69.24
12.5001 091	Draperies, filter light control	LF			62.55	62.55
12.5001 101	Draperies, flameproof	LF			67.20	67.20
12.5001 111	Draperies, velour grand	LF			73.05	73.05
12.5001 121	Blinds, vertical	SF			6.28	6.28
12.8000 000	**OFFICE LANDSCAPE, FURNITURE BY STATION:**					
12.8000 001	Note: The following prices are based on quantities of 100 or more stations.					
12.8001 000	OFFICE LANDSCAPE, FURNITURE BY STATION:					
12.8001 001	Note: The following workstations include: Reception:........................... horseshoe counter and secretarial chair. Secretarial:........................... desk with return and secretarial chair. Engineering:........ desk, chair, drafting table and 35" x 80" partition with shelving. Managerial:... desk, 3 chairs, table, 70" x 80" partition with shelving, and file storage. Executive:............ desk, 5 chairs, table and work table with over/under storage.					
12.8001 011	Furniture, reception station, good	EA			3,170.88	3,170.88
12.8001 021	Furniture, reception station, better	EA			6,332.89	6,332.89
12.8001 031	Furniture, reception station, best	EA			15,310.14	15,310.14
12.8001 051	Furniture, secretarial station, good	EA			846.59	846.59
12.8001 061	Furniture, secretarial station, better	EA			2,199.84	2,199.84
12.8001 071	Furniture, secretarial station, best	EA			5,177.43	5,177.43
12.8001 101	Furniture, engineering station, good	EA			6,110.69	6,110.69
12.8001 111	Furniture, engineering station, better	EA			8,112.84	8,112.84
12.8001 121	Furniture, engineering station, best	EA			9,617.14	9,617.14
12.8001 151	Furniture, managerial station, good	EA			5,766.31	5,766.31
12.8001 161	Furniture, managerial station, better	EA			7,379.52	7,379.52
12.8001 171	Furniture, managerial station, best	EA			11,465.95	11,465.95
12.8001 201	Furniture, executive station, good	EA			4,599.72	4,599.72
12.8001 211	Furniture, executive station, better	EA			7,999.46	7,999.46
12.8001 221	Furniture, executive station, best	EA			13,399.12	13,399.12

FURNISHINGS - 12

Division	Description	Unit	Labor Hours	Labor Cost	Material Cost	Total Cost
12.8002 000	ADDERS & DEDUCTORS FOR OFFICE LANDSCAPE:					
12.8002 011	Partitioning, 5'6" high, good	LF			195.54	195.54
12.8002 021	Partitioning system, work surface, good	STA			3,135.56	3,135.56
12.8002 031	Desk, good	EA			457.73	457.73
12.8002 041	Desk, better	EA			1,326.57	1,326.57
12.8002 051	Desk, best	EA			3,199.81	3,199.81
12.8002 061	Add for task light per SF cover	SF			105.83	105.83
12.8002 071	Secretarial desk with return, good	EA			686.64	686.64
12.8002 081	Secretarial desk with return, better	EA			1,831.02	1,831.02
12.8002 091	Secretarial desk with return, best	EA			4,577.48	4,577.48
12.8002 101	Drafting table, good	EA			733.29	733.29
12.8002 111	Drafting table, better	EA			1,304.35	1,304.35
12.8002 121	Drafting table, best	EA			1,579.37	1,579.37
12.8002 131	Desk chair, good	EA			530.11	530.11
12.8002 141	Desk chair, better	EA			924.58	924.58
12.8002 151	Desk chair, best	EA			1,588.23	1,588.23
12.8002 161	Secretarial chair, good	EA			263.02	263.02
12.8002 171	Secretarial chair, better	EA			530.11	530.11
12.8002 181	Secretarial chair, best	EA			924.58	924.58
12.8002 191	Drafting stool, good	EA			206.60	206.60
12.8002 201	Drafting stool, better	EA			411.08	411.08
12.8002 211	Drafting stool, best	EA			686.64	686.64
12.8002 221	Side chair, good	EA			183.31	183.31
12.8002 231	Side chair, better	EA			342.22	342.22
12.8002 241	Side chair, best	EA			593.27	593.27
12.8002 251	Credenza, 66" long, good	EA			522.19	522.19
12.8002 261	Credenza, 66" long, better	EA			1,417.70	1,417.70
12.8002 271	Credenza, 66" long, best	EA			2,517.60	2,517.60
12.8002 281	Bookcase, 72" high, 36" wide, good	EA			251.09	251.09
12.8002 291	Bookcase, 72" high, 36" wide, better	EA			433.28	433.28
12.8002 301	Bookcase, 72" high, 36" wide, best	EA			628.83	628.83
12.8002 311	File cabinet, 4 drawer, good	EA			228.88	228.88
12.8002 321	File cabinet, 4 drawer, better	EA			297.77	297.77
12.8002 331	File cabinet, 4 drawer, best	EA			457.73	457.73
12.8002 341	Conference table, 8', good	EA			548.81	548.81
12.8002 351	Conference table, 8', better	EA			1,944.32	1,944.32
12.8002 361	Conference table, 8', best	EA			4,577.48	4,577.48
12.8002 371	Conference chair, good	EA			206.60	206.60
12.8002 381	Conference chair, better	EA			433.28	433.28
12.8002 391	Conference chair, best	EA			688.85	688.85

13 - SPECIAL CONSTRUCTION

Division	Description	Unit	Labor Hours	Labor Cost	Material Cost	Total Cost
13.0000 000	**SPECIAL CONSTRUCTION:**					
13.1001 000	**AUDIOMETRIC ROOMS:**					
13.1001 011	Audiometric rooms, 500 SF & over	SFSA	0.1783	19.24	59.36	78.60
13.1001 021	Audiometric masking system, 5000 SF	SFSA	0.0061	0.66	2.00	2.66
13.1002 000	**BOWLING ALLEYS:**					
13.1002 011	Bowling lanes, complete, auto	LANE			72,714.64	72,714.64
13.1002 021	Automatic scorer, 4 lane	EA			45,424.85	45,424.85
13.1003 000	**BROADCASTING STUDIOS:**					
13.1003 011	Sound wall, 4" short wall, record	SFSA	0.1523	16.43	55.43	71.86
13.1003 021	Sound wall, 4" wall, echo chamber	SFSA	0.1270	13.70	38.08	51.78
13.1004 000	**INSULATED ROOMS:**					
13.1004 001	*Note: The floating floor item listed below includes jack-up neoprene mounts, 3/4" perimeter board, 6 ml polyethylene, raising floors to operating position and grouting jack-screw holes flush. Concrete, repair & caulking are not included.*					
13.1004 011	Blast absorption chamber, 4" perforated metal	SF			69.85	69.85
13.1004 021	Sound deadening enclosure, 4"	SF			46.56	46.56
13.1004 031	Floating floor sound isolate system	SF	0.0644	6.95	7.32	14.27
13.1005 000	**GREENHOUSES:**					
13.1005 001	*Note: The following items are 500 sf or less, foundations & stubwalls not included.*					
13.1005 011	Greenhouses, to 1000 SF	SF			39.36	39.36
13.1005 021	Greenhouses, to 4000 SF	SF			29.35	29.35
13.1005 031	Greenhouse, to 10,000 SF	SF			25.61	25.61
13.1006 000	**INCINERATORS:**					
13.1006 011	Incinerator, 50#/hour, no scrubber	EA			18,710.18	18,710.18
13.1006 021	Incinerator, 100#/hour, no scrubber	EA			23,387.72	23,387.72
13.1006 031	Incinerator, 200#/hour, no scrubber	EA			26,985.86	26,985.86
13.1006 041	Incinerator, 500#/hour, no scrubber	EA			55,770.84	55,770.84
13.1006 051	Incinerator, 1000#/hour, no scrubber	EA			71,962.34	71,962.34
13.1007 000	**INTEGRATED CEILINGS:**					
13.1007 011	Suspended ceiling, T bar, 100 foot candles	SF			19.16	19.16
13.1007 021	Suspended ceiling, T bar, 85 foot candles	SF			18.25	18.25
13.1007 031	Suspended ceiling, T bar, 70 foot candles	SF			17.31	17.31
13.1007 041	Suspended ceiling, T bar, 55 foot candles	SF			15.44	15.44
13.1007 051	T bar ceiling, 80% plastic, 100 foot candles	SF			23.38	23.38
13.1007 061	T bar ceiling, 70% plastic, 85 foot candles	SF			21.70	21.70
13.1007 071	T bar ceiling, 60% plastic, 70 foot candles	SF			20.78	20.78
13.1007 081	T bar ceiling, 50% plastic, 55 foot candles	SF			19.14	19.14
13.1008 000	**PEDESTAL FLOORS:**					
13.1008 011	Pedestal floor, vinyl tile, gridless	SF			22.61	22.61
13.1008 021	Pedestal floor, vinyl tile, grid	SF			27.15	27.15
13.1008 031	Pedestal floor, perma kleen, grid	SF			29.59	29.59
13.1008 041	Pedestal floor, carpeted system	SF			30.56	30.56
13.1008 051	Pedestal floor, ramps	SF			32.28	32.28
13.1008 061	Add for cutouts	EA			256.29	256.29
13.1008 071	Add for floor grills	EA			113.95	113.95
13.1008 081	Add for seismic bracing	SF			0.07	0.07
13.1008 091	Add for sheet metal trim & casing	LF			126.47	126.47
13.1008 101	Add for CO2 fire system (smoke detector)	SF			7.24	7.24
13.1008 111	Add for automatic fire alarm	SF			0.48	0.48
13.1100 000	**PREFABRICATED STRUCTURES:**					
13.1100 001	*Note: See also section 05.1108.*					
13.1101 000	**PREFABRICATED WALL PANELS:**					
13.1101 011	Prefab wall panel, 84" high	LF	0.1860	20.07	83.61	103.68
13.1101 021	Prefab wall panel, 96" high	LF	0.1953	21.07	93.85	114.93
13.1101 031	Prefab wall panel, 108" high	LF	0.2046	22.07	104.07	126.14
13.1101 041	Prefab wall panel, 120" high	LF	0.2139	23.08	114.94	138.02

SPECIAL CONSTRUCTION - 13

Division	Description	Unit	Labor Hours	Labor Cost	Material Cost	Total Cost
13.1101 051	Prefab wall panel, 132" high	LF	0.2232	24.08	127.17	151.25
13.1101 061	Prefab wall panel, 144" high	LF	0.2325	25.08	136.06	161.15
13.1102 000	**PREFABRICATED STRUCTURES, CORNER POSTS:**					
13.1102 011	Prefab, corner posts, 96" high	EA	0.1860	20.07	65.28	85.35
13.1102 021	Prefab, corner posts, 108" high	EA	0.1860	20.07	80.24	100.31
13.1102 031	Prefab, corner posts, 120" high	EA	0.1860	20.07	95.24	115.31
13.1102 041	Prefab, corner posts, 132" high	EA	0.1860	20.07	110.14	130.21
13.1102 051	Prefab, corner posts, 144" high	EA	0.1860	20.07	125.12	145.19
13.1103 000	**PREFABRICATED STRUCTURES, WALL STARTS:**					
13.1103 011	Prefab, wall starts, 96" high	EA			32.62	32.62
13.1103 021	Prefab, wall starts, 108" high	EA			36.70	36.70
13.1103 031	Prefab, wall starts, 120" high	EA			40.79	40.79
13.1103 041	Prefab, wall starts, 132" high	EA			44.88	44.88
13.1103 051	Prefab, wall starts, 144" high	EA			48.92	48.92
13.1104 000	**PREFABRICATED STRUCTURES, DOORS & WINDOWS:**					
13.1104 011	Prefab, door with threshold	EA			517.06	517.06
13.1104 021	Prefab, door closer	EA			163.23	163.23
13.1104 031	Door lites, 20x30, 1/8" tempered, fixed	EA			87.04	87.04
13.1104 041	Prefab, lock sets	EA			40.79	40.79
13.1104 051	Prefab, window 3'-4' fixed lite, 1/8" tempered	EA			217.61	217.61
13.1104 061	Prefab, window 3-4' side slide & 1 hung, 1/8"	EA			340.12	340.12
13.1104 071	Prefab, window 2' fixed lites, 1/8" tempered	EA			163.23	163.23
13.1105 000	**PREFABRICATED STRUCTURES, ROOF PANELS:**					
13.1105 011	Prefab, roof panel	SF			13.42	13.42
13.1105 021	Prefab, roof fascia	LF			3.35	3.35
13.1105 031	Prefab, ridge beam	LF			16.22	16.22
13.1105 041	Angle to attach to each wall	LF			8.12	8.12
13.1106 000	**PREFABRICATED STRUCTURES, ELECTRICAL:**					
13.1106 011	Prefab, switch or duplex outlet	EA			40.79	40.79
13.1106 021	Prefab, floor fixture, #4 tube	EA			141.49	141.49
13.1106 031	Prefab, balance stems, 2 fixture	SET			40.79	40.79
13.1106 041	Prefab, circuit breaker, 4 circuit	EA			81.61	81.61
13.1106 051	Prefab, circuit breaker, 6 circuit	EA			108.85	108.85
13.1106 061	Prefab, circuit breaker, 8 circuit	EA			157.77	157.77
13.1106 071	Prefab, circuit breaker, 10 circuit	EA			239.42	239.42
13.1107 000	**PREFABRICATED STRUCTURES, HVAC:**					
13.1107 011	Prefab HVAC, heater, 100 watt, baseboard	EA			122.43	122.43
13.1107 021	Thermostat for above	EA			51.68	51.68
13.1107 031	Prefab HVAC, heater, 3200-5600w, 220v, fan force	EA			435.42	435.42
13.1107 041	Prefab, fan, exhaust, 160 CFM	EA			195.86	195.86
13.1107 051	Prefab, fan, exhaust, 350 CFM	EA			342.85	342.85
13.1107 061	Prefab, air conditioner, panel prepared	EA			81.61	81.61
13.1107 071	Caulking, silicone, per tube	EA			13.59	13.59
13.1108 000	**PREFABRICATED STRUCTURES, COMPLETE, IN-PLACE:**					
13.1108 011	Prefab, in place, 64-144 SF	SF			106.23	106.23
13.1108 021	Prefab, in place, 145-288 SF	SF			59.63	59.63
13.1108 031	Prefab, in place, 289-576 SF	SF			57.12	57.12
13.1108 041	Prefab, in place, 577-1132 SF	SF			55.73	55.73
13.1108 051	Prefab, in place, 1133-2264 SF	SF			55.09	55.09
13.1201 000	**RADIATION PROTECTION:**					
13.1201 011	Lead lined lath, 2#	SF			10.25	10.25
13.1201 021	Lead lined lath, 4#	SF			15.16	15.16
13.1201 031	Lead lined lath, 6#	SF			18.16	18.16
13.1201 041	Lead lined lath, 8#	SF			23.41	23.41
13.1201 051	Lead glass windows with lead frames	SF			380.79	380.79
13.1201 061	Lead lined door, to 4#	SF			39.67	39.67
13.1201 071	Lead lined door frames	EA			519.09	519.09

13 - SPECIAL CONSTRUCTION

Division	Description	Unit	Labor Hours	Labor Cost	Material Cost	Total Cost
13.1201 081	Lead lined window frame	SF			66.13	66.13
13.1201 111	Lead lined gypsum wall board, 2#	SF			6.44	6.44
13.1201 121	Lead lined gypsum wall board, 2-1/2#	SF			7.20	7.20
13.1201 131	Lead lined gypsum wall board, 4#	SF			13.83	13.83
13.1201 141	Add for second layer lead lining, 2#	SF			5.00	5.00
13.1201 151	Add for corner angle, 2" x 2", 8'	LF			31.75	31.75
13.1202 000	**CHIMNEY, INSULATING REFRACTORY, NO FOUNDATIONS:**					
13.1202 001	Note: The following item includes 28 ga aluminized steel jacket, floor support, tee, clean out, straight sections and spark screen. It is 24" deep with a 30' stack.					
13.1202 011	Chimney, insulated 1800 - 2000 degrees F	LF			407.95	407.95
13.1203 000	**SWIMMING POOLS:**					
13.1203 001	Note: The following items include filters, chlorinators, heaters, and gunite or concrete. Gunite is listed separately in section 03.8001. For plaster see section 09.1400. Pool deck or flatwork is not included.					
13.1203 011	Pool, residential	SF			73.74	73.74
13.1203 021	Pool, multiple residence	SF			83.27	83.27
13.1203 031	Pool, community	SF			88.94	88.94
13.1203 041	Pool, hotel, resort	SF			95.86	95.86
13.1203 051	Pool, school, 42' x 75'	SF			98.23	98.23
13.1203 061	Pool, school, 42' x 165'	SF			101.98	101.98
13.1203 071	Pool, school, 30' x 30'	SF			106.29	106.29
13.1203 081	Pool deck concrete	SF			7.94	7.94
13.1203 091	Pool cool deck	SF			11.65	11.65
13.1204 000	**SPA:**					
13.1204 011	Spa with pool, 4' diameter	EA			11,777.03	11,777.03
13.1204 021	Spa, recreational, 8' diameter	EA			28,264.82	28,264.82
13.1205 000	**DOMES, IN PLACE:**					
13.1205 001	Note: The following prices are for net area covered, including roofing.					
13.1205 011	Dome, corrugated metal, belem truss	SF			27.60	27.60
13.1205 021	Dome, steel deck, cable suspended	SF			34.06	34.06
13.1205 031	Dome, fabric cover, cable suspended	SF			38.89	38.89
13.1205 041	Dome, air floated fabric cover, cable suspended	SF			31.49	31.49
13.1205 051	Dome, fink, steel, corrugated cover	SF			34.06	34.06
13.1205 061	Dome, aluminum, geodesic	SF			43.01	43.01
13.1205 071	Dome, glu-lams, decking, ribbed	SF			30.56	30.56
13.1205 081	Dome, glu-lam, decking, tridesic	SF			34.43	34.43
13.1205 091	Dome, thin-shell concrete, hyperbolic paraboloid	SF			41.70	41.70
13.1205 101	Dome, steel rib frame, skylight cover	SF			62.91	62.91
13.1205 111	Dome, steel rib frame, skylight, partial opening	SF			125.80	125.80
13.1205 121	Dome, aluminum skin, revolve, 10-40' base	SF			259.86	259.86
13.1206 000	**AIR STRUCTURE, FABRIC COMPLETE:**					
13.1206 011	Foundation and floor concrete	SF	0.0491	4.70	3.73	8.43
13.1206 021	Fabric structure, 3,000 SF, inflated	SF	0.0048	0.46	20.03	20.49
13.1206 031	Fabric structure, 5,000 SF, inflated	SF	0.0048	0.46	17.61	18.07
13.1206 041	Fabric structure, 10,000 SF, inflated	SF	0.0048	0.46	15.27	15.73
13.1206 051	Air curtain, doorway, 3' x 8'	EA	8.0410	720.31	1,201.24	1,921.56
13.1206 061	Air curtain, doorway, 5' x 10'	EA	12.0615	1,080.47	1,766.56	2,847.03
13.1207 000	**BULLET RESISTANT CONSTRUCTION:**					
13.1207 001	Note: Fiberglass matting is usually built into regular stud walls. See Division 9, for other wall components.					
13.1207 011	Bullet resistant fiberglass matting	SF	0.0381	4.11	34.89	39.00
13.1207 021	Bullet resistant service window, voice	EA	1.6250	175.32	3,122.23	3,297.56
13.1207 031	Bullet resistant service window, speaker	EA	1.8750	202.29	4,183.41	4,385.70

CONVEYING EQUIPMENT - 14

Division	Description	Unit	Labor Hours	Labor Cost	Material Cost	Total Cost
14.0000 000	**CONVEYING SYSTEMS:**					
14.1001 000	**DUMBWAITER, TRAY, FOOD OR RECORD:**					
14.1001 011	Dumbwaiter, 2 station, manual, 200#, 24x24x36	EA			11,275.50	11,275.50
14.1001 021	Add for additional stop	STOP			1,931.28	1,931.28
14.1001 031	Dumbwaiter, 2 station, electric, 300#, 30x30x36	EA			28,046.78	28,046.78
14.1001 041	Add for additional stop	STOP			3,890.94	3,890.94
14.1002 000	**DUMBWAITER, CART, FLOOR LEVEL, ELECTRIC:**					
14.1002 011	Dumbwaiter, 2 station, electric, 500#, 23x56x48	EA			58,834.32	58,834.32
14.1002 021	Add for additional stop	STOP			4,672.07	4,672.07
14.1003 000	**ELEVATOR, HOMELIFT, 450#:**					
14.1003 011	Elevator, 450#, 2 stop	EA			24,240.91	24,240.91
14.1003 021	Elevator, 450#, 3 stop	EA			27,095.29	27,095.29
14.1003 031	Chairlift	FLOOR			11,502.70	11,502.70
14.1005 000	**ELEVATOR, HYDRAULIC, 125'/MIN, 2000#, 5'X6' CAB, AUTO EXIT:**					
14.1005 001	Note: For 3 stops or more, elevator must be at least 2500# capacity with a platform at least 7' x 5'.					
14.1005 011	Elevator, 2000#, 2 stop	EA			77,611.96	77,611.96
14.1005 021	Elevator, 2000#, 4 stop	EA			101,342.05	101,342.05
14.1005 031	Elevator, 2000#, 5 stop	EA			116,851.45	116,851.45
14.1005 041	Add for additional stop	STOP			15,721.81	15,721.81
14.1005 051	Add for premium 2 stop basic	EA			65,324.19	65,324.19
14.1006 000	**ELEVATOR, HYDRAULIC, 126'/MIN, 2500#, 7' X 5' CAB:**					
14.1006 011	Elevator, 2500#, 2 stop	EA			79,867.60	79,867.60
14.1006 021	Elevator, 2500#, 3 stop	EA			92,055.58	92,055.58
14.1006 031	Elevator, 2500#, 4 stop	EA			104,286.58	104,286.58
14.1006 041	Elevator, 2500#, 5 stop	EA			120,253.68	120,253.68
14.1006 051	Add for each additional stop	EA			16,178.05	16,178.05
14.1006 061	Add for premium 2 stop basic	EA			67,227.50	67,227.50
14.1007 000	**ELEVATOR, HYDRAULIC, 125'-150'/MIN, 4000#, 8'X 6' CAB:**					
14.1007 011	Elevator, 4000#, 3 stop	EA			107,667.36	107,667.36
14.1008 000	**ELEVATOR, HYDRAULIC, 100'/MIN, 20,000#, 10X16 CAB, AUTO EXIT:**					
14.1008 011	Elevator, 20,000#, 3 stop	EA			204,567.98	204,567.98
14.1008 021	Elevator, 20,000#, 4 stop	EA			226,101.45	226,101.45
14.1009 000	**ELEVATOR, GEAR, 350'/MIN, 3500#, 5' X 8' CAB:**					
14.1009 011	Elevator, 3500#, 10 stop	EA			268,149.82	268,149.82
14.1009 021	Elevator, 3500#, 15 stop	EA			319,717.12	319,717.12
14.1009 031	Add for additional stop	STOP			10,313.45	10,313.45
14.1010 000	**ELEVATOR, GEAR, 350'/MIN, 4000#, 8'X 6' CAB:**					
14.1010 011	Elevator, 4000#, 5 stop	EA			344,535.52	344,535.52
14.1011 000	**ELEVATOR, GEAR, 200'/MIN, 4000#, 6'X 8' CAB:**					
14.1011 011	Elevator, 4000#, 3 stop	EA			301,468.58	301,468.58
14.1012 000	**ELEVATOR, GEAR, 350'/MIN, 4500#, 6' X 9' CAB:**					
14.1012 011	Elevator, 4500#, 10 stop	EA			307,074.82	307,074.82
14.1012 021	Elevator, 4500#, 15 stop	EA			377,938.27	377,938.27
14.1012 031	Add for additional stop	STOP			15,353.71	15,353.71
14.1013 000	**ELEVATOR, GEAR, 100'/MIN, 8' X 10' CAB:**					
14.1013 011	Elevator, 6000#, 3 stop	EA			323,002.06	323,002.06
14.1014 000	**ELEVATOR, GEARLESS, 500'/MIN, 3500#, 6' X 9' CAB:**					
14.1014 011	Elevator, gearless, 3500#, 10 stop	EA			377,111.48	377,111.48
14.1014 021	Elevator, gearless, 3500#, 15 stop	EA			450,100.85	450,100.85
14.1014 031	Add for additional stop	STOP			18,247.30	18,247.30
14.1015 000	**ELEVATOR, GEARLESS, 700'/MIN, 4500#, 6' X 9' CAB:**					
14.1015 011	Elevator, gearless, 4500#, 10 stop	EA			401,331.61	401,331.61
14.1015 021	Elevator, gearless, 4500#, 15 stop	EA			463,581.53	463,581.53
14.1015 031	Elevator, gearless, 4500#, 20 stop	EA			525,831.52	525,831.52
14.1015 041	Add for additional stop	STOP			12,747.41	12,747.41

14 - CONVEYING EQUIPMENT

Division	Description	Unit	Labor Hours	Labor Cost	Material Cost	Total Cost
14.1016 000	**ELEVATOR, GEARLESS, 1200'/MIN, 4500#, 6'X9' CAB:**					
14.1016 011	Elevator, gearless, 4500#, 20 stop	EA			748,834.41	748,834.41
14.1016 021	Add for additional stop	STOP			14,105.76	14,105.76
14.1017 000	**ELEVATOR, GEAR, FREIGHT, 100'/MIN, 4500#, MANUAL DOOR:**					
14.1017 011	Elevator, freight, 2 stop	EA			159,342.86	159,342.86
14.1017 021	Add for additional stop	STOP			13,703.43	13,703.43
14.1017 031	Elevator, sidewalk, 2500#, 2 stop	EA			63,478.03	63,478.03
14.1018 000	**ELEVATOR CAB, INCLUDED WITH ABOVE PASSENGER ELEVATOR:**					
14.1018 011	Elevator cab, hydraulic, 4' x 5'	CAR			7,435.95	7,435.95
14.1018 021	Elevator cab, hydraulic, 5' x 6'	CAR			8,710.75	8,710.75
14.1018 031	Elevator cab, geared, 5' x 8'	CAR			9,773.03	9,773.03
14.1018 041	Elevator cab, geared, 6' x 9'	CAR			13,809.71	13,809.71
14.1018 051	Elevator cab, gearless, 6' x 9'	CAR			13,809.71	13,809.71
14.1018 061	Add for glass finish	EA			43,066.95	43,066.95
14.1018 071	Add for stainless steel finish	EA			16,150.06	16,150.06
14.1018 081	Add for grouting sills	STOP			652.30	652.30
14.1019 000	**INCLINE ELEVATORS:**					
14.1019 011	Incline elevator, 3 stop, 100 fpm	EA			34,587.98	34,587.98
14.1019 021	Add for additional stops	EA			17,910.09	17,910.09
14.1019 031	Incline elevator, 5 stop, 250 fpm	EA			487,589.21	487,589.21
14.1019 041	Add for additional stops	EA			21,776.89	21,776.89
14.1020 000	**ESCALATORS:**					
14.1020 001	*Note: The following prices are based on 12' floor to floor height.*					
14.1020 011	Escalator, 24"	FLOOR			183,901.11	183,901.11
14.1020 021	Escalator, 32"	FLOOR			187,327.87	187,327.87
14.1020 031	Escalator, 36"	FLOOR			192,129.67	192,129.67
14.1020 041	Escalator, 40"	FLOOR			192,711.89	192,711.89
14.1020 051	Escalator, 44"	FLOOR			196,975.94	196,975.94
14.1020 061	Escalator, 48"	FLOOR			205,326.51	205,326.51
14.1020 081	Add for baked enamel sides	FLOOR			7,651.91	7,651.91
14.1020 091	Add for glass sides	FLOOR			7,881.45	7,881.45
14.1020 101	Add for stainless steel sides	FLOOR			10,508.64	10,508.64
14.1021 000	**AIR HOIST:**					
14.1021 011	Air hoist, 2000#, 30' lift, rail	EA	8.0204	755.68	11,411.64	12,167.32
14.1021 021	Air hoist, 4000#, 30' lift, rail	EA	8.0204	755.68	12,425.93	13,181.61
14.1022 000	**ELECTRIC HOIST:**					
14.1022 011	Electric hoist, 1000#, 30' lift, rail	EA	8.0204	755.68	7,426.58	8,182.26
14.1022 021	Electric hoist, 500#, 30' lift, rail	EA	8.0204	755.68	6,448.44	7,204.12
14.1023 000	**CRANES:**					
14.1023 011	Crane, hydraulic, 2000#, portable	EA			4,067.56	4,067.56
14.1023 021	Crane, gantry, 4000#, 10-20' range	EA			4,227.59	4,227.59
14.1024 000	**CRANE, MONORAIL, OVERHEAD:**					
14.1024 011	Crane, monorail, 200#/LF, manual, channel	LF	0.3151	29.69	30.08	59.77
14.1024 021	Crane, monorail, 100#/LF, manual, channel	LF	0.2619	24.68	12.36	37.04
14.1024 031	Rail system	LF			106.17	106.17
14.1024 041	Crane, 1/2 ton	EA			4,249.13	4,249.13
14.1025 000	**BRIDGE CRANES WITHOUT RAILS:**					
14.1025 011	Crane, bridge, 1 ton	EA			10,622.82	10,622.82
14.1025 021	Crane, bridge, 3 ton	EA			21,245.69	21,245.69
14.1025 031	Crane, bridge, 5 ton	EA			31,868.57	31,868.57
14.1026 000	**MANLIFTS:**					
14.1026 001	*Note: Be sure to check local codes about the use of the following items. Their use may be restricted or illegal.*					
14.1026 011	Manlift, 2 stop	EA			25,561.58	25,561.58
14.1026 021	Manlift, 3 stop	EA			28,060.95	28,060.95
14.1026 031	Manlift, 4 stop	EA			30,560.35	30,560.35
14.1026 041	Add for additional stop	EA			2,499.32	2,499.32

CONVEYING EQUIPMENT - 14

Division	Description	Unit	Labor Hours	Labor Cost	Material Cost	Total Cost
14.1027 000	**LIFT, SCISSOR TYPE, PORTABLE:**					
14.1027 001	*Note: For garage lift, see section 11.1801.*					
14.1027 011	Lift, scissor, 2000#, 4' lift	EA			4,197.72	4,197.72
14.1027 021	Lift, scissor, 1000#, 3' lift	EA			3,749.00	3,749.00
14.1028 000	**CONVEYORS, BELT TYPE:**					
14.1028 011	Conveyor, belt, horizontal, 24"	LF			880.41	880.41
14.1028 021	Conveyor, belt, elevated & descending, 24"	FLIGH			9,713.40	9,713.40
14.1028 031	Add for direction changes	EA			8,733.47	8,733.47
14.1028 041	Add for starter section	EA			3,652.43	3,652.43
14.1029 000	**CONVEYORS, AUTO/SELECTIVE/COLLECTIVE:**					
14.1029 011	Horizontal	LF			3,621.22	3,621.22
14.1029 021	Vertical	LF			4,856.69	4,856.69
14.1030 000	**MAIL CONVEYORS, AUTOMATIC, ELECTRONIC:**					
14.1030 011	Mail conveyor, horizontal	LF			1,959.64	1,959.64
14.1030 021	Mail conveyor, vertical, 12' high	FLOOR			26,811.21	26,811.21
14.1031 000	**MOVING SIDEWALKS:**					
14.1031 011	Moving sidewalk, horizontal, 48"	LF			2,568.28	2,568.28
14.1031 021	Moving sidewalk, horizontal, 72"	LF			3,056.87	3,056.87
14.1031 031	Moving sidewalk, elevating, 48"	LF			4,947.58	4,947.58
14.1032 000	**BAGGAGE CAROUSELS:**					
14.1032 011	Baggage carousel, round, 20'	EA	166.5823	15,695.38	44,461.13	60,156.51
14.1032 021	Baggage carousel, round, 25'	EA	208.2284	19,619.28	57,354.81	76,974.09
14.1032 031	Baggage carousel, rectangular 75' x 30'	EA			194,836.57	194,836.57
14.1032 041	Baggage carousel, average cost/LF	LF			922.98	922.98
14.1033 000	**PNEUMATIC TUBE SYSTEMS:**					
14.1033 011	Pneumatic tube, twin, 3", 2 station	TOTAL			28,344.97	28,344.97
14.1033 021	Add for additional station	STATION			11,815.11	11,815.11
14.1033 031	Pneumatic tube, twin, 4", 2 station	TOTAL			29,253.82	29,253.82
14.1033 041	Add for additional station	STATION			12,269.55	12,269.55
14.1033 051	Pneumatic tube, twin, 4" x 7", 2 station	TOTAL			45,550.79	45,550.79
14.1033 061	Add for additional station	STATION			17,177.40	17,177.40
14.1033 071	Pneumatic tube, twin, electric, 4", 2 station	TOTAL			59,507.48	59,507.48
14.1033 081	Add for additional station	STATION			23,465.54	23,465.54
14.1033 091	Pneumatic tube, twin, electric, 6", 2 station	TOTAL			78,525.35	78,525.35
14.1033 101	Add for additional station	STATION			26,993.06	26,993.06
14.1033 111	Pneumatic tube, twin, electric, 4"x7", 2 station	TOTAL			87,028.86	87,028.86
14.1033 121	Add for additional station	STATION			33,741.38	33,741.38
14.1033 131	Pneumatic tube, twin, electric, 4x12, 2 station	TOTAL			115,530.08	115,530.08
14.1033 141	Add for additional station	STATION			48,771.57	48,771.57
14.1033 151	Pneumatic tube, pharmacy, 8", positive interlock	STATION			43,324.09	43,324.09

Plumbing

Table of Contents

15.0000.000	**MECHANICAL WORK - PLUMBING**	**161**
15.1000.000	Equipment	161
15.1200.000	Fixtures	162
15.1300.000	Piping	164
15.1400.000	Valves & Specialties	168
15.1500.000	Insulation, Piping	171
15.1600.000	Miscellaneous Plumbing Specialties	172
15.1700.000	Medical & Laboratory Equipment & Pipe	174
15.1800.000	Fees, Permits & Sterilization	176
15.1900.000	Industrial Piping	176
15.2000.000	Gate, Globe & Check Valves, Cast Steel	178
15.2100.000	Industrial Piping Insulation	179

MECHANICAL - PLUMBING - 15

Division	Description	Unit	Labor Hours	Labor Cost	Material Cost	Total Cost
15.0000 000	**PLUMBING:**					
15.0000 001	**MECHANICAL WORK:**					
15.0001 000	**PLUMBING, IN PLACE COST PER FIXTURE:**					
15.0001 001	*Note: The following prices include piping to 5' beyond building line. Gas lines and storm system piping are not included.*					
15.0001 011	Plumbing, residential, tract housing	FIX	4.4415	390.85	824.26	1,215.11
15.0001 021	Plumbing, residential, multi, 2 story	FIX	5.1816	455.98	961.56	1,417.54
15.0001 031	Plumbing, residential, multi, 3 story	FIX	6.4860	570.77	1,201.15	1,771.92
15.0001 041	Plumbing, residential, custom, good to better	FIX	7.0853	623.51	1,317.35	1,940.86
15.0001 051	Plumbing, custom residential, best	FIX	10.2224	899.57	1,902.11	2,801.68
15.0001 061	Plumbing, commercial, frame construction	FIX	15.0162	1,321.43	2,012.64	3,334.07
15.0001 071	Plumbing, commercial, type 1 building	FIX	16.8845	1,485.84	2,258.00	3,743.84
15.0001 081	Plumbing, commercial, high rise	FIX	21.8547	1,923.21	2,930.10	4,853.32
15.0001 091	Plumbing, institutional, schools, elementary	FIX	20.5152	1,805.34	2,767.45	4,572.79
15.0001 101	Plumbing, institutional, high schools	FIX	25.1328	2,211.69	3,392.06	5,603.75
15.0001 111	Plumbing, institutional, public structures	FIX	27.3182	2,404.00	3,686.39	6,090.39
15.0001 121	Plumbing, institutional, hospital, 1 & 2 story	FIX	34.2272	3,011.99	4,619.66	7,631.65
15.0001 131	Plumbing, institutional, hospital, high rise	FIX	39.2678	3,455.57	5,190.45	8,646.02
15.1000 000	**EQUIPMENT:**					
15.1000 001	*Note: The following prices do not include valving, auxiliary equipment or supports, unless otherwise noted. For boilers see section 15.3200.*					
15.1001 000	**WATER HEATER, GLASS LINED, RESIDENTIAL, ELECTRIC:**					
15.1001 011	Heater, hot water, electric, 30 gallon, 5 year, glass lined	EA	2.2990	279.24	639.49	918.72
15.1001 021	Heater, hot water, electric, 40 gallon, 5 year, glass lined	EA	2.2990	279.24	705.04	984.28
15.1001 031	Heater, hot water, electric, 82 gallon, 5 year, glass lined	EA	3.0653	372.31	1,164.76	1,537.07
15.1001 041	Heater, hot water, electric, 119 gallon, 5 year, glass lined	EA	3.8316	465.39	1,737.44	2,202.82
15.1001 051	Heater, hot water, electric, 30 gallon, 10 year, glass lined	EA	2.2990	279.24	673.79	953.02
15.1001 061	Heater, hot water, electric, 40 gallon, 10 year, glass lined	EA	2.2990	279.24	739.43	1,018.67
15.1001 071	Heater, hot water, electric, 82 gallon, 10 year, glass lined	EA	3.0653	372.31	1,199.05	1,571.36
15.1001 081	Heater, hot water, electric, 119 gallon, 10 year, glass lined	EA	3.8316	465.39	1,772.92	2,238.30
15.1002 000	**WATER HEATERS, GLASS LINED, RESIDENTIAL:**					
15.1002 011	Heater, hot water, gas, 30 gallon, 5 year, glass lined	EA	2.6822	325.78	605.22	931.00
15.1002 021	Heater, hot water, gas, 40 gallon, 5 year, glass lined	EA	2.6822	325.78	682.30	1,008.08
15.1002 031	Heater, hot water, gas, 75 gallon, 5 year, glass lined	EA	3.4485	418.85	1,378.92	1,797.77
15.1002 041	Heater, hot water, gas, 100 gallon, 5 year, glass lined	EA	4.2148	511.93	2,235.42	2,747.35
15.1002 051	Heater, hot water, gas, 30 gallon, 10 year, glass lined	EA	2.6822	325.78	679.45	1,005.23
15.1002 061	Heater, hot water, gas, 40 gallon, 10 year, glass lined	EA	2.6822	325.78	719.42	1,045.20
15.1002 071	Heater, hot water, gas, 50 gallon, 10 year, glass lined	EA	3.0653	372.31	876.55	1,248.86
15.1002 081	Heater, hot water, gas, 100g, 10 year, glass lined	EA	4.2150	511.95	2,309.38	2,821.33
15.1003 000	**WATER HEATERS, COMMERCIAL, 3 YEAR, WITH CONNECTIONS:**					
15.1003 011	Heater, hot water, electric, 6 gallon, 17 GPM, with connection	EA	2.0000	242.92	719.42	962.34
15.1003 021	Heater, hot water, electric, 50 gallon, 100 GPH, with connection	EA	4.9811	605.00	3,985.46	4,590.46
15.1003 031	Heater, hot water, electric, 85 gallon, 200 GPH, with connection	EA	4.9811	605.00	5,010.36	5,615.36
15.1003 041	Heater, hot water, electric, 120 gallon, 221 GPH, with connection	EA	6.1306	744.62	5,664.15	6,408.77
15.1003 051	Heater, hot water, gas, 20 gallon, 100 GPH, with connection	EA	5.3643	651.55	1,721.88	2,373.43
15.1003 061	Heater, hot water, gas, 50 gallon, 100 GPH, with connection	EA	5.3643	651.55	2,041.28	2,692.83
15.1003 071	Heater, hot water, gas, 75 gallon, 300 GPH, with connection	EA	5.7474	698.08	5,327.28	6,025.36
15.1003 081	Heater, hot water, gas, 85 gallon, 168 GPH, with connection	EA	5.7474	698.08	6,906.05	7,604.13
15.1003 091	Heater, hot water, 100 gallon, 235 GPH, with connection	EA	6.1306	744.62	7,134.40	7,879.02
15.1004 000	**INTERCEPTORS, CAST IRON:**					
15.1004 011	Grease intercept, cast iron, 4 GPM, 8#	EA	1.9158	232.69	807.90	1,040.60
15.1004 021	Grease intercept, cast iron, 10 GPM, 20#	EA	2.2990	279.24	1,321.83	1,601.07
15.1004 031	Grease intercept, cast iron, 20 GPM, 40#	EA	3.0653	372.31	2,398.11	2,770.42
15.1004 041	Grease intercept, cast iron, 50 GPM, 100#	EA	9.1959	1,116.93	4,408.01	5,524.95
15.1004 051	Hair intercept, cast iron, small	EA	1.9158	232.69	302.98	535.68
15.1004 061	Hair intercept, cast iron, large	EA	2.6822	325.78	557.86	883.64
15.1004 071	Plaster intercept, cast iron, small	EA	1.9158	232.69	459.68	692.37

15 - PLUMBING - MECHANICAL

Division	Description	Unit	Labor Hours	Labor Cost	Material Cost	Total Cost
15.1004 081	Plaster intercept, cast iron, large	EA	2.6822	325.78	1,175.76	1,501.54
15.1005 000	**PUMPS, CIRCULATING, IN LINE, FLANGED, IRON BODY:**					
15.1005 011	Pump, circulating, iron body, 3/4"-1-1/2", flanged, 1/12 hp	EA	1.4561	176.86	709.69	886.55
15.1005 021	Pump, circulating, iron body, to 2", flanged, 1/6 hp	EA	1.6860	204.78	858.78	1,063.56
15.1005 031	Pump, circulating, iron body, to 2-1/2", flanged, 1/4 hp	EA	2.1457	260.62	2,261.08	2,521.70
15.1005 041	Pump, circulating, iron body, to 3", flanged, 1/3 hp	EA	2.3756	288.54	2,884.11	3,172.65
15.1006 000	**PUMPS, SEWAGE EJECTOR, WITH TANKS & FITTINGS:**					
15.1006 011	Pump, sewage ejector, single, 1/2 hp, 2", tank	EA	8.5828	1,042.47	5,183.68	6,226.15
15.1006 021	Pump, sewage ejector, single, 1 hp, 3", tank, fittings	EA	10.3454	1,256.55	12,507.28	13,763.83
15.1006 031	Pump, sewage ejector, single, 2 hp, 4", tank, fittings	EA	11.8780	1,442.70	13,844.95	15,287.65
15.1006 041	Pump, sewage ejector, duplex, 2 hp, tank, fittings	EA	16.3227	1,982.56	20,603.60	22,586.15
15.1006 051	Pump, sewage ejector, duplex, 3 hp, tank, fittings	EA	19.7711	2,401.40	20,800.94	23,202.34
15.1006 061	Pump, sewage ejector, duplex, 5 hp, tank, fittings	EA	23.2195	2,820.24	24,713.43	27,533.67
15.1007 000	**PUMPS, SUMP, ELECTRIC, WITH IRON GUARD ACCESSORIES:**					
15.1007 001	*Note: For bronze body, add 67% to the material costs.*					
15.1007 011	Sump, 1/4 hp, 2' deep, 1-1/4" outlet	EA	1.5327	186.16	1,344.95	1,531.11
15.1007 021	Sump, 1/3 hp, 3' deep, 1-1/2" outlet	EA	1.9158	232.69	2,177.01	2,409.71
15.1007 031	Sump, 1/2 hp, 6' deep, 2" outlet	EA	2.6822	325.78	2,743.69	3,069.47
15.1008 000	**TANKS, SEPTIC, STEEL, INCLUDE BURY:**					
15.1008 011	Septic tank, steel, 200 gallon, buried	EA	3.4485	328.92	1,160.21	1,489.13
15.1008 021	Septic tank, steel, 500 gallon, buried	EA	5.7474	548.19	1,654.02	2,202.20
15.1008 031	Septic tank, steel, 1,000 gallon, buried	EA	8.2380	785.74	3,642.85	4,428.59
15.1008 041	Septic tank, steel, 10,000 gallon, buried	EA	31.4192	2,996.76	24,869.43	27,866.20
15.1009 000	**COMPRESSOR, AIR, TANK MOUNTED, CONTROL PANEL WITH ACCESS:**					
15.1009 011	Compressor, simplex, 5 hp, reciprocating, tank mounted	EA	16.0000	1,943.36	9,538.62	11,481.98
15.1009 021	Compressor, simplex, 7.5 hp, reciprocating, tank mounted	EA	18.0000	2,186.28	13,219.67	15,405.95
15.1009 031	Compressor, simplex, 10 hp, tank mounted	EA	20.0000	2,429.20	17,292.26	19,721.46
15.1009 041	Compressor, simplex, 15 hp, reciprocating, tank mounted	EA	21.0000	2,550.66	23,732.00	26,282.66
15.1009 051	Compressor, simplex, 20 hp, reciprocating, tank mounted	EA	24.0000	2,915.04	28,788.73	31,703.77
15.1010 000	**TANKS, WATER, GLASS LINED, ASME:**					
15.1010 011	Tank, water, glass lined, 140 gallon, ASME	EA	4.9811	605.00	2,691.39	3,296.39
15.1010 021	Tank, water, glass lined, 200 gallon, ASME	EA	6.8586	833.05	3,247.72	4,080.77
15.1010 031	Tank, water, glass lined, 350 gallon, ASME	EA	9.7706	1,186.74	4,398.83	5,585.57
15.1010 041	Tank, water, glass lined, 400 gallon, ASME	EA	10.9584	1,331.01	4,426.22	5,757.23
15.1011 000	**WATER SOFTENERS, WITH BRINETANK & START-UP:**					
15.1011 011	Water softener, 8 GPM, brinetank & start-up	EA	4.5000	546.57	2,122.08	2,668.65
15.1011 021	Water softener, 25 GPM, brinetank & start-up	EA	6.1306	744.62	4,050.48	4,795.10
15.1011 031	Water softener, 50 GPM, brinetank & start-up	EA	11.1117	1,349.63	7,382.36	8,731.99
15.1012 000	**PUMP, PRESSURE BOOSTER SYSTEM:**					
15.1012 001	*Note: The following prices include pumps, valves, controls. All items are pre-piped and skid-mounted.*					
15.1012 011	2 pump system, 100 GPM @ 50 PSI	EACH	24.0000	2,915.04	25,386.00	28,301.04
15.1012 021	2 pump system, 320 GPM @ 50 PSI	EACH	24.0000	2,915.04	36,698.13	39,613.17
15.1012 031	3 pump system, 300 GPM @ 100 PSI	EACH	32.0000	3,886.72	39,436.78	43,323.50
15.1012 041	3 pump system, 950 GPM @ 100 PSI	EACH	32.0000	3,886.72	76,543.81	80,430.53
15.1200 000	**FIXTURES:**					
15.1200 001	*Note: The following prices include trim with stops, hangars and supports. For rough in at fixtures see section 15.1300.*					
15.1201 000	**FIXTURES, ECONOMY GRADE:**					
15.1201 011	Bath tub, steel, with shower	EA	2.6822	325.78	464.81	790.59
15.1201 021	Bath tub, steel, with out shower	EA	2.2990	279.24	397.72	676.96
15.1201 031	Tub, fiberglass, integral walls	EA	2.6822	325.78	911.55	1,237.33
15.1201 041	Bidets, floor mounted	EA	1.5327	186.16	631.38	817.55
15.1201 051	Lavatory, steel, wall hung	EA	1.5327	186.16	260.37	446.54
15.1201 061	Lavatory, steel, vanity mounted	EA	1.5327	186.16	203.52	389.68
15.1201 071	Service sink	EA	1.9158	232.69	481.39	714.09
15.1201 081	Shower & drain receptor, 32" square	EA	2.2990	279.24	374.77	654.01

MECHANICAL - PLUMBING - 15

Division	Description	Unit	Labor Hours	Labor Cost	Material Cost	Total Cost
15.1201 091	Shower cabinet, with door, 32" square	EA	3.4485	418.85	797.05	1,215.91
15.1201 101	Sink, porcelain on steel, counter, single	EA	1.1495	139.62	177.49	317.11
15.1201 111	Sink, porcelain on cast iron, counter, single	EA	1.1495	139.62	252.46	392.07
15.1201 121	Sink, stainless steel, counter, single	EA	1.1495	139.62	264.34	403.96
15.1201 131	Sink, porcelain on steel, counter, double	EA	1.1495	139.62	195.97	335.59
15.1201 141	Sink, porcelain on cast iron, counter, double	EA	1.1495	139.62	268.26	407.88
15.1201 151	Sink, stainless steel, counter double	EA	1.1495	139.62	302.97	442.59
15.1201 161	Sink, bar	EA	1.1495	139.62	256.87	396.49
15.1201 171	Sink, floor	EA	0.7664	93.09	189.31	282.40
15.1201 181	Urinal, floor, with flush valve	EA	1.5327	186.16	571.78	757.94
15.1201 191	Urinal, wall, with flush valve, carrier	EA	1.5327	186.16	603.98	790.14
15.1201 201	Water closet, floor, with tank	EA	1.5327	186.16	587.88	774.04
15.1201 211	Water closet, wall, with flush valve, carrier	EA	1.5327	186.16	623.41	809.57
15.1202 000	**FIXTURES, STANDARD GRADE:**					
15.1202 011	Bath tub, porcelain enamel on cast iron, with shower	EA	2.6822	325.78	1,549.87	1,875.65
15.1202 021	Bath tub, porcelain enamel on cast iron, no shower	EA	2.2990	279.24	1,535.79	1,815.03
15.1202 031	Tub, fiberglass, with integral wall	EA	2.6822	325.78	1,193.66	1,519.44
15.1202 041	Bidet, floor mounted	EA	1.5327	186.16	994.17	1,180.33
15.1202 051	Lavatory, wall hung	EA	1.5327	186.16	737.57	923.74
15.1202 061	Lavatory, vanity mounted	EA	1.5327	186.16	525.73	711.90
15.1202 071	Service sink	EA	1.9158	232.69	1,297.16	1,529.85
15.1202 081	Shower & drain receptor, 32" square	EA	2.2990	279.24	953.16	1,232.40
15.1202 091	Shower, cabinet with door, 32" square	EA	3.4485	418.85	1,574.56	1,993.42
15.1202 101	Sink, porcelain on cast iron, counter, single	EA	1.1495	139.62	868.19	1,007.80
15.1202 111	Sink, stainless steel, counter, single	EA	1.1495	139.62	890.10	1,029.72
15.1202 121	Sink, porcelain on cast iron, counter, double	EA	1.1495	139.62	981.12	1,120.74
15.1202 131	Sink, stainless steel, counter, double	EA	1.1495	139.62	1,208.60	1,348.22
15.1202 141	Sink, bar	EA	1.1495	139.62	619.28	758.90
15.1202 151	Sink, floor	EA	0.7664	93.09	262.88	355.97
15.1202 161	Urinal, trough, with flush valve	EA	1.5327	186.16	845.59	1,031.75
15.1202 171	Urinal, wall, flush valve & carrier	EA	1.5327	186.16	934.20	1,120.36
15.1202 181	Water closet, floor, with tank	EA	1.5327	186.16	652.31	838.47
15.1202 191	Water closet, wall, with flush valve & carrier	EA	1.5327	186.16	684.51	870.68
15.1202 201	Water closet, floor, with flush valve, handicap	EA	1.5327	186.16	801.70	987.87
15.1203 000	**FIXTURES, INSTITUTIONAL GRADE:**					
15.1203 011	Bath tub, with shower	EA	3.0653	372.31	2,037.77	2,410.08
15.1203 021	Bath tub, with out shower	EA	2.6822	325.78	2,023.69	2,349.47
15.1203 031	Bidet, floor mounted	EA	1.9158	232.69	1,085.54	1,318.23
15.1203 041	Lavatory, wall hung	EA	1.9158	232.69	964.67	1,197.36
15.1203 051	Lavatory, vanity mounted	EA	1.9158	232.69	761.08	993.77
15.1203 061	Service sink	EA	1.9158	232.69	1,537.25	1,769.94
15.1203 071	Shower & drain receptor, 36" square	EA	2.6822	325.78	1,077.89	1,403.67
15.1203 081	Shower cabinet, with door, 36" square	EA	3.8316	465.39	1,784.95	2,250.34
15.1203 091	Shower cabinet, corner, with door & trim	EACH	3.8316	465.39	5,468.48	5,933.86
15.1203 101	Shower cabinet, handicap, with door, trim	EACH	5.5111	669.38	5,359.85	6,029.22
15.1203 111	Sink, clinic	EA	1.9158	232.69	1,613.23	1,845.92
15.1203 121	Sink, counter, single, porcelain on cast iron	EA	1.5327	186.16	996.01	1,182.17
15.1203 131	Sink, counter, single, stainless steel	EA	1.5327	186.16	1,024.31	1,210.47
15.1203 141	Sink, counter, double, porcelain on cast iron	EA	1.5327	186.16	1,121.25	1,307.41
15.1203 151	Sink, counter, double, stainless steel	EA	1.5327	186.16	1,403.41	1,589.57
15.1203 161	Sink, bar	EA	1.5327	186.16	652.07	838.23
15.1203 171	Sink, floor	EA	1.1495	139.62	518.11	657.73
15.1203 181	Sitz bath	EA	2.6822	325.78	2,102.48	2,428.26
15.1203 191	Urinal, floor, with flush valve	EA	1.9158	232.69	1,195.54	1,428.24
15.1203 201	Urinal, wall, flush valve & carrier	EA	1.9158	232.69	1,372.26	1,604.95
15.1203 211	Urinal, trough 4 deg, valve, carrier	EA	3.0653	372.31	565.41	937.72
15.1203 221	Water closet, floor, with flush valve	EA	1.9158	232.69	1,001.37	1,234.07

15 - PLUMBING - MECHANICAL

Division	Description	Unit	Labor Hours	Labor Cost	Material Cost	Total Cost
15.1203 231	Water closet, wall, with flush valve & carrier	EA	1.9158	232.69	978.89	1,211.58
15.1203 241	Water closet, floor, with bed pan flush	EA	2.2990	279.24	1,161.20	1,440.43
15.1203 251	Water closet, wall, bed pan flush, flush valve	EA	2.2990	279.24	2,104.80	2,384.04
15.1203 261	Water closet lavatory, hospital, standard	EA	3.3719	409.55	3,437.09	3,846.64
15.1203 271	Water closet lavatory, hospital, deluxe	EA	3.3719	409.55	4,371.73	4,781.28
15.1204 000	**SPECIALTY FIXTURES:**					
15.1204 001	Note: For stainless steel, add 30% to the material costs. For enameled steel, add 18%.					
15.1204 011	Column showers, 2 head	FIX	4.2148	511.93	973.70	1,485.63
15.1204 021	Column showers, 3 head	FIX	4.5213	549.16	1,189.80	1,738.96
15.1204 031	Column showers, 6 head	FIX	4.9811	605.00	1,693.84	2,298.85
15.1204 041	Drinking fountain, cast iron, wall	FIX	0.7664	93.09	482.02	575.11
15.1204 051	Drinking fountain, stainless steel, wall	FIX	0.7664	93.09	624.71	717.80
15.1204 061	Water cooler, elec, enamel, semi-recess	FIX	1.1495	139.62	952.05	1,091.67
15.1204 071	Water cooler, elec, stainless steel, semi-recess	FIX	1.1495	139.62	1,084.43	1,224.04
15.1204 081	Water cooler, elec, enamel, wheel chair	EA	1.1495	139.62	1,297.37	1,436.99
15.1204 091	Water cooler, elec, stainless steel, wheel chair	EA	1.1495	139.62	1,486.84	1,626.46
15.1204 101	Water cooler, elec, stainless steel, dual purpose	FIX	1.1495	139.62	1,965.77	2,105.39
15.1204 111	Water fountain, granite 36" half circle	FIX	3.8316	465.39	3,980.70	4,446.09
15.1204 121	Water fountain, granite 54" half circle	FIX	4.2148	511.93	4,909.54	5,421.47
15.1204 131	Water fountain, granite 36" full circle	FIX	3.8316	465.39	4,511.48	4,976.86
15.1204 141	Water fountain, granite 54" full circle	FIX	4.2148	511.93	5,838.40	6,350.33
15.1204 171	Eye-wash fountain	FIX	1.5327	186.16	479.66	665.83
15.1204 181	Eye-wash & shower	FIX	4.5980	558.47	1,214.05	1,772.52
15.1204 191	Shower, head drench	FIX	3.8316	465.39	568.39	1,033.77
15.1204 201	Water closet, stainless steel, park & rec	FIX	2.6822	325.78	1,107.01	1,432.79
15.1204 251	Urinal, stainless steel, park & rec	FIX	2.6822	325.78	1,429.03	1,754.81
15.1204 261	Lavatory basin, stainless steel, park & rec	FIX	2.2990	279.24	908.26	1,187.50
15.1204 271	Decontamination shower, walk-thru, eye-wash	FIX	9.1959	1,116.93	5,647.60	6,764.53
15.1204 281	Bathing pool (tub), 5' x 42"	EA	2.6822	325.78	2,116.28	2,442.06
15.1204 291	Bathing pool (tub), 6' x 36"	EA	2.6822	325.78	2,943.48	3,269.26
15.1204 301	Recess bath with whirlpool	EA	3.8316	465.39	6,999.02	7,464.41
15.1204 311	Water closet & lavatory, stainless steel, jail, wall	FIX	2.6822	325.78	2,582.56	2,908.34
15.1204 321	Water closet & lavatory, stainless steel, jail, floor	FIX	2.6822	325.78	4,546.42	4,872.20
15.1204 331	Water closet, in floor, stainless steel, jail	FIX	2.2990	279.24	3,287.44	3,566.68
15.1204 341	Shower, wall mounted unit, h & c	FIX	3.8316	465.39	1,248.26	1,713.65
15.1204 351	Water closet, stainless steel, wall, jail	FIX	2.2990	279.24	1,385.60	1,664.84
15.1204 361	Lavatory, stainless steel, oval, wall, jail	FIX	2.2990	279.24	1,231.16	1,510.40
15.1204 371	Shower stall, jail, 36" x 36"	FIX	4.5980	558.47	7,467.58	8,026.05
15.1204 401	Add for paddle control for sink	EA			340.80	340.80
15.1204 411	Add for flush sensor	EA			292.11	292.11
15.1300 000	**PIPING:**					
15.1301 000	**ROUGH-IN FOR FIXTURES:**					
15.1301 001	Note: Use the following prices for schematic and preliminary estimates only, where piping takeoff is not possible. Unit prices include allowances for all piping and valving from fixture to 5' beyond building perimeter. Gas lines, roof drains and off-site work are not included.					
15.1301 011	Tract housing	FIX	1.7273	209.80	386.97	596.77
15.1301 021	Custom housing	FIX	3.2076	389.60	740.82	1,130.41
15.1301 031	Apartment building	FIX	2.3970	291.14	536.27	827.40
15.1301 041	Industrial building	FIX	4.1243	500.94	988.13	1,489.06
15.1301 051	Commercial building	FIX	4.4766	543.73	1,211.37	1,755.10
15.1301 061	Institutional structures	FIX	15.3335	1,862.41	2,706.69	4,569.09
15.1301 071	Schools	FIX	11.0684	1,344.37	1,956.03	3,300.39
15.1301 081	Hospital, 1 or 2 story	FIX	18.5763	2,256.28	3,290.05	5,546.33
15.1301 091	Hospital, high rise	FIX	21.5022	2,611.66	3,809.04	6,420.70
15.1301 101	Office building, high rise	FIX	14.8752	1,806.74	2,609.15	4,415.89

MECHANICAL - PLUMBING - 15

Division	Description	Unit	Labor Hours	Labor Cost	Material Cost	Total Cost
15.1302 000	**ROUGH-IN AT FIXTURES:**					
15.1302 001	Note: The following prices include fittings and valving required to connect fixtures to the waste line, vent riser and water runs. Straight run piping and risers should be taken off separately. Use these prices only when you are not making a detailed takeoff of connecting piping and fittings at fixtures. Do not use for residential work.					
15.1302 011	Rough-in at bath tubs	EA	2.6822	325.78	381.11	706.89
15.1302 021	Rough-in at fountains & coolers	EA	1.7243	209.43	222.58	432.01
15.1302 031	Rough-in at lavatory	EA	2.4906	302.51	314.42	616.93
15.1302 041	Rough-in at shower	EA	2.6822	325.78	349.99	675.77
15.1302 051	Rough-in at sinks	EA	2.4906	302.51	315.58	618.09
15.1302 061	Rough-in at urinal	EA	2.4906	302.51	246.91	549.42
15.1302 071	Rough-in at washing machine	EA	2.2990	279.24	329.34	608.58
15.1302 081	Rough-in at water closet	EA	3.2186	390.93	397.04	787.97
15.1302 091	Rough-in at wash fountain	EA	3.8316	465.39	718.31	1,183.70
15.1303 000	**CAST IRON PIPE, SOIL, SERVICE WEIGHT, SINGLE HUB:**					
15.1303 011	Cast iron pipe, soil, 1 hub, 2", service weight	LF	0.1839	22.34	11.33	33.67
15.1303 021	Cast iron pipe, soil, 1 hub, 3", service weight	LF	0.1954	23.73	13.90	37.63
15.1303 031	Cast iron pipe, soil, 1 hub, 4", service weight	LF	0.2108	25.60	18.49	44.09
15.1303 041	Cast iron pipe, soil, 1 hub, 5", service weight	LF	0.2376	28.86	26.19	55.05
15.1303 051	Cast iron pipe, soil, 1 hub, 6", service weight	LF	0.2721	33.05	32.12	65.17
15.1303 061	Cast iron pipe, soil, 1 hub, 8", service weight	LF	0.2989	36.30	61.95	98.26
15.1303 071	Cast iron pipe, soil, 1 hub, 10", service weight	LF	0.3460	42.03	81.33	123.36
15.1304 000	**CAST IRON PIPE, SOIL, SERVICE WEIGHT, HUBBELL:**					
15.1304 011	Cast iron pipe, soil, 1-1/2", no hub	LF	0.1700	20.65	11.60	32.25
15.1304 021	Cast iron pipe, soil, 2", no hub, service weight	LF	0.1700	20.65	11.96	32.61
15.1304 031	Cast iron pipe, soil, 3", no hub, service weight	LF	0.1860	22.59	13.51	36.10
15.1304 041	Cast iron pipe, soil, 4", no hub, service weight	LF	0.1957	23.77	18.06	41.83
15.1304 051	Cast iron pipe, soil, 5", no hub, service weight	LF	0.2245	27.27	28.12	55.39
15.1304 061	Cast iron pipe, soil, 6", no hub, service weight	LF	0.2578	31.31	33.14	64.46
15.1304 071	Cast iron pipe, soil, 8", no hub, service weight	LF	0.2912	35.37	59.50	94.87
15.1304 081	Cast iron pipe, soil, 10", no hub, service weight	LF	0.3303	40.12	85.65	125.77
15.1305 000	**CAST IRON PIPE, SOIL, EXTRA HEAVY, SINGLE HUB:**					
15.1305 011	Cast iron pipe, soil, extra heavy, 1 hub, 2"	LF	0.1839	22.34	13.10	35.44
15.1305 021	Cast iron pipe, soil, extra heavy, 1 hub, 3"	LF	0.1954	23.73	16.17	39.90
15.1305 031	Cast iron pipe, soil, extra heavy, 1 hub, 4"	LF	0.2108	25.60	21.38	46.99
15.1305 041	Cast iron pipe, soil, extra heavy, 1 hub, 5"	LF	0.2376	28.86	28.28	57.14
15.1305 051	Cast iron pipe, soil, extra heavy, 1 hub, 6"	LF	0.2721	33.05	37.13	70.18
15.1305 061	Cast iron pipe, soil, extra heavy, 1 hub, 8"	LF	0.2989	36.30	71.55	107.85
15.1305 071	Cast iron pipe, soil, extra heavy, 1 hub, 10"	LF	0.3460	42.03	93.92	135.94
15.1305 081	Cast iron pipe, soil, extra heavy, 1 hub, 12"	LF	0.3908	47.47	135.26	182.73
15.1306 000	**CAST IRON PIPE, DUR IRON:**					
15.1306 011	Cast iron pipe, dur iron, 2"	LF	0.1763	21.41	60.72	82.13
15.1306 021	Cast iron pipe, dur iron, 3"	LF	0.1954	23.73	76.07	99.81
15.1306 031	Cast iron pipe, dur iron, 4"	LF	0.2077	25.23	106.14	131.37
15.1306 041	Cast iron pipe, dur iron, 6"	LF	0.2683	32.59	168.49	201.07
15.1306 051	Cast iron pipe, dur iron, 8"	LF	0.2902	35.25	317.76	353.00
15.1307 000	**COPPER PIPE, "K", UNDERGROUND, WITH TRENCHING:**					
15.1307 011	Pipe, copper, 'K', soft, 1/2" coils, underground	LF	0.0632	7.68	6.65	14.32
15.1307 021	Pipe, copper, 'K', soft, 3/4" coils, underground	LF	0.0728	8.84	11.81	20.65
15.1307 031	Pipe, copper, 'K', soft, 1" coils, underground	LF	0.0728	8.84	15.87	24.72
15.1307 041	Pipe, copper, 'K', soft, 1-1/4" coil	LF	0.1302	15.81	20.80	36.62
15.1307 051	Pipe, copper, 'K', hard, 1-1/2", underground	LF	0.1342	16.30	26.20	42.50
15.1307 061	Pipe, copper, 'K', hard, 2", underground	LF	0.1609	19.54	40.48	60.02
15.1307 071	Pipe, copper, 'K', hard, 2-1/2", underground	LF	0.1609	19.54	57.71	77.25
15.1307 081	Pipe, copper, 'K', hard, 3", underground	LF	0.1648	20.02	80.38	100.39
15.1307 091	Pipe, copper, 'K', hard, 4", underground	LF	0.1835	22.29	132.92	155.20
15.1307 101	Pipe, copper, 'K', hard, 5", underground	LF	0.2414	29.32	292.54	321.86

15 - PLUMBING - MECHANICAL

Division	Description	Unit	Labor Hours	Labor Cost	Material Cost	Total Cost
15.1307 111	Pipe, copper, 'K', hard, 6", underground	LF	0.2644	32.11	387.26	419.37
15.1308 000	**COPPER PIPE, "L", IN BUILDING, WITH FITTINGS & SUPPORTS:**					
15.1308 011	Pipe, copper, 'L', 1/2", in building, with fittings & supports	LF	0.0791	9.61	5.31	14.91
15.1308 021	Pipe, copper, 'L', 3/4", in building, with fittings & supports	LF	0.0962	11.68	8.25	19.94
15.1308 031	Pipe, copper, 'L', 1", in building, with fittings & supports	LF	0.1142	13.87	11.94	25.81
15.1308 041	Pipe, copper, 'L', 1-1/4", in building, with fittings & supports	LF	0.1233	14.98	17.10	32.07
15.1308 051	Pipe, copper, 'L', 1-1/2", in building, with fittings & supports	LF	0.1311	15.92	21.32	37.24
15.1308 061	Pipe, copper, 'L', 2", in building, with fittings & supports	LF	0.1581	19.20	26.44	45.64
15.1308 071	Pipe, copper, 'L', 2-1/2", in building, with fittings & supports	LF	0.1752	21.28	46.61	67.89
15.1308 081	Pipe, copper, 'L', 3", in building, with fittings & supports	LF	0.2012	24.44	64.12	88.56
15.1308 091	Pipe, copper, 'L', 4", in building, with fittings & supports	LF	0.2452	29.78	107.68	137.46
15.1308 101	Pipe, copper, 'L', 5", in building, with fittings & supports	LF	0.3073	37.32	256.74	294.07
15.1308 111	Pipe, copper, 'L', 6", in building, with fittings & supports	LF	0.3762	45.69	302.66	348.36
15.1309 000	**COPPER PIPE, "M", IN BUILDING, WITH FITTINGS & SUPPORTS:**					
15.1309 011	Pipe, copper, 'M', 1/2", in building, with fittings & supports	LF	0.0791	9.61	4.37	13.98
15.1309 021	Pipe, copper, 'M', 3/4", in building, with fittings & supports	LF	0.0962	11.68	6.85	18.54
15.1309 031	Pipe, copper, 'M', 1", in building, with fittings & supports	LF	0.1142	13.87	10.23	24.10
15.1309 041	Pipe, copper, 'M', 1-1/4", in building, with fittings & supports	LF	0.1233	14.98	15.37	30.35
15.1309 051	Pipe, copper, 'M', 1-1/2", in building, with fittings & supports	LF	0.1311	15.92	20.45	36.37
15.1309 061	Pipe, copper, 'M', 2", in building, with fittings & supports	LF	0.1581	19.20	32.21	51.41
15.1309 071	Pipe, copper, 'M', 3", in building, with fittings & supports	LF	0.2012	24.44	57.64	82.08
15.1309 081	Pipe, copper, 'M', 4", in building, with fittings & supports	LF	0.2452	29.78	103.56	133.34
15.1309 091	Pipe, copper, 'M', 5", in building, with fittings & supports	LF	0.3142	38.16	245.47	283.63
15.1309 101	Pipe, copper, 'M', 6", in building, with fittings & supports	LF	0.3051	37.06	315.30	352.36
15.1310 000	**COPPER 'DWV' DRAINAGE TUBE:**					
15.1310 011	Tube, copper, 'DWV', 1-1/4"	LF	0.1233	14.98	26.42	41.39
15.1310 021	Tube, copper, 'DWV', 1-1/2"	LF	0.1311	15.92	28.38	44.30
15.1310 031	Tube, copper, 'DWV', 2"	LF	0.1581	19.20	41.11	60.31
15.1310 041	Tube, copper, 'DWV', 3"	LF	0.2012	24.44	55.08	79.52
15.1310 051	Tube, copper, 'DWV', 4"	LF	0.2452	29.78	119.54	149.32
15.1310 061	Tube, copper, 'DWV', 5"	LF	0.3142	38.16	257.62	295.79
15.1311 000	**PVC, SCH 40, IN BUILDING, WITH FITTINGS & SUPPORTS:**					
15.1311 001	*Note: For schedule 80, add 31% to the material costs.*					
15.1311 011	Pipe, PVC, sch 40, 1/2", in building, with fittings & supports	LF	0.0556	6.75	1.63	8.38
15.1311 021	Pipe, PVC, sch 40, 3/4", in building, with fittings & supports	LF	0.0621	7.54	1.90	9.45
15.1311 031	Pipe, PVC, sch 40, 1", in building, with fittings & supports	LF	0.0621	7.54	2.51	10.05
15.1311 041	Pipe, PVC, sch 40, 1-1/4", building, with fittings & supports	LF	0.1073	13.03	3.20	16.23
15.1311 051	Pipe, PVC, sch 40, 1-1/2", in building, with fittings & supports	LF	0.1226	14.89	3.44	18.33
15.1311 061	Pipe, PVC, sch 40, 2", in building, with fittings & supports	LF	0.1594	19.36	4.45	23.81
15.1311 071	Pipe, PVC, sch 40, 3", in building, with fittings & supports	LF	0.1924	23.37	8.65	32.02
15.1311 081	Pipe, PVC, sch 40, 4", in building, with fittings & supports	LF	0.2130	25.87	13.28	39.15
15.1311 091	Pipe, PVC, sch 40, 5", in building, with fittings & supports	LF	0.2663	32.34	22.38	54.73
15.1311 101	Pipe, PVC, sch 40, 6", in building, with fittings & supports	LF	0.3198	38.84	26.43	65.27
15.1312 000	**ACID WASTE PIPE, POLYPROPYLENE:**					
15.1312 011	Pipe, polypropylene, 2", acid waste	LF	0.0505	6.13	25.21	31.35
15.1312 021	Pipe, polypropylene, 3", acid waste	LF	0.0758	9.21	34.78	43.99
15.1312 031	Pipe, polypropylene, 4", acid waste	LF	0.0912	11.08	46.75	57.82
15.1312 041	Pipe, polypropylene, 6", acid waste	LF	0.1318	16.01	87.17	103.17
15.1313 000	**PLASTIC 'DWV', ABS:**					
15.1313 011	Pipe, plastic, 'DWV', 1-1/2"	LF	0.1226	14.89	9.92	24.82
15.1313 021	Pipe, plastic, 'DWV', 2"	LF	0.1594	19.36	12.84	32.20
15.1313 031	Pipe, plastic, 'DWV', 3"	LF	0.1924	23.37	19.21	42.58
15.1313 041	Pipe, plastic, 'DWV', 4"	LF	0.2130	25.87	32.66	58.53
15.1313 051	Pipe, plastic, 'DWV', 6"	LF	0.2184	26.53	123.46	149.99
15.1314 000	**PYREX GLASS:**					
15.1314 011	Pipe, Pyrex, 1"	LF	0.2088	25.36	28.18	53.54
15.1314 021	Pipe, Pyrex, 1-1/2"	LF	0.2298	27.91	39.84	67.75

MECHANICAL - PLUMBING - 15

Division	Description	Unit	Labor Hours	Labor Cost	Material Cost	Total Cost
15.1314 031	Pipe, Pyrex, 2"	LF	0.2508	30.46	48.85	79.32
15.1314 041	Pipe, Pyrex, 3"	LF	0.3231	39.24	53.58	92.82
15.1314 051	Pipe, Pyrex, 4"	LF	0.4375	53.14	93.38	146.52
15.1315 000	**STEEL PIPE, BLACK, WELD, SCH 40, A-120, SCREWED:**					
15.1315 001	*Note: The following prices include malleable iron fittings and supports.*					
15.1315 011	Black steel pipe, weld, A-120, sch 40, 1/2", screwed	LF	0.0752	9.13	2.34	11.47
15.1315 021	Black steel pipe, weld, A-120, sch 40, 3/4", screwed	LF	0.0931	11.31	2.75	14.06
15.1315 031	Black steel pipe, weld, A-120, sch 40, 1", screwed	LF	0.1122	13.63	3.56	17.19
15.1315 041	Black steel pipe, weld, A-120, sch 40, 1-1/4", screwed	LF	0.1211	14.71	4.94	19.65
15.1315 051	Black steel pipe, weld, A-120, sch 40, 1-1/2", screwed	LF	0.1402	17.03	5.57	22.60
15.1315 061	Black steel pipe, weld, A-120, sch 40, 2", screwed	LF	0.1908	23.17	7.41	30.59
15.1315 071	Black steel pipe, weld, A-120, sch 40, 2-1/2", screwed	LF	0.2291	27.83	11.38	39.21
15.1315 081	Black steel pipe, weld, A-120, sch 40, 3", screwed	LF	0.2674	32.48	15.37	47.85
15.1315 091	Black steel pipe, weld, A-120, sch 40, 4", screwed	LF	0.3487	42.35	26.82	69.17
15.1315 101	Black steel pipe, weld, A-120, sch 40, 5", screwed	LF	0.3771	45.80	34.55	80.35
15.1315 111	Black steel pipe, weld, A-120, sch 40, 6", screwed	LF	0.4215	51.20	39.62	90.81
15.1316 000	**STEEL PIPE, GALVANIZED, WELD, SCH 40, A-120, SCREWED:**					
15.1316 001	*Note: The following prices include gmi fittings and supports.*					
15.1316 011	Galvanized steel pipe, A-120, sch 40, 1/2", screwed	LF	0.0752	9.13	2.93	12.07
15.1316 021	Galvanized steel pipe, A-120, sch 40, 3/4", screwed	LF	0.0931	11.31	3.47	14.78
15.1316 031	Galvanized steel pipe, A-120, sch 40, 1", screwed	LF	0.1122	13.63	4.19	17.82
15.1316 041	Galvanized steel pipe, A-120, sch 40, 1-1/4", screwed	LF	0.1211	14.71	5.90	20.61
15.1316 051	Galvanized steel pipe, A-120, sch 40, 1-1/2", screwed	LF	0.1402	17.03	6.44	23.47
15.1316 061	Galvanized steel pipe, A-120, sch 40, 2", screwed	LF	0.1908	23.17	8.48	31.66
15.1316 071	Galvanized steel pipe, A-120, sch 40, 2-1/2", screwed	LF	0.2291	27.83	14.20	42.03
15.1316 081	Galvanized steel pipe, A-120, sch 40, 3", screwed	LF	0.2674	32.48	19.07	51.55
15.1316 091	Galvanized steel pipe, A-120, sch 40, 4", screwed	LF	0.3485	42.33	30.37	72.70
15.1316 101	Galvanized steel pipe, A-120, sch 40, 5", screwed	LF	0.3771	45.80	45.21	91.01
15.1316 111	Galvanized steel pipe, A-120, sch 40, 6", screwed	LF	0.4215	51.20	62.30	113.49
15.1317 000	**STEEL PIPE, BLACK, WELDED, SCH 40, A-53:**					
15.1317 011	Black steel pipe, welded, A-53, sch 40, 2"	LF	0.1924	23.37	16.31	39.68
15.1317 021	Black steel pipe, welded, A-53, sch 40, 2-1/2"	LF	0.2452	29.78	21.45	51.24
15.1317 031	Black steel pipe, welded, A-53, sch 40, 3"	LF	0.2790	33.89	26.63	60.52
15.1317 041	Black steel pipe, welded, A-53, sch 40, 4"	LF	0.3583	43.52	37.26	80.78
15.1318 000	**STEEL PIPE, SEAMLESS, SCH 40, A-53, WELDED:**					
15.1318 011	Steel pipe, seamless, A-53, sch 40, 2", welded	LF	0.1924	23.37	19.40	42.77
15.1318 021	Steel pipe, seamless, A-53, sch 40, 2-1/2", welded	LF	0.2489	30.23	21.80	52.03
15.1318 031	Steel pipe, seamless, A-53, sch 40, 3", welded	LF	0.2785	33.83	26.63	60.46
15.1318 041	Steel pipe, seamless, A-53, sch 40, 4", welded	LF	0.3583	43.52	37.84	81.36
15.1318 051	Steel pipe, seamless, A-53, sch 40, 5", welded	LF	0.3985	48.40	45.50	93.90
15.1318 061	Steel pipe, seamless, A-53, sch 40, 6", welded	LF	0.4398	53.42	47.96	101.38
15.1318 071	Steel pipe, seamless, A-53, sch 40, 8", welded	LF	0.4904	59.56	73.55	133.12
15.1318 081	Steel pipe, seamless, A-53, sch 40, 10", welded	LF	0.5517	67.01	138.62	205.63
15.1318 091	Steel pipe, seamless, A-53, sch 40, 12", welded	LF	0.7203	87.49	210.66	298.15
15.1319 000	**STEEL PIPE, SEAMLESS, SCH 80, A-53, WELDED:**					
15.1319 011	Steel pipe, seamless, A-53, sch 80, 2", welded	LF	0.2207	26.81	20.80	47.61
15.1319 021	Steel pipe, seamless, A-53, sch 80, 2-1/2", welded	LF	0.2862	34.76	24.47	59.23
15.1319 031	Steel pipe, seamless, A-53, sch 80, 3", welded	LF	0.3196	38.82	30.11	68.92
15.1319 041	Steel pipe, seamless, A-53, sch 80, 4", welded	LF	0.4115	49.98	44.16	94.14
15.1319 051	Steel pipe, seamless, A-53, sch 80, 5", welded	LF	0.4583	55.67	53.64	109.31
15.1319 061	Steel pipe, seamless, A-53, sch 80, 6", welded	LF	0.5058	61.43	59.12	120.56
15.1319 071	Steel pipe, seamless, A-53, sch 80, 8", welded	LF	0.5640	68.50	90.74	159.24
15.1319 081	Steel pipe, seamless, A-53, sch 80, 10", welded	LF	0.6345	77.07	179.21	256.28
15.1319 091	Steel pipe, seamless, A-53, sch 80, 12", welded	LF	0.8277	100.53	266.98	367.51
15.1320 000	**STEEL PIPE, BLACK, WELDED, SCH 40, A-120:**					
15.1320 001	*Note: The following prices include victaulic couplings with victaulic fittings.*					
15.1320 011	Black steel pipe, weld, A-120, sch 40, 2"	LF	0.1016	12.34	19.17	31.51

15 - PLUMBING - MECHANICAL

Division	Description	Unit	Labor Hours	Labor Cost	Material Cost	Total Cost
15.1320 021	Black steel pipe, weld, A-120, sch 40, 2-1/2"	LF	0.1314	15.96	21.45	37.41
15.1320 031	Black steel pipe, weld, A-120, sch 40, 3"	LF	0.1471	17.87	26.63	44.50
15.1320 041	Black steel pipe, weld, A-120, sch 40, 4"	LF	0.1892	22.98	34.38	57.36
15.1320 051	Black steel pipe, weld, A-120, sch 40, 5"	LF	0.2104	25.56	44.80	70.35
15.1320 061	Black steel pipe, weld, A-120, sch 40, 6"	LF	0.2322	28.20	46.92	75.12
15.1320 071	Black steel pipe, weld, A-120, sch 40, 8"	LF	0.2587	31.42	78.10	109.52
15.1320 081	Black steel pipe, weld, A-120, sch 40, 10"	LF	0.3016	36.63	136.43	173.07
15.1320 091	Black steel pipe, weld, A-120, sch 40, 12"	LF	0.4878	59.25	207.88	267.12
15.1321 000	**FLANGES, CAST IRON, SCREWED, BLACK, 125#:**					
15.1321 011	Cast iron flange, black, screwed, 125#, 1-1/2"	EA	0.7281	88.44	32.19	120.63
15.1321 021	Cast iron flange, black, screwed, 125#, 2"	EA	0.7664	93.09	40.21	133.30
15.1321 031	Cast iron flange, black, screwed, 125#, 2-1/2"	EA	0.8430	102.39	48.05	150.44
15.1321 041	Cast iron flange, black, screwed, 125#, 3"	EA	0.9426	114.49	53.08	167.57
15.1321 051	Cast iron flange, black, screwed, 125#, 4"	EA	1.1495	139.62	70.55	210.17
15.1321 061	Cast iron flange, black, screwed, 125#, 5"	EA	1.2874	156.37	84.75	241.12
15.1321 071	Cast iron flange, black, screwed, 125#, 6"	EA	1.5095	183.34	83.97	267.31
15.1321 081	Cast iron flange, black, screwed, 125#, 8"	EA	1.6093	195.47	144.72	340.18
15.1321 091	Cast iron flange, black, screwed, 125#, 10"	EA	1.8085	219.66	235.45	455.11
15.1322 000	**FLANGES, SLIP-ON, 150#:**					
15.1322 011	Steel flange, slip-on, 150#, 2"	EA	0.8430	102.39	20.94	123.33
15.1322 021	Steel flange, slip-on, 150#, 3"	EA	1.2262	148.93	30.11	179.04
15.1322 031	Steel flange, slip-on, 150#, 4"	EA	1.6093	195.47	38.28	233.74
15.1322 041	Steel flange, slip-on, 150#, 5"	EA	1.9925	242.01	54.09	296.10
15.1322 051	Steel flange, slip-on, 150#, 6"	EA	2.3756	288.54	62.84	351.38
15.1322 061	Steel flange, slip-on, 150#, 8"	EA	3.2952	400.23	95.57	495.80
15.1322 071	Steel flange, slip-on, 150#, 10"	EA	4.0615	493.31	172.77	666.08
15.1322 081	Steel flange, slip-on, 150#, 12"	EA	4.9811	605.00	359.07	964.07
15.1400 000	**VALVES & SPECIALTIES:**					
15.1401 000	**VACUUM BREAKERS, ANTI-SIPHON:**					
15.1401 011	Vacuum breaker, brass, 1/2", anti-siphon	EA	0.2553	31.01	24.88	55.89
15.1401 021	Vacuum breaker, brass, 3/4", anti-siphon	EA	0.3064	37.22	29.38	66.60
15.1401 031	Vacuum breaker, brass, 1", anti-siphon	EA	0.3090	37.53	46.01	83.54
15.1401 041	Vacuum breaker, brass, 1-1/4", anti-siphon	EA	0.3827	46.48	76.74	123.22
15.1401 051	Vacuum breaker, brass, 1-1/2", anti-siphon	EA	0.4589	55.74	89.68	145.41
15.1401 061	Vacuum breaker, brass, 2", anti-siphon	EA	0.5506	66.88	139.84	206.71
15.1401 071	Vacuum breaker, brass, 2-1/2", anti-siphon	EA	1.8751	227.75	402.27	630.02
15.1401 081	Vacuum breaker, brass, 3", anti-siphon	EA	1.6198	196.74	534.78	731.52
15.1402 000	**STRAINERS, 'Y', WITH STAINLESS SCREENS:**					
15.1402 011	Y strainer, 1/2" screwed, 250# cast iron, stainless steel screen	EA	0.2451	29.77	27.50	57.27
15.1402 021	Y strainer, 3/4" screwed, 250# cast iron, stainless steel screen	EA	0.2936	35.66	31.80	67.46
15.1402 031	Y strainer, 1" screwed, 250# cast iron, stainless steel screen	EA	0.3060	37.17	39.72	76.89
15.1402 041	Y strainer, 1-1/4" screwed, 250# cast iron, screen	EA	0.3918	47.59	60.51	108.10
15.1402 051	Y strainer, 1-1/2" screwed, 250# cast iron, screen	EA	0.4528	55.00	67.84	122.84
15.1402 061	Y strainer, 2" screwed, 250# cast iron, stainless steel screen	EA	0.5403	65.62	102.82	168.45
15.1402 071	Y strainer, 2-1/2" flange, 150# cast iron, screen	EA	1.6146	196.11	339.36	535.47
15.1402 081	Y strainer, 3" flange, 150# cast iron, stainless steel screen	EA	1.8637	226.37	403.27	629.64
15.1402 091	Y strainer, 4" flange, 150# cast iron, stainless steel screen	EA	2.7407	332.89	714.81	1,047.70
15.1402 101	Y strainer, 6" flange, 150# cast iron, stainless steel screen	EA	3.3041	401.32	1,474.14	1,875.46
15.1402 111	Y strainer, 8" flange, 150# cast iron, stainless steel screen	EA	4.5056	547.25	2,350.34	2,897.59
15.1402 121	Y strainer, 10" flange, 150# cast iron, stainless steel screen	EA	4.9000	595.15	4,094.29	4,689.45
15.1402 131	Y strainer 12" flange, 150# cast iron, stainless steel screen	EA	5.4000	655.88	6,141.43	6,797.31
15.1403 000	**GATE, GLOBE & CHECK VALVES, BRASS, 125#, SCREWED:**					
15.1403 011	Valves, brass, 125#, screwed, 1/2"	EA	0.2560	31.09	13.31	44.40
15.1403 021	Valves, brass, 125#, screwed, 3/4"	EA	0.3066	37.24	17.30	54.54
15.1403 031	Valves, brass, 125#, screwed, 1"	EA	0.3196	38.82	24.72	63.54
15.1403 041	Valves, brass, 125#, screwed, 1-1/4"	EA	0.4092	49.70	35.23	84.93
15.1403 051	Valves, brass, 125#, screwed, 1-1/2"	EA	0.6729	81.73	46.34	128.08

MECHANICAL - PLUMBING - 15

Division	Description	Unit	Labor Hours	Labor Cost	Material Cost	Total Cost
15.1403 061	Valves, brass, 125#, screwed, 2"	EA	1.1400	138.46	70.15	208.61
15.1403 071	Valves, brass, 125#, screwed, 2-1/2"	EA	1.5500	188.26	126.23	314.50
15.1403 081	Valves, brass, 125#, screwed, 3"	EA	1.9200	233.20	175.97	409.17
15.1403 091	Valves, brass, 125#, screwed, 4"	EA	2.7000	327.94	379.95	707.89
15.1404 000	**GATE, GLOBE, & CHECK VALVES, PVC:**					
15.1404 011	Valve, gate, globe & check, PVC, 1/2", solvent joint	EA	0.2000	24.29	40.26	64.55
15.1404 021	Valve, gate, globe & check, PVC, 3/4", solvent joint	EA	0.2500	30.37	53.93	84.30
15.1404 031	Valve, gate, globe & check, PVC, 1", solvent joint	EA	0.2800	34.01	67.52	101.53
15.1404 041	Valve, gate, globe & check, PVC, 1-1/4", solvent joint	EA	0.3300	40.08	81.17	121.25
15.1404 051	Valve, gate, globe & check, PVC, 1-1/2", solvent joint	EA	0.5500	66.80	94.76	161.57
15.1404 061	Valve, gate, globe & check, PVC, 2", solvent joint	EA	0.8800	106.88	186.85	293.73
15.1404 071	Valve, gate, globe & check, PVC, 2-1/2", solvent joint	EA	1.3500	163.97	279.02	442.99
15.1404 081	Valve, gate, globe & check, PVC, 3", solvent joint	EA	1.7200	208.91	461.44	670.35
15.1404 091	Valve, gate, globe & check, PVC, 4", solvent joint	EA	1.9500	236.85	643.84	880.69
15.1405 000	**GATE, GLOBE & CHECK VALVES, IRON, FLANGED, 125#:**					
15.1405 001	Note: The following prices do not include companion flanges or bolt and gaskets sets.					
15.1405 011	Valve, gate, globe & check, iron body, flange, 125#, 2"	EA	1.4116	171.45	442.36	613.81
15.1405 021	Valve, gate, globe & check, iron body, flange, 125#, 2-1/2"	EA	1.6860	204.78	483.55	688.33
15.1405 031	Valve, gate, globe & check, iron body, flange, 125#, 3"	EA	1.9604	238.11	551.77	789.88
15.1405 041	Valve, gate, globe & check, iron body, flange, 125#, 4"	EA	2.8618	347.59	809.68	1,157.28
15.1405 051	Valve, gate, globe & check, iron body, flange, 125#, 5-6"	EA	3.8000	461.55	1,403.97	1,865.52
15.1405 061	Valve, gate, globe & check, iron body, flange, 125#, 8"	EA	5.6000	680.18	2,772.98	3,453.16
15.1405 071	Valve, gate, globe & check, iron body, flange, 125#, 10"	EA	7.5000	910.95	4,582.30	5,493.25
15.1405 081	Valve, gate, globe & check, iron body, flange, 125#, 12"	EA	9.9000	1,202.45	6,121.23	7,323.69
15.1405 091	Valve, gate, globe & check, iron body, flange, 125#, 14"	EA	12.3000	1,493.96	10,025.04	11,518.99
15.1406 000	**GAS SERVICE COCKS, BRASS, SCREWED:**					
15.1406 011	Gas cock, brass, screwed, 1/2"	EA	0.3525	42.81	32.68	75.50
15.1406 021	Gas cock, brass, screwed, 3/4"	EA	0.3726	45.26	38.28	83.53
15.1406 031	Gas cock, brass, screwed, 1"	EA	0.5217	63.37	55.08	118.44
15.1406 041	Gas cock, brass, screwed, 1-1/4"	EA	0.6077	73.81	64.72	138.53
15.1406 051	Gas cock, brass, screwed, 1-1/2"	EA	0.6708	81.48	107.39	188.87
15.1406 061	Gas cock, brass, screwed, 2"	EA	0.9681	117.59	168.74	286.32
15.1406 071	Gas cock, iron, flanged, 2-1/2"	EA	1.3000	157.90	287.88	445.78
15.1406 081	Gas cock, iron, flanged, 3"	EA	1.7000	206.48	456.14	662.62
15.1406 091	Gas cock, iron, flanged, 4"	EA	2.1000	255.07	706.44	961.50
15.1407 000	**HOSE GATE VALVE, WITH BRASS CAP & HOSE BIBB:**					
15.1407 011	Hose gate valve, 1", with cap & bibb	EA	0.3648	44.31	164.90	209.21
15.1407 021	Hose gate valve, 1-1/2", cap, bibb	EA	0.4944	60.05	257.60	317.65
15.1407 031	Hose gate valve, 2", with cap & bibb	EA	0.7302	88.69	366.86	455.55
15.1407 041	Hose bibb, brass, 3/4"	EA	0.3046	37.00	23.77	60.77
15.1407 051	Hose bibb, with wall box, 3/4"	EA	0.5073	61.62	74.88	136.50
15.1407 061	Wall hydrant, auto drain, wall to 14" thick	EA	0.7500	91.10	258.33	349.43
15.1407 071	Wall hydrant with box, wall to 14" thick	EACH	0.9500	115.39	456.73	572.12
15.1408 000	**PRESSURE REDUCING VALVE, IRON, BRONZE TRIM, 125, 0-100:**					
15.1408 011	Pressure reducing valve, brass, 125#, 0-100, 3/4"	EA	0.5000	60.73	145.31	206.04
15.1408 021	Pressure reducing valve, brass, 125#, 0-100, 1"	EA	0.6000	72.88	226.37	299.25
15.1408 031	Pressure reducing valve, iron body, bronze trim, 125#, 0-100, 1-1/2"	EA	1.2296	149.35	660.46	809.81
15.1408 041	Pressure reducing valve, iron body, bronze trim, 125#, 0-100, 2"	EA	1.3915	169.01	1,358.20	1,527.21
15.1408 051	Pressure reducing valve, iron body, bronze trim, 125#, 0-100, 3"	EA	1.6388	199.05	1,714.17	1,913.22
15.1408 061	Pressure reducing valve, iron body, bronze trim, 125#, 0-100, 4"	EA	2.7041	328.44	2,107.29	2,435.73
15.1408 071	Pressure reducing valve, iron body, bronze trim, 125#, 0-100, 6"	EA	3.3599	408.09	3,148.38	3,556.48
15.1408 081	Pressure reducing valve, iron body, bronze trim, 125#, 0-100, 8"	EA	4.4245	537.40	5,275.93	5,813.33
15.1408 091	Pressure reducing valve, iron body, bronze trim, 125#, 0-100, 10"	EA	6.0639	736.52	10,527.80	11,264.32
15.1409 000	**RELIEF VALVES, BRONZE, SCREWED, HOT WATER:**					
15.1409 011	Valve, relief, hot water, screwed, bronze, 3/4"	EA	0.3362	40.83	58.18	99.02
15.1409 021	Valve, relief, hot water, screwed, bronze, 1"	EA	0.4099	49.79	76.51	126.29

15 - PLUMBING - MECHANICAL

Division	Description	Unit	Labor Hours	Labor Cost	Material Cost	Total Cost
15.1409 031	Valve, relief, hot water, screwed, bronze, 1-1/2"	EA	0.5568	67.63	247.68	315.31
15.1409 041	Valve, relief, hot water, screwed, bronze, 2"	EA	0.9014	109.48	278.84	388.32
15.1410 000	**GAS REGULATOR, SCREWED, WITH AUTO SHUTOFF & RELIEF:**					
15.1410 011	Gas regulator valve, screwed, 1", shut off, relief	EA	0.4099	49.79	295.00	344.78
15.1410 021	Gas regulator valve, screwed, 1-1/2", shut off, relief	EA	0.5568	67.63	356.48	424.11
15.1410 031	Gas regulator valve, screwed, 2", shut off, relief	EA	0.9014	109.48	668.88	778.37
15.1410 041	Gas regulator valve, quake, shut off, CA approved, 3/4"	EA	0.3462	42.05	727.13	769.18
15.1410 051	Gas regulator valve, quake, shut off, CA approved, 2"	EA	0.9114	110.70	6,657.20	6,767.90
15.1411 000	**STEAM TRAPS, CAST IRON, SCREWED, WITH STAINLESS STEEL BUCKET:**					
15.1411 011	Steam trap, cast iron, screwed, 1/2", bucket	EA	0.3282	39.86	265.43	305.29
15.1411 021	Steam trap, cast iron, screwed, 3/4", bucket	EA	0.4529	55.01	425.73	480.74
15.1411 031	Steam trap, cast iron, screwed, 1", stainless steel bucket	EA	0.5732	69.62	846.56	916.18
15.1411 041	Steam trap, cast iron, screwed, 1-1/2", bucket	EA	0.7374	89.56	1,313.85	1,403.41
15.1412 000	**THERMOSTATIC MIXING VALVES:**					
15.1412 011	Thermostatic mix valve, cab, 3/4x3/4	EA	1.2500	151.82	2,539.52	2,691.34
15.1412 021	Thermostatic mix valve, cab, 3/4 x 1	EA	1.3300	161.54	2,978.69	3,140.24
15.1413 000	**VALVES, SOLENOID:**					
15.1413 011	Valve, solenoid, water or air, 1/2"	EA	0.5000	60.73	184.91	245.64
15.1413 021	Valve, solenoid, water or air, 3/4"	EA	0.5000	60.73	230.31	291.04
15.1413 031	Valve, solenoid, water or air, 1"	EA	0.6000	72.88	362.79	435.67
15.1413 051	Valve, solenoid, water or air, 1-1/2"	EA	0.8000	97.17	673.27	770.44
15.1413 061	Valve, solenoid, water or air, 2"	EA	1.0000	121.46	997.70	1,119.16
15.1413 071	Valve, solenoid, gas, 1/2"	EA	0.5000	60.73	177.85	238.58
15.1413 081	Valve, solenoid, gas, 3/4"	EA	0.5000	60.73	207.15	267.88
15.1413 091	Valve, solenoid, gas, 1"	EA	0.6000	72.88	288.85	361.73
15.1414 000	**BACKFLOW PREVENTERS, WITH SUPPORTS:**					
15.1414 011	Backflow preventer, 3/4" screwed	EA	0.8971	108.96	1,288.47	1,397.44
15.1414 021	Backflow preventer, 1" screwed	EA	1.0110	122.80	1,304.52	1,427.31
15.1414 031	Backflow preventer, 1-1/2" screwed	EA	1.8810	228.47	2,056.31	2,284.78
15.1414 041	Backflow preventer, 2" screwed	EA	2.1101	256.29	4,928.70	5,185.00
15.1414 051	Backflow preventer, 3" flanged	EA	3.0210	366.93	6,445.27	6,812.20
15.1414 061	Backflow preventer, 4" flanged	EA	5.8750	713.58	7,650.03	8,363.60
15.1414 071	Backflow preventer, 6" flanged	EA	6.4340	781.47	11,261.54	12,043.01
15.1414 081	Backflow preventer, 8" flanged	EA	8.4660	1,028.28	19,203.98	20,232.26
15.1414 091	Backflow preventer, 10" flanged	EA	11.6140	1,410.64	21,086.55	22,497.18
15.1415 000	**VICTAULIC COUPLINGS:**					
15.1415 011	Victaulic coupling, 2"	EA	0.1226	14.89	41.18	56.07
15.1415 021	Victaulic coupling, 3"	EA	0.1226	14.89	55.42	70.31
15.1415 031	Victaulic coupling, 4"	EA	0.1226	14.89	78.57	93.46
15.1415 041	Victaulic coupling, 6"	EA	0.1533	18.62	136.55	155.17
15.1415 051	Victaulic coupling, 8"	EA	0.1533	18.62	215.14	233.76
15.1415 061	Victaulic coupling, 10"	EA	0.1533	18.62	329.58	348.20
15.1415 071	Victaulic coupling, 12"	EA	0.1533	18.62	368.91	387.53
15.1416 000	**WATER METER, TURBINE, AWWA:**					
15.1416 001	Note: The following prices include flanges and bolt-ups.					
15.1416 011	Water meter, turbine, 3/4"	EA	0.9330	113.32	324.68	438.00
15.1416 021	Water meter, turbine, 1-1/4"	EA	1.2020	145.99	663.08	809.08
15.1416 031	Water meter, turbine, with 2 bolt & gaskets, 2"	EA	1.0722	130.23	1,453.79	1,584.02
15.1416 041	Water meter, turbine, with 2 bolt & gaskets, 3"	EA	2.6204	318.27	2,138.80	2,457.08
15.1416 051	Water meter, turbine, with 2 bolt & gaskets, 4"	EA	4.0700	494.34	4,256.87	4,751.21
15.1416 061	Water meter, turbine, with 2 bolt & gaskets, 6"	EA	11.2400	1,365.21	6,388.97	7,754.18
15.1416 071	Water meter, turbine, with 2 bolt & gaskets, 8"	EA	5.4000	655.88	10,414.68	11,070.56
15.1416 081	Add remote reader	EA	2.0200	245.35	11,239.03	11,484.38
15.1417 000	**VALVE, EARTHQUAKE ACTUATED, GAS SHUT OFF:**					
15.1417 011	Valve, earthquake actuated, gas shut off, 3/4"	EA	0.3462	42.05	387.80	429.85
15.1417 021	Valve, earthquake actuated, gas shut off, 1"	EA	0.4500	54.66	416.25	470.91

MECHANICAL - PLUMBING - 15

Division	Description	Unit	Labor Hours	Labor Cost	Material Cost	Total Cost
15.1417 031	Valve, earthquake actuated, gas shut off, 1-1/4"	EA	0.6200	75.31	442.04	517.35
15.1417 041	Valve, earthquake actuated, gas shut off, 1-1/2"	EA	0.7560	91.82	470.46	562.29
15.1417 051	Valve, earthquake actuated, gas shut off, flanged, 2"	EA	0.9114	110.70	3,102.20	3,212.90
15.1417 061	Valve, earthquake actuated, gas shut off, flanged, 3"	EA	1.5000	182.19	5,935.59	6,117.78
15.1417 071	Valve, earthquake actuated, gas shut off, flanged, 4"	EA	2.0000	242.92	7,714.19	7,957.11
15.1418 000	**EXPANSION JOINTS:**					
15.1418 011	Expansion joints, neoprene, flanged, 6" length, to 1-1/2"	EA	0.8003	97.20	457.17	554.37
15.1418 021	Expansion joints, neoprene, flanged, 6" length, to 3"	EA	1.4882	180.76	554.96	735.71
15.1418 031	Expansion joints, neoprene, flanged, 6" length, to 4"	EA	2.0193	245.26	583.40	828.67
15.1418 041	Expansion joints, neoprene, flanged, 6" length, to 6"	EA	2.4730	300.37	742.24	1,042.61
15.1418 051	Expansion joints, neoprene, flanged, 6" length, to 8"	EA	3.1408	381.48	848.26	1,229.74
15.1418 061	Expansion joints, neoprene, flanged, 6" length, to 10"	EA	3.3917	411.96	1,198.08	1,610.03
15.1418 071	Expansion joints, neoprene, flanged, 6" length, to 12"	EA	3.6869	447.81	1,350.67	1,798.48
15.1418 111	Expansion joints, neoprene, flanged, 10" length, to 3"	EA	1.5358	186.54	796.04	982.58
15.1418 121	Expansion joints, neoprene, flanged, 10" length, to 4"	EA	2.1110	256.40	886.52	1,142.92
15.1418 131	Expansion joints, neoprene, flanged, 10" length, to 6"	EA	2.8150	341.91	1,061.35	1,403.26
15.1418 141	Expansion joints, neoprene, flanged, 10" length, to 8"	EA	3.3776	410.24	1,253.61	1,663.86
15.1418 151	Expansion joints, neoprene, flanged, 10" length, to 10"	EA	3.6719	445.99	1,377.30	1,823.29
15.1418 161	Expansion joints, neoprene, flanged, 10" length, to 12"	EA	4.2351	514.40	1,569.50	2,083.89
15.1418 171	Expansion joints, neoprene, flanged, 10" length, to 16"	EA	5.8234	707.31	2,258.12	2,965.43
15.1418 181	Expansion joints, neoprene, flanged, 10" length, to 20"	EA	8.0419	976.77	2,665.23	3,641.99
15.1418 191	Expansion joints, neoprene, flanged, 10" length, to 24"	EA	9.3824	1,139.59	3,091.98	4,231.56
15.1418 211	Expansion joints, neoprene, flanged, 10" length, to 30"	EA	14.2719	1,733.46	3,816.95	5,550.41
15.1418 221	Expansion joints, neoprene, flanged, 10" length, to 36"	EA	16.9427	2,057.86	4,664.77	6,722.63
15.1500 000	**INSULATION, PIPING:**					
15.1500 001	Note: Prices include insulation allowance at pipe runs, valves and fittings.					
15.1501 000	**INSULATION, 1-1/2" CALCIUM SILICATE:**					
15.1501 001	Note: For 1" calsil on pipe sizes 1" to 6" .... add 28% to material cost For 2" calsil on sizes 1" to 6" .... add 56% For 2" calsil on sizes 8" to 14" .... add 41% For 2 1/2" calsil on sizes 1" to 6" .... add 94% For 2 1/2" calsil on sizes 8" to 14" .... add 80%					
15.1501 011	Insulation, 1-1/2" calcium silicate, 1/2" pipe	LF	0.0261	3.17	4.76	7.93
15.1501 021	Insulation, 1-1/2" calcium silicate, 3/4" pipe	LF	0.0261	3.17	4.85	8.02
15.1501 031	Insulation, 1-1/2" calcium silicate, 1" pipe	LF	0.0261	3.17	5.21	8.38
15.1501 041	Insulation, 1-1/2" calcium silicate, 1-1/4" pipe	LF	0.0261	3.17	5.41	8.58
15.1501 051	Insulation, 1-1/2" calcium silicate, 1-1/2" pipe	LF	0.0261	3.17	5.87	9.04
15.1501 061	Insulation, 1-1/2" calcium silicate, 2" pipe	LF	0.0384	4.66	6.49	11.15
15.1501 071	Insulation, 1-1/2" calcium silicate, 2-1/2" pipe	LF	0.0384	4.66	7.04	11.70
15.1501 081	Insulation, 1-1/2" calcium silicate, 3" pipe	LF	0.0514	6.24	7.40	13.65
15.1501 091	Insulation, 1-1/2" calcium silicate, 4" pipe	LF	0.0514	6.24	8.53	14.77
15.1501 101	Insulation, 1-1/2" calcium silicate, 5" pipe	LF	0.0514	6.24	9.59	15.84
15.1501 111	Insulation, 1-1/2" calcium silicate, 6" pipe	LF	0.0514	6.24	9.89	16.13
15.1501 121	Insulation, 1-1/2" calcium silicate, 8" pipe	LF	0.0644	7.82	13.42	21.24
15.1501 131	Insulation, 1-1/2" calcium silicate, 10" pipe	LF	0.0767	9.32	17.90	27.22
15.1501 141	Insulation, 1-1/2" calcium silicate, 12" pipe	LF	0.0897	10.89	21.12	32.02
15.1501 151	Insulation, 1-1/2" calcium silicate, 14" pipe	LF	0.0897	10.89	24.02	34.92
15.1502 000	**INSULATION, 3" CALCIUM SILICATE:**					
15.1502 011	Insulation, 3" calcium silicate, 3" pipe	LF	0.0767	9.32	16.73	26.05
15.1502 021	Insulation, 3" calcium silicate, 4" pipe	LF	0.0767	9.32	21.92	31.24
15.1502 031	Insulation, 3" calcium silicate, 5" pipe	LF	0.0767	9.32	24.94	34.25
15.1502 041	Insulation, 3" calcium silicate, 6" pipe	LF	0.0767	9.32	26.87	36.19
15.1502 061	Insulation, 3" calcium silicate, 8" pipe	LF	0.1035	12.57	31.94	44.51
15.1502 071	Insulation, 3" calcium silicate, 10" pipe	LF	0.1150	13.97	38.51	52.47
15.1502 081	Insulation, 3" calcium silicate, 12" pipe	LF	0.1380	16.76	42.65	59.42
15.1502 101	Insulation, 3" calcium silicate, 14" pipe	LF	0.1590	19.31	47.86	67.17
15.1503 000	**INSULATION, 1" FIBERGLASS WITH ALUMINUM JACKET:**					

15 - PLUMBING - MECHANICAL

Division	Description	Unit	Labor Hours	Labor Cost	Material Cost	Total Cost
15.1503 001	Note: For 1-1/2" fiberglass, sizes 1/2" to 1-1/2"... add 100% to matl					
	For sizes 2" to 3"..................... add 70%					
	For sizes 4" to 6"..................... add 45%					
	For sizes 8" to 12"..................... add 25%					
15.1503 011	Insulation, 1" fiberglass, 1/2" pipe, with aluminum jacket	LF	0.0384	4.66	2.56	7.22
15.1503 021	Insulation, 1" fiberglass, 3/4" pipe, with aluminum jacket	LF	0.0384	4.66	2.74	7.40
15.1503 031	Insulation, 1" fiberglass, 1" pipe, with aluminum jacket	LF	0.0384	4.66	2.91	7.57
15.1503 041	Insulation, 1" fiberglass, 1-1/4" pipe, with aluminum jacket	LF	0.0384	4.66	3.58	8.24
15.1503 051	Insulation, 1" fiberglass, 1-1/2" pipe, with aluminum jacket	LF	0.0384	4.66	3.62	8.29
15.1503 061	Insulation, 1" fiberglass, 2" pipe, with aluminum jacket	LF	0.0514	6.24	3.93	10.17
15.1503 071	Insulation, 1" fiberglass, 2-1/2" pipe, with aluminum jacket	LF	0.0514	6.24	4.34	10.59
15.1503 081	Insulation, 1" fiberglass, 3" pipe, with aluminum jacket	LF	0.0644	7.82	4.85	12.67
15.1503 091	Insulation, 1" fiberglass, 4" pipe, with aluminum jacket	LF	0.0644	7.82	6.34	14.16
15.1503 101	Insulation, 1" fiberglass, 5" pipe, with aluminum jacket	LF	0.0644	7.82	7.20	15.02
15.1503 111	Insulation, 1" fiberglass, 6" pipe, with aluminum jacket	LF	0.0644	7.82	7.79	15.61
15.1503 121	Insulation, 1" fiberglass, 8" pipe, with aluminum jacket	LF	0.1035	12.57	11.21	23.78
15.1503 131	Insulation, 1" fiberglass, 10" pipe, with aluminum jacket	LF	0.1410	17.13	13.20	30.33
15.1503 141	Insulation, 1" fiberglass, 12" pipe, with aluminum jacket	LF	0.1533	18.62	15.12	33.74
15.1504 000	**INSULATION, 2" FIBERGLASS WITH ALUMINUM JACKET:**					
15.1504 011	Insulation, 2" fiberglass, 1/2" pipe, with aluminum jacket	LF	0.0514	6.24	8.46	14.70
15.1504 021	Insulation, 2" fiberglass, 3/4" pipe, with aluminum jacket	LF	0.0514	6.24	8.69	14.93
15.1504 031	Insulation, 2" fiberglass, 1" pipe, with aluminum jacket	LF	0.0514	6.24	9.16	15.40
15.1504 041	Insulation, 2" fiberglass, 1-1/4" pipe with aluminum jacket	LF	0.0514	6.24	9.71	15.95
15.1504 051	Insulation, 2" fiberglass, 1-1/2" pipe, with aluminum jacket	LF	0.0514	6.24	10.22	16.47
15.1504 061	Insulation, 2" fiberglass, 2" pipe, with aluminum jacket	LF	0.0644	7.82	10.60	18.42
15.1504 071	Insulation, 2" fiberglass, 2-1/2" pipe, with aluminum jacket	LF	0.0644	7.82	11.52	19.34
15.1504 081	Insulation, 2" fiberglass, 3" pipe, with aluminum jacket	LF	0.0767	9.32	12.23	21.54
15.1504 091	Insulation, 2" fiberglass, 4" pipe, with aluminum jacket	LF	0.0767	9.32	14.30	23.62
15.1504 101	Insulation, 2" fiberglass, 5" pipe, with aluminum jacket	LF	0.0767	9.32	16.12	25.44
15.1504 111	Insulation, 2" fiberglass, 6" pipe, with aluminum jacket	LF	0.0767	9.32	16.43	25.75
15.1504 121	Insulation, 2" fiberglass, 8" pipe, with aluminum jacket	LF	0.1150	13.97	20.41	34.38
15.1504 131	Insulation, 2" fiberglass, 10" pipe, with aluminum jacket	LF	0.1533	18.62	24.32	42.94
15.1504 141	Insulation, 2" fiberglass, 12" pipe, with aluminum jacket	LF	0.1663	20.20	27.16	47.36
15.1505 000	**INSULATION, VALVES, FIBERGLASS WITH ALUMINUM JACKET:**					
15.1505 011	Insulation, fiberglass, 1" valve, with aluminum jacket	EA	0.3525	42.81	10.51	53.32
15.1505 021	Insulation, fiberglass, 2" valve, with aluminum jacket	EA	0.5671	68.88	28.16	97.04
15.1505 031	Insulation, fiberglass, 3" valve, with aluminum jacket	EA	0.7281	88.44	36.05	124.49
15.1505 041	Insulation, fiberglass, 4" valve, with aluminum jacket	EA	0.9119	110.76	42.20	152.96
15.1505 051	Insulation, fiberglass, 6" valve, with aluminum jacket	EA	1.3564	164.75	65.93	230.68
15.1505 061	Insulation, fiberglass, 8" valve, with aluminum jacket	EA	1.7778	215.93	75.02	290.95
15.1505 071	Insulation, fiberglass, 10" valve, with aluminum jacket	EA	2.3296	282.95	84.34	367.29
15.1505 081	Insulation, fiberglass, 12" valve, with aluminum jacket	EA	3.0806	374.17	93.79	467.96
15.1600 000	**MISCELLANEOUS PLUMBING SPECIALTIES:**					
15.1601 000	**ACCESS DOORS:**					
15.1601 001	Note: For fire rating, add 300% to the material costs. Stainless steel, add 200%.					
15.1601 011	Access door, painted steel, 8" x 8"	EA	0.3066	37.24	85.29	122.53
15.1601 021	Access door, painted steel, 12" x 12"	EA	0.3066	37.24	95.38	132.62
15.1601 031	Access door, painted steel, 18" x 18"	EA	0.3832	46.54	136.30	182.84
15.1601 041	Access door, 24" x 24", painted steel	EA	0.3832	46.54	194.26	240.80
15.1601 051	Access door, painted steel, 36" x 36"	EA	0.5365	65.16	378.23	443.39
15.1602 000	**CLEANOUTS, CAST IRON, FLOOR & WALL:**					
15.1602 011	Floor cleanout, cast iron, 2" & 3"	EA	0.8047	97.74	137.75	235.49
15.1602 021	Floor cleanout, cast iron, 4"	EA	0.8199	99.59	187.30	286.89
15.1602 031	Floor cleanout, cast iron, 6"	EA	0.8430	102.39	326.46	428.86
15.1602 041	Floor cleanout, cast iron, 8"	EA	0.9579	116.35	336.06	452.40
15.1602 051	Cleanout, cast iron, 4", to grade	EA	0.8430	102.39	291.22	393.62
15.1602 061	Cleanout, cast iron, 6", to grade	EA	0.9579	116.35	476.62	592.96

MECHANICAL - PLUMBING - 15

Division	Description	Unit	Labor Hours	Labor Cost	Material Cost	Total Cost
15.1602 071	Cleanout, cast iron, 8", to grade	EA	1.0729	130.31	544.00	674.32
15.1602 081	Wall cleanout, cast iron, 2"	EA	0.7664	93.09	40.40	133.48
15.1602 091	Wall cleanout, cast iron, 4"	EA	0.8047	97.74	51.44	149.18
15.1602 101	Wall cleanout, cast iron, 6"	EA	0.8199	99.59	88.68	188.26
15.1603 000	**DRAINS, CAST IRON, AREA & FLOOR:**					
15.1603 011	Area drain, 2", cast iron, with 5" x 5" strainer	EA	1.1495	139.62	120.17	259.79
15.1603 021	Area drain, 3", cast iron, with 6" x 6" strainer	EA	1.1495	139.62	139.26	278.88
15.1603 031	Area drain, 4", cast iron, with 8" x 8" strainer	EA	1.1495	139.62	248.65	388.27
15.1603 041	Floor drain, 2"-4", cast iron, with trap	EA	1.1495	139.62	142.66	282.27
15.1603 051	Floor drain, 6", cast iron, with trap	EA	1.3411	162.89	362.46	525.35
15.1604 000	**DRAINS, ROOF, CAST IRON, WITH ALUMINUM DOMES:**					
15.1604 001	*Note: For best quality roof drains, add 75% to the material costs.*					
15.1604 011	Roof drain, cast iron, 2"-4", aluminum dome	EA	1.4561	176.86	191.37	368.23
15.1604 021	Roof drain, cast iron, 5"-6", aluminum dome	EA	1.8392	223.39	273.07	496.46
15.1604 031	Roof drain, cast iron, 8", aluminum dome	EA	2.1457	260.62	354.98	615.59
15.1605 000	**DRAINS, SHOWER, WITH NICALOY STRAINER:**					
15.1605 011	Shower drain, 1-1/2", with 4" strainer	EA	1.2262	148.93	107.05	255.98
15.1605 021	Shower drain, 2", with 7" strainer	EA	1.2262	148.93	166.21	315.14
15.1605 031	Shower drain, 3", with 7" strainer	EA	1.3794	167.54	181.05	348.59
15.1605 041	Shower drain, 4", with 8" strainer	EA	1.6093	195.47	203.28	398.75
15.1605 051	Shower drain, 2", with 5" x 5" strainer	EA	1.2262	148.93	183.75	332.69
15.1605 061	Shower drain, 3", with 6" x 6" strainer	EA	1.3794	167.54	203.19	370.73
15.1605 071	Shower drain, 4", with 8" x 8" strainer	EA	1.6093	195.47	285.13	480.59
15.1606 000	**MISCELLANEOUS ITEMS:**					
15.1606 011	Roof jacks, galvanized iron, 4"	EA	0.3066	37.24	64.11	101.35
15.1606 021	Roof jacks, galvanized iron, 6"	EA	0.3832	46.54	98.97	145.51
15.1606 031	Water hammer arrestor, 1-11 FU, 3/4"	EACH	0.4600	55.87	136.62	192.49
15.1606 041	Water hammer arrestor, 12-32 FU, 1"	EA	0.4800	58.30	273.66	331.96
15.1606 051	Water hammer arrestor, 33-60 FU, 1"	EA	0.4800	58.30	409.53	467.83
15.1606 061	Water hammer arrestor, 61-113 FU, 1"	EA	0.4800	58.30	1,027.03	1,085.33
15.1607 000	**FIRE BARRIER PENETRATION SYSTEM, PLASTIC PIPE:**					
15.1607 001	*Note: The following systems are rated for 4 hours.*					
15.1607 011	Fire barrier, water, coupling, fitting with plug, 1/2-1-1/2"	EA	0.0780	9.47	2.48	11.95
15.1607 021	Fire barrier, water, coupling, fitting with plug, 2"	EA	0.0903	10.97	6.39	17.36
15.1607 031	Fire barrier, water, coupling, fitting with plug, 2-1/2-3"	EA	0.1103	13.40	7.55	20.95
15.1607 041	Fire barrier, water, coupling, fitting with plug, 4-5"	EA	0.1247	15.15	10.92	26.07
15.1607 051	Fire barrier, water, coupling, fitting with plug, 6"	EA	0.1352	16.42	24.14	40.56
15.1607 061	Fire barrier, water, coupling, fitting with plug, 8"	EA	0.1488	18.07	36.88	54.95
15.1607 071	Fire barrier, DWV, coupling, fitting with plug, 2"	EA	0.0999	12.13	15.09	27.23
15.1607 081	Fire barrier, DWV, coupling, fitting with plug, 3"	EA	0.1299	15.78	21.73	37.51
15.1607 091	Fire barrier, DWV, coupling, fitting with plug, 4"	EA	0.1500	18.22	27.62	45.84
15.1607 101	Fire barrier, DWV, coupling, fitting with plug, 6"	EA	0.1786	21.69	41.86	63.56
15.1607 111	Fire barrier, DWV, coupling, fitting with plug, 8"	EA	0.2012	24.44	47.73	72.17
15.1607 121	Fire barrier, caulk, wrap/strip, 1/2" - 1"	EA	0.0999	12.13	1.99	14.13
15.1607 131	Fire barrier, caulk, wrap/strip, 2"	EA	0.1200	14.58	2.85	17.43
15.1607 141	Fire barrier, caulk, wrap/strip, 3"	EA	0.1398	16.98	5.49	22.47
15.1607 151	Fire barrier, caulk, wrap/strip, 4"	EA	0.1499	18.21	6.97	25.17
15.1607 161	Fire barrier, caulk, wrap/strip, 5"	EA	0.1700	20.65	7.44	28.09
15.1607 171	Fire barrier, caulk, wrap/strip, 6"	EA	0.1899	23.07	12.48	35.55
15.1607 181	Fire barrier, caulk, wrap/strip, 8"	EA	0.2099	25.49	22.34	47.83
15.1607 191	Fire barrier, batt with smoke seal, 1/2" - 1"	EA	0.0931	11.31	0.03	11.34
15.1607 201	Fire barrier, batt with smoke seal, 2"	EA	0.1090	13.24	0.11	13.35
15.1607 211	Fire barrier, batt with smoke seal, 3"	EA	0.1233	14.98	0.30	15.27
15.1607 221	Fire barrier, batt with smoke seal, 4"	EA	0.1450	17.61	0.49	18.10
15.1607 231	Fire barrier, batt with smoke seal, 5"	EA	0.1607	19.52	0.63	20.15
15.1607 241	Fire barrier, batt with smoke seal, 6"	EA	0.1852	22.49	1.20	23.70
15.1607 251	Fire barrier, batt with smoke seal, 8"	EA	0.2028	24.63	2.05	26.68

15 - PLUMBING - MECHANICAL

Division	Description	Unit	Labor Hours	Labor Cost	Material Cost	Total Cost
15.1700 000	**MEDICAL & LABORATORY EQUIPMENT & PIPE:**					
15.1701 000	**MANIFOLDS, OXYGEN OR NITROUS OXIDE:**					
15.1701 011	Manifold, 4 cylinder	EA	4.9045	595.70	10,631.83	11,227.53
15.1701 021	Manifold, 6 cylinder	EA	6.8969	837.70	11,179.82	12,017.51
15.1701 031	Manifold, 12 cylinder	EA	9.1959	1,116.93	12,309.66	13,426.59
15.1701 041	Manifold, 24 cylinder	EA	16.0000	1,943.36	16,187.85	18,131.21
15.1701 051	Zone valve, with box, 1 @ 1/2"	EA	0.5748	69.82	514.17	583.99
15.1701 061	Zone valve, with box, 2 @ 1/2"	EA	0.9196	111.69	731.58	843.27
15.1701 071	Zone valve, with box 3 @ 1/2"	EA	1.5111	183.54	1,037.24	1,220.78
15.1701 081	Zone valve, with box 5 @ 1/2	EA	2.1000	255.07	1,899.69	2,154.75
15.1701 091	Shut-off valve, with out box, 1"	EA	0.3449	41.89	197.30	239.19
15.1701 101	Shut-off valve, with out box, 1-1/4"	EA	0.4598	55.85	288.41	344.25
15.1701 111	Shut-off valve, with out box, 1-1/2"	EA	0.5365	65.16	403.50	468.66
15.1701 121	Gas outlet, wall	EA	0.4750	57.69	136.95	194.64
15.1702 000	**ALARMS:**					
15.1702 011	Alarm, line press, local, 1 gas	EA	0.5888	71.52	3,739.74	3,811.25
15.1702 021	Alarm, line press, local, 2 gas	EA	0.7664	93.09	4,634.42	4,727.51
15.1702 031	Alarm, line press, local, 3 gas, liquid	EA	1.1495	139.62	5,806.68	5,946.29
15.1702 041	Alarm, line press, local, 5 gas	EA	1.9458	236.34	6,850.06	7,086.40
15.1702 051	Emergency inlet connection, liquid	EA	2.0110	244.26	2,622.65	2,866.91
15.1702 061	Alarm, master, 15 signal	EA	2.5100	304.86	5,044.73	5,349.59
15.1702 071	Switch, vacuum	EA	0.7664	93.09	647.92	741.01
15.1702 081	Switch, nitrogen	EA	0.7664	93.09	539.92	633.01
15.1702 091	Alarm, pressure switch, liquid	EA	0.7664	93.09	539.92	633.01
15.1703 000	**MEDICAL VACUUM PUMPS WITH ALL RELATED ACCESSORIES:**					
15.1703 011	Vacuum pump, duplex, 30 CFM, 5 hp	EA	22.9896	2,792.32	35,790.61	38,582.92
15.1703 021	Vacuum pump, duplex, 60 CFM, 10 hp	EA	27.5876	3,350.79	42,456.29	45,807.08
15.1703 031	Vacuum pump, duplex, 210 CFM, 15 hp	EA	30.6528	3,723.09	45,979.44	49,702.53
15.1704 000	**MEDICAL AIR COMPRESSOR WITH ACCESSORIES:**					
15.1704 011	Air compressor, duplex, 2 @ 3 hp	EA	12.2612	1,489.25	19,517.32	21,006.57
15.1704 021	Air compressor, duplex, 2 @ 5 hp	EA	16.1000	1,955.51	49,895.15	51,850.66
15.1704 031	Air compressor, duplex, 2 @ 10 hp	EA	22.9896	2,792.32	78,988.73	81,781.05
15.1705 000	**MEDICAL GAS PIPE, COPPER, 'L':**					
15.1705 001	Note: The following prices include purging, sterilizing, etc.					
15.1705 011	Medical gas pipe, 1/2", copper, 'L'	LF	0.1582	19.21	13.18	32.39
15.1705 021	Medical gas pipe, 3/4", copper, 'L'	LF	0.1922	23.34	17.90	41.25
15.1705 031	Medical gas pipe, 1", copper, 'L'	LF	0.2282	27.72	23.41	51.13
15.1705 041	Medical gas pipe, 1-1/4", copper, 'L'	LF	0.2464	29.93	31.09	61.02
15.1705 051	Medical gas pipe, 1-1/2", copper, 'L'	LF	0.2621	31.83	37.28	69.11
15.1705 061	Medical gas pipe, 2", copper, 'L'	LF	0.3162	38.41	55.44	93.85
15.1705 071	Medical gas pipe, 2-1/2", copper, 'L'	LF	0.3503	42.55	82.52	125.07
15.1705 081	Medical gas pipe, 3", copper, 'L'	LF	0.4024	48.88	105.30	154.17
15.1706 000	**STAINLESS STEEL PIPE, NON-SPOOLED, '304', SCH 10, SEAMLESS:**					
15.1706 001	Note: The following items are normally used in labs, manufacturing or chemical plants. Note that all pricing by user should be P.O.A. from suppliers.					
15.1706 011	Stainless steel pipe, seamless, '304', sch 10, 1/2", non-spooled	LF	0.0903	10.09	11.38	21.47
15.1706 021	Stainless steel pipe, seamless, '304', sch 10, 3/4", non-spooled	LF	0.1118	12.49	12.96	25.45
15.1706 031	Stainless steel pipe, seamless, '304', sch 10, 1", non-spooled	LF	0.1346	15.04	18.08	33.12
15.1706 041	Stainless steel pipe, seamless, '304', sch 10, 1-1/4", non-spooled	LF	0.1454	16.24	22.23	38.48
15.1706 051	Stainless steel pipe, seamless, '304', sch 10, 1-1/2", non-spooled	LF	0.1683	18.80	24.93	43.73
15.1706 061	Stainless steel pipe, seamless, '304', sch 10, 2", non-spooled	LF	0.2290	25.58	28.26	53.84
15.1706 071	Stainless steel pipe, seamless, '304', sch 10, 2-1/2", non-spooled	LF	0.2751	30.73	38.68	69.41
15.1706 081	Stainless steel pipe, seamless, '304', sch 10, 3", non-spooled	LF	0.3210	35.86	48.30	84.16
15.1706 091	Stainless steel pipe, seamless, '304', sch 10, 4", non-spooled	LF	0.4185	46.75	59.94	106.69
15.1707 000	**STAINLESS STEEL PIPE, '304', SCH 40, SEAMLESS:**					
15.1707 011	Stainless steel pipe, seamless, '304', sch 40, 1/2"	LF	0.0947	10.58	13.59	24.17
15.1707 021	Stainless steel pipe, seamless, '304', sch 40, 3/4"	LF	0.1173	14.25	16.04	30.29

MECHANICAL - PLUMBING - 15

Division	Description	Unit	Labor Hours	Labor Cost	Material Cost	Total Cost
15.1707 031	Stainless steel pipe, seamless, '304', sch 40, 1"	LF	0.1414	17.17	20.17	37.34
15.1707 041	Stainless steel pipe, seamless, '304', sch 40, 1-1/4"	LF	0.1527	18.55	24.27	42.82
15.1707 051	Stainless steel pipe, seamless, '304', sch 40, 1-1/2"	LF	0.1767	21.46	28.32	49.78
15.1707 061	Stainless steel pipe, seamless, '304', sch 40, 2"	LF	0.2405	29.21	36.05	65.26
15.1707 071	Stainless steel pipe, seamless, '304', sch 40, 2-1/2"	LF	0.2887	35.07	52.92	87.99
15.1707 081	Stainless steel pipe, seamless, '304', sch 40, 3"	LF	0.3370	40.93	68.05	108.98
15.1707 091	Stainless steel pipe, seamless, '304', sch 40, 4"	LF	0.4394	53.37	90.25	143.62
15.1707 101	Stainless steel pipe, seamless, '304', sch 40, 6"	LF	0.5311	64.51	154.88	219.39
15.1707 111	Stainless steel pipe, seamless, '304', sch 40, 8"	LF	0.5734	69.65	249.92	319.57
15.1708 000	**STAINLESS STEEL PIPE, '304', SCH 80, SEAMLESS:**					
15.1708 011	Stainless steel pipe, seamless, '304', sch 80, 1/2"	LF	0.0996	12.10	17.82	29.92
15.1708 021	Stainless steel pipe, seamless, '304', sch 80, 3/4"	LF	0.1233	14.98	22.47	37.45
15.1708 031	Stainless steel pipe, seamless, '304', sch 80, 1"	LF	0.1484	18.02	29.52	47.55
15.1708 041	Stainless steel pipe, seamless, '304', sch 80, 1-1/4"	LF	0.1604	19.48	37.47	56.96
15.1708 051	Stainless steel pipe, seamless, '304', sch 80, 1-1/2"	LF	0.1767	21.46	42.95	64.41
15.1708 061	Stainless steel pipe, seamless, '304', sch 80, 2"	LF	0.2648	32.16	52.87	85.03
15.1708 071	Stainless steel pipe, seamless, '304', sch 80, 2-1/2"	LF	0.3435	41.72	108.16	149.88
15.1708 081	Stainless steel pipe, seamless, '304', sch 80, 3"	LF	0.3836	46.59	135.23	181.82
15.1708 091	Stainless steel pipe, seamless, '304', sch 80, 4"	LF	0.4939	59.99	201.55	261.54
15.1708 101	Stainless steel pipe, seamless, '304', sch 80, 6"	LF	0.5500	66.80	471.00	537.80
15.1708 111	Stainless steel pipe, seamless, '304', sch 80, 8"	LF	0.6070	73.73	788.12	861.85
15.1709 000	**STAINLESS STEEL PIPE, '316' EXTRA LOW CARBON, SCH 10, SEAMLESS:**					
15.1709 011	Stainless steel pipe, seamless, '316', extra low carbon, sch 10, 1/2"	LF	0.0903	10.97	13.01	23.98
15.1709 021	Stainless steel pipe, seamless, '316', extra low carbon, sch 10, 3/4"	LF	0.1118	13.58	15.04	28.61
15.1709 031	Stainless steel pipe, seamless, '316', extra low carbon, sch 10, 1"	LF	0.1346	16.35	21.45	37.80
15.1709 041	Stainless steel pipe, seamless, '316', extra low carbon, sch 10, 1-1/4"	LF	0.1454	17.66	25.66	43.32
15.1709 051	Stainless steel pipe, seamless, '316', extra low carbon, sch 10, 1-1/2"	LF	0.1683	20.44	29.05	49.49
15.1709 061	Stainless steel pipe, seamless, '316', extra low carbon, sch 10, 2"	LF	0.2290	27.81	36.75	64.57
15.1709 071	Stainless steel pipe, seamless, '316', extra low carbon, sch 10 2-1/2"	LF	0.2751	33.41	45.60	79.01
15.1709 081	Stainless steel pipe, seamless, '316', extra low carbon, sch 10, 3"	LF	0.3210	38.99	59.41	98.40
15.1709 091	Stainless steel pipe, seamless, '316', extra low carbon, sch 10, 4"	LF	0.4185	50.83	72.63	123.47
15.1710 000	**STAINLESS STEEL PIPE, '316' EXTRA LOW CARBON, SCH 40, SEAMLESS:**					
15.1710 011	Stainless steel pipe, seamless, '316', extra low carbon, sch 40, 1/2"	LF	0.0947	11.50	18.42	29.92
15.1710 021	Stainless steel pipe, seamless, '316', extra low carbon, sch 40, 3/4"	LF	0.1173	14.25	21.19	35.44
15.1710 031	Stainless steel pipe, seamless, '316', extra low carbon, sch 40, 1"	LF	0.1414	17.17	28.60	45.78
15.1710 041	Stainless steel pipe, seamless, '316', extra low carbon, sch 40, 1-1/4"	LF	0.1527	18.55	34.35	52.89
15.1710 051	Stainless steel pipe, seamless, '316', extra low carbon, sch 40, 1-1/2"	LF	0.1767	21.46	40.49	61.95
15.1710 061	Stainless steel pipe, seamless, '316', extra low carbon, sch 40, 2"	LF	0.2405	29.21	52.51	81.72
15.1710 071	Stainless steel pipe, seamless, '316', extra low carbon, sch 40, 2-1/2"	LF	0.2887	35.07	83.72	118.78
15.1710 081	Stainless steel pipe, seamless, '316', extra low carbon, sch 40, 3"	LF	0.3370	40.93	112.10	153.04
15.1710 091	Stainless steel pipe, seamless, '316', extra low carbon, sch 40, 4"	LF	0.4394	53.37	148.96	202.33
15.1710 101	Stainless steel pipe, seamless, '316', extra low carbon, sch 40, 6"	LF	0.5311	64.51	275.36	339.87
15.1710 111	Stainless steel pipe, seamless, '316', extra low carbon, sch 40, 8"	LF	0.5734	69.65	492.66	562.30
15.1711 000	**STAINLESS STEEL PIPE, '316', SCH 80, SEAMLESS:**					
15.1711 011	Stainless steel pipe, seamless, '316', extra low carbon, sch 80, 1/2"	LF	0.0996	12.10	23.41	35.51
15.1711 021	Stainless steel pipe, seamless, '316', extra low carbon, sch 80, 3/4"	LF	0.1233	14.98	27.29	42.26
15.1711 031	Stainless steel pipe, seamless, '316', extra low carbon, sch 80, 1"	LF	0.1484	18.02	35.85	53.87
15.1711 041	Stainless steel pipe, seamless, '316', extra low carbon, sch 80, 1-1/4"	LF	0.1604	19.48	48.11	67.59
15.1711 051	Stainless steel pipe, seamless, '316', extra low carbon, sch 80, 1-1/2"	LF	0.1767	21.46	53.79	75.26
15.1711 061	Stainless steel pipe, seamless, '316', extra low carbon, sch 80, 2"	LF	0.2648	32.16	67.84	100.01
15.1711 071	Stainless steel pipe, seamless, '316', extra low carbon, sch 80, 2-1/2"	LF	0.3435	41.72	166.44	208.16
15.1711 081	Stainless steel pipe, seamless, '316', extra low carbon, sch 80, 3"	LF	0.3836	46.59	189.32	235.91
15.1711 091	Stainless steel pipe, seamless, '316', extra low carbon, sch 80, 4"	LF	0.4939	59.99	251.77	311.76
15.1711 101	Stainless steel pipe, seamless, '316', extra low carbon, sch 80, 6"	LF	0.5500	66.80	476.75	543.56
15.1711 111	Stainless steel pipe, seamless, '316', extra low carbon, sch 80, 8"	LF	0.6070	73.73	832.89	906.61

15 - PLUMBING - MECHANICAL

Division	Description	Unit	Labor Hours	Labor Cost	Material Cost	Total Cost
15.1800 000	**FEES, PERMITS & STERILIZATION:**					
15.1800 001	Note: The fees listed below represent typical installation costs for Contra Costa County, California. It is recommended that you contact the local utility district for charges in your area. Permits: Under $2,000 .... 2.00% $2,000 to 10,000 .... 1.00% $10,000 to 50,000 .... 0.50% $50,000 to 100,000 .... 0.25%					
15.1801 000	**FEES, WATER METER (WHERE APPLICABLE):**					
15.1801 011	Water meter fee, 3/4" connection	EA			1,506.27	1,506.27
15.1801 021	Water meter fee, 1" connection	EA			2,317.87	2,317.87
15.1801 031	Water meter fee, 1-1/2" connect	EA			4,464.41	4,464.41
15.1801 041	Water meter fee, 2" connection	EA			6,916.43	6,916.43
15.1801 051	Water meter fee, 3" connection	EA			12,725.06	12,725.06
15.1801 061	Water meter fee, 4" connection	EA			20,190.10	20,190.10
15.1801 071	Water meter fee, 6" connection	EA			40,247.38	40,247.38
15.1802 000	**SEWER CONNECTION FEE, NO PLANT OR LINE CHARGE:**					
15.1802 011	Fee, sewer connection, average	FIX			346.67	346.67
15.1803 000	**STERILIZATION, TESTING & CLEANING:**					
15.1803 021	Testing & cleaning, per fixture	FIX			94.55	94.55
15.1900 000	**INDUSTRIAL PIPING:**					
15.1900 001	Note: This section deals with piping and related specialties for general industrial applications. It is presented with two different approaches - complex systems and straight run pipe. All prices include equipment time, small tools, plumber's assistant and equipment operators. Prices are complete unless otherwise noted.					
15.1901 000	**PIPE, CHROME-MOLY, SCH 120, 2-1/2" INSULATED, ALUMINUM SHEATH:**					
15.1901 001	Note: The following prices are for complex flanged prefabricated spool pieces, and include all hangers, fittings, pre-heating, stress- relieving testing and site insulation.					
15.1901 011	Pipe, chrome-moly, '120', 1", with insulation	LF	1.7817	216.41	62.03	278.44
15.1901 021	Pipe, chrome-moly, '120', 2", insulation, aluminum sheath	LF	2.2320	271.10	133.30	404.40
15.1901 031	Pipe, chrome-moly, '120', 4", insulation, aluminum sheath	LF	3.2282	392.10	222.78	614.88
15.1901 041	Pipe, chrome-moly, '120', 6", insulation, aluminum sheath	LF	3.6113	438.63	368.87	807.50
15.1901 051	Pipe, chrome-moly, '120', 10", insulation, aluminum sheath	LF	4.4447	539.85	973.50	1,513.36
15.1901 061	Pipe, chrome-moly, '120', 12", insulation, aluminum sheath	LF	6.1210	743.46	1,512.75	2,256.21
15.1902 000	**PIPE, CHROME-MOLY, SCH 80, 2-1/2" INSULATED, ALUMINUM SHEATH:**					
15.1902 001	Note: The following prices are for complex flanged prefabricated spool pieces and include all hangers, fittings, pre-heating, stress- relieving testing and site insulation.					
15.1902 011	Pipe, chrome-moly, '80', 1", insulation, aluminum sheath	LF	1.5040	182.68	23.47	206.15
15.1902 021	Pipe, chrome-moly, '80', 2", insulation, aluminum sheath	LF	1.8967	230.37	48.98	279.35
15.1902 031	Pipe, chrome-moly, '80', 3", insulation, aluminum sheath	LF	2.1074	255.96	81.93	337.89
15.1902 041	Pipe, chrome-moly, '80', 4", insulation, aluminum sheath	LF	2.7780	337.42	146.33	483.75
15.1902 051	Pipe, chrome-moly, '80', 6", insulation, aluminum sheath	LF	2.9408	357.19	231.57	588.76
15.1902 061	Pipe, chrome-moly, '80', 8", insulation, aluminum sheath	LF	3.3335	404.89	356.25	761.14
15.1902 071	Pipe, chrome-moly, '80', 10", insulation, aluminum sheath	LF	4.3489	528.22	511.12	1,039.34
15.1902 081	Pipe, chrome-moly, '80', 12", insulation, aluminum sheath	LF	5.3834	653.87	689.61	1,343.48
15.1903 000	**STEEL PIPE, A-53 GRADE-B ELECTRIC RESISTANCE WELDED, STD WALL THICKNESS:**					
15.1903 001	Note: The following prices are for straight run pipe (simply supported). Prices do not include fittings, insulation or painting.					
15.1903 011	Steel pipe, weld, A-53, '40', grade-B, electric resistance welded, 2"	LF	0.3014	36.61	5.78	42.38
15.1903 021	Steel pipe, weld, A-53, '40', grade-B, electric resistance welded, 3"	LF	0.3168	38.48	10.19	48.67
15.1903 031	Steel pipe, weld, A-53, '40', grade-B, electric resistance welded, 4"	LF	0.4584	55.68	17.43	73.11
15.1903 041	Steel pipe, weld, A-53, '40', grade-B, electric resistance welded, 6"	LF	0.5474	66.49	36.21	102.70
15.1903 051	Steel pipe, weld, A-53, '40', grade-B, electric resistance welded, 8"	LF	0.7170	87.09	41.49	128.57

MECHANICAL - PLUMBING - 15

Division	Description	Unit	Labor Hours	Labor Cost	Material Cost	Total Cost
15.1903 061	Steel pipe, weld, A-53, '40', grade-B, electric resistance welded, 10"	LF	0.8104	98.43	68.37	166.80
15.1903 071	Steel pipe, weld, A-53, .375, grade-B, electric resistance welded, 12"	LF	0.8884	107.91	70.90	178.81
15.1903 081	Steel pipe, weld, A-53, '30', grade-B, electric resistance welded, 14"	LF	1.1040	134.09	92.68	226.77
15.1903 091	Steel pipe, weld, A-53, '30', grade-B, electric resistance welded, 16"	LF	1.2081	146.74	102.05	248.79
15.1903 101	Steel pipe, weld, A-53, .375, grade-B, electric resistance welded, 18"	LF	1.2903	156.72	131.39	288.11
15.1903 111	Steel pipe, weld, A-53, '20', grade-B, electric resistance welded, 20"	LF	1.4290	173.57	161.05	334.62
15.1903 121	Steel pipe, weld, A-53, '20', grade-B, electric resistance welded, 24"	LF	1.5884	192.93	207.38	400.31
15.1904 000	**STEEL PIPE, A-53 GRADE-B ELECTRIC RESISTANCE WELDED, STD WALL THICKNESS:**					
15.1904 001	*Note: The following prices are for complex field installations and include fittings and simple supports. Prices do not include insulation or painting. Prices are based on 500 feet of 20 foot random pipe 'lengths and include 13 bends, 2 tee pieces and 11 S.O. flanges. Make allowance for the complexity of your own job when using these prices. For weld neck flanges, add 3% to the material costs.*					
15.1904 011	Steel pipe, weld, A-53, '40', grade-B, electric resistance welded, 2"	LF	0.4272	51.89	8.38	60.27
15.1904 021	Steel pipe, weld, A-53, '40', grade-B, electric resistance welded, 4"	LF	0.6160	74.82	19.00	93.82
15.1904 031	Steel pipe, weld, A-53, '40', grade-B, electric resistance welded, 6"	LF	0.8103	98.42	38.80	137.22
15.1904 041	Steel pipe, weld, A-53, '40', grade-B, electric resistance welded, 8"	LF	1.1742	142.62	47.18	189.80
15.1904 051	Steel pipe, weld, A-53, '40', grade-B, electric resistance welded, 10"	LF	1.3202	160.35	77.29	237.64
15.1904 061	Steel pipe, weld, A-53, .0375, grade-B, electric resistance welded, 12"	LF	1.5719	190.92	83.77	274.70
15.1904 071	Steel pipe, weld, A-53, '30', grade-B, electric resistance welded, 16"	LF	2.3951	290.91	125.33	416.24
15.1904 081	Steel pipe, weld, A-53, '20', grade-B, electric resistance welded, 20"	LF	3.1410	381.51	203.38	584.89
15.1905 000	**STEEL PIPE, A-53 GRADE-B, SEAMLESS:**					
15.1905 001	*Note: The following prices are for straight run field erected pipe. Prices do not include insulation or painting.*					
15.1905 011	Steel pipe, seamless, A-53, sch 40, grade-B, 2"	LF	0.3014	36.61	11.14	47.75
15.1905 021	Steel pipe, seamless, A-53, sch 40, grade-B, 3"	LF	0.3168	38.48	17.85	56.33
15.1905 031	Steel pipe, seamless, A-53, sch 40, grade-B, 4"	LF	0.4584	55.68	24.27	79.95
15.1905 041	Steel pipe, seamless, A-53, sch 40, grade-B, 6"	LF	0.5474	66.49	47.07	113.55
15.1905 051	Steel pipe, seamless, A-53, sch 40, grade-B, 8"	LF	0.7170	87.09	61.60	148.69
15.1905 061	Steel pipe, seamless, A-53, sch 40, grade-B, 10"	LF	0.8104	98.43	91.50	189.93
15.1905 071	Steel pipe, seamless, A-53, .375, grade-B, 12"	LF	0.8884	107.91	106.76	214.67
15.1905 081	Steel pipe, seamless, A-53, sch 30, grade-B, 14"	LF	1.1040	134.09	136.95	271.04
15.1905 091	Steel pipe, seamless, A-53, sch 30, grade-B, 16"	LF	1.2081	146.74	147.96	294.70
15.1905 101	Steel pipe, seamless, A-53, .375, grade-B, 18"	LF	1.2903	156.72	168.49	325.21
15.1905 111	Steel pipe, seamless, A-53, sch 20, grade-B, 20"	LF	1.4290	173.57	185.04	358.60
15.1905 121	Steel pipe, seamless, A-53, sch 20, grade-B, 24"	LF	1.5884	192.93	227.00	419.93
15.1906 000	**STEEL PIPE, A-53 GRADE-B, SEAMLESS:**					
15.1906 001	*Note: The following prices are for complex fabricated and field erected piping. Prices include all fittings and simple support. For weld neck flanges, add 3% to the material costs.*					
15.1906 011	Steel pipe, seamless, A-53, sch 40, grade-B, 2"	LF	0.4272	51.89	16.79	68.68
15.1906 021	Steel pipe, seamless, A-53, sch 40, grade-B, 3"	LF	0.5753	69.88	24.12	94.00
15.1906 031	Steel pipe, seamless, A-53, sch 40, grade-B, 4"	LF	0.6160	74.82	31.22	106.04
15.1906 041	Steel pipe, seamless, A-53, sch 40, grade-B, 6"	LF	0.8103	98.42	57.79	156.21
15.1906 051	Steel pipe, seamless, A-53, sch 40, grade-B, 8"	LF	1.1742	142.62	77.25	219.87
15.1906 061	Steel pipe, seamless, A-53, sch 40, grade-B, 10"	LF	1.3202	160.35	114.67	275.02
15.1906 071	Steel pipe, seamless, A-53, .375, grade-B, 12"	LF	1.5719	190.92	138.45	329.37
15.1906 081	Steel pipe, seamless, A-53, sch 30, grade-B, 14"	LF	1.9837	240.94	195.22	436.16
15.1906 091	Steel pipe, seamless, A-53, sch 30, grade-B, 16"	LF	2.3955	290.96	226.33	517.28
15.1906 101	Steel pipe, seamless, A-53, .375, grade-B, 18"	LF	2.7684	336.25	257.70	593.95
15.1906 111	Steel pipe, seamless, A-53, sch 20, grade-B, 20"	LF	3.1410	381.51	280.74	662.24
15.1906 121	Steel pipe, seamless, A-53, sch 20, grade-B, 24"	LF	3.5146	426.88	347.26	774.15
15.1907 000	**STAINLESS STEEL PIPE, '316', SCH 10, SEAMLESS, STRAIGHT RUNS:**					
15.1907 011	Stainless steel pipe, seamless, '316', '10', 1/2", straight run	LF	0.2410	29.27	16.35	45.63
15.1907 021	Stainless steel pipe, seamless, '316', '10', 3/4", straight run	LF	0.2410	29.27	18.86	48.14
15.1907 031	Stainless steel pipe, seamless, '316', '10', 1", straight run	LF	0.2517	30.57	29.44	60.01
15.1907 041	Stainless steel pipe, seamless, '316', '10', 1-1/4", straight run	LF	0.2574	31.26	35.74	67.01

15 - PLUMBING - MECHANICAL

Division	Description	Unit	Labor Hours	Labor Cost	Material Cost	Total Cost
15.1907 051	Stainless steel pipe, seamless, '316', '10', 1-1/2", straight run	LF	0.2626	31.90	40.19	72.09
15.1907 061	Stainless steel pipe, seamless, '316', '10', 2", straight run	LF	0.2687	32.64	46.34	78.98
15.1907 071	Stainless steel pipe, seamless, '316', '10', 2-1/2", straight run	LF	0.2790	33.89	57.59	91.47
15.1907 081	Stainless steel pipe, seamless, '316', '10', 3", straight run	LF	0.2960	35.95	65.19	101.14
15.1907 091	Stainless steel pipe, seamless, '316', '10', 4", straight run	LF	0.3282	39.86	90.88	130.74
15.1908 000	**STAINLESS STEEL PIPE, '316', SCH 10, SEAMLESS, COMPLEX SYSTEM:**					
15.1908 011	Stainless steel pipe, seamless, '316', '10', 1/2", complex	LF	0.8034	97.58	24.50	122.08
15.1908 021	Stainless steel pipe, seamless, '316', '10', 3/4", complex	LF	0.8034	97.58	28.33	125.91
15.1908 031	Stainless steel pipe, seamless, '316', '10', 1", complex	LF	0.8391	101.92	44.18	146.10
15.1908 041	Stainless steel pipe, seamless, '316', '10', 1-1/4", complex	LF	0.8580	104.21	53.71	157.93
15.1908 051	Stainless steel pipe, seamless, '316', '10', 1-1/2", complex	LF	0.8753	106.31	60.34	166.65
15.1908 061	Stainless steel pipe, seamless, '316', '10', 2", complex	LF	0.8956	108.78	69.59	178.37
15.1908 071	Stainless steel pipe, seamless, '316', '10', 2-1/2", complex	LF	0.9299	112.95	86.38	199.32
15.1908 081	Stainless steel pipe, seamless, '316', '10', 3", complex	LF	0.9867	119.84	97.82	217.67
15.1908 091	Stainless steel pipe, seamless, '316', '10', 4", complex	LF	1.0950	133.00	136.27	269.27
15.1909 000	**STAINLESS STEEL PIPE, '316', SCH 40, SEAMLESS, STRAIGHT RUNS:**					
15.1909 011	Stainless steel pipe, seamless, '316', '40', 1/2", straight run	LF	0.4969	60.35	20.48	80.83
15.1909 021	Stainless steel pipe, seamless, '316', '40', 3/4", straight run	LF	0.4969	60.35	26.77	87.12
15.1909 031	Stainless steel pipe, seamless, '316', '40', 1", straight run	LF	0.5280	64.13	34.23	98.36
15.1909 041	Stainless steel pipe, seamless, '316', '40', 1-1/4", straight run	LF	0.5438	66.05	45.51	111.56
15.1909 051	Stainless steel pipe, seamless, '316', '40', 1-1/2", straight run	LF	0.5596	67.97	51.93	119.89
15.1909 061	Stainless steel pipe, seamless, '316', '40', 2", straight run	LF	0.5931	72.04	63.36	135.40
15.1909 071	Stainless steel pipe, seamless, '316', '40', 2-1/2", straight run	LF	0.6247	75.88	90.99	166.87
15.1909 081	Stainless steel pipe, seamless, '316', '40', 3", straight run	LF	0.6869	83.43	109.67	193.10
15.1909 091	Stainless steel pipe, seamless, '316', '40', 4", straight run	LF	0.8128	98.72	156.27	254.99
15.1909 101	Stainless steel pipe, seamless, '316', '40', 6", straight run	LF	0.9646	117.16	274.71	391.87
15.1909 111	Stainless steel pipe, seamless, '316', '40', 8", straight run	LF	1.1946	145.10	470.07	615.17
15.1910 000	**STAINLESS STEEL PIPE, '316', SCH 40, SEAMLESS, COMPLEX SYSTEM:**					
15.1910 011	Stainless steel pipe, seamless, '316', '40', 1/2", complex	LF	1.6562	201.16	30.75	231.91
15.1910 021	Stainless steel pipe, seamless, '316', '40', 3/4", complex	LF	1.6562	201.16	40.19	241.35
15.1910 031	Stainless steel pipe, seamless, '316', '40', 1", complex	LF	1.7600	213.77	51.36	265.13
15.1910 041	Stainless steel pipe, seamless, '316', '40', 1-1/4", complex	LF	1.8126	220.16	68.31	288.47
15.1910 051	Stainless steel pipe, seamless, '316', '40', 1-1/2", complex	LF	1.8654	226.57	77.91	304.48
15.1910 061	Stainless steel pipe, seamless, '316', '40', 2", complex	LF	1.9771	240.14	95.03	335.17
15.1910 071	Stainless steel pipe, seamless, '316', '40', 2-1/2", complex	LF	2.0825	252.94	136.54	389.48
15.1910 081	Stainless steel pipe, seamless, '316', '40', 3", complex	LF	2.2894	278.07	164.58	442.65
15.1910 091	Stainless steel pipe, seamless, '316', '40', 4", complex	LF	2.7092	329.06	234.39	563.45
15.1910 101	Stainless steel pipe, seamless, '316', '40', 6", complex	LF	3.2150	390.49	412.12	802.61
15.1910 111	Stainless steel pipe, seamless, '316', '40', 8", complex	LF	3.9820	483.65	705.18	1,188.83
15.1911 000	**CRYOGENIC PIPING, VACUUM JACKETED:**					
15.1911 001	*Note: The following prices are for inbar schedule 5 internal pipe and 304 stainless steel external pipe. The pipe diameter shown below is for the internal pipe. The external pipe is 2" larger in diameter. For complex systems, add 25% to the material costs.*					
15.1911 011	Vacuum jacket inbar/304 stainless steel, 1"	LF	0.4109	49.91	126.47	176.38
15.1911 021	Vacuum jacket inbar/304 stainless steel, 1-1/2"	LF	0.4483	54.45	137.01	191.46
15.1911 031	Vacuum jacket inbar/304 stainless steel, 2"	LF	0.4857	58.99	141.40	200.39
15.1911 041	Vacuum jacket inbar/304 stainless steel, 2-1/2"	LF	0.5105	62.01	157.70	219.71
15.1911 051	Vacuum jacket inbar/304 stainless steel, 3"	LF	0.5479	66.55	171.31	237.85
15.1911 061	Vacuum jacket inbar/304 stainless steel, 4"	LF	0.5853	71.09	189.01	260.10
15.2000 000	**GATE, GLOBE & CHECK VALVES, CAST STEEL:**					
15.2000 001	*Note: The following prices include site handling and bolt-ups. Prices do not include insulation.*					
15.2001 000	**VALVES, CAST STEEL, CLASS 150:**					
15.2001 011	Valve, steel, 150#, flanged, 2"	EA	2.0011	243.05	609.84	852.89
15.2001 021	Valve, steel, 150#, flanged, 2-1/2"	EA	2.3177	281.51	819.92	1,101.42
15.2001 031	Valve, steel, 150#, flanged, 3"	EA	2.5100	304.86	726.14	1,031.00

MECHANICAL - PLUMBING - 15

Division	Description	Unit	Labor Hours	Labor Cost	Material Cost	Total Cost
15.2001 041	Valve, steel, 150#, flanged, 4"	EA	3.4012	413.11	1,067.29	1,480.40
15.2001 051	Valve, steel, 150#, flanged, 6"	EA	4.2000	510.13	1,673.25	2,183.38
15.2001 061	Valve, steel, 150#, flanged, 8"	EA	4.9000	595.15	2,838.45	3,433.60
15.2001 071	Valve, steel, 150#, flanged, 10"	EA	5.9000	716.61	5,291.90	6,008.51
15.2001 081	Valve, steel, 150#, flanged, 12"	EA	6.6000	801.64	7,003.03	7,804.67
15.2002 000	**VALVES, CAST STEEL, CLASS 300:**					
15.2002 001	Note: For motorized valves man-hours remain as listed for valve items. Use the adders at the end of the section to increase the material prices.					
15.2002 011	Valve, steel, 300#, flanged, 2"	EA	3.0000	364.38	841.36	1,205.74
15.2002 021	Valve, steel, 300#, flanged, 2-1/2"	EA	3.4000	412.96	1,034.62	1,447.58
15.2002 031	Valve, steel, 300#, flanged, 3"	EA	3.8000	461.55	1,105.19	1,566.74
15.2002 041	Valve, steel, 300#, flanged, 4"	EA	4.2000	510.13	1,468.99	1,979.13
15.2002 051	Valve, steel, 300#, flanged, 6"	EA	5.4000	655.88	2,422.89	3,078.77
15.2002 061	Valve, steel, 300#, flanged, 8"	EA	6.8000	825.93	4,001.21	4,827.14
15.2002 071	Valve, steel, 300#, flanged, 10"	EA	8.3000	1,008.12	8,527.57	9,535.69
15.2002 081	Valve, steel, 300#, flanged, 12"	EA	9.5000	1,153.87	11,507.60	12,661.47
15.2002 091	Add for motorized valve, 1"-3"	EA	2.0000	242.92	1,160.59	1,403.51
15.2002 101	Add for motorized valve, 4"-6"	EA	4.0000	485.84	4,642.33	5,128.17
15.2002 111	Add for motorized valve, 8"-10"	EA	6.0000	728.76	8,356.20	9,084.96
15.2003 000	**GLOBE VALVES, CRYOGENIC, VACUUM JACKETED, FLANGED:**					
15.2003 011	Globe valve, cryogenic, vacuum jacket, flange, 1"	EA	1.8928	229.90	3,347.02	3,576.92
15.2003 021	Globe valve, cryogenic, vacuum jacket, flange, 1-1/2"	EA	1.8928	229.90	4,554.73	4,784.63
15.2003 031	Globe valve, cryogenic, vacuum jacket, flange, 2"	EA	3.1630	384.18	5,417.33	5,801.51
15.2003 041	Globe valve, cryogenic, vacuum jacket, flange, 3"	EA	3.8479	467.37	13,112.17	13,579.53
15.2003 051	Globe valve, cryogenic, vacuum jacket, flange, 4"	EA	5.1180	621.63	15,872.61	16,494.24
15.2100 000	**INDUSTRIAL PIPING INSULATION:**					
15.2101 000	**PIPE INSULATION, 1-1/2" FIBERGLASS, ALUMINUM SHEATH:**					
15.2101 001	Note: The following prices are for straight run pipe. For insulation without sheath, deduct 38% from the material costs. For complex piping, add 30% to the material costs.					
15.2101 011	1-1/2" insulation with aluminum sheath, 1" pipe	LF	0.0493	5.99	6.96	12.94
15.2101 021	1-1/2" insulation with aluminum sheath, 2" pipe	LF	0.0551	6.69	8.58	15.28
15.2101 031	1-1/2" insulation with aluminum sheath, 3" pipe	LF	0.0626	7.60	9.55	17.15
15.2101 041	1-1/2" insulation with aluminum sheath, 4" pipe	LF	0.0741	9.00	11.02	20.02
15.2101 051	1-1/2" insulation with aluminum sheath, 6" pipe	LF	0.0836	10.15	13.10	23.25
15.2101 061	1-1/2" insulation with aluminum sheath, 8" pipe	LF	0.0969	11.77	16.53	28.29
15.2101 071	1-1/2" insulation with aluminum sheath, 10" pipe	LF	0.1063	12.91	19.20	32.11
15.2101 081	1-1/2" insulation with aluminum sheath, 12" pipe	LF	0.1215	14.76	22.86	37.62
15.2101 091	1-1/2" insulation with aluminum sheath, 14" pipe	LF	0.1367	16.60	25.38	41.99
15.2101 101	1-1/2" insulation with aluminum sheath, 16" pipe	LF	0.1462	17.76	30.50	48.25
15.2101 111	1-1/2" insulation with aluminum sheath, 20" pipe	LF	0.1651	20.05	34.20	54.25
15.2101 121	1-1/2" insulation with aluminum sheath, 24" pipe	LF	0.1879	22.82	41.03	63.85
15.2102 000	**PIPE INSULATION, 3" FIBERGLASS, ALUMINUM SHEATH:**					
15.2102 001	Note: For insulation without sheath, deduct 38% from the material costs. For complex piping, add 30% to the material costs.					
15.2102 011	3" fiberglass insulation, aluminum sheath, 1" pipe	LF	0.0493	5.99	15.87	21.86
15.2102 021	3" fiberglass insulation, aluminum sheath, 2" pipe	LF	0.0551	6.69	17.22	23.92
15.2102 031	3" fiberglass insulation, aluminum sheath, 3" pipe	LF	0.0626	7.60	19.70	27.30
15.2102 041	3" fiberglass insulation, aluminum sheath, 4" pipe	LF	0.0741	9.00	22.81	31.81
15.2102 051	3" fiberglass insulation, aluminum sheath, 6" pipe	LF	0.0836	10.15	27.53	37.68
15.2102 061	3" fiberglass insulation, aluminum sheath, 8" pipe	LF	0.0969	11.77	33.61	45.38
15.2102 071	3" fiberglass insulation, aluminum sheath, 10" pipe	LF	0.1063	12.91	38.78	51.69
15.2102 081	3" fiberglass insulation, aluminum sheath, 12" pipe	LF	0.1215	14.76	43.43	58.19
15.2102 091	3" fiberglass insulation, aluminum sheath, 14" pipe	LF	0.1367	16.60	48.12	64.72
15.2102 101	3" fiberglass insulation, aluminum sheath, 16" pipe	LF	0.1462	17.76	58.38	76.14
15.2102 111	3" fiberglass insulation, aluminum sheath, 20" pipe	LF	0.1651	20.05	62.62	82.67
15.2102 121	3" fiberglass insulation, aluminum sheath, 24" pipe	LF	0.1879	22.82	73.92	96.74

15 - PLUMBING - MECHANICAL

Division	Description	Unit	Labor Hours	Labor Cost	Material Cost	Total Cost
15.2103 000	**VALVE INSULATION, 3" FIBERGLASS, ALUMINUM BOX:**					
15.2103 011	3" insulation, fiberglass, aluminum box, 1"-2" valve	EA	0.0071	0.86	25.72	26.58
15.2103 021	3" insulation, fiberglass, aluminum box, 3"-4" valve	EA	1.1400	138.46	38.54	177.00
15.2103 031	3" insulation, aluminum box, 1" to 2" valve	EA	0.0071	0.86	25.72	26.58
15.2103 041	3" insulation, fiberglass, aluminum box, 6" valve	EA	1.6955	205.94	60.21	266.15
15.2103 051	3" insulation, aluminum box, 3" to 4" valve	EA	1.1400	138.46	38.54	177.00
15.2103 061	3" insulation, fiberglass, aluminum box, 8" valve	EA	2.2224	269.93	68.52	338.45
15.2103 071	3" insulation, aluminum box, 6" valve	EA	1.6955	205.94	60.21	266.15
15.2103 081	3" insulation, fiberglass, aluminum box, 10" valve	EA	2.9121	353.70	77.05	430.75
15.2103 091	3" insulation, aluminum box, 8" valve	EA	2.2224	269.93	68.52	338.45
15.2103 101	3" insulation, fiberglass, aluminum box, 12" valve	EA	3.8508	467.72	85.66	553.38
15.2103 111	3" insulation, aluminum box, 10" valve	EA	2.9121	353.70	77.05	430.75
15.2103 121	3" insulation, fiberglass, aluminum box, 14"-16" valve	EA	5.7666	700.41	159.04	859.45
15.2103 131	3" insulation, aluminum box, 12" valve	EA	3.8508	467.72	85.66	553.38
15.2103 141	3" insulation, fiberglass, aluminum box, 18" valve	EA	6.7916	824.91	173.45	998.36
15.2103 151	3" insulation, aluminum box, 14" to 16" valve	EA	5.7666	700.41	159.04	859.45
15.2103 161	3" insulation, fiberglass, aluminum box, 20" valve	EA	7.8069	948.23	192.71	1,140.94
15.2103 171	3" insulation, aluminum box, 18" valve	EA	6.7916	824.91	173.45	998.36
15.2103 181	3" insulation, fiberglass, aluminum sheath, 16" pipe	LF	0.1462	17.76	58.38	76.14
15.2103 191	3" insulation, aluminum box, 20" valve	EA	7.8069	948.23	192.71	1,140.94
15.2103 201	3" insulation, fiberglass, aluminum sheath, 20" pipe	LF	0.1651	20.05	62.62	82.67
15.2103 211	3" insulation, fiberglass, aluminum sheath, 24" pipe	LF	0.1879	22.82	73.92	96.74

Heating, Ventilation & Air Conditioning

Table of Contents

15.3000 000	**MECHANICAL WORK - HVAC**	**182**
15.3100 000	Equipment, Furnaces	182
15.3200 000	Equipment, Hot Water & Steam Boilers	183
15.3300 000	Equipment, Cooling	184
15.3400 000	Equipment, Heating & Cooling Combinations	186
15.3500 000	Auxiliary Heating & Cooling Combinations	187
15.3600 000	Air Handling Equipment, Primary	188
15.3700 000	Distribution, Terminal Equipment	190
15.3800 000	Miscellaneous Equipment	192
15.3900 000	Controls	193
15.4000 000	Duct Work, Grills & Registers	193
15.4100 000	Piping & Insulation	195
15.4200 000	Fittings	197
15.4300 000	Valves & Specialties	198
15.4400 000	Insulation, Piping	200
15.5500 000	Fire Protection Systems	201

15 - HVAC - MECHANICAL

Division	Description	Unit	Labor Hours	Labor Cost	Material Cost	Total Cost
15.3000 000	**HVAC:**					
15.3001 000	**HVAC, IN PLACE COSTS:**					
15.3001 011	HVAC, auditoriums and theaters	SF	0.1479	18.70	15.09	33.79
15.3001 021	HVAC, banks	SF	0.1123	14.20	11.72	25.92
15.3001 031	HVAC, colleges, classroom & administration	SF	0.1541	19.49	16.85	36.33
15.3001 041	HVAC, dormitories	SF	0.0616	7.79	6.50	14.29
15.3001 051	Ventilation, parking garages	SF	0.0040	0.51	0.37	0.87
15.3001 061	Heating & ventilating, tract residential	SF	0.0190	2.40	1.90	4.30
15.3001 071	HVAC, tract housing	SF	0.0339	4.29	2.97	7.25
15.3001 081	HVAC, custom housing	SF	0.0425	5.37	3.25	8.63
15.3001 091	Heating & ventilating, multi residential	SF	0.0203	2.57	2.13	4.70
15.3001 101	HVAC, multiple residence	SF	0.0316	4.00	2.84	6.84
15.3001 111	HVAC, hospitals	SF	0.2657	33.60	32.42	66.02
15.3001 121	HVAC, institutional	SF	0.1253	15.84	27.78	43.62
15.3001 131	Heating and ventilating, manufacturing	SF	0.0172	2.17	13.13	15.31
15.3001 141	HVAC, medical clinics	SF	0.1746	22.08	19.38	41.46
15.3001 151	Heating, small office building	SF	0.0660	8.35	9.03	17.38
15.3001 161	HVAC, small office building	SF	0.0949	12.00	10.07	22.07
15.3001 171	HVAC, high rise office building	SF	0.1210	15.30	18.14	33.44
15.3001 181	HVAC, schools	SF	0.1209	15.29	16.97	32.26
15.3001 191	HVAC, wet laboratories	SF	0.3986	50.40	46.30	96.70
15.3100 000	**EQUIPMENT, FURNACES:**					
15.3101 000	**FURNACES, GAS FIRED, RESIDENTIAL, WITH FLUE & VALVING:**					
15.3101 011	Wall furnace, single, 25 MBTU, manual	EA	2.5433	308.91	437.62	746.53
15.3101 021	Wall furnace, single, 35 MBTU, manual	EA	2.8134	341.72	495.96	837.68
15.3101 031	Wall furnace, single, 25 MBTU, thermostat	EA	3.4634	420.66	522.72	943.38
15.3101 041	Wall furnace, single, 35 MBTU, thermostat	EA	3.7876	460.04	581.09	1,041.13
15.3101 051	Wall furnace, dual, 25 MBTU, manual	EA	2.7593	335.14	488.68	823.83
15.3101 061	Wall furnace, dual, 35 MBTU, manual	EA	2.8676	348.30	522.72	871.02
15.3101 071	Wall furnace, dual, 25 MBTU, thermostat	EA	3.6792	446.88	573.81	1,020.68
15.3101 081	Wall furnace, dual, 35 MBTU, thermostat	EA	3.8417	466.61	593.21	1,059.82
15.3101 091	Wall furnace, dual, 50 MBTU, thermostat	EA	4.7075	571.77	746.44	1,318.22
15.3101 101	Wall furnace, dual, 60 MBTU, thermostat	EA	5.0326	611.26	826.69	1,437.95
15.3101 111	Floor furnace, 32 MBTU, thermostat	EA	4.5451	552.05	1,539.11	2,091.15
15.3101 121	Floor furnace, 45 MBTU, thermostat	EA	5.0858	617.72	1,626.36	2,244.09
15.3101 131	Floor furnace, 65 MBTU, thermostat	EA	5.6809	690.00	1,893.44	2,583.44
15.3102 000	**FURNACE, FORCED AIR, GAS FIRE, FLUE, VALVE, THERMO:**					
15.3102 001	*Note: For A/C preparation, add 12% to the material costs.*					
15.3102 011	Furnace, up flow, 50 MBTU, gas fired	EA	3.4083	413.97	758.55	1,172.52
15.3102 021	Furnace, up flow, 80 MBTU, gas fired	EA	3.6792	446.88	816.94	1,263.81
15.3102 031	Furnace, up flow, 100 MBTU, gas fired	EA	4.0584	492.93	899.55	1,392.48
15.3102 041	Furnace, up flow, 120 MBTU, gas fired	EA	4.9776	604.58	962.81	1,567.39
15.3102 051	Furnace, up flow, 150 MBTU, gas fired	EA	6.1141	742.62	1,429.66	2,172.28
15.3102 061	Furnace, horizontal flow, 80 MBTU, gas fired	EA	4.5451	552.05	946.74	1,498.79
15.3102 071	Furnace, horizontal flow, 100 MBTU, gas fired	EA	5.1400	624.30	1,039.95	1,664.25
15.3102 081	Furnace, counter flow, 80 MBTU, gas fired	EA	3.6792	446.88	1,678.34	2,125.21
15.3102 091	Furnace, counter flow, 100 MBTU, gas fired	EA	4.0584	492.93	1,735.66	2,228.59
15.3102 101	Furnace, counter flow, 120 MBTU, gas fired	EA	4.9776	604.58	1,864.83	2,469.41
15.3102 111	Furnace, counter flow, 160 MBTU, gas fired	EA	6.1141	742.62	1,950.86	2,693.48
15.3103 000	**UNIT HEATERS, GAS FIRED, WITH FLUE & VALVE:**					
15.3103 011	Unit heater, suspended, 50 MBTU, gas fired	EA	3.7876	460.04	1,095.44	1,555.48
15.3103 021	Unit heater, suspended, 75 MBTU, gas fired	EA	4.1036	498.42	1,188.22	1,686.64
15.3103 031	Unit heater, suspended, 125 MBTU, gas fired	EA	4.9597	602.41	1,423.86	2,026.27
15.3103 041	Unit heater, suspended, 175 MBTU, gas fired	EA	6.5827	799.53	1,900.17	2,699.70
15.3103 051	Unit heater, suspended, 225 MBTU, gas fired	EA	7.0792	859.84	2,045.56	2,905.40
15.3103 061	Unit heater, suspended, 300 MBTU, gas fired	EA	8.0257	974.80	2,607.09	3,581.89

MECHANICAL - HVAC - 15

Division	Description	Unit	Labor Hours	Labor Cost	Material Cost	Total Cost
15.3103 071	Unit heater, suspended, 400 MBTU, gas fired	EA	9.4236	1,144.59	3,231.24	4,375.83
15.3104 000	**DUCT HEATERS, GAS FIRED, WITH FLUE & VALVE:**					
15.3104 011	Duct heater, indoor, 100 MBTU, gas fired	EA	3.9680	481.95	1,617.61	2,099.56
15.3104 021	Duct heater, indoor, 125 MBTU, gas fired	EA	4.3398	527.11	1,743.78	2,270.90
15.3104 031	Duct heater, indoor, 175 MBTU, gas fired	EA	4.7480	576.69	2,039.03	2,615.72
15.3104 041	Duct heater, indoor, 225 MBTU, gas fired	EA	5.1605	626.79	2,412.55	3,039.35
15.3104 051	Duct heater, indoor, 250 MBTU, gas fired	EA	5.3615	651.21	2,755.68	3,406.89
15.3104 061	Duct heater, indoor, 300 MBTU, gas fired	EA	6.0732	737.65	2,944.98	3,682.63
15.3104 071	Duct heater, indoor, 350 MBTU, gas fired	EA	6.5421	794.60	3,318.50	4,113.10
15.3104 081	Duct heater, indoor, 400 MBTU, gas fired	EA	7.0110	851.56	3,626.41	4,477.97
15.3104 091	Duct heater, roof, 100 MBTU, gas fired	EA	3.7510	455.60	2,425.12	2,880.72
15.3104 101	Duct heater, roof, 125 MBTU, gas fired	EA	4.4432	539.67	2,662.35	3,202.02
15.3104 111	Duct heater, roof, 175 MBTU, gas fired	EA	5.8274	707.80	3,045.95	3,753.75
15.3104 121	Duct heater, roof, 225 MBTU, gas fired	EA	7.1003	862.40	3,351.31	4,213.71
15.3104 131	Duct heater, roof, 250 MBTU, gas fired	EA	7.6810	932.93	3,858.51	4,791.45
15.3104 141	Duct heater, roof, 300 MBTU, gas fired	EA	8.1719	992.56	4,093.20	5,085.76
15.3104 151	Duct heater, roof, 350 MBTU, gas fired	EA	9.2212	1,120.01	4,396.00	5,516.01
15.3104 161	Duct heater, roof, 400 MBTU, gas fired	EA	10.2705	1,247.45	5,021.92	6,269.38
15.3105 000	**ELECTRIC FURNACES, UNIT, DUCT, BASEBOARD HEATERS:**					
15.3105 001	*Note: The following prices include supports, relays & thermostats.*					
15.3105 011	Furnace, up flow, 68 MBTU/hr	EA	3.0661	372.41	1,220.48	1,592.89
15.3105 021	Furnace, up flow, 85 MBTU/hr	EA	3.2917	399.81	1,511.05	1,910.86
15.3105 031	Furnace, up flow, 102 MBTU/hr	EA	3.5619	432.63	1,797.51	2,230.14
15.3105 041	Unit heater, 10 MBTU/hr, 3kw, 240v	EA	1.8939	230.03	546.52	776.56
15.3105 051	Unit heater, 18 MBTU/hr, 5kw, 240v	EA	2.1194	257.42	572.18	829.60
15.3105 061	Unit heater, 34 MBTU/hr, 10kw, 240v	EA	2.8403	344.98	806.97	1,151.95
15.3105 071	Unit heater, 40 MBTU/hr, 12kw, 240v	EA	3.2917	399.81	930.86	1,330.67
15.3105 081	Duct heater, 3.4 MBTU/hr, 1kw, no enclosure	EA	0.7804	94.79	216.22	311.01
15.3105 091	Duct heater, 10 MBTU/hr, 3kw, no enclosure	EA	0.9537	115.84	273.55	389.39
15.3105 101	Duct heater, 18 MBTU/hr, 5kw, no enclosure	EA	1.0842	131.69	323.34	455.02
15.3105 111	Duct heater, 25 MBTU/hr, 75kw, no enclosure	EA	1.4883	180.77	421.84	602.60
15.3105 121	Duct heater, 34 MBTU/hr, 10kw, no enclosure	EA	1.8035	219.05	483.40	702.45
15.3105 131	Duct heater, 40 MBTU/hr, 12kw, no enclosure	EA	2.1646	262.91	601.94	864.86
15.3105 151	Baseboard heater, hot water, 1030 BTU	LF	0.1157	14.05	37.93	51.99
15.3105 161	Baseboard heater, hot water, 2 row	LF	0.1487	18.06	54.26	72.32
15.3105 171	Baseboard heater, elec, 240/160w, aluminum fin	LF	0.1039	12.62	55.12	67.74
15.3105 181	Baseboard heater, elec, 240/140w	LF	0.1039	12.62	34.41	47.03
15.3200 000	**EQUIPMENT, HOT WATER & STEAM BOILERS:**					
15.3200 001	*Note: The following prices do not include stack and breeching.* *For oil firing .... add 2% to the material costs* *For gas, oil firing ...... add 8%*					
15.3201 000	**BOILERS, GAS FIRED, CAST IRON, 15# STEAM, 30# WATER:**					
15.3201 001	*Note: The following prices include burner, fire control and trim.* *For copper water tube to 200 MBH ...... deduct 4% from material* *For copper water tube over 200 MBH .... deduct 30%* *For forced draft 335 MBH and above .... add 50% to material costs*					
15.3201 011	Boiler, cast iron, gas, 4hp, 134 MBTU/hr, 15#/30#	EA	18.3917	2,233.86	4,171.81	6,405.66
15.3201 021	Boiler, cast iron, gas, 6hp, 200 MBTU/hr, 15#/30#	EA	18.3917	2,233.86	5,084.31	7,318.17
15.3201 031	Boiler, cast iron, gas, 8hp, 268 MBTU/hr, 15#/30#	EA	19.1580	2,326.93	6,002.15	8,329.08
15.3201 041	Boiler, cast iron, gas, 10hp, 335 MBTU/hr, 15#/30#	EA	21.4570	2,606.17	8,845.77	11,451.93
15.3201 051	Boiler, cast iron, gas, 12hp, 402 MBTU/hr, 15#/30#	EA	24.5223	2,978.48	9,755.85	12,734.33
15.3201 061	Boiler, cast iron, gas, 15hp, 502 MBTU/hr, 15#/30#	EA	29.1202	3,536.94	10,661.04	14,197.98
15.3201 071	Boiler, cast iron, gas, 20hp, 670 MBTU/hr, 15#/30#	EA	31.4192	3,816.18	14,228.86	18,045.04
15.3201 081	Boiler, cast iron, gas, 25hp, 838 MBTU/hr, 15#/30#	EA	45.9792	5,584.63	16,119.91	21,704.54
15.3201 091	Boiler, cast iron, gas, 30hp, 1000 MBTU/hr, 15#/30#	EA	49.8108	6,050.02	18,013.79	24,063.81
15.3201 101	Boiler, cast iron, gas, 35hp, 1170 MBTU/hr, 15#/30#	EA	53.6424	6,515.41	21,075.42	27,590.83

15 - HVAC - MECHANICAL

Division	Description	Unit	Labor Hours	Labor Cost	Material Cost	Total Cost
15.3202 000	**BOILERS, GAS FIRED, STEEL WATER TUBE, 15# STEAM, 30# WATER:**					
15.3202 001	*Note: The following prices include burner, fire control and trim. For steel fire tube 35 HP to 100 HP … add 50% to material cost.*					
15.3202 011	Boiler, steel, gas, 10hp, 335 MBTU/hr, 15#/30#	EA	11.1117	1,349.63	14,384.11	15,733.74
15.3202 021	Boiler, steel, gas, 20hp, 670 MBTU/hr, 15#/30#	EA	11.8780	1,442.70	16,794.36	18,237.06
15.3202 031	Boiler, steel, gas, 40hp, 1340 MBTU/hr, 15#/30#	EA	13.7938	1,675.39	20,722.68	22,398.07
15.3202 041	Boiler, steel, gas, 60hp, 2010 MBTU/hr, 15#/30#	EA	16.8591	2,047.71	22,898.23	24,945.94
15.3202 051	Boiler, steel, gas, 100hp, 3350 MBTU/hr, 15/30#	EA	27.5876	3,350.79	33,840.33	37,191.12
15.3202 061	Add for forced draft	EA			6,224.90	6,224.90
15.3203 000	**BOILERS, GAS, STEEL TUBE, 15# STEAM 30# WATER, FORCED DRAFT:**					
15.3203 001	*Note: The following prices include burner, fire control and trim.*					
15.3203 011	Boiler, steel, 150hp, 5020 MBTU/hr, 15#/30#	EA	33.7181	4,095.40	77,117.07	81,212.47
15.3203 021	Boiler, steel, 200hp, 6700 MBTU/hr, 15#/30#	EA	38.3160	4,653.86	102,253.63	106,907.49
15.3203 031	Boiler, steel, 300hp, 10050 MBTU/hr, 15#/30#	EA	59.0067	7,166.95	126,749.20	133,916.15
15.3203 041	Boiler, steel, 400hp, 13400 MBTU/hr, 15#/30#	EA	60.6900	7,371.41	148,402.20	155,773.61
15.3203 051	Boiler, steel, 600hp, 20100 MBTU/hr, 15#/30#	EA	76.6320	9,307.72	199,507.76	208,815.48
15.3204 000	**BOILERS, STEEL TUBE, GAS FIRED, 150# STEAM:**					
15.3204 001	*Note: The following prices include burner, fire control and trim.*					
15.3204 011	Boiler, steel, 10hp, 335 MBTU/hr, 150# steam	EA	11.1117	1,349.63	19,591.97	20,941.60
15.3204 021	Boiler, steel, 20hp, 670 MBTU/hr, 150# steam	EA	11.8780	1,442.70	27,124.59	28,567.29
15.3204 031	Boiler, steel, 40hp, 1340 MBTU/hr, 150# steam	EA	13.7938	1,675.39	41,704.47	43,379.86
15.3204 041	Boiler, steel, 70hp, 2345 MBTU/hr, 150# steam	EA	16.8591	2,047.71	55,378.04	57,425.75
15.3204 051	Boiler, steel, 100hp, 3350 MBTU/hr, 150# steam	EA	27.5876	3,350.79	66,410.27	69,761.06
15.3204 061	Boiler, steel, 150hp, 5020 MBTU/hr, 150# steam	EA	33.7181	4,095.40	82,008.03	86,103.43
15.3204 071	Boiler, steel, 200hp, 6700 MBTU/hr, 150# steam	EA	38.3160	4,653.86	99,101.44	103,755.30
15.3204 081	Boiler, steel, 300hp, 10,050 MBTU/hr, 150# steam	EA	59.0067	7,166.95	142,262.48	149,429.43
15.3204 091	Boiler, steel, 400hp, 13,400 MBTU/hr, 150# steam	EA	65.1372	7,911.56	164,785.15	172,696.72
15.3204 101	Boiler, steel, 500hp, 16,740 MBTU/hr, 150# steam	EA	76.6320	9,307.72	191,926.08	201,233.80
15.3204 111	Boiler, steel, 600hp, 20,100 MBTU/hr, 150# steam	EA	91.9584	11,169.27	230,073.88	241,243.15
15.3205 000	**BLACK IRON FOR STACKS & BREECHING:**					
15.3205 011	Black iron, rectangular, 10 ga - 3/16"	#	0.0151	1.98	2.70	4.69
15.3205 021	Black iron, rectangular, 1/4" & over	#	0.0130	1.71	2.46	4.17
15.3205 031	Black iron, round, 10 ga - 3/16"	#	0.0151	1.98	2.61	4.60
15.3205 041	Black iron, round, 1/4" & over	#	0.0130	1.71	2.30	4.01
15.3206 000	**STACKS & BREECHING, INSULATED, 2000 DEGREE:**					
15.3206 011	Stack, 12" dia, insulated, 2,000 degree	LF	0.4241	55.74	65.77	121.51
15.3206 021	Stack, 18" dia, insulated, 2,000 degree	LF	0.6126	80.52	102.18	182.70
15.3206 031	Stack, 24" dia, insulated, 2,000 degree	LF	0.9430	123.95	156.64	280.58
15.3206 041	Stack, 30" dia, insulated, 2,000 degree	LF	1.2253	161.05	204.23	365.28
15.3206 051	Stack, 36" dia, insulated, 2,000 degree	LF	1.5548	204.36	229.91	434.28
15.3206 061	Breeching, exhaust duct, insulated	SF	0.0988	12.99	5.74	18.73
15.3300 000	**EQUIPMENT, COOLING:**					
15.3301 000	**CHILLERS, WITH STARTER & TRIM, ELECTRIC DRIVEN:**					
15.3301 001	*Note: The hot water absorption chiller prices for chillers above 300 tons include duct coil section, remote condenser, controls and accessories. For steam absorption, add 9% to the material costs.*					
15.3301 011	Chiller, reciprocating, air cool, 20 ton	EA	38.6226	4,691.10	25,176.97	29,868.07
15.3301 021	Chiller, reciprocating, air cool, 50 ton	EA	55.1750	6,701.56	52,982.63	59,684.19
15.3301 031	Chiller, reciprocating, air cool, 100 ton	EA	96.5563	11,727.73	86,667.83	98,395.56
15.3301 041	Chiller, reciprocating, air cool, 150 ton	EA	110.3501	13,403.12	116,987.01	130,390.14
15.3301 051	Chiller, reciprocating, water cool, 20 ton	EA	38.6226	4,691.10	24,846.19	29,537.29
15.3301 061	Chiller, reciprocating, water cool, 40 ton	EA	45.9792	5,584.63	40,244.39	45,829.02
15.3301 071	Chiller, reciprocating, water cool, 70 ton	EA	68.9688	8,376.95	63,164.64	71,541.59
15.3301 081	Chiller, reciprocating, water cool, 100 ton	EA	96.5563	11,727.73	73,094.13	84,821.86
15.3301 091	Chiller, reciprocating, water cool, 150 ton	EA	110.3501	13,403.12	116,687.05	130,090.17
15.3301 101	Chiller, reciprocating, water cool, 200 ton	EA	183.9168	22,338.53	147,530.83	169,869.36
15.3301 111	Chiller, centrifugal, water cool, 40 ton	EA	36.7834	4,467.71	47,374.84	51,842.55

MECHANICAL - HVAC - 15

Division	Description	Unit	Labor Hours	Labor Cost	Material Cost	Total Cost
15.3301 121	Chiller, centrifugal, water cool, 60 ton	EA	45.9792	5,584.63	61,930.68	67,515.31
15.3301 131	Chiller, centrifugal, water cool, 80 ton	EA	64.3709	7,818.49	74,948.49	82,766.98
15.3301 141	Chiller, centrifugal, water cool, 100 ton	EA	78.1646	9,493.87	85,713.71	95,207.58
15.3301 151	Chiller, centrifugal, water cool, 150 ton	EA	105.7522	12,844.66	116,007.00	128,851.66
15.3301 161	Chiller, centrifugal, water cool, 200 ton	EA	137.9376	16,753.90	149,859.03	166,612.93
15.3301 171	Chiller, centrifugal, water cool, 250 ton	EA	160.9272	19,546.22	153,872.59	173,418.81
15.3301 181	Chiller, centrifugal, water cool, 300 ton	EA	183.9168	22,338.53	184,647.55	206,986.09
15.3301 191	Chiller, centrifugal, water cool, 400 ton	EA	224.1486	27,225.09	235,987.08	263,212.17
15.3301 201	Chiller, centrifugal, water cool, 600 ton	EA	249.0540	30,250.10	322,490.19	352,740.29
15.3301 211	Chiller, centrifugal, water cool, 1000 ton	EA	348.6756	42,350.14	445,699.85	488,049.99
15.3301 221	Chiller, absorption, hot water, 50 ton	EA	55.1750	6,701.56	80,282.03	86,983.58
15.3301 231	Chiller, absorption, hot water, 100 ton	EA	78.1646	9,493.87	160,561.82	170,055.69
15.3301 241	Chiller, absorption, hot water, 150 ton	EA	105.7522	12,844.66	240,843.86	253,688.52
15.3301 251	Chiller, absorption, hot water, 200 ton	EA	137.9376	16,753.90	294,364.40	311,118.30
15.3301 261	Chiller, absorption, hot water, 250 ton	EA	160.9272	19,546.22	367,957.22	387,503.44
15.3301 271	Chiller, absorption, hot water, 300 ton	EA	183.9168	22,338.53	401,407.84	423,746.37
15.3301 281	Chiller, absorption, hot water, 350 ton	EA	193.1126	23,455.46	468,309.19	491,764.65
15.3301 291	Chiller, absorption, hot water, 400 ton	EA	206.9064	25,130.85	535,210.51	560,341.36
15.3301 301	Chiller, absorption, hot water, 650 ton	EA	243.6898	29,598.56	835,100.24	864,698.80
15.3301 311	Chiller, absorption hot water, 1000 ton	EA	321.8544	39,092.44	1,070,420.96	1,109,513.39
15.3302 000	**COOLING TOWERS:**					
15.3302 011	Cooling tower, 20 ton, compressor chiller	EA	9.1958	1,116.92	9,367.16	10,484.08
15.3302 021	Cooling tower, 40 ton, compressor chiller	EA	13.7938	1,675.39	18,732.16	20,407.55
15.3302 031	Cooling tower, 60 ton, compressor chiller	EA	18.3917	2,233.86	28,099.28	30,333.13
15.3302 041	Cooling tower, 80 ton, compressor chiller	EA	22.9896	2,792.32	32,110.59	34,902.90
15.3302 051	Cooling tower, 100 ton, compressor chiller	EA	27.5875	3,350.78	40,139.84	43,490.62
15.3302 061	Cooling tower, 150 ton, compressor chiller	EA	34.0247	4,132.64	60,209.87	64,342.51
15.3302 071	Cooling tower, 200 ton, compressor chiller	EA	40.4617	4,914.48	67,436.45	72,350.93
15.3302 081	Cooling tower, 250 ton, compressor chiller	EA	46.8989	5,696.34	84,295.55	89,991.89
15.3302 091	Cooling tower, 300 ton, compressor chiller	EA	55.1750	6,701.56	101,154.68	107,856.24
15.3302 101	Cooling tower, 400 ton, compressor chiller	EA	68.9688	8,376.95	134,014.44	142,391.39
15.3302 111	Cooling tower, 600 ton, compressor chiller	EA	91.9584	11,169.27	202,311.60	213,480.87
15.3302 121	Cooling tower, 1000 ton, compressor chiller	EA	128.7418	15,636.98	307,745.14	323,382.12
15.3302 131	Cooling tower, 1600 ton, compressor chiller	EA	165.5251	20,104.68	449,577.07	469,681.75
15.3302 141	Cooling tower, 50 ton, absorption chiller	EA	18.3917	2,233.86	18,476.69	20,710.54
15.3302 151	Cooling tower, 100 ton, absorption chiller	EA	26.6680	3,239.10	22,221.34	25,460.44
15.3302 161	Cooling tower, 150 ton, absorption chiller	EA	35.8638	4,356.02	32,578.65	36,934.67
15.3302 171	Cooling tower, 200 ton, absorption chiller	EA	43.2205	5,249.56	41,491.91	46,741.47
15.3302 181	Cooling tower, 250 ton, absorption chiller	EA	49.6576	6,031.41	45,404.78	51,436.20
15.3302 191	Cooling tower, 300 ton, absorption chiller	EA	59.7730	7,260.03	46,678.03	53,938.06
15.3302 201	Cooling tower, 350 ton, absorption chiller	EA	67.1297	8,153.57	51,345.88	59,499.45
15.3302 211	Cooling tower, 400 ton, absorption chiller	EA	71.7276	8,712.03	57,051.34	65,763.37
15.3302 221	Cooling tower, 600 ton, absorption chiller	EA	102.9935	12,509.59	83,370.46	95,880.05
15.3302 231	Cooling tower, 1000 ton, absorption chiller	EA	141.6160	17,200.68	138,792.18	155,992.86
15.3302 241	Cooling tower, 1600 ton, absorption chiller	EA	200.4694	24,349.01	210,663.18	235,012.20
15.3302 251	Cooling tower, 100 ton, closed circuit unit	EA	29.8865	3,630.01	38,898.42	42,528.44
15.3302 261	Cooling tower, 200 ton, closed circuit unit	EA	45.9792	5,584.63	73,906.98	79,491.61
15.3303 000	**AIR CONDITIONERS (D-X), PACKAGE UNIT:**					
15.3303 011	Air conditioner, 8 ton, roof mounted, D-X	EA	9.1959	1,116.93	14,702.51	15,819.44
15.3303 021	Air conditioner, 10 ton, roof mounted, D-X	EA	10.7285	1,303.08	19,199.04	20,502.13
15.3303 031	Air conditioner, 15 ton, roof mounted, D-X	EA	12.2612	1,489.25	23,324.16	24,813.41
15.3303 041	Air conditioner, 20 ton, roof mounted, D-X	EA	13.7938	1,675.39	27,014.43	28,689.82
15.3303 051	Air conditioner, 2 ton, thru-wall, D-X	EA	4.5980	558.47	2,428.07	2,986.54
15.3303 061	Air conditioner, 3 ton, thru-wall, D-X	EA	5.3643	651.55	3,495.80	4,147.35
15.3303 071	Air conditioner, 4 ton, thru-wall, D-X	EA	6.1306	744.62	4,826.92	5,571.54
15.3303 081	Air conditioner, 5 ton, thru-wall, D-X	EA	7.6632	930.77	5,964.96	6,895.74
15.3304 000	**ELECTRIC HEATERS, FOR USE WITH ABOVE D-X UNITS:**					

15 - HVAC - MECHANICAL

Division	Description	Unit	Labor Hours	Labor Cost	Material Cost	Total Cost
15.3304 011	Electric heater, 9.6 kw, 327 MBTU/hr	EA	0.7664	93.09	607.89	700.98
15.3304 021	Electric heater, 19.2 kw, 655 MBTU/hr	EA	0.8430	102.39	1,021.49	1,123.88
15.3304 031	Electric heater, 28.8 kw, 983 MBTU/hr	EA	1.0346	125.66	1,246.91	1,372.57
15.3305 000	**SPLIT SYSTEM, FOR USE WITH FORCED AIR FURNACE:**					
15.3305 011	Air conditioner, residential, 2 ton, split system	EA	3.4485	418.85	3,368.53	3,787.39
15.3305 021	Air conditioner, residential, 3 ton, split system	EA	3.6401	442.13	4,267.12	4,709.25
15.3305 031	Air conditioner, residential, 4 ton, split system	EA	5.3643	651.55	5,647.27	6,298.81
15.3305 041	Air conditioner, residential, 5 ton, split system	EA	6.5138	791.17	6,570.12	7,361.28
15.3305 101	Air conditioner, commercial, standard, (Title-24), 3 ton, split system	EA	5.4400	660.74	4,912.74	5,573.48
15.3305 111	Air conditioner, commercial, standard, (Title-24), 4 ton, split system	EA	6.8600	833.22	5,452.92	6,286.14
15.3305 121	Air conditioner, commercial, standard, (Title-24), 5 ton, split system	EA	8.0100	972.89	6,095.92	7,068.81
15.3305 131	Air conditioner, commercial, standard, (Title-24), 6 ton, split system	EA	8.8800	1,078.56	7,063.05	8,141.61
15.3305 141	Air conditioner, commercial, standard, (Title-24), 7-1/2 ton, split system	EA	9.9600	1,209.74	9,089.87	10,299.61
15.3305 151	Air conditioner, commercial, standard, (Title-24), 8-1/2 ton, split system	EA	11.2400	1,365.21	10,982.99	12,348.20
15.3305 161	Air conditioner, commercial, standard, (Title-24), 10 ton, split system	EA	12.8400	1,559.55	11,644.03	13,203.58
15.3305 171	Air conditioner, commercial, standard, (Title-24), 12-1/2 ton, split system	EA	14.0200	1,702.87	13,261.86	14,964.73
15.3305 181	Air conditioner, commercial, standard, (Title-24), 20 ton, split system	EA	12.6443	1,535.78	19,629.14	21,164.92
15.3305 191	Air conditioner, commercial, standard, (Title-24), 30 ton, split system	EA	24.5223	2,978.48	29,488.10	32,466.57
15.3305 193	Air conditioner, commercial, standard, (Title-24), 50 ton, split system	EA	29.8865	3,630.01	42,504.87	46,134.88
15.3305 195	Air conditioner, commercial, standard, (Title-24), 70 ton, split system	EA	38.3160	4,653.86	52,893.31	57,547.18
15.3305 201	Air conditioner, commercial, high efficiency, 3 ton, split system	EA	5.4400	660.74	7,253.41	7,914.16
15.3305 211	Air conditioner, commercial, high efficiency, 4 ton, split system	EA	6.8600	833.22	8,590.86	9,424.07
15.3305 221	Air conditioner, commercial, high efficiency, 5 ton, split system	EA	8.0100	972.89	9,311.09	10,283.99
15.3305 231	Air conditioner, commercial, high efficiency, 6 ton, split system	EA	8.8800	1,078.56	10,257.63	11,336.19
15.3305 241	Air conditioner, commercial, high efficiency, 7-1/2 ton, split system	EA	9.9600	1,209.74	12,546.83	13,756.57
15.3305 251	Air conditioner, commercial, high efficiency, 8-1/2 ton, split system	EA	11.2400	1,365.21	14,120.97	15,486.18
15.3305 261	Air conditioner, commercial, high efficiency, 10 ton, split system	EA	12.8400	1,559.55	15,822.39	17,381.94
15.3305 271	Air conditioner, commercial, high efficiency, 12-1/2 ton, split system	EA	14.0200	1,702.87	17,523.90	19,226.77
15.3305 401	Economizer, add to above, 3 ton to 6 ton unit	EA			1,944.55	1,944.55
15.3305 411	Economizer, add to above, 7-1/2 ton to 12-1/2 ton unit	EA			2,722.34	2,722.34
15.3306 000	**COMPUTER ROOM AIR CONDITIONING UNITS:**					
15.3306 011	Air cooled, computer room, air conditioning unit, 5 ton	EA	24.5223	2,978.48	19,029.74	22,008.22
15.3306 021	Air cooled, computer room, air conditioning unit, 8 ton	EA	26.0549	3,164.63	27,554.74	30,719.37
15.3306 031	Air cooled, computer room, air conditioning unit, 10 ton	EA	49.0445	5,956.94	34,812.29	40,769.23
15.3306 041	Air cooled, computer room, air conditioning unit, 15 ton	EA	68.9688	8,376.95	44,802.86	53,179.81
15.3306 051	Air cooled, computer room, air conditioning unit, 20 ton	EA	73.5668	8,935.42	56,743.51	65,678.93
15.3306 061	Water cooled, computer room, air conditioning unit, 5 ton	EA	22.9896	2,792.32	14,068.66	16,860.98
15.3306 071	Water cooled, computer room, air conditioning unit, 8 ton	EA	24.5223	2,978.48	20,614.78	23,593.26
15.3306 081	Water cooled, computer room, air conditioning unit, 10 ton	EA	45.9792	5,584.63	25,412.92	30,997.55
15.3306 091	Water cooled, computer room, air conditioning unit, 15 ton	EA	65.9036	8,004.65	34,998.40	43,003.05
15.3306 101	Water cooled, computer room, air conditioning unit, 20 ton	EA	70.5015	8,563.11	41,939.63	50,502.74
15.3400 000	**EQUIPMENT, HEATING & COOLING COMBINATIONS:**					
15.3401 000	**GAS HEAT AND D-X COOLING:**					
15.3401 001	Note: The following items are roof mounted, single zone, with disposable filters, isolation dampers, curbs, controls and valving. For multi-zoned units, 2nd zone . . . add 8% to material costs For 3rd to 8th zones . . . . add 4% to material costs					
15.3401 011	Air conditioner, 2 ton, 80 MBTU/hr	EA	9.9622	1,210.01	5,205.92	6,415.93
15.3401 021	Air conditioner, 3 ton, 100 MBTU/hr	EA	10.7285	1,303.08	7,656.50	8,959.59
15.3401 031	Air conditioner, 4 ton, 140 MBTU/hr	EA	11.4948	1,396.16	10,258.78	11,654.93
15.3401 041	Air conditioner, 5 ton, 140 MBTU/hr	EA	13.0275	1,582.32	10,932.48	12,514.80
15.3401 051	Air conditioner, 8 ton, 200 MBTU/hr	EA	15.3264	1,861.54	17,516.48	19,378.02
15.3401 061	Air conditioner, 10 ton, 260 MBTU/hr	EA	19.9244	2,420.02	21,130.08	23,550.10
15.3401 071	Air conditioner, 15 ton, 300 MBTU/hr	EA	22.9896	2,792.32	27,561.19	30,353.51
15.3401 081	Air conditioner, 20 ton, 400 MBTU/hr	EA	30.6528	3,723.09	48,999.10	52,722.19
15.3401 091	Air conditioner, 30 ton, 550 MBTU/hr	EA	45.9792	5,584.63	68,902.60	74,487.23
15.3401 101	Air conditioner, 40 ton, 760 MBTU/hr	EA	65.1372	7,911.56	88,807.75	96,719.31

MECHANICAL - HVAC - 15

Division	Description	Unit	Labor Hours	Labor Cost	Material Cost	Total Cost
15.3401 111	Air conditioner, 60 ton, 1000 MBTU/hr	EA	84.2952	10,238.49	125,555.43	135,793.93
15.3402 000	**HEAT PUMPS:**					
15.3402 011	Heat pump, 2 ton, 26,000 BTU, thru-wall	EA	6.1306	744.62	2,376.57	3,121.19
15.3402 021	Heat pump, 3 ton, 37,000 BTU, thru-wall	EA	7.6632	930.77	3,521.33	4,452.10
15.3402 031	Heat pump, 4 ton, 52,000 BTU, thru-wall	EA	9.1959	1,116.93	4,840.25	5,957.19
15.3402 041	Heat pump, 5 ton, 61,000 BTU, thru-wall	EA	10.7285	1,303.08	5,826.22	7,129.30
15.3402 051	Heat pump, 8 ton, 90,000 BTU, roof, duct	EA	15.3264	1,861.54	14,147.94	16,009.49
15.3402 061	Heat pump, 10 ton, 120,000 BTU, roof, duct	EA	19.9244	2,420.02	17,763.32	20,183.33
15.3402 071	Heat pump, 15 ton, 180,000 BTU, roof, duct	EA	22.9896	2,792.32	26,029.75	28,822.07
15.3402 081	Heat pump, 20 ton, 240,000 BTU, roof, duct	EA	30.6528	3,723.09	33,685.59	37,408.67
15.3402 091	Heat pump, 1 ton, 13,000 BTU, plenum	EA	4.5980	558.47	3,069.17	3,627.64
15.3402 101	Heat pump, 2 ton, 26,000 BTU, plenum	EA	5.3643	651.55	3,981.02	4,632.57
15.3402 111	Heat pump, 3 ton, 37,000 BTU, plenum	EA	5.7474	698.08	5,206.09	5,904.16
15.3402 121	Heat pump, 4 ton, 52,000 BTU, plenum	EA	6.1306	744.62	6,737.40	7,482.02
15.3403 000	**D-X COOLING & ELECTRIC HEAT AIR CONDITIONER:**					
15.3403 001	*Note: The following items are packaged units, electric heat, including controls and mounting. Skid-mounted, pre-piped, and prewired.*					
15.3403 011	Air conditioner, 8 ton D-X, 58 MBTU/hr electric	EA	7.6632	930.77	18,260.71	19,191.48
15.3403 021	Air conditioner, 10 ton D-X, 50 MBTU/hr electric	EA	9.1959	1,116.93	24,161.72	25,278.66
15.3404 000	**PACKAGED HYDRONIC HEATING & COOLING UNIT, BOILER CLOSED:**					
15.3404 001	*Note: The following items include boiler, closed circuit evaporative cooler, pumps, controls, and are skid mounted, pre-piped and pre-wired.*					
15.3404 011	Hydronic unit, 50 ton, 300 MBTU/hr	EA	16.0000	1,943.36	40,074.02	42,017.38
15.3404 021	Hydronic unit, 100 ton, 600 MBTU/hr	EA	24.0000	2,915.04	71,516.74	74,431.78
15.3404 031	Hydronic unit, 150 ton, 900 MBTU/hr	EA	30.0000	3,643.80	91,247.25	94,891.05
15.3404 041	Hydronic unit, 200 ton, 1200 MBTU/hr	EA	32.0000	3,886.72	110,974.07	114,860.79
15.3405 000	**INFRA-RED HEATERS, GAS FIRED:**					
15.3405 011	Infra-red heater, 30,000 BTU, gas fired	EA	2.2442	272.58	1,212.40	1,484.98
15.3405 021	Infra-red heater, 45,000 BTU, gas fired	EA	2.6975	327.64	1,923.73	2,251.37
15.3405 031	Infra-red heater, 60,000 BTU, gas fired	EA	3.3718	409.54	2,187.56	2,597.09
15.3405 041	Infra-red heater, 80,000 BTU, gas fired	EA	3.9908	484.72	2,346.48	2,831.21
15.3406 000	**RADIANT HEAT PANELS, CEILING:**					
15.3406 011	Radiant heat panel, 24x24, 357 w	EA	0.7664	93.09	383.00	476.08
15.3406 021	Radiant heat panel, 24x48, 500 w	EA	0.7664	93.09	427.12	520.21
15.3500 000	**AUXILIARY HEATING & COOLING EQUIPMENT:**					
15.3501 000	**HUMIDIFIERS, STEAM OR ELECTRIC:**					
15.3501 011	Humidifier, steam, 50#/hr	EA	6.1306	744.62	1,064.52	1,809.14
15.3501 021	Humidifier, steam, 100#/hr	EA	9.1959	1,116.93	1,580.59	2,697.53
15.3501 031	Humidifier, steam, 150#/hr	EA	10.7285	1,303.08	1,881.64	3,184.72
15.3501 041	Humidifier, steam, 300#/hr	EA	14.3685	1,745.20	2,913.88	4,659.07
15.3501 051	Humidifier, electric, 50#/hr	EA	4.5980	558.47	5,270.78	5,829.25
15.3501 061	Humidifier, electric, 100#/hr	EA	6.8969	837.70	7,212.69	8,050.38
15.3501 071	Humidifier, electric, 150#/hr	EA	9.1959	1,116.93	9,154.52	10,271.46
15.3502 000	**PUMPS, CONDENSATE, DUPLEX:**					
15.3502 011	Pump, condensate, duplex, 15 GPM	EA	9.1959	1,116.93	5,036.34	6,153.27
15.3502 021	Pump, condensate, duplex, 25 GPM	EA	10.7285	1,303.08	6,176.06	7,479.14
15.3502 031	Pump, condensate, duplex, 40 GPM	EA	15.3264	1,861.54	7,670.08	9,531.63
15.3502 041	Pump, condensate, duplex, 50 GPM	EA	16.8591	2,047.71	8,409.35	10,457.06
15.3502 051	Pump, condensate, duplex, 60 GPM	EA	19.9244	2,420.02	8,686.62	11,106.64
15.3502 061	Pump, condensate, duplex, 75 GPM	EA	22.9896	2,792.32	8,994.61	11,786.93
15.3502 071	Pump, condensate, duplex, 100 GPM	EA	26.0549	3,164.63	11,397.33	14,561.96
15.3502 081	Pump, condensate, duplex, 120 GPM	EA	28.3539	3,443.86	11,745.24	15,189.10
15.3503 000	**PUMPS, WATER SERVICE:**					
15.3503 001	*Note: For 100 foot head, add 30% to the material costs. For 150 foot head, add 40%. For 200 foot head, add 70%.*					
15.3503 011	Pump, 7-1/2 GPM, 50' head	EA	3.0653	372.31	1,215.31	1,587.62
15.3503 021	Pump, 20 GPM, 50' head	EA	3.4485	418.85	1,416.98	1,835.83

15 - HVAC - MECHANICAL

Division	Description	Unit	Labor Hours	Labor Cost	Material Cost	Total Cost
15.3503 031	Pump, 30 GPM, 50' head	EA	4.5980	558.47	1,925.23	2,483.71
15.3503 041	Pump, 45 GPM, 50' head	EA	5.3643	651.55	1,971.44	2,622.99
15.3503 051	Pump, 60 GPM, 50' head	EA	7.6632	930.77	2,016.24	2,947.01
15.3503 061	Pump, 90 GPM, 50' head	EA	8.4296	1,023.86	2,110.11	3,133.97
15.3503 071	Pump, 120 GPM, 50' head	EA	9.9622	1,210.01	2,341.04	3,551.05
15.3503 081	Pump, 150 GPM, 50' head	EA	10.7285	1,303.08	2,922.92	4,226.00
15.3503 091	Pump, 200 GPM, 50' head	EA	13.0275	1,582.32	3,409.55	4,991.87
15.3503 101	Pump, 250 GPM, 50' head	EA	14.5601	1,768.47	4,261.92	6,030.39
15.3503 111	Pump, 300 GPM, 50' head	EA	17.6254	2,140.78	4,601.01	6,741.79
15.3503 121	Pump, 400 GPM, 50' head	EA	19.1580	2,326.93	4,995.25	7,322.18
15.3503 131	Pump, 600 GPM, 50' head	EA	21.4570	2,606.17	6,372.76	8,978.93
15.3503 141	Pump, 800 GPM, 50' head	EA	24.5223	2,978.48	6,572.71	9,551.18
15.3503 151	Pump, 1000 GPM, 50' head	EA	27.5876	3,350.79	6,818.07	10,168.86
15.3503 161	Pump, 1500 GPM, 50' head	EA	28.7370	3,490.40	7,892.80	11,383.19
15.3503 171	Pump, 2000 GPM, 50' head	EA	33.7181	4,095.40	15,232.38	19,327.78
15.3503 181	Pump, 2500 GPM, 50' head	EA	35.2508	4,281.56	15,515.74	19,797.31
15.3503 191	Pump, 3000 GPM, 50' head	EA	37.5497	4,560.79	16,055.45	20,616.23
15.3503 201	Pump, 3500 GPM, 50' head	EA	39.8487	4,840.02	20,076.25	24,916.27
15.3503 211	Pump, 4000 GPM, 50' head	EA	42.1476	5,119.25	20,375.81	25,495.06
15.3504 000	**PUMPS, FUEL OIL:**					
15.3504 001	Note: The following prices are based on #2 oil at 100 pounds pressure, with explosion-proof motor.					
15.3504 011	Pump, fuel oil, 1 GPM, 1/4 hp	EA	4.9811	605.00	1,205.80	1,810.80
15.3504 021	Pump, fuel oil, 5 GPM, 1/4 hp	EA	7.6632	930.77	1,394.85	2,325.62
15.3504 031	Pump, fuel oil, 10 GPM, 1/4 hp	EA	8.4296	1,023.86	1,544.29	2,568.15
15.3504 041	Pump, fuel oil, 20 GPM, 1/4 hp	EA	9.9622	1,210.01	1,721.87	2,931.88
15.3504 051	Pump, fuel oil, 50 GPM, 5 hp	EA	15.3264	1,861.54	3,934.27	5,795.82
15.3600 000	**AIR HANDLING EQUIPMENT, PRIMARY:**					
15.3600 001	Note: The following prices include attachments and accessories.					
15.3601 000	**AIR HANDLERS, CENTRAL STATION:**					
15.3601 001	Note: The following items have 2 rows heating, 6 rows cooling coils, with disposable filter and isolation package. For multi-zone use, add 5% to the material costs For exterior weather tight, add 15%.					
15.3601 011	Air handler, 1600 CFM, 3 ton	EA	6.1306	805.81	6,798.36	7,604.16
15.3601 021	Air handler, 2000 CFM, 5 ton	EA	7.2801	956.90	7,641.64	8,598.53
15.3601 031	Air handler, 2500 CFM, 8 ton	EA	9.5790	1,259.06	9,552.12	10,811.19
15.3601 041	Air handler, 4000 CFM, 10 ton	EA	10.3454	1,359.80	16,542.07	17,901.87
15.3601 051	Air handler, 5200 CFM, 12 ton	EA	12.2612	1,611.61	21,504.69	23,116.31
15.3601 061	Air handler, 6250 CFM, 15 ton	EA	14.5601	1,913.78	25,846.97	27,760.75
15.3601 071	Air handler, 7500 CFM, 20 ton	EA	16.8591	2,215.96	31,016.38	33,232.34
15.3601 081	Air handler, 10,000 CFM, 25 ton	EA	19.9244	2,618.86	40,697.89	43,316.75
15.3601 091	Air handler, 15,000 CFM, 30 ton	EA	22.9896	3,021.75	66,117.23	69,138.98
15.3601 101	Air handler, 18,000 CFM, 35 ton	EA	30.6528	4,029.00	73,963.92	77,992.92
15.3601 111	Air handler, 22,000 CFM, 50 ton	EA	38.3160	5,036.26	90,175.65	95,211.91
15.3601 121	Air handler, 28,000 CFM, 75 ton	EA	49.8108	6,547.13	114,769.73	121,316.86
15.3601 131	Air handler, 32,000 CFM, 100 ton	EA	57.4740	7,554.38	131,091.51	138,645.89
15.3602 000	**AIR HANDLERS, CENTRAL STATION:**					
15.3602 001	Note: The following items are single zone, without coils. Prices include disposable filters and isolation package. See following sections for components. For multi-zone use, add 3% to the material.					
15.3602 011	Air handler, 1600 CFM, 3 ton	EA	4.5980	604.36	4,519.28	5,123.64
15.3602 021	Air handler, 2000 CFM, 5 ton	EA	5.3643	705.08	5,135.62	5,840.70
15.3602 031	Air handler, 2500 CFM, 8 ton	EA	6.1306	805.81	5,648.22	6,454.02
15.3602 041	Air handler, 4000 CFM, 10 ton	EA	6.8969	906.53	9,658.51	10,565.03
15.3602 051	Air handler, 5200 CFM, 12 ton	EA	7.6632	1,007.25	11,186.12	12,193.37
15.3602 061	Air handler, 6250 CFM, 15 ton	EA	9.1959	1,208.71	13,218.97	14,427.68
15.3602 071	Air handler, 7500 CFM, 20 ton	EA	12.2612	1,611.61	17,795.76	19,407.37

MECHANICAL - HVAC - 15

Division	Description	Unit	Labor Hours	Labor Cost	Material Cost	Total Cost
15.3602 081	Air handler, 10,000 CFM, 25 ton	EA	14.5601	1,913.78	21,351.33	23,265.11
15.3602 091	Air handler, 15,000 CFM, 30 ton	EA	16.8591	2,215.96	28,470.58	30,686.54
15.3602 101	Air handler, 18,000 CFM, 35 ton	EA	24.5223	3,223.21	40,668.30	43,891.51
15.3602 111	Air handler, 22,000 CFM, 50 ton	EA	30.6528	4,029.00	50,837.24	54,866.24
15.3602 121	Air handler, 28,000 CFM, 75 ton	EA	36.7834	4,834.81	66,086.50	70,921.31
15.3602 131	Air handler, 32,000 CFM, 100 ton	EA	42.9140	5,640.62	81,338.80	86,979.42
15.3603 000	**COILS, CHILLED WATER, 6 ROW, VALVE, TRAP & DRAIN PAN:**					
15.3603 001	*Note: The following items are for use with the air handlers listed above. Costs for exterior weather tight, add 15%.*					
15.3603 011	Coils, chilled water, 44 SF, 22,000 CFM	EA	13.4873	1,638.17	15,434.49	17,072.66
15.3603 021	Coils, chilled water, 48 SF, 24,000 CFM	EA	14.7134	1,787.09	16,536.56	18,323.64
15.3603 031	Coils, chilled water, 56 SF, 28,000 CFM	EA	17.1656	2,084.93	19,908.86	21,993.79
15.3603 041	Coils, chilled water, 64 SF, 32,000 CFM	EA	18.3917	2,233.86	22,114.10	24,347.95
15.3603 051	Use 1 SF of coil at each 500 CFM	SF	0.3066	37.24	340.08	377.32
15.3604 000	**COILS, HOT WATER, 2 ROW, VALVE, TRAP & DRAIN PAN:**					
15.3604 001	*Note: The following items are for use with the air handlers listed above.*					
15.3604 011	Coils, hot water, 4 SF, 4000 CFM	EA	1.2262	148.93	1,053.77	1,202.70
15.3604 021	Coils, hot water, 6 SF, 6000 CFM	EA	1.8392	223.39	1,347.73	1,571.12
15.3604 031	Coils, hot water, 8 SF, 8000 CFM	EA	2.4523	297.86	1,623.34	1,921.20
15.3604 041	Coils, hot water, 10 SF, 10,000 CFM	EA	3.0653	372.31	1,843.87	2,216.18
15.3604 051	Coils, hot water, 12 SF, 12,000 CFM	EA	3.6784	446.78	2,101.15	2,547.92
15.3604 061	Coils, hot water, 14 SF, 14,000 CFM	EA	4.2914	521.23	2,315.12	2,836.35
15.3604 071	Coils, hot water, 16 SF, 16,000 CFM	EA	4.9045	595.70	2,548.60	3,144.30
15.3604 081	Coils, hot water, 18 SF, 18,000 CFM	EA	5.5176	670.17	2,719.28	3,389.45
15.3604 091	Coils, hot water, 20 SF, 20,000 CFM	EA	6.1306	744.62	2,928.74	3,673.36
15.3604 101	Coils, hot water, 22 SF, 22,000 CFM	EA	6.7437	819.09	3,029.69	3,848.78
15.3604 111	Coils, hot water, 24 SF, 24,000 CFM	EA	7.3567	893.54	3,270.57	4,164.11
15.3604 121	Coils, hot water, 28 SF, 28,000 CFM	EA	8.5828	1,042.47	4,401.22	5,443.68
15.3604 131	Coils, hot water, 32 SF, 32,000 CFM	EA	9.8089	1,191.39	5,071.20	6,262.59
15.3605 000	**FILTERS & FRAMES, 32,000 CFM:**					
15.3605 011	Filter & frame, throw away	M/CFM	0.0613	8.06	18.27	26.32
15.3605 021	High pressure filter frame, up to 90%, roll & bag	M/CFM	1.3028	171.24	574.30	745.54
15.3605 031	Active carbon, filter unit, complete	M/CFM	1.6860	221.61	907.14	1,128.75
15.3605 041	High efficiency ceiling lay-in	EA	0.9196	120.87	1,005.55	1,126.42
15.3606 000	**FANS, SUPPLY, LOW PRESSURE, 1-1/2" UTILITY SET:**					
15.3606 001	*Note: The following items include vibration mounts.*					
15.3606 011	Fan, low pressure, 600 CFM, 1-1/2" utility	EA	2.8538	375.10	2,067.09	2,442.19
15.3606 021	Fan, low pressure, 1000 CFM, 1-1/2" utility	EA	4.1290	542.72	2,602.97	3,145.69
15.3606 031	Fan, low pressure, 2000 CFM, 1-1/2" utility	EA	5.8296	766.24	3,214.17	3,980.41
15.3606 041	Fan, low pressure, 4000 CFM, 1-1/2" utility	EA	6.6801	878.03	3,350.15	4,228.18
15.3606 051	Fan, low pressure, 6000 CFM, 1-1/2" utility	EA	9.6514	1,268.58	4,440.26	5,708.84
15.3606 061	Fan, low pressure, 10,000 CFM, 1-1/2" utility	EA	10.6050	1,393.92	5,204.93	6,598.85
15.3606 071	Fan, low pressure, 16,000 CFM, 1-1/2" utility	EA	14.0011	1,840.30	6,737.14	8,577.44
15.3606 081	Fan, low pressure, 20,000 CFM, 1-1/2" utility	EA	16.2058	2,130.09	8,880.88	11,010.97
15.3606 091	Fan, low pressure, 30,000 CFM, 1-1/2" utility	EA	19.8987	2,615.49	11,025.33	13,640.82
15.3606 101	Fan, low pressure, 40,000 CFM, 1-1/2" utility	EA	26.7504	3,516.07	15,158.17	18,674.24
15.3606 111	Fan, low pressure, 60,000 CFM, 1-1/2" utility	EA	36.1044	4,745.56	25,417.29	30,162.85
15.3606 121	Fan, low pressure, 80,000 CFM, 1-1/2" utility	EA	46.0542	6,053.36	30,065.45	36,118.81
15.3607 000	**FANS, SUPPLY, HIGH PRESSURE, 3-1/2" UTILITY SET:**					
15.3607 001	*Note: The following items include vibration mounts.*					
15.3607 011	Fan, high pressure, 2000 CFM, 3-1/2" utility	EA	5.8984	775.29	3,712.89	4,488.17
15.3607 021	Fan, high pressure, 4000 CFM, 3-1/2" utility	EA	6.8510	900.50	5,062.26	5,962.76
15.3607 031	Fan, high pressure, 10,000 CFM, 3-1/2" utility	EA	11.2009	1,472.25	7,467.01	8,939.25
15.3607 041	Fan, high pressure, 16,000 CFM, 3-1/2" utility	EA	14.9539	1,965.54	9,370.20	11,335.74
15.3607 051	Fan, high pressure, 20,000 CFM, 3-1/2" utility	EA	17.4569	2,294.53	11,631.16	13,925.70
15.3607 061	Fan, high pressure, 30,000 CFM, 3-1/2" utility	EA	21.0310	2,764.31	13,177.89	15,942.20
15.3607 071	Fan, high pressure, 40,000 CFM, 3-1/2" utility	EA	28.0023	3,680.62	18,935.79	22,616.41

15 - HVAC - MECHANICAL

Division	Description	Unit	Labor Hours	Labor Cost	Material Cost	Total Cost
15.3607 081	Fan, high pressure, 60,000 CFM, 3-1/2" utility	EA	37.3555	4,910.01	31,698.53	36,608.53
15.3607 091	Fan, high pressure, 80,000 CFM, 3-1/2" utility	EA	47.3052	6,217.80	46,681.19	52,898.99
15.3700 000	**DISTRIBUTION, TERMINAL EQUIPMENT:**					
15.3701 000	**FAN, EXHAUST, CEILING AND WALL, RESIDENTIAL:**					
15.3701 011	Exhaust fan, wall, 60 CFM	EA	0.6457	84.87	96.61	181.48
15.3701 021	Exhaust fan, wall, 100 CFM	EA	0.6457	84.87	113.52	198.39
15.3701 031	Exhaust fan, wall, 200 CFM	EA	1.3406	176.21	123.12	299.32
15.3701 041	Exhaust fan, wall, 300 CFM	EA	1.7870	234.88	193.22	428.10
15.3702 000	**FAN, EXHAUST, ROOF MOUNTED, BELT DRIVE:**					
15.3702 001	*Note: The following prices include curb, hood and bird screen.*					
15.3702 011	Exhaust fan, roof, 600 CFM, 1/2" static pressure	EA	3.5253	463.37	1,376.06	1,839.43
15.3702 021	Exhaust fan, roof, 1600 CFM, 1/2" static pressure	EA	3.8728	509.04	1,525.76	2,034.80
15.3702 031	Exhaust fan, roof, 2000 CFM, 1/2" static pressure	EA	6.2064	815.77	2,328.06	3,143.83
15.3702 041	Exhaust fan, roof, 4000 CFM, 1/2" static pressure	EA	8.8375	1,161.60	3,420.37	4,581.97
15.3702 051	Exhaust fan, roof, 6000 CFM, 1/2" static pressure	EA	14.5475	1,912.12	5,847.49	7,759.61
15.3702 061	Exhaust fan, roof, 10,000 CFM, 1/2" static pressure	EA	15.5898	2,049.12	6,165.56	8,214.68
15.3702 071	Exhaust fan, roof, 16,000 CFM, 1/2" static pressure	EA	17.1289	2,251.42	6,564.43	8,815.85
15.3703 000	**FAN, EXHAUST, UTILITY SET, WITH VIBRATION MOUNTS:**					
15.3703 011	Exhaust fan, 600 CFM, 3/4" static pressure, vibration mounts	EA	2.0852	274.08	1,053.22	1,327.30
15.3703 021	Exhaust fan, 1000 CFM, 3/4" static pressure, vibration mounts	EA	3.1276	411.09	1,540.64	1,951.73
15.3703 031	Exhaust fan, 2000 CFM, 3/4" static pressure, vibration mounts	EA	4.4188	580.81	2,231.59	2,812.39
15.3703 041	Exhaust fan, 4000 CFM, 3/4" static pressure, vibration mounts	EA	4.9153	646.07	2,429.71	3,075.78
15.3703 051	Exhaust fan, 10,000 CFM, 3/4" static pressure, vibration mounts	EA	8.2915	1,089.83	4,134.71	5,224.55
15.3703 061	Exhaust fan, 16,000 CFM, 3/4" static pressure, vibration mounts	EA	8.8375	1,161.60	4,460.59	5,622.19
15.3703 071	Exhaust fan, 20,000 CFM, 3/4" static pressure, vibration mounts	EA	10.8731	1,429.16	5,594.64	7,023.80
15.3703 081	Exhaust fan, 30,000 CFM, 3/4" static pressure, vibration mounts	EA	15.5898	2,049.12	8,514.51	10,563.63
15.3703 091	Exhaust fan, 40,000 CFM, 3/4" static pressure, vibration mounts	EA	20.7530	2,727.77	11,350.89	14,078.67
15.3703 101	Exhaust fan, 60,000 CFM, 3/4" static pressure, vibration mounts	EA	28.5480	3,752.35	17,026.45	20,778.80
15.3703 111	Exhaust fan, 80,000 CFM, 3/4" static pressure, vibration mounts	EA	36.3429	4,776.91	22,297.84	27,074.75
15.3704 000	**FANS, EXHAUST, PROPELLER BELT DRIVEN, WALL SHUTTER:**					
15.3704 011	Fan, exhaust, propeller belt driven, 5,000 CFM	EA	5.6000	736.06	1,181.63	1,917.70
15.3704 021	Fan, exhaust, propeller belt driven, 6,000 CFM	EA	6.1000	801.78	1,263.14	2,064.92
15.3704 031	Fan, exhaust, propeller belt driven, 8,000 CFM	EA	6.8000	893.79	1,494.01	2,387.80
15.3704 041	Fan, exhaust, propeller belt driven, 10,000 CFM	EA	7.1000	933.22	1,825.40	2,758.63
15.3704 051	Fan, exhaust, propeller belt driven, 15,000 CFM	EA	8.3000	1,090.95	2,580.58	3,671.53
15.3705 000	**RETURN FANS, VANE AXIAL, WITH VIBRATION MOUNTS, ADJUSTABLE PITCH:**					
15.3705 001	*Note: For controllable pitch, add 60% to the material costs.*					
15.3705 011	Return fan, 10,000 CFM, 5 hp, vibration mounts	EA	8.2763	1,087.84	8,816.89	9,904.72
15.3705 021	Return fan, 12,500 CFM, 5 hp, vibration mounts	EA	8.2763	1,087.84	9,246.96	10,334.80
15.3705 031	Return fan, 17,500 CFM, 5 hp, vibration mounts	EA	8.8127	1,158.34	9,806.14	10,964.48
15.3705 041	Return fan, 21,000 CFM, 7.5 hp, vibration mounts	EA	10.7285	1,410.15	15,805.89	17,216.05
15.3705 051	Return fan, 31,000 CFM, 10 hp, vibration mounts	EA	15.3264	2,014.50	19,999.31	22,013.82
15.3705 061	Return fan, 115,000 CFM, 50 hp, vibration mounts	EA	38.3160	5,036.26	29,226.32	34,262.58
15.3705 071	Return fan, 150,000 CFM, 75 hp, vibration mounts	EA	45.9792	6,043.51	38,193.50	44,237.00
15.3705 081	Return fan, 178,000 CFM, 200 hp, vibration mounts	EA	52.1098	6,849.31	56,459.96	63,309.27
15.3706 000	**FAN, CENTRIFUGAL IN-LINE WITH VIBRATION ISOLATOR:**					
15.3706 011	Fan, centrifugal in-line, 3,000 CFM	EA	4.4890	545.23	3,505.26	4,050.50
15.3706 021	Fan, centrifugal in-line, 4,000 CFM	EA	4.9000	595.15	4,093.43	4,688.59
15.3706 031	Fan, centrifugal in-line, 6,000 CFM	EA	6.1800	750.62	5,955.96	6,706.58
15.3706 041	Fan, centrifugal in-line, 10,000 CFM	EA	8.2910	1,007.02	7,392.47	8,399.50
15.3706 051	Fan, centrifugal in-line, 15,000 CFM	EA	8.8127	1,070.39	10,048.22	11,118.61
15.3706 061	Fan, centrifugal in-line, 20,000 CFM	EA	10.1111	1,228.09	12,010.22	13,238.31
15.3706 071	Fan, centrifugal in-line, 30,000 CFM	EA	15.5898	1,893.54	14,547.42	16,440.96
15.3706 081	Fan, centrifugal in-line, 35,000 CFM	EA	18.1010	2,198.55	17,688.42	19,886.97
15.3706 091	Fan, centrifugal, in-line, 45,000 CFM	EA	22.7351	2,761.41	21,293.00	24,054.41

MECHANICAL - HVAC - 15

Division	Description	Unit	Labor Hours	Labor Cost	Material Cost	Total Cost
15.3707 000	**CABINET BLOWERS:**					
15.3707 011	Cabinet blower, 8,000 CFM, 1" static pressure	EA	8.2915	1,089.83	2,832.26	3,922.09
15.3707 021	Cabinet blower, 10,000 CFM, 1" static pressure	EA	12.4615	1,637.94	4,723.02	6,360.96
15.3707 031	Cabinet blower, 20,000 CFM, 1" static pressure	EA	22.8381	3,001.84	9,445.99	12,447.83
15.3708 000	**FAN COIL UNITS, DUCT MOUNTED, 2 PIPE, SINGLE COIL:**					
15.3708 011	Fan coil unit, duct mounted, 400 CFM, 2 pipe, 1 coil	EA	1.6230	213.33	1,429.50	1,642.83
15.3708 021	Fan coil unit, duct mounted, 600 CFM, 2 pipe, 1 coil	EA	1.8939	248.93	1,567.18	1,816.12
15.3708 031	Fan coil unit, duct mounted, 1000 CFM, 2 pipe, 1 coil	EA	2.8403	373.33	2,284.54	2,657.87
15.3708 041	Fan coil unit, duct mounted, 1500 CFM, 2 pipe, 1 coil	EA	3.5167	462.24	2,756.86	3,219.10
15.3709 000	**FAN COIL UNITS, WITH CABINETS, 2 PIPE, SINGLE COIL:**					
15.3709 001	*Note: For 3 pipe, 2 coil, add 25% to the total costs. For 4 pipe, 2 coil, add 35%.*					
15.3709 011	Fan coil unit, with cabinet, 400 CFM, 2 pipe, 1 coil	EA	1.8939	248.93	1,514.48	1,763.41
15.3709 021	Fan coil unit, with cabinet, 600 CFM, 2 pipe, 1 coil	EA	2.3445	308.16	1,652.17	1,960.33
15.3709 031	Fan coil unit, with cabinet, 1000 CFM, 2 pipe, 1 coil	EA	3.2014	420.79	3,028.95	3,449.74
15.3709 041	Fan coil unit, with cabinet, 1500 CFM, 2 pipe, 1 coil	EA	3.7876	497.84	3,403.52	3,901.36
15.3710 000	**INDUCTION UNITS:**					
15.3710 011	Induction unit, 250 CFM @ outlet	EA	2.6154	343.77	676.61	1,020.37
15.3710 021	Induction unit, 500 CFM @ outlet	EA	3.0661	403.01	725.75	1,128.76
15.3710 031	Induction unit, 800 CFM @ outlet	EA	3.1563	414.86	772.18	1,187.05
15.3710 041	Induction unit, 1000 CFM @ outlet	EA	3.3820	444.53	878.05	1,322.58
15.3710 051	Induction unit, 1200 CFM @ outlet	EA	3.6070	474.10	942.63	1,416.73
15.3711 000	**MIXING BOXES, CONSTANT VOLUME:**					
15.3711 001	*Note: For variable volume, add 14% to the total costs.*					
15.3711 011	Mixing box, constant volume, 200 CFM #4	EA	1.9841	260.79	1,007.18	1,267.98
15.3711 021	Mixing box, constant volume, 550 CFM #6	EA	2.5250	331.89	1,087.23	1,419.12
15.3711 031	Mixing box, constant volume, 850 CFM #8	EA	2.8403	373.33	1,247.38	1,620.71
15.3711 041	Mixing box, constant volume, 1400 CFM #10	EA	3.2917	432.66	1,433.35	1,866.01
15.3711 051	Mixing box, constant volume, 2500 CFM #12	EA	3.7876	497.84	1,836.20	2,334.04
15.3711 061	Mixing box, constant volume, 3200 CFM #14	EA	4.1480	545.21	2,153.89	2,699.10
15.3711 071	Mixing box, constant volume, 5000 CFM #16	EA	4.4640	586.75	2,396.67	2,983.42
15.3712 000	**AIR TERMINAL UNITS, WITH RE-HEAT COILS:**					
15.3712 011	Air terminal unit, #7, 4-650 CFM, 1" static pressure, 1.1 SF coil	EA	2.0744	272.66	857.17	1,129.83
15.3712 021	Air terminal unit, #8, 6-850 CFM, 1" static pressure, 1.6 SF coil	EA	2.3445	308.16	928.09	1,236.25
15.3712 031	Air terminal unit, #9, 8-1050 CFM, 1" static pressure, 2.8 SF coil	EA	2.6605	349.70	956.90	1,306.59
15.3712 041	Air terminal unit, #10, 6-1000 CFM, 1" static pressure, 2 SF coil	EA	2.6605	349.70	1,023.29	1,372.98
15.3712 051	Air terminal unit, #12, 12-1800 CFM, 1" static pressure, 2.1 SF coil	EA	2.8403	373.33	1,120.75	1,494.08
15.3712 061	Air terminal unit, #14, 2-2800 CFM, 1" static pressure, 3.4 SF coil	EA	3.5167	462.24	1,258.13	1,720.36
15.3713 000	**AIR TERMINAL UNITS, CONSTANT VOLUME:**					
15.3713 001	*Note: The following items are single duct, without coils.*					
15.3713 011	Air terminal unit, constant volume, 350 CFM	EA	1.7870	234.88	371.39	606.27
15.3713 021	Air terminal unit, constant volume, 500 CFM	EA	1.8866	247.97	414.11	662.09
15.3713 031	Air terminal unit, constant volume, 800 CFM	EA	2.2837	300.17	456.84	757.00
15.3713 041	Air terminal unit, constant volume, 1600 CFM	EA	2.6812	352.42	588.34	940.76
15.3713 051	Air terminal unit, constant volume, 2400 CFM	EA	3.1276	411.09	706.67	1,117.76
15.3713 061	Air terminal unit, constant volume, 3000 CFM	EA	3.9717	522.04	861.14	1,383.18
15.3714 000	**AIR TERMINAL UNITS, VARIABLE VOLUME:**					
15.3714 011	Air terminal unit, variable volume, 150 CFM	EA	1.4896	195.79	478.70	674.49
15.3714 021	Air terminal unit, variable volume, 350 CFM	EA	1.9376	254.68	715.64	970.32
15.3714 031	Air terminal unit, variable volume, 500 CFM	EA	2.2447	295.04	833.39	1,128.44
15.3714 041	Air terminal unit, variable volume, 800 CFM	EA	2.6631	350.04	988.75	1,338.78
15.3714 051	Air terminal unit, variable volume, 1000 CFM	EA	2.9253	384.50	1,086.02	1,470.52
15.3714 061	Air terminal unit, variable volume, 1600 CFM	EA	3.5763	470.07	1,327.71	1,797.78
15.3714 071	Air terminal unit, variable volume, 2400 CFM	EA	4.0075	526.75	1,487.78	2,014.52
15.3714 081	Air terminal unit, variable volume, 3200 CFM	EA	5.0546	664.38	1,950.74	2,615.12
15.3715 000	**RE-HEAT COILS, 2 ROW:**					
15.3715 011	Re-heat coil, 2 row, 1/2 SF	EA	0.7022	92.30	286.49	378.79
15.3715 021	Re-heat coil, 2 row, 1 SF	EA	0.7437	97.75	358.14	455.89

15 - HVAC - MECHANICAL

Division	Description	Unit	Labor Hours	Labor Cost	Material Cost	Total Cost
15.3715 031	Re-heat coil, 2 row, 1-1/2 SF	EA	0.7847	103.14	639.54	742.68
15.3715 041	Re-heat coil, 2 row, 2 SF	EA	0.8262	108.60	690.71	799.31
15.3715 051	Re-heat coil, 2 row, 2-1/2 SF	EA	0.8671	113.97	729.05	843.02
15.3715 061	Re-heat coil, 2 row, 3 SF	EA	0.9086	119.43	936.28	1,055.71
15.3715 071	Re-heat coil, 2 row, 3-1/2 SF	EA	0.9501	124.88	1,005.37	1,130.26
15.3715 081	Re-heat coil, 2 row, 4 SF	EA	0.9911	130.27	1,092.32	1,222.59
15.3800 000	**MISCELLANEOUS EQUIPMENT:**					
15.3801 000	**EXPANSION TANKS:**					
15.3801 011	Expansion tank, 16 gallon, chilled water	EA	1.3629	165.54	1,020.99	1,186.53
15.3801 021	Expansion tank, 44 gallon, chilled water	EA	1.5692	190.60	1,446.98	1,637.57
15.3801 031	Expansion tank, 55 gallon, ASME code	EA	5.6081	681.16	2,951.85	3,633.01
15.3801 041	Expansion tank, 88 gallon, ASME code	EA	6.8600	833.22	4,378.04	5,211.26
15.3801 051	Expansion tank, 132 gallon, ASME code	EA	9.1200	1,107.72	5,625.93	6,733.64
15.3801 061	Expansion tank, 150 gallon, ASME code	EA	11.6300	1,412.58	6,408.45	7,821.03
15.3801 071	Expansion tank, 250 gallon, ASME code	EA	13.5600	1,647.00	8,436.56	10,083.56
15.3801 081	Expansion tank, 375 gallon, ASME code	EA	16.0200	1,945.79	12,034.36	13,980.15
15.3802 000	**SEPARATORS-AIR ELIMINATION, 150#, WITH STRAINER:**					
15.3802 001	*Note: The following prices do not include companion flanges or bolt and gasket sets.*					
15.3802 011	Air separator, steel, 150#, 2", strainer	EA	1.5110	183.53	1,202.10	1,385.62
15.3802 021	Air separator, steel, 150#, 2-1/2", strainer	EA	1.7225	209.21	1,421.43	1,630.65
15.3802 031	Air separator, steel, 150#, 3", strainer	EA	1.9700	239.28	2,080.93	2,320.21
15.3802 041	Air separator, steel, 150#, 4", strainer	EA	2.4000	291.50	2,998.67	3,290.17
15.3802 051	Air separator, steel, 150#, 5", strainer	EA	3.2500	394.75	3,800.03	4,194.78
15.3802 061	Air separator, steel, 150#, 6", strainer	EA	4.1500	504.06	4,562.66	5,066.72
15.3802 071	Air separator, steel, 150#, 8", strainer	EA	5.6000	680.18	6,811.68	7,491.86
15.3802 081	Air separator, steel, 150#, 10", strainer	EA	6.2000	753.05	10,728.08	11,481.13
15.3802 091	Air separator, steel, 150#, 12", strainer	EA	7.6120	924.55	15,381.25	16,305.80
15.3803 000	**WATER PURIFICATION & TREATMENT:**					
15.3803 011	Water purification, organic, 5m grain, 1 GPM	EA	6.8965	837.65	1,007.18	1,844.83
15.3804 000	**RELIEF VENT:**					
15.3804 011	Relief vent, 750 CFM, aluminum	EA	2.3714	288.03	1,404.52	1,692.55
15.3804 021	Relief vent, 1,500 CFM, aluminum	EA	2.9511	358.44	1,821.56	2,180.01
15.3804 031	Relief vent, 3,000 CFM, aluminum	EA	4.2792	519.75	2,528.38	3,048.13
15.3804 041	Relief vent, 6,000 CFM, aluminum	EA	6.5827	799.53	3,951.46	4,751.00
15.3804 051	Relief vent, 12,000 CFM, aluminum	EA	10.6241	1,290.40	6,491.16	7,781.56
15.3804 061	Relief vent, 20,000 CFM, aluminum	EA	14.2845	1,735.00	8,642.19	10,377.18
15.3804 071	Relief vent, 30,000 CFM, aluminum	EA	24.4579	2,970.66	15,091.40	18,062.06
15.3804 081	Relief vent, 40,000 CFM, aluminum	EA	28.9314	3,514.01	19,187.63	22,701.64
15.3804 091	Relief vent, 50,000 CFM, aluminum	EA	32.9014	3,996.20	20,408.00	24,404.21
15.3805 000	**VIBRATION ISOLATORS:**					
15.3805 011	Vibration isolator, pad mounted, neoprene, to 400 lbs	EA			65.69	65.69
15.3805 021	Vibration isolator, pad mounted, neoprene, to 1,200 lbs	EA			156.05	156.05
15.3805 031	Vibration isolator, pad mounted, neoprene, to 4,000 lbs	EA			352.58	352.58
15.3805 111	Vibration isolator, pad mounted, spring, to 300 lbs	EA			215.38	215.38
15.3805 121	Vibration isolator, pad mounted, spring, to 1,000 lbs	EA			304.58	304.58
15.3805 131	Vibration isolator, pad mounted, spring, to 1,300 lbs	EA			341.94	341.94
15.3805 141	Vibration isolator, pad mounted, spring, to 1,800 lbs	EA			488.75	488.75
15.3805 151	Vibration isolator, pad mounted, spring, to 2,600 lbs	EA			544.95	544.95
15.3805 161	Vibration isolator, pad mounted, spring, to 4,000 lbs	EA			626.82	626.82

MECHANICAL - HVAC - 15

Division	Description	Unit	Labor Hours	Labor Cost	Material Cost	Total Cost
15.3900 000	**CONTROLS:**					
15.3900 001	Note: Controls are expressed as a percentage of equipment, since it is impossible to classify all possible control systems available for HVAC. Controls for built-up systems: $100,000 or Less............28.00% $100,000 to 200,000...........26.00% $200,000 to 500,000...........24.00% $500,000 to 1,000,000........22.00% $1,000,000 and Up............20.00% Controls for package systems: $100,000 or Less.............11.00% $100,000 to 200,000............9.00% $200,000 to 500,000............7.00%					
15.3901 000	**COMPUTERIZED ENERGY MANAGEMENT SYSTEM ONLY:**					
15.3901 011	Control computer, under 300 points	EA			60,212.94	60,212.94
15.3901 021	Control computer, 500-1,000 data control points	EA			182,789.29	182,789.29
15.3901 031	Add for each point up to 100	EA			2,150.45	2,150.45
15.3901 041	Add per point, 100-200	EA			1,612.83	1,612.83
15.3901 051	Add per point, 200-500	EA			1,290.25	1,290.25
15.3901 061	Add per point over 500	EA			1,075.24	1,075.24
15.3902 000	**PNEUMATIC CONTROLS:**					
15.3902 011	Air compressor station, complete	EA	22.9896	2,792.32	28,816.20	31,608.52
15.3903 000	**PNEUMATIC ZONE CONTROLS:**					
15.3903 001	Note: The two variable air volume items listed below include thermostat and tubing, with motor by box manufacturer.					
15.3903 011	Reheat zone control, complete	EA			645.17	645.17
15.3903 021	Variable air volume zone	EA			580.63	580.63
15.3903 031	Variable air volume, complete	EA			806.36	806.36
15.3904 000	**MISCELLANEOUS ELECTRICAL CONTROLS:**					
15.3904 011	Thermostat with fan switch, heat only	EA	0.7185	87.27	103.69	190.96
15.3904 021	Thermostat with fan switch, cool only	EA	0.7185	87.27	103.69	190.96
15.3904 031	Thermostat with fan switch, heat & cool	EA	1.1974	145.44	103.84	249.28
15.4000 000	**DUCT WORK, GRILLS & REGISTERS:**					
15.4000 001	Note: The following prices include supports, joints, fittings and duct tape. Insulation is not included.					
15.4001 000	**ALUMINUM DUCTWORK, WITH SUPPORTS & ACCESSORIES:**					
15.4001 011	Duct, aluminum, rectangular, under 4,000#	#	0.0880	11.57	17.64	29.20
15.4001 021	Duct, aluminum, rectangular, 4,000-8,000#	#	0.0820	10.78	16.63	27.41
15.4001 031	Duct, aluminum, rectangular, over 8,000#	#	0.0759	9.98	15.18	25.16
15.4001 041	Duct, aluminum, round, under 4,000#	#	0.0934	12.28	18.85	31.13
15.4001 051	Duct, aluminum, round, over 4,000#	#	0.0893	11.74	16.22	27.95
15.4002 000	**FIBERGLASS DUCT WITH SUPPORTS & ACCESSORIES:**					
15.4002 001	Note: The following items do not require insulation.					
15.4002 011	Duct, fiberglass, rectangular	SF	0.0545	7.16	7.63	14.80
15.4002 021	Duct, fiberglass, round	SF	0.0162	2.13	9.08	11.21
15.4003 000	**GALVANIZED IRON DUCT WITH SUPPORTS & ACCESSORIES:**					
15.4003 001	Note: The following prices are based on medium pressure, to 3.5# Static pressure. The following is a list of nominal weights by largest side dimension with waste and laps: To 12" 26 GA.......MINIMUM 1.0#/SF 13"-30" 24 GA...............1.3#/SF 31"-54" 22 GA...............1.6#/SF 55"-84" 20 GA...............2.0#/SF 85" + 18 GA.................2.4#/SF					
15.4003 021	Duct, galvanized iron, rectangular, under 10,000 #, shop fab	#	0.0474	6.23	7.24	13.47
15.4003 031	Duct, galvanized iron, rectangular, 10,000-20,000 #, shop fab	#	0.0399	5.24	6.73	11.97
15.4003 041	Duct, galvanized iron, rectangular, over 20,000 #, shop fab	#	0.0350	4.60	5.98	10.58
15.4003 051	Duct, galvanized iron, round, under 10,000 #, shop fab	#	0.0397	5.22	6.82	12.04
15.4003 061	Duct, galvanized iron, round, 10,000-20,000 #, shop fab	#	0.0356	4.68	6.50	11.18

15 - HVAC - MECHANICAL

Division	Description	Unit	Labor Hours	Labor Cost	Material Cost	Total Cost
15.4003 071	Duct, galvanized iron, round, over 20,000 #, shop fab	#	0.0333	4.38	5.64	10.02
15.4003 081	Duct, galvanized iron, spiral, 3", 26 ga, with fittings & supports	LF	0.0120	1.58	4.09	5.67
15.4003 091	Duct, galvanized iron, spiral, 4", 26 ga, with fittings & supports	LF	0.0142	1.87	5.00	6.86
15.4003 101	Duct, galvanized iron, spiral, 6", 26 ga, with fittings & supports	LF	0.0196	2.58	6.81	9.38
15.4003 111	Duct, galvanized iron, spiral, 8", 26 ga, with fittings & supports	LF	0.0263	3.46	9.67	13.13
15.4003 121	Duct, galvanized iron, spiral, 10", 24 ga, with fittings & supports	LF	0.0417	5.48	14.85	20.33
15.4003 131	Duct, galvanized iron, spiral, 12", 24 ga, with fittings & supports	LF	0.0492	6.47	17.09	23.55
15.4003 141	Duct, galvanized iron, spiral, 16", 24 ga, with fittings & supports	LF	0.0632	8.31	23.06	31.36
15.4003 151	Duct, galvanized iron, spiral, 20", 24 ga, with fittings & supports	LF	0.0780	10.25	29.17	39.42
15.4003 161	Duct, galvanized iron, spiral, 24", 22 ga, with fittings & supports	LF	0.0928	12.20	38.53	50.73
15.4003 171	Duct, galvanized iron, spiral, 30", 22 ga, with fittings & supports	LF	0.1270	16.69	53.98	70.67
15.4003 181	Duct, galvanized iron, spiral, 36", 22 ga, with fittings & supports	LF	0.1707	22.44	64.10	86.53
15.4003 191	Duct, galvanized iron, spiral, 42", 20 ga, with fittings & supports	LF	0.2385	31.35	83.54	114.89
15.4003 201	Duct, galvanized iron, spiral, 48", 20 ga, with fittings & supports	LF	0.2722	35.78	95.99	131.77
15.4004 000	**STAINLESS STEEL DUCT (304) WITH SUPPORTS & ACCESSORIES:**					
15.4004 011	Duct, stainless steel, rectangular	#	0.0437	5.74	31.50	37.25
15.4004 021	Duct, stainless steel, round	#	0.0465	6.11	32.91	39.03
15.4005 000	**FLEXIBLE DUCT, WITH CLAMPS:**					
15.4005 011	Duct, flexible, 4", clamps	LF	0.0511	6.72	4.39	11.11
15.4005 021	Duct, flexible, 6", clamps	LF	0.0632	8.31	5.92	14.23
15.4005 031	Duct, flexible, 8", clamps	LF	0.0913	12.00	8.43	20.44
15.4005 041	Duct, flexible, 10", clamps	LF	0.1201	15.79	8.80	24.59
15.4005 051	Duct, flexible, 12", clamps	LF	0.1688	22.19	10.80	32.98
15.4005 061	Duct, flexible, 14", clamps	LF	0.2072	27.23	11.92	39.15
15.4005 071	Duct, flexible, 16", clamps	LF	0.2400	29.15	15.18	44.34
15.4005 081	Duct, flexible, 18", clamps	LF	0.2909	35.33	18.50	53.83
15.4005 091	Duct, flexible, 20", clamps	LF	0.3429	41.65	19.80	61.45
15.4006 000	**FIBERGLASS REINFORCED PLASTIC DUCT, BURIED BELOW BUILDING:**					
15.4006 011	Fiberglass reinforced plastic duct, 8", buried	LF	0.1370	18.01	41.19	59.19
15.4006 021	Fiberglass reinforced plastic duct, 10", buried	LF	0.1639	21.54	49.37	70.91
15.4006 031	Fiberglass reinforced plastic duct, 12", buried	LF	0.1754	23.05	52.81	75.86
15.4006 041	Fiberglass reinforced plastic duct, 14", buried	LF	0.2144	28.18	64.50	92.68
15.4006 051	Fiberglass reinforced plastic duct, 16", buried	LF	0.2459	32.32	74.06	106.38
15.4006 061	Fiberglass reinforced plastic duct, 18", buried	LF	0.2844	37.38	85.72	123.10
15.4006 071	Fiberglass reinforced plastic duct, 20", buried	LF	0.2992	39.33	90.12	129.45
15.4006 081	1-1/4" PVC terminal adapters	EA			4.07	4.07
15.4007 000	**FIRE DAMPERS:**					
15.4007 011	Fire damper, in wall, sleeve to 1 SF	EA	0.8441	110.95	219.47	330.42
15.4007 021	Fire damper, in wall, sleeve 1-2 SF	EA	1.0426	137.04	289.75	426.79
15.4007 031	Fire damper, in wall, sleeve 2-4 SF	EA	1.5392	202.31	370.69	573.00
15.4007 041	Fire damper, in wall, sleeve 4-6 SF	EA	1.7870	234.88	704.39	939.27
15.4007 051	Fire damper, in wall, sleeve 6-10 SF	EA	2.5817	339.34	760.87	1,100.21
15.4007 061	Fire damper, in wall, sleeve 10-16 SF	EA	3.1276	411.09	939.70	1,350.79
15.4007 071	Fire damper, in wall, sleeve 16-25 SF	EA	4.3875	576.69	1,144.34	1,721.03
15.4007 081	Fire damper, in wall, sleeve 25-30 SF	EA	5.2650	692.03	1,835.03	2,527.06
15.4007 091	Fire damper, in wall, sleeve 30-35 SF	EA	5.4425	715.36	2,525.41	3,240.77
15.4007 101	Fire damper, in wall, sleeve 35-40 SF	EA	6.2200	817.56	2,679.01	3,496.57
15.4007 111	Fire damper, in wall, sleeve 40-45 SF	EA	6.5475	860.60	3,203.09	4,063.70
15.4007 121	Fire damper, in wall, sleeve 45-55 SF	EA	8.0025	1,051.85	4,147.04	5,198.89
15.4007 131	Fire damper, in duct, to 1 SF	EA	0.3970	52.18	120.97	173.15
15.4007 141	Fire damper, in duct, 1-2 SF	EA	0.5461	71.78	159.28	231.06
15.4007 151	Fire damper, in duct, 2-4 SF	EA	0.7445	97.86	208.69	306.55
15.4007 161	Fire damper, in duct, 4-6 SF	EA	0.8441	110.95	291.13	402.08
15.4007 171	Fire damper, in duct, 6-10 SF	EA	1.3406	176.21	428.47	604.68
15.4007 181	Fire damper, in duct, 10-16 SF	EA	1.5392	202.31	587.80	790.11
15.4008 000	**CEILING RETURN OR EXHAUST REGISTER:**					
15.4008 011	Ceiling return, to 10" large dimension	EA	0.5461	71.78	64.83	136.61

MECHANICAL - HVAC - 15

Division	Description	Unit	Labor Hours	Labor Cost	Material Cost	Total Cost
15.4008 021	Ceiling return, 12-18" large dimension	EA	0.6457	84.87	90.87	175.74
15.4008 031	Ceiling return, 20-30" large dimension	EA	0.8441	110.95	125.77	236.72
15.4009 000	**CEILING EXHAUST GRILL:**					
15.4009 011	Ceiling exhaust grill, to 10" large dimension	EA	0.4966	65.27	55.40	120.67
15.4009 021	Ceiling exhaust grill, 12-18" large dimension	EA	0.5962	78.36	76.07	154.44
15.4009 031	Ceiling exhaust grill, 20-30" large dimension	EA	0.7946	104.44	102.80	207.24
15.4010 000	**WALL EXHAUST REGISTER:**					
15.4010 011	Wall exhaust, to 10" large dimension	EA	0.4966	65.27	64.83	130.10
15.4010 021	Wall exhaust, 12-18" large dimension	EA	0.6457	84.87	90.87	175.74
15.4010 031	Wall exhaust, 20-30" large dimension	EA	0.7946	104.44	125.77	230.22
15.4011 000	**WALL EXHAUST GRILL:**					
15.4011 011	Wall exhaust grill, to 10" large dimension	EA	0.3970	52.18	55.40	107.58
15.4011 021	Wall exhaust grill, 12-18" large dimension	EA	0.5461	71.78	76.07	147.85
15.4011 031	Wall exhaust grill, 20-30" large dimension	EA	0.7445	97.86	102.80	200.65
15.4012 000	**CEILING DIFFUSER, RECTANGULAR, TWO WAY:**					
15.4012 011	Ceiling diffuser, to 12", louver, rectangular, 2 way	EA	0.5962	78.36	90.87	169.23
15.4012 021	Ceiling diffuser, 14-20", louver, rectangular, 2 way	EA	0.7445	97.86	130.82	228.67
15.4012 031	Ceiling diffuser, 25-32", louver, rectangular, 2 way	EA	0.9430	123.95	197.40	321.35
15.4012 041	Ceiling diffuser, to 12", perforated, rectangular, 2 way	EA	0.5962	78.36	79.87	158.23
15.4012 051	Ceiling diffuser, 14-20", perforated, rectangular, 2 way	EA	0.7445	97.86	110.09	207.94
15.4012 061	Ceiling diffuser, 24-30", perforated, rectangular, 2 way	EA	0.9430	123.95	160.24	284.18
15.4013 000	**LINEAR DIFFUSER, ALUMINUM:**					
15.4013 011	Linear diffuser, aluminum, ceiling or wall 2" wide	EA	0.2644	34.75	58.85	93.60
15.4013 021	Linear diffuser, aluminum, ceiling or wall 3" wide	EA	0.2872	37.75	70.90	108.65
15.4013 031	Linear diffuser, aluminum, ceiling or wall 4" wide	EA	0.3219	42.31	82.90	125.21
15.4013 041	Linear diffuser, aluminum, ceiling or wall 6" wide	EA	0.3524	46.32	107.24	153.56
15.4013 051	Linear diffuser, aluminum, ceiling or wall 8" wide	EA	0.3850	50.60	125.29	175.90
15.4013 061	Linear diffuser, aluminum, ceiling or wall 10" wide	EA	0.4231	55.61	139.49	195.10
15.4013 071	Linear diffuser, aluminum, ceiling or wall 12" wide	EA	0.4687	61.61	156.02	217.62
15.4014 000	**WALL SUPPLY REGISTERS:**					
15.4014 011	Wall supply register, to 10" large dimension	EA	0.4966	65.27	60.81	126.08
15.4014 021	Wall supply register, 12-15" large dimension	EA	0.6457	84.87	85.44	170.31
15.4014 031	Wall supply register, 18-24" large dimension	EA	0.7445	97.86	111.46	209.32
15.4014 041	Wall supply register, 30-36" large dimension	EA	0.8441	110.95	140.31	251.25
15.4015 000	**MISCELLANEOUS DUCT ITEMS, INSULATION:**					
15.4015 011	Insulation, duct, internal, 1"	SFCA	0.0230	3.02	2.54	5.57
15.4015 021	Insulation, duct, external, 1", plain	SFCA	0.0077	1.01	1.24	2.25
15.4015 031	Insulation, duct, exterior, 1", vapor barrier	SFCA	0.0153	2.01	1.87	3.88
15.4015 041	Insulation, duct, exterior, 1", rigid board	SFCA	0.0384	5.05	3.32	8.37
15.4016 000	**SOUND ATTENUATORS:**					
15.4016 011	Attenuator, 3' l x 24" w x 24" h	EA	2.2990	302.18	1,873.79	2,175.97
15.4016 021	Attenuator, 3' l x 36" w x 24" h	EA	2.6822	352.55	2,066.77	2,419.31
15.4016 031	Attenuator, 3' l x 48" w x 28" h	EA	2.6822	352.55	3,456.52	3,809.07
15.4016 041	Attenuator, 5' l x 36" w x 24" h	EA	3.0653	402.90	3,085.41	3,488.31
15.4016 051	Attenuator, 5' l x 48" w x 24" h	EA	3.4485	453.27	4,071.27	4,524.54
15.4016 061	Attenuator, 5' l x 72" w x 28" h	EA	3.8316	503.63	5,576.75	6,080.38
15.4016 071	Attenuator, 7' l x 20" w x 18" h	EA	4.5980	604.36	2,886.41	3,490.77
15.4016 081	Attenuator, 7' l x 32" w x 20" h	EA	4.7895	629.53	3,723.80	4,353.33
15.4016 091	Attenuator, 7' l x 48" w x 20" h	EA	4.9811	654.72	5,576.75	6,231.47
15.4100 000	**PIPING & INSULATION:**					
15.4101 000	**COPPER "L" IN BUILDING, WITH FITTINGS & SUPPORTS:**					
15.4101 001	Note: The following prices include fittings, hangers and supports. Prices do not include valves or insulation. For piping not covered here, see section 15.2100.					
15.4101 011	Pipe, copper, 'L', 1/2", in building, with fittings & supports	LF	0.0841	10.21	5.33	15.55
15.4101 021	Pipe, copper 'L', 3/4", in building, with fittings & supports	LF	0.1012	12.29	8.31	20.61
15.4101 031	Pipe, copper, 'L', 1", in building, with fittings & supports	LF	0.1192	14.48	12.11	26.59

15 - HVAC - MECHANICAL

Division	Description	Unit	Labor Hours	Labor Cost	Material Cost	Total Cost
15.4101 041	Pipe, copper, 'L', 1-1/4", in building, with fittings & supports	LF	0.1283	15.58	17.25	32.83
15.4101 051	Pipe, copper, 'L', 1-1/2", in building, with fittings & supports	LF	0.1361	16.53	21.56	38.09
15.4101 061	Pipe, copper, 'L', 2", in building, with fittings & supports	LF	0.1631	19.81	26.79	46.60
15.4102 000	**STEEL PIPE, BLACK, WELDED, SCH 40, A-120, SCREWED:**					
15.4102 001	*Note: The following prices include malleable iron fittings & supports.*					
15.4102 011	Black steel pipe, weld, A-120, '40', 1/2", screwed	LF	0.0752	9.13	2.21	11.35
15.4102 021	Black steel pipe, weld, A-120, '40', 3/4", screwed	LF	0.0931	11.31	2.62	13.93
15.4102 031	Black steel pipe, weld, A-120, '40', 1", screwed	LF	0.1122	13.63	3.38	17.01
15.4102 041	Black steel pipe, weld, A-120, '40', 1-1/4", screwed	LF	0.1211	14.71	4.62	19.33
15.4102 051	Black steel pipe, weld, A-120, '40', 1-1/2", screwed	LF	0.1402	17.03	5.25	22.28
15.4102 061	Black steel pipe, weld, A-120, '40', 2", screwed	LF	0.1908	23.17	7.04	30.21
15.4102 071	Black steel pipe, weld, A-120, '40', 2-1/2", screwed	LF	0.2291	27.83	10.84	38.67
15.4102 081	Black steel pipe, weld, A-120, '40', 3", screwed	LF	0.2674	32.48	14.61	47.09
15.4102 091	Black steel pipe, weld, A-120, '40', 4", screwed	LF	0.3487	42.35	25.49	67.84
15.4102 101	Black steel pipe, weld, A-120, '40', 5", screwed	LF	0.3771	45.80	36.29	82.10
15.4102 111	Black steel pipe, weld, A-120, '40', 6", screwed	LF	0.4215	51.20	46.45	97.64
15.4103 000	**STEEL PIPE, BLACK, WELDED, SCH 40, A-53, WELD CONSTRUCTION:**					
15.4103 001	*Note: The following prices include black welded fittings & supports.*					
15.4103 011	Black steel pipe, weld, A-53, '40', 2", welded	LF	0.1924	23.37	15.47	38.84
15.4103 021	Black steel pipe, weld, A-53, '40', 2-1/2", welded	LF	0.2452	29.78	20.39	50.17
15.4103 031	Black steel pipe, weld, A-53, '40', 3", welded	LF	0.2790	33.89	25.26	59.15
15.4103 041	Black steel pipe, weld, A-53, '40', 4", welded	LF	0.3583	43.52	35.42	78.94
15.4103 051	Black steel pipe, weld, A-53, '40', 5", welded	LF	0.3985	48.40	42.47	90.87
15.4103 061	Black steel pipe, weld, A-53, '40', 6", welded	LF	0.4398	53.42	44.64	98.06
15.4103 071	Black steel pipe, weld, A-53, '40', 8", welded	LF	0.4904	59.56	74.25	133.81
15.4104 000	**STEEL PIPE, SEAMLESS, SCH 40, A-53, WELDED CONSTRUCTION:**					
15.4104 001	*Note: The following prices include black welded fittings & supports.*					
15.4104 011	Steel pipe, seamless, A-53, '40', 2", welded	LF	0.1924	23.37	18.42	41.79
15.4104 021	Steel pipe, seamless, A-53, '40', 2-1/2", welded	LF	0.2489	30.23	20.72	50.95
15.4104 031	Steel pipe, seamless, A-53, '40', 3", welded	LF	0.2785	33.83	25.29	59.12
15.4104 041	Steel pipe, seamless, A-53, '40', 4", welded	LF	0.3583	43.52	35.95	79.47
15.4104 051	Steel pipe, seamless, A-53, '40', 5", welded	LF	0.3985	48.40	43.27	91.68
15.4104 061	Steel pipe, seamless, A-53, '40', 6", welded	LF	0.4398	53.42	45.60	99.02
15.4104 071	Steel pipe, seamless, A-53, '40', 8", welded	LF	0.4904	59.56	69.87	129.44
15.4104 081	Steel pipe, seamless, A-53, '40', 10", welded	LF	0.5517	67.01	131.73	198.74
15.4105 000	**STEEL PIPE, SEAMLESS, SCH 80:**					
15.4105 001	*Note: The following prices include black welded fittings & supports.*					
15.4105 011	Steel pipe, seamless, A-53, '80', 2"	LF	0.2207	26.81	19.80	46.61
15.4105 021	Steel pipe, seamless, A-53, '80', 2-1/2"	LF	0.2862	34.76	23.20	57.96
15.4105 031	Steel pipe, seamless, A-53, '80', 3"	LF	0.3196	38.82	28.60	67.42
15.4105 041	Steel pipe, seamless, A-53, '80', 4"	LF	0.4115	49.98	41.99	91.97
15.4105 051	Steel pipe, seamless, A-53, '80', 5"	LF	0.4583	55.67	50.99	106.65
15.4105 061	Steel pipe, seamless, A-53, '80', 6"	LF	0.5058	61.43	56.14	117.58
15.4105 071	Steel pipe, seamless, A-53, '80', 8"	LF	0.5640	68.50	86.24	154.74
15.4105 081	Steel pipe, seamless, A-53, '80', 10"	LF	0.6345	77.07	170.34	247.41
15.4106 000	**STEEL PIPE, BLACK, WELDED, SCH 40, A-120, VICTAULIC COUPLINGS:**					
15.4106 001	*Note: The following prices include victaulic fittings and supports.*					
15.4106 011	Black steel pipe, weld, A-120, '40', 2", victaulic	LF	0.1016	12.34	18.26	30.60
15.4106 021	Black steel pipe, weld, A-120, '40', 2-1/2", victaulic	LF	0.1314	15.96	20.39	36.35
15.4106 031	Black steel pipe, weld, A-120, '40', 3", victaulic	LF	0.1471	17.87	25.26	43.12
15.4106 041	Black steel pipe, weld, A-120, '40', 4", victaulic	LF	0.1892	22.98	32.68	55.66
15.4106 051	Black steel pipe, weld, A-120, '40', 5", victaulic	LF	0.2104	25.56	42.47	68.03
15.4106 061	Black steel pipe, weld, A-120, '40', 6", victaulic	LF	0.2322	28.20	44.64	72.84
15.4106 071	Black steel pipe, weld, A-120, '40', 8", victaulic	LF	0.2589	31.45	74.25	105.70
15.4106 081	Black steel pipe, weld, A-120, '40', 10", victaulic	LF	0.3016	36.63	94.64	131.27
15.4106 091	Black steel pipe, weld, A-120, '40', 12", victaulic	LF	0.4878	59.25	108.68	167.92
15.4106 101	Black steel pipe, weld, A-120, '40', 14", victaulic	LF	0.5007	60.82	123.99	184.80

MECHANICAL - HVAC - 15

Division	Description	Unit	Labor Hours	Labor Cost	Material Cost	Total Cost
15.4106 111	Black steel pipe, weld, A-120, '40', 16", victaulic	LF	0.5335	64.80	148.25	213.05
15.4106 121	Black steel pipe, weld, A-120, '40', 18", victaulic	LF	0.6566	79.75	197.50	277.25
15.4106 131	Black steel pipe, weld, A-120, '40', 20", victaulic	LF	0.7462	90.63	255.86	346.49
15.4106 141	Black steel pipe, weld, A-120, '40', 24", victaulic	LF	0.8812	107.03	321.67	428.70
15.4200 000	**FITTINGS:**					
15.4201 000	**FLANGES, CAST IRON., SCREWED, BLACK, 125#:**					
15.4201 011	Cast iron flange, black, screwed, 125#, 1-1/2"	EA	0.5477	66.52	32.19	98.72
15.4201 021	Cast iron flange, black, screwed, 125#, 2"	EA	0.6547	79.52	40.21	119.73
15.4201 031	Cast iron flange, black, screwed, 125#, 2-1/2"	EA	0.7126	86.55	45.50	132.05
15.4201 041	Cast iron flange, black, screwed, 125#, 3"	EA	0.9426	114.49	48.05	162.54
15.4201 051	Cast iron flange, black, screwed, 125#, 4"	EA	1.1495	139.62	53.08	192.70
15.4201 061	Cast iron flange, black, screwed, 125#, 5"	EA	1.2874	156.37	70.55	226.92
15.4201 071	Cast iron flange, black, screwed, 125#, 6"	EA	1.5096	183.36	83.97	267.32
15.4201 081	Cast iron flange, black, screwed, 125#, 8"	EA	1.6093	195.47	144.72	340.18
15.4201 091	Cast iron flange, black, screwed, 125#, 10"	EA	1.8085	219.66	235.45	455.11
15.4202 000	**FLANGES, STEEL, SLIP-ON, FORGED STEEL 150#:**					
15.4202 001	Note: For weld neck flange, add 20% to the material costs.					
15.4202 011	Steel flange, slip-on, 2", forged steel 150#	EA	0.5726	69.55	20.94	90.49
15.4202 021	Steel flange, slip-on, 2-1/2", forged steel 150	EA	0.6786	82.42	29.75	112.17
15.4202 031	Steel flange, slip-on, 3", forged steel 150#	EA	0.8355	101.48	30.11	131.59
15.4202 041	Steel flange, slip-on, 4", forged steel 150#	EA	1.0977	133.33	38.28	171.60
15.4202 051	Steel flange, slip-on, 5", forged steel 150#	EA	1.3327	161.87	54.09	215.96
15.4202 061	Steel flange, slip-on, 6", forged steel 150#	EA	1.6150	196.16	62.84	258.99
15.4202 071	Steel flange, slip-on, 8", forged steel 150#	EA	2.2348	271.44	95.57	367.00
15.4202 081	Steel flange, slip-on, 10", forged steel 150#	EA	2.7443	333.32	172.77	506.10
15.4202 091	Steel flange, slip-on, 12", forged steel 150#	EA	3.3870	411.39	359.07	770.45
15.4202 101	Steel flange, slip-on, 14", forged steel 150#	EA	3.9207	476.21	415.93	892.14
15.4202 111	Steel flange, slip-on, 16", forged steel 150#	EA	4.6267	561.96	466.83	1,028.79
15.4202 121	Steel flange, slip-on, 18", forged steel 150#	EA	5.3711	652.37	618.32	1,270.69
15.4202 131	Steel flange, slip-on, 20", forged steel 150#	EA	6.4687	785.69	736.13	1,521.82
15.4202 141	Steel flange, slip-on, 24", forged steel 150#	EA	8.0765	980.97	966.33	1,947.30
15.4203 000	**VICTAULIC COUPLINGS, WITH GROOVING:**					
15.4203 011	Victaulic coupling, 2", grooving	EA	0.8033	97.57	38.56	136.13
15.4203 021	Victaulic coupling, 3", grooving	EA	0.9014	109.48	52.97	162.45
15.4203 031	Victaulic coupling, 4", grooving	EA	0.9752	118.45	110.42	228.87
15.4203 041	Victaulic coupling, 6", grooving	EA	1.1063	134.37	144.37	278.75
15.4203 051	Victaulic coupling, 8", grooving	EA	1.6388	199.05	236.08	435.13
15.4203 061	Victaulic coupling, 10", grooving	EA	2.0486	248.82	333.31	582.13
15.4203 071	Victaulic coupling, 12", grooving	EA	2.4176	293.64	378.92	672.56
15.4203 081	Victaulic coupling, 14", grooving	EA	2.7859	338.38	445.04	783.42
15.4203 091	Victaulic coupling, 16", grooving	EA	3.1140	378.23	572.78	951.00
15.4300 000	**VALVES & SPECIALTIES:**					
15.4300 001	Note: The following prices include 2 appropriate connections to pipe or equipment.					
15.4301 000	**GATE, GLOBE & CHECK VALVES, BRASS, 125#, SCREWED:**					
15.4301 011	Valve, gate, globe & check, brass, 125#, screwed, 1/2"	EA	0.2816	34.20	10.99	45.19
15.4301 021	Valve, gate, globe & check, brass, 125#, screwed, 3/4"	EA	0.3373	40.97	14.33	55.29
15.4301 031	Valve, gate, globe & check, brass, 125#, screwed, 1"	EA	0.3516	42.71	20.39	63.09
15.4301 041	Valve, gate, globe & check, brass, 125#, screwed, 1-1/4"	EA	0.4501	54.67	29.13	83.80
15.4301 051	Valve, gate, globe & check, brass, 125#, screwed, 1-1/2"	EA	0.6729	81.73	38.31	120.04
15.4301 061	Valve, gate, globe & check, brass, 125#, screwed, 2"	EA	1.2540	152.31	57.93	210.24
15.4301 071	Valve, gate, globe & check, brass, 125#, screwed, 2-1/2"	EA	1.6500	200.41	104.28	304.68
15.4301 081	Valve, gate, globe & check, brass, 125#, screwed, 3"	EA	1.9200	233.20	145.36	378.56
15.4301 091	Valve, gate, globe & check, brass, 125#, screwed, 4"	EA	2.7000	327.94	314.06	642.01
15.4302 000	**GATE, GLOBE & CHECK VALVES, BRONZE, 200#, SCREW:**					
15.4302 011	Valve, gate, globe & check, bronze, 200#, screwed, 1/2"	EA	0.2816	34.20	73.21	107.41
15.4302 021	Valve, gate, globe & check, bronze, 200#, screwed, 3/4"	EA	0.3373	40.97	94.64	135.61

15 - HVAC - MECHANICAL

Division	Description	Unit	Labor Hours	Labor Cost	Material Cost	Total Cost
15.4302 031	Valve, gate, globe & check, bronze, 200#, screwed, 1"	EA	0.3516	42.71	131.99	174.69
15.4302 041	Valve, gate, globe & check, bronze, 200#, screwed, 1-1/4"	EA	0.4501	54.67	197.33	252.00
15.4302 051	Valve, gate, globe & check, bronze, 200#, screwed, 1-1/2"	EA	0.0402	4.88	227.14	232.02
15.4302 061	Valve, gate, globe & check, bronze, 200#, screwed, 2"	EA	1.2540	152.31	345.43	497.74
15.4302 071	Valve, gate, globe & check, bronze, 200#, screwed, 2-1/2"	EA	1.6500	200.41	753.88	954.29
15.4303 000	**GATE, GLOBE & CHECK VALVES, IRON, FLANGE, 125#, BOLT & GASKET:**					
15.4303 001	*Note: The following prices do not include companion flanges.*					
15.4303 011	Valve, gate, globe & check, iron body, flange, 125#, 2", bolt & gaskets	EA	1.2100	146.97	365.56	512.52
15.4303 021	Valve, gate, globe & check, iron body, flange, 125#, 2-1/2", bolt & gaskets	EA	1.3500	163.97	399.63	563.60
15.4303 031	Valve, gate, globe & check, iron body, flange, 125#, 3", bolt & gaskets	EA	1.5000	182.19	456.01	638.20
15.4303 041	Valve, gate, globe & check, iron body, flange, 125#, 4", bolt & gaskets	EA	2.4000	291.50	669.18	960.68
15.4303 051	Valve, gate, globe & check, iron body, flange, 125#, 5-6", bolt & gaskets	EA	3.5000	425.11	1,160.35	1,585.46
15.4303 061	Valve, gate, globe & check, iron body, flange, 125#, 8", bolt & gaskets	EA	4.6000	558.72	2,291.76	2,850.48
15.4303 071	Valve, gate, globe & check, iron body, flange, 125#, 10", bolt & gaskets	EA	6.6000	801.64	3,787.00	4,588.64
15.4303 081	Valve, gate, globe & check, iron body, flange, 125#, 12", bolt & gaskets	EA	7.5000	910.95	5,058.85	5,969.80
15.4303 091	Valve, gate, globe & check, iron body, flange, 125#, 14", bolt & gaskets	EA	12.3000	1,493.96	8,285.18	9,779.14
15.4303 101	Valve, gate, globe & check, iron body, flange, 125#, 16", bolt & gaskets	EA	16.1000	1,955.51	16,347.76	18,303.27
15.4303 111	Valve, gate, globe & check, iron body, flange, 125#, 18", bolt & gaskets	EA	19.7000	2,392.76	19,861.12	22,253.88
15.4304 000	**VALVE, TRIPLE DUTY, IRON, FLANGE, WITH BOLT & GASKET SET:**					
15.4304 001	*Note: The following prices do not include companion flanges.*					
15.4304 011	Valve, 3 duty, iron body, flange .3", bolt & gaskets	EA	1.7000	206.48	729.75	936.23
15.4304 021	Valve, 3 duty, iron body, flange, 4", bolt & gaskets	EA	2.9000	352.23	1,547.45	1,899.68
15.4304 031	Valve, 3 duty, iron body, flange, 5", bolt & gaskets	EA	3.8000	461.55	1,820.04	2,281.59
15.4304 041	Valve, 3 duty, iron body, flange, 6", bolt & gaskets	EA	4.1000	497.99	2,497.07	2,995.06
15.4304 051	Valve, 3 duty, iron body, flange, 8", bolt & gaskets	EA	5.6000	680.18	3,385.12	4,065.29
15.4305 000	**PRESSURE RED VALVES, IRON, BRONZE TRIM, 125, 9-100:**					
15.4305 011	Pressure reducing valve, iron body, bronze trim, 125#, 1-1/2", 9-100	EA	1.2296	149.35	798.36	947.71
15.4305 021	Pressure reducing valve, iron body, bronze trim, 125#, 2", 9-100	EA	1.3915	169.01	903.05	1,072.06
15.4305 031	Pressure reducing valve, iron body, bronze trim, 125#, 3", 9-100	EA	1.9120	232.23	1,278.50	1,510.74
15.4305 041	Pressure reducing valve, iron body, bronze trim, 125#, 4", 9-100	EA	2.8500	346.16	1,814.01	2,160.17
15.4305 051	Pressure reducing valve, iron body, bronze trim, 125#, 6", 9-100	EA	3.8000	461.55	2,879.09	3,340.64
15.4305 061	Pressure reducing valve, iron body, bronze trim, 125#, 8", 9-100	EA	5.6000	680.18	4,831.40	5,511.57
15.4305 071	Pressure reducing valve, iron body, bronze trim, 125#, 10", 9-100	EA	7.5000	910.95	6,625.67	7,536.62
15.4306 000	**VALVES, PRESSURE RELIEF, ASME:**					
15.4306 011	Valve, pressure relief, bronze, 3/4"	EA	0.3378	41.03	53.37	94.40
15.4306 021	Valve, pressure relief, bronze, 1"	EA	0.3516	42.71	100.38	143.08
15.4306 031	Valve, pressure relief, bronze, 1-1/2"	EA	0.6229	75.66	376.80	452.46
15.4306 041	Valve, pressure relief, bronze, 2"	EA	1.2540	152.31	419.43	571.74
15.4307 000	**STEAM TRAPS, CAST IRON, SCREWED, WITH STAINLESS STEEL BUCKET:**					
15.4307 011	Steam trap, cast iron, 1/2", screwed, bucket	EA	0.3282	39.86	166.33	206.20
15.4307 021	Steam trap, cast iron, 3/4", screwed, bucket	EA	0.4529	55.01	237.14	292.15
15.4307 031	Steam trap, cast iron, 1", screwed, bucket	EA	0.5732	69.62	399.18	468.80
15.4307 041	Steam trap, cast iron, 1-1/2", screwed, bucket	EA	0.7374	89.56	721.23	810.80
15.4308 000	**VALVES, BUTTERFLY, IRON BODY & DISCONNECT, NYLON COAT, WAFER:**					
15.4308 001	*Note: The following prices include buna seat, handles and wafer body.*					
15.4308 011	Butterfly valve, iron body, nylon, 2", wafer body	EA	1.0587	128.59	167.80	296.39
15.4308 021	Butterfly valve, iron body, nylon, 2-1/2", wafer	EA	1.2645	153.59	177.01	330.60
15.4308 031	Butterfly valve, iron body, nylon, 3", wafer body	EA	1.4703	178.58	190.78	369.36
15.4308 041	Butterfly valve, iron body, nylon, 4", wafer body	EA	2.1464	260.70	239.09	499.79
15.4308 051	Butterfly valve, iron body, nylon, 5", wafer body	EA	2.8500	346.16	342.50	688.66
15.4308 061	Butterfly valve, iron body, nylon, 6", wafer body	EA	2.8500	346.16	418.33	764.49
15.4308 071	Butterfly valve, iron body, nylon, 8", wafer, chain	EA	4.2000	510.13	625.28	1,135.41
15.4308 081	Butterfly valve, iron body, nylon, 10", wafer, chain	EA	5.6250	683.21	866.60	1,549.82
15.4308 091	Butterfly valve, iron body, nylon, 12", wafer, chain	EA	7.4250	901.84	1,333.28	2,235.12

MECHANICAL - HVAC - 15

Division	Description	Unit	Labor Hours	Labor Cost	Material Cost	Total Cost
15.4309 000	**VALVES, BUTTERFLY, IRON BODY & DISCONNECT, NYLON COAT, LUG TYPE:**					
15.4309 001	*Note: The following prices include buna seat and handles.*					
15.4309 011	Butterfly valve, iron body, nylon, 2", lug type	EA	1.0587	128.59	199.96	328.55
15.4309 021	Butterfly valve, iron body, nylon, 2-1/2", lug type	EA	1.2645	153.59	204.61	358.20
15.4309 031	Butterfly valve, iron body, nylon, 3", lug type	EA	1.4703	178.58	227.58	406.16
15.4309 041	Butterfly valve, iron body, nylon, 4", lug type	EA	2.1464	260.70	285.05	545.75
15.4309 051	Butterfly valve, iron body, nylon, 5", lug type	EA	2.8500	346.16	434.44	780.60
15.4309 061	Butterfly valve, iron body, nylon, 6", lug type	EA	2.8500	346.16	485.06	831.22
15.4309 071	Butterfly valve, iron body, nylon, 8", lug, chain	EA	4.2000	510.13	673.50	1,183.63
15.4309 081	Butterfly valve, iron body, nylon, 10", lug, chain	EA	5.6250	683.21	947.08	1,630.29
15.4309 091	Butterfly valve, iron body, nylon, 12", lug, chain	EA	7.4250	901.84	1,498.77	2,400.61
15.4310 000	**VALVES, FLOW CONTROL:**					
15.4310 011	Valve, flow control, straight angle, 3/4"	EA	0.3066	37.24		37.24
15.4310 021	Valve, flow control, straight angle, 1"	EA	0.3196	38.82		38.82
15.4310 031	Valve, flow control, straight angle, 1-1/4"	EA	0.4092	49.70		49.70
15.4310 041	Valve, flow control, straight angle, 1-1/2"	EA	0.6729	81.73		81.73
15.4310 051	Valve, flow control, straight angle, 2"	EA	1.1400	138.46		138.46
15.4310 061	Valve, flow control, straight, 2-1/2"	EA	1.5500	188.26		188.26
15.4310 071	Valve, flow control, straight, 3"	EA	1.9200	233.20		233.20
15.4310 081	Valve, flow control, straight, 4"	EA	2.7000	327.94		327.94
15.4311 000	**SUCTION DIFFUSER, ANGLE BODY:**					
15.4311 001	*Note: The following prices include inlet, vanes, strainer and permanent magnet.*					
15.4311 011	Suction diffuser, angle body, 3 x 3	EA	1.7500	212.56	632.46	845.02
15.4311 021	Suction diffuser, angle body, 4 x 3	EA	2.5700	312.15	746.73	1,058.88
15.4311 031	Suction diffuser, angle body, 4 x 4	EA	2.5700	312.15	868.74	1,180.89
15.4311 041	Suction diffuser, angle body, 6 x 4	EA	3.4500	419.04	1,028.71	1,447.74
15.4311 051	Suction diffuser, angle body, 6 x 6	EA	3.4500	419.04	1,280.14	1,699.18
15.4311 061	Suction diffuser, angle body, 8 x 6	EA	5.1000	619.45	1,379.28	1,998.73
15.4311 071	Suction diffuser, angle body, 8 x 8	EA	5.1510	625.64	2,385.09	3,010.73
15.4311 081	Suction diffuser, angle body, 10 x 10	EA	7.0700	858.72	3,230.95	4,089.68
15.4312 000	**VALVES, BALANCING, CIRCUIT SETTER:**					
15.4312 011	Valve, balancing, circuit setter, 1/2"	EA	0.2560	31.09	89.93	121.02
15.4312 013	Valve, balancing, circuit setter, 3/4"	EA	0.3066	37.24	109.72	146.96
15.4312 021	Valve, balancing, circuit setter, 1"	EA	0.3196	38.82	126.96	165.77
15.4312 041	Valve, balancing, circuit setter, 1-1/4"	EA	0.4092	49.70	156.47	206.17
15.4312 051	Valve, balancing, circuit setter, 1-1/2"	EA	0.6729	81.73	168.62	250.35
15.4312 061	Valve, balancing, circuit setter, 2"	EA	1.1400	138.46	277.30	415.77
15.4312 071	Valve, balancing, circuit setter, 2-1/2"	EA	1.5500	188.26	473.41	661.67
15.4312 081	Valve, balancing, circuit setter, 3"	EA	1.9200	233.20	703.23	936.43
15.4312 091	Valve, balancing, circuit setter, 4"	EA	2.7000	327.94	999.55	1,327.49
15.4313 000	**FLEXIBLE CONNECTOR, WITH NEOPRENE COVER, BELLOWS, 125#, FLANGE:**					
15.4313 001	*Note: The following prices do not include companion flanges or bolt and gasket sets.*					
15.4313 011	Flexible connector, flange, 125#, 2", cover, bellows	EA	0.6800	82.59	167.05	249.65
15.4313 021	Flexible connector, flange, 125#, 2-1/2", cover, bellows	EA	0.7500	91.10	205.57	296.67
15.4313 031	Flexible connector, flange, 125#, 3", cover, bellows	EA	1.1000	133.61	221.70	355.30
15.4313 041	Flexible connector, flange, 125#, 4", cover, bellows	EA	1.5500	188.26	298.70	486.96
15.4313 051	Flexible connector, flange, 125#, 5", cover, bellows	EA	1.9900	241.71	327.71	569.42
15.4313 061	Flexible connector, flange, 125#, 6", cover, bellows	EA	2.5100	304.86	411.21	716.08
15.4314 000	**METER, FLOW, CIRCUIT SENSOR:**					
15.4314 011	Meter, flow, circuit sensor, 6"	EA	3.5000	425.11	322.87	747.98
15.4314 021	Meter, flow, circuit sensor, 8"	EA	4.6000	558.72	472.20	1,030.92
15.4314 031	Meter, flow, circuit sensor, 10"	EA	6.0600	736.05	543.85	1,279.90
15.4314 041	Meter, flow, circuit sensor, 12"	EA	7.5000	910.95	906.38	1,817.33

15 - HVAC - MECHANICAL

Division	Description	Unit	Labor Hours	Labor Cost	Material Cost	Total Cost
15.4400 000	**INSULATION, PIPING:**					
15.4400 001	Note: The following prices include insulation allowance at pipe runs, valves and fittings.					
15.4401 000	**INSULATION, 1-1/2" CALCIUM SILICATE:**					
15.4401 001	Note: For 1" calsil on pipe sizes 1" to 6"... deduct 28% from material					
	For 2" calsil on pipe sizes 1" to 6"... add 56%					
	For 2" calsil on sizes 8" to 14"...... add 41%					
	For 2 1/2" calsil on sizes 1" to 6".... add 94%					
	For 2 1/2" calsil on sizes 8" to 14"... add 80%					
15.4401 011	Insulation, 1-1/2" calcium silicate, 1/2" pipe	LF	0.0261	3.17	4.33	7.50
15.4401 021	Insulation, 1-1/2" calcium silicate, 3/4" pipe	LF	0.0261	3.17	4.44	7.61
15.4401 031	Insulation, 1-1/2" calcium silicate, 1" pipe	LF	0.0261	3.17	4.79	7.96
15.4401 041	Insulation, 1-1/2" calcium silicate, 1-1/4" pipe	LF	0.0261	3.17	5.00	8.17
15.4401 051	Insulation, 1-1/2" calcium silicate, 1-1/2" pipe	LF	0.0261	3.17	5.36	8.53
15.4401 061	Insulation, 1-1/2" calcium silicate, 2" pipe	LF	0.0384	4.66	5.94	10.60
15.4401 071	Insulation, 1-1/2" calcium silicate, 2-1/2" pipe	LF	0.0384	4.66	6.45	11.12
15.4401 081	Insulation, 1-1/2" calcium silicate, 3" pipe	LF	0.0514	6.24	6.78	13.03
15.4401 091	Insulation, 1-1/2" calcium silicate, 4" pipe	LF	0.0514	6.24	7.82	14.06
15.4401 101	Insulation, 1-1/2" calcium silicate, 5" pipe	LF	0.0514	6.24	8.79	15.03
15.4401 111	Insulation, 1-1/2" calcium silicate, 6" pipe	LF	0.0514	6.24	9.12	15.37
15.4401 121	Insulation, 1-1/2" calcium silicate, 8" pipe	LF	0.0644	7.82	12.33	20.15
15.4401 131	Insulation, 1-1/2" calcium silicate, 10" pipe	LF	0.0767	9.32	16.45	25.76
15.4401 141	Insulation, 1-1/2" calcium silicate, 12" pipe	LF	0.0897	10.89	19.39	30.29
15.4401 151	Insulation, 1-1/2" calcium silicate, 14" pipe	LF	0.0897	10.89	22.03	32.92
15.4402 000	**INSULATION, 3" CALCIUM SILICATE:**					
15.4402 011	Insulation, 3" calcium silicate, 1" pipe	LF	0.0510	6.19	11.35	17.54
15.4402 021	Insulation, 3" calcium silicate, 2" pipe	LF	0.0610	7.41	13.31	20.71
15.4402 031	Insulation, 3" calcium silicate, 3" pipe	LF	0.0767	9.32	15.32	24.64
15.4402 041	Insulation, 3" calcium silicate, 4" pipe	LF	0.0767	9.32	20.15	29.46
15.4402 051	Insulation, 3" calcium silicate, 5" pipe	LF	0.0926	11.25	22.92	34.17
15.4402 061	Insulation, 3" calcium silicate, 6" pipe	LF	0.0926	11.25	24.63	35.88
15.4402 071	Insulation, 3" calcium silicate, 8" pipe	LF	0.1035	12.57	29.29	41.86
15.4402 081	Insulation, 3" calcium silicate, 10" pipe	LF	0.1150	13.97	35.33	49.30
15.4402 091	Insulation, 3" calcium silicate, 12" pipe	LF	0.1380	16.76	39.16	55.92
15.4402 101	Insulation, 3" calcium silicate, 14" pipe	LF	0.1590	19.31	43.93	63.24
15.4403 000	**INSULATION, 1" FIBERGLASS WITH ALUMINUM JACKET:**					
15.4403 001	Note: For 1-1/2" fiberglass on sizes 1/2" to 1-1/2"... add 100% to material					
	For sizes 2" to 3"........................ add 70%					
	For sizes 4" to 6"........................ add 45%					
	For sizes 8" to 12"....................... add 25%					
15.4403 011	Insulation, 1" fiberglass, 1/2" pipe, with aluminum jacket	LF	0.0384	4.66	2.28	6.94
15.4403 021	Insulation, 1" fiberglass, 3/4" pipe, with aluminum jacket	LF	0.0384	4.66	2.51	7.17
15.4403 031	Insulation, 1" fiberglass, 1" pipe, with aluminum jacket	LF	0.0384	4.66	2.69	7.36
15.4403 041	Insulation, 1" fiberglass, 1-1/4" pipe, with aluminum jacket	LF	0.0384	4.66	3.24	7.91
15.4403 051	Insulation, 1" fiberglass, 1-1/2" pipe, with aluminum jacket	LF	0.0384	4.66	3.29	7.95
15.4403 061	Insulation, 1" fiberglass, 2" pipe, with aluminum jacket	LF	0.0514	6.24	3.59	9.83
15.4403 071	Insulation, 1" fiberglass, 2-1/2" pipe, with aluminum jacket	LF	0.0514	6.24	4.02	10.27
15.4403 081	Insulation, 1" fiberglass, 3" pipe, with aluminum jacket	LF	0.0644	7.82	4.44	12.26
15.4403 091	Insulation, 1" fiberglass, 4" pipe, with aluminum jacket	LF	0.0644	7.82	5.81	13.63
15.4403 101	Insulation, 1" fiberglass, 5" pipe, with aluminum jacket	LF	0.0811	9.85	6.67	16.52
15.4403 111	Insulation, 1" fiberglass, 6" pipe, with aluminum jacket	LF	0.0811	9.85	7.14	16.99
15.4403 121	Insulation, 1" fiberglass, 8" pipe, with aluminum jacket	LF	0.1035	12.57	10.31	22.89
15.4403 131	Insulation, 1" fiberglass, 10" pipe, with aluminum jacket	LF	0.1410	17.13	12.15	29.27
15.4403 141	Insulation, 1" fiberglass, 12" pipe, with aluminum jacket	LF	0.1533	18.62	13.89	32.51
15.4404 000	**INSULATION, 2" FIBERGLASS WITH ALUMINUM JACKET:**					
15.4404 011	Insulation, 2" fiberglass, 1/2" pipe, with aluminum jacket	LF	0.0514	6.24	7.78	14.02
15.4404 021	Insulation, 2" fiberglass, 3/4" pipe, with aluminum jacket	LF	0.0514	6.24	7.96	14.21
15.4404 031	Insulation, 2" fiberglass, 1" pipe, with aluminum jacket	LF	0.0514	6.24	8.41	14.65

MECHANICAL - FIRE PROTECTION - 15

Division	Description	Unit	Labor Hours	Labor Cost	Material Cost	Total Cost
15.4404 041	Insulation, 2" fiberglass, 1-1/4" pipe, with aluminum jacket	LF	0.0514	6.24	8.90	15.15
15.4404 051	Insulation, 2" fiberglass, 1-1/2" pipe, with aluminum jacket	LF	0.0514	6.24	9.37	15.62
15.4404 061	Insulation, 2" fiberglass, 2" pipe, with aluminum jacket	LF	0.0644	7.82	9.73	17.55
15.4404 071	Insulation, 2" fiberglass, 2-1/2" pipe, with aluminum jacket	LF	0.0644	7.82	10.57	18.39
15.4404 081	Insulation, 2" fiberglass, 3" pipe, with aluminum jacket	LF	0.0767	9.32	11.25	20.57
15.4404 091	Insulation, 2" fiberglass, 4" pipe, with aluminum jacket	LF	0.0767	9.32	13.11	22.43
15.4404 101	Insulation, 2" fiberglass, 5" pipe, with aluminum jacket	LF	0.0767	9.32	14.83	24.15
15.4404 111	Insulation, 2" fiberglass, 6" pipe, with aluminum jacket	LF	0.0767	9.32	15.10	24.42
15.4404 121	Insulation, 2" fiberglass, 8" pipe, with aluminum jacket	LF	0.1150	13.97	18.73	32.69
15.4404 131	Insulation, 2" fiberglass, 10" pipe, with aluminum jacket	LF	0.1533	18.62	22.38	41.00
15.4404 141	Insulation, 2" fiberglass, 12" pipe, with aluminum jacket	LF	0.1663	20.20	24.95	45.15
15.4405 000	**INSULATION, VALVES, FIBERGLASS WITH ALUMINUM JACKET:**					
15.4405 011	Insulation, fiberglass, 1" valve, aluminum jacket	EA	0.3525	42.81	9.65	52.46
15.4405 021	Insulation, fiberglass, 2" valve, aluminum jacket	EA	0.5671	68.88	25.82	94.70
15.4405 031	Insulation, fiberglass, 3" valve, aluminum jacket	EA	0.7281	88.44	33.09	121.52
15.4405 041	Insulation, fiberglass, 4" valve, aluminum jacket	EA	0.9119	110.76	38.71	149.47
15.4405 051	Insulation, fiberglass, 6" valve, aluminum jacket	EA	1.3564	164.75	60.49	225.23
15.4405 061	Insulation, fiberglass, 8" valve, aluminum jacket	EA	1.7778	215.93	68.82	284.75
15.4405 071	Insulation, fiberglass, 10" valve, aluminum jacket	EA	2.3296	282.95	77.41	360.37
15.4405 081	Insulation, fiberglass, 12" valve, aluminum jacket	EA	3.0806	374.17	86.04	460.21
15.4406 000	**PERMITS, TEST AND BALANCE:**					
15.4406 001	Note: Permits will equal approximately 3% of the HVAC. Testing will be approximately 5% of the equipment costs.					
15.4406 031	Balance	REG	0.7664	93.09		93.09
15.5500 000	**FIRE PROTECTION SYSTEMS:**					
15.5500 001	Note: Remember to add alarm and valve riser for each fire sprinkler system.					
15.5501 000	**FIRE PROTECTION, EXPOSED SYSTEM, WET, PLACED, NORMAL HAZARD:**					
15.5501 011	Fire protection sprinklers, exposed, wet, 5,000 SF, normal hazard	SF	0.0261	2.73	2.38	5.11
15.5501 021	Fire protection sprinklers, exposed, wet, 6,000-15,000 SF, normal hazard	SF	0.0190	1.98	1.98	3.97
15.5501 031	Fire protection sprinklers, exposed, wet, over 15,000 SF, normal hazard	SF	0.0180	1.88	1.90	3.78
15.5502 000	**FIRE PROTECTION,, CONCEALED SYSTEM, WET, PLACED, NORMAL HAZARD:**					
15.5502 001	Note: For high hazard, add 30% to the material costs and 15% to the labor costs For light hazard deduct 15% from the material costs For concealed systems in rooms less than 2,500 SF, add 40% to the total costs					
15.5502 011	Fire protection sprinklers, concealed, wet, 5,000 SF, normal hazard	SF	0.0283	2.96	3.82	6.77
15.5502 021	Fire protection sprinklers, concealed, wet, 6,000-15,000 SF, normal hazard	SF	0.0210	2.19	3.35	5.54
15.5502 031	Fire protection sprinklers, concealed, wet, over 15,000 SF, normal hazard	SF	0.0197	2.06	3.15	5.21
15.5502 041	Add for testing systems	HEAD	0.0261	2.73	3.23	5.96
15.5503 000	**FIRE PROTECTION, EXPOSED SYSTEM, WET, PER HEAD:**					
15.5503 011	Fire protection sprinklers, exposed, wet, small	HEAD	2.0863	217.93	172.37	390.31
15.5503 021	Fire protection sprinklers, exposed, wet, medium	HEAD	1.9560	204.32	159.58	363.91
15.5503 031	Fire protection sprinklers, exposed, wet, large	HEAD	1.7779	185.72	145.09	330.80
15.5504 000	**FIRE PROTECTION, CONCEALED SYSTEM, WET, PER HEAD:**					
15.5504 011	Fire protection sprinklers, concealed, wet, small	HEAD	2.3471	245.18	191.57	436.75
15.5504 021	Fire protection sprinklers, concealed, wet, medium	HEAD	2.2167	231.56	180.86	412.42
15.5504 031	Fire protection sprinklers, concealed, wet, large	HEAD	2.0390	212.99	166.41	379.41
15.5505 000	**FIRE WATER CONNECTION FEES (WHERE APPLICABLE):**					
15.5505 011	Fire water, 4" connection	EA	53.2226	5,559.63	3,821.11	9,380.74
15.5505 021	Fire water, 6" connection	EA	61.6387	6,438.78	4,426.14	10,864.92
15.5505 031	Fire water, 8" connection	EA	79.5390	8,308.64	5,709.83	14,018.47
15.5506 000	**STANDPIPE, DRY, 6" DIAMETER, HOOK UP:**					
15.5506 011	Standpipe connection, pumper, 6"	EA	5.3106	554.75	1,934.64	2,489.39
15.5506 021	Standpipe connection, pumper, 4"	EA	5.1715	540.21	1,842.82	2,383.04
15.5506 031	Gate valve, 3 x 2-1/2 PB	EA	1.0706	111.83	319.76	431.60
15.5506 041	Roof manifold, with valves	EA	6.8954	720.29	1,181.08	1,901.37

15 - FIRE PROTECTION - MECHANICAL

Division	Description	Unit	Labor Hours	Labor Cost	Material Cost	Total Cost
15.5507 000	**STANDPIPE, WET, HOSE RACK STATION:**					
15.5507 001	*Note: The following prices include 75' hose, cabinet, angle valves, flow switch and pipe.*					
15.5507 011	Standpipe, wet, 2-1/2", per floor	EA	7.6709	801.30	2,545.51	3,346.81
15.5507 021	Standpipe, 4", wet, per floor	EA	8.4927	887.15	2,727.19	3,614.34
15.5507 031	Standpipe, 6", wet, per floor	EA	9.0933	949.89	3,127.92	4,077.80
15.5508 000	**ALARM & VALVE RISERS:**					
15.5508 001	*Note: The following items should be added to each fire sprinkler system.*					
15.5508 011	Alarm & valve riser, 4"	EA	21.6007	2,256.41	2,817.06	5,073.47
15.5508 021	Alarm & valve riser, 6"	EA	22.1275	2,311.44	3,342.07	5,653.51
15.5508 031	Alarm & valve riser, 8"	EA	23.7081	2,476.55	4,029.39	6,505.94
15.5509 000	**FIRE PUMPS, ELECTRIC:**					
15.5509 011	Fire pump, electric, 250 GPM @ 40 PSI	EA	45.0000	4,700.70	20,621.28	25,321.98
15.5509 021	Fire pump, electric, 500 GPM @ 100 PSI	EA	48.0000	5,014.08	27,799.71	32,813.79
15.5509 031	Fire pump, electric, 750 GPM @ 100 PSI	EA	54.0000	5,640.84	30,036.19	35,677.03
15.5509 041	Fire pump, electric, 1000 GPM @ 150 PSI	EA	60.0000	6,267.60	57,234.67	63,502.27
15.5509 051	Fire pump, electric, 1500 GPM @ 150 PSI	EA	80.0000	8,356.80	59,904.05	68,260.85
15.5509 061	Fire pump, electric, 2000 GPM @ 150 PSI	EA	104.0000	10,863.84	63,282.77	74,146.61
15.5509 071	Fire pump, electric, 3000 GPM @ 150 PSI	EA	124.0000	12,953.04	84,120.52	97,073.56
15.5509 081	Jockey pump, 10 hp	EA	8.0000	835.68	2,849.72	3,685.40
15.5510 000	**FIRE PUMPS, DIESEL:**					
15.5510 011	Fire pump, diesel, 500 GPM @ 100 PSI	EA	50.0000	5,223.00	57,499.16	62,722.16
15.5510 021	Fire pump, diesel, 750 GPM @ 100 PSI	EA	56.0000	5,849.76	65,423.03	71,272.79
15.5510 031	Fire pump, diesel, 1000 GPM @ 150 PSI	EA	66.0000	6,894.36	93,727.77	100,622.13
15.5510 041	Fire pump, diesel, 1500 GPM @ 150 PSI	EA	82.0000	8,565.72	105,956.25	114,521.97
15.5510 051	Fire pump, diesel, 2000 GPM @ 150 PSI	EA	108.0000	11,281.68	110,900.58	122,182.26
15.5510 061	Fire pump, diesel, 3000 GPM @ 150 PSI	EA	142.0000	14,833.32	201,764.19	216,597.51
15.5511 000	**BACKFLOW PREVENTER, OUTSIDE STEM & YOKE, UL APPROVED:**					
15.5511 011	Back flow preventer, outside stem & yoke, UL approved, 2-1/2"	EA	6.2011	647.77	5,352.50	6,000.27
15.5511 021	Back flow preventer, outside stem & yoke, UL approved, 3"	EA	7.4762	780.96	5,684.76	6,465.73
15.5511 031	Back flow preventer, outside stem & yoke, UL approved, 4"	EA	11.8184	1,234.55	7,369.56	8,604.11
15.5511 041	Back flow preventer, outside stem & yoke, UL approved, 6"	EA	13.8711	1,448.98	10,004.24	11,453.21
15.5511 051	Back flow preventer, outside stem & yoke, UL approved, 8"	EA	16.0112	1,672.53	15,686.87	17,359.40
15.5511 061	Back flow preventer, outside stem & yoke, UL approved, 10"	EA	24.1313	2,520.76	18,086.21	20,606.96
15.5512 000	**FIRE PROTECTION, HALON SYSTEM, 8' CEILING & 1" RAISED FLOOR:**					
15.5512 011	Fire protection, halon, 200 SF, 8' ceiling, 1" floor	SF			115.62	115.62
15.5512 021	Fire protection, halon, 500 SF, 8' ceiling, 1" floor	SF			42.77	42.77
15.5512 031	Fire protection, halon, 600 SF, 8' ceiling, 1" floor	SF			41.65	41.65
15.5512 041	Fire protection, halon, 700 SF, 8' ceiling, 1" floor	SF			39.33	39.33
15.5512 051	Fire protection, halon, 1,000 SF, 8' ceiling, 1" floor	SF			34.64	34.64
15.5512 061	Fire protection, halon, 2,000 SF, 8' ceiling, 1" floor	SF			26.55	26.55
15.5512 071	Fire protection, halon, 2,500 SF, 8' ceiling, 1" floor	SF			23.68	23.68
15.5512 081	Fire protection, halon, 3,000 SF, 8' ceiling, 1" floor	SF			21.37	21.37
15.5512 091	Fire protection, halon, 4,000 SF, 8' ceiling, 1" floor	SF			19.67	19.67
15.5512 101	Fire protection, halon, 5,000 SF, 8' ceiling, 1" floor	SF			17.90	17.90
15.5512 111	Fire protection, halon, 10,000 SF, 8' ceiling, 1" floor	SF			16.19	16.19
15.5512 121	Fire protection, halon, 20,000 SF, 8' ceiling, 1" floor	SF			13.90	13.90
15.5513 000	**FIRE PROTECTION, CHEMETRON SYSTEM, FM-200:**					
15.5513 011	Fire protection, chemetron, raised floor, to 5,000 SF	CF			7.12	7.12
15.5513 021	Fire protection, chemetron, raised floor, to 10,000 SF	CF			6.12	6.12
15.5513 031	Fire protection, chemetron, non raised floor, to 10,000 SF	CF			5.10	5.10
15.5515 000	**FIRE PROTECTION, IN PLACE, TYPICAL PER SQUARE FOOT:**					
15.5515 011	Fire protection, housing high rise	SF	0.0169	1.77	2.10	3.86
15.5515 021	Fire protection, auditorium/theater	SF	0.0261	2.73	3.14	5.87
15.5515 031	Fire protection, college/school	SF	0.0189	1.97	2.26	4.23
15.5515 041	Fire protection, garage, underground	SF	0.0126	1.32	1.54	2.85
15.5515 051	Fire protection, hospital	SF	0.0273	2.85	3.32	6.18

MECHANICAL - FIRE PROTECTION - 15

Division	Description	Unit	Labor Hours	Labor Cost	Material Cost	Total Cost
15.5515 061	Fire protection, government building	SF	0.0223	2.33	2.74	5.07
15.5515 071	Fire protection, manufacturing	SF	0.0142	1.48	1.66	3.15
15.5515 081	Fire protection, medical clinic	SF	0.0185	1.93	2.26	4.19
15.5515 091	Fire protection, office, low rise	SF	0.0180	1.88	2.21	4.09
15.5515 111	Fire protection, office, high rise	SF	0.0199	2.08	2.42	4.50
15.5515 121	Fire protection, warehouse	SF	0.0126	1.32	1.54	2.85
15.5515 131	Fire protection, shopping center, 1 story	SF	0.0164	1.71	1.95	3.66
15.5515 141	Fire protection, shopping center, quality	SF	0.0185	1.93	2.26	4.19
15.5515 151	Fire protection, hotel, low rise	SF	0.0196	2.05	2.35	4.40
15.5515 161	Fire protection, hotel, mid rise	SF	0.0206	2.15	2.52	4.67
15.5515 171	Fire protection, hotel, high rise	SF	0.0205	2.14	2.48	4.62
15.5515 181	Fire protection, hotel, high, deluxe	SF	0.0216	2.26	2.58	4.83
15.5515 191	Fire protection, hotel, high, luxury	SF	0.0227	2.37	2.74	5.11

Electrical

Table of Contents

16.0000.000	**ELECTRICAL WORK**	**205**
16.0100.000	Total Electrical Work, Buildings	205
16.1000.000	Electrical Costs, In-Place, Preliminary Estimates	205
16.1100.000	Main Switchboards, 600V, Service & Distribution	206
16.1200.000	Distribution Panels to 600V	206
16.1300.000	Transformers	207
16.1400.000	Raceway & Wire, Combined	208
16.1500.000	Underfloor Distribution Systems	208
16.1600.000	Lighting Fixtures, In-Place	209
16.1700.000	Branch Circuit Runs, Special Purpose Conduit & Wire	210
16.1800.000	Signal & Communications Systems	210
16.1900.000	Branch Circuit Outlets & Devices	211
16.2000.000	Equipment, Unit Substations	211
16.2100.000	Equipment, Switchgear & Transformers	212
16.2200.000	Equipment, High Voltage Transformers	212
16.2300.000	Service Sections	213
16.2400.000	Combination Service & Distribution Switchboards	215
16.3000.000	Motor Control Centers	218
16.4000.000	Panelboards, 600V Max, Bolt-on Breakers	220
16.4100.000	Transformers, Dry, Low Voltage	220
16.4200.000	Panelboards for Bolt-on Breakers, 120/240V, 1PH, 3W	221
16.4300.000	Load Centers, Main Lug & Circuit Breaker Types, 240V Max	221
16.4400.000	Plug-in Circuit Breakers, Type QO	222
16.4500.000	Special Gear	223
16.5000.000	PVC, RSC, IMC & Aluminum Raceway	223
16.5100.000	PVC, RSC, IMC & Alum Conduit Terminals, Elbows & Fittings	226
16.5200.000	EMT Raceway, Terminations & Elbows	227
16.5300.000	EMT, MI Cable & Terminations	229
16.5400.000	Specialty Fittings, Explosion Proof	229
16.5500.000	Underfloor & Flush Trench Duct, Cable Tray	230
16.5600.000	Steel Gutters, Pull Boxes, Unistrut Hangers	231
16.5700.000	Special Raceway Assembly Systems	231
16.5800.000	Conductor Only	232
16.5900.000	Busways	235
16.5950.000	Raceway & Wire Combined	237
16.6000.000	Lighting Fixtures	238
16.7000.000	Electric & Signal Devices	242
16.7100.000	Communication, Intercom, Public Address	244
16.7200.000	Special Hospital Systems	246
16.7300.000	Prison Cell Door Control Systems	247
16.7500.000	Soft Wire Systems, 3 Wire	247
16.7600.000	Energy & Building Management Systems	248
16.7700.000	Testing	248

ELECTRICAL - 16

Division	Description	Unit	Labor Hours	Labor Cost	Material Cost	Total Cost
16.0000 000	**ELECTRICAL WORK:**					
16.0101 000	**TOTAL ELECTRICAL WORK, BUILDINGS:**					
16.0101 011	Electrical work, Commercial stores	SF	0.0798	8.91	4.41	13.33
16.0101 015	Electrical work, Commercial stores, quality retail	SF	0.1277	14.27	9.40	23.67
16.0101 021	Electrical work, Market buildings	SF	0.0829	9.26	4.67	13.93
16.0101 031	Electrical work, Recreational buildings	SF	0.0932	10.41	15.53	25.94
16.0101 041	Electrical work, Schools, elementary & high	SF	0.2186	24.42	12.78	37.20
16.0101 051	Electrical work, College buildings	SF	0.1378	15.39	14.56	29.95
16.0101 055	Electrical work, College dormitory	SF	0.0875	9.77	7.73	17.50
16.0101 061	Electrical work, Clinical-mob	SF	0.2260	25.25	13.07	38.31
16.0101 071	Electrical work, Hospitals, full service	SF	0.3191	35.65	22.00	57.64
16.0101 081	Electrical work, Office buildings, high rise, 4 - 7 story	SF	0.1364	15.24	7.68	22.92
16.0101 085	Electrical work, Office buildings, high rise, 8 - 30 story	SF	0.1100	12.29	6.79	19.08
16.0101 091	Electrical work, Warehouses	SF	0.0432	4.83	2.47	7.30
16.0101 101	Electrical work, Parking lots, 2.5 foot candles	SF	0.0128	1.43	1.51	2.94
16.0101 111	Electrical work, Garages, 5 foot candles	SF	0.0193	2.16	2.15	4.30
16.0101 121	Electrical work, Assembly buildings, light industrial	SF	0.0844	9.43	8.58	18.01
16.0101 131	Electrical work, Laboratory buildings, wet science	SF	0.2469	27.58	27.39	54.97
16.0101 141	Electrical work, City hall	SF	0.1959	21.88	17.95	39.84
16.0101 151	Electrical work, Hotel, low rise	SF	0.0733	8.19	4.59	12.78
16.0101 161	Electrical work, Hotel, mid rise	SF	0.0891	9.95	6.10	16.05
16.0101 171	Electrical work, Hotel, high rise, first class	SF	0.0984	10.99	8.24	19.23
16.0101 181	Electrical work, Hotel, high rise, deluxe	SF	0.1094	12.22	9.40	21.63
16.0101 191	Electrical work, Hotel, high rise, luxury	SF	0.1280	14.30	12.79	27.09
16.0101 201	Electrical work, Housing, low rise	SF	0.0447	4.99	2.50	7.49
16.0101 211	Electrical work, Housing, high rise	SF	0.0639	7.14	4.69	11.83
16.1000 000	**ELECTRICAL COST, IN-PLACE, PRELIMINARY ESTIMATES:**					
16.1001 000	**SELECTOR SWITCHES:**					
16.1001 011	Selector switch, 3-way, high voltage	EA	15.9588	1,782.76	9,175.59	10,958.35
16.1001 021	Selector switch, 4-way, high voltage	EA	17.5234	1,957.54	9,850.30	11,807.84
16.1002 000	**HIGH VOLTAGE UNIT SUBSTATIONS:**					
16.1002 011	Unit substation, high voltage, 150kva	EA	52.3776	5,851.10	56,497.09	62,348.20
16.1002 021	Unit substation, high voltage, 225kva	EA	65.4720	7,313.88	66,540.99	73,854.86
16.1002 031	Unit substation, high voltage, 300kva	EA	73.6560	8,228.11	79,095.86	87,323.97
16.1002 041	Unit substation, high voltage, 500kva	EA	98.2080	10,970.82	95,417.28	106,388.10
16.1002 051	Unit substation, high voltage, 750kva	EA	130.9440	14,627.75	112,994.13	127,621.89
16.1002 061	Unit substation, high voltage, 1000kva	EA	163.6800	18,284.69	125,549.08	143,833.78
16.1002 071	Unit substation, high voltage, 1500kva	EA	196.4160	21,941.63	144,381.40	166,323.03
16.1002 081	Unit substation, high voltage, 2000kva	EA	261.8880	29,255.51	163,213.77	192,469.28
16.1003 000	**DISTRIBUTION TRANSFORMERS, PRIMARY:**					
16.1003 011	Transformer, high voltage, 112.5kva	EA	26.1888	2,925.55	20,656.30	23,581.85
16.1003 021	Transformer, high voltage, 150kva	EA	29.4624	3,291.24	24,098.89	27,390.13
16.1003 031	Transformer, high voltage, 225kva	EA	44.1936	4,936.87	35,158.78	40,095.64
16.1003 041	Transformer, high voltage, 300kva	EA	49.1040	5,485.41	44,190.63	49,676.04
16.1003 051	Transformer, high voltage, 500kva	EA	57.2880	6,399.64	56,897.20	63,296.84
16.1003 061	Transformer, high voltage, 750kva	EA	65.4720	7,313.88	65,806.93	73,120.81
16.1003 071	Transformer, high voltage, 1000kva	EA	81.8400	9,142.35	81,422.55	90,564.90
16.1003 081	Transformer, high voltage, 1500kva	EA	98.2080	10,970.82	93,685.23	104,656.04
16.1003 091	Transformer, high voltage, 2000kva	EA	114.5760	12,799.28	104,833.15	117,632.43
16.1003 101	Transformer, high voltage, 2500kva	EA	130.9440	14,627.75	120,997.69	135,625.44
16.1003 111	Transformer, high voltage, 3000kva	EA	147.3120	16,456.22	143,850.88	160,307.10
16.1004 000	**MAIN SWITCHGEAR, TO 600V, LIGHT COMMERCIAL:**					
16.1004 001	*Note: The following prices are based on average service amps.*					
16.1004 011	Service enclosure with metering	AMP	0.0058	0.65	1.28	1.93
16.1004 021	Pull section	AMP	0.0026	0.29	0.59	0.88
16.1004 031	Fire alarm circuit breaker	AMP	0.0015	0.17	0.24	0.41

16 - ELECTRICAL

Division	Description	Unit	Labor Hours	Labor Cost	Material Cost	Total Cost
16.1004 041	Main disconnect	AMP	0.0054	0.60	1.27	1.87
16.1004 051	Distribution circuit breaker to 150% service	AMP	0.0173	1.93	4.01	5.94
16.1004 061	Total average	AMP	0.0322	3.60	7.54	11.14
16.1005 000	**MAIN SWITCHGEAR, TO 600V, COMMERCIAL & SMALL INSTITUTIONAL:**					
16.1005 001	*Note: The following prices are based on average service amps.*					
16.1005 011	Service enclosure with metering	AMP	0.0218	2.44	5.07	7.51
16.1005 021	Pull section	AMP	0.0054	0.60	1.28	1.89
16.1005 031	Fire alarm circuit breaker	AMP	0.0024	0.27	0.37	0.64
16.1005 041	Main disconnect	AMP	0.0082	0.92	1.92	2.83
16.1005 051	Distribution circuit breaker to 150% service	AMP	0.0229	2.56	4.51	7.06
16.1005 061	Total average	AMP	0.0587	6.56	13.70	20.26
16.1006 000	**MAIN SWITCHGEAR, TO 600V, LARGE COMMERCIAL & INSTITUTIONAL:**					
16.1006 001	*Note: The following prices are based on average service amps.*					
16.1006 011	Service enclosure with metering	AMP	0.0184	2.06	4.47	6.53
16.1006 021	Pull section	AMP	0.0035	0.39	0.83	1.22
16.1006 031	Fire alarm circuit breaker	AMP	0.0009	0.10	0.07	0.17
16.1006 041	Main disconnect	AMP	0.0382	4.27	9.45	13.72
16.1006 051	Distribution circuit breaker to 150% service	AMP	0.0246	2.75	9.31	12.06
16.1006 061	Ground fault system	AMP	0.0026	0.29	0.69	0.98
16.1006 071	Customer option metering	AMP	0.0024	0.27	0.24	0.51
16.1006 081	Total average	AMP	0.1024	11.44	22.43	33.86
16.1100 000	**MAIN SWITCHBOARDS, 600V, SERVICE & DISTRIBUTION:**					
16.1101 000	**MAIN SWITCHBOARDS, 600V, COMMERCIAL & LIGHT INSTITUTIONAL:**					
16.1101 001	*Note: The following prices include service & distribution sides with average distribution and feeder breakers.*					
16.1101 011	Main switchboard, 600v, 400a	EA	32.7360	3,656.94	2,024.37	5,681.30
16.1101 021	Main switchboard, 600v, 600a	EA	39.2832	4,388.33	3,373.97	7,762.30
16.1101 031	Main switchboard, 600v, 800a	EA	52.3776	5,851.10	6,747.88	12,598.98
16.1101 041	Main switchboard, 600v, 1200a	EA	58.9248	6,582.49	9,447.07	16,029.56
16.1101 051	Main switchboard, 600v, 1600a	EA	72.0192	8,045.26	11,808.88	19,854.15
16.1101 061	Main switchboard, 600v, 2000a	EA	90.0240	10,056.58	13,495.89	23,552.47
16.1101 071	Main switchboard, 600v, 2500a	EA	114.5760	12,799.28	15,182.88	27,982.16
16.1101 081	Main switchboard, 600v, 3000a	EA	163.6800	18,284.69	18,556.91	36,841.60
16.1102 000	**MAIN PANELS, 250V, LIGHT COMMERCIAL:**					
16.1102 011	Main panel, 250v, 100a	EA	11.4576	1,279.93	570.99	1,850.92
16.1102 021	Main panel, 250v, 150a	EA	14.7312	1,645.62	856.54	2,502.16
16.1102 031	Main panel, 250v, 225a	EA	19.6416	2,194.16	1,142.13	3,336.29
16.1102 041	Main panel, 250v, 400a	EA	29.4624	3,291.24	1,713.12	5,004.36
16.1102 051	Main panel, 250v, 600a	EA	34.3728	3,839.79	2,855.20	6,694.99
16.1102 061	Main panel, 250v, 800a	EA	39.2832	4,388.33	4,282.84	8,671.16
16.1102 071	Main panel, 250v, 1000a	EA	49.1040	5,485.41	5,710.46	11,195.87
16.1200 000	**DISTRIBUTION PANELS TO 600V:**					
16.1201 000	**DISTRIBUTION, POWER PANELS, TO 600V, WITHOUT MAINS:**					
16.1201 001	*Note: The following prices include breakers installed in typical distribution by connected breaker load.*					
16.1201 011	Power panel, 600v, 100a & circuit breaker	EA	10.1485	1,133.69	1,489.68	2,623.37
16.1201 021	Power panel, 600v, 225a & circuit breaker	EA	12.2760	1,371.35	2,383.47	3,754.82
16.1201 031	Power panel, 600v, 400a & circuit breaker	EA	16.3680	1,828.47	3,575.24	5,403.71
16.1201 041	Power panel, 600v, 800a & circuit breaker	EA	26.1888	2,925.55	4,766.96	7,692.51
16.1201 051	Power panel, 600v, 1200a & circuit breaker	EA	39.2832	4,388.33	5,958.75	10,347.08
16.1201 061	Power panel, 600v, 1600a & circuit breaker	EA	52.3776	5,851.10	7,448.42	13,299.52
16.1201 071	Power panel, 600v, 2000a & circuit breaker	EA	65.4720	7,313.88	8,938.10	16,251.97
16.1201 081	Power panel, 600v, 3000a & circuit breaker	EA	90.0240	10,056.58	13,407.20	23,463.78
16.1201 091	Power panel, 600v, 4000a & circuit breaker	EA	114.5760	12,799.28	16,386.62	29,185.91
16.1202 000	**BRANCH CIRCUIT PANELS, TO 600V, COMMERCIAL:**					

ELECTRICAL - 16

Division	Description	Unit	Labor Hours	Labor Cost	Material Cost	Total Cost
16.1202 011	Panelboard, 600v, 100a & circuit breaker	EA	8.1840	914.23	868.89	1,783.13
16.1202 021	Panelboard, 600v, 150a & circuit breaker	EA	9.8208	1,097.08	1,216.50	2,313.58
16.1202 031	Panelboard, 600v, 225a & circuit breaker	EA	11.4576	1,279.93	1,477.25	2,757.17
16.1202 041	Panelboard, 600v, 400a & circuit breaker	EA	13.0944	1,462.78	2,085.53	3,548.30
16.1203 000	**BRANCH CIRCUIT PANELS, TO 600V, INSTITUTIONAL:**					
16.1203 011	Panelboard, 600v, 100a & circuit breaker	EA	9.8208	1,097.08	1,564.10	2,661.19
16.1203 021	Panelboard, 600v, 150a & circuit breaker	EA	11.4576	1,279.93	2,033.41	3,313.34
16.1203 031	Panelboard, 600v, 225a & circuit breaker	EA	13.9128	1,554.20	2,398.36	3,952.56
16.1203 041	Panelboard, 600v, 400a & circuit breaker	EA	16.3680	1,828.47	3,441.18	5,269.65
16.1300 000	**TRANSFORMERS:**					
16.1301 000	**DISTRIBUTION TRANSFORMERS, LIGHT/POWER TO 600V:**					
16.1301 011	Transformer, 600v, 9kva	EA	4.9104	548.54	2,400.44	2,948.98
16.1301 021	Transformer, 600v, 15kva	EA	7.3656	822.81	3,249.14	4,071.95
16.1301 031	Transformer, 600v, 30kva	EA	14.4000	1,608.62	4,155.22	5,763.84
16.1301 041	Transformer, 600v, 45 kva	EA	16.0000	1,787.36	4,446.64	6,234.00
16.1301 051	Transformer, 600v, 75kva	EA	22.9152	2,559.86	6,118.61	8,678.46
16.1301 061	Transformer, 600v, 112.5kva	EA	26.1888	2,925.55	8,142.06	11,067.61
16.1301 071	Transformer, 600v, 150kva	EA	27.8256	3,108.40	10,636.15	13,744.55
16.1301 081	Transformer, 600v, 225kva	EA	40.9200	4,571.17	14,181.61	18,752.78
16.1301 091	Transformer, 600v, 300kva	EA	52.3776	5,851.10	18,184.41	24,035.52
16.1301 101	Transformer, 600v, 500kva	EA	63.0168	7,039.61	28,820.67	35,860.28
16.1301 111	Transformer, 600v, 750kva	EA	65.4720	7,313.88	46,833.64	54,147.52
16.1301 121	Transformer, 600v, 1000kva	EA	81.8400	9,142.35	56,497.72	65,640.06
16.1301 131	Transformer, 600v, 1500kva	EA	98.2080	10,970.82	86,189.69	97,160.51
16.1301 141	Transformer, 600v, 2000kva	EA	114.5760	12,799.28	100,268.52	113,067.81
16.1301 151	Transformer, 600v, 2500kva	EA	130.9440	14,627.75	119,154.71	133,782.47
16.1301 161	Transformer, 600v, 3000kva	EA	147.3120	16,456.22	136,323.97	152,780.19
16.1302 000	**MOTOR CONTROL CENTERS, 600V, 22,000 AMPERE INTERRUPT CAPACITY:**					
16.1302 011	Motor control center, 100a section	EA	4.9104	548.54	1,740.64	2,289.18
16.1302 021	Motor control center, 225a section	EA	7.3656	822.81	2,320.85	3,143.66
16.1302 031	Motor control center, 400a section	EA	8.1840	914.23	2,901.09	3,815.33
16.1302 041	Motor control center, 800a section	EA	9.8208	1,097.08	3,481.32	4,578.41
16.1302 051	Motor control center, 1200a section	EA	13.0944	1,462.78	4,641.78	6,104.55
16.1302 061	Motor control center, 1600a section	EA	16.3680	1,828.47	5,775.24	7,603.71
16.1302 071	Combination starter, size 1	EA	4.2966	479.97	1,387.88	1,867.85
16.1302 081	Combination starter, size 2	EA	6.1380	685.68	1,718.88	2,404.56
16.1302 091	Combination starter, size 3	EA	10.2300	1,142.79	2,319.68	3,462.47
16.1302 101	Combination starter, size 4	EA	16.3680	1,828.47	4,241.56	6,070.03
16.1302 111	Combination starter, size 5	EA	26.1888	2,925.55	14,924.50	17,850.05
16.1302 112	Combination starter, size 6	EA	36.5641	4,084.58	4,965.04	9,049.62
16.1302 113	Combination starter, size 7	EA	47.6633	5,324.47	6,472.25	11,796.72
16.1303 000	**EMERGENCY GENERATORS:**					
16.1303 011	Emergency generator, to 30kw	EA	32.7360	3,656.94	35,639.70	39,296.64
16.1303 021	Emergency generator, to 60kw	EA	49.1040	5,485.41	42,532.98	48,018.39
16.1303 031	Emergency generator, to 100kw	EA	65.4720	7,313.88	60,761.46	68,075.34
16.1303 041	Emergency generator, to 150kw	EA	81.8400	9,142.35	72,913.78	82,056.12
16.1303 051	Emergency generator, to 200kw	EA	98.2080	10,970.82	85,066.05	96,036.87
16.1303 061	Emergency generator, to 400kw	EA	114.5760	12,799.28	176,208.20	189,007.49
16.1303 071	Emergency generator, to 600kw	EA	130.9440	14,627.75	273,426.62	288,054.37
16.1303 081	Emergency generator, to 750kw	EA	163.6800	18,284.69	382,797.22	401,081.91
16.1303 091	Emergency generator, to 1000kw	EA	196.4160	21,941.63	440,094.68	462,036.31
16.1304 000	**AUTO TRANSFER SWITCHES:**					
16.1304 011	Auto transfer switch, to 30a	EA	3.2736	365.69	8,374.68	8,740.37
16.1304 021	Auto transfer switch, to 70a	EA	4.9104	548.54	8,471.04	9,019.58
16.1304 031	Auto transfer switch, to 100a	EA	6.5472	731.39	9,263.55	9,994.94
16.1304 041	Auto transfer switch, to 150a	EA	8.1840	914.23	9,866.29	10,780.53

16 - ELECTRICAL

Division	Description	Unit	Labor Hours	Labor Cost	Material Cost	Total Cost
16.1304 051	Auto transfer switch, to 225a	EA	9.8208	1,097.08	16,037.64	17,134.72
16.1304 061	Auto transfer switch, to 400a	EA	13.0944	1,462.78	17,990.43	19,453.21
16.1304 071	Auto transfer switch, to 800a	EA	16.3680	1,828.47	33,014.70	34,843.17
16.1304 081	Auto transfer switch, to 1200a	EA	21.2784	2,377.01	56,618.55	58,995.56
16.1304 091	Auto transfer switch, to 1600a	EA	26.1888	2,925.55	70,288.03	73,213.58
16.1304 101	Auto transfer switch to 2000a	EA	29.4600	3,290.98	73,211.33	76,502.31
16.1400 000	**RACEWAY & WIRE, COMBINED:**					
16.1401 000	**PVC & COPPER WIRE:**					
16.1401 011	PVC & copper wire, to 30a	LF	0.0345	3.85	4.36	8.21
16.1401 021	PVC & copper wire, to 60a	LF	0.0459	5.13	8.99	14.12
16.1401 031	PVC & copper wire, to 100a	LF	0.0983	10.98	17.03	28.01
16.1401 041	PVC & copper wire, to 150a	LF	0.1064	11.89	27.03	38.92
16.1401 051	PVC & copper wire, to 225a	LF	0.1310	14.63	54.65	69.28
16.1401 061	PVC & copper wire, to 400a	LF	0.1790	20.00	112.84	132.84
16.1401 071	PVC & copper wire to 800a	LF	0.1977	22.09	251.75	273.83
16.1402 000	**EMT & COPPER WIRE:**					
16.1402 011	EMT & copper wire, to 30a	LF	0.0655	7.32	3.78	11.09
16.1402 021	EMT & copper wire, to 60a	LF	0.0819	9.15	11.22	20.37
16.1402 031	EMT & copper wire, to 100a	LF	0.0942	10.52	26.45	36.97
16.1402 041	EMT & copper wire, to 150a	LF	0.1083	12.10	36.20	48.29
16.1402 051	EMT & copper wire, to 225a	LF	0.1362	15.21	69.86	85.08
16.1403 000	**RSC & COPPER WIRE:**					
16.1403 011	RSC & copper wire, to 30a	LF	0.0807	9.01	7.22	16.24
16.1403 021	RSC & copper wire, to 60a	LF	0.0990	11.06	13.79	24.85
16.1403 031	RSC & copper wire, to 100a	LF	0.1225	13.68	32.90	46.59
16.1403 041	RSC & copper wire, to 150a	LF	0.1392	15.55	45.69	61.24
16.1403 051	RSC & copper wire, to 225a	LF	0.1801	20.12	87.46	107.58
16.1403 061	RSC & copper wire, to 400a	LF	0.2777	31.02	175.67	206.69
16.1403 071	RSC & copper wire to 800a	LF	0.3530	39.43	393.60	433.03
16.1404 000	**PVC & ALUMINUM WIRE:**					
16.1404 001	Note: Aluminum wire not recommended for under 100 amps.					
16.1404 011	PVC & aluminum wire, to 100a	LF	0.0901	10.07	16.73	26.79
16.1404 021	PVC & aluminum wire, to 150a	LF	0.1024	11.44	23.78	35.22
16.1404 031	PVC & aluminum wire, to 225a	LF	0.1228	13.72	33.37	47.08
16.1404 041	PVC & aluminum wire, to 400a	LF	0.1474	16.47	42.89	59.35
16.1405 000	**EMT & ALUMINUM WIRE:**					
16.1405 011	EMT & aluminum wire, to 100a	LF	0.0737	8.23	19.86	28.09
16.1405 021	EMT & aluminum wire, to 150a	LF	0.0901	10.07	29.05	39.11
16.1405 031	EMT & aluminum wire, to 225a	LF	0.1064	11.89	39.91	51.79
16.1405 041	EMT & aluminum wire, to 400a	LF	0.1310	14.63	48.36	63.00
16.1406 000	**RSC & ALUMINUM WIRE:**					
16.1406 011	RSC & aluminum wire, to 100a	LF	0.0983	10.98	28.46	39.44
16.1406 021	RSC & aluminum wire, to 150a	LF	0.1310	14.63	40.37	55.00
16.1406 031	RSC & aluminum wire, to 225a	LF	0.1637	18.29	52.24	70.53
16.1406 041	RSC & aluminum wire, to 400a	LF	0.2456	27.44	69.32	96.76
16.1500 000	**UNDERFLOOR DISTRIBUTION SYSTEMS:**					
16.1501 000	**UNDERFLOOR DUCT & ACCESSORIES:**					
16.1501 011	Underfloor duct, blank standard	LF	0.0615	6.87	6.08	12.95
16.1501 021	Underfloor duct, blank jumbo	LF	0.0901	10.07	13.51	23.57
16.1501 031	Underfloor duct, insert standard	LF	0.0615	6.87	8.54	15.41
16.1501 041	Underfloor duct, insert jumbo	LF	0.0901	10.07	21.51	31.58
16.1501 051	Underfloor duct, J-box, 1 way	EA	2.0460	228.56	253.21	481.77
16.1501 061	Underfloor duct, J-box, 2 way	EA	2.8644	319.98	382.33	702.32
16.1501 071	Underfloor duct, J-box, 3 way	EA	4.0920	457.12	642.45	1,099.56
16.1501 081	Underfloor duct, panel riser, standard	EA	3.2736	365.69	117.75	483.45
16.1501 091	Underfloor duct, panel riser, jumbo	EA	4.0920	457.12	189.61	646.73

ELECTRICAL - 16

Division	Description	Unit	Labor Hours	Labor Cost	Material Cost	Total Cost
16.1502 000	**CELLULAR FLOOR SYSTEMS:**					
16.1502 011	Bond-seal floor cell joints	SF			0.24	0.24
16.1502 021	Flush trench duct, to 12"	LF	0.6548	73.15	119.70	192.85
16.1502 031	Flush trench duct, to 24"	LF	0.9003	100.57	168.21	268.78
16.1502 041	Flush trench duct, to 36"	LF	1.6368	182.85	255.64	438.49
16.1502 051	Flush trench ell, to 12"	EA	0.8184	91.42	398.07	489.49
16.1502 061	Flush trench ell, to 24"	EA	1.4322	159.99	647.29	807.28
16.1502 071	Flush trench ell, to 36"	EA	2.0460	228.56	1,067.99	1,296.55
16.1502 081	Flush trench tee, to 12"	EA	1.2276	137.14	398.07	535.21
16.1502 091	Flush trench tee, to 24"	EA	1.8414	205.70	647.29	852.99
16.1502 101	Flush trench tee, to 36"	EA	2.8644	319.98	1,067.99	1,387.97
16.1502 111	Flush trench riser, to 12"	EA	1.6368	182.85	453.08	635.92
16.1502 121	Flush trench riser, to 24"	EA	2.4552	274.27	527.54	801.81
16.1502 131	Flush trench riser, to 36"	EA	3.2736	365.69	679.66	1,045.35
16.1502 141	Flush trench duct, 3" grommet	EA	0.0013	0.15	6.41	6.56
16.1502 151	Flush trench duct, 6" grommet	EA	0.0030	0.34	9.67	10.01
16.1503 000	**FLOOR SYSTEM OUTLETS:**					
16.1503 011	Core drill, add afterset insert	EA	0.2865	32.00	12.92	44.92
16.1503 021	Flush outlet at insert, power	EA	0.2865	32.00	59.79	91.79
16.1503 031	Flush outlet at insert, signal	EA	0.1637	18.29	51.75	70.03
16.1503 041	Surface outlet at insert, power	EA	0.4092	45.71	98.70	144.41
16.1503 051	Surface outlet at insert, signal	EA	0.3274	36.57	88.95	125.52
16.1503 061	Preset power/signal access box	EA	0.2046	22.86	53.65	76.51
16.1503 071	Add 110v duplex at access box	EA	0.2046	22.86	17.70	40.56
16.1600 000	**LIGHTING FIXTURES, IN-PLACE:**					
16.1601 000	**FIXTURES BY BUILDING TYPE:**					
16.1601 011	Fixtures, college, classroom, 70 foot candles	SF	0.0156	1.74	3.93	5.67
16.1601 021	Fixtures, store, commercial	SF	0.0193	2.16	2.97	5.13
16.1601 031	Fixtures, garage, commercial, 5 foot candles	SF	0.0013	0.15	0.05	0.19
16.1601 041	Fixtures, hospital, general space 70 foot candles	SF	0.0156	1.74	3.93	5.67
16.1601 051	Fixtures, hospital, wards & rooms 50 foot candles	SF	0.0099	1.11	1.76	2.86
16.1601 061	Fixtures, market, 70 foot candles	SF	0.0156	1.74	2.62	4.37
16.1601 071	Fixtures, light manufacturing, 100 foot candles	SF	0.0153	1.71	2.97	4.68
16.1601 081	Fixtures, office, general & high rise	SF	0.0156	1.74	3.93	5.67
16.1601 091	Fixtures, office, drafting, 100 foot candles	SF	0.0206	2.30	6.07	8.37
16.1601 101	Fixtures, elementary & high school, 70 foot candles	SF	0.0136	1.52	3.28	4.80
16.1601 111	Fixtures, recreational facility, 40 foot candles	SF	0.0124	1.39	3.13	4.52
16.1601 121	Fixtures, warehouse, 20 foot candles	SF	0.0025	0.28	0.18	0.46
16.1601 131	Fixtures, churches	SF	0.0152	1.70	5.49	7.19
16.1601 141	Fixtures, banks, 50 foot candles	SF	0.0200	2.23	3.28	5.52
16.1601 151	Fixtures, libraries, 70 foot candles	SF	0.0194	2.17	5.15	7.32
16.1601 161	Fixtures, motels & hotels	SF	0.0095	1.06	1.03	2.09
16.1602 000	**FIXTURES, COMMERCIAL, AVERAGE:**					
16.1602 011	Incandescent, surface, 100w	EA	0.6957	77.72	119.47	197.19
16.1602 021	Incandescent, surface, 150w	EA	0.7775	86.85	126.13	212.98
16.1602 031	Incandescent, surface, 200w	EA	0.9003	100.57	146.02	246.60
16.1602 041	Incandescent, recessed, 100w	EA	1.0230	114.28	159.27	273.55
16.1602 051	Incandescent, recessed, 150w	EA	1.0640	118.86	172.55	291.41
16.1602 061	Incandescent, recessed, 200w	EA	1.2276	137.14	199.18	336.31
16.1602 071	Mercury vapor, 100w fixture	EA	1.1458	128.00	199.18	327.18
16.1602 081	Mercury vapor, 175w fixture	EA	1.3095	146.28	252.33	398.61
16.1602 091	Mercury vapor, 250w fixture	EA	1.6368	182.85	332.00	514.84
16.1602 101	Mercury vapor, 400w fixture	EA	2.4552	274.27	531.21	805.48
16.1602 111	High pressure sodium/lucalux fixture, 100w	EA	1.2276	137.14	371.84	508.98
16.1602 121	High pressure sodium/lucalux fixture, 250w	EA	2.0460	228.56	663.98	892.54
16.1602 131	High pressure sodium/lucalux fixture, 400w	EA	3.2736	365.69	863.23	1,228.93
16.1602 141	Fluorescent strip, surface, 2 lamp, 4'	EA	0.6138	68.57	74.35	142.91

16 - ELECTRICAL

Division	Description	Unit	Labor Hours	Labor Cost	Material Cost	Total Cost
16.1602 151	Fluorescent strip, surface, 2 lamp, 8'	EA	0.9003	100.57	90.22	190.79
16.1602 161	Fluorescent, lens, surface, 2 lamp, 4'	EA	0.8184	91.42	100.86	192.28
16.1602 171	Fluorescent, lens, surface, 2 lamp, 8'	LF	1.1458	128.00	199.18	327.18
16.1602 181	Fluorescent, lens, recessed, 2'x2', 2 lamp	EA	0.8184	91.42	153.99	245.42
16.1602 191	Fluorescent, lens, recessed, 2'x4', 2 lamp	EA	0.8184	91.42	164.70	256.13
16.1602 201	Fluorescent, lens, recessed, 2'x4', 3 lamp	EA	0.8184	91.42	172.55	263.97
16.1602 211	Fluorescent, lens, recessed, 2'x4', 4 lamp	EA	0.8184	91.42	212.47	303.89
16.1602 221	Fluorescent, industrial, suspended, 2 lamp, 4'	EA	1.2276	137.14	212.47	349.60
16.1602 231	Fluorescent, industrial, suspended, 2 lamp, 8'	EA	1.6368	182.85	345.25	528.10
16.1602 241	Exit signs, surface mounted	EA	0.8184	91.42	159.27	250.70
16.1602 251	Exit signs, recessed	EA	1.2276	137.14	212.47	349.60
16.1700 000	**BRANCH CIRCUIT RUNS, SPECIAL PURPOSE CONDUIT & WIRE:**					
16.1701 000	LIGHTING OUTLETS & DEVICES WITHOUT WIRE & CONDUIT:					
16.1701 011	Fixture outlets	EA	0.2046	22.86	8.88	31.74
16.1701 021	Fixture junction boxes	EA	0.1637	18.29	8.88	27.17
16.1701 031	Fixture flex assemblies	EA	0.0819	9.15	12.05	21.20
16.1701 041	Fixture switch outlets	EA	0.6400	71.49	30.71	102.20
16.1701 051	Fixture switch & pilot lights	EA	0.7900	88.25	21.08	109.34
16.1701 061	Fixture switch outlets, waterproof	EA	0.7900	88.25	49.98	138.23
16.1701 071	Fixture switch, two gang	EA	0.7900	88.25	36.53	124.78
16.1702 000	LIGHTING CIRCUITS:					
16.1702 011	Lighting circuits, EMT & wire	LF	0.0492	5.50	1.59	7.09
16.1702 021	Lighting circuits, RSC & wire	LF	0.0615	6.87	3.60	10.47
16.1800 000	**SIGNAL & COMMUNICATIONS SYSTEMS:**					
16.1801 000	TELEPHONE SYSTEMS:					
16.1801 011	Telephone main terminal	EA	2.4552	274.27	107.85	382.12
16.1801 021	Telephone auxiliary terminal	EA	1.6368	182.85	77.01	259.86
16.1801 031	Telephone riser raceway	LF	0.1188	13.27	13.36	26.63
16.1801 041	Telephone riser sleeve	EA	0.4092	45.71	19.99	65.70
16.1801 051	Telephone outlet, wall	EA	0.2456	27.44	16.86	44.29
16.1801 061	Telephone outlet, floor box	EA	0.4092	45.71	126.74	172.45
16.1801 071	Telephone outlet raceway	LF	0.0440	4.92	1.72	6.64
16.1802 000	FIRE ALARM SYSTEMS:					
16.1802 011	Fire alarm main panel	EA	9.8208	1,097.08	3,952.90	5,049.98
16.1802 021	Fire alarm annunciator	EA	8.1840	914.23	2,900.32	3,814.55
16.1802 031	Fire alarm power supply	EA	4.9104	548.54	2,289.71	2,838.25
16.1802 041	Fire alarm terminal cabinet	EA	2.4552	274.27	457.88	732.15
16.1802 051	Fire alarm stations	EA	1.0230	114.28	167.84	282.11
16.1802 061	Fire alarm horns	EA	0.8184	91.42	152.63	244.05
16.1802 071	Fire alarm bells	EA	0.8184	91.42	137.33	228.76
16.1802 081	Fire alarm chimes	EA	0.8184	91.42	167.84	259.26
16.1802 091	Fire alarm smoke detector, ceiling	EA	1.2276	137.14	220.80	357.93
16.1802 101	Fire alarm smoke detector, duct	EA	2.0460	228.56	641.11	869.67
16.1802 111	Fire alarm ionization detector	EA	1.0230	114.28	374.72	489.00
16.1802 121	Fire alarm heat detector	EA	0.6138	68.57	213.38	281.95
16.1802 131	Fire alarm flow switch connect	EA	1.0230	114.28	213.38	327.66
16.1802 141	Fire alarm door hold assembly	EA	1.6368	182.85	201.43	384.28
16.1802 151	Fire alarm door release assembly	EA	1.6368	182.85	201.43	384.28
16.1802 161	Fire alarm distribution feeders	LF	0.1801	20.12	10.91	31.03
16.1802 171	Fire alarm device circuits	LF	0.0573	6.40	3.71	10.11
16.1803 000	COMMUNICATION, INTERCOM, PUBLIC ADDRESS:					
16.1803 011	Main amplifier panel power supply	EA	6.5472	731.39	3,897.39	4,628.78
16.1803 021	Auxiliary terminal cabinet	EA	3.2736	365.69	389.73	755.42
16.1803 031	Speaker enclosure, surface	EA	0.2865	32.00	16.59	48.60
16.1803 041	Speaker enclosure, flush	EA	0.4092	45.71	38.75	84.46
16.1803 051	Microphone outlet, wall	EA	0.2865	32.00	16.88	48.88

ELECTRICAL - 16

Division	Description	Unit	Labor Hours	Labor Cost	Material Cost	Total Cost
16.1803 061	Microphone outlet, floor	EA	0.4092	45.71	49.30	95.01
16.1803 071	Intercom outlet, wall	EA	0.2865	32.00	16.88	48.88
16.1803 081	Intercom outlet, floor	EA	0.4092	45.71	49.30	95.01
16.1803 091	System feeder raceway	LF	0.1188	13.27	3.32	16.59
16.1803 101	System feeder cable	LF	0.0164	1.83	9.07	10.90
16.1803 111	System device raceway	LF	0.0287	3.21	2.41	5.62
16.1803 121	System device cable	LF	0.0066	0.74	3.03	3.76
16.1900 000	**BRANCH CIRCUIT OUTLETS & DEVICES:**					
16.1901 000	**POWER OUTLETS, AVERAGE, WITH CONDUIT & WIRE:**					
16.1901 011	Duplex outlet	EA	0.7600	84.90	24.78	109.68
16.1901 021	Duplex outlet, waterproof	EA	0.8100	90.49	57.27	147.76
16.1901 031	Duplex outlet, ground fault interrupter	EA	0.9200	102.77	51.49	154.27
16.1901 041	Duplex outlet, ground fault interrupter, waterproof	EA	1.1800	131.82	97.05	228.87
16.1901 051	Double duplex outlet	EA	0.5310	59.32	150.24	209.56
16.1901 061	Outlet, 30 amp	EA	0.9200	102.77	32.33	135.10
16.1901 071	Outlet, 50 amp	EA	1.0160	113.50	41.72	155.22
16.1901 081	Outlet, 60 amp, welding receptacle	EA	1.4000	156.39	320.70	477.09
16.1901 091	Outlet, 100 amp, welding receptacle	EA	1.7500	195.49	482.50	677.99
16.1901 101	Motor connect, 1 ph, fractional hp	EA	1.5200	169.80	30.89	200.69
16.1901 111	Motor connection, 3 ph	EA	1.8000	201.08	38.69	239.77
16.1901 121	Clock outlet	EA	0.9200	102.77	57.27	160.04
16.1901 131	J-box outlet	EA	0.3200	35.75	15.04	50.79
16.1901 141	J-box outlet & equipment connection	EA	0.9500	106.12	30.89	137.02
16.1901 151	Floor box with flush 110v outlet	EA	0.6138	68.57	105.37	173.93
16.1901 161	Floor box with surface 110v outlet	EA	1.0230	114.28	119.82	234.10
16.1901 171	Floor box with equipment connection	EA	1.4322	159.99	119.82	279.81
16.1901 181	Outlet circuit, EMT/wire to 20a	LF	0.0459	5.13	1.44	6.57
16.1901 191	Outlet circuit, RSC/wire to 20a	LF	0.0655	7.32	3.47	10.78
16.1901 201	Outlet circuit, PVC/wire to 20a	LF	0.0369	4.12	2.52	6.64
16.1950 000	**FEES, PERMITS, TESTING:**					
16.1951 000	**FEES, INSPECTION:**					
16.1951 011	Fee, inspection, per fixture	EA			2.10	2.10
16.1951 021	Fee, inspection, per outlet	EA			0.68	0.68
16.1951 031	Fee, inspection, per amp	EA			0.12	0.12
16.1951 041	Fee, inspection, per horsepower	EA			4.23	4.23
16.1952 000	**TESTING:**					
16.1952 011	Testing, per fixture	EA			4.91	4.91
16.1952 021	Testing, per outlet	EA			2.80	2.80
16.1952 031	Testing, per amp, service switchgear	EA			0.34	0.34
16.1952 041	Testing, per horsepower	EA			21.06	21.06
16.2000 000	**EQUIPMENT, UNIT SUBSTATIONS:**					
16.2001 000	**UNIT SUB-STATION, HIGH VOLTAGE, CUSTOMER OWNED:**					
16.2001 011	Substation, 4.16kv, 75,000 va, 1200a	EA	35.9664	4,017.81	22,210.71	26,228.52
16.2001 021	Substation, 4.16kv, 250,000 va, 2000a	EA	69.6821	7,784.19	34,306.34	42,090.52
16.2001 031	Substation, 4.16kv, 350,000 va, 1200a	EA	57.3587	6,407.54	36,075.40	42,482.94
16.2001 041	Substation, 4.16kv, 350,000 va, 3000a	EA	102.6012	11,461.58	65,059.66	76,521.24
16.2001 051	Substation, 138kv, 500,000 va, 1200a	EA	61.6852	6,890.85	37,708.28	44,599.13
16.2001 061	Substation, 13.8kv, 500,000 va, 2000a	EA	89.2400	9,969.00	41,639.42	51,608.42
16.2001 071	Substation, 13.8kv, 750,000 va, 1200a	EA	97.2011	10,858.33	46,764.89	57,623.22
16.2001 081	Substation, 13.8kv, 1,000,000 va, 1200a	EA	119.6326	13,364.16	64,560.70	77,924.85
16.2001 091	Substation, 138kv, 1,000,000 va, 3000a	EA	148.0522	16,538.91	99,335.90	115,874.81
16.2002 000	**UNIT SUB-STATION, 2400V OR 4160V, 3 PH, 120/208-240, 4 WIRE:**					
16.2002 011	Substation, 112.5kva, 5kv, 120/208v, 3ph, 4 wire	EA	28.6953	3,205.55	39,739.76	42,945.31
16.2002 021	Substation, 150kva, 5kv, 120/208v, 3ph, 4 wire	EA	30.0456	3,356.39	43,206.96	46,563.35
16.2002 031	Substation, 225kva, 5kv, 120/208v, 3ph, 4 wire	EA	42.1989	4,714.04	50,274.78	54,988.82
16.2002 041	Substation, 300kva, 5kv, 120/208v, 3ph, 4 wire	EA	45.2372	5,053.45	59,596.31	64,649.76

16 - ELECTRICAL

Division	Description	Unit	Labor Hours	Labor Cost	Material Cost	Total Cost
16.2002 051	Substation, 500kva, 5kv, 120/208v, 3ph, 4 wire	EA	50.3010	5,619.12	85,547.26	91,166.38
16.2003 000	**UNIT SUB-STATION, 2400V OR 4160V, 3 PH, 277/480, 4 WIRE:**					
16.2003 011	Substation, 112.5kva, 5kv, 277/480v, 3ph, 4 wire	EA	28.6953	3,205.55	36,831.33	40,036.88
16.2003 021	Substation, 150kva, 5kv, 277/480v, 3ph, 4 wire	EA	30.0456	3,356.39	39,752.40	43,108.79
16.2003 031	Substation, 225kva, 5kv, 277/480v, 3ph, 4 wire	EA	42.1989	4,714.04	45,467.64	50,181.68
16.2003 041	Substation, 300kva, 5kv, 277/480v, 3ph, 4 wire	EA	45.2372	5,053.45	53,748.33	58,801.78
16.2003 051	Substation, 500kva, 5kv, 277/480v, 3ph, 4 wire	EA	50.3010	5,619.12	77,104.59	82,723.72
16.2004 000	**UNIT SUB-STATION, 12000-13800V, 3 PH, 120/208-240, 4 WIRE:**					
16.2004 011	Substation, 112.5kva, 15kv, 120/208, 3ph, 4 wire	EA	28.4927	3,182.92	51,728.31	54,911.23
16.2004 021	Substation, 150kva, 15kv, 120/208v, 3ph, 4 wire	EA	29.8430	3,333.76	57,482.62	60,816.38
16.2004 031	Substation, 225kva, 15kv, 120/208v, 3ph, 4 wire	EA	42.1989	4,714.04	68,884.52	73,598.56
16.2004 041	Substation, 300kva, 15kv, 120/208v, 3ph, 4 wire	EA	45.3046	5,060.98	83,713.59	88,774.57
16.2004 051	Substation, 500kva, 15kv, 120/208v, 3ph, 4 wire	EA	48.9507	5,468.28	103,556.80	109,025.09
16.2005 000	**UNIT SUB-STATION, 12000V-13800V, 3 PH, 277/408, 4 WIRE:**					
16.2005 011	Substation, 112.5kva, 15kv, 277/480, 3ph, 4 wire	EA	28.4927	3,182.92	50,694.89	53,877.81
16.2005 021	Substation, 150kva, 15kv, 277/480v, 3ph, 4 wire	EA	29.8430	3,333.76	56,749.21	60,082.97
16.2005 031	Substation, 225kva, 15kv, 277/480v, 3ph, 4 wire	EA	42.1989	4,714.04	66,617.40	71,331.44
16.2005 041	Substation, 300kva, 15kv, 277/480v, 3ph, 4 wire	EA	45.3046	5,060.98	81,886.60	86,947.58
16.2005 051	Substation, 500kva, 15kv, 277/480v, 3ph, 4 wire	EA	48.9507	5,468.28	101,223.12	106,691.40
16.2006 000	**UNIT SUB-STATION, HIGH VOLTAGE SECTION OPTIONS:**					
16.2006 011	Single feed, 3 conductors pothead	EA	4.0511	452.55	1,809.29	2,261.84
16.2006 021	Lightning arrest, 3-3 kv	EA	1.6881	188.58	961.46	1,150.04
16.2006 031	Lightning arrest, 3-4.5 kv	EA	1.6881	188.58	1,373.53	1,562.11
16.2006 041	Lightning arrest, 3-6 kv	EA	1.6881	188.58	1,373.53	1,562.11
16.2006 051	Lightning arrest, 3-9 kv	EA	1.6881	188.58	1,716.87	1,905.45
16.2006 061	Lightning arrest, 3-15 kv	EA	1.6881	188.58	2,589.43	2,778.01
16.2006 101	Switch, ram, oil, 15kv, 400a, 3ph, 4 wire	EA	48.0000	5,362.08	46,278.47	51,640.55
16.2006 111	Switch, fused, oil, 15kv, 100a	EA	29.5000	3,295.45	15,374.72	18,670.17
16.2006 121	Switch, fused, oil, 15kv, 600a	EA	39.0000	4,356.69	29,206.87	33,563.56
16.2006 131	Switch, interrupt, gas, 15kv, 4 pole	EA	43.0000	4,803.53	32,485.43	37,288.96
16.2007 000	**UNIT SUB-STATION, LOW VOLTAGE SECTION OPTIONS:**					
16.2007 011	Kirk key interlock	EA			462.28	462.28
16.2007 021	Indicating watt meter	EA			3,342.95	3,342.95
16.2007 031	One phase indicating ammeter	EA			1,102.47	1,102.47
16.2007 041	One phase indicating voltmeter	EA			1,102.47	1,102.47
16.2007 051	Indicating watt-hour meter	EA			2,044.83	2,044.83
16.2007 061	Demand watt-hour meter	EA			497.78	497.78
16.2007 071	Selector switch with volt or ammeter	EA			248.86	248.86
16.2007 081	Potential transformer	EA			1,066.86	1,066.86
16.2007 091	Current transformer, 5-800a	EA			800.10	800.10
16.2007 101	Current transformer, to 1500a	EA			1,315.81	1,315.81
16.2100 000	**EQUIPMENT, SWITCHGEAR & TRANSFORMERS:**					
16.2100 001	Note: Main switchgear consists of two sides service & distribution. Both sides must be listed to complete the switchgear. These are the component parts: SERVICE SIDE: DISTRIBUTION SIDE: Equipment Enclosure... Switchboard Enclosures Current Transformers.........Feeder Breakers Volt Meter...................Fused Switches Amp Meter Main Breaker Fire Alarm Breaker Pull Section Light commercial applications, a volt meter, amp meter, fire alarm breaker or current transformers may not be required. See diagram at beginning of this section.					
16.2200 000	**EQUIPMENT, HIGH VOLTAGE TRANSFORMERS:**					
16.2200 001	Note: Items are based on 150 degree rise. Primary voltages are 2400 delta, 4160 delta, and 4160/2400 delta. Secondary voltages are 120/208, 277/480, 240 delta and 600 delta.					

ELECTRICAL - 16

Division	Description	Unit	Labor Hours	Labor Cost	Material Cost	Total Cost
16.2201 000	TRANSFORMERS, DRY, 150 DEGREE RISE, HIGH VOLTAGE, 3PH, 2.4KV & 5KV:					
16.2201 011	Transformer, to 5kv(p)/norm(s) 45kva	EA	17.8924	1,998.76	6,992.06	8,990.82
16.2201 021	Transformer, to 5kv(p)/norm(s) 75kva	EA	19.9179	2,225.03	8,708.01	10,933.04
16.2201 031	Transformer, to 5kv(p)/norm(s) 112.5kva	EA	27.6825	3,092.41	10,388.27	13,480.69
16.2201 041	Transformer, to 5kv(p)/norm(s) 150kva	EA	29.0328	3,243.25	12,575.29	15,818.54
16.2201 051	Transformer, to 5kv(p)/norm(s) 225kva	EA	45.2372	5,053.45	28,718.18	33,771.63
16.2201 061	Transformer, to 5kv(p)/norm(s) 300kva	EA	49.2882	5,505.98	34,157.26	39,663.25
16.2201 071	Transformer, to 5kv(p)/norm(s) 500kva	EA	56.7152	6,335.65	45,035.43	51,371.09
16.2201 081	Transformer, to 5kv(p)/norm(s) 750kva	EA	60.7662	6,788.19	59,176.97	65,965.17
16.2201 091	Transformer, to 5kv(p)/norm(s) 1000kva	EA	63.4670	7,089.90	69,511.27	76,601.17
16.2201 101	Transformer, to 5kv(p)/norm(s) 1500kva	EA	67.5180	7,542.44	80,933.29	88,475.73
16.2201 111	Transformer, to 5kv(p)/norm(s) 2000kva	EA	84.3976	9,428.06	95,074.87	104,502.92
16.2201 121	Transformer, to 5kv(p)/norm(s) 2500kva	EA	100.6019	11,238.24	108,128.60	119,366.84
16.2201 131	Transformer, to 5kv(p)/norm(s) 3000kva	EA	108.0288	12,067.90	129,426.77	141,494.67
16.2202 000	TRANSFORMERS, DRY, 150 DEGREE RISE, HIGH VOLTAGE, 3PH, 15KV, 480/277V:					
16.2202 011	Transformer, 15kv/277-480, 3ph, 4 wire, 45kva	EA	17.5547	1,961.04	10,574.88	12,535.92
16.2202 021	Transformer, 15kv/277-480, 3ph, 4 wire, 75kva	EA	19.9179	2,225.03	13,597.64	15,822.67
16.2202 031	Transformer, 15kv/277-480, 3ph, 112.5kva	EA	27.6825	3,092.41	15,642.42	18,734.84
16.2202 041	Transformer, 15kv/277-480, 3ph, 4 wire, 150kva	EA	29.0328	3,243.25	18,393.99	21,637.25
16.2202 051	Transformer, 15kv/277-480, 3ph, 4 wire, 225kva	EA	45.2372	5,053.45	34,307.94	39,361.39
16.2202 061	Transformer, 15kv/277-480, 3ph, 4 wire, 300kva	EA	49.2882	5,505.98	43,121.26	48,627.25
16.2202 071	Transformer, 15kv/277-480, 3ph, 4 wire, 500kva	EA	56.7152	6,335.65	55,520.29	61,855.95
16.2202 081	Transformer, 15kv/277-480, 3ph, 4 wire, 750kva	EA	60.7662	6,788.19	69,661.89	76,450.08
16.2202 091	Transformer, 15kv/277-480, 3ph, 4 wire, 1000kva	EA	63.4670	7,089.90	79,452.21	86,542.11
16.2202 101	Transformer, 15kv/277-480, 3ph, 4 wire, 1500kva	EA	67.5180	7,542.44	91,418.19	98,960.62
16.2202 111	Transformer, 15kv/277/480, 3ph, 4 wire, 2000kva	EA	84.3976	9,428.06	102,296.25	111,724.31
16.2202 121	Transformer, 15kv/277-480, 3ph, 4 wire, 2500kva	EA	100.6019	11,238.24	118,069.59	129,307.82
16.2202 131	Transformer, 15kv/277-480, 3ph, 4 wire, 3000kva	EA	108.0288	12,067.90	140,369.74	152,437.64
16.2300 000	SERVICE SECTIONS:					
16.2301 000	SERVICE SECTION, AIR CIRCUIT BREAKER MAIN, 480V:					
16.2301 001	Note: The following prices include air circuit main breaker, horizontal bus, underground pull section and Class I gear to 600 volts.					
16.2301 011	Service section, 1200a, 3ph, 480v, 3 wire, air circuit breaker	EA	19.6816	2,198.63	21,593.45	23,792.08
16.2301 021	Service section, 1600a, 3ph, 480v, 3 wire, air circuit breaker	EA	24.3065	2,715.28	22,870.34	25,585.62
16.2301 031	Service section, 2000a, 3ph, 480v, 3 wire, air circuit breaker	EA	34.4343	3,846.66	52,655.13	56,501.79
16.2301 041	Service section, 3000a, 3ph, 480v, 3 wire, air circuit breaker	EA	47.2626	5,279.71	62,156.69	67,436.39
16.2301 051	Service section, 4000a, 3ph, 480v, 3 wire, air circuit breaker	EA	56.0400	6,260.23	87,554.20	93,814.43
16.2302 000	SERVICE SECTION, AIR CIRCUIT BREAKER MAIN, 120/208V:					
16.2302 001	Note: The following prices include air circuit main breaker, horizontal bus, underground pull section and Class I gear to 600 volts.					
16.2302 011	Service section, 1200a, 3ph, 120/208v, air circuit breaker	EA	21.3696	2,387.20	23,126.99	25,514.19
16.2302 021	Service section, 1600a, 3ph, 120/208v, air circuit breaker	EA	26.3321	2,941.56	24,877.29	27,818.85
16.2302 031	Service section, 2000a, 3ph, 120/208v, air circuit breaker	EA	36.7974	4,110.64	55,492.28	59,602.91
16.2302 041	Service section, 3000a, 3ph, 120/208v, air circuit breaker	EA	50.6386	5,656.84	64,527.11	70,183.95
16.2302 051	Service section, 4000a, 3ph, 120/208v, air circuit breaker	EA	60.0911	6,712.78	90,007.89	96,720.67
16.2303 000	SERVICE SECTION, AIR CIRCUIT BREAKER MAIN, 277/480V:					
16.2303 001	Note: The following prices include air circuit main breaker, horizontal bus, underground pull section and Class I gear to 600 volts.					
16.2303 011	Service section, 1200a, 3ph, 277/480v, air circuit breaker	EA	21.3696	2,387.20	22,700.25	25,087.45
16.2303 021	Service section, 1600a, 3ph, 277/480v, air circuit breaker	EA	26.3321	2,941.56	24,453.91	27,395.47
16.2303 031	Service section, 2000a, 3ph, 277/480v, air circuit breaker	EA	36.7974	4,110.64	55,068.90	59,179.54
16.2303 041	Service section, 3000a, 3ph, 277/480v, air circuit breaker	EA	50.6386	5,656.84	64,103.74	69,760.58
16.2303 051	Service section, 4000a, 3ph, 277/480v, air circuit breaker	EA	60.0911	6,712.78	89,751.25	96,464.02
16.2304 000	SERVICE SECTION, AIR CIRCUIT BREAKER MAIN, OPTIONS:					
16.2304 011	Add for ground fault protection	EA	14.1988	1,586.15	6,403.56	7,989.71

16 - ELECTRICAL

Division	Description	Unit	Labor Hours	Labor Cost	Material Cost	Total Cost
16.2305 000	**SERVICE SECTION, BOLTED PRESSURE SWITCH MAIN, 480V:**					
16.2305 001	*Note: The following prices include horizontal bus, underground pull section and Class I gear to 600 volts.*					
16.2305 011	Service section, 1200a, 3ph, 480v, 3 wire, bolted pressure switch	EA	20.3568	2,274.06	16,055.83	18,329.88
16.2305 021	Service section, 1600a, 3ph, 480v, 3 wire, bolted pressure switch	EA	25.6569	2,866.13	18,046.17	20,912.30
16.2305 031	Service section, 2000a, 3ph, 480v, 3 wire, bolted pressure switch	EA	31.7336	3,544.96	20,496.59	24,041.55
16.2305 041	Service section, 2500a, 3ph, 480v, 3 wire, bolted pressure switch	EA	36.4598	4,072.92	25,277.35	29,350.27
16.2305 051	Service section, 3000a, 3ph, 480v, 3 wire, bolted pressure switch	EA	41.1861	4,600.90	34,238.82	38,839.72
16.2305 061	Service section, 4000a, 3ph, 480v, 3 wire, bolted pressure switch	EA	49.2882	5,505.98	44,917.21	50,423.19
16.2306 000	**SERVICE SECTION, BOLTED PRESSURE SWITCH MAIN, 277/480V:**					
16.2306 001	*Note: The following prices include horizontal bus, underground pull section and Class I gear to 600 volts.*					
16.2306 011	Service section, 1200a, 3ph, 277/480v, bolted pressure switch	EA	22.6186	2,526.72	16,986.03	19,512.76
16.2306 021	Service section, 1600a, 3ph, 277/480v, bolted pressure switch	EA	28.3576	3,167.83	19,106.42	22,274.25
16.2306 031	Service section, 2000a, 3ph, 277/480v, bolted pressure switch	EA	34.7718	3,884.36	21,720.14	25,604.50
16.2306 041	Service section, 2500a, 3ph, 277/480v, bolted pressure switch	EA	39.8357	4,450.05	26,637.57	31,087.62
16.2306 051	Service section, 3000a, 3ph, 277/480v, bolted pressure switch	EA	45.2372	5,053.45	35,389.04	40,442.48
16.2306 061	Service section, 4000a, 3ph, 277/480v, bolted pressure switch	EA	53.3393	5,958.53	46,770.82	52,729.35
16.2307 000	**SERVICE SECTION, BOLTED PRESSURE SWITCH MAIN, OPTIONS:**					
16.2307 011	Add for ground fault protection	EA			5,984.59	5,984.59
16.2308 000	**SERVICE SECTION, MOLDED CASE MAIN CIRCUIT BREAKER, 120/240V:**					
16.2308 001	*Note: The following prices include molded case main circuit breaker, horizontal bus, underground pull section and Class I gear to 600 volts.*					
16.2308 011	Service section, 225a, 1ph, 120/240v, molded case main breaker	EA	12.3897	1,384.05	3,117.12	4,501.17
16.2308 021	Service section, 400a, 1ph, 120/240v, molded case main breaker	EA	13.8076	1,542.45	4,780.71	6,323.15
16.2308 031	Service section, 800a, 1ph, 120/240v, molded case main breaker	EA	17.0822	1,908.25	9,194.70	11,102.95
16.2308 041	Service section, 1000a, 1ph, 120/240v, molded case main breaker	EA	20.1880	2,255.20	11,061.71	13,316.91
16.2308 051	Service section, 1200a, 1ph, 120/240v, molded case main breaker	EA	20.6944	2,311.77	13,105.40	15,417.17
16.2309 000	**SERVICE SECTION, MOLDED CASE MAIN CIRCUIT BREAKER, 277/480V:**					
16.2309 001	*Note: The following prices include molded case main circuit breaker, horizontal bus, underground pull section and Class I gear to 600 volts.*					
16.2309 011	Service section, 225a, 3ph, 277/480v, molded case main breaker	EA	12.8960	1,440.61	3,705.33	5,145.94
16.2309 021	Service section, 400a, 3ph, 277/480v, molded case main breaker	EA	14.6178	1,632.95	5,448.41	7,081.36
16.2309 031	Service section, 600a, 3ph, 277/480v, molded case main breaker	EA	17.4197	1,945.95	8,312.38	10,258.33
16.2309 041	Service section, 800a, 3ph, 277/480v, molded case main breaker	EA	18.4325	2,059.09	8,525.17	10,584.26
16.2309 051	Service section, 1000a, 3ph, 277/480v, molded case main breaker	EA	22.3823	2,500.33	12,979.82	15,480.14
16.2309 061	Service section, 1200a, 3ph, 277/480v, molded case main breaker	EA	22.5511	2,519.18	14,510.22	17,029.40
16.2309 071	Service section, 1600a, 3ph, 277/480v, molded case main breaker	EA	26.3321	2,941.56	18,256.83	21,198.39
16.2309 081	Service section, 2000a, 3ph, 277/480v, molded case main breaker	EA	36.7974	4,110.64	22,797.32	26,907.96
16.2309 091	Service section, 2500a, 3ph, 277/480v, molded case main breaker	EA	44.5619	4,978.01	31,922.56	36,900.57
16.2310 000	**SERVICE SECTION, MOLDED CASE MAIN CIRCUIT BREAKER, OPTIONS:**					
16.2310 011	Add for ground fault protection	EA	14.1988	1,586.15	5,984.59	7,570.74
16.2311 000	**SERVICE SECTION, FUSIBLE SWITCH, 1 PH, 120/240V:**					
16.2311 001	*Note: The following prices include fusible switch, horizontal bus, underground pull section and Class I gear to 600 volts.*					
16.2311 011	Service section, 225a, 1ph, 120/240v, fusible switch	EA	9.9927	1,116.28	1,500.12	2,616.41
16.2311 021	Service section, 400a, 1ph, 120/240v, fusible switch	EA	12.2209	1,365.20	3,737.17	5,102.37
16.2311 031	Service section, 600a, 1ph, 120/240v, fusible switch	EA	15.2255	1,700.84	4,627.38	6,328.22
16.2311 041	Service section, 800a, 1ph, 120/240v, fusible switch	EA	16.2382	1,813.97	7,147.74	8,961.71
16.2311 051	Service section, 1200a, 1ph, 120/240v, fusible switch	EA	18.8376	2,104.35	8,864.70	10,969.05
16.2312 000	**SERVICE SECTION, FUSIBLE SWITCH, 3 PH, 600V:**					
16.2312 001	*Note: The following prices include fusible switch, horizontal bus, underground pull section and Class I gear to 600 volts.*					
16.2312 011	Service section, 225a, 3ph, 600v, 3 wire, fusible switch	EA	9.9927	1,116.28	3,807.22	4,923.51
16.2312 021	Service section, 400a, 3ph, 600v, 3 wire, fusible switch	EA	12.2209	1,365.20	4,844.03	6,209.23

ELECTRICAL - 16

Division	Description	Unit	Labor Hours	Labor Cost	Material Cost	Total Cost
16.2312 031	Service section, 600a, 3ph, 600v, 3 wire, fusible switch	EA	15.2255	1,700.84	7,064.36	8,765.20
16.2312 041	Service section, 800a, 3ph, 600v, 3 wire, fusible switch	EA	16.2382	1,813.97	8,487.96	10,301.92
16.2312 051	Service section, 1200a, 3ph, 600v, 3 wire, fusible switch	EA	18.8376	2,104.35	9,918.21	12,022.56
16.2313 000	**SERVICE SECTION, FUSIBLE SWITCH, 3 PH, 120/240V:**					
16.2313 001	Note: The following prices include fusible switch, horizontal bus, underground pull section and Class I gear to 600 volts.					
16.2313 011	Service section, 225a, 3ph, 120/208v, fusible switch	EA	12.5584	1,402.90	3,740.17	5,143.06
16.2313 021	Service section, 400a, 3ph, 120/208v, fusible switch	EA	15.2592	1,704.61	4,489.53	6,194.13
16.2313 031	Service section, 600a, 3ph, 120/208v, fusible switch	EA	17.5547	1,961.04	7,162.97	9,124.01
16.2313 041	Service section, 800a, 3ph, 120/208v, fusible switch	EA	19.3440	2,160.92	8,512.41	10,673.33
16.2313 051	Service section, 1200a, 3ph, 120/208v, fusible switch	EA	22.3485	2,496.55	10,195.18	12,691.73
16.2314 000	**SERVICE SECTION, FUSIBLE SWITCH, 3 PH, 277/480V:**					
16.2314 001	Note: The following prices include fusible switch, horizontal bus, underground pull section and Class I gear to 600 volts.					
16.2314 011	Service section, 225a, 3ph, 277/480v, fusible switch	EA	12.5584	1,402.90	4,110.52	5,513.42
16.2314 021	Service section, 400a, 3ph, 277/480v, fusible switch	EA	15.2592	1,704.61	5,080.72	6,785.33
16.2314 031	Service section, 600a, 3ph, 277/480v, fusible switch	EA	17.5547	1,961.04	7,687.82	9,648.85
16.2314 041	Service section, 800a, 3ph, 277/480v, fusible switch	EA	19.3440	2,160.92	9,274.69	11,435.61
16.2314 051	Service section, 1200a, 3ph, 277/480v, fusible switch	EA	22.3485	2,496.55	19,016.35	21,512.90
16.2315 000	**SERVICE SECTION, FUSIBLE SWITCH, OPTIONS:**					
16.2315 011	Add for ground fault protection	EA	14.1988	1,586.15	5,984.59	7,570.74
16.2400 000	**COMBINATION SERVICE & DISTRIBUTION SWITCHBOARDS:**					
16.2400 001	Note: The following prices include main switch and distribution circuit breakers.					
16.2401 000	**COMBINATION SWITCHBOARDS WITH FUSIBLE MAIN:**					
16.2401 011	Switchboard, 225a, 1ph, 120/240v, fusible	EA	7.3259	818.38	3,113.74	3,932.12
16.2401 021	Switchboard, 400a, 1ph, 120/240v, fusible	EA	8.7437	976.76	4,157.30	5,134.06
16.2401 031	Switchboard, 600a, 1ph, 120/240v, fusible	EA	10.1616	1,135.15	5,360.81	6,495.96
16.2401 041	Switchboard, 225a, 3ph, 480v, fusible, 3 wide	EA	7.3259	818.38	4,960.18	5,778.55
16.2401 051	Switchboard, 400a, 3ph, 480v, fusible, 3 wide	EA	8.7437	976.76	7,533.20	8,509.96
16.2401 061	Switchboard, 600a, 3ph, 480v, fusible, 3 wide	EA	10.1616	1,135.15	9,615.98	10,751.13
16.2401 071	Switchboard, 200a, 3ph, 120/208v, fusible	EA	7.8321	874.92	4,361.64	5,236.57
16.2401 081	Switchboard, 400a, 3ph, 120/208v, fusible	EA	9.5539	1,067.27	6,230.92	7,298.18
16.2401 091	Switchboard, 600a, 3ph, 120/208v, fusible	EA	11.1744	1,248.29	8,376.78	9,625.07
16.2401 101	Switchboard, 225a, 3ph, 277/480v, fusible	EA	7.8321	874.92	4,575.13	5,450.05
16.2401 111	Switchboard, 400a, 3ph, 277/480v, fusible	EA	9.4526	1,055.95	5,590.91	6,646.86
16.2401 121	Switchboard, 600a, 3ph, 277/480v, fusible	EA	11.1744	1,248.29	7,851.22	9,099.51
16.2402 000	**COMBINATION SWITCHBOARD WITH CIRCUIT BREAKER MAIN:**					
16.2402 011	Switchboard, 225a, 1ph, 120/240v, circuit breaker main	EA	7.3259	818.38	3,584.50	4,402.87
16.2402 021	Switchboard, 400a, 1ph, 120/240v, circuit breaker main	EA	8.7437	976.76	5,733.78	6,710.53
16.2402 031	Switchboard, 600a, 1ph, 120/240v, circuit breaker main	EA	10.1616	1,135.15	6,794.48	7,929.63
16.2402 041	Switchboard, 800a, 1ph, 120/240v, circuit breaker main	EA	10.4992	1,172.87	9,542.49	10,715.35
16.2402 051	Switchboard, 1000a, 1ph, 120/240v, circuit breaker main	EA	12.1533	1,357.65	11,856.44	13,214.08
16.2402 061	Switchboard, 1200a, 1ph, 120/240v, circuit breaker main	EA	13.4362	1,500.96	14,604.29	16,105.24
16.2402 071	Switchboard, 225a, 3ph, 480v, circuit breaker main, 3 wire	EA	7.3259	818.38	4,930.75	5,749.13
16.2402 081	Switchboard, 400a, 3ph, 480v, circuit breaker main, 3 wire	EA	8.7437	976.76	8,081.20	9,057.96
16.2402 091	Switchboard, 600a, 3ph, 480v, circuit breaker main, 3 wire	EA	10.1616	1,135.15	10,214.90	11,350.06
16.2402 101	Switchboard, 800a, 3ph, 480v, circuit breaker main, 3 wire	EA	10.4992	1,172.87	13,518.78	14,691.65
16.2402 111	Switchboard, 1000a, 3ph, 480v, circuit breaker main, 3 wire	EA	12.1533	1,357.65	16,862.65	18,220.30
16.2402 121	Switchboard, 1200a, 3ph, 480v, circuit breaker main, 3 wire	EA	13.4362	1,500.96	19,953.22	21,454.18
16.2402 131	Switchboard, 225a, 3ph, 120/208v, circuit breaker main	EA	7.8321	874.92	4,604.02	5,478.94
16.2402 141	Switchboard, 400a, 3ph, 120/208v, circuit breaker main	EA	9.5539	1,067.27	7,241.13	8,308.40
16.2402 151	Switchboard, 600a, 3ph, 120/208v, circuit breaker main	EA	11.1744	1,248.29	9,104.77	10,353.06
16.2402 161	Switchboard, 800a, 3ph, 120/208v, circuit breaker main	EA	11.8495	1,323.71	12,008.55	13,332.26
16.2402 171	Switchboard, 1000a, 3ph, 120/208v, circuit breaker main	EA	13.5712	1,516.04	15,185.69	16,701.72
16.2402 181	Switchboard, 1200a, 3ph, 120/208v, circuit breaker main	EA	14.2802	1,595.24	17,126.04	18,721.28
16.2402 191	Switchboard, 225a, 3ph, 277/480v, circuit breaker main	EA	7.8321	874.92	4,604.02	5,478.94
16.2402 201	Switchboard, 400a, 3ph, 277/480v, circuit breaker main	EA	9.5539	1,067.27	7,241.13	8,308.40

16 - ELECTRICAL

Division	Description	Unit	Labor Hours	Labor Cost	Material Cost	Total Cost
16.2402 211	Switchboard, 600a, 3ph, 277/480v, circuit breaker main	EA	11.1744	1,248.29	9,104.77	10,353.06
16.2402 221	Switchboard, 800a, 3ph, 277/480v, circuit breaker main	EA	11.8495	1,323.71	12,008.55	13,332.26
16.2402 231	Switchboard, 1000a, 3ph, 277/480v, circuit breaker main	EA	13.5712	1,516.04	23,260.35	24,776.39
16.2402 241	Switchboard, 1200a, 3ph, 277/480v, circuit breaker main	EA	14.2802	1,595.24	25,220.68	26,815.92
16.2403 000	**COMBINATION SERVICE & DISTRIBUTION SWITCHBOARD OPTIONS:**					
16.2403 011	Add for underground pull section	EA	3.2072	358.28	389.98	748.26
16.2403 021	Add for ground fault protection	EA	14.1988	1,586.15	6,403.56	7,989.71
16.2404 000	**DISTRIBUTION SWITCHBOARD, 36", ENCLOSURE ONLY:**					
16.2404 011	Switchboard enclosure, 36", 225a, 1ph, 120/240v	EA	4.5576	509.13	1,076.75	1,585.88
16.2404 021	Switchboard enclosure, 36", 400a, 1ph, 120/240v	EA	4.5576	509.13	1,173.42	1,682.55
16.2404 031	Switchboard enclosure, 36", 600a, 1ph, 120/240v	EA	4.5576	509.13	1,343.42	1,852.55
16.2404 041	Switchboard enclosure, 36", 800a, 1ph, 120/240v	EA	4.8952	546.84	1,480.13	2,026.98
16.2404 051	Switchboard enclosure, 36", 1200a, 1ph, 120/240v	EA	5.7391	641.11	1,846.84	2,487.95
16.2404 061	Switchboard enclosure, 36", 1600a, 1ph, 120/240v	EA	7.4271	829.68	2,183.64	3,013.32
16.2404 071	Switchboard enclosure, 36", 2000a, 1ph, 120/240v	EA	11.4782	1,282.23	2,517.06	3,799.29
16.2404 081	Switchboard enclosure, 36", 225a, 3ph, 120/208v	EA	4.7264	527.99	1,210.12	1,738.11
16.2404 091	Switchboard enclosure, 36", 400a, 3ph, 120/208v	EA	4.7264	527.99	1,310.11	1,838.10
16.2404 101	Switchboard enclosure, 36", 600a, 3ph, 120/208v	EA	4.7264	527.99	1,510.13	2,038.11
16.2404 111	Switchboard enclosure, 36", 800a, 3ph, 120/208v	EA	4.7264	527.99	1,730.17	2,258.16
16.2404 121	Switchboard enclosure, 36", 1200a, 3ph, 120/208v	EA	6.0767	678.83	2,200.29	2,879.12
16.2404 131	Switchboard enclosure, 36", 1600a, 3ph, 120/208v	EA	8.7774	980.52	2,587.05	3,567.57
16.2404 141	Switchboard enclosure, 36", 2000a, 3ph, 120/208v	EA	8.7774	980.52	3,020.41	4,000.94
16.2404 151	Switchboard enclosure, 36", 2500a, 3ph, 120/208v	EA	14.8540	1,659.34	4,463.95	6,123.29
16.2404 161	Switchboard enclosure, 36", 3000a, 3ph, 120/208v	EA	16.2044	1,810.19	5,204.08	7,014.27
16.2404 171	Switchboard enclosure, 36", 4000a, 3ph, 120/208v	EA	20.9307	2,338.17	6,280.94	8,619.11
16.2404 181	Switchboard enclosure, 36", 225a, 3ph, 277/480v	EA	4.5576	509.13	1,089.01	1,598.14
16.2404 191	Switchboard enclosure, 36", 400a, 3ph, 277/480v	EA	4.5576	509.13	1,243.38	1,752.51
16.2404 201	Switchboard enclosure, 36", 600a, 3ph, 277/480v	EA	4.5576	509.13	1,413.53	1,922.66
16.2404 211	Switchboard enclosure, 36", 800a, 3ph, 277/480v	EA	4.8952	546.84	1,546.84	2,093.69
16.2404 221	Switchboard enclosure, 36", 1200a, 3ph, 277/480v	EA	5.7391	641.11	1,916.90	2,558.01
16.2404 231	Switchboard enclosure, 36", 1600a, 3ph, 277/480v	EA	7.4271	829.68	2,250.29	3,079.97
16.2404 241	Switchboard enclosure, 36", 2000a, 3ph, 277/480v	EA	11.4782	1,282.23	2,603.69	3,885.92
16.2404 251	Switchboard enclosure, 36", 2500a, 3ph, 277/480v	EA	14.1789	1,583.92	4,030.60	5,614.53
16.2404 261	Switchboard enclosure, 36", 3000a, 3ph, 277/480v	EA	14.8540	1,659.34	4,784.02	6,443.36
16.2404 271	Switchboard enclosure, 36", 4000a, 3ph, 277/480v	EA	18.2300	2,036.47	5,204.08	7,240.55
16.2405 000	**CIRCUIT BREAKER FOR DISTRIBUTION SWITCHBOARDS, BOLT-ON, 240/480V:**					
16.2405 011	Circuit breaker, 15/1-50/1, 240v, 10k ampere interrupt capacity, bolt on	EA	0.1892	21.14	21.44	42.58
16.2405 021	Circuit breaker, 15/2-50/2, 240v, 10k ampere interrupt capacity, bolt on	EA	0.3377	37.72	42.37	80.09
16.2405 031	Circuit breaker, 15/3-50/3, 240v, 10k, ampere interrupt capacity, bolt on	EA	0.4052	45.26	145.61	190.87
16.2405 041	Circuit breaker, 70/3-100/3, 240v, 10k ampere interrupt capacity, bolt on	EA	0.4727	52.81	211.83	264.64
16.2405 051	Circuit breaker, 15/1-50/1, 240v, 20k ampere interrupt capacity, bolt on	EA	0.1892	21.14	45.05	66.18
16.2405 061	Circuit breaker, 15/2-50/2, 240v, 20k ampere interrupt capacity, bolt on	EA	0.3782	42.25	94.00	136.25
16.2405 071	Circuit breaker, 15/3-50/3, 240v, 20k ampere interrupt capacity, bolt on	EA	0.5672	63.36	177.40	240.76
16.2405 081	Circuit breaker, 70/1-100/1, 240v, 20k ampere interrupt capacity, bolt on	EA	0.5200	58.09	58.28	116.36
16.2405 091	Circuit breaker, 70/2-100/2, 240v, 20k ampere interrupt capacity, bolt on	EA	0.7200	80.43	209.18	289.62
16.2405 101	Circuit breaker, 70/3-100/3, 240v, 20k ampere interrupt capacity, bolt on	EA	1.0128	113.14	268.11	381.25
16.2405 111	Circuit breaker, 125/2-225/2, 240v, 10k ampere interrupt capacity, bolt on	EA	1.2492	139.55	280.69	420.24
16.2405 121	Circuit breaker, 125/3-225/3, 240v, 10k ampere interrupt capacity, bolt on	EA	2.0000	223.42	722.90	946.32
16.2405 131	Circuit breaker, 15/1-50/1, 480v, 20k ampere interrupt capacity, bolt on	EA	0.1892	21.14	94.03	115.17
16.2405 141	Circuit breaker, 15/2-50/2, 480v, 20k ampere interrupt capacity, bolt on	EA	0.3782	42.25	328.36	370.61
16.2405 151	Circuit breaker, 15/3-50/3, 480v, 20k ampere interrupt capacity, bolt on	EA	0.5672	63.36	423.30	486.67
16.2405 161	Circuit breaker, 70/2-100/2, 480v, 20k ampere interrupt capacity, bolt on	EA	0.7200	80.43	428.87	509.30
16.2405 171	Circuit breaker, 70/3-100/3, 480v, 20k ampere interrupt capacity, bolt on	EA	1.0128	113.14	503.10	616.24
16.2405 181	Circuit breaker, 70/2-225/2 600v, 25k ampere interrupt capacity, bolt on	EA	1.2800	142.99	812.67	955.66
16.2405 191	Circuit breaker, 70/3-225/3, 600v, 25k ampere interrupt capacity, bolt on	EA	2.0000	223.42	1,005.23	1,228.65
16.2405 201	Circuit breaker, 125/2-400/2, 600v, 50k ampere interrupt capacity, bolt on	EA	1.8800	210.01	1,445.97	1,655.99

ELECTRICAL - 16

Division	Description	Unit	Labor Hours	Labor Cost	Material Cost	Total Cost
16.2405 211	Circuit breaker, 125/3-400/3, 600v, 50k ampere interrupt capacity, bolt on	EA	2.8000	312.79	1,726.62	2,039.41
16.2405 221	Circuit breaker, 500/2-600/2, 600v, 50k ampere interrupt capacity, bolt on	EA	2.7684	309.26	2,399.04	2,708.30
16.2405 231	Circuit breaker, 500/3-600/3, 600v, 50k ampere interrupt capacity, bolt on	EA	4.1525	463.88	3,042.05	3,505.93
16.2405 241	Circuit breaker, 700/2-800/2, 600v, 50k ampere interrupt capacity, bolt on	EA	2.7684	309.26	3,156.37	3,465.63
16.2405 251	Circuit breaker, 700/3-800/3, 600v, 50k ampere interrupt capacity, bolt on	EA	4.1525	463.88	3,998.51	4,462.39
16.2405 261	Circuit breaker, 1000/2, 600v, 50k ampere interrupt capacity, bolt-on	EA	3.7811	422.39	4,226.97	4,649.35
16.2405 271	Circuit breaker, 1000/3, 600v, 50k ampere interrupt capacity, bolt-on	EA	5.6716	633.57	5,062.62	5,696.19
16.2405 281	Circuit breaker, 1200/2, 600v, 50k ampere interrupt capacity, bolt-on	EA	3.9161	437.47	6,129.95	6,567.42
16.2405 291	Circuit breaker, 1220/3, 600v, 50k ampere interrupt capacity, bolt-on	EA	5.8741	656.20	7,001.55	7,657.74
16.2405 301	Circuit breaker, 2000/3, 600v, 50k ampere interrupt capacity, bolt-on	EA	7.0895	791.97	10,125.31	10,917.28
16.2406 000	MAIN/BRANCH FUSED SWITCH FOR DISTRIBUTION SWITCHBOARD, 240/480V:					
16.2406 011	Fused switch, 30/30 twin, 240v, 2p	EA	0.6753	75.44	387.57	463.01
16.2406 021	Fused switch, 60/60 twin, 240v, 2p	EA	1.0803	120.68	387.57	508.25
16.2406 031	Fused switch, 100a, 240v, 2p	EA	0.7428	82.98	300.28	383.26
16.2406 041	Fused switch, 200a, 240v, 2p	EA	1.6881	188.58	653.02	841.60
16.2406 051	Fused switch, 400a, 240v, 2p	EA	2.7008	301.71	1,459.89	1,761.60
16.2406 061	Fused switch, 600a, 240v, 2p	EA	3.5110	392.21	1,899.83	2,292.05
16.2406 071	Fused switch, 30/30 twin, 480v, 3p	EA	1.0128	113.14	495.82	608.96
16.2406 081	Fused switch, 60/60 twin, 480v, 3p	EA	1.6205	181.03	523.83	704.85
16.2406 091	Fused switch, 100a, 480v, 3p	EA	1.1142	124.47	405.06	529.53
16.2406 101	Fused switch, 200a, 480v, 3p	EA	2.5321	282.86	897.46	1,180.32
16.2406 111	Fused switch, 400a, 480v, 3p	EA	4.0511	452.55	2,078.03	2,530.58
16.2406 121	Fused switch, 600a, 480v, 3p	EA	5.2665	588.32	2,514.63	3,102.95
16.2407 000	DISTRIBUTION SWITCHBOARDS FOR CURRENT LIMITING BREAKERS, 240/480V:					
16.2407 001	Note: The following prices are based on 36" wide enclosures only. Components must be added.					
16.2407 011	Distribution switchboard, 225a, 120/240v, 1ph	EA	4.5576	509.13	1,302.64	1,811.77
16.2407 021	Distribution switchboard, 400a, 120/240v, 1ph	EA	4.5576	509.13	1,407.43	1,916.56
16.2407 031	Distribution switchboard, 600a, 120/240v, 1ph	EA	4.5576	509.13	1,582.02	2,091.15
16.2407 041	Distribution switchboard, 800a, 120/240v, 1ph	EA	4.8952	546.84	1,795.09	2,341.93
16.2407 051	Distribution switchboard, 1200a, 120/240v, 1ph	EA	5.7391	641.11	2,217.79	2,858.91
16.2407 061	Distribution switchboard, 1600a, 120/240v, 1ph	EA	7.4271	829.68	2,636.88	3,466.56
16.2407 071	Distribution switchboard, 2000a, 120/240v, 1ph	EA	11.4782	1,282.23	3,042.00	4,324.23
16.2407 081	Distribution switchboard, 225a, 277/480v, 3ph	EA	4.5576	509.13	1,311.29	1,820.42
16.2407 091	Distribution switchboard, 400a, 277/480v, 3ph	EA	4.5576	509.13	1,393.82	1,902.95
16.2407 101	Distribution switchboard, 600a, 277/480v, 3ph	EA	4.5576	509.13	1,707.85	2,216.98
16.2407 111	Distribution switchboard, 800a, 277/480v, 3ph	EA	4.8952	546.84	1,934.89	2,481.73
16.2407 121	Distribution switchboard, 1200a, 277/480v, 3ph	EA	5.7391	641.11	2,374.89	3,016.01
16.2407 131	Distribution switchboard, 1600a, 277/480v, 3ph	EA	7.4271	829.68	2,815.00	3,644.68
16.2407 141	Distribution switchboard, 2000a, 277/480v, 3ph	EA	11.4782	1,282.23	3,272.50	4,554.73
16.2407 151	Distribution switchboard, 3000a, 277/480v, 3ph	EA	14.8540	1,659.34	5,207.38	6,866.72
16.2407 161	Distribution switchboard, 4000a, 277/480v, 3ph	EA	18.2300	2,036.47	7,316.93	9,353.40
16.2408 000	MAIN/BRANCH CURRENT LIMITING BREAKERS, 240 OR 480V:					
16.2408 011	Current limiting circuit breaker, 15-100a, 2p, 600v	EA	0.4862	54.31	1,845.09	1,899.41
16.2408 021	Current limiting circuit breaker, 15-100a, 3p, 600v	EA	0.7293	81.47	2,224.49	2,305.96
16.2408 031	Current limiting circuit breaker, 125-225a, 2p, 600v	EA	0.7563	84.49	3,437.26	3,521.75
16.2408 041	Current limiting circuit breaker, 125-225a, 3p, 600v	EA	1.1344	126.72	4,313.89	4,440.62
16.2408 051	Current limiting circuit breaker, 250-400a, 2p, 600v	EA	1.3910	155.39	3,895.45	4,050.84
16.2408 061	Current limiting circuit breaker, 250-400a, 3p, 600v	EA	2.0864	233.07	4,958.37	5,191.45
16.2408 071	Current limiting circuit breaker, 600-800a, 2p, 600v	EA	2.3632	263.99	5,587.28	5,851.27
16.2408 081	Current limiting circuit breaker, 600-800a, 3p, 600v	EA	3.5449	396.00	6,981.92	7,377.92
16.2408 091	Current limiting circuit breaker, 800-1600a, 2p, 600v	EA	3.1734	354.50	14,288.48	14,642.98
16.2408 101	Current limiting circuit breaker, 800-1600a, 3p, 600v	EA	4.7601	531.75	17,368.09	17,899.84
16.2408 111	Current limiting circuit breaker, 1800-2000a, 2p, 600v	EA	4.7264	527.99	17,368.09	17,896.08
16.2408 121	Current limiting circuit breaker, 1800-2000a, 3p, 600v	EA	7.0895	791.97	18,442.00	19,233.96

16 - ELECTRICAL

Division	Description	Unit	Labor Hours	Labor Cost	Material Cost	Total Cost
16.2409 000	**ENCLOSED CIRCUIT BREAKERS:**					
16.2409 011	Circuit breaker to 30/3, 600v, with enclosure	EA	2.0000	223.42	811.76	1,035.18
16.2409 021	Circuit breaker to 60/3, 600v, with enclosure	EA	2.2000	245.76	811.76	1,057.52
16.2409 031	Circuit breaker to 100/3, 600v, with enclosure	EA	2.6000	290.45	909.09	1,199.53
16.2409 041	Circuit breaker to 225/3, 600v, with enclosure	EA	5.6000	625.58	3,207.56	3,833.13
16.2409 051	Circuit breaker to 400/3, 600v, with enclosure	EA	9.6000	1,072.42	4,172.81	5,245.23
16.2409 061	Circuit breaker to 600/3, 600v, with enclosure	EA	11.2000	1,251.15	5,288.20	6,539.35
16.2409 071	Circuit breaker to 800/3, 600v, with enclosure	EA	12.8000	1,429.89	6,566.92	7,996.81
16.2501 000	**ENCLOSED CIRCUIT BREAKERS, EXPLOSION PROOF:**					
16.2501 011	Circuit breaker to 30/3, 600v, explosion proof	EA	3.6500	407.74	1,294.70	1,702.44
16.2501 021	Circuit breaker to 60/3, 600v, explosion proof	EA	4.2500	474.77	1,387.65	1,862.42
16.2501 031	Circuit breaker to 100/3, 600v, explosion proof	EA	5.6300	628.93	1,603.74	2,232.67
16.2501 041	Circuit breaker to 225/3, 600v, explosion proof	EA	10.7500	1,200.88	3,233.47	4,434.35
16.2501 051	Circuit breaker to 400/3, 600v, explosion proof	EA	18.0200	2,013.01	6,495.06	8,508.08
16.2501 061	Circuit breaker to 600/3, 600v, explosion proof	EA	24.9600	2,788.28	11,306.44	14,094.72
16.2501 071	Circuit breaker to 800/3, 600v, explosion proof	EA	28.9400	3,232.89	15,075.97	18,308.86
16.3000 000	**MOTOR CONTROL CENTERS:**					
16.3001 000	**MOTOR CONTROL CENTER ENCLOSURE, ADD COMPONENTS:**					
16.3001 011	Motor control center enclosure only, 600v, 225a section	EA	7.1789	801.95	1,643.33	2,445.28
16.3001 021	Motor control center enclosure only, 600v, 400a section	EA	7.1789	801.95	2,738.85	3,540.80
16.3001 031	Motor control center enclosure only, 600v, 600a section	EA	7.1789	801.95	2,860.92	3,662.88
16.3001 041	Motor control center enclosure only, 600v, 800a section	EA	7.1789	801.95	2,957.99	3,759.95
16.3001 051	Motor control center enclosure only, 600v, 1200a section	EA	10.0170	1,119.00	5,477.80	6,596.80
16.3001 061	Motor control center enclosure only, 600v, 1600a section	EA	10.0170	1,119.00	8,237.26	9,356.26
16.3002 000	**COMPONENTS FOR MOTOR CONTROL CENTERS:**					
16.3002 011	Start-stop push button	EA	0.4052	45.26	175.84	221.11
16.3002 021	HOA selector switch	EA	0.5402	60.35	175.84	236.19
16.3002 031	Pilot light	EA	0.2701	30.17	219.12	249.29
16.3002 041	Electric interlock	EA	0.2701	30.17	128.96	159.13
16.3002 051	Ammeter in cover	EA	0.5402	60.35	472.83	533.18
16.3002 061	Ammeter & switch	EA	0.8103	90.52	2,026.93	2,117.45
16.3002 071	Voltmeter	EA	0.2701	30.17	1,337.01	1,367.18
16.3002 081	Voltmeter & switch	EA	0.5402	60.35	2,026.93	2,087.28
16.3002 091	Control transformers for starters, standard	EA	0.5402	60.35	149.32	209.66
16.3002 101	Combination starter, size 00, full voltage non reversing, to fractional hp	EA	4.2966	479.97	1,010.35	1,490.32
16.3002 111	Combination starter, size 0, full voltage non reversing, to 5hp	EA	4.2966	479.97	1,081.05	1,561.02
16.3002 121	Combination starter, size 1, full voltage non reversing, to 10hp	EA	4.2966	479.97	1,350.23	1,830.20
16.3002 131	Combination starter, size 2, full voltage non reversing, to 25hp	EA	6.1380	685.68	1,672.30	2,357.98
16.3002 141	Combination starter, size 3, full voltage non reversing, to 50hp	EA	10.2300	1,142.79	2,256.74	3,399.53
16.3002 151	Combination starter, size 4, full voltage non reversing, to 100hp	EA	16.3680	1,828.47	4,126.59	5,955.06
16.3002 161	Combination starter, size 5, full voltage non reversing, to 200hp	EA	26.1888	2,925.55	14,519.89	17,445.44
16.3002 171	Combination starter, size 6, full voltage non reversing, to 400hp	EA	36.5641	4,084.58	35,807.90	39,892.48
16.3002 181	Combination starter, size 7, full voltage non reversing, to 600hp	EA	47.6633	5,324.47	55,489.33	60,813.80
16.3002 191	Combination starter, size 8, full voltage non reversing, to 900hp	EA	57.1959	6,389.35	83,392.26	89,781.61
16.3002 201	Machine tool control transformer, 100va	EA	0.2701	30.17	130.51	160.68
16.3002 211	Machine tool control transformer, 300va	EA	0.2701	30.17	227.09	257.27
16.3002 221	Machine tool control transformer, 500va	EA	0.2701	30.17	245.76	275.94
16.3002 231	Machine tool control transformer, 1000va	EA	0.3377	37.72	381.08	418.80
16.3003 000	**AC CONTROL RELAYS:**					
16.3003 011	Relay, 20a, 2 pole, 600v	EA	0.2701	30.17	124.63	154.80
16.3003 021	Relay, 20a, 3 pole, 600v	EA	0.4052	45.26	145.41	190.68
16.3003 031	Relay, 20a, 6 pole, 600v	EA	0.8103	90.52	207.69	298.21
16.3003 041	Relay, 20a, 8 pole, 600v	EA	1.0803	120.68	249.30	369.98
16.3003 051	Relay, 20a, 10 pole, 600v	EA	1.3504	150.85	290.81	441.66
16.3003 061	Relay, 20a, 12 pole, 600v	EA	1.6205	181.03	332.39	513.42
16.3004 000	**MOTOR STARTERS, 600V, WITH ENCLOSURE:**					
16.3004 011	Fractional hp manual starter	EA	0.5879	65.67	68.04	133.71

ELECTRICAL - 16

Division	Description	Unit	Labor Hours	Labor Cost	Material Cost	Total Cost
16.3004 021	Manual starter, 1ph, to 5hp, with enclosure	EA	1.4000	156.39	192.13	348.53
16.3004 031	Magnetic starter, 3ph, to 10hp, with enclosure	EA	2.8000	312.79	373.99	686.78
16.3004 041	Magnetic starter, 3ph, to 25hp, with enclosure	EA	5.6000	625.58	747.98	1,373.55
16.3004 051	Magnetic starter, 3ph, to 50hp, with enclosure	EA	7.2000	804.31	1,246.58	2,050.90
16.3004 061	Magnetic starter, 3ph, to 100hp, with enclosure	EA	12.0000	1,340.52	2,929.48	4,270.00
16.3005 000	**COMBINATION MOTOR STARTERS, 600V, WITH ENCLOSURE:**					
16.3005 011	Combination starter, to size 0, full voltage non reversing to 5hp	EA	1.7500	195.49	1,496.74	1,692.23
16.3005 021	Combination starter, size 1, full voltage non reversing to 10hp	EA	2.2500	251.35	1,557.03	1,808.38
16.3005 031	Combination starter, size 2, full voltage non reversing, 25hp	EA	3.5000	390.99	2,203.89	2,594.88
16.3005 041	Combination starter, size 3, full voltage non reversing to 50hp	EA	4.5000	502.70	3,205.68	3,708.38
16.3005 051	Combination starter, size 4, full voltage non reversing to 100hp	EA	5.7500	642.33	7,035.98	7,678.31
16.3005 061	Combination starter, size 5, full voltage non reversing, to 200hp	EA	8.0000	893.68	16,317.15	17,210.83
16.3005 071	Combination starter, size 6, full voltage non reversing, to 400hp	EA	12.5000	1,396.38	40,690.71	42,087.08
16.3005 081	Combination starter, size 7, full voltage non reversing, to 600hp	EA	15.0000	1,675.65	63,056.01	64,731.66
16.3006 000	**SWITCHES, FUSIBLE DISCONNECT, 240/480V:**					
16.3006 011	Switch, fusible disconnect, single pole, 30a, 240/480v	EA	2.0000	223.42	295.33	518.75
16.3006 021	Switch, fusible disconnect, 3 pole, 30a, 240/480v	EA	2.2000	245.76	344.88	590.64
16.3006 031	Switch, fusible disconnect, 3 pole, 60a, 240/480v	EA	2.8000	312.79	395.96	708.74
16.3006 041	Switch, fusible disconnect, 3 pole, 100a, 240/480v	EA	3.2000	357.47	714.41	1,071.88
16.3006 051	Switch, fusible disconnect, 3 pole, 200a, 240/480v	EA	5.0000	558.55	1,042.74	1,601.29
16.3006 061	Switch, fusible disconnect, 3 pole, 400a, 240/480v	EA	8.0000	893.68	2,674.59	3,568.27
16.3006 071	Switch, fusible disconnect, 3 pole, 600a, 240/480v	EA	11.2000	1,251.15	4,379.03	5,630.18
16.3007 000	**SWITCHES, NONFUSED DISCONNECT, 240/480V:**					
16.3007 011	Switch, nonfuse disconnect, single pole, 30a, 240/480v	EA	1.3500	150.81	83.17	233.98
16.3007 021	Switch, nonfuse disconnect, 3 pole, 30a, 240/480v	EA	1.5000	167.57	140.21	307.78
16.3007 031	Switch, nonfuse disconnect, 3 pole, 60a, 240/480v	EA	2.0000	223.42	266.13	489.55
16.3007 041	Switch, nonfuse disconnect, 3 pole, 100a, 240/480v	EA	2.5000	279.27	452.94	732.21
16.3007 051	Switch, nonfuse disconnect, 3 pole, 200a, 240/480v	EA	4.5000	502.70	699.09	1,201.79
16.3007 061	Switch, nonfuse disconnect, 3 pole, 400a, 240/480v	EA	7.6000	849.00	1,448.87	2,297.87
16.3007 071	Switch, nonfuse disconnect, 3 pole, 600a, 240/480v	EA	10.4000	1,161.78	2,579.28	3,741.07
16.3007 081	Switch, nonfuse, explosion proof, single pole, 30a, 240/480v	EA	3.2730	365.63	524.55	890.17
16.3007 091	Switch, nonfuse, explosion proof, 3 pole, 30a, 240/480v	EA	3.8510	430.20	1,091.85	1,522.04
16.3007 101	Switch, nonfuse, explosion proof, 3 pole, 60a, 240/480v	EA	4.5220	505.15	1,159.42	1,664.58
16.3007 111	Switch, nonfuse, explosion proof, 3 pole, 100a, 240/480v	EA	5.8540	653.95	1,447.65	2,101.60
16.3007 121	Switch, nonfuse, explosion proof, 3 pole, 200a, 240/480v	EA	10.4630	1,168.82	3,049.56	4,218.39
16.3007 131	Switch, nonfuse, explosion proof, 3 pole, 400a, 240/480v	EA	16.9210	1,890.24	6,065.48	7,955.73
16.3007 141	Switch, nonfuse, explosion proof, 3 pole, 600a, 240/480v	EA	23.5450	2,630.21	10,758.88	13,389.09
16.3008 000	**MOTOR CONNECTIONS:**					
16.3008 011	Motor connection, 1 ph, to 5hp	EA	1.5200	169.80	41.37	211.17
16.3008 021	Motor connection, 3 ph, to 5hp	EA	1.6000	178.74	49.69	228.43
16.3008 031	Motor connection, 3 ph, to 15hp	EA	2.2500	251.35	82.79	334.14
16.3008 041	Motor connection, 3 ph, to 40hp	EA	5.5000	614.41	151.90	766.31
16.3008 051	Motor connection, 3 ph, to 75hp	EA	10.0000	1,117.10	220.96	1,338.06
16.3008 061	Motor connection, 3 ph, to 100hp	EA	15.0000	1,675.65	262.30	1,937.95
16.3008 071	Motor connection, 3 ph, to 150hp	EA	21.0000	2,345.91	345.23	2,691.14
16.3008 081	Motor connection, 3 ph, to 250hp	EA	25.0000	2,792.75	690.47	3,483.22
16.3008 091	Motor connection, 3 ph, to 500hp	EA	40.0000	4,468.40	1,104.78	5,573.18
16.3008 101	Start/stop panelboard station, NEMA 1	EA	0.4000	44.68	74.16	118.84
16.3008 111	On/off pilot light, panelboard station, NEMA 1	EA	0.4500	50.27	181.39	231.66
16.3008 121	3 position panelboard station NEMA 1	EA	0.4000	44.68	113.33	158.01
16.3008 131	Start/stop panelboard explosion proof	EA	2.3000	256.93	330.45	587.38
16.3008 141	On/off pilot, panelboard, explosion proof	EA	2.3000	256.93	696.83	953.76
16.3008 151	3 position panelboard station explosion proof	EA	2.6000	290.45	674.36	964.80
16.3009 000	**CAPACITORS:**					
16.3009 011	Capacitor, 240v, 3ph to 5kvar	EA	3.2736	365.69	1,465.20	1,830.89
16.3009 021	Capacitor, 240v, 3ph to 15kvar	EA	4.0920	457.12	3,587.04	4,044.15
16.3009 031	Capacitor, 240v, 3ph to 25kvar	EA	5.3196	594.25	4,573.79	5,168.04

16 - ELECTRICAL

Division	Description	Unit	Labor Hours	Labor Cost	Material Cost	Total Cost
16.3009 041	Capacitor, 480v, 3ph to 5kvar	EA	3.2736	365.69	1,805.11	2,170.80
16.3009 051	Capacitor, 480v, 3ph to 15kvar	EA	4.0920	457.12	2,059.15	2,516.27
16.3009 061	Capacitor, 480v, 3ph to 40kvar	EA	4.9104	548.54	3,593.65	4,142.19
16.3009 071	Capacitor, 480v, 3ph to 50kvar	EA	6.1380	685.68	4,141.44	4,827.11
16.4000 000	**PANELBOARDS, 600V MAX, BOLT-ON BREAKERS:**					
16.4000 001	Note: The following prices include circuit breakers for branch circuits.					
16.4001 000	**PANELBOARDS, BOLT-ON CIRCUIT BREAKERS, 277/480V, 3 PH, 4 WIRE, MAIN CIRCUIT BREAKER:**					
16.4001 001	Note: The following items include main circuit breaker.					
16.4001 011	Panel, 18 circuit, 100a, 277/480v, main circuit breaker	EA	4.4563	497.81	2,504.62	3,002.43
16.4001 021	Panel, 24 circuit, 100a, 277/480v, main circuit breaker	EA	4.4563	497.81	3,708.30	4,206.11
16.4001 031	Panel, 30 circuit, 225a, 277/480v, main circuit breaker	EA	10.4000	1,161.78	3,933.46	5,095.25
16.4001 041	Panel, 42 circuit, 225a, 277/480v, main circuit breaker	EA	14.8800	1,662.24	5,140.43	6,802.68
16.4001 051	Panel, 30 circuit, 400a, 277/480v, main circuit breaker	EA	13.2000	1,474.57	4,217.39	5,691.96
16.4001 061	Panel, 42 circuit, 400a, 277/480v, main circuit breaker	EA	16.4800	1,840.98	5,390.11	7,231.09
16.4001 071	Panel, 42 circuit, 600a, 277/480v, main circuit breaker	EA	18.0800	2,019.72	5,618.46	7,638.18
16.4002 000	**PANELBOARDS, BOLT-ON BREAKERS, 277/480V, 3 PH, 4 WIRE, MAIN LUG ONLY:**					
16.4002 001	Note: The following prices are main lug only.					
16.4002 011	Panel, 18 circuit, 100a, 277/480v, main lug only	EA	4.2800	478.12	2,428.93	2,907.05
16.4002 021	Panel, 24 circuit, 100a, 277/480v, main lug only	EA	6.3468	709.00	3,067.51	3,776.51
16.4002 031	Panel, 30 circuit, 100a, 277/480v, main lug only	EA	10.4000	1,161.78	4,316.85	5,478.63
16.4002 041	Panel, 30 circuit, 400a, 277/480v, main lug only	EA	14.0000	1,563.94	6,855.39	8,419.33
16.4002 051	Panel, 30 circuit, 225a, 277/480v, main lug only	EA	11.2000	1,251.15	5,468.70	6,719.85
16.4002 061	Panel, 42 circuit, 225a, 277/480v, main lug only	EA	15.6800	1,751.61	6,657.73	8,409.34
16.4002 071	Panel, 42 circuit, 400a, 277/480v, main lug only	EA	17.2800	1,930.35	8,068.94	9,999.29
16.4100 000	**TRANSFORMERS, DRY, LOW VOLTAGE:**					
16.4101 000	**TRANSFORMERS, LOW VOLTAGE, 1 PH, 240/480V PRIMARY, 120/240V SECOND:**					
16.4101 011	Transformer, 0.5kva, 240-480/120-240v, 1ph	EA	2.4000	268.10	192.41	460.51
16.4101 021	Transformer, 1 kva, 240-480/120-240v, 1ph	EA	2.8000	312.79	298.69	611.48
16.4101 031	Transformer, 1.5kva, 240-480/120-240v, 1ph	EA	2.8000	312.79	357.30	670.09
16.4101 041	Transformer, 2kva, 240-480/120-240v, 1ph	EA	3.2000	357.47	445.03	802.51
16.4101 051	Transformer, 3kva, 240-480/120-240v, 1ph	EA	3.4000	379.81	569.33	949.14
16.4101 061	Transformer, 5kva, 240-480/120-240v, 1ph	EA	4.2000	469.18	771.76	1,240.95
16.4101 071	Transformer, 7.5kva, 240-480/120-240v, 1ph	EA	6.4000	714.94	1,084.52	1,799.47
16.4101 081	Transformer, 10kva, 240-480/120-240v, 1ph	EA	7.2000	804.31	1,345.10	2,149.41
16.4101 091	Transformer, 15kva, 240-480/120-240v, 1ph	EA	9.6000	1,072.42	1,804.16	2,876.58
16.4101 101	Transformer, 25kva, 240-480/120-240v, 1ph	EA	12.0000	1,340.52	2,345.43	3,685.95
16.4101 111	Transformer, 38kva, 240-480/120-240v, 1ph	EA	14.0000	1,563.94	3,127.23	4,691.17
16.4101 121	Transformer, 50kva, 240-480/120-240v, 1ph	EA	16.0000	1,787.36	3,804.79	5,592.15
16.4101 131	Transformer, 75kva, 240-480/120-240v, 1ph	EA	18.0000	2,010.78	5,159.98	7,170.76
16.4101 141	Transformer, 100kva, 240-480/120-240v, 1ph	EA	20.0000	2,234.20	6,671.45	8,905.65
16.4102 000	**TRANSFORMERS, LOW VOLTAGE, 3 PH, 480V PRIME, 4 WIRE, 120/208V SECOND:**					
16.4102 011	Transformer, 3kva, 480/120-208v, 3ph	EA	6.4000	714.94	1,171.88	1,886.82
16.4102 021	Transformer, 6kva, 480/120-208v, 3ph	EA	8.0000	893.68	1,343.27	2,236.95
16.4102 031	Transformer, 9kva, 480/120-208v, 3ph	EA	9.6000	1,072.42	1,794.21	2,866.63
16.4102 041	Transformer, 15kva, 480/120-208v, 3ph	EA	11.2000	1,251.15	2,698.40	3,949.55
16.4102 051	Transformer, 30kva, 480/120-208v, 3ph	EA	14.4000	1,608.62	4,051.02	5,659.64
16.4102 061	Transformer, 45kva, 480/120-208v, 3ph	EA	16.0000	1,787.36	4,335.21	6,122.57
16.4102 071	Transformer, 75kva, 480/120-208v, 3ph	EA	18.0000	2,010.78	6,533.36	8,544.14
16.4102 081	Transformer, 112.5kva, 480/120-208v, 3ph	EA	20.0000	2,234.20	8,693.96	10,928.16
16.4102 091	Transformer, 150kva, 480/120-208v, 3ph	EA	21.6000	2,412.94	11,357.11	13,770.05
16.4102 101	Transformer, 225kva, 480/120-208v, 3ph	EA	28.0000	3,127.88	15,142.79	18,270.67
16.4102 111	Transformer, 300kva, 480/120-208v, 3ph	EA	32.0000	3,574.72	19,416.98	22,991.70
16.4102 121	Transformer, 500kva, 480/120-208v, 3ph	EA	40.0000	4,468.40	30,774.19	35,242.59

ELECTRICAL - 16

Division	Description	Unit	Labor Hours	Labor Cost	Material Cost	Total Cost
16.4102 131	Transformer, 750kva, 480/120-208v, 3ph	EA	48.0000	5,362.08	50,008.00	55,370.08
16.4102 141	Transformer, 1000kva, 480/120-208v, 3ph	EA	52.0000	5,808.92	60,327.12	66,136.04
16.4103 000	**TRANSFORMERS, LOW VOLTAGE, 3 PH, 480V PRIME, 240V SECOND:**					
16.4103 011	Transformer, 6kva, 480/240v, 3ph	EA	8.0000	893.68	1,221.21	2,114.89
16.4103 021	Transformer, 9kva, 480/240v, 3ph	EA	9.6000	1,072.42	1,631.10	2,703.52
16.4103 031	Transformer, 15kva, 480/240v, 3ph	EA	11.2000	1,251.15	2,453.09	3,704.24
16.4103 041	Transformer, 30kva, 480/240v, 3ph	EA	14.4000	1,608.62	3,275.02	4,883.64
16.4103 051	Transformer, 45kva, 480/240v, 3ph	EA	16.0000	1,787.36	3,941.09	5,728.45
16.4103 061	Transformer, 75kva, 480/240v, 3ph	EA	18.0000	2,010.78	5,939.43	7,950.21
16.4103 071	Transformer, 112.5kva, 480/240v, 3ph	EA	20.0000	2,234.20	7,903.60	10,137.80
16.4103 081	Transformer, 150kva, 480/240v, 3ph	EA	21.6000	2,412.94	10,324.65	12,737.59
16.4103 091	Transformer, 225kva, 480/240v, 3ph	EA	28.0000	3,127.88	13,765.56	16,893.44
16.4103 101	Transformer, 300kva, 480/240v, 3ph	EA	32.0000	3,574.72	17,651.81	21,226.53
16.4200 000	**PANELBOARDS FOR BOLT-ON BREAKERS, 1240/240V, 1PH, 3 WIRE:**					
16.4200 001	*Note: The following prices include circuit breakers for branch circuits.*					
16.4201 000	**PANELBOARDS, BOLT-ON BEAKERS, 120/240V, 1 PH, 3 WIRE, MAIN LUG ONLY:**					
16.4201 001	*Note: The following items are main lug only.*					
16.4201 011	Panel, 12 circuit, 100a, 120/240v, main lug only	EA	4.2800	478.12	836.69	1,314.81
16.4201 021	Panel, 24 circuit, 100a, 120/240v, main lug only	EA	6.8000	759.63	1,114.44	1,874.06
16.4201 031	Panel, 30 circuit, 225a, 120/240v, main lug only	EA	10.4000	1,161.78	1,463.19	2,624.97
16.4201 041	Panel, 42 circuit, 225a, 120/240v, main lug only	EA	14.8800	1,662.24	1,850.60	3,512.85
16.4201 051	Panel, 30 circuit, 400a, 120/240v, main lug only	EA	13.2000	1,474.57	1,773.08	3,247.65
16.4201 061	Panel, 42 circuit, 400a, 120/240v, main lug only	EA	16.4800	1,840.98	2,158.82	3,999.80
16.4202 000	**PANELBOARDS, BOLT-ON BREAKERS, 120/240V, 1 PH, 3 WIRE, MAIN CIRCUIT BREAKER:**					
16.4202 001	*Note: The following prices include main circuit breaker.*					
16.4202 011	Panel, 12 circuit, 100a, 120/240v, main circuit breaker	EA	4.2800	478.12	1,140.13	1,618.25
16.4202 021	Panel, 24 circuit, 100a, 120/240v, main circuit breaker	EA	7.2000	804.31	1,416.16	2,220.48
16.4202 031	Panel, 30 circuit, 225a, 120/240v, main circuit breaker	EA	11.2000	1,251.15	2,710.72	3,961.87
16.4202 041	Panel, 42 circuit, 225a, 120/240v, main circuit breaker	EA	15.6800	1,751.61	3,094.83	4,846.44
16.4202 051	Panel, 30 circuit, 400a, 120/240v, main circuit breaker	EA	14.0000	1,563.94	3,998.78	5,562.72
16.4202 061	Panel, 42 circuit, 400a, 120/240v, main circuit breaker	EA	17.2800	1,930.35	4,391.02	6,321.37
16.4203 000	**PANELBOARDS, BOLT-ON BREAKERS, 120/208V, 3 PH, 4 WIRE, MAIN LUG ONLY:**					
16.4203 001	*Note: The following items are main lug only.*					
16.4203 011	Panel, 12 circuit, 100a, 120/208v, main lug only	EA	4.2800	478.12	900.05	1,378.17
16.4203 021	Panel, 24 circuit, 100a, 120/208v, main lug only	EA	6.8000	759.63	1,306.83	2,066.46
16.4203 031	Panel, 30 circuit, 100a, 120/208v, main lug only	EA	9.8000	1,094.76	1,498.92	2,593.68
16.4203 041	Panel, 30 circuit, 225a, 120/208v, main lug only	EA	10.4000	1,161.78	1,552.40	2,714.18
16.4203 051	Panel, 30 circuit, 400a, 120/208v, main lug only	EA	13.2000	1,474.57	1,891.50	3,366.07
16.4203 061	Panel, 42 circuit, 225a, 120/208v, main lug only	EA	14.3800	1,606.39	2,069.96	3,676.35
16.4203 071	Panel, 42 circuit, 400a, 120/208v, main lug only	EA	16.4800	1,840.98	2,277.27	4,118.25
16.4204 000	**PANELBOARDS, BOLT-ON BREAKERS, 120/208V, 3 PH, 4 WIRE, MAIN CIRCUIT BREAKER:**					
16.4204 001	*Note: The following prices include main circuit breaker.*					
16.4204 011	Panel, 12 circuit, 100a, 120/208v, main circuit breaker	EA	3.8824	433.70	1,313.66	1,747.36
16.4204 021	Panel, 24 circuit, 100a, 120/208v, main circuit breaker	EA	8.9600	1,000.92	1,720.52	2,721.45
16.4204 031	Panel, 30 circuit, 100a, 120/208v, main circuit breaker	EA	10.6000	1,184.13	1,912.88	3,097.00
16.4204 041	Panel, 42 circuit, 225a, 120/208v, main circuit breaker	EA	15.6800	1,751.61	3,450.12	5,201.73
16.4204 051	Panel, 30 circuit, 400a, 120/208v, main circuit breaker	EA	13.8000	1,541.60	4,420.58	5,962.17
16.4204 061	Panel, 42 circuit, 400a, 120/208v, main circuit breaker	EA	17.2800	1,930.35	4,811.20	6,741.54
16.4300 000	**LOAD CENTERS, MAIN LUG & CIRCUIT BREAKER TYPES, 240V MAX**					
16.4300 001	*Note: The following items are for plug-in breakers and include ground bus and cover with door. Prices do not include breakers.*					
16.4301 000	**LOAD CENTERS, 120/240V, 1 PH, 3W, NEMA 1, INDOOR:**					

16 - ELECTRICAL

Division	Description	Unit	Labor Hours	Labor Cost	Material Cost	Total Cost
16.4301 001	Note: The following items are main lug only.					
16.4301 011	Panel, 12 circuit, 100a, 120/240, NEMA 1	EA	2.2500	251.35	94.60	345.95
16.4301 021	Panel, 16 circuit, 125a, 120/240, NEMA 1	EA	2.4983	279.09	110.66	389.74
16.4301 031	Panel, 24 circuit, 125a, 120/240, NEMA 1	EA	4.0174	448.78	134.70	583.48
16.4301 041	Panel, 24 circuit, 150a, 120/240, NEMA 1	EA	4.0174	448.78	239.08	687.87
16.4301 051	Panel, 30 circuit, 150a, 120/240, NEMA 1	EA	4.0174	448.78	258.71	707.50
16.4301 061	Panel, 16 circuit, 225a, 120/240, NEMA 1	EA	2.4983	279.09	240.83	519.91
16.4301 071	Panel, 24 circuit, 225a, 120/240, NEMA 1	EA	4.0174	448.78	269.37	718.15
16.4301 081	Panel, 30 circuit, 225a, 120/240, NEMA 1	EA	5.5000	614.41	296.17	910.58
16.4301 091	Panel, 42 circuit, 225a, 120/240, NEMA 1	EA	6.5000	726.12	497.86	1,223.97
16.4302 000	LOAD CENTER, 120/240V, 1 PH, 3 WIRE, SOLID NEUTRAL , NEMA 3R, OUTDOOR, MAIN LUG ONLY:					
16.4302 001	Note: The following items are main lug only.					
16.4302 011	Panel, 3 circuit, 60a, 120/208v, NEMA 3R	EA	1.5000	167.57	62.39	229.95
16.4302 021	Panel, 12 circuit, 125a, 120/208v, NEMA 3R	EA	3.5000	390.99	215.87	606.86
16.4302 031	Panel, 12 circuit, 200a, 120/208v, NEMA 3R	EA	4.0000	446.84	292.46	739.30
16.4302 041	Panel, 20 circuit, 125a, 120/208v, NEMA 3R	EA	4.5000	502.70	306.73	809.42
16.4302 051	Panel, 30 circuit, 225a, 120/208v, NEMA 3R	EA	4.4000	491.52	426.41	917.93
16.4302 061	Panel, 42 circuit, 225a, 120/208v, NEMA 3R	EA	5.2000	580.89	583.50	1,164.39
16.4303 000	LOAD CENTER, 120/240V, 1 PH, 3 WIRE, SOLID NEUTRAL, NEMA 1, INDOOR:					
16.4303 001	Note: The following prices include main circuit breaker.					
16.4303 011	Panel, 16 circuit, 100a, 120/240v, main circuit breaker	EA	2.9600	330.66	212.37	543.04
16.4303 021	Panel, 20 circuit, 100a, 120/240v, main circuit breaker	EA	3.3600	375.35	256.06	631.40
16.4303 031	Panel, 24 circuit, 125a, 120/240v, main circuit breaker	EA	4.2560	475.44	486.29	961.73
16.4303 041	Panel, 30 circuit, 150a, 120/240v, main circuit breaker	EA	4.2560	475.44	506.76	982.20
16.4303 051	Panel, 30 circuit, 225a, 120/240v, main circuit breaker	EA	5.0000	558.55	626.30	1,184.85
16.4303 061	Panel, 42 circuit, 225a, 120/240v, main circuit breaker	EA	5.1990	580.78	735.26	1,316.04
16.4304 000	LOAD CENTER, 120/208V, 3 PH, 4 WIRE, NEMA 1, INDOOR:					
16.4304 001	Note: The following prices include main circuit breaker.					
16.4304 011	Panel, 30 circuit, 125a, 120/208v, main circuit breaker	EA	4.9104	548.54	1,208.04	1,756.58
16.4304 021	Panel, 30 circuit, 150a, 120/208v, main circuit breaker	EA	4.9104	548.54	1,208.04	1,756.58
16.4304 031	Panel, 30 circuit, 225a, 120/208v, main circuit breaker	EA	4.9104	548.54	1,208.04	1,756.58
16.4304 041	Panel, 42 circuit, 150a, 120/208v, main circuit breaker	EA	4.9104	548.54	1,325.80	1,874.35
16.4304 051	Panel, 42 circuit, 225a, 120/208v, main circuit breaker	EA	4.9104	548.54	1,325.80	1,874.35
16.4400 000	PLUG-IN CIRCUIT BREAKERS, TYPE QO:					
16.4401 000	PLUG-IN CIRCUIT BREAKER, MAX 240V:					
16.4401 011	Plug-in circuit breaker, to 30/1, 10k ampere interrupt capacity	EA	0.2800	31.28	19.22	50.50
16.4401 021	Plug-in circuit breaker, to 40/1, 10k ampere interrupt capacity	EA	0.3000	33.51	19.22	52.74
16.4401 031	Plug-in circuit breaker, to 50/1, 10k ampere interrupt capacity	EA	0.3200	35.75	19.22	54.97
16.4401 041	Plug-in circuit breaker, to 30/1, with ground fault interrupter	EA	0.2800	31.28	150.61	181.89
16.4401 051	Plug-in circuit breaker, to 30/2, 10k ampere interrupt capacity	EA	0.4400	49.15	43.73	92.88
16.4401 061	Plug-in circuit breaker, to 40/2, 10k ampere interrupt capacity	EA	0.4600	51.39	43.73	95.12
16.4401 071	Plug-in circuit breaker, to 50/2, 10k ampere interrupt capacity	EA	0.4800	53.62	43.73	97.35
16.4401 081	Plug-in circuit breaker, to 60/2, 10k ampere interrupt capacity	EA	0.6000	67.03	43.73	110.76
16.4401 091	Plug-in circuit breaker, to 70/2, 10k ampere interrupt capacity	EA	0.6000	67.03	88.90	155.93
16.4401 101	Plug-in circuit breaker, to 100/2, 10k ampere interrupt capacity	EA	0.7200	80.43	125.27	205.70
16.4401 111	Plug-in circuit breaker, to 125/2, 10k ampere interrupt capacity	EA	0.8000	89.37	266.76	356.12
16.4401 121	Plug-in circuit breaker, to 150/2, 10k ampere interrupt capacity	EA	0.8800	98.30	266.76	365.06
16.4401 131	Plug-in circuit breaker, to 30/3, 10k ampere interrupt capacity	EA	0.4400	49.15	150.61	199.76
16.4401 141	Plug-in circuit breaker, to 40/3, 10k ampere interrupt capacity	EA	0.4600	51.39	150.61	202.00
16.4401 151	Plug-in circuit breaker, to 50/3, 10k ampere interrupt capacity	EA	0.4800	53.62	150.61	204.23
16.4401 161	Plug-in circuit breaker, to 60/3, 10k ampere interrupt capacity	EA	0.6078	67.90	150.61	218.51
16.4401 171	Plug-in circuit breaker, to 70/3, 10k ampere interrupt capacity	EA	0.6078	67.90	194.20	262.10
16.4401 181	Plug-in circuit breaker, to 90/3, 10k ampere interrupt capacity	EA	0.6800	75.96	318.88	394.85
16.4401 191	Plug-in circuit breaker, to 100/3, 10k ampere interrupt capacity	EA	0.7200	80.43	298.01	378.44

ELECTRICAL - 16

Division	Description	Unit	Labor Hours	Labor Cost	Material Cost	Total Cost
16.4402 000	PLUG-IN CIRCUIT BREAKER, MAX 240V, 65K AMPERE INTERRUPT CAPACITY:					
16.4402 011	Plug-in circuit breaker, to 30/1, 65k ampere interrupt capacity	EA	0.2800	31.28	69.87	101.15
16.4402 021	Plug-in circuit breaker, to 30/2, 65k ampere interrupt capacity	EA	0.4400	49.15	172.99	222.14
16.4402 031	Plug-in circuit breaker, to 30/3, 65k ampere interrupt capacity	EA	0.4400	49.15	302.21	351.36
16.4500 000	SPECIAL GEAR:					
16.4501 000	EMERGENCY GENERATORS WITH ACCESSORIES:					
16.4501 011	Emergency generator to 30 kw	EA	32.7360	3,656.94	35,639.70	39,296.64
16.4501 021	Emergency generator to 60 kw	EA	49.1040	5,485.41	42,532.98	48,018.39
16.4501 031	Emergency generator to 100 kw	EA	65.4720	7,313.88	60,761.46	68,075.34
16.4501 041	Emergency generator to 150 kw	EA	81.8400	9,142.35	72,913.78	82,056.12
16.4501 051	Emergency generator to 200 kw	EA	98.2080	10,970.82	85,066.05	96,036.87
16.4501 061	Emergency generator to 400 kw	EA	114.5760	12,799.28	176,208.20	189,007.49
16.4501 071	Emergency generator to 600 kw	EA	130.9440	14,627.75	273,426.62	288,054.37
16.4501 081	Emergency generator to 750 kw	EA	163.6800	18,284.69	382,797.22	401,081.91
16.4501 091	Emergency generator to 1000 kw	EA	196.4160	21,941.63	440,094.68	462,036.31
16.4502 000	TRANSFER SWITCHES, AUTOMATIC:					
16.4502 011	Auto transfer switch to 30a	EA	3.2736	365.69	8,374.68	8,740.37
16.4502 021	Auto transfer switch to 70a	EA	4.9104	548.54	8,485.67	9,034.21
16.4502 031	Auto transfer switch to 100a	EA	6.5472	731.39	9,263.55	9,994.94
16.4502 041	Auto transfer switch to 150a	EA	8.1840	914.23	9,866.29	10,780.53
16.4502 051	Auto transfer switch to 225a	EA	9.8208	1,097.08	16,037.64	17,134.72
16.4502 061	Auto transfer switch to 260a	EA	11.4576	1,279.93	16,037.64	17,317.57
16.4502 071	Auto transfer switch to 400a	EA	13.0944	1,462.78	17,990.43	19,453.21
16.4502 081	Auto transfer switch to 600a	EA	14.7312	1,645.62	27,623.27	29,268.89
16.4502 091	Auto transfer switch to 800a	EA	16.3680	1,828.47	33,014.70	34,843.17
16.4502 101	Auto transfer switch to 1000a	EA	18.0048	2,011.32	54,620.29	56,631.60
16.4502 111	Auto transfer switch to 1200a	EA	19.6416	2,194.16	56,618.55	58,812.72
16.4502 121	Auto transfer switch to 1600a	EA	24.5520	2,742.70	70,287.89	73,030.59
16.4502 131	Auto transfer switch to 2000a	EA	29.4624	3,291.24	73,211.33	76,502.58
16.4503 000	TRANSFER SWITCHES, AUTOMATIC WITH BYPASS:					
16.4503 011	Auto transfer switch to 150a, bypass	EA	8.1800	913.79	32,718.66	33,632.45
16.4503 021	Auto transfer switch to 260a, bypass	EA	11.4500	1,279.08	36,271.19	37,550.27
16.4503 031	Auto transfer switch to 400a, bypass	EA	13.0900	1,462.28	49,368.37	50,830.65
16.4503 041	Auto transfer switch to 600a, bypass	EA	14.7300	1,645.49	68,076.09	69,721.58
16.4503 051	Auto transfer switch to 800a, bypass	EA	16.3600	1,827.58	80,652.29	82,479.87
16.4503 061	Auto transfer switch to 1000a, bypass	EA	18.0000	2,010.78	123,393.81	125,404.59
16.4503 071	Auto transfer switch to 1200a, bypass	EA	19.6400	2,193.98	139,052.86	141,246.84
16.4503 081	Auto transfer switch to 1600a, bypass	EA	24.5500	2,742.48	179,681.89	182,424.37
16.4503 091	Auto transfer switch to 2000a, bypass	EA	29.4600	3,290.98	200,923.40	204,214.38
16.4504 000	UNINTERRUPTED POWER SYSTEMS:					
16.4504 011	Uninterrupted power supply with battery bank, 37.5kva	EA	128.6116	14,367.20	148,098.72	162,465.93
16.4504 021	Uninterrupted power supply with battery bank, 50kva	EA	171.4794	19,155.96	164,636.30	183,792.26
16.4504 031	Uninterrupted power supply with battery bank, 75kva	EA	194.8611	21,767.93	183,109.32	204,877.25
16.4504 041	Uninterrupted power supply with battery bank, 100kva	EA	245.5200	27,427.04	223,689.75	251,116.79
16.4504 051	Uninterrupted power supply with battery bank, 125kva	EA	286.0000	31,949.06	230,055.34	262,004.40
16.4504 061	Uninterrupted power supply with battery bank, 225 kva	EA	346.0000	38,651.66	275,045.30	313,696.96
16.5000 000	PVC, RSC, IMC & ALUMINUM RACEWAY:					
16.5001 000	PVC CONDUIT, SCHEDULE 40, CONCRETE ENCASED, TRENCH & BURIED:					
16.5001 011	PVC conduit, '40', trench & buried in concrete, 1-2"	LF	0.0504	5.63	31.01	36.64
16.5001 021	PVC conduit, '40', trench & buried in concrete, 2-2"	LF	0.0619	6.91	38.17	45.09
16.5001 031	PVC conduit, '40', trench & buried in concrete, 3-2"	LF	0.0718	8.02	39.57	47.59
16.5001 041	PVC conduit, '40', trench & buried in concrete, 4-2"	LF	0.0829	9.26	39.57	48.83
16.5001 051	PVC conduit, '40', trench & buried in concrete, 1-3"	LF	0.0664	7.42	33.22	40.63
16.5001 061	PVC conduit, '40', trench & buried in concrete, 2-3"	LF	0.0830	9.27	40.26	49.54
16.5001 071	PVC conduit, '40', trench & buried in concrete, 3-3"	LF	0.0995	11.12	48.04	59.15

16 - ELECTRICAL

Division	Description	Unit	Labor Hours	Labor Cost	Material Cost	Total Cost
16.5001 081	PVC conduit, '40', trench & buried in concrete, 4-3"	LF	0.1105	12.34	50.89	63.24
16.5001 091	PVC conduit, '40', trench & buried in concrete, 1-4"	LF	0.1105	12.34	39.84	52.18
16.5001 101	PVC conduit, '40', trench & buried in concrete, 2-4"	LF	0.1436	16.04	48.33	64.37
16.5001 111	PVC conduit, '40', trench & buried in concrete, 3-4"	LF	0.1659	18.53	58.04	76.58
16.5001 121	PVC conduit, '40', trench & buried in concrete, 4-4"	LF	0.1824	20.38	62.56	82.94
16.5001 131	PVC conduit, '40', trench & buried in concrete, 1-5"	LF	0.1271	14.20	51.46	65.66
16.5001 141	PVC conduit, '40', trench & buried in concrete, 2-5"	LF	0.1769	19.76	61.36	81.12
16.5001 151	PVC conduit, '40', trench & buried in concrete, 3-5"	LF	0.2046	22.86	65.63	88.49
16.5001 161	PVC conduit, '40', trench & buried in concrete, 4-5"	LF	0.2323	25.95	69.24	95.19
16.5001 171	PVC conduit, '40', trench & buried in concrete, 1-6"	LF	0.1328	14.84	59.43	74.27
16.5001 181	PVC conduit, '40', trench & buried in concrete, 2-6"	LF	0.1837	20.52	69.25	89.77
16.5001 191	PVC conduit, '40', trench & buried in concrete, 4-6"	LF	0.2818	31.48	106.56	138.04
16.5001 201	PVC conduit, '40', trench & buried in concrete, 6-6"	LF	0.3765	42.06	142.37	184.43
16.5001 211	PVC conduit, '40', trench & buried in concrete, 8-6"	LF	0.5186	57.93	195.98	253.91
16.5001 221	PVC conduit, '40', trench & buried in concrete, 10-6"	LF	0.6053	67.62	228.73	296.35
16.5001 231	PVC conduit, '40', trench & buried in concrete, 12-6"	LF	0.6947	77.60	262.53	340.13
16.5002 000	**PVC CONDUIT, SCHEDULE 40, EMBEDDED:**					
16.5002 011	PVC conduit, 1/2", underground embed in slab	LF	0.0320	3.57	1.10	4.67
16.5002 021	PVC conduit, 3/4", underground embed in slab	LF	0.0320	3.57	1.46	5.03
16.5002 031	PVC conduit, 1", underground embed in slab	LF	0.0320	3.57	2.22	5.79
16.5002 041	PVC conduit, 1-1/4", underground embed in slab	LF	0.0320	3.57	3.04	6.61
16.5002 051	PVC conduit, 1-1/2", underground embed in slab	LF	0.0400	4.47	3.71	8.18
16.5002 061	PVC conduit, 2", underground embed in slab	LF	0.0480	5.36	4.64	10.01
16.5002 071	PVC conduit, 2-1/2", underground embed in slab	LF	0.0720	8.04	7.48	15.52
16.5002 081	PVC conduit, 3", underground embed in slab	LF	0.0960	10.72	9.39	20.12
16.5002 091	PVC conduit, 3-1/2", underground embed in slab	LF	0.1120	12.51	11.88	24.39
16.5002 101	PVC conduit, 4", underground embed in slab	LF	0.1280	14.30	13.34	27.64
16.5002 111	PVC conduit, 5", underground embed in slab	LF	0.1760	19.66	19.04	38.70
16.5002 121	PVC conduit, 6", underground embed in slab	LF	0.2400	26.81	24.69	51.50
16.5003 000	**PVC CONDUIT TERMINATIONS & ELBOWS:**					
16.5003 011	1/2" PVC terminal adapters	EA	0.1000	11.17	1.05	12.22
16.5003 021	3/4" PVC terminal adapters	EA	0.1200	13.41	1.95	15.36
16.5003 031	1" PVC terminal adapters	EA	0.1500	16.76	2.44	19.19
16.5003 041	1-1/4" PVC terminal adapters	EA	0.1700	18.99	3.19	22.18
16.5003 051	1-1/2" PVC terminal adapters	EA	0.2000	22.34	3.81	26.15
16.5003 061	2" PVC terminal adapters	EA	0.2500	27.93	5.53	33.46
16.5003 071	2-1/2" terminal adapters	EA	0.3000	33.51	9.45	42.96
16.5003 081	3" PVC terminal adapters	EA	0.3500	39.10	13.67	52.77
16.5003 091	3-1/2" PVC terminal adapters	EA	0.4000	44.68	17.95	62.64
16.5003 101	4" PVC terminal adapters	EA	0.4500	50.27	23.45	73.72
16.5003 111	5" PVC terminal adapters	EA	0.4845	54.12	46.33	100.45
16.5003 121	6" PVC terminal adapters	EA	0.5815	64.96	55.78	120.74
16.5003 131	5" PVC elbow with 1 coupling	EA	1.7445	194.88	175.48	370.36
16.5003 141	6" PVC elbow with 1 coupling	EA	2.0352	227.35	278.39	505.74
16.5003 151	1/2" PVC elbow with 1 coupling	EA	0.1800	20.11	3.70	23.80
16.5003 161	3/4" PVC elbow with 1 coupling	EA	0.2200	24.58	4.18	28.76
16.5003 171	1" PVC elbow with 1 coupling	EA	0.2600	29.04	6.54	35.58
16.5003 181	1-1/4" PVC elbow with 1 coupling	EA	0.3300	36.86	9.06	45.92
16.5003 191	1-1/2" PVC elbow with 1 coupling	EA	0.4500	50.27	12.27	62.54
16.5003 201	2" PVC elbow with 1 coupling	EA	0.5600	62.56	17.50	80.06
16.5003 211	2-1/2" PVC elbow with 1 coupling	EA	0.8200	91.60	31.79	123.40
16.5003 221	3" PVC elbow with 1 coupling	EA	1.2000	134.05	54.54	188.60
16.5003 231	3-1/2" PVC elbow with 1 coupling	EA	1.5000	167.57	72.37	239.94
16.5003 241	4" PVC elbow with 1 coupling	EA	1.6767	187.30	92.22	279.52
16.5004 000	**RIGID STEEL CONDUIT, EMBEDDED:**					
16.5004 011	RSC, 1/2", embedded	LF	0.0320	3.57	2.53	6.10
16.5004 021	RSC, 3/4", embedded	LF	0.0400	4.47	3.17	7.63

ELECTRICAL - 16

Division	Description	Unit	Labor Hours	Labor Cost	Material Cost	Total Cost
16.5004 031	RSC, 1", embedded	LF	0.0480	5.36	4.39	9.75
16.5004 041	RSC, 1-1/4", embedded	LF	0.0600	6.70	5.63	12.33
16.5004 051	RSC, 1-1/2", embedded	LF	0.0720	8.04	7.12	15.16
16.5004 061	RSC, 2", embedded	LF	0.0880	9.83	9.21	19.04
16.5004 071	RSC, 2-1/2", embedded	LF	0.0960	10.72	15.54	26.26
16.5004 081	RSC, 3", embedded	LF	0.1200	13.41	18.97	32.38
16.5004 091	RSC, 3-1/2", embedded	LF	0.1440	16.09	23.72	39.81
16.5004 101	RSC, 4", embedded	LF	0.1680	18.77	27.99	46.76
16.5004 111	RSC, 5", embedded	LF	0.2240	25.02	63.95	88.97
16.5004 121	RSC, 6", embedded	LF	0.2880	32.17	80.48	112.65
16.5005 000	**RIGID STEEL CONDUIT, CONCEALED:**					
16.5005 011	RSC, 1/2", concealed	LF	0.0440	4.92	2.63	7.55
16.5005 021	RSC, 3/4", concealed	LF	0.0520	5.81	3.32	9.12
16.5005 031	RSC, 1", concealed	LF	0.0640	7.15	4.59	11.74
16.5005 041	RSC, 1-1/4", concealed	LF	0.0800	8.94	5.93	14.86
16.5005 051	RSC, 1-1/2", concealed	LF	0.0960	10.72	7.43	18.15
16.5005 061	RSC, 2", concealed	LF	0.1120	12.51	9.69	22.20
16.5005 071	RSC, 2-1/2", concealed	LF	0.1280	14.30	16.27	30.57
16.5005 081	RSC, 3", concealed	LF	0.1600	17.87	19.85	37.72
16.5005 091	RSC, 3-1/2", concealed	LF	0.1920	21.45	24.91	46.36
16.5005 101	RSC, 4", concealed	LF	0.2240	25.02	29.40	54.43
16.5005 111	RSC, 5", concealed	LF	0.3040	33.96	67.82	101.78
16.5005 121	RSC, 6", concealed	LF	0.3840	42.90	84.50	127.40
16.5006 000	**RIGID STEEL CONDUIT, EXPOSED:**					
16.5006 011	RSC, 1/2", exposed	LF	0.0541	6.04	2.73	8.77
16.5006 021	RSC, 3/4", exposed	LF	0.0660	7.37	3.37	10.75
16.5006 031	RSC, 1", exposed	LF	0.0800	8.94	4.70	13.64
16.5006 041	RSC, 1-1/4", exposed	LF	0.1000	11.17	6.12	17.29
16.5006 051	RSC, 1-1/2", exposed	LF	0.1200	13.41	7.65	21.05
16.5006 061	RSC, 2", exposed	LF	0.1440	16.09	9.97	26.06
16.5006 071	RSC, 2-1/2", exposed	LF	0.1600	17.87	16.81	34.68
16.5006 081	RSC, 3", exposed	LF	0.2000	22.34	20.46	42.80
16.5006 091	RSC, 3-1/2", exposed	LF	0.2400	26.81	25.61	52.42
16.5006 101	RSC, 4", exposed	LF	0.2800	31.28	31.69	62.97
16.5006 111	RSC, 5", exposed	LF	0.3760	42.00	73.24	115.24
16.5006 121	RSC, 6", exposed	LF	0.4800	53.62	91.25	144.87
16.5007 000	**INTERMEDIATE METAL CONDUIT, EMBEDDED:**					
16.5007 011	Intermediate metal conduit, 1/2", embedded	LF	0.0315	3.52	1.96	5.48
16.5007 021	Intermediate metal conduit, 3/4", embedded	LF	0.0414	4.62	2.32	6.95
16.5007 031	Intermediate metal conduit, 1", embedded	LF	0.0534	5.97	3.26	9.22
16.5007 041	Intermediate metal conduit, 1-1/4", embedded	LF	0.0534	5.97	4.16	10.12
16.5007 051	Intermediate metal conduit, 1-1/2", embedded	LF	0.0652	7.28	4.81	12.09
16.5007 061	Intermediate metal conduit, 2", embedded	LF	0.0720	8.04	6.60	14.64
16.5007 071	Intermediate metal conduit, 2-1/2", embedded	LF	0.0800	8.94	13.61	22.55
16.5007 081	Intermediate metal conduit, 3", embedded	LF	0.1274	14.23	17.05	31.28
16.5007 091	Intermediate metal conduit, 3-1/2", embedded	LF	0.1600	17.87	19.88	37.76
16.5007 101	Intermediate metal conduit, 4", embedded	LF	0.1687	18.85	23.56	42.40
16.5008 000	**INTERMEDIATE METAL CONDUIT, CONCEALED:**					
16.5008 011	Intermediate metal conduit, 1/2", concealed	LF	0.0389	4.35	2.07	6.41
16.5008 021	Intermediate metal conduit, 3/4", concealed	LF	0.0562	6.28	2.36	8.63
16.5008 031	Intermediate metal conduit, 1", concealed	LF	0.0623	6.96	3.37	10.33
16.5008 041	Intermediate metal conduit, 1-1/4", concealed	LF	0.0623	6.96	4.33	11.29
16.5008 051	Intermediate metal conduit, 1-1/2", concealed	LF	0.0762	8.51	5.04	13.55
16.5008 061	Intermediate metal conduit, 2", concealed	LF	0.0960	10.72	7.00	17.73
16.5008 071	Intermediate metal conduit, 2-1/2", concealed	LF	0.1040	11.62	14.30	25.92
16.5008 081	Intermediate metal conduit, 3", concealed	LF	0.1280	14.30	17.91	32.21
16.5008 091	Intermediate metal conduit, 3-1/2", concealed	LF	0.1440	16.09	20.91	37.00

16 - ELECTRICAL

Division	Description	Unit	Labor Hours	Labor Cost	Material Cost	Total Cost
16.5008 101	Intermediate metal conduit, 4", concealed	LF	0.1977	22.09	24.72	46.81
16.5009 000	**INTERMEDIATE METAL CONDUIT, EXPOSED:**					
16.5009 011	Intermediate metal conduit, 1/2", exposed	LF	0.0520	5.81	2.14	7.95
16.5009 021	Intermediate metal conduit, 3/4", exposed	LF	0.0700	7.82	2.41	10.23
16.5009 031	Intermediate metal conduit, 1", exposed	LF	0.0754	8.42	3.47	11.89
16.5009 041	Intermediate metal conduit, 1-1/4", exposed	LF	0.0754	8.42	4.47	12.89
16.5009 051	Intermediate metal conduit, 1-1/2", exposed	LF	0.1000	11.17	5.15	16.32
16.5009 061	Intermediate metal conduit, 2", exposed	LF	0.1200	13.41	7.14	20.55
16.5009 071	Intermediate metal conduit, 2-1/2", exposed	LF	0.1280	14.30	14.70	28.99
16.5009 081	Intermediate metal conduit, 3", exposed	LF	0.1600	17.87	18.40	36.28
16.5009 091	Intermediate metal conduit, 3-1/2", exposed	LF	0.1800	20.11	21.51	41.62
16.5009 101	Intermediate metal conduit, 4", exposed	LF	0.2259	25.24	25.42	50.65
16.5010 000	**ALUMINUM CONDUIT, CONCEALED:**					
16.5010 011	Aluminum conduit, 1/2", concealed	LF	0.0384	4.29	2.25	6.54
16.5010 021	Aluminum conduit, 3/4", concealed	LF	0.0520	5.81	3.00	8.81
16.5010 031	Aluminum conduit, 1", concealed	LF	0.0600	6.70	4.27	10.98
16.5010 041	Aluminum conduit, 1-1/4" concealed	LF	0.0720	8.04	5.58	13.62
16.5010 051	Aluminum conduit, 1-1/2", concealed	LF	0.0880	9.83	7.01	16.84
16.5010 061	Aluminum conduit, 2", concealed	LF	0.0640	7.15	9.34	16.48
16.5010 071	Aluminum conduit, 2-1/2", concealed	LF	0.0880	9.83	14.75	24.58
16.5010 081	Aluminum conduit, 3", concealed	LF	0.1120	12.51	19.41	31.92
16.5010 091	Aluminum conduit, 3-1/2", concealed	LF	0.1280	14.30	23.10	37.39
16.5010 101	Aluminum conduit, 4", concealed	LF	0.1520	16.98	27.57	44.55
16.5010 111	Aluminum conduit, 5", concealed	LF	0.1920	21.45	39.45	60.90
16.5010 121	Aluminum conduit, 6", concealed	LF	0.2400	26.81	52.01	78.82
16.5011 000	**ALUMINUM CONDUIT, EXPOSED:**					
16.5011 011	Aluminum conduit, 1/2", exposed	LF	0.0520	5.81	2.40	8.21
16.5011 021	Aluminum conduit, 3/4", exposed	LF	0.0600	6.70	3.26	9.96
16.5011 031	Aluminum conduit, 1", exposed	LF	0.0740	8.27	4.53	12.80
16.5011 041	Aluminum conduit, 1-1/4", exposed	LF	0.0920	10.28	6.01	16.29
16.5011 051	Aluminum conduit, 1-1/2", exposed	LF	0.1120	12.51	7.43	19.94
16.5011 061	Aluminum conduit, 2", exposed	LF	0.0800	8.94	9.97	18.91
16.5011 071	Aluminum conduit, 2-1/2", exposed	LF	0.1080	12.06	15.80	27.87
16.5011 081	Aluminum conduit, 3", exposed	LF	0.1400	15.64	20.77	36.41
16.5011 091	Aluminum conduit, 3-1/2" exposed	LF	0.1600	17.87	24.71	42.59
16.5011 101	Aluminum conduit, 4", exposed	LF	0.1920	21.45	29.50	50.94
16.5011 111	Aluminum conduit, 5", exposed	LF	0.2400	26.81	42.23	69.04
16.5011 121	Aluminum conduit, 6", exposed	LF	0.2960	33.07	55.64	88.71
16.5100 000	**PVC, RSC, IMC & ALUMINUM CONDUIT TERMINALS, ELBOWS & FITTINGS:**					
16.5101 000	**CONDUIT TERMINATION FOR RSC, IMC & ALUMINUM:**					
16.5101 011	1/2" double locknut & ground bushing	EA	0.2000	22.34	5.63	27.97
16.5101 021	3/4" double locknut & ground bushing	EA	0.2500	27.93	6.53	34.46
16.5101 031	1" double locknut & ground bushing	EA	0.3000	33.51	9.70	43.22
16.5101 041	1-1/4" double locknut & ground bushing	EA	0.3500	39.10	11.56	50.66
16.5101 051	1-1/2" double locknut & ground bushing	EA	0.5000	55.86	14.31	70.17
16.5101 061	2" double locknut & ground bushing	EA	0.6000	67.03	20.20	87.22
16.5101 071	2-1/2" double locknut & ground bushing	EA	1.0000	111.71	35.03	146.74
16.5101 081	3" double locknut & ground bushing	EA	1.3000	145.22	44.92	190.14
16.5101 091	3-1/2" double locknut & ground bushing	EA	1.5000	167.57	65.96	233.52
16.5101 101	4" double locknut & ground bushing	EA	1.8000	201.08	84.33	285.41
16.5101 111	5" double locknut & ground busing	EA	2.8000	312.79	96.98	409.77
16.5101 121	6" double locknut & ground bushing	EA	4.0000	446.84	152.22	599.06
16.5102 000	**CONDUIT HUBS:**					
16.5102 011	1/2" conduit hub	EA	0.2500	27.93	8.32	36.25
16.5102 021	3/4" conduit hub	EA	0.3000	33.51	9.53	43.04

ELECTRICAL - 16

Division	Description	Unit	Labor Hours	Labor Cost	Material Cost	Total Cost
16.5102 031	1" conduit hub	EA	0.3500	39.10	12.02	51.11
16.5102 041	1-1/4" conduit hub	EA	0.4000	44.68	13.75	58.43
16.5102 051	1-1/2" conduit hub	EA	0.5000	55.86	14.33	70.18
16.5102 061	2" conduit hub	EA	0.7500	83.78	19.63	103.41
16.5102 071	2-1/2" conduit hub	EA	1.0000	111.71	38.78	150.49
16.5102 081	3" conduit hub	EA	1.5000	167.57	54.82	222.39
16.5102 091	3-1/2" conduit hub	EA	2.0000	223.42	72.32	295.74
16.5102 101	4" conduit hub	EA	2.8000	312.79	90.39	403.18
16.5102 111	5" conduit hub	EA	4.0000	446.84	166.76	613.60
16.5102 121	6" conduit hub	EA	5.0000	558.55	223.63	782.18
16.5103 000	**RIGID CONDUIT ELBOWS:**					
16.5103 011	1/2" galvanized rigid conduit elbow with 1 coupling	EA	0.2500	27.93	9.65	37.57
16.5103 021	3/4" galvanized rigid conduit elbow with 1 coupling	EA	0.3000	33.51	11.69	45.20
16.5103 031	1" galvanized rigid conduit elbow with 1 coupling	EA	0.3500	39.10	17.11	56.21
16.5103 041	1-1/4" galvanized rigid conduit elbow with 1 coupling	EA	0.5500	61.44	23.89	85.33
16.5103 051	1-1/2" galvanized rigid conduit elbow with 1 coupling	EA	0.7000	78.20	29.92	108.12
16.5103 061	2" galvanized rigid conduit elbow with 1 coupling	EA	0.9000	100.54	43.00	143.54
16.5103 071	2-1/2" galvanized rigid conduit elbow with 1 coupling	EA	1.4000	156.39	77.03	233.42
16.5103 081	3" galvanized rigid conduit elbow with 1 coupling	EA	2.0000	223.42	112.71	336.13
16.5103 091	3-1/2" galvanized rigid conduit elbow with 1 coupling	EA	2.5000	279.27	184.68	463.96
16.5103 101	4" galvanized rigid conduit elbow with 1 coupling	EA	3.2580	363.95	210.42	574.37
16.5103 111	5" galvanized rigid conduit elbow with 1 coupling	EA	4.0000	446.84	457.58	904.42
16.5103 121	6" galvanized rigid conduit elbow with 1 coupling	EA	5.0000	558.55	723.14	1,281.69
16.5104 000	**INTERMEDIATE METAL CONDUIT ELBOWS:**					
16.5104 011	1/2" intermediate metal conduit elbow	EA	0.2000	22.34	6.02	28.36
16.5104 021	3/4" intermediate metal conduit elbow	EA	0.2500	27.93	7.90	35.83
16.5104 031	1" intermediate metal conduit elbow	EA	0.3000	33.51	11.37	44.88
16.5104 041	1-1/4" intermediate metal conduit elbow	EA	0.5000	55.86	15.92	71.78
16.5104 051	1-1/2" intermediate metal conduit elbow	EA	0.6500	72.61	20.17	92.78
16.5104 061	2" intermediate metal conduit elbow	EA	0.8500	94.95	29.10	124.06
16.5104 071	2-1/2" intermediate metal conduit elbow	EA	1.2500	139.64	50.29	189.93
16.5104 081	3" intermediate metal conduit elbow	EA	1.7500	195.49	76.97	272.46
16.5104 091	3-1/2" intermediate metal conduit elbow	EA	2.0000	223.42	135.91	359.33
16.5104 101	4" intermediate metal conduit elbow	EA	2.7500	307.20	157.67	464.87
16.5105 000	**ALUMINUM ELBOWS:**					
16.5105 011	1/2" aluminum elbow	EA	0.1500	16.76	6.37	23.12
16.5105 021	3/4" aluminum elbow	EA	0.2000	22.34	8.69	31.03
16.5105 031	1" aluminum elbow	EA	0.2580	28.82	12.07	40.89
16.5105 041	1-1/4" aluminum elbow	EA	0.4500	50.27	19.21	69.48
16.5105 051	1-1/2" aluminum elbow	EA	0.6000	67.03	25.57	92.59
16.5105 061	2" aluminum elbow	EA	0.8000	89.37	37.62	126.99
16.5105 071	2-1/2" aluminum elbow	EA	1.2000	134.05	63.51	197.56
16.5105 081	3" aluminum elbow	EA	1.5000	167.57	98.06	265.63
16.5105 091	3-1/2" aluminum elbow	EA	1.7500	195.49	152.93	348.42
16.5105 101	4" aluminum elbow	EA	2.5000	279.27	181.39	460.66
16.5105 111	5" aluminum elbow	EA	3.2500	363.06	495.07	858.13
16.5105 121	6" aluminum elbow	EA	4.0000	446.84	685.26	1,132.10
16.5200 000	**EMT RACEWAY, TERMINATIONS & ELBOWS:**					
16.5201 000	**EMT CONDUIT, CONCEALED:**					
16.5201 011	1/2" EMT conduit, concealed	LF	0.0350	3.91	0.87	4.78
16.5201 021	3/4" EMT conduit, concealed	LF	0.0400	4.47	1.09	5.55
16.5201 031	1" EMT conduit, concealed	LF	0.0450	5.03	1.71	6.74
16.5201 041	1-1/4" EMT conduit, concealed	LF	0.0550	6.14	2.73	8.87
16.5201 051	1-1/2" EMT conduit, concealed	LF	0.0650	7.26	3.25	10.51
16.5201 061	2" EMT conduit, concealed	LF	0.0750	8.38	3.75	12.13
16.5201 071	2-1/2" EMT conduit, concealed	LF	0.0900	10.05	10.54	20.59
16.5201 081	3" EMT conduit, concealed	LF	0.1100	12.29	13.19	25.48

16 - ELECTRICAL

Division	Description	Unit	Labor Hours	Labor Cost	Material Cost	Total Cost
16.5201 091	4" EMT conduit, concealed	LF	0.1400	15.64	20.25	35.89
16.5202 000	**EMT CONDUIT, EXPOSED:**					
16.5202 011	1/2" EMT conduit, exposed	LF	0.0400	4.47	0.91	5.38
16.5202 021	3/4" EMT conduit, exposed	LF	0.0450	5.03	1.14	6.17
16.5202 031	1" EMT conduit, exposed	LF	0.0500	5.59	1.86	7.45
16.5202 041	1-1/4" EMT conduit, exposed	LF	0.0650	7.26	2.99	10.25
16.5202 051	1-1/2" EMT conduit, exposed	LF	0.0750	8.38	3.44	11.82
16.5202 061	2" EMT conduit, exposed	LF	0.0850	9.50	4.41	13.91
16.5202 071	2-1/2" EMT conduit, exposed	LF	0.1000	11.17	11.73	22.90
16.5202 081	3" EMT conduit, exposed	LF	0.1200	13.41	13.19	26.60
16.5202 091	4" EMT conduit, exposed	LF	0.1600	17.87	22.62	40.50
16.5203 000	**EMT CONDUIT TERMINATIONS:**					
16.5203 011	1/2" EMT set-screw connectors	EA	0.1000	11.17	1.51	12.68
16.5203 021	3/4" EMT set-screw connectors	EA	0.1000	11.17	2.51	13.68
16.5203 031	1" EMT set-screw connectors	EA	0.1500	16.76	4.10	20.86
16.5203 041	1-1/4" EMT set-screw connectors	EA	0.1500	16.76	7.42	24.17
16.5203 051	1-1/2" EMT set-screw connectors	EA	0.2000	22.34	10.80	33.14
16.5203 061	2" EMT set-screw connectors	EA	0.2500	27.93	15.63	43.56
16.5203 071	2-1/2" EMT set-screw connectors	EA	0.3500	39.10	69.54	108.64
16.5203 081	3" EMT set-screw connectors	EA	0.5000	55.86	82.13	137.99
16.5203 091	3-1/2" EMT set-screw connectors	EA	0.6000	67.03	110.13	177.15
16.5203 101	4" EMT set-screw connectors	EA	0.6000	67.03	120.15	187.18
16.5204 000	**EMT ELBOWS:**					
16.5204 011	1" EMT elbow with 2 couplings	EA	0.1500	16.76	12.58	29.34
16.5204 021	1-1/4" elbow with 2 couplings	EA	0.2000	22.34	21.79	44.13
16.5204 031	1-1/2" EMT elbow with 2 couplings	EA	0.3000	33.51	30.27	63.78
16.5204 041	2" EMT elbow with 2 couplings	EA	0.3500	39.10	43.13	82.23
16.5204 051	2-1/2" EMT elbow with 2 couplings	EA	0.6000	67.03	105.76	172.79
16.5204 061	3" EMT elbow with 2 couplings	EA	0.9000	100.54	137.48	238.02
16.5204 071	3-1/2" EMT elbow with 2 couplings	EA	1.4000	156.39	172.20	328.60
16.5204 081	4" EMT elbow with 2 couplings	EA	1.4000	156.39	200.03	356.43
16.5205 000	**CONDUIT BODIES WITH COVER & GASKET:**					
16.5205 011	1/2" LB form 8, cover, gasket	EA	0.4000	44.68	13.88	58.56
16.5205 021	3/4" LB form 8, cover, gasket	EA	0.4000	44.68	16.64	61.32
16.5205 031	1" LB form 8, cover, gasket	EA	0.5000	55.86	23.56	79.41
16.5205 041	1-1/4" LB form 8, cover, gasket	EA	0.8000	89.37	32.98	122.35
16.5205 051	1-1/2" LB form 8, cover, gasket	EA	1.0000	111.71	39.94	151.65
16.5205 061	2" LB form 8, cover, gasket	EA	1.2000	134.05	64.17	198.22
16.5205 071	2-1/2" LB form 8, cover, gasket	EA	2.0000	223.42	131.95	355.37
16.5205 081	3" LB form 8, cover, gasket	EA	2.6000	290.45	166.76	457.21
16.5205 091	3-1/2" LB form 8, cover, gasket	EA	3.0000	335.13	263.96	599.09
16.5205 101	4" LB form 8, cover, gasket	EA	3.6000	402.16	295.70	697.85
16.5205 111	1/2" T form 8, cover, gasket	EA	0.6000	67.03	15.69	82.72
16.5205 121	3/4" T form 8, cover, gasket	EA	0.6000	67.03	19.34	86.37
16.5205 131	1" T form 8, cover, gasket	EA	0.7500	83.78	26.33	110.11
16.5205 141	1-1/4" T form 8, cover, gasket	EA	1.2000	134.05	38.09	172.14
16.5205 151	1-1/2" T form 8, cover, gasket	EA	1.5000	167.57	47.52	215.08
16.5205 161	2" T form 8, cover, gasket	EA	1.8000	201.08	73.24	274.31
16.5205 171	2-1/2" T form 8, cover, gasket	EA	3.0000	335.13	132.26	467.39
16.5205 181	3" T form 8, cover, gasket	EA	3.9000	435.67	181.56	617.23
16.5205 191	3-1/2" T form 8, cover, gasket	EA	4.5000	502.70	315.36	818.06
16.5205 201	4" T form 8, cover, gasket	EA	5.4000	603.23	344.10	947.34
16.5205 211	1/2" X form 8, cover, gasket	EA	0.7500	83.78	19.88	103.67
16.5205 221	3/4" X form 8, cover, gasket	EA	0.7500	83.78	23.56	107.34
16.5205 231	1" X form 8, cover, gasket	EA	0.9000	100.54	31.69	132.23
16.5205 241	1-1/4" X form 8, cover, gasket	EA	1.5000	167.57	45.07	212.63
16.5205 251	1-1/2" X form 8, cover, gasket	EA	1.8000	201.08	54.87	255.94

ELECTRICAL - 16

Division	Description	Unit	Labor Hours	Labor Cost	Material Cost	Total Cost
16.5205 261	2" X form 8, cover, gasket	EA	2.0000	223.42	92.27	315.69
16.5206 000	**HIGH VOLTAGE CONDUIT FITTINGS:**					
16.5206 011	2" LBD condulet	EA	1.2000	134.05	109.38	243.43
16.5206 021	2-1/2" LBD condulet	EA	2.0000	223.42	323.81	547.23
16.5206 031	3" LBD condulet	EA	2.6000	290.45	323.81	614.25
16.5206 041	3-1/2" LBD condulet	EA	3.0000	335.13	600.96	936.09
16.5206 051	4" LBD condulet	EA	3.6000	402.16	596.82	998.98
16.5300 000	**EMT, MI CABLE & TERMINATIONS:**					
16.5301 000	**EMT CONDUIT:**					
16.5301 011	1/2" EMT conduit, concealed	LF	0.0222	2.48	0.64	3.12
16.5301 021	3/4" EMT conduit, concealed	LF	0.0244	2.73	0.91	3.64
16.5301 031	1" EMT conduit, concealed	LF	0.0267	2.98	1.43	4.42
16.5302 000	**EMT TERMINATIONS:**					
16.5302 011	1/2" EMT quick connect term	EA	0.0583	6.51	0.67	7.18
16.5302 021	3/4" EMT quick connect term	EA	0.0666	7.44	1.57	9.01
16.5302 031	1" EMT quick connect terminator	EA	0.0750	8.38	1.49	9.87
16.5303 000	**MI CABLE**					
16.5303 011	MI cable with 3 #12 solid copper wire	LF	0.0776	8.67	7.05	15.72
16.5303 021	MI cable with 4 #12 solid copper wire	LF	0.0776	8.67	8.28	16.95
16.5304 011	3 #12 termination	EA	0.7406	82.73	13.08	95.81
16.5304 021	4 #12 termination	EA	0.8888	99.29	19.57	118.86
16.5400 000	**SPECIALTY FITTINGS, EXPLOSION PROOF:**					
16.5401 000	**CONDUIT BODIES, EXPLOSION PROOF:**					
16.5401 011	1/2" EYS, explosion proof	EA	1.0000	111.71	17.13	128.84
16.5401 021	3/4" EYS, explosion proof	EA	1.2000	134.05	20.78	154.84
16.5401 031	1" EYS, explosion proof	EA	1.5000	167.57	27.22	194.78
16.5401 041	1-1/4" EYS, explosion proof	EA	2.4000	268.10	33.01	301.11
16.5401 051	1-1/2" EYS, explosion proof	EA	2.5000	279.27	48.97	328.25
16.5401 061	2" EYS, explosion proof	EA	2.6000	290.45	63.37	353.82
16.5401 071	2-1/2" EYS, explosion proof	EA	2.7000	301.62	98.87	400.49
16.5401 081	3" EYS, explosion proof	EA	2.8000	312.79	124.57	437.36
16.5401 091	1/2" UNY, explosion proof	EA	0.2000	22.34	11.63	33.98
16.5401 101	3/4" UNY, explosion proof	EA	0.2500	27.93	16.52	44.45
16.5401 111	1" UNY, explosion proof	EA	0.3000	33.51	28.11	61.62
16.5401 121	1-1/4" UNY, explosion proof	EA	0.3500	39.10	43.16	82.26
16.5401 131	1-1/2" UNY, explosion proof	EA	0.4500	50.27	55.99	106.26
16.5401 141	2" UNY, explosion proof	EA	0.5500	61.44	72.24	133.68
16.5401 151	2-1/2" UNY, explosion proof	EA	0.6000	67.03	117.23	184.26
16.5401 161	3" UNY, explosion proof	EA	0.6500	72.61	159.17	231.78
16.5401 171	1/2" 2-hub GUA, explosion proof	EA	0.4000	44.68	37.16	81.84
16.5401 181	3/4" 2-hub GUA, explosion proof	EA	0.4000	44.68	38.92	83.61
16.5401 191	1" 2-hub GUA, explosion proof	EA	0.5000	55.86	49.58	105.43
16.5401 201	1-1/4" 2-hub GUA, explosion proof	EA	0.8000	89.37	83.76	173.13
16.5401 211	1-1/2" 2-hub GUA, explosion proof	EA	1.0000	111.71	166.69	278.40
16.5401 221	2" 2-hub GUA, explosion proof	EA	1.2000	134.05	177.04	311.10
16.5401 231	1/2" 3-hub GUA, explosion proof	EA	0.6000	67.03	38.92	105.95
16.5401 241	3/4" 3-hub GUA, explosion proof	EA	0.6000	67.03	42.48	109.51
16.5401 251	1" 3-hub GUA, explosion proof	EA	0.7500	83.78	52.21	135.99
16.5401 261	1-1/4" 3-hub GUA, explosion proof	EA	1.2000	134.05	89.10	223.15
16.5401 271	1-1/2" 3-hub GUA, explosion proof	EA	1.5000	167.57	182.90	350.47
16.5401 281	2" 3-hub GUA, explosion proof	EA	1.8000	201.08	188.84	389.92
16.5401 291	1/2" 4-hub GUA, explosion proof	EA	0.8000	89.37	41.57	130.94
16.5401 301	3/4" 4-hub GUA, explosion proof	EA	0.8000	89.37	46.01	135.37
16.5401 311	1" 4-hub GUA, explosion proof	EA	1.0000	111.71	58.10	169.81
16.5401 321	1-1/4" 4-hub GUA, explosion proof	EA	1.6000	178.74	94.71	273.45
16.5401 331	1-1/2" 4-hub GUA, explosion proof	EA	2.0000	223.42	188.84	412.26

16 - ELECTRICAL

Division	Description	Unit	Labor Hours	Labor Cost	Material Cost	Total Cost
16.5401 341	2" 4-hub GUA, explosion proof	EA	2.4000	268.10	199.16	467.26
16.5403 000	**FLEXIBLE COUPLINGS, EXPLOSION PROOF:**					
16.5403 011	Flex coupling, 1/2", to 15", explosion proof	EA	0.5600	62.56	111.52	174.08
16.5403 021	Flex coupling, 3/4", to 15", explosion proof	EA	0.6400	71.49	141.17	212.66
16.5403 031	Flex coupling, 1", to 15", explosion proof	EA	0.8000	89.37	237.98	327.34
16.5403 041	Flex coupling, 1-1/4", to 15", explosion proof	EA	0.9600	107.24	365.40	472.64
16.5403 051	Flex coupling, 1-1/2", to 15", explosion proof	EA	1.1200	125.12	478.87	603.99
16.5403 061	Flex coupling, 2", to 15", explosion proof	EA	1.3600	151.93	616.11	768.03
16.5403 071	Flex coupling, 2-1/2", to 15", explosion proof	EA	1.8414	205.70	1,310.31	1,516.01
16.5403 081	Flex coupling, 3", to 15", explosion proof	EA	2.1279	237.71	1,747.05	1,984.76
16.5403 091	Flex coupling, 4", to 15", explosion proof	EA	2.8644	319.98	2,083.98	2,403.96
16.5404 000	**CONDUIT UNIONS, EXPLOSION PROOF**					
16.5404 011	Conduit union, explosion proof, 1/2"	EA	0.4500	50.27	10.26	60.53
16.5404 021	Conduit union, explosion proof, 3/4"	EA	0.4500	50.27	13.77	64.04
16.5404 031	Conduit union, explosion proof, 1"	EA	0.4500	50.27	25.20	75.47
16.5404 041	Conduit union, explosion proof, 1-1/4"	EA	0.4500	50.27	37.70	87.97
16.5404 051	Conduit union, explosion proof, 1-1/2"	EA	0.4500	50.27	48.41	98.68
16.5404 061	Conduit union, explosion proof, 2"	EA	0.4500	50.27	63.21	113.48
16.5404 071	Conduit union, explosion proof, 2-1/2"	EA	0.4500	50.27	91.34	141.61
16.5404 081	Conduit union, explosion proof, 3"	EA	0.4500	50.27	130.53	180.80
16.5404 091	Conduit union, explosion proof, 3-1/2"	EA	0.4500	50.27	194.04	244.31
16.5404 101	Conduit union, explosion proof, 4"	EA	0.4500	50.27	219.73	270.00
16.5500 000	**UNDERFLOOR & FLUSH TRENCH DUCT, CABLE TRAY:**					
16.5501 000	**STEEL UNDERFLOOR DUCT:**					
16.5501 011	Underfloor duct, blank, standard	LF	0.0609	6.80	9.90	16.70
16.5501 021	Underfloor duct, blank, jumbo	LF	0.0879	9.82	18.39	28.21
16.5501 031	Underfloor duct, junction box, 1 duct	EA	2.0460	228.56	252.12	480.68
16.5501 041	Underfloor duct, junction box, 2 duct	EA	2.8644	319.98	351.43	671.41
16.5501 051	Underfloor duct, junction box, 3 duct	EA	4.0920	457.12	381.97	839.08
16.5501 061	Underfloor duct, panel riser, standard	EA	3.2736	365.69	106.83	472.53
16.5501 071	Underfloor duct, panel riser, jumbo	EA	4.0920	457.12	172.01	629.12
16.5502 000	**FLUSH TRENCH DUCT:**					
16.5502 011	Flush trench duct, to 12"	LF	0.6548	73.15	107.28	180.43
16.5502 021	Flush trench duct, to 24"	LF	0.9003	100.57	152.63	253.20
16.5502 031	Flush trench duct, to 36"	LF	1.6368	182.85	231.44	414.28
16.5502 041	Flush trench duct, ell, to 12"	EA	3.6000	402.16	360.75	762.91
16.5502 051	Flush trench duct, ell, to 24"	EA	4.8000	536.21	588.90	1,125.11
16.5502 061	Flush trench duct, ell, to 36"	EA	6.4000	714.94	972.19	1,687.14
16.5502 071	Flush trench duct, tee, to 12"	EA	4.8000	536.21	360.75	896.96
16.5502 081	Flush trench duct, tee, to 24"	EA	6.0000	670.26	588.90	1,259.16
16.5502 091	Flush trench duct, tee, to 36"	EA	7.6000	849.00	972.19	1,821.19
16.5502 101	Flush trench duct, riser, to 12"	EA	3.5200	393.22	414.29	807.51
16.5502 111	Flush trench duct, riser, t0 24"	EA	4.8000	536.21	479.07	1,015.28
16.5502 121	Flush trench duct, riser, to 36"	EA	6.2000	692.60	617.10	1,309.70
16.5502 131	Flush trench duct, 3" grommet	EA	0.0013	0.15	5.83	5.98
16.5502 141	Flush trench duct, 6" grommet	EA	0.0030	0.34	8.77	9.10
16.5502 151	Flush trench duct, add barrier	LF	0.0819	9.15	3.21	12.36
16.5502 161	Core drill, add afterset insert	EA	0.2046	22.86	5.83	28.69
16.5502 171	Flush outlet @ insert, 110v	EA	0.2865	32.00	54.28	86.28
16.5502 181	Flush outlet @ insert, phone	EA	0.1637	18.29	46.95	65.24
16.5502 191	Surface outlet @ insert, 110v	EA	0.4092	45.71	80.71	126.42
16.5502 201	Surface outlet @ insert, phone	EA	0.3274	36.57	80.71	117.29
16.5502 211	Preset power/signal access box	EA	0.2046	22.86	48.73	71.59
16.5502 221	Add 110v duplex @ access box	EA	0.2046	22.86	16.06	38.92
16.5503 000	**CABLE TRAY:**					
16.5503 011	Cable tray, ladder, to 12"	LF	0.1120	12.51	19.42	31.93
16.5503 021	Cable tray, ladder, to 24"	LF	0.1280	14.30	21.26	35.56

ELECTRICAL - 16

Division	Description	Unit	Labor Hours	Labor Cost	Material Cost	Total Cost
16.5503 031	Cable tray trough to 12"	LF	0.1375	15.36	23.87	39.23
16.5503 041	Cable tray trough to 24"	LF	0.1576	17.61	27.47	45.08
16.5503 051	Cable tray horizontal, 90 elbow 12"	LF	2.4000	268.10	226.27	494.38
16.5503 061	Cable tray horizontal 90 elbow 24"	EA	3.2000	357.47	269.63	627.10
16.5503 071	Cable tray vertical outside ell-12"	EA	1.8500	206.66	238.33	445.00
16.5503 081	Cable tray vertical outside ell-24"	EA	2.4666	275.54	260.00	535.54
16.5503 091	Cable tray inside ell-12"	EA	1.9000	212.25	238.33	450.58
16.5503 101	Cable tray vertical inside ell-24"	EA	2.5333	282.99	260.00	542.99
16.5503 111	Cable tray tee fitting-to 12"	EA	3.6000	402.16	418.88	821.03
16.5503 121	Cable tray tee fitting-to 24"	LF	4.4000	491.52	532.00	1,023.52
16.5503 131	Cable tray end plate to 12"	LF	2.1500	240.18	29.84	270.02
16.5503 141	Cable tray end plate to 24"	LF	2.8595	319.43	38.98	358.42
16.5503 151	Cable tray drop out to 12"	EA	2.5500	284.86	20.46	305.32
16.5503 161	Cable tray drop out to 24"	EA	3.3500	374.23	24.77	399.00
16.5503 171	Cable tray/cover, ell, to 36"	EA	4.6400	518.33	532.25	1,050.59
16.5503 181	Cable tray/cover, tee, to 12"	EA	3.7600	420.03	379.22	799.25
16.5503 191	Cable tray/cover, tee, to 24"	EA	5.4400	607.70	532.25	1,139.95
16.5503 201	Cable tray/cover, tee, to 36"	EA	7.6000	849.00	728.88	1,577.87
16.5503 211	Cable tray/cover, drop, to 12"	EA	2.9600	330.66	330.26	660.93
16.5503 221	Cable tray/cover, drop, to 24"	EA	5.0400	563.02	420.06	983.08
16.5503 231	Cable tray/cover, drop, to 36"	EA	6.8000	759.63	587.85	1,347.47
16.5600 000	**STEEL GUTTERS, PULL BOXES & UNISTRUT HANGERS:**					
16.5601 000	**STEEL GUTTERS:**					
16.5601 011	Steel gutter, 4" x 4"	LF	0.1637	18.29	15.31	33.60
16.5601 021	Steel gutter, 6" x 6"	LF	0.2456	27.44	31.30	58.73
16.5601 031	Steel gutter, 8" x 8"	LF	0.2865	32.00	52.27	84.27
16.5601 041	Steel gutter, 10" x 10"	LF	0.3274	36.57	61.48	98.05
16.5601 051	Steel gutter, 12" x 12"	LF	0.4092	45.71	71.02	116.73
16.5602 000	**PULL BOXES:**					
16.5602 011	Pull box, 6" x 6" x 6"	EA	0.8000	89.37	25.22	114.59
16.5602 021	Pull box, 8" x 8" x 6"	EA	0.8800	98.30	34.39	132.70
16.5602 031	Pull box, 12" x 12" x 8"	EA	1.4000	156.39	77.48	233.87
16.5602 041	Pull box, 18" x 18" x 8"	EA	1.8000	201.08	183.49	384.57
16.5602 051	Pull box, 24" x 24" x 8"	EA	2.4000	268.10	206.22	474.32
16.5602 061	Pull box, 24" x 24" x 12"	EA	2.8000	312.79	423.90	736.69
16.5603 000	**UNISTRUT CONDUIT HANGERS:**					
16.5603 011	Unistrut hangers, 12" x 24"	EA	0.4092	45.71	16.27	61.98
16.5603 021	Unistrut hangers, 18" x 24"	EA	0.4911	54.86	18.44	73.30
16.5603 031	Unistrut hangers, 24" x 24"	EA	0.4911	54.86	18.44	73.30
16.5603 041	Unistrut hangers, 36" x 24"	EA	0.8184	91.42	24.91	116.33
16.5700 000	**SPECIAL RACEWAY ASSEMBLY SYSTEMS:**					
16.5701 000	**WIREMOLD SURFACE RACEWAY:**					
16.5701 011	Wiremold, to size 1000	LF	0.0819	9.15	12.92	22.07
16.5701 021	Wiremold, to size 2000	LF	0.1024	11.44	16.58	28.02
16.5701 031	Wiremold, to size 3000	LF	0.1433	16.01	22.46	38.47
16.5701 041	Wiremold, to size 4000	LF	0.1842	20.58	24.87	45.45
16.5701 051	Wiremold, to size 5000	LF	0.2456	27.44	35.51	62.95
16.5701 061	Wiremold, to size 6000	LF	0.3274	36.57	53.33	89.90
16.5702 000	**PLUGMOLD:**					
16.5702 011	Plugmold, 2000, 1 ft with 4 outlets	EA	0.4092	45.71	52.53	98.24
16.5702 021	Plugmold, 2000, 3 ft with 6 outlets	EA	0.8184	91.42	55.18	146.60
16.5702 031	Plugmold, 2000, 5 ft with 10 outlets	EA	1.2276	137.14	83.10	220.24
16.5702 041	Plugmold, 2000, 6 ft with 12 outlets	EA	1.4322	159.99	84.41	244.40
16.5702 051	Telephone-power pole, hangar assembly 10'8"	EA	0.8800	98.30	528.51	626.81
16.5703 000	**FLAT WIRE UNDER-CARPET POWER SYSTEMS:**					
16.5703 011	Under-carpet system, 3 wire, low density	SF			3.37	3.37

16 - ELECTRICAL

Division	Description	Unit	Labor Hours	Labor Cost	Material Cost	Total Cost
16.5703 021	Under-carpet system, 3 wire, high density	SF			6.93	6.93
16.5704 000	**WIREMOLD OVERHEAD DISTRIBUTION SYSTEMS (ODS):**					
16.5704 011	Wiremold overhead distribution system	SF			3.52	3.52
16.5704 021	Add for light fixture	EA	0.6957	77.72	179.16	256.87
16.5704 031	Add for power & telephone pole	EA	0.9821	109.71	230.62	340.33
16.5705 000	**FLEXIBLE STEEL CONDUIT:**					
16.5705 011	Flex conduit, 1/2", wood	LF	0.0320	3.57	1.77	5.34
16.5705 021	Flex conduit, 1/2", in ceiling	LF	0.0400	4.47	1.77	6.24
16.5705 031	Flex conduit, 3/4", wood	LF	0.0400	4.47	2.10	6.57
16.5705 041	Flex conduit, 3/4", in ceiling	LF	0.0520	5.81	2.10	7.91
16.5705 051	Flex conduit, 1", wood	LF	0.0640	7.15	5.01	12.16
16.5705 061	Flex conduit, 1", in ceiling	LF	0.0800	8.94	5.01	13.95
16.5705 071	Flex conduit, 1-1/4", wood	LF	0.0880	9.83	5.63	15.46
16.5705 081	Flex conduit, 1-1/4", in ceiling	LF	0.1080	12.06	6.41	18.48
16.5705 091	Flex conduit, 1-1/2", wood	LF	0.1200	13.41	7.78	21.18
16.5705 101	Flex conduit, 1-1/2", in ceiling	LF	0.1520	16.98	8.84	25.82
16.5705 111	Flex conduit, 2", wood	LF	0.1600	17.87	10.18	28.05
16.5705 121	Flex conduit, 2", in ceiling	LF	0.2000	22.34	11.32	33.66
16.5705 131	Flex conduit, 3", in ceiling	LF	0.4000	44.68	16.81	61.49
16.5706 000	**ARMORED CABLE, COPPER, BX, 600V, WITH CONNECTORS:**					
16.5706 011	Copper BX, 14/2 solid, 600v, wood frame, connector	LF	0.0280	3.13	2.00	5.13
16.5706 021	Copper BX, 12/2 solid, 600v, wood frame, connector	LF	0.0320	3.57	2.17	5.75
16.5706 031	Copper BX, 10/2 solid, 600v, wood frame, connector	LF	0.0360	4.02	3.67	7.70
16.5706 041	Copper BX, 8/2 solid, 600v, wood frame, connector	LF	0.0480	5.36	5.30	10.67
16.5706 051	Copper BX, 14/3 solid, 600v, wood frame, connector	LF	0.0320	3.57	2.44	6.01
16.5706 061	Copper BX, 12/3 solid, 600v, wood frame, connector	LF	0.0360	4.02	3.07	7.09
16.5706 071	Copper BX, 10/3 solid, 600v, wood frame, connector	LF	0.0400	4.47	4.47	8.94
16.5706 081	Copper BX, 8/3 solid, 600v, wood frame, connector	LF	0.0560	6.26	6.53	12.78
16.5706 091	Copper BX, 14/4 solid, 600v, wood frame, connector	LF	0.0360	4.02	3.12	7.14
16.5706 101	Copper BX, 12/4 solid, 600v, wood frame, connector	LF	0.0400	4.47	3.96	8.43
16.5706 111	Copper BX, 10/4 solid, 600v, wood frame, connector	LF	0.0440	4.92	6.41	11.33
16.5706 121	Copper BX, 8/4 solid, 600v, wood frame, connector	LF	0.0640	7.15	10.48	17.63
16.5800 000	**CONDUCTOR ONLY:**					
16.5801 000	**COPPER WIRE, THW, THHN, 600V:**					
16.5801 011	Copper wire, solid, 600v, THW #14	LF	0.0050	0.56	0.21	0.77
16.5801 021	Copper wire, solid, 600v, THW #12	LF	0.0059	0.66	0.31	0.97
16.5801 031	Copper wire, solid, 600v, THW #10	LF	0.0076	0.85	0.44	1.29
16.5801 041	Copper wire, solid, 600v, THHN #14	LF	0.0050	0.56	0.21	0.77
16.5801 051	Copper wire, solid, 600v, THHN #12	LF	0.0059	0.66	0.31	0.97
16.5801 061	Copper wire, solid, 600v, THHN #10	LF	0.0076	0.85	0.44	1.29
16.5801 071	Copper wire, stranded, 600v, THHN #12	LF	0.0059	0.66	0.34	0.99
16.5801 081	Copper wire, stranded, 600v, THHN #10	LF	0.0076	0.85	0.58	1.43
16.5801 091	Copper wire, stranded, 600v, THHN #8	LF	0.0094	1.05	0.91	1.96
16.5801 101	Copper wire, stranded, 600v, THHN #6	LF	0.0098	1.09	1.20	2.30
16.5801 111	Copper wire, stranded, 600v, THHN #4	LF	0.0118	1.32	1.83	3.14
16.5801 121	Copper wire, stranded, 600v, THHN #3	LF	0.0128	1.43	2.37	3.80
16.5801 131	Copper wire, stranded, 600v, THHN #2	LF	0.0144	1.61	3.07	4.68
16.5801 141	Copper wire, stranded, 600v, THHN #1	LF	0.0160	1.79	3.95	5.74
16.5801 151	Copper wire, stranded, 600v, THW with THHN 1/0	LF	0.0180	2.01	4.84	6.85
16.5801 161	Copper wire, stranded, 600v, THW with THHN 2/0	LF	0.0200	2.23	5.71	7.94
16.5801 171	Copper wire, stranded, 600v, THW with THHN 3/0	LF	0.0224	2.50	7.01	9.52
16.5801 181	Copper wire, stranded, 600v, THW with THHN 4/0	LF	0.0248	2.77	8.76	11.53
16.5801 191	Copper wire, stranded, 600v, THW with THHN 250	LF	0.0272	3.04	10.81	13.85
16.5801 201	Copper wire, stranded, 600v, THW with THHN 300	LF	0.0296	3.31	12.57	15.88
16.5801 211	Copper wire, stranded, 600v, THW with THHN 350	LF	0.0320	3.57	14.66	18.24
16.5801 221	Copper wire, stranded, 600v, THW with THHN 400	LF	0.0352	3.93	17.43	21.37
16.5801 231	Copper wire, stranded, 600v, THW with THHN 500	LF	0.0392	4.38	20.81	25.19

ELECTRICAL - 16

Division	Description	Unit	Labor Hours	Labor Cost	Material Cost	Total Cost
16.5801 241	Copper wire, stranded, 600v, THW with THHN 600	LF	0.0432	4.83	28.83	33.65
16.5801 251	Copper wire, stranded, 600v, THW with THHN 750	LF	0.0480	5.36	35.41	40.77
16.5801 261	Copper wire, stranded, 600v, THHN 1000	LF	0.0560	6.26	54.21	60.46
16.5802 000	**ALUMINUM WIRE, THHN, 600V:**					
16.5802 011	Aluminum wire, stranded, 600v, THHN #8	LF	0.0072	0.80	0.62	1.43
16.5802 021	Aluminum wire, stranded, 600v, THHN #6	LF	0.0080	0.89	0.73	1.62
16.5802 031	Aluminum wire, stranded, 600v, THHN #4	LF	0.0092	1.03	1.03	2.06
16.5802 041	Aluminum wire, stranded, 600v, THHN #2	LF	0.0120	1.34	1.41	2.75
16.5802 051	Aluminum wire, stranded, 600v, THHN #1	LF	0.0136	1.52	2.07	3.59
16.5802 061	Aluminum wire, stranded, 600v, THHN 1/0	LF	0.0152	1.70	2.32	4.02
16.5802 071	Aluminum wire, stranded, 600v, THHN 2/0	LF	0.0168	1.88	2.76	4.64
16.5802 081	Aluminum wire, stranded, 600v, THHN 3/0	LF	0.0184	2.06	3.34	5.39
16.5802 091	Aluminum wire, stranded, 600v, THHN 4/0	LF	0.0200	2.23	3.85	6.08
16.5802 101	Aluminum wire, stranded, 600v, THHN 250	LF	0.0216	2.41	4.60	7.01
16.5802 111	Aluminum wire, stranded, 600v, THHN 300	LF	0.0240	2.68	5.94	8.62
16.5802 121	Aluminum wire, stranded, 600v, THHN 350	LF	0.0264	2.95	6.57	9.52
16.5802 131	Aluminum wire, stranded, 600v, THHN 400	LF	0.0288	3.22	7.07	10.29
16.5802 141	Aluminum wire, stranded, 600v, THHN 500	LF	0.0312	3.49	8.32	11.80
16.5802 151	Aluminum wire, stranded, 600v, THHN 600	LF	0.0344	3.84	9.92	13.77
16.5802 161	Aluminum wire, stranded, 600v, THHN 750	LF	0.0384	4.29	11.65	15.94
16.5802 171	Aluminum wire, stranded, 600v, THHN 1000	LF	0.0432	4.83	17.67	22.49
16.5803 000	**COPPER WIRE, RHH-RHW; AERIAL-DIRECT BURIAL, 600V:**					
16.5803 001	Note: The following prices do not include trenching.					
16.5803 011	Copper wire, buried, solid, 600v, #12	LF	0.0047	0.53	0.58	1.10
16.5803 021	Copper wire, buried, solid, 600v, #10	LF	0.0057	0.64	0.73	1.36
16.5803 031	Copper wire, buried, solid, 600v, #8	LF	0.0064	0.71	1.20	1.92
16.5803 041	Copper wire, buried, stranded, 600v, #14	LF	0.0040	0.45	0.61	1.06
16.5803 051	Copper wire, buried, stranded, 600v, #12	LF	0.0047	0.53	0.62	1.15
16.5803 061	Copper wire, buried, stranded, 600v, #10	LF	0.0060	0.67	0.75	1.42
16.5803 071	Copper wire, buried, stranded, 600v, #8	LF	0.0064	0.71	1.17	1.88
16.5803 081	Copper wire, buried, stranded, 600v, #6	LF	0.0076	0.85	1.39	2.24
16.5803 091	Copper wire, buried, stranded, 600v, #4	LF	0.0089	0.99	2.21	3.20
16.5803 101	Copper wire, buried, stranded, 600v, #2	LF	0.0115	1.28	3.34	4.62
16.5803 111	Copper wire, buried, stranded, 600v, #1	LF	0.0128	1.43	4.36	5.79
16.5803 121	Copper wire, buried, stranded, 600v, 1/0	LF	0.0144	1.61	5.27	6.88
16.5803 131	Copper wire, buried, stranded, 600v, 2/0	LF	0.0160	1.79	6.24	8.03
16.5803 141	Copper wire, buried, stranded, 600v, 3/0	LF	0.0179	2.00	7.72	9.72
16.5803 151	Copper wire, buried, stranded, 600v, 4/0	LF	0.0198	2.21	9.37	11.58
16.5803 161	Copper wire, buried, stranded, 600v, 250	LF	0.0217	2.42	11.66	14.08
16.5803 171	Copper wire, buried, stranded, 600v, 300	LF	0.0236	2.64	13.56	16.20
16.5803 181	Copper wire, buried, stranded, 600v, 350	LF	0.0256	2.86	15.62	18.48
16.5803 191	Copper wire, buried, stranded, 600v, 400	LF	0.0281	3.14	18.59	21.73
16.5803 201	Copper wire, buried, stranded, 600v, 500	LF	0.0313	3.50	21.59	25.09
16.5803 211	Copper wire, buried, stranded, 600v, 600	LF	0.0345	3.85	31.39	35.24
16.5803 221	Copper wire, buried, stranded, 600v, 750	LF	0.0384	4.29	37.72	42.01
16.5803 231	Trench, 1" w x 2' d, backfill, tamp	LF	0.0104	1.16	1.66	2.83
16.5804 000	**COPPER WIRE, BARE, GROUNDING:**					
16.5804 011	Copper wire, bare, ground, soft #14 solid	LF	0.0039	0.44	0.21	0.64
16.5804 021	Copper wire, bare, ground soft #12 solid	LF	0.0047	0.53	0.31	0.84
16.5804 031	Copper wire, bare, ground soft #10 solid	LF	0.0057	0.64	0.44	1.08
16.5804 041	Copper wire, bare, ground, soft #8 stranded	LF	0.0064	0.71	0.84	1.56
16.5804 051	Copper wire, bare, ground soft #6 stranded	LF	0.0076	0.85	1.17	2.02
16.5804 061	Copper wire, bare, ground soft #4 stranded	LF	0.0089	0.99	1.83	2.82
16.5804 071	Copper wire, bare, ground soft #2 stranded	LF	0.0015	0.17	3.05	3.22
16.5804 081	Copper wire, bare, ground, soft #1 stranded	LF	0.0128	1.43	3.92	5.35
16.5804 091	Copper wire, bare, ground soft #1/0 stranded	LF	0.0144	1.61	4.71	6.32
16.5804 101	Copper wire, bare, ground soft #2/0 stranded	LF	0.0160	1.79	5.57	7.36

16 - ELECTRICAL

Division	Description	Unit	Labor Hours	Labor Cost	Material Cost	Total Cost
16.5804 111	Copper wire, bare, ground soft #3/0 stranded	LF	0.0179	2.00	6.93	8.93
16.5804 121	Copper wire, bare, ground soft #4/0 stranded	LF	0.0198	2.21	8.61	10.82
16.5805 000	COPPER CONTROL CABLE, MULTICONDUCTOR, #14:					
16.5805 011	Copper control cable, #14, 2 conductors	LF	0.0089	0.99	0.65	1.64
16.5805 021	Copper control cable, #14, 3 conductors	LF	0.0130	1.45	1.03	2.48
16.5805 031	Copper control cable, #14, 4 conductors	LF	0.0176	1.97	1.32	3.28
16.5805 041	Copper control cable, #14, 8 conductors	LF	0.0312	3.49	2.80	6.28
16.5805 051	Copper control cable, #14, 10 conductors	LF	0.0387	4.32	3.48	7.80
16.5805 061	Copper control cable, #14, 12 conductors	LF	0.0468	5.23	4.18	9.41
16.5805 071	Copper control cable, #14, 16 conductors	LF	0.0562	6.28	5.49	11.77
16.5805 081	Copper control cable, #14, 20 conductors	LF	0.0703	7.85	7.01	14.87
16.5805 091	Copper control cable, #14, 24 conductors	LF	0.0792	8.85	8.61	17.45
16.5805 101	Copper control cable, #14, 36 conductors	LF	0.1115	12.46	13.02	25.48
16.5806 000	COPPER POWER CABLE, 5 KV, SHIELDED, PULLED AND SPLICED:					
16.5806 011	Copper wire cable, 5kv, #8, shielded, pulled & spliced	LF	0.0138	1.54	4.16	5.70
16.5806 021	Copper power cable, 5kv, #6, shielded, pulled & spliced	LF	0.0143	1.60	4.49	6.09
16.5806 031	Copper power cable, 5kv, #4, shielded, pulled & spliced	LF	0.0182	2.03	5.13	7.16
16.5806 041	Copper power cable, 5kv, #2, shielded, pulled & spliced	LF	0.0205	2.29	6.29	8.58
16.5806 051	Copper power cable, 5kv, #1, shielded, pulled & spliced	LF	0.0234	2.61	8.49	11.11
16.5806 061	Copper power cable, 5kv, 1/0, shielded, pulled & spliced	LF	0.0257	2.87	10.70	13.57
16.5806 071	Copper power cable, 5kv, 2/0, shielded, pulled & spliced	LF	0.0286	3.19	13.86	17.06
16.5806 081	Copper power cable, 5kv, 4/0, shielded, pulled & spliced	LF	0.0308	3.44	14.50	17.94
16.5806 091	Copper power cable, 5kv, 250, shielded, pulled & spliced	LF	0.0386	4.31	21.24	25.55
16.5806 101	Copper power cable, 5kv, 350, shielded, pulled & spliced	LF	0.0458	5.12	26.25	31.37
16.5806 111	Copper power cable, 5kv, 500, shielded, pulled & spliced	LF	0.0558	6.23	33.63	39.87
16.5806 121	Copper power cable, 5kv, 750, shielded, pulled & spliced	LF	0.0686	7.66	51.86	59.53
16.5807 000	ALUMINUM POWER CABLE, 5 KV, SHIELDED, PULLED AND SPLICED:					
16.5807 011	Aluminum power cable, 5kv, #8, shielded, pulled & spliced	LF	0.0107	1.20	1.37	2.57
16.5807 021	Aluminum power cable, 5kv, #6, shielded, pulled & spliced	LF	0.0116	1.30	2.68	3.98
16.5807 031	Aluminum power cable, 5kv, #4, shielded, pulled & spliced	LF	0.0146	1.63	3.08	4.72
16.5807 041	Aluminum power cable, 5kv, #2, shielded, pulled & spliced	LF	0.0164	1.83	4.10	5.93
16.5807 051	Aluminum power cable, 5kv, #1, shielded, pulled & spliced	LF	0.0186	2.08	4.93	7.01
16.5807 061	Aluminum power cable, 5kv, 1/0, shielded, pulled & spliced	LF	0.0205	2.29	6.07	8.36
16.5807 071	Aluminum power cable, 5kv, 2/0, shielded, pulled & spliced	LF	0.0221	2.47	6.80	9.27
16.5807 081	Aluminum power cable, 5kv, 4/0, shielded, pulled & spliced	LF	0.0270	3.02	9.18	12.20
16.5807 091	Aluminum power cable, 5kv, 250, shielded, pulled & spliced	LF	0.0308	3.44	10.70	14.14
16.5807 101	Aluminum power cable, 5kv, 350, shielded, pulled & spliced	LF	0.0367	4.10	13.63	17.73
16.5807 111	Aluminum power cable, 5kv, 500, shielded, pulled & spliced	LF	0.0446	4.98	17.33	22.31
16.5807 121	Aluminum power cable, 5kv, 750, shielded, pulled & spliced	LF	0.0541	6.04	25.88	31.92
16.5808 000	COPPER POWER CABLE, 15 KV, SHIELD, GROUND, PULLED:					
16.5808 011	Copper power cable, 15kv, #4, ground, shielded & pulled	LF	0.0140	1.56	3.81	5.38
16.5808 021	Copper power cable, 15kv, #2, ground, shielded & pulled	LF	0.0158	1.77	6.39	8.15
16.5808 031	Copper power cable, 15 kv, 1/0, ground, shielded & pulled	LF	0.0198	2.21	7.95	10.16
16.5808 041	Copper power cable, 15kv, 2/0, ground, shielded & pulled	LF	0.0220	2.46	7.95	10.41
16.5808 051	Copper power cable, 15kv, 4/0, ground, shielded & pulled	LF	0.0260	2.90	11.82	14.72
16.5808 061	Copper power cable, 15kv, 250, ground, shielded & pulled	LF	0.0237	2.65	14.10	16.74
16.5808 071	Copper power cable, 15kv, ground, 350, shielded & pulled	LF	0.0282	3.15	16.76	19.91
16.5808 081	Copper power cable, 15kv, 500, ground, shielded & pulled	LF	0.0342	3.82	20.72	24.54
16.5808 091	Copper power cable, 15kv, 750, ground, shielded & pulled	LF	0.0416	4.65	26.39	31.03
16.5809 000	ALUMINUM POWER CABLE, 15 KV, SHIELD, GROUND, PULLED:					
16.5809 011	Aluminum power cable, 15kv, #4, ground, shielded & pulled	LF	0.0112	1.25	3.23	4.49
16.5809 021	Aluminum power cable, 15kv, #2, ground, shielded & pulled	LF	0.0126	1.41	4.03	5.44
16.5809 031	Aluminum power cable, 15kv, 1/0, ground, shielded & pulled	LF	0.0158	1.77	5.74	7.51
16.5809 041	Aluminum power cable, 15kv, 2/0, ground, shielded & pulled	LF	0.0176	1.97	6.63	8.60
16.5809 051	Aluminum power cable, 15kv, 4/0, ground, shielded & pulled	LF	0.0208	2.32	8.47	10.79
16.5809 061	Aluminum power cable, 15kv, 250, ground, shielded & pulled	LF	0.0237	2.65	9.42	12.06
16.5809 071	Aluminum power cable, 15kv, 350, ground, shielded & pulled	LF	0.0282	3.15	11.82	14.97

ELECTRICAL - 16

Division	Description	Unit	Labor Hours	Labor Cost	Material Cost	Total Cost
16.5809 081	Aluminum power cable, 15kv, 500, ground, shielded & pulled	LF	0.0342	3.82	15.47	19.29
16.5809 091	Aluminum power cable, 15kv, 750, ground, shielded & pulled	LF	0.0416	4.65	20.93	25.58
16.5900 000	**BUSWAYS:**					
16.5901 000	**ALUMINUM BUSWAY, PLUG-IN, 600V, 3 P, PLUG-INS:**					
16.5901 011	Aluminum busway, plug-in, 600v, 3 pole, 400a	LF	0.2400	26.81	201.77	228.58
16.5901 021	Aluminum busway, plug-in, 600v, 3 pole, 600a	LF	0.2800	31.28	232.87	264.15
16.5901 031	Aluminum busway, plug-in, 600v, 3 pole, 800a	LF	0.3200	35.75	253.77	289.52
16.5901 041	Aluminum busway, plug-in, 600v, 3 pole, 1000a	LF	0.3600	40.22	284.93	325.14
16.5901 051	Aluminum busway, plug-in, 600v, 3 pole, 1350a	LF	0.4000	44.68	420.15	464.83
16.5901 061	Aluminum busway, plug-in, 600v, 3 pole, 1600a	LF	0.4400	49.15	513.72	562.87
16.5901 071	Aluminum busway, plug-in, 600v, 3 pole, 2000a	LF	0.4800	53.62	609.38	663.00
16.5901 081	Aluminum busway, plug-in, 600v, 3 pole, 2500a	LF	0.5600	62.56	734.18	796.74
16.5901 091	Aluminum busway, plug-in, 600v, 3 pole, 3000a	LF	0.7200	80.43	840.29	920.72
16.5902 000	**ALUMINUM BUSWAY, PLUG-IN, 600V, 4 POLE:**					
16.5902 011	Aluminum busway, plug-in, 600v, 4 pole, 400a	LF	0.2520	28.15	230.78	258.93
16.5902 021	Aluminum busway, plug-in, 600v, 4 pole, 600a	LF	0.2940	32.84	291.19	324.03
16.5902 031	Aluminum busway, plug-in, 600v, 4 pole, 800a	LF	0.3360	37.53	301.61	339.15
16.5902 041	Aluminum busway, plug-in, 600v, 4 pole, 1000a	LF	0.3780	42.23	359.77	402.00
16.5902 051	Aluminum busway, plug-in, 600v, 4 pole, 1350a	LF	0.4200	46.92	507.50	554.42
16.5902 061	Aluminum busway, plug-in, 600v, 4 pole, 1600a	LF	0.4620	51.61	598.95	650.56
16.5902 071	Aluminum busway, plug-in, 600v, 4 pole, 2000a	LF	0.5040	56.30	734.18	790.48
16.5902 081	Aluminum busway, plug-in, 600v, 4 pole, 2500a	LF	0.5880	65.69	877.65	943.34
16.5902 091	Aluminum busway, plug-in, 600v, 4 pole, 3000a	LF	0.7560	84.45	1,037.85	1,122.30
16.5903 000	**COPPER BUSWAYS, PLUG-IN, 600V, 3 POLE:**					
16.5903 011	Copper busway, plug-in, 600v, 3 pole, 400a	LF	0.3200	35.75	284.93	320.68
16.5903 021	Copper busway, plug-in, 600v, 3 pole, 600a	LF	0.3600	40.22	343.19	383.41
16.5903 031	Copper busway, plug-in, 600v, 3 pole, 800a	LF	0.4000	44.68	405.53	450.22
16.5903 041	Copper busway, plug-in, 600v, 3 pole, 1000a	LF	0.4400	49.15	422.21	471.36
16.5903 051	Copper busway, plug-in, 600v, 3 pole, 1350a	LF	0.4800	53.62	623.97	677.59
16.5903 061	Copper busway, plug-in, 600v, 3 pole, 1600a	LF	0.5600	62.56	734.18	796.74
16.5903 071	Copper busway, plug-in, 600v, 3 pole, 2000a	LF	0.6400	71.49	913.07	984.57
16.5903 081	Copper busway, plug-in, 600v, 3 pole, 2500a	LF	0.7200	80.43	1,131.49	1,211.92
16.5903 091	Copper busway, plug-in, 600v, 3 pole, 3000a	LF	0.8800	98.30	1,441.26	1,539.56
16.5904 000	**COPPER BUSWAYS, PLUG-IN, 600V, 4 POLE:**					
16.5904 011	Copper busway, plug-in, 600v, 4 pole, 400a	LF	0.3360	37.53	412.37	449.91
16.5904 021	Copper busway, plug-in, 600v, 4 pole, 600a	LF	0.3780	42.23	445.64	487.86
16.5904 031	Copper busway, plug-in, 600v, 4 pole, 800a	LF	0.4200	46.92	560.93	607.84
16.5904 041	Copper busway, plug-in, 600v, 4 pole, 1000a	LF	0.4620	51.61	638.58	690.19
16.5904 051	Copper busway, plug-in, 600v, 4 pole, 1350a	LF	0.5040	56.30	851.41	907.71
16.5904 061	Copper busway, plug-in, 600v, 4 pole, 1600a	LF	0.5880	65.69	1,024.34	1,090.03
16.5904 071	Copper busway, plug-in, 600v, 4 pole, 2000a	LF	0.6720	75.07	1,235.00	1,310.06
16.5904 081	Copper busway, plug-in, 600v, 4 pole, 2500a	LF	0.7560	84.45	1,514.41	1,598.87
16.5904 091	Copper busway, plug-in, 600v, 4 pole, 3000a	LF	0.9240	103.22	1,909.04	2,012.26
16.5905 000	**ALUMINUM BUSWAY, TAP BOX, 600V, 3 POLE:**					
16.5905 011	Aluminum busway tap box, 600v, 400a, 3 pole	EA	5.3196	594.25	3,316.37	3,910.62
16.5905 021	Aluminum busway tap box, 600v, 600a, 3 pole	EA	6.5472	731.39	3,316.37	4,047.75
16.5905 031	Aluminum busway tap box, 600v, 800a, 3 pole	EA	7.7748	868.52	3,464.79	4,333.31
16.5905 041	Aluminum busway tap box, 600v, 1000a, 3 pole	EA	9.8208	1,097.08	3,594.71	4,691.79
16.5905 051	Aluminum busway tap box, 600v, 1350a, 3 pole	EA	11.4576	1,279.93	4,510.20	5,790.12
16.5905 061	Aluminum busway tap box, 600v, 1600a, 3 pole	EA	13.0944	1,462.78	4,968.37	6,431.14
16.5905 071	Aluminum busway tap box, 600v, 2000a, 3 pole	EA	14.7312	1,645.62	5,373.61	7,019.23
16.5905 081	Aluminum busway tap box, 600v, 2500a, 3 pole	EA	18.8232	2,102.74	9,448.34	11,551.08
16.5905 091	Aluminum busway tap box, 600v, 3000a, 3 pole	EA	21.2784	2,377.01	10,683.37	13,060.38
16.5906 000	**ALUMINUM BUSWAY, TAP BOX, 600V, 4 POLE:**					
16.5906 011	Aluminum busway tap box, 600v, 400a, 4 pole	EA	6.1380	685.68	3,316.37	4,002.04
16.5906 021	Aluminum busway tap box, 600v, 600a, 4 pole	EA	7.3656	822.81	3,316.37	4,139.18
16.5906 031	Aluminum busway tap box, 600v, 800a, 4 pole	EA	9.0024	1,005.66	4,631.77	5,637.43

16 - ELECTRICAL

Division	Description	Unit	Labor Hours	Labor Cost	Material Cost	Total Cost
16.5906 041	Aluminum busway tap box, 600v, 1000a, 4 pole	EA	11.4576	1,279.93	4,906.06	6,185.99
16.5906 051	Aluminum busway tap box, 600v, 1350a, 4 pole	EA	13.0944	1,462.78	5,199.05	6,661.83
16.5906 061	Aluminum busway tap box, 600v, 1600a, 4 pole	EA	14.7312	1,645.62	5,738.31	7,383.93
16.5906 071	Aluminum busway tap box, 600v, 2000a, 4 pole	EA	17.1864	1,919.89	6,361.68	8,281.57
16.5906 081	Aluminum busway tap box, 600v, 2500a, 4 pole	EA	19.6416	2,194.16	9,861.26	12,055.42
16.5906 091	Aluminum busway tap box, 600v, 3000a, 4 pole	EA	24.5520	2,742.70	12,323.72	15,066.43
16.5907 000	**COPPER BUSWAY, TAP BOX, 600V, 3 POLE:**					
16.5907 011	Copper busway tap box, 600v, 400a, 3 pole	EA	4.9100	548.50	3,535.36	4,083.86
16.5907 021	Copper busway tap box, 600v, 600a, 3 pole	EA	6.1300	684.78	3,535.36	4,220.14
16.5907 031	Copper busway tap box, 600v, 800a, 3 pole	EA	6.9500	776.38	3,693.61	4,469.99
16.5907 041	Copper busway tap box, 600v, 1000a, 3 pole	EA	8.5900	959.59	3,832.10	4,791.68
16.5907 051	Copper busway tap box, 600v, 1350a, 3 pole	EA	10.2300	1,142.79	4,808.02	5,950.81
16.5907 061	Copper busway tap box, 600v, 1600a, 3 pole	EA	12.2700	1,370.68	5,296.53	6,667.21
16.5907 071	Copper busway tap box, 600v, 2000a, 3 pole	EA	14.7300	1,645.49	5,728.47	7,373.96
16.5907 081	Copper busway tap box, 600v, 2500a, 3 pole	EA	17.1800	1,919.18	10,072.28	11,991.46
16.5907 091	Copper busway tap box, 600v, 3000a, 3 pole	EA	19.6400	2,193.98	102.59	2,296.58
16.5908 000	**COPPER BUSWAY, TAP BOX, 600V, 4 POLE:**					
16.5908 011	Copper busway tap box, 600v, 400a, 4 pole	EA	5.7288	639.96	3,535.36	4,175.32
16.5908 021	Copper busway tap box, 600v, 600a, 4 pole	EA	6.9564	777.10	3,535.36	4,312.46
16.5908 031	Copper busway tap box, 600v, 800a, 4 pole	EA	8.5932	959.95	4,937.68	5,897.63
16.5908 041	Copper busway tap box, 600v, 1000a, 4 pole	EA	10.2300	1,142.79	5,229.98	6,372.77
16.5908 051	Copper busway tap box, 600v, 1350a, 4 pole	EA	12.2760	1,371.35	5,542.41	6,913.76
16.5908 061	Copper busway tap box, 600v, 1600a, 4 pole	EA	14.7312	1,645.62	6,117.24	7,762.87
16.5908 071	Copper busway tap box, 600v, 2000a, 4 pole	EA	17.1864	1,919.89	6,781.79	8,701.68
16.5908 081	Copper busway tap box, 600v, 2500a, 4 pole	EA	19.6416	2,194.16	10,512.48	12,706.65
16.5908 091	Copper busway tap box, 600v, 3000a, 4 pole	EA	22.0968	2,468.43	10,512.48	12,980.92
16.5909 000	**PLUG-IN UNIT, CIRCUIT BREAKER:**					
16.5909 011	Plug-in circuit breaker, to 60a, 3 pole, 600v	EA	1.4000	156.39	1,283.78	1,440.18
16.5909 021	Plug-in circuit breaker, 100a, 3 pole, 600v	EA	1.8000	201.08	1,416.86	1,617.93
16.5909 031	Plug-in circuit breaker, 225a, 3 pole, 600v	EA	4.2000	469.18	3,070.98	3,540.17
16.5909 041	Plug-in circuit breaker, 400a, 3 pole, 600v	EA	9.2000	1,027.73	6,248.37	7,276.11
16.5909 051	Plug-in circuit breaker, 600a, 3 pole, 600v	EA	12.8000	1,429.89	10,208.43	11,638.32
16.5909 061	Plug-in circuit breaker, 800a, 3 pole, 600v	EA	20.4000	2,278.88	12,026.65	14,305.54
16.5909 071	Plug-in circuit breaker, 1000a, 3 pole, 600v	EA	25.6000	2,859.78	13,669.63	16,529.41
16.5909 081	Plug-in circuit breaker, 1200a, 3 pole, 600v	EA	28.0000	3,127.88	22,404.73	25,532.61
16.5909 091	Plug-in circuit breaker, 1400a, 3 pole, 600v	EA	28.0000	3,127.88	22,404.73	25,532.61
16.5909 101	Plug-in circuit breaker, 1600a, 3 pole, 600v	EA	34.4000	3,842.82	22,404.73	26,247.55
16.5909 111	Plug-in circuit breaker, to 60a, 3 pole, 240v	EA	1.4000	156.39	968.92	1,125.32
16.5909 121	Plug-in circuit breaker, 100a, 3 pole, 240v	EA	1.8000	201.08	1,135.19	1,336.26
16.5909 131	Plug-in circuit breaker, to 60a, 3 pole, 480v	EA	1.4000	156.39	1,181.80	1,338.20
16.5909 141	Plug-in circuit breaker, 100a, 3 pole, 480v	EA	1.8000	201.08	1,290.44	1,491.52
16.5909 151	Plug-in circuit breaker, to 60a, 4 pole, 277/480v	EA	1.4000	156.39	1,314.85	1,471.25
16.5909 161	Plug-in circuit breaker, 100a, 4 pole, 277/480v	EA	1.8000	201.08	1,412.28	1,613.36
16.5909 171	Plug-in circuit breaker, 225a, 4 pole, 277/480v	EA	4.2000	469.18	3,290.49	3,759.67
16.5909 181	Plug-in circuit breaker, 400a, 4 pole, 277/480v	EA	9.2000	1,027.73	6,574.22	7,601.96
16.5909 191	Plug-in circuit breaker, 600a, 4 pole, 277/480v	EA	12.8000	1,429.89	9,277.28	10,707.17
16.5909 201	Plug-in circuit breaker, 800a, 4 pole, 277/480v	EA	20.4000	2,278.88	13,044.39	15,323.27
16.5909 211	Plug-in circuit breaker, 1000a, 4 pole, 277/480v	EA	25.6000	2,859.78	13,044.39	15,904.16
16.5909 221	Plug-in circuit breaker, 1200a, 4 pole, 277/480v	EA	28.0000	3,127.88	20,250.65	23,378.53
16.5909 231	Plug-in circuit breaker, 1400a, 4 pole, 277/480v	EA	28.0000	3,127.88	20,250.65	23,378.53
16.5909 241	Plug-in circuit breaker, 1600a, 4 pole, 277/480v	EA	34.4000	3,842.82	6,580.15	10,422.98
16.5910 000	**BUSWAY PLUG-IN, 600V, 3 POLE, FUSED:**					
16.5910 011	Bus plug, 600v, fused, 30a, 3 pole	EA	0.9791	109.38	698.39	807.76
16.5910 021	Bus plug, 600v, fused, 60a, 3 pole	EA	0.9791	109.38	743.15	852.52
16.5910 031	Bus plug, 600v, fused, 100a, 3 pole	EA	1.1479	128.23	1,016.32	1,144.56
16.5910 041	Bus plug, 600v, fused, 200a, 3 pole	EA	2.8358	316.79	1,721.48	2,038.27
16.5910 051	Bus plug, 600v, fused, 400a, 3 pole	EA	5.5366	618.49	4,365.28	4,983.78

ELECTRICAL - 16

Division	Description	Unit	Labor Hours	Labor Cost	Material Cost	Total Cost
16.5910 061	Bus plug, 600v, fused, 600a, 3 pole	EA	8.2373	920.19	6,250.20	7,170.39
16.5911 000	**BUSWAY PLUG-IN, 600V, 4 POLE, FUSED:**					
16.5911 011	Bus plug, 600v, fused, 30a, 4 pole	EA	1.0466	116.92	794.63	911.54
16.5911 021	Bus plug, 600v, fused, 60a, 4 pole	EA	1.0466	116.92	826.03	942.95
16.5911 031	Bus plug, 600v, fused, 100a, 4 pole	EA	1.2154	135.77	1,164.08	1,299.85
16.5911 041	Bus plug, 600v, fused, 200a, 4 pole	EA	2.9033	324.33	1,922.94	2,247.27
16.5911 051	Bus plug, 600v, fused, 400a, 4 pole	EA	5.6041	626.03	4,696.63	5,322.67
16.5911 061	Bus plug, 600v, fused, 600a, 4 pole	EA	8.3048	927.73	6,836.76	7,764.49
16.5912 000	**BUSWAY PLUG-IN, 600V, BOLTED FUSE:**					
16.5912 011	Bus plug, 800a, 3 pole, hi-cap fuse	EA	11.6131	1,297.30	10,770.30	12,067.60
16.5912 021	Bus plug, 1000a, 3 pole, hi-cap fuse	EA	15.6642	1,749.85	12,771.04	14,520.88
16.5912 031	Bus plug, 1200a, 3 pole, hi-cap fuse	EA	21.7409	2,428.68	20,203.94	22,632.62
16.5912 041	Bus plug, 1600a, 3 pole, hi-cap fuse	EA	25.7920	2,881.22	20,203.94	23,085.16
16.5912 051	Bus plug, 800a, 4 pole, hi-cap fuse	EA	11.6807	1,304.85	11,189.74	12,494.59
16.5912 061	Bus plug, 1000a, 4 pole, hi-cap fuse	EA	15.7318	1,757.40	13,194.92	14,952.32
16.5912 071	Bus plug, 1200a, 4 pole, hi-cap fuse	EA	21.8084	2,436.22	20,484.35	22,920.57
16.5912 081	Bus plug, 1600a, 4 pole, hi-cap fuse	EA	25.8598	2,888.80	20,484.35	23,373.15
16.5913 000	**BUSWAY PLUG-IN, 600V, HI-CAP CIRCUIT BREAKER:**					
16.5913 011	Bus plug, 30a, 3 pole, hi-cap circuit breaker	EA	0.8441	94.29	1,489.97	1,584.26
16.5913 021	Bus plug, 60a, 3 pole, hi-cap circuit breaker	EA	0.8441	94.29	1,489.97	1,584.26
16.5913 031	Bus plug, 100a, 3 pole, hi-cap circuit breaker	EA	0.3350	37.42	547.05	584.47
16.5913 041	Bus plug, 225a, 3 pole, hi-cap circuit breaker	EA	1.0128	113.14	1,676.11	1,789.25
16.5913 051	Bus plug, 400a, 3 pole, hi-cap circuit breaker	EA	5.4015	603.40	7,689.38	8,292.78
16.5913 061	Bus plug, 600a, 3 pole, hi-cap circuit breaker	EA	8.1022	905.10	11,024.25	11,929.35
16.5913 071	Bus plug, 800a, 3 pole, hi-cap circuit breaker	EA	8.2046	916.54	13,345.56	14,262.10
16.5913 081	Bus plug, 1200a, 3 pole, hi-cap circuit breaker	EA	15.5292	1,734.77	15,896.62	17,631.39
16.5913 091	Bus plug, 1400a, 3 pole, hi-cap circuit breaker	EA	21.6058	2,413.58	20,922.28	23,335.86
16.5913 101	Bus plug, 1600a, 3 pole, hi-cap circuit breaker	EA	23.6314	2,639.86	23,532.64	26,172.51
16.5950 000	**RACEWAY & WIRE COMBINED:**					
16.5951 000	**PVC & COPPER WIRE:**					
16.5951 011	PVC & copper wire to 30a	LF	0.0345	3.85	3.95	7.81
16.5951 021	PVC & copper wire to 60a	LF	0.0459	5.13	8.17	13.30
16.5951 031	PVC & copper wire to 100a	LF	0.0983	10.98	15.47	26.45
16.5951 041	PVC & copper wire to 150a	LF	0.1064	11.89	24.60	36.48
16.5951 051	PVC & copper wire to 200a	LF	0.1187	13.26	39.72	52.98
16.5951 061	PVC & copper wire to 225a	LF	0.1310	14.63	49.68	64.31
16.5951 071	PVC & copper wire to 250a	LF	0.1418	15.84	56.10	71.94
16.5951 081	PVC & copper wire to 300a	LF	0.1499	16.75	79.87	96.61
16.5951 091	PVC & copper wire to 400a	LF	0.1790	20.00	102.57	122.57
16.5951 101	PVC & copper wire to 600a	LF	0.1883	21.03	180.31	201.35
16.5951 111	PVC & copper wire to 800a	LF	0.1977	22.09	228.87	250.96
16.5951 121	PVC & copper wire to 1000a	LF	0.2831	31.63	252.37	284.00
16.5951 131	PVC & copper wire to 1200a	LF	0.3685	41.17	362.68	403.85
16.5951 141	PVC & copper wire to 1600a	LF	0.4707	52.58	515.82	568.40
16.5951 151	PVC & copper wire to 2000a	LF	0.5412	60.46	631.97	692.43
16.5952 000	**EMT & COPPER WIRE:**					
16.5952 011	EMT & copper wire to 30a	LF	0.0655	7.32	3.44	10.76
16.5952 021	EMT & copper wire to 60a	LF	0.0819	9.15	10.18	19.33
16.5952 031	EMT & copper wire to 100a	LF	0.0942	10.52	24.07	34.59
16.5952 041	EMT & copper wire to 150a	LF	0.1083	12.10	32.90	45.00
16.5952 051	EMT & copper wire to 200a	LF	0.1225	13.68	44.98	58.66
16.5952 061	EMT & copper wire to 225a	LF	0.1362	15.21	63.52	78.74
16.5952 071	EMT & copper wire to 250a	LF	0.1576	17.61	75.37	92.98
16.5952 081	EMT & copper wire to 300a	LF	0.1683	18.80	100.47	119.27
16.5952 091	EMT & copper wire to 400a	LF	0.1790	20.00	127.57	147.57
16.5952 101	EMT & copper wire to 600a	LF	0.1860	20.78	228.44	249.22
16.5952 111	EMT & copper wire to 800a	LF	0.1930	21.56	285.40	306.96

16 - ELECTRICAL

Division	Description	Unit	Labor Hours	Labor Cost	Material Cost	Total Cost
16.5952 121	EMT & copper wire to 1000a	LF	0.2612	29.18	334.22	363.39
16.5952 131	EMT & copper wire to 1200a	LF	0.3295	36.81	474.16	510.97
16.5952 141	EMT & copper wire to 1600a	LF	0.4142	46.27	657.88	704.15
16.5952 151	EMT & copper wire to 2000a	LF	0.5540	61.89	787.82	849.71
16.5953 000	**IMC & COPPER WIRE**					
16.5953 011	Intermediate metal conduit & copper wire to 30a	LF	0.0740	8.27	4.60	12.86
16.5953 021	Intermediate metal conduit & copper wire to 60a	LF	0.0925	10.33	11.14	21.47
16.5953 031	Intermediate metal conduit & copper wire to 100a	LF	0.1064	11.89	28.36	40.25
16.5953 041	Intermediate metal conduit & copper wire to 150a	LF	0.1223	13.66	39.69	53.35
16.5953 051	Intermediate metal conduit & copper wire to 200a	LF	0.1384	15.46	55.88	71.34
16.5953 061	Intermediate metal conduit & copper wire to 225a	LF	0.1539	17.19	71.56	88.75
16.5953 071	Intermediate metal conduit & copper wire to 250a	LF	0.1780	19.88	88.81	108.69
16.5953 081	Intermediate metal conduit & copper wire to 300a	LF	0.1901	21.24	114.51	135.74
16.5953 091	Intermediate metal conduit & copper wire to 400a	LF	0.2022	22.59	154.14	176.73
16.5953 101	Intermediate metal conduit & copper wire to 600a	LF	0.2101	23.47	260.15	283.62
16.5953 111	Intermediate metal conduit & copper wire to 800a	LF	0.2180	24.35	342.34	366.69
16.5953 121	Intermediate metal conduit & copper wire to 1000a	LF	0.2951	32.97	382.28	415.24
16.5953 131	Intermediate metal conduit & copper wire to 1200a	LF	0.3723	41.59	538.56	580.15
16.5953 141	Intermediate metal conduit & copper wire to 1600a	LF	0.4680	52.28	738.37	790.65
16.5953 151	Intermediate metal conduit & copper wire to 2000a	LF	0.5601	62.57	895.14	957.71
16.5954 000	**RSC & COPPER WIRE**					
16.5954 011	RSC & copper wire to 30a	LF	0.0807	9.01	6.54	15.55
16.5954 021	RSC & copper wire to 60a	LF	0.0990	11.06	12.56	23.62
16.5954 031	RSC & copper wire to 100a	LF	0.1225	13.68	29.92	43.61
16.5954 041	RSC & copper wire to 150a	LF	0.1392	15.55	41.53	57.08
16.5954 051	RSC & copper wire to 200a	LF	0.1516	16.94	59.22	76.16
16.5954 061	RSC & copper wire to 225a	LF	0.1801	20.12	79.51	99.63
16.5954 071	RSC & copper wire to 250a	LF	0.2289	25.57	92.22	117.79
16.5954 081	RSC & copper wire to 300a	LF	0.2533	28.30	131.22	159.52
16.5954 091	RSC & copper wire to 400a	LF	0.2777	31.02	159.69	190.71
16.5954 101	RSC & copper wire to 600a	LF	0.3153	35.22	298.78	334.00
16.5954 111	RSC & copper wire to 800a	LF	0.3530	39.43	357.84	397.28
16.5954 121	RSC & copper wire to 1000a	LF	0.3873	43.27	423.88	467.15
16.5954 131	RSC & copper wire to 1200a	LF	0.4217	47.11	609.65	656.76
16.5954 141	RSC & copper wire to 1600a	LF	0.4866	54.36	837.94	892.30
16.5954 151	RSC & copper wire to 2000a	LF	0.5698	63.65	1,017.62	1,081.27
16.6000 000	**LIGHTING FIXTURES:**					
16.6001 000	**INCANDESCENT FIXTURES, COMMERCIAL:**					
16.6001 011	Incandescent fixture, surface, with lens, 100w	EA	0.4092	45.71	58.89	104.60
16.6001 021	Incandescent fixture, surface, with lens, 150w	EA	0.4298	48.01	68.95	116.96
16.6001 031	Incandescent fixture, surface, with lens, 200w	EA	0.4502	50.29	81.86	132.15
16.6001 041	Incandescent fixture, surface, with lens, 300w	EA	0.4911	54.86	93.54	148.40
16.6001 051	Incandescent fixture, recessed, open, 100w	EA	0.8184	91.42	70.88	162.30
16.6001 061	Incandescent fixture, recessed, open, 150w	EA	0.9003	100.57	79.37	179.94
16.6001 071	Incandescent fixture, recessed, open, 200w	EA	1.0230	114.28	91.05	205.33
16.6001 081	Incandescent fixture, recessed, open, 300w	EA	1.2276	137.14	103.00	240.13
16.6001 091	Incandescent fixture, recessed, with reflector, 100w	EA	1.0230	114.28	87.51	201.78
16.6001 101	Incandescent fixture, recessed, with reflector, 150w	EA	1.1049	123.43	98.25	221.68
16.6001 111	Incandescent fixture, recessed, with reflector, 200w	EA	1.2276	137.14	112.71	249.85
16.6001 121	Incandescent fixture, recessed, with reflector, 300w	EA	1.4322	159.99	127.53	287.52
16.6001 131	Incandescent fixture, recessed, with lens, 100w	EA	1.0640	118.86	99.54	218.40
16.6001 141	Incandescent fixture, recessed, with lens, 150w	EA	1.1458	128.00	114.34	242.34
16.6001 151	Incandescent fixture, recessed, with lens, 200w	EA	1.3504	150.85	129.47	280.32
16.6001 161	Incandescent fixture, recessed, with lens, 300w	EA	1.6368	182.85	144.62	327.47
16.6001 171	Keyless light fixture, utility	EA	0.2800	31.28	11.63	42.91
16.6001 181	Recessed shower light, 60w	EA	0.4041	45.14	27.36	72.50

ELECTRICAL - 16

Division	Description	Unit	Labor Hours	Labor Cost	Material Cost	Total Cost
16.6002 000	**FLUORESCENT FIXTURES, COMMERCIAL**					
16.6002 011	Fluorescent fixture, surface, 2-9w lamps	EA	0.4092	45.71	262.68	308.39
16.6002 021	Fluorescent fixture surface, 2-13w lamps	EA	0.4502	50.29	303.08	353.37
16.6002 031	Fluorescent fixture, recessed, 2-9w lamps	EA	0.8184	91.42	144.32	235.75
16.6002 041	Fluorescent fixture, recessed, 2-13w lamps	EA	1.0230	114.28	165.57	279.85
16.6002 051	Fluorescent fixture recessed, quad tube	EA	1.2276	137.14	422.84	559.98
16.6002 061	Fluorescent, recessed 7" shower light, 1-13w lamp	EA	1.0290	114.95	103.65	218.59
16.6002 071	Fluorescent drum light, surf, 2-13w lamps	EA	0.9101	101.67	303.08	404.75
16.6002 081	Fluorescent drum light, pendant, 2-13w lamps	EA	0.9101	101.67	335.58	437.25
16.6002 091	Emergency ballast for above	EA	0.5000	55.86	263.68	319.54
16.6003 000	**ELECTRIC DISCHARGE LIGHTING FIXTURES:**					
16.6003 011	Mercury vapor fixture, indoor, 75w	EA	1.2276	137.14	308.75	445.89
16.6003 021	Mercury vapor fixture, indoor, 100w	EA	1.4322	159.99	308.75	468.74
16.6003 031	Mercury vapor fixture, indoor, 175w	EA	1.5550	173.71	318.68	492.39
16.6003 041	Mercury vapor fixture, indoor, 250w	EA	1.8414	205.70	321.46	527.16
16.6003 051	Mercury vapor fixture, indoor, 400w	EA	2.0460	228.56	439.02	667.57
16.6003 061	Mercury vapor fixture, outdoor, 75w	EA	1.4322	159.99	330.02	490.01
16.6003 071	Mercury vapor fixture, outdoor, 100w	EA	1.5550	173.71	330.02	503.73
16.6003 081	Mercury vapor fixture, outdoor, 175w	EA	1.7187	192.00	444.64	636.64
16.6003 091	Mercury vapor fixture, outdoor, 250w	EA	2.0460	228.56	454.54	683.10
16.6003 101	Mercury vapor fixture, outdoor, 400w	EA	2.4552	274.27	618.30	892.57
16.6003 111	High pressure sodium/lucalux fixture, indoor, 75w	EA	1.2276	137.14	381.05	518.19
16.6003 121	High pressure sodium/lucalux fixture, indoor, 100w	EA	1.4322	159.99	389.45	549.44
16.6003 131	High pressure sodium/lucalux fixture, indoor, 175w	EA	1.5550	173.71	423.53	597.24
16.6003 141	High pressure sodium/lucalux fixture, indoor, 250w	EA	1.8414	205.70	560.82	766.53
16.6003 151	High pressure sodium/lucalux fixture, indoor, 400w	EA	2.0460	228.56	608.95	837.51
16.6003 161	High pressure sodium/lucalux fixture, outdoor, 75w	EA	1.4322	159.99	506.93	666.92
16.6003 171	High pressure sodium/lucalux fixture, outdoor, 100w	EA	1.5550	173.71	506.93	680.64
16.6003 181	High pressure sodium/lucalux fixture, outdoor, 175w	EA	1.7187	192.00	538.18	730.17
16.6003 191	High pressure sodium/lucalux fixture, outdoor, 250w	EA	2.0460	228.56	572.10	800.66
16.6003 201	High pressure sodium/lucalux fixture, outdoor, 400w	EA	2.4552	274.27	606.15	880.42
16.6003 211	High pressure sodium/lucalux fixture, outdoor, 1000w	EA	2.7500	307.20	804.79	1,112.00
16.6004 000	**FLUORESCENT FIXTURES, COMMERCIAL:**					
16.6004 011	Fluorescent strip, 1 lamp, to 4'	EA	0.7520	84.01	60.16	144.16
16.6004 021	Fluorescent strip, 2 lamp, to 4'	EA	0.8000	89.37	75.06	164.43
16.6004 031	Fluorescent strip, 4 lamp, to 4'	EA	0.9000	100.54	144.43	244.97
16.6004 041	Fluorescent strip, 1 lamp, to 8'	EA	0.9520	106.35	92.57	198.91
16.6004 051	Fluorescent strip, 2 lamp, to 8'	EA	1.0320	115.28	107.35	222.64
16.6004 061	Fluorescent strip, 4 lamp, to 8'	EA	1.5000	167.57	245.80	413.36
16.6004 071	Fluorescent, with reflector, 2 lamp, to 4'	EA	0.6548	73.15	117.52	190.67
16.6004 081	Fluorescent, with reflector, 4 lamp, to 4'	EA	0.8184	91.42	203.39	294.81
16.6004 091	Fluorescent, with reflector, 2 lamp, to 8'	EA	0.8594	96.00	200.25	296.26
16.6004 101	Fluorescent, with reflector, 4 lamp, to 8'	EA	1.1049	123.43	325.74	449.16
16.6004 111	Fluorescent, baffle type, 2 lamp, to 4'	EA	0.6957	77.72	146.69	224.41
16.6004 121	Fluorescent, baffle type, 4 lamp, to 4'	EA	0.9003	100.57	266.79	367.36
16.6004 131	Fluorescent, baffle type, 2 lamp, to 8'	EA	0.9412	105.14	290.89	396.03
16.6004 141	Fluorescent, RLM reflector, 4 lamp, to 8'	EA	1.2276	137.14	322.62	459.75
16.6004 151	Fluorescent, 1 piece lens, 1 lamp, to 4'	EA	0.8000	89.37	143.30	232.66
16.6004 161	Fluorescent, 1 piece lens, 2 lamp, to 4'	EA	0.9100	101.66	156.84	258.49
16.6004 171	Fluorescent, 1 piece lens, 1 lamp, to 8'	EA	1.4000	156.39	229.68	386.07
16.6004 181	Fluorescent, 1 piece lens, 2 lamp, to 8'	EA	1.6000	178.74	251.25	429.99
16.6004 191	Fluorescent, surface, lens, 2 lamp, 4'	EA	0.9100	101.66	138.07	239.73
16.6004 201	Fluorescent, surface, lens, 4 lamp, 4'	EA	1.1000	122.88	242.85	365.73
16.6004 211	Fluorescent, surface, lens, 2 lamp, 8'	EA	1.8000	201.08	254.06	455.14
16.6004 221	Fluorescent, surface, lens, 4 lamp, 8'	EA	1.9000	212.25	362.96	575.21
16.6005 000	**FLUORESCENT FIXTURES, RECESSED, COMMERCIAL:**					
16.6005 011	2' x 2' lay-in, recessed, 2 lamp, economy	EA	0.6500	72.61	139.47	212.08

16 - ELECTRICAL

Division	Description	Unit	Labor Hours	Labor Cost	Material Cost	Total Cost
16.6005 021	2'x 2' lay-in, recessed, 3 lamp, medium	EA	0.7000	78.20	160.38	238.58
16.6005 031	2'x 2' lay-in, recessed, 4 lamp, custom	EA	0.7500	83.78	184.44	268.22
16.6005 041	1'x 4' lay-in, recessed, 2 lamp, economy	EA	0.7000	78.20	155.76	233.96
16.6005 051	1'x 4' lay-in, recessed, 3 lamp, medium	EA	0.7500	83.78	206.00	289.78
16.6005 061	1'x 4' lay-in, recessed, 4 lamp, custom	EA	0.8000	89.37	217.36	306.73
16.6005 071	2'x 4' lay-in, recessed, 2 lamp, economy	EA	0.8000	89.37	177.04	266.41
16.6005 081	2'x 4' lay-in, recessed, 3 lamp, medium	EA	0.8500	94.95	203.51	298.47
16.6005 091	2'x 4' lay-in, recessed, 4 lamp, custom	EA	0.9000	100.54	228.73	329.27
16.6005 101	2'x 2' spline, recessed, 2 lamp, economy	EA	0.7500	83.78	182.77	266.56
16.6005 111	2'x 2' spline, recessed, 3 lamp, medium	EA	0.8000	89.37	210.17	299.53
16.6005 121	2'x 2' spline, recessed, 4 lamp, custom	EA	0.8500	94.95	241.68	336.64
16.6005 131	1'x 4' spline, recessed, 2 lamp, economy	EA	0.8000	89.37	182.43	271.80
16.6005 141	1'x 4' spline, recessed, 3 lamp, medium	EA	0.8500	94.95	232.64	327.59
16.6005 151	1'x 4' spline, recessed, 4 lamp, custom	EA	0.9000	100.54	267.53	368.07
16.6005 161	2'x 4' spline, recessed, 2 lamp, economy	EA	0.9000	100.54	204.00	304.54
16.6005 171	2'x 4' spline, recessed, 3 lamp, medium	EA	0.9500	106.12	234.58	340.70
16.6005 181	2'x 4' spline, recessed, 4 lamp, custom	EA	1.0000	111.71	255.58	367.29
16.6005 191	4'x 4' lay-in, recessed, 4 lamp, economy	EA	1.5000	167.57	453.18	620.75
16.6005 201	4'x 4' lay-in, recessed, 6 lamp, medium	EA	1.6000	178.74	515.48	694.21
16.6005 211	4'x 4' lay-in, recessed, 8 lamp, custom	EA	1.7000	189.91	615.34	805.25
16.6005 221	4'x 4' spline, recessed, 4 lamp, economy	EA	2.0000	223.42	494.31	717.73
16.6005 231	4'x 4' spline, recessed, 6 lamp, medium	EA	2.0500	229.01	556.58	785.59
16.6005 241	4'x 4' spline, recessed, 8 lamp, custom	EA	2.1000	234.59	656.41	891.01
16.6005 251	1x4 lay-in, recessed, 3 lamp, master-slave	EA	1.6787	187.53	287.93	475.46
16.6005 261	2x4 lay-in, recessed, 3 lamp, master-slave	EA	1.6787	187.53	299.59	487.12
16.6006 000	**FLUORESCENT FIXTURES, RECESSED, INSTITUTIONAL:**					
16.6006 011	2'x 2' lay-in, recessed, 2 lamp, economy	EA	0.7000	78.20	144.80	223.00
16.6006 021	2'x 2' lay-in, recessed, 3 lamp, medium	EA	0.7500	83.78	160.38	244.17
16.6006 031	2'x 2' lay-in, recessed, 4 lamp, custom	EA	0.8000	89.37	193.16	282.53
16.6006 041	1'x 4' lay-in, recessed, 2 lamp, economy	EA	0.7500	83.78	155.76	239.54
16.6006 051	1'x 4' lay-in, recessed, 3 lamp, medium	EA	0.8000	89.37	206.00	295.36
16.6006 061	1'x 4' lay-in, recessed, 4 lamp, custom	EA	0.8500	94.95	217.36	312.32
16.6006 071	2'x 4' lay-in, recessed, 2 lamp, economy	EA	0.8500	94.95	177.04	272.00
16.6006 081	2'x 4' lay-in, recessed, 3 lamp, medium	EA	0.9000	100.54	203.51	304.05
16.6006 091	2'x 4' lay-in, recessed, 4 lamp, custom	EA	0.9500	106.12	228.73	334.86
16.6006 101	2'x 2' spline, recessed, 2 lamp, economy	EA	0.8000	89.37	182.77	272.14
16.6006 111	2'x 2' spline, recessed, 3 lamp, medium	EA	0.8500	94.95	217.31	312.26
16.6006 121	2'x 2' spline, recessed, 4 lamp, custom	EA	0.9000	100.54	253.54	354.08
16.6006 131	1'x 4' spline, recessed, 2 lamp, economy	EA	0.8500	94.95	182.43	277.38
16.6006 141	1'x 4' spline, recessed, 3 lamp, medium	EA	0.9000	100.54	232.64	333.18
16.6006 151	1'x 4' spline, recessed, 4 lamp, custom	EA	0.9500	106.12	267.53	373.65
16.6006 161	2'x 4' spline, recessed, 2 lamp, economy	EA	0.9500	106.12	205.26	311.38
16.6006 171	2'x 4' spline, recessed, 3 lamp, medium	EA	1.0000	111.71	253.54	365.25
16.6006 181	2'x 4' spline, recessed, 4 lamp, custom	EA	1.0500	117.30	289.75	407.04
16.6006 191	4'x 4' lay-in, recessed, 4 lamp, economy	EA	1.6000	178.74	453.18	631.92
16.6006 201	4'x 4' lay-in, recessed, 6 lamp, medium	EA	1.7000	189.91	515.48	705.38
16.6006 211	4'x 4' lay-in, recessed, 8 lamp, custom	EA	1.8000	201.08	615.34	816.42
16.6006 221	4'x 4' spline, recessed, 4 lamp, economy	EA	2.2100	246.88	494.31	741.19
16.6006 231	4'x 4' spline, recessed, 6 lamp, medium	EA	2.2000	245.76	556.58	802.34
16.6006 241	4'x 4' spline, recessed, 8 lamp, custom	EA	2.3000	256.93	651.94	908.88
16.6006 251	Add for fluorescent battery unit	EA			584.51	584.51
16.6006 261	Add for 1 x 4 parabolic louvers	EA			50.06	50.06
16.6006 271	Add for 2 x 4 parabolic louvers	EA			97.68	97.68
16.6007 000	**FLOODLIGHT FIXTURES, INDUSTRIAL:**					
16.6007 011	Fluorescent hi-bay, reflector, 4', 2 lamp, high output	EA	1.2800	142.99	181.08	324.06
16.6007 021	Fluorescent hi-bay, reflector, 4', 4 lamp, high output	EA	1.4800	165.33	231.78	397.11
16.6007 031	Fluorescent hi-bay, reflector, 8', 4 lamp, high output	EA	2.0460	228.56	386.33	614.89

ELECTRICAL - 16

Division	Description	Unit	Labor Hours	Labor Cost	Material Cost	Total Cost
16.6007 041	Mercury vapor hi-bay, 250w	EA	2.0800	232.36	422.53	654.89
16.6007 051	Mercury vapor hi-bay, 400w	EA	2.4000	268.10	482.91	751.01
16.6007 061	Mercury vapor hi-bay, 750w	EA	2.4000	268.10	784.76	1,052.87
16.6007 071	Mercury vapor hi-bay, 1000w	EA	2.4000	268.10	1,086.58	1,354.68
16.6007 081	Mercury vapor hi-bay, 1500w	EA	2.8000	312.79	1,146.97	1,459.76
16.6007 091	High pressure sodium hi-bay, 250w	EA	2.0800	232.36	474.40	706.76
16.6007 101	High pressure sodium hi-bay, 400w	EA	2.4000	268.10	524.00	792.11
16.6007 111	High pressure sodium hi-bay, 750w	EA	2.8000	312.79	864.81	1,177.60
16.6007 121	High pressure sodium hi-bay, 1000w	EA	2.8000	312.79	890.74	1,203.53
16.6007 131	High pressure sodium hi-bay, 1500w	EA	2.8000	312.79	1,255.62	1,568.41
16.6007 141	Metal halide hi-bay, 250w, economy	EA	2.0700	231.24	310.99	542.23
16.6007 151	Metal halide hi-bay, 250w, custom	EA	2.0700	231.24	477.38	708.62
16.6007 161	Metal halide hi-bay, 400w, custom	EA	2.3884	266.81	682.05	948.86
16.6007 171	Metal halide hi-bay, 400w complete assembly	EA	4.4082	492.44	661.91	1,154.35
16.6007 181	Metal halide hi-bay, 400w with stand by	EA	4.4082	492.44	805.15	1,297.59
16.6008 000	**OUTDOOR WALL PACK FIXTURES:**					
16.6008 011	Wall pack, 250w high pressure sodium	EA	1.8326	204.72	545.68	750.40
16.6008 021	Wall pack, 400w high pressure sodium	EA	1.8326	204.72	589.29	794.01
16.6008 031	Wall pack, 1000w high pressure sodium	EA	1.8326	204.72	804.79	1,009.51
16.6009 000	**SPECIALTY LIGHTING:**					
16.6009 011	Neon tubing baked enamel finish	LF	0.0950	10.61	22.48	33.10
16.6009 021	Cold cathode transformer	EA	0.2354	26.30	45.15	71.45
16.6009 031	Tivoli lighting	LF	0.0950	10.61	63.80	74.41
16.6009 041	Transformer 120vp, 24vs @ 24' (3" OC)	EA	0.2354	26.30	27.28	53.57
16.6009 051	Fluorescent aisle fixture, louvered, faceplate	EA	0.8500	94.95	87.40	182.35
16.6009 061	Incandescent aisle fixture, louver, faceplate	EA	0.8500	94.95	77.87	172.82
16.6009 071	Perimeter lighting, flour strip 1-lamp	LF	0.2500	27.93	61.37	89.30
16.6009 081	Cove lighting, indirect fluorescent, 1-lamp	LF	0.2500	27.93	119.21	147.13
16.6009 091	Valance lighting, indirect fluorescent, 1-l	LF	0.2500	27.93	203.11	231.04
16.6009 101	Architectural custom perimeter lighting	LF	0.2500	27.93	477.38	505.31
16.6009 111	Surface track, single circuit	LF	0.2456	27.44	15.97	43.40
16.6009 121	Surface track, 2-circuit	LF	0.2456	27.44	17.17	44.60
16.6009 131	Track cylinder, 50w R-20	EA	0.1600	17.87	80.87	98.75
16.6009 141	Track sphere, 6" dia, 50w	EA	0.1600	17.87	68.73	86.60
16.6009 151	Track spot, 200w, par-46	EA	0.1600	17.87	145.51	163.39
16.6009 161	Recessed shower light, 60w	EA	0.4091	45.70	47.91	93.61
16.6009 171	6" dia tube light fixture, pendulum mounted	EA	0.7469	83.44	167.42	250.86
16.6009 181	Linear lighting, uplight, pendant mounted	EA	0.3000	33.51	79.11	112.62
16.6009 191	Linear lighting, up-down light, pendant mounted	EA	0.3500	39.10	96.71	135.81
16.6009 201	Linear lighting, custom, pendant mounted	EA	0.3500	39.10	105.48	144.58
16.6009 211	Linear lighting, uplight, wall mounted	EA	0.2000	22.34	70.33	92.67
16.6009 221	Linear lighting, custom, wall mounted	EA	0.2000	22.34	96.71	119.06
16.6009 311	Dock light, dual arm, 300w flood	EA	2.0460	228.56	450.13	678.69
16.6010 000	**EXIT SIGNS, FLUORESCENT & INCANDESCENT:**					
16.6010 011	Exit sign, incandescent, surface, 1 circuit	EA	0.6138	68.57	133.65	202.22
16.6010 021	Exit sign, incandescent, surface, 2 circuit	EA	0.7366	82.29	187.57	269.85
16.6010 031	Exit sign, incandescent, recessed, 1 circuit	EA	1.0230	114.28	122.13	236.41
16.6010 041	Exit sign, incandescent, recessed, 2 circuit	EA	1.1049	123.43	176.22	299.65
16.6010 051	Exit sign, fluorescent, surface, 1 circuit	EA	0.6138	68.57	133.65	202.22
16.6010 061	Exit sign, fluorescent, surface, 2 circuit	EA	0.7366	82.29	147.88	230.17
16.6010 071	Exit sign, fluorescent, recessed, 1 circuit	EA	1.0230	114.28	134.05	248.33
16.6010 081	Exit sign, fluorescent, recessed, 2 circuit	EA	1.1000	122.88	158.44	281.32
16.6010 091	Add for emergency battery unit	EA			266.97	266.97
16.6010 101	Exit sign, self luminous, 1 side	EA	0.3000	33.51	592.41	625.92
16.6010 111	Exit sign, self luminous, 2 sides	EA	0.3000	33.51	1,040.66	1,074.17
16.6011 000	**EMERGENCY LIGHT FIXTURES:**					
16.6011 011	Emergency light, economy, 1 head	EA	1.6368	182.85	196.40	379.24

16 - ELECTRICAL

Division	Description	Unit	Labor Hours	Labor Cost	Material Cost	Total Cost
16.6011 021	Emergency light, economy, 2 head	EA	1.6368	182.85	227.24	410.09
16.6011 031	Emergency light, lead acid battery, custom	EA	1.6368	182.85	811.17	994.02
16.6011 041	Emergency light, nickel cadmium battery, custom	EA	1.6368	182.85	1,214.82	1,397.67
16.6011 051	Remote emergency light, 1 head	EA	0.6138	68.57	64.40	132.97
16.6011 061	Remote emergency light, 2 head	EA	0.7366	82.29	132.63	214.92
16.6011 071	Emergency light, square, battery, commercial	EA	0.9100	101.66	297.50	399.15
16.6012 000	**CHANDELIERS, INCANDESCENT**					
16.6012 011	Chandelier, incandescent, 6 lamp, 15 lbs.	EA	10.6218	1,186.56	720.80	1,907.36
16.6012 021	Chandelier, incandescent, 8 lamp, 25 lbs.	EA	13.1011	1,463.52	954.03	2,417.55
16.6012 031	Chandelier, incandescent, 12 lamp, 50 lbs.	EA	16.5416	1,847.86	1,188.56	3,036.42
16.6012 041	Chandelier, incandescent, 16 lamp, 75 lbs.	EA	19.6394	2,193.92	1,656.45	3,850.37
16.6012 051	Chandelier, incandescent, 20 lamp, 100 lbs.	EA	24.4146	2,727.35	2,364.00	5,091.35
16.6012 061	Chandelier, incandescent, 24 lamp, 150 lbs.	EA	28.9112	3,229.67	3,099.13	6,328.80
16.6012 071	Chandelier, incandescent, 30 lamp, 200 lbs.	EA	34.4552	3,848.99	4,557.74	8,406.73
16.6012 081	Chandelier, incandescent, 36 lamp, 300 lbs.	EA	38.6950	4,322.62	5,314.54	9,637.16
16.6013 000	**EXPLOSION PROOF FIXTURES**					
16.6013 011	Explosion proof, incandescent, 100 watt	EA	1.6974	189.62	352.65	542.27
16.6013 021	Explosion proof, incandescent, 200 watt	EA	2.0016	223.60	576.93	800.53
16.6013 031	Explosion proof, incandescent, 300 watt	EA	2.4165	269.95	775.02	1,044.97
16.6013 041	Explosion proof, fluorescent, 1' x 4'	EA	2.9611	330.78	2,959.93	3,290.72
16.6013 051	Explosion proof, fluorescent, 2' x 2'	EA	3.0691	342.85	3,265.20	3,608.04
16.6013 061	Explosion proof, fluorescent, 2' x 4'	EA	4.0864	456.49	3,829.20	4,285.69
16.6013 071	Explosion proof, H. I. D. 175 watt	EA	2.6978	301.37	1,448.22	1,749.59
16.6013 081	Explosion proof, H. I. D. 250 watt	EA	3.0203	337.40	1,568.59	1,905.98
16.6013 091	Explosion proof, H. I. D. 400 watt	EA	3.6786	410.94	2,081.68	2,492.62
16.6013 101	Explosion proof, H. I. D. 1,000 watt	EA	4.4213	493.90	2,764.33	3,258.24
16.6013 111	Explosion proof, exit sign, 1 face	EA	1.7621	196.84	602.82	799.66
16.6013 121	Explosion proof, exit sign, 2 face	EA	1.9926	222.59	683.02	905.62
16.6014 000	**VANDAL PROOF FIXTURES**					
16.6014 011	Vandal proof, max security, fluorescent, 1'x4', 2 tube	EA	2.1043	235.07	465.38	700.45
16.6014 021	Vandal proof, max security, fluorescent, 2'x4', 2 tube	EA	2.5168	281.15	538.25	819.40
16.6014 031	Vandal proof, max security, fluorescent, 2'x4', 3 tube	EA	2.9304	327.35	606.63	933.99
16.6014 041	Vandal proof, max security, fluorescent, 2'x4', 4 tube	EA	3.3442	373.58	673.21	1,046.79
16.6014 051	Vandal proof, max security, fluorescent, 2' strip, 1 tube	EA	1.7726	198.02	428.01	626.02
16.6014 061	Vandal proof, max security, fluorescent, 4' strip, 2 tube	EA	2.1298	237.92	455.04	692.96
16.6014 071	Vandal proof, med security, fluorescent, 1'x4', 2 tube	EA	1.8486	206.51	413.02	619.53
16.6014 081	Vandal proof, med security, fluorescent, 2'x4', 2 tube	EA	2.2141	247.34	443.15	690.49
16.6014 091	Vandal proof, med security, fluorescent, 2'x4', 3 tube	EA	2.5781	288.00	506.96	794.96
16.6014 101	Vandal proof, med security, fluorescent, 2'x4', 4 tube	EA	2.9429	328.75	562.23	890.98
16.6014 111	Vandal proof, med security, fluorescent, 2' strip, 1 tube	EA	1.6529	184.65	393.01	577.66
16.6014 121	Vandal proof, med security, fluorescent, 4' strip, 2 tube	EA	1.8736	209.30	409.45	618.75
16.7000 000	**ELECTRIC & SIGNAL DEVICES:**					
16.7001 000	**LIGHTING DEVICES & OUTLETS:**					
16.7001 011	Fixture outlet, commercial	EA	0.2046	22.86	8.21	31.07
16.7001 021	Fixture junction box, commercial	EA	0.1637	18.29	8.21	26.50
16.7001 031	Fixture flex conduit/wire assembly	EA	0.1481	16.54	14.08	30.63
16.7001 041	Switch/pilot light outlet	EA	0.8100	90.49	23.55	114.03
16.7001 051	Switch, 20a, 1 pole, commercial	EA	0.6400	71.49	27.90	99.40
16.7001 061	Switch, 20a, 2-gang, commercial	EA	0.7900	88.25	32.33	120.58
16.7001 071	Switch 20a, 3-gang, commercial	EA	0.8999	100.53	78.38	178.91
16.7001 081	Switch, 20a, 4-gang, commercial	EA	1.0498	117.27	103.89	221.16
16.7001 091	Switch, 20a, 5-gang, commercial	EA	1.2000	134.05	152.40	286.45
16.7001 101	Switch, 20a, 3-way, commercial	EA	0.8100	90.49	31.16	121.64
16.7001 111	Switch, 20a, 4-way, commercial	EA	0.8900	99.42	66.87	166.29
16.7001 121	Switch, 20a, waterproof, commercial	EA	0.7900	88.25	46.65	134.90
16.7001 131	Incandescent dimmer, 600w	EA	0.8400	93.84	53.99	147.83
16.7001 141	Incandescent dimmer, 1000w	EA	0.9400	105.01	101.88	206.88

ELECTRICAL - 16

Division	Description	Unit	Labor Hours	Labor Cost	Material Cost	Total Cost
16.7001 151	Incandescent dimmer, 1500w	EA	1.0400	116.18	192.98	309.15
16.7001 161	Incandescent dimmer, 2000w	EA	1.0400	116.18	248.29	364.47
16.7001 171	Switch 20a, tamperproof keyed, 1 pole	EA	0.6400	71.49	50.27	121.76
16.7001 181	Switch 20a, explosion proof, 1 pole	EA	0.8953	100.01	228.27	328.29
16.7001 191	Fluorescent dimmer, 2-12 lamps	EA	0.6400	71.49	143.75	215.24
16.7001 201	Fluorescent dimmer, 4-220 lamps	EA	0.7900	88.25	179.67	267.92
16.7001 211	Fluorescent dimmer, 6-30 lamps	EA	0.8100	90.49	358.29	448.78
16.7001 221	Fluorescent dimmer, 8-40 lamps	EA	0.8100	90.49	465.22	555.70
16.7001 231	Time switch, 7 day calendar	EA	2.2000	245.76	183.12	428.88
16.7001 241	Time switch, 1 pole, 24 hr, dial	EA	1.8000	201.08	126.82	327.90
16.7001 251	Time switch, 24 hr, with reserve power	EA	2.0000	223.42	608.31	831.73
16.7001 261	Time switch, photoelectric cell, k-1100	EA	0.8000	89.37	77.51	166.88
16.7001 271	Time switch, photoelectric cell, k-1900	EA	0.8000	89.37	135.26	224.62
16.7001 281	Lighting circuits, EMT/wire	LF	0.0459	5.13	1.47	6.59
16.7001 291	Lighting circuits, RSC/wire	LF	0.0573	6.40	3.28	9.68
16.7001 301	Lighting circuits, modular boxes	EA	0.0655	7.32	40.02	47.34
16.7001 311	Lighting circuits, modular flex	LF	0.0124	1.39	1.43	2.82
16.7002 000	**OCCUPANCY SENSOR SYSTEM**					
16.7002 011	Wall switch, 120v or 277v	EA	0.6400	71.49	166.74	238.23
16.7002 021	Sensor, 1-way, relay & transformer or switch pack	EA	1.2276	137.14	230.88	368.02
16.7002 031	Sensor, room, 2-way, relay & transformer or switchpack	EA	1.3000	145.22	340.25	485.47
16.7002 041	Sensor, warehouse, relay & transformer or switchpack	EA	1.3000	145.22	345.09	490.31
16.7002 051	Remote control relay	EA	0.2701	30.17	48.62	78.79
16.7002 061	Transformer, 120v	EA	0.2354	26.30	26.98	53.27
16.7002 071	Transformer, 277v	EA	0.2354	26.30	38.88	65.17
16.7002 081	Switchpack, 120/277v	EA	0.0100	1.12	53.42	54.54
16.7002 091	Circuit timer, 120/277v	EA	0.8000	89.37	54.01	143.38
16.7003 000	**CONTACTOR, LIGHTING:**					
16.7003 011	Contactor, lighting to 30a, 4 pole, enclosed	EA	2.4552	274.27	347.57	621.84
16.7003 021	Contactor, lighting, to 60a, 4 pole, enclosed	EA	3.0690	342.84	710.95	1,053.78
16.7003 031	Contactor, lighting, to 100a, 4 pole, enclosed	EA	3.6828	411.41	1,169.04	1,580.44
16.7003 041	Contactor, lighting, to 200a, 4 pole, enclosed	EA	5.3100	593.18	2,969.98	3,563.16
16.7004 000	**BRANCH CIRCUIT DEVICES/OUTLETS:**					
16.7004 011	Duplex receptacle, commercial	EA	0.7600	84.90	21.99	106.89
16.7004 021	Double duplex receptacle, commercial	EA	0.9200	102.77	37.77	140.54
16.7004 031	Duplex receptacle, waterproof	EA	0.8100	90.49	50.82	141.31
16.7004 041	Duplex receptacle, ground fault interrupter	EA	0.9200	102.77	39.60	142.38
16.7004 051	Duplex receptacle, ground fault interrupter, waterproof, enclosed	EA	0.8100	90.49	68.45	158.94
16.7004 061	Receptacle, 30a, commercial	EA	0.9200	102.77	28.70	131.47
16.7004 071	Receptacle, 50a, commercial	EA	1.0160	113.50	32.43	145.93
16.7004 081	Tamperproof receptacle, 20a	EA	0.7600	84.90	33.00	117.90
16.7004 091	Isolated ground duplex receptacle	EA	0.9200	102.77	28.14	130.92
16.7004 093	Surge suppression duplex receptacle	EA	0.8900	99.42	89.79	189.21
16.7004 101	60a welding receptacle	EA	1.4000	156.39	284.39	440.78
16.7004 111	100a welding receptacle	EA	1.7500	195.49	432.16	627.66
16.7004 121	Duplex receptacle, hospital grade	EA	0.9200	102.77	50.82	153.60
16.7004 131	Single receptacle, commercial	EA	0.6666	74.47	19.33	93.79
16.7004 135	Single receptacle, explosion proof, 20a	EA	1.1100	124.00	306.56	430.55
16.7004 136	Single receptacle, explosion proof, 30a	EA	1.2500	139.64	517.06	656.70
16.7004 137	Single receptacle, explosion proof, 60a	EA	1.9100	213.37	735.68	949.05
16.7004 141	Power cord reel, 10" dia, 50' cord	EA	0.5000	55.86	390.20	446.06
16.7004 151	Branch circuit junction box, indoor	EA	0.3200	35.75	12.30	48.05
16.7004 161	Branch circuit junction box, water proof	EA	0.5600	62.56	23.31	85.87
16.7004 171	Branch circuit junction box, equipment connector	EA	0.9500	106.12	27.46	133.59
16.7004 181	Branch circuit junction box, equipment connector, water proof	EA	1.3600	151.93	34.32	186.25
16.7004 191	Motor-rated toggle switch	EA	0.6400	71.49	34.56	106.05
16.7004 201	Motor connection, 20a, 1 ph	EA	1.5200	169.80	27.46	197.26

16 - ELECTRICAL

Division	Description	Unit	Labor Hours	Labor Cost	Material Cost	Total Cost
16.7004 211	Motor connection, 20a, 3 ph	EA	1.8000	201.08	34.32	235.40
16.7004 221	Floor box & flush outlet	EA	0.6138	68.57	93.47	162.03
16.7004 231	Floor box & surface outlet/enclosure	EA	1.0230	114.28	106.31	220.59
16.7004 241	Floor box & equipment flex connector	EA	1.4322	159.99	106.31	266.30
16.7004 251	Outlet circuits, EMT/wire	LF	0.0459	5.13	1.27	6.40
16.7004 261	Outlet circuits, RSC/wire	LF	0.0655	7.32	3.03	10.34
16.7004 271	Outlet circuits, PVC/wire	LF	0.0246	2.75	2.24	4.99
16.7005 000	**TELEPHONE SYSTEMS:**					
16.7005 011	Phone backboards, 4' x 8' ply	EA	1.6368	182.85	84.03	266.87
16.7005 021	Phone cabinets, average	EA	2.0460	228.56	240.31	468.87
16.7005 031	Phone cabinets, large	EA	3.2736	365.69	432.55	798.24
16.7005 041	Phone raceway, 2" EMT, average	LF	0.0983	10.98	3.73	14.71
16.7005 051	Phone raceway, 3" EMT, average	LF	0.1188	13.27	10.41	23.68
16.7005 061	Phone raceway, 4" EMT average	LF	0.1801	20.12	17.50	37.62
16.7005 071	Phone riser sleeves, to 4"	EA	0.4092	45.71	40.70	86.41
16.7005 081	Phone outlet, wall	EA	0.2456	27.44	15.39	42.83
16.7005 091	Phone outlet, floor box	EA	0.4092	45.71	115.85	161.56
16.7005 101	Phone/data outlet, wall	EA	0.3100	34.63	26.61	61.24
16.7005 111	Phone raceway, 3/4" EMT, average	LF	0.0477	5.33	1.27	6.60
16.7005 121	Phone raceway, 1" EMT, average	LF	0.0573	6.40	1.47	7.87
16.7005 131	Phone raceway, 1" RSC, average	LF	0.0655	7.32	3.17	10.48
16.7005 141	Phone raceway, 1" PVC, average	LF	0.0295	3.30	1.19	4.49
16.7005 151	Phone raceway, 1" PVC, average	LF	0.0295	3.30	1.19	4.49
16.7006 000	**TELEPHONE SYSTEM:**					
16.7006 011	Telephone cable, 2 pair	LF	0.0079	0.88	0.09	0.97
16.7006 021	Telephone cable, 4 pair	LF	0.0098	1.09	0.18	1.28
16.7006 031	Telephone cable, 25 pair	LF	0.0197	2.20	1.21	3.41
16.7007 000	**FIRE ALARM SYSTEMS, LIFE SAFETY:**					
16.7007 011	Fire alarm main panel, 10 zone	EA	10.8000	1,206.47	3,707.80	4,914.27
16.7007 021	Fire alarm, main panel, 30 zone	EA	14.4000	1,608.62	7,191.00	8,799.62
16.7007 031	Fire alarm, main panel, 50 zone	EA	18.0000	2,010.78	11,235.95	13,246.73
16.7007 041	Fire alarm annunciator, 10 zone	EA	4.9104	548.54	1,544.90	2,093.44
16.7007 051	Fire alarm annunciator, 30 zone	EA	8.1840	914.23	2,668.52	3,582.76
16.7007 061	Fire alarm annunciator, 50 zone	EA	13.0944	1,462.78	4,353.98	5,816.76
16.7007 071	Fire alarm power supply, 10 zone	EA	3.2736	365.69	1,685.34	2,051.04
16.7007 081	Fire alarm power supply, 30 zone	EA	4.9104	548.54	3,160.01	3,708.55
16.7007 091	Fire alarm power supply, 50 zone	EA	6.5472	731.39	4,845.43	5,576.82
16.7007 101	Fire alarm, terminal cabinet, average	EA	2.0460	228.56	280.86	509.42
16.7007 111	Fire alarm, 4 zone, with battery backup	EA	3.6000	402.16	2,363.49	2,765.65
16.7007 121	Fire alarm, manual stations	EA	1.0230	114.28	162.72	277.00
16.7007 131	Fire alarm horn	EA	0.8184	91.42	107.87	199.30
16.7007 141	Fire alarm bells	EA	0.8184	91.42	149.89	241.32
16.7007 151	Fire alarm chimes	EA	0.8184	91.42	141.66	233.08
16.7007 161	Fire alarm visual alarm/horn	EA	1.3000	145.22	195.84	341.06
16.7007 171	Fire alarm visual alarm/signal	EA	1.3000	145.22	195.33	340.56
16.7007 181	Fire alarm, smoke detect, ceiling	EA	1.2276	137.14	208.48	345.62
16.7007 191	Fire alarm, smoke detector, duct	EA	2.0460	228.56	603.67	832.23
16.7007 201	Fire alarm, ionization detector	EA	1.0230	114.28	379.91	494.19
16.7007 211	Fire alarm, heat detector	EA	0.6138	68.57	87.05	155.62
16.7007 221	Fire alarm, flow switch connector	EA	1.0230	114.28	123.23	237.51
16.7007 231	Fire alarm, door hold assembly	EA	1.6368	182.85	167.67	350.52
16.7007 241	Fire alarm, door release assembly	EA	1.6368	182.85	167.67	350.52
16.7007 251	Fire alarm, phone jacks	EA	0.4937	55.15	84.33	139.48
16.7007 261	Handsets for phone jack	EA			120.49	120.49
16.7007 271	Fire alarm, distribution circuits, 10 zone	LF	0.0819	9.15	5.15	14.30
16.7007 281	Fire alarm, distribution circuits, 30 zone	LF	0.1801	20.12	10.11	30.23
16.7007 291	Fire alarm, distribution circuits, 50 zone	LF	0.2743	30.64	18.49	49.13

ELECTRICAL - 16

Division	Description	Unit	Labor Hours	Labor Cost	Material Cost	Total Cost
16.7007 301	Fire alarm circuits, EMT/wire	LF	0.0573	6.40	3.35	9.75
16.7100 000	**COMMUNICATION, INTERCOM, PUBLIC ADDRESS:**					
16.7101 000	COMMUNICATION, INTERCOM, PUBLIC ADDRESS:					
16.7101 011	Public address/intercom, main amp, power supply	EA	6.5472	731.39	3,511.22	4,242.61
16.7101 021	Public address/intercom, auxiliary terminal cabinet	EA	3.2736	365.69	351.07	716.77
16.7101 031	Public address/intercom, speaker enclosure, surface	EA	0.2865	32.00	14.94	46.94
16.7101 041	Public address/intercom, speaker enclosures, flush	EA	0.4092	45.71	34.93	80.64
16.7101 051	Public address/intercom, microphone outlet, wall	EA	0.2865	32.00	14.94	46.94
16.7101 061	Public address/intercom, microphone outlet, floor	EA	0.4092	45.71	44.33	90.04
16.7101 071	Public address/intercom, intercom outlet, wall	EA	0.2865	32.00	14.94	46.94
16.7101 081	Public address/intercom, intercom outlet, floor	EA	0.4092	45.71	44.33	90.04
16.7101 091	Public address/intercom, television outlet, wall	EA	0.2865	32.00	14.94	46.94
16.7101 101	Public address/intercom, television outlet, floor	EA	0.4092	45.71	44.33	90.04
16.7101 111	Public address/intercom, speakers, flush	EA	0.4092	45.71	55.50	101.21
16.7101 121	Public address/intercom, speakers, horn	EA	0.4092	45.71	97.15	142.86
16.7101 131	Volume control	EA	0.5000	55.86	105.01	160.86
16.7101 141	Public address/intercom, distribution raceway	LF	0.1188	13.27	2.92	16.19
16.7101 151	Public address/intercom device raceway	LF	0.0287	3.21	2.16	5.37
16.7101 161	Public address/intercom device cable	LF	0.0066	0.74	0.91	1.65
16.7101 171	Intercom station - all master desk	EA	0.6600	73.73	589.60	663.33
16.7101 181	Intercom power supply, 1 for 3 master station	EA	0.2000	22.34	227.97	250.31
16.7101 191	Intercom 1 station master, wall mounted, flush	EA	0.4800	53.62	781.82	835.44
16.7101 201	Intercom 1 station remote, wall mounted flush	EA	0.4800	53.62	210.06	263.68
16.7102 000	**TELEPHONE TIE-IN FOR PAGING:**					
16.7102 011	6 line tie-in module	EA	3.9000	435.67	2,505.35	2,941.02
16.7102 021	19" relay rack, 84" high	EA	1.2000	134.05	1,020.97	1,155.02
16.7103 000	**TELEVISION ANTENNA SYSTEMS:**					
16.7103 011	TV antenna mast, lead-in box	EA	6.5472	731.39	179.56	910.95
16.7103 021	TV main control cabinet, average	EA	3.2736	365.69	1,879.20	2,244.89
16.7103 031	TV main control cabinet, large	EA	5.3196	594.25	3,261.04	3,855.29
16.7103 041	TV auxiliary terminal cabinets	EA	3.2736	365.69	276.37	642.06
16.7103 051	TV distribution raceway	LF	0.0819	9.15	3.63	12.78
16.7103 061	TV outlet	EA	0.2046	22.86	14.90	37.76
16.7103 071	TV outlet & jack assembly	EA	0.4092	45.71	26.54	72.25
16.7103 081	TV outlet raceway	LF	0.0492	5.50	1.21	6.71
16.7103 091	TV outlet coax cable	LF	0.0066	0.74	0.12	0.85
16.7104 000	**CLOSED-CIRCUIT TELEVISION SYSTEM:**					
16.7104 011	Commercial camera & monitor	EA	3.2736	365.69	2,215.18	2,580.87
16.7104 021	Commercial camera station only	EA	1.6000	178.74	1,256.88	1,435.61
16.7104 031	Industrial camera & monitor	EA	5.7288	639.96	9,811.65	10,451.61
16.7104 041	Industrial camera station only	EA	1.6368	182.85	5,621.87	5,804.72
16.7104 051	Add for low light	EA	1.4322	159.99	3,147.58	3,307.57
16.7104 061	Add for weatherproof station	EA	0.6138	68.57	1,726.08	1,794.65
16.7104 071	Add for pan & tilt	EA	2.4000	268.10	3,407.24	3,675.35
16.7104 081	Add for zoom - remote control	EA	1.2686	141.72	3,935.23	4,076.95
16.7104 091	Closed circuit TV monitor with rack	EA	2.0000	223.42	1,718.77	1,942.19
16.7104 101	Closed circuit TV switcher	EA	2.0000	223.42	1,010.27	1,233.69
16.7104 111	Add for automatic iris	EA	1.2686	141.72	3,567.33	3,709.04
16.7104 121	Video cable	LF	0.0089	0.99	7.41	8.40
16.7104 131	EMT & video cable, 3/4" average	LF	0.0417	4.66	8.35	13.01
16.7104 141	RSC & video cable, 3/4" average	LF	0.0595	6.65	10.32	16.96
16.7105 000	**SECURITY SYSTEM:**					
16.7105 011	Master control panel	EA	12.0000	1,340.52	698.47	2,038.99
16.7105 021	Intruder alarm panel	EA	6.0000	670.26	290.00	960.26
16.7105 031	Security starter kit	EA	6.0000	670.26	702.59	1,372.85
16.7105 041	Passive infrared motion detector	EA	0.7000	78.20	343.75	421.94
16.7105 051	Sound discriminator detector	EA	0.7000	78.20	241.83	320.03

16 - ELECTRICAL

Division	Description	Unit	Labor Hours	Labor Cost	Material Cost	Total Cost
16.7105 061	Security conduit/wire	EA	0.0350	3.91	3.10	7.01
16.7106 000	**MISCELLANEOUS ALARMS:**					
16.7106 011	Alarm panel, 4 zone	EA	2.8100	313.91	1,258.63	1,572.54
16.7106 021	Alarm panel, 20 zone	EA	9.0000	1,005.39	7,127.25	8,132.64
16.7106 031	20 zone annunciator	EA	4.5000	502.70	634.30	1,137.00
16.7106 041	Door switch	EA	0.3310	36.98	21.63	58.60
16.7106 051	Hold up button	EA	0.4200	46.92	255.71	302.63
16.7106 061	Foot switch	EA	0.4850	54.18	501.46	555.64
16.7108 000	**SEISMIC MONITORING SYSTEMS**					
16.7108 011	Accelerograph, seismic monitor, digital	EA	9.5000	1,061.24	8,086.20	9,147.44
16.7108 021	Accelerograph, seismic monitor, analog	EA	11.8000	1,318.18	9,096.98	10,415.16
16.7200 000	**SPECIAL HOSPITAL SYSTEMS:**					
16.7201 000	**NURSE CALL SYSTEMS-AUDIO VISUAL:**					
16.7201 011	Nurse call, audio visual station, single	EA	0.8184	91.42	360.16	451.59
16.7201 021	Nurse call, 20 station desk master	EA	6.5472	731.39	2,514.59	3,245.97
16.7201 031	Nurse call, 30 station desk master	EA	9.8208	1,097.08	3,007.17	4,104.26
16.7201 041	Nurse call, 40 station desk master	EA	13.0944	1,462.78	3,508.36	4,971.13
16.7201 051	Nurse call, master control panel with power supply	EA	14.2500	1,591.87	3,214.56	4,806.42
16.7201 061	Staff locator station	EA	0.8184	91.42	317.97	409.39
16.7201 071	Pillow speaker	EA	0.2000	22.34	233.29	255.63
16.7201 081	Code blue panel, 50 lights	EA	9.5000	1,061.24	674.02	1,735.27
16.7201 091	Code blue control panel	EA	9.5000	1,061.24	1,309.08	2,370.32
16.7201 101	Code blue push button	EA	0.8184	91.42	165.91	257.33
16.7201 111	Emergency pull cord or push button	EA	0.8184	91.42	88.37	179.80
16.7201 121	Dome light, single	EA	0.6138	68.57	88.37	156.94
16.7201 131	Dome light, double	EA	0.6957	77.72	103.03	180.75
16.7201 141	Dome light, quad	EA	0.7438	83.09	128.84	211.93
16.7201 151	Nurse call system junction box	EA	0.4092	45.71	11.74	57.45
16.7201 161	Nurse call system feeder	LF	0.2046	22.86	10.84	33.69
16.7201 171	Nurse call system device circuit	LF	0.0573	6.40	3.64	10.04
16.7202 000	**RADIO PAGING SYSTEMS:**					
16.7202 011	Master station & power supply	EA	6.5472	731.39	3,683.35	4,414.73
16.7202 021	Personal call pocket receiver	EA	0.8184	91.42	368.32	459.74
16.7203 000	**DOCTORS REGISTRY SYSTEMS:**					
16.7203 011	Doctors registry display, to 50	EA	9.8208	1,097.08	4,515.24	5,612.33
16.7203 021	Doctors registry display, to 100	EA	16.3680	1,828.47	6,554.41	8,382.88
16.7203 031	Doctors registry display, to 200	EA	26.1888	2,925.55	14,565.47	17,491.02
16.7204 000	**ROOM OCCUPANCY SYSTEMS:**					
16.7204 011	Room occupancy display, to 50	EA	9.8208	1,097.08	3,495.66	4,592.74
16.7204 021	Room occupancy display, to 100	EA	16.3680	1,828.47	5,243.53	7,072.00
16.7204 031	Room occupancy display, to 200	EA	26.1888	2,925.55	7,282.71	10,208.26
16.7205 000	**MEDICAL CARE SYSTEMS:**					
16.7205 011	Bedside patient consoles, normal	EA	6.5472	731.39	3,641.33	4,372.71
16.7205 021	Bedside patient consoles, intensive care unit	EA	9.8208	1,097.08	6,554.41	7,651.49
16.7206 000	**EQUIPOTENTIAL GROUNDING SYSTEMS:**					
16.7206 011	Equipotential ground, operating room	SF	0.0410	4.58	5.78	10.36
16.7206 021	Equipotential ground, ward area	SF	0.0206	2.30	3.13	5.43
16.7206 031	Equipotential ground, private room	SF	0.0246	2.75	3.85	6.60
16.7206 041	Equipotential ground, intensive care unit ward	SF	0.0287	3.21	3.59	6.80
16.7206 051	Equipotential ground, intensive care unit room	SF	0.0369	4.12	4.31	8.43
16.7206 061	Ground modules, wall outlet	EA	1.6368	182.85	189.32	372.17
16.7206 071	Master ground module	EA	6.5472	731.39	786.56	1,517.95
16.7206 081	Equipotential circuits, RSC/wire	LF	0.0655	7.32	6.40	13.72
16.7207 000	**HOSPITAL ISOLATING PANELS:**					
16.7207 011	Power/light panel, 8 circuit, 3kva, operating room	EA	11.0484	1,234.22	8,090.06	9,324.28
16.7207 021	Power/light panel, 8 circuit, 5kva, operating room	EA	12.2760	1,371.35	8,365.63	9,736.98
16.7207 031	Power/light panel, 8 circuit, 75kva, operating room	EA	13.0944	1,462.78	8,867.61	10,330.39

ELECTRICAL - 16

Division	Description	Unit	Labor Hours	Labor Cost	Material Cost	Total Cost
16.7207 041	Power/light panel, 8 circuit, 10kva, operating room	EA	14.7312	1,645.62	9,753.60	11,399.22
16.7207 051	Intensive care unit/coronary care unit area panel, 8 circuit, 3 kva	EA	11.0484	1,234.22	9,677.23	10,911.45
16.7207 061	Intensive care unit/coronary care unit area panel, 8 circuit, 5 kva	EA	12.2760	1,371.35	9,884.78	11,256.13
16.7207 071	Intensive care unit/coronary care unit area panel, 8 circuit, 75 kva	EA	13.0944	1,462.78	10,774.22	12,237.00
16.7207 081	Intensive care unit/coronary care unit area panel, 8 circuit, 10 kva	EA	14.7312	1,645.62	11,776.10	13,421.72
16.7207 091	X-ray panel, 15 kva, 277/208v	EA	14.7312	1,645.62	19,598.23	21,243.85
16.7207 101	X-ray panel, 25 kva, 277/208v	EA	18.0048	2,011.32	20,069.70	22,081.02
16.7207 111	X-ray outlets, 60 a	EA	0.6000	67.03	1,168.63	1,235.66
16.7208 000	**CLOCK AND PROGRAM SYSTEMS:**					
16.7208 011	Clock main panel/generator	EA	6.5472	731.39	3,224.19	3,955.58
16.7208 021	Clock terminal cabinet	EA	2.0460	228.56	280.30	508.85
16.7208 031	Clock distribution raceway	LF	0.0819	9.15	3.63	12.78
16.7208 041	Clock distribution cable	LF	0.0066	0.74	6.41	7.15
16.7208 051	Clock outlets	EA	0.2046	22.86	14.90	37.76
16.7208 061	Elapsed timer, 15 seconds to 2 hour	EA	0.8184	91.42	186.24	277.66
16.7208 071	Clock, standard, wall mounted	EA	0.8184	91.42	45.82	137.24
16.7208 081	Clock, standard, flush mounted	EA	1.6368	182.85	157.48	340.33
16.7208 091	Clock circuits, EMT/wire	LF	0.0573	6.40	2.99	9.39
16.7209 000	**LOW VOLTAGE SWITCHING SYSTEMS:**					
16.7209 011	Low voltage relays, one pole	EA	0.0819	9.15	49.00	58.15
16.7209 021	Low voltage relays, multi-pole	EA	0.3683	41.14	124.34	165.48
16.7209 031	Low voltage relay cabinet, 18 pole	EA	2.8644	319.98	232.55	552.53
16.7209 041	Low voltage switch, single	EA	0.3274	36.57	38.29	74.86
16.7209 051	Low voltage switch, master	EA	1.2276	137.14	436.59	573.72
16.7209 061	Low voltage transformer	EA	0.2046	22.86	45.82	68.68
16.7209 071	Low voltage push-button	EA	0.2046	22.86	10.46	33.31
16.7209 081	Low voltage door switch	EA	0.4092	45.71	31.70	77.41
16.7209 091	Low voltage raceway, EMT	LF	0.0492	5.50	1.21	6.71
16.7209 101	Low voltage conductor #14	LF	0.0050	0.56	0.07	0.63
16.7300 000	**PRISON CELL DOOR CONTROL SYSTEMS:**					
16.7300 011	Door, prison cell, locking relay cabinet	EA	10.5673	1,180.47	1,319.12	2,499.59
16.7300 021	Door, prison cell, gang control cabinet	EA	4.2501	474.78	203.35	678.13
16.7300 031	Door, prison cell, gang control panel	EA	7.5241	840.52	628.58	1,469.10
16.7300 041	Door, prison cell, cable gutter	LF	0.1814	20.26	25.74	46.01
16.7300 051	Door, prison cell, junction box	EA	0.1274	14.23	8.25	22.48
16.7300 061	Door, prison cell, door connections	EA	0.2513	28.07	14.85	42.92
16.7300 071	Door, prison cell, 2-2" concealed, RSG conduit	LF	0.2130	23.79	19.48	43.27
16.7300 081	Door, prison cell, 2-2" concealed, RSG conduit	LF	0.2828	31.59	29.20	60.79
16.7300 091	Door, prison cell, 2-2" concealed, RSG conduit	LF	0.0652	7.28	5.03	12.31
16.7300 101	Door, prison cell, #14 wire	LF	0.0031	0.35	0.10	0.45
16.7300 111	Door, prison cell, #16 wire	LF	0.0019	0.21	0.05	0.26
16.7500 000	**SOFT WIRE SYSTEMS, 3 WIRE:**					
16.7501 000	**LIGHTING, 15 FIXTURE GROUP:**					
16.7501 001	*Note: The following prices include wiring only. Fixtures must be added separately.*					
16.7501 011	Fixtures on 6' centers	GR	1.1867	132.57	58.91	191.48
16.7501 021	Fixtures on 8' centers	GR	1.1867	132.57	62.39	194.95
16.7501 031	Fixtures on 10' centers	GR	1.1867	132.57	65.80	198.36
16.7501 041	Fixtures on 12' centers	GR	1.1867	132.57	68.63	201.19
16.7501 051	Fixtures on 16' centers	GR	1.1867	132.57	76.66	209.22
16.7501 061	Fixtures on 20' centers	GR	1.1867	132.57	83.25	215.82
16.7502 000	**SWITCHING:**					
16.7502 011	Single pole - single level	EA	0.4092	45.71	106.80	152.51
16.7502 021	Single pole - double level	EA	0.4092	45.71	147.25	192.96
16.7502 031	Three-way, per pair of switches	EA	0.9003	100.57	172.14	272.72
16.7502 041	Four-way, per switch drop	EA	0.4092	45.71	127.17	172.88
16.7503 000	**OUTLETS, 9 OUTLET GROUP:**					

16 - ELECTRICAL

Division	Description	Unit	Labor Hours	Labor Cost	Material Cost	Total Cost
16.7503 011	Outlets on 4' centers	GR	3.3146	370.27	469.49	839.77
16.7503 021	Outlets on 8' centers	GR	3.3146	370.27	476.40	846.68
16.7503 031	Outlets on 10' centers	GR	0.3315	37.03	479.50	516.53
16.7503 041	Outlets on 12' centers	GR	0.3315	37.03	482.42	519.45
16.7503 051	Outlets on 16' centers	GR	0.3315	37.03	490.29	527.32
16.7504 000	**POWER COLUMNS, 9 COLUMN GROUP:**					
16.7504 011	Power columns on 10' centers	GR	14.7312	1,645.62	1,854.10	3,499.73
16.7504 021	Power columns on 12' centers	GR	14.7312	1,645.62	1,857.27	3,502.89
16.7504 031	Power columns on 16' centers	GR	14.7312	1,645.62	1,860.61	3,506.23
16.7504 041	Power columns on 20' centers	GR	14.7312	1,645.62	1,868.40	3,514.02
16.7504 051	Power columns on 24' centers	GR	14.7312	1,645.62	1,882.44	3,528.07
16.7504 061	Soft wire poke-thru receptacle	EA	1.1949	133.48	185.79	319.27
16.7600 000	**ENERGY & BUILDING MANAGEMENT SYSTEMS:**					
16.7601 000	**ENERGY & BUILDING MANAGEMENT SYSTEM:**					
16.7601 011	Relay panel	EA	1.1294	126.17	414.88	541.05
16.7601 021	Monitor point, 3" pipe	EA	0.9684	108.18	192.19	300.37
16.7601 031	Monitor point, 4" pipe	EA	0.9821	109.71	259.55	369.26
16.7601 041	Monitor point, 5" pipe	EA	0.9957	111.23	382.27	493.50
16.7601 051	Monitor point, 6" pipe	EA	1.0093	112.75	463.45	576.20
16.7601 061	Monitor point, 8" pipe	EA	1.0230	114.28	558.44	672.72
16.7601 071	Monitor point, 10" pipe	EA	1.0366	115.80	584.31	700.11
16.7601 081	Monitor point, 12" pipe	LF	1.0502	117.32	805.65	922.97
16.7601 091	Monitor EMT & wire, 3/4" average	LF	0.0459	5.13	1.29	6.42
16.7601 101	Monitor RSC & wire, 3/4" average	LF	0.0573	6.40	2.76	9.16
16.7601 111	Monitor point termination	EA	0.1364	15.24	0.12	15.35
16.7601 121	Master control panel	EA	6.9974	781.68	2,420.36	3,202.04
16.7700 000	**TESTING:**					
16.7701 000	**TESTING:**					
16.7701 011	Testing, light fixtures	EA			8.75	8.75
16.7701 021	Testing, wiring devices	EA			6.19	6.19
16.7701 031	Testing, motors & control by hp	EA			36.36	36.36

Assembly Costs

Table of Contents

1.1	**DEMOLITION**	**252**
01.1105	Building	252
01.1110	Wall	252
01.1115	Roof	252
01.1120	Ceiling	252
01.1200	Foundation	252
01.1305	Paving, Curbs & Walls	252
01.1400	Utilities	253
1.21	**SITEWORK**	**254**
01.2100	Clear & Grub	254
01.2200	Mass Excavation	254
01.2300	Site Fill	254
01.2400	Site Paving, Asphalt	254
01.2500	Site Paving, Concrete	254
01.2600	Site Other, Surface	254
01.2700	Site Utilities	255
01.2800	Site Lighting	256
2.1	**SUBSTRUCTURE**	**257**
02.1100	Excavation for Buildings	257
02.1200	Piles	257
02.1300	Foundations and Pile Caps	257
02.1400	Walls below Grade	258
02.1500	Waterproofing	258
02.1600	Substructure Drainage	258
02.1700	Slabs on Grade	258
3.0	**STRUCTURE**	**259**
03.0105	Columns, Cast In Place	259
03.0110	Columns, Precast	259
03.0115	Columns, Steel	259
03.0120	Columns, Wood	260
03.0205	Beams & Girders, Cast in Place	260
03.0210	Beams & Girders, Precast	260
03.0215	Beams & Girders, Steel	260
03.0220	Beams & Girders, Wood	260
03.0305	Floors, Cast In Place	261
03.0310	Floors, Precast	261
03.0315	Floors, Concrete on Steel Joists	261
03.0320	Floors, Wood	261
03.0325	Roofs, Metal Deck on Steel Joists	261
03.0330	Roofs, Wood	262
03.0400	Floor, Topping	262
03.0500	Fireproofing	262
4.1	**ENCLOSURE, VERTICAL**	**263**
04.1105	Walls, Cast In Place	263
04.1110	Walls, Tilt-Up Concrete Panel	263
04.1115	Walls, Precast Concrete Panels	263
04.1120	Walls, Concrete Block	264
04.1125	Walls, Brick Veneer with Block Back-Up	264
04.1130	Walls, Brick	264
04.1135	Walls, Stone Veneer with Block Back-Up	264
04.1140	Walls, Brick Veneer on Stud Frame	265
04.1145	Walls, Siding on Stud Frame	266
04.1150	Walls, Stone Veneer on Stud Frame	266
04.1155	Walls, Stucco on Stud Frame	267
04.1160	Walls, Metal Siding	267
04.1165	Curtain Walls	267
04.1170	Walls, Exterior Coating	268
04.1175	Interior Surface of Exterior Walls	269
04.1200	Fenestration	269
04.1300	Exterior Doors	270

Assembly Costs

Table of Contents

04.1400	Store Front Systems	270
04.1500	Special Doors	271

4.2 ENCLOSURE, HORIZONTAL 272

04.2100	Roof & Roof Materials	272
04.2200	Rigid Insulation	273

4.3 SUPPORT ITEMS 272

04.3100	Miscellaneous Iron	274
04.3200	Sheetmetal	274
04.3300	Skylights	275
04.3400	Insulation	275
04.3500	Caulking & Sealants	276

5.1 INTERNALS, VERTICAL 277

05.1105	Gypsum Board on Stud	277
05.1110	Lath and Plaster on Stud	278
05.1115	Concrete Block	279
05.1120	Structural Tile	280
05.1125	Wall Finishes	280
05.1200	Doors, Including Hardware	281
05.1300	Internal Fenestration	282

5.2 INTERNALS, HORIZONTAL 283

05.2100	Ceiling	283
05.2200	Floor, Resilient	283

5.3 FINISHES, SPECIAL 284

05.3100	Tile	284
05.3200	Terrazzo	284
05.3300	Hardwood Floors	284
05.3400	Wall Covering	284
05.3500	Wood	284

5.4 INTERIORS 285

05.4100	Cabinets & Tops	285
05.4300	Carpet	286

6.0 SPECIALTIES 287

06.0100	Chalk & Tack Boards	287
06.0200	Toilet Partitions	287
06.0300	Demountable Partitions	287
06.0400	Toilet Accessories	288

7.0 EQUIPMENT 289

07.0100	Bank Equipment	289
07.0200	Educational Equipment	289
07.0300	Food Service Equipment	290
07.0400	Gymnasium & Playground Equipment	291
07.0500	Industrial Equipment	292
07.0600	Parking Lot Equipment	293
07.0700	Material Handling Equipment	293
07.0800	Laboratory Equipment	293
07.0900	Library Equipment	294
07.1000	Hospital Equipment	294
07.1100	Dental Equipment	296
07.1200	Stage Equipment	296
07.1300	Garbage Compactors	298

8.0 SPECIAL CONSTRUCTION 299

08.0100	Raised Floors	299
08.0200	X-Ray Room Construction	299
08.0300	Pools	299

9.0 CONVEYING 300

09.0100	Stairs	300
09.0205	Elevators, Hydraulic	300
09.0210	Elevators, Electric Gear	300

Assembly Costs

Table of Contents

09.0215	Elevators, Electric Gearless	300
09.0220	Hoists and Cranes	300
09.0225	Manlift	301
09.0300	Dumbwaiters	301
09.0400	Escalators	301
09.0500	Pneumatic Systems	301

10.1 PLUMBING & FIRE PROTECTION 302

10.1100	Equipment	302
10.1205	Fixtures, Economy Grade	302
10.1210	Fixtures, Standard Grade	303
10.1215	Fixtures, Institutional Grade	303
10.1305	Rough-Ins	303
10.1310	Pipe, Cast Iron	304
10.1315	Pipe, Copper	304
10.1320	Pipe, PVC	305
10.1325	Pipe, Polypropylene	305
10.1330	Pipe, Plastic	305
10.1335	Pipe, Pyrex Unit Cost	305
10.1340	Pipe, Steel Unit Cost	305
10.1400	Miscellaneous Plumbing Specialties	305
10.1500	Medical Gases, Accessories	306
10.1515	Medical Gas Piping	306
10.1600	Fees, Permits, Sterilization	306
10.1700	Fire Protection Systems	306

10.2 HEAT, VENT & AIR CONDITIONING 308

10.2105	Furnaces	308
10.2110	Boilers	308
10.2115	Chillers	308
10.2120	Cooling Towers	308
10.2125	Air Conditioners	308
10.2130	Split Systems	309
10.2135	Computer Room Air Conditioning	309
10.2140	Roof Mounted Units	309
10.2145	Heat Pumps	309
10.2150	Hydronic Systems	309
10.2155	Infrared and Radiant Systems	309
10.2160	Humidifiers	309
10.2165	Pumps	309
10.2170	Air Handlers	310
10.2175	Coils	310
10.2180	Fans	310
10.2185	Tanks	310
10.2200	Controls	310
10.2300	Ductwork	311
10.2400	Piping	311

11.0 ELECTRICAL 313

11.0100	Switchgear	313
11.0200	Emergency Systems	313
11.0300	Feeder Conduit & Wire	313
11.0400	Fixtures	314
11.0500	Signal & Communications	315
11.0600	Devices	315

1.1 - DEMOLITION

CSI#	Description	Unit	Material Cost	Install Cost	Total Cost
	The demolition costs in this section include disposal. However, no allowances are included for dumping charges or special handling of toxic or hazardous waste. Other conditions, such as excessive haul distances, unusual work hours, high voltage lines, or limited access must also be accounted for by increased costs or additional allowances. Consult local authorities for special conditions and legislation which may affect the demolition and disposal costs.				
01.1105	**DEMOLITION, BUILDING**				
01.1105 100	Demolition, frame building, one story	SF	0.18	4.24	4.42
01.1105 105	Demolition, frame building, two story	SF	0.18	4.10	4.29
01.1105 110	Demolition, frame building, three story	SF	0.18	4.02	4.20
01.1105 115	Demolition, concrete block building, two story	SF	0.88	8.35	9.22
01.1105 120	Demolition, concrete block building, three story	SF	0.81	8.07	8.87
01.1105 125	Demolition, concrete building, monolithic, large	SF	3.96	9.41	13.38
01.1105 130	Demolition, concrete building, precast panel, frame roof	SF	3.04	8.63	11.67
01.1105 135	Demolition, steel frame building, no fireproofing	SF	1.73	10.02	11.75
01.1105 140	Demolition, steel frame building, with fireproofing	SF	2.25	13.03	15.27
01.1105 145	Demolition, steel frame building, salvage steel	SF	2.59	15.03	17.62
01.1110	**DEMOLITION, WALL**				
01.1110 105	Demolition, remove wall, concrete to 10" thick, reinforced	SF	2.92	4.16	7.07
01.1110 110	Demolition, remove wall, concrete block, reinforced	SF	1.56	2.99	4.55
01.1110 115	Demolition, remove wall, stucco/plaster, metal stud	SF	0.02	1.10	1.12
01.1110 120	Demolition, remove wall, gypsum board, metal stud	SF	0.02	0.81	0.83
01.1110 125	Demolition, remove wall, brick veneer, overlaid	SF	0.27	7.51	7.78
01.1110 130	Demolition, remove wall, 8" solid brick, 10" cavity	SF	0.45	7.15	7.60
01.1110 135	Demolition, remove wall, 10"-12" solid brick, reinforced, grout	SF	0.90	9.85	10.75
01.1110 140	Demolition, remove wall, metal siding, no save	SF	0.33	4.88	5.21
01.1110 145	Demolition, remove curtain wall, save glass	SF	0.61	6.33	6.94
01.1115	**DEMOLITION, ROOF**				
01.1115 100	Demolition, remove built-up roof, on plywood	SQ	2.16	62.47	64.62
01.1115 105	Demolition, remove built-up roof, on metal deck	SQ	3.17	46.50	49.67
01.1115 110	Demolition, remove built-up roof, on gypsum plank	SQ	2.16	47.54	49.70
01.1115 115	Demolition, remove built-up roof, on concrete	SQ	2.16	54.60	56.76
01.1115 120	Demolition, remove roof, asphalt shingles	SQ	2.16	40.48	42.63
01.1115 125	Demolition, remove roof, wood shingles	SQ	2.16	38.33	40.49
01.1120	**DEMOLITION, CEILING**				
01.1120 100	Demolition, remove ceiling, plaster/lath/frame	SF	0.03	2.13	2.17
01.1120 105	Demolition, remove ceiling, plaster suspended grid	SF	0.03	1.60	1.64
01.1120 110	Demolition, remove ceiling, acoustic suspended grid	SF	0.03	0.49	0.53
01.1120 115	Demolition, remove ceiling, acoustic salvage	SF	0.03	0.93	0.96
01.1120 120	Demolition, remove ceiling, area light, salvage	SF	0.03	1.72	1.76
01.1200	**DEMOLITION, FOUNDATION**				
01.1200 100	Demolition, remove concrete foundation, no rebar	CY	50.37	73.92	124.29
01.1200 105	Demolition, remove concrete foundation, with rebar	CY	58.89	109.19	168.08
01.1305	**DEMOLITION, PAVING, CURBS & WALKS**				
01.1305 100	Demolition, remove pavement, asphaltic concrete, short haul, 5,000 to 25,000 SF	SF	0.63	0.32	0.95
01.1305 105	Demolition, remove pavement, asphaltic concrete, short haul, 25,000 to 50,000 SF	SF	0.40	0.32	0.72
01.1305 110	Demolition, remove pavement, asphaltic concrete, short haul, over 50,000 SF	SF	0.15	1.14	1.29
01.1305 115	Demolition, remove pavement, concrete slab, 5" thick, no rebar	SF	0.59	0.85	1.44
01.1305 120	Demolition, remove pavement, concrete slab, 5" thick, with rebar	SF	0.66	1.10	1.76
01.1305 125	Demolition, remove pavement, concrete slab, 9"- 12" thick, with rebar	SF	3.17	4.78	7.95
01.1305 130	Demolition, remove pavement, concrete slab, 13"- 18" thick, with rebar	SF	4.59	8.60	13.19
01.1305 135	Demolition, remove pavement, concrete curb and gutter, no sawing	LF	1.09	1.88	2.98
01.1305 140	Demolition, remove pavement, concrete curb-planter & batt	SF	1.36	2.65	4.01
01.1305 145	Demolition, remove pavement, concrete drive with curb & retaining wall	LF	0.63	1.11	1.74
01.1305 150	Demolition, remove pavement, concrete sidewalk	SF	0.48	0.98	1.46
01.1305 155	Demolition, remove pavement, concrete catch basin, sump and dry well	EACH	124.76	201.40	326.16

DEMOLITION - 1.1

CSI#	Description	Unit	Material Cost	Install Cost	Total Cost
01.1400	**DEMOLITION, UTILITIES**				
01.1400 100	Demolition, remove hydrant, with reset thrust blocks	EACH	136.37	450.89	587.26
01.1400 105	Demolition, remove storm drain line, dispose	LF	3.24	7.18	10.42
01.1400 110	Demolition, remove sanitary line, dispose	LF	3.76	7.07	10.83

1.21 - SITEWORK

CSI#	Description	Unit	Material Cost	Install Cost	Total Cost
	The sitework costs in this section are typical of costs associated with the construction of a building. They include allowances for materials, labor, overhead, and profit for the subcontractor. There are no allowances for the markup or profit for the general contractor.				
	The costs include all costs associated with completion of the installation. For example, the costs for underground utilities and site lighting have allowances for trenching and backfill included. However, there are no built-in allowances for adverse weather conditions nor are there any legislation allowances for unusual soil or rock conditions.				
01.2100	**CLEAR & GRUB**				
01.2100 100	Clear & grub, brush, turf, roots, with disposal	SF	0.06	0.09	0.15
01.2100 105	Clear & grub, large area, without disposal	SF	0.03	0.09	0.13
01.2100 110	Clear & grub, large products	CY	1.37	1.34	2.72
01.2100 115	Strip and stockpile 6" deep	CY	1.81	1.60	3.41
01.2100 120	Scarify and compact 6" topping	SF	0.07	0.22	0.29
01.2100 125	Rough grade machine	SF	0.03	0.09	0.13
01.2100 130	Fine grade machine	SF	0.05	0.17	0.22
01.2100 135	Fine grade hand	SF		0.85	0.85
01.2200	**MASS EXCAVATION**				
01.2200 100	Site cut and fill, earth. 5M - 20M	CY	2.72	2.69	5.41
01.2200 105	Site cut and fill, earth. 20M - 50M	CY	1.91	1.88	3.79
01.2200 110	Site cut and fill, rock and earth mixture	CY	6.63	5.93	12.55
01.2300	**SITE FILL**				
01.2300 100	Engineered fill, 1M - 2.5M, imported one mile	CY	11.71	1.42	13.12
01.2300 105	Engineered fill, 2.5M - 10M, imported one mile	CY	11.37	0.85	12.22
01.2300 110	Engineered fill, 10M - 25M, imported one mile	CY	7.02	1.10	8.12
01.2300 115	Engineered fill, 25M - 50M, imported one mile	CY	6.51	0.97	7.49
01.2300 120	Engineered fill, 100 - 2.5M	CY	8.08	0.55	8.63
01.2300 125	Engineered fill, 2.5M - 25M	CY	5.22	0.36	5.58
01.2300 130	Engineered fill, 25M - 50M	CY	4.46	0.19	4.65
01.2300 135	Engineered fill, compaction, by roller	CY	0.83	1.19	2.02
01.2300 140	Engineered fill, compaction, by sheepsfoot	CY	1.08	1.52	2.60
01.2300 145	Engineered fill, dozer spread, no material	CY	0.65	0.82	1.46
01.2300 150	Backfill, machine, no compaction	CY	3.81	12.84	16.65
01.2300 155	Backfill, hand, no compaction	CY	6.51	71.27	77.78
01.2300 160	Backfill, select import, compacted	CY	15.71	34.77	50.48
01.2300 165	Backfill, site material, compacted	CY	7.87	22.39	30.26
01.2400	**SITE PAVING, ASPHALT**				
01.2400 100	Pavement, asphalt, driveway, 2" thick, 4" base	SF	2.14	1.70	3.84
01.2400 105	Pavement, asphalt, parking lot, 2" thick, 6" base	SF	2.43	2.11	4.54
01.2400 110	Pavement, asphalt, truck & ramp, 3" thick, 8" base	SF	2.98	2.67	5.66
01.2400 115	Pavement, asphalt, street, 3" thick, 8" base, 10" sub-base	SF	4.10	3.68	7.78
01.2400 120	Pavement, asphalt, armor coat, 2 shot, 4" base	SF	0.96	1.17	2.12
01.2400 125	Add or deduct for each additional 1" of baserock	SF	0.22		0.22
01.2400 130	Add or deduct for each additional 1" of asphaltic concrete	SF	0.84		0.84
01.2500	**SITE PAVING, CONCRETE**				
01.2500 100	Pavement, concrete, driveway, 2" thick, 4" base	SF	2.14	1.70	3.84
01.2500 105	Pavement, concrete, parking lot, 2" thick, 6" base	SF	2.43	2.11	4.54
01.2500 110	Pavement, concrete, truck & ramp, 3" thick, 8" base	SF	2.98	2.67	5.66
01.2500 115	Pavement, concrete, street, 3" thick, 8" base, 10" sub-base	SF	4.10	3.68	7.78
01.2500 120	Bumper block, parking, precast, 3' long	EACH	29.01	17.92	46.92
01.2500 125	Bumper block, parking, precast, 6' long	EACH	46.58	21.92	68.50
01.2500 130	Striping, parking stall, one line per stall, 1 coat	EACH	0.05	11.29	11.34
01.2500 135	Street sign with pole	EACH	147.52	139.87	287.39
01.2600	**SITE OTHER, SURFACE**				
01.2600 100	Sidewalk, concrete walk 4", 100% seeded color	SF	4.64	4.93	9.57
01.2600 105	Sidewalk, concrete walk 4", 80 - 100% seeded aggregate	SF	4.51	4.80	9.30
01.2600 110	Sidewalk, concrete walk 4", exposed aggregate wash	SF	3.64	2.90	6.54

SITEWORK - 1.21

CSI#	Description	Unit	Material Cost	Install Cost	Total Cost
01.2600 115	Sidewalk, concrete walk 4", broom finish	SF	3.54	1.92	5.46
01.2600 120	Curb and gutter, 4' sidewalk, mono	LF	16.74	8.21	24.95
01.2600 125	Curb 6", asphaltic concrete	LF	2.78	7.77	10.55
01.2600 130	Curb and gutter, machine work	LF	9.33	4.44	13.78
01.2600 140	Curb forms fabricacted edge 6" x 12"	LF	1.44	2.81	4.25
01.2600 145	Pavements miscellaneous, tennis court, bases, AC, color, seal	SF	7.35		7.35
01.2600 150	Pavements miscellaneous, tennis court, 7200 SF, fence/stripe	SF	77,057.49		77,057.49
01.2600 155	Pavements miscellaneous, parking gravel, 6" of 1 1/2" rock	SF	0.61		0.61
01.2600 160	Sidewalk and concrete finishes, precast pavers, grouted	SF	3.62	7.65	11.27
01.2600 165	Sidewalk and concrete finishes, precast pavers, sand bed, 1" compact	SF	3.00	5.02	8.02
01.2600 170	Sidewalk and concrete finishes, brick pavers, grouted	SF	4.15	8.25	12.39
01.2600 175	brick pavers, sand bed, 1" compact	SF	3.66	6.11	9.78
01.2600 180	Sidewalk and concrete finishes, concrete hardner, iron base	SF	0.38	0.59	0.97
01.2600 185	Sidewalk and concrete finishes, concrete hardner, granolithic	SF	0.01	0.30	0.31
01.2600 190	Sidewalk and concrete finishes, concrete hardner, chemical	SF	0.10	0.30	0.41
01.2600 195	finish broom	SF		0.81	0.81
01.2600 200	Sidewalk and concrete finishes, finish steel trowel	SF		0.81	0.81
01.2700	**SITE UTILITIES**				
01.2700 100	Underground service, concrete encased, 100 Amp, 600 V, 4-THHN #2 copper wire, grounded	LF	34.28	11.04	45.33
01.2700 105	Underground service, concrete encased, 175 Amp, 600 V, 3-THHN 2/0 copper wire, grounded	LF	39.11	11.31	50.42
01.2700 110	Underground service, concrete encased, 400 Amp, 600 V, 3-500 MCM copper wire, grounded	LF	84.41	17.75	102.16
01.2700 115	Underground service, concrete encased, 800 Amp, 600 V, 6-500 MCM copper wire, grounded	LF	146.83	30.88	177.72
01.2700 120	Underground service, concrete encased, 1200 Amp, 600 V, 12-350 MCM copper wire, grounded	LF	197.92	47.51	245.43
01.2700 125	Underground service, fiber conduit, 175 Amp, 600 V, 3-THW/THHN 2/0 copper wire, grounded	LF	19.45	16.42	35.87
01.2700 130	Underground service, fiber conduit, 300 Amp, 600 V, 4-THHN 350 MCM copper wire, grounded	LF	60.97	24.02	84.99
01.2700 135	Underground service, rigid steel conduit, 30 Amp, 600 V, 1-THHN #10 copper wire	LF	10.45	17.24	27.69
01.2700 140	Underground service, rigid steel conduit, 60 Amp, 600 V, 4-THHN #6 copper wire	LF	15.81	21.12	36.93
01.2700 145	Underground service, rigid steel conduit, 100 Amp, 600 V, 4-THHN #2 copper wire	LF	23.34	24.23	47.57
01.2700 150	Underground service, rigid steel conduit, 400 Amp, 600 V, 4-THHN 500 MCM copper wire	LF	98.25	49.39	147.64
01.2700 152	Underground service, rigid steel conduit, 600 Amp, 600 V, 6-THHN 350 MCM copper wire	LF	128.99	81.29	210.28
01.2700 155	Utility pole, with transformer, 250 KVA,	EACH	24,504.67	13,289.59	37,794.26
01.2700 160	Utility pole, with transformer, 100 KVA,	EACH	8,578.69	10,470.32	19,049.01
01.2700 165	Utility pole, with transformer, 75 KVA,	EACH	9,105.39	7,865.51	16,970.89
01.2700 170	Utility pole, with transformer, 25 KVA,	EACH	4,433.79	6,169.22	10,603.01
01.2700 175	Site drainage, underdrain, 4" polyethelene pipe	LF	9.30	26.15	35.45
01.2700 180	Site drainage, underdrain, 6" polyethelene pipe	LF	10.12	26.53	36.65
01.2700 185	Site drainage, underdrain, 10" polyethelene pipe	LF	13.71	27.23	40.94
01.2700 190	Sanitary sewer, cast iron pipe, 4" Diameter	LF	17.96	19.85	37.81
01.2700 192	Sanitary sewer, cast iron pipe, 6" Diameter	LF	31.07	26.74	57.81
01.2700 195	Sanitary sewer, cast iron pipe, 10" Diameter	LF	77.29	32.14	109.43
01.2700 200	Storm drainage, reinforced concrete pipe, 12" Diameter	LF	22.98	26.54	49.52
01.2700 205	Storm drainage, reinforced concrete pipe, 15" Diameter	LF	32.56	32.64	65.20
01.2700 210	Storm drainage, reinforced concrete pipe, 18" Diameter	LF	42.04	40.56	82.60
01.2700 215	Storm drainage, reinforced concrete pipe, 36" Diameter	LF	122.17	67.93	190.10
01.2700 220	Storm drainage, corrugated metal pipe, 8" Diameter	LF	9.30	18.49	27.79
01.2700 225	Storm drainage, corrugated metal pipe, 15" Diameter	LF	12.69	24.78	37.47
01.2700 230	Sanitary sewer, PVC pipe, 4" Diameter	LF	10.37	20.55	30.92
01.2700 235	Sanitary sewer, PVC pipe, 8" Diameter	LF	16.18	25.47	41.65
01.2700 240	Sanitary sewer, PVC pipe, 12" Diameter	LF	30.01	46.14	76.15
01.2700 245	Water service, copper pipe, 2" Diameter	LF	49.20	18.80	68.00
01.2700 250	Water service, copper pipe, 3" Diameter	LF	88.99	21.46	110.45

1.21 - SITEWORK

CSI#	Description	Unit	Material Cost	Install Cost	Total Cost
01.2700 255	Water service, copper pipe, 4" Diameter	LF	137.06	20.46	157.52
01.2700 260	Water service, PVC MUNI C900, 4" Diameter	LF	27.42	15.73	43.14
01.2700 265	Water service, PVC MUNI C900, 6" Diameter	LF	33.28	17.70	50.98
01.2700 270	Water service, PVC MUNI C900, 8" Diameter	LF	43.96	19.53	63.50
01.2700 275	Water service, ductile iron pipe, 3" Diameter	LF	19.79	23.49	43.27
01.2700 280	Water service, ductile iron pipe, 4" Diameter	LF	20.80	23.49	44.29
01.2700 285	Water service, ductile iron pipe, 6" Diameter	LF	22.47	29.57	52.04
01.2700 290	Water service, ductile iron pipe, 8" Diameter	LF	24.86	29.57	54.43
01.2700 295	Gas/Steam service, black steel pipe, wrapped, 1" Diameter	LF	7.46	17.83	25.30
01.2700 300	Gas/Steam service, black steel pipe, wrapped, 2" Diameter	LF	21.03	22.19	43.22
01.2700 305	Gas/Steam service, black steel pipe, wrapped, 3" Diameter	LF	35.19	27.34	62.53
01.2700 310	Gas/Steam service, black steel pipe, wrapped, 4" Diameter	LF	49.09	34.38	83.48
01.2700 315	Fire hydrant, 4 outlets, 2 valves	EACH	3,643.37	2,162.47	5,805.85
01.2700 320	Catch basin, with grate, 2'X2'X2'	EACH	628.47	500.75	1,129.21
01.2700 325	Catch basin, with grate, 2'X2'X4'	EACH	794.20	779.33	1,573.53
01.2700 330	Drop inlet, precast with grate, 12"X12"X4"	EACH	396.75	321.03	717.78
01.2700 335	Drop inlet, precast with grate, 16"X16"X5"	EACH	521.49	401.28	922.77
01.2700 340	Manhole, with lid, 4' diameter X 6'-8' deep	EACH	1,791.56	1,348.08	3,139.64
01.2700 345	Manhole, with lid, 4' diameter X 6'-8' deep	EACH	1,791.56	1,348.08	3,139.64
01.2700 350	Fuel storage tank, underground, 500 gallon	EACH	9,669.08	730.92	10,400.00
01.2700 355	Fuel storage tank, underground, 1000 gallon	EACH	12,313.92	913.65	13,227.56
01.2700 360	Fuel storage tank, underground, 2000 gallon	EACH	15,798.97	998.44	16,797.41
01.2800	**SITE LIGHTING**				
01.2800 100	Site lighting, underground wire, 30' pole, mercury vapor, 175 watt	EACH	1,376.77	2,446.16	3,822.93
01.2800 105	Site lighting, underground wire, 30' pole, mercury vapor, 1000 watt	EACH	2,173.47	2,535.18	4,708.65
01.2800 110	Site lighting, underground wire, 42" bollards incandescent, 150 watt	EACH	912.22	311.67	1,223.89
01.2800 115	Site lighting, underground wire, 42" bollards metal halide, 175 watt	EACH	1,171.26	311.67	1,482.93
01.2800 120	Site lighting, parking lot, small	SF	0.78	0.85	1.63
01.2800 125	Site lighting, parking lot, large	SF	0.51	0.49	1.00

SUBSTRUCTURE - 2.1

CSI#	Description	Unit	Material Cost	Install Cost	Total Cost
	The costs in this section include materials, labor, equipment rental, supervision, and subcontractor overhead and profit. There are no allowances for the general contractor.				
	These costs are typical of those associated with the construction of a building.				
	Costs are complete, and represent normal conditions related to weather and soil. For cast in place concrete items, allowances for forms and finishing are included. Pile costs include equipment costs and below grade walls include waterproofing and an allowance for excavation.				
	Foundation wall and footing costs are combined as one assembly and include trenching and backfill. The costs, for these items, should be adjusted for excessive climate conditions and non-standard construction practices.				
02.1100	**EXCAVATION FOR BUILDINGS**				
02.1100 100	Excavation & backfill, sand or gravel, 4' deep, no removal	CY	14.34	39.78	54.12
02.1100 105	Excavation & backfill, sand or gravel, 4' deep, remove excess	CY	17.63	42.73	60.36
02.1100 110	Excavation & backfill, clay, 4' deep, gravel backfill	CY	14.34	39.78	54.12
02.1100 115	Excavation & backfill, sand or gravel, 8' deep, no removal	CY	17.04	42.69	59.73
02.1100 120	Excavation & backfill, sand or gravel, 8' deep, remove excess	CY	20.33	45.64	65.97
02.1100 125	Excavation & backfill, clay, 8' deep, gravel backfill	CY	17.04	42.69	59.73
02.1100 130	Excavation & backfill, sand or gravel, 16' deep, no removal	CY	17.04	42.69	59.73
02.1100 135	Excavation & backfill, sand or gravel, 16' deep, remove excess	CY	20.33	45.64	65.97
02.1100 140	Excavation & backfill, clay, 16' deep, gravel backfill	CY	14.34	39.78	54.12
02.1200	**PILES**				
02.1200 100	Piles, precast concrete, square 10"	LF	35.15	17.99	53.14
02.1200 105	Piles, precast concrete, square 12"	LF	40.89	20.53	61.42
02.1200 110	Piles, precast concrete, square 16"	LF	41.95	21.92	63.87
02.1200 115	Piles, precast concrete, square 18"	LF	45.29	23.07	68.36
02.1200 120	Piles, steel H section, 8"x8"x36#	LF	28.42	22.48	50.89
02.1200 125	Piles, steel H section, 10"x10"x57#	LF	44.63	25.27	69.90
02.1200 130	Piles, steel H section, 12"x12"x74#	LF	54.78	26.63	81.41
02.1200 135	Piles, steel H section, 14"x14"x89#	LF	63.88	28.06	91.94
02.1200 140	Piles, pipe, 12", concrete filled	LF	18.40	23.57	41.97
02.1200 145	Piles, pipe, 12", unfilled	LF	15.83	21.03	36.86
02.1200 150	Piles, pipe, 16", concrete filled	LF	26.58	25.27	51.85
02.1200 155	Piles, pipe, 16", unfilled	LF	22.72	22.49	45.21
02.1200 160	Piles, steel step tapered, concrete filled, 12" butt	LF	15.99	16.54	32.53
02.1200 165	Piles, steel step tapered, concrete filled, 16" butt	LF	25.83	18.74	44.57
02.1200 170	Piles, wood, untreated to 39', 12" butt	LF	5.42	16.54	21.96
02.1200 175	Piles, wood, untreated to 70', 13" butt	LF	5.82	17.99	23.81
02.1200 180	Piles, wood, treated to 39', 12" butt	LF	7.66	16.54	24.20
02.1200 185	Piles, wood, treated to 70', 13" butt	LF	8.53	17.99	26.52
02.1200 190	Piles, soldier, steel, recovered, no lagging, 15'max, pulled	LF	22.45	13.51	35.96
02.1200 195	Piles, soldier, steel, recovered, no lagging, 20'max, pulled	LF	14.43	10.84	25.28
02.1200 200	Piles, soldier, steel, recovered, no lagging, 30'max, pulled	LF	13.85	10.08	23.93
02.1200 205	Piles, soldier, steel, recovered, no lagging, 50'max, pulled	LF	14.91	11.03	25.94
02.1300	**FOUNDATIONS AND PILE CAPS**				
02.1300 100	Foundation, residential, continuous footing, base 8"x16", wall 8" wide, 20" deep	LF	22.18	41.44	63.62
02.1300 105	Foundation, commercial, continuous footing, base 12"x24", wall 12" wide, 24" deep	LF	36.11	66.94	103.05
02.1300 110	Foundation, institutional, continuous footing, base 18"x36", wall 18" wide, 30" deep	LF	94.12	199.61	293.73
02.1300 115	Foundation, continuous footing, wall 8" wide, add for each additional foot deep	LF	22.18	41.44	63.62
02.1300 120	Foundation, continuous footing, wall 12" wide, add for each additional foot deep	LF	11.37	31.64	43.01
02.1300 125	Foundation, continuous footing, wall 18" wide, add for each additional foot deep	LF	15.22	33.60	48.81
02.1300 130	Foundation, spread footing, 2'x2'x12" deep, 3000 psi	EA	41.07	125.05	166.12
02.1300 135	Foundation, spread footing, 3'x3'x12" deep, 3000 psi	EA	84.62	196.10	280.72
02.1300 140	Foundation, spread footing, 4'x4'x12" deep, 3000 psi	EA	153.31	275.38	428.69
02.1300 145	Foundation, spread footing, 6'x6'x12" deep, 3000 psi	EA	323.88	450.10	773.98
02.1300 155	Foundation, spread footing, 6'x6'x18" deep, 3000 psi	EA	473.02	665.41	1,138.43
02.1300 160	Foundation, spread footing, 8'x8'x18" deep, 3000 psi	EA	813.55	955.44	1,768.99

2.1 - SUBSTRUCTURE

CSI#	Description	Unit	Material Cost	Install Cost	Total Cost
02.1300 165	Foundation, spread footing, 10'x10'x24" deep, 3000 psi	EA	1,681.21	1,728.73	3,409.94
02.1300 170	Foundation, spread footing, 12'x12'x24" deep, 3000 psi	EA	2,374.08	2,197.03	4,571.11
02.1300 175	Foundation, pilecap, 6'x4'x24" deep, cap for two piles, 3000 psi	EA	435.54	535.14	970.68
02.1300 180	Foundation, pilecap, 6'x4'x36" deep, cap for two piles, 3000 psi	EA	634.90	799.73	1,434.63
02.1300 185	Foundation, pilecap, 6'x6'x24" deep, cap for 4 piles, 3000 psi	EA	630.86	687.72	1,318.58
02.1300 195	Foundation, pilecap, 6'x6'x36" deep, cap for 4 piles, 3000 psi	EA	940.83	1,028.88	1,969.71
02.1300 200	Foundation, pilecap, 8'x6'x36" deep, cap for 6 piles, 3000 psi	EA	1,241.65	1,256.97	2,498.62
02.1300 205	Foundation, pilecap, 8'x8'x36" deep, cap for 8 piles, 3000 psi	EA	1,631.15	1,521.67	3,152.83
02.1300 210	Foundation, pilecap, 12'x8'x42" deep, cap for 10 piles, 3000 psi	EA	2,420.37	2,073.75	4,494.12
02.1300 215	Foundation, underpinning, 6' deep, complete	LF	311.63	456.76	768.39
02.1300 220	Foundation, underpinning, each additional foot deep, complete	LF	42.18	69.76	111.93
02.1300 225	Foundation, cast in place concrete pier, poured neat, 16"	LF	19.83	12.49	32.32
02.1300 230	Foundation, cast in place concrete pier, poured neat, 24"	LF	39.13	21.28	60.41
02.1300 235	Foundation, cast in place concrete pier, poured neat, 36"	LF	83.42	40.99	124.41
02.1400	**WALLS BELOW GRADE**				
02.1400 100	Walls below grade, reinforced concrete, finished one side, 8" thick	SF	11.74	30.02	41.76
02.1400 105	Walls below grade, reinforced concrete, finished one side, 12" thick	SF	13.16	33.65	46.81
02.1400 110	Walls below grade, concrete block, reinforced, filled, 8" thick	SF	14.70	15.84	30.54
02.1400 115	Walls below grade, concrete block, reinforced, filled, 12" thick	SF	14.98	16.15	31.13
02.1500	**WATERPROOFING**				
02.1500 100	Waterproofing, hot coatings, vertical surfaces, bitumals, walls, per coat	SF	0.05	1.14	1.19
02.1500 105	Waterproofing, hot coatings, vertical surfaces, hot mop, 15# felt, 2 ply	SF	0.27	1.60	1.87
02.1500 110	Waterproofing, hot coatings, vertical surfaces, hot mop, 30# felt, 2 ply	SF	0.27	1.60	1.87
02.1500 115	Waterproofing, hot coatings, vertical surfaces, hot mop, glass fiber, 2 ply	SF	0.94	1.60	2.54
02.1500 120	Waterproofing, cold application, verticals, waterproof, 1/32" butyl	SF	0.80	1.97	2.77
02.1500 125	Waterproofing, cold application, verticals, waterproof, 1/16" butyl	SF	1.17	1.97	3.15
02.1500 130	Waterproofing, cold application, verticals, waterproof, 1/16" neoprene	SF	1.21	1.97	3.18
02.1500 135	Waterproofing, cold application, verticals, elastic polyurethane, 60 ML, vertical	SF	1.50	2.62	4.12
02.1500 140	Waterproofing, emulsion, verticals, waterproof, asphalt mastic, 1/8"	SF	0.33	1.22	1.56
02.1500 145	Waterproofing, emulsion, verticals, waterproof, asphalt paint, brush/coat	SF	0.12	0.47	0.59
02.1500 150	Waterproofing, emulsion, verticals, waterproof, silicone, spray, 2 coat	SF	0.28	0.47	0.75
02.1500 155	Waterproofing, specialties, verticals, volclay bent panels, 3/16"	SF	1.63	1.26	2.88
02.1500 160	Waterproofing, specialties, verticals, volclay bent panels, 9/16"	SF	2.08	1.26	3.34
02.1500 165	Waterproofing, specialties, verticals, bentonite panel, 3/8"	SF	2.80	1.39	4.19
02.1500 170	Waterproofing, specialties, verticals, exterior, cement pargetting, 1/2", 2 coat	SF	0.53	1.95	2.48
02.1500 175	Waterproofing, specialties, verticals, ironite, 2 coats	SF	0.82	3.22	4.04
02.1600	**SUBSTRUCTURE DRAINAGE**				
02.1600 100	Underdrain, corrugated polyethylene pipe, perforated, 3" diameter	LF	0.47	8.92	9.40
02.1600 105	Underdrain, corrugated polyethylene pipe, perforated, 4" diameter	LF	0.63	9.15	9.78
02.1600 110	Underdrain, corrugated polyethylene pipe, perforated, 6" diameter	LF	1.44	9.51	10.95
02.1600 115	Underdrain, corrugated polyethylene pipe, perforated, 8" diameter	LF	2.58	9.78	12.36
02.1600 120	Underdrain, corrugated polyethylene pipe, perforated, 10" diameter	LF	4.98	10.19	15.17
02.1600 125	Underdrain, corrugated polyethylene pipe, perforated, 12" diameter	LF	7.10	10.50	17.59
02.1700	**SLABS ON GRADE**				
02.1700 100	Slab on grade, reinforced concrete, 4" thick	SF	3.48	3.20	6.68
02.1700 105	Slab on grade, reinforced concrete, 5" thick	SF	4.00	3.32	7.32
02.1700 110	Slab on grade, reinforced concrete, 6" thick	SF	4.85	3.79	8.64
02.1700 115	Slab on grade, reinforced concrete, kalman floor, 6" thick	SF	5.44	5.92	11.36

STRUCTURE - 3.0

CSI#	Description	Unit	Material Cost	Install Cost	Total Cost
	The costs in this section include materials, labor, equipment rental, supervision, and subcontractor overhead and profit. There are no allowances for the general contractor.				
	These costs are typical of those associated with the construction of a building.				
	Costs are complete, and represent normal conditions related to weather and soil. For cast in place concrete items, allowances for forms and finishing are included.				
	Concrete columns include allowances for forms and reinforcing. All column costs have a pro-rated allowance for column base. Steel members include allowances for welding where appropriate.				
	Roof structure costs do not include roof cover or insulation. These costs are found in section 4.2, Horizontal Enclosure.				
03.0105	**COLUMNS, CAST IN PLACE**				
03.0105 100	Concrete Column cast in place, round, 12" diameter	LF	28.14	30.90	59.04
03.0105 105	Concrete Column, cast in place, round, 14" diameter	LF	37.37	36.43	73.80
03.0105 110	Concrete Column, cast in place, round, 16" diameter	LF	47.37	42.40	89.78
03.0105 115	Concrete Column, cast in place, round, 20" diameter	LF	73.64	59.59	133.23
03.0105 120	Concrete Column, cast in place, round, 24" diameter	LF	101.50	77.42	178.92
03.0105 125	Concrete Column, cast in place, round, 30" diameter	LF	155.07	111.79	266.86
03.0105 130	Concrete Column, cast in place, round, 36" diameter	LF	213.37	151.93	365.30
03.0105 135	Concrete Column, cast in place, square, 10"x10"	LF	19.94	48.95	68.89
03.0105 140	Concrete Column, cast in place, square, 12"x12"	LF	26.39	60.21	86.60
03.0105 145	Concrete Column, cast in place, square, 16"x16"	LF	28.26	77.75	106.01
03.0105 150	Concrete Column, cast in place, square, 20"x20"	LF	59.05	110.05	169.10
03.0105 155	Concrete Column, cast in place, square, 24"x24"	LF	80.61	139.48	220.09
03.0105 160	Concrete Column, cast in place, square, 30"x30"	LF	119.21	187.88	307.10
03.0105 165	Concrete Column, cast in place, square, 36"x36"	LF	165.66	242.18	407.84
03.0110	**COLUMNS, PRECAST**				
03.0110 100	Column, Precast concrete, 12"x12",	LF	92.49	13.22	105.71
03.0110 105	Column, Precast concrete, 14"x14"	LF	129.40	18.50	147.90
03.0110 110	Column, Precast concrete, 16"x16"	LF	157.53	22.52	180.05
03.0110 115	Column, Precast concrete, 18"x18"	LF	182.29	26.06	208.35
03.0110 120	Column, Precast concrete, 20"x20"	LF	196.91	28.15	225.06
03.0115	**COLUMNS, STEEL**				
03.0115 100	Steel column, wide flange, W6x20	LF	74.44	14.36	88.80
03.0115 105	Steel column, wide flange, W8x24	LF	85.53	16.03	101.56
03.0115 110	Steel column, wide flange, W10x45	LF	143.80	24.77	168.56
03.0115 115	Steel column, wide flange, W12x72	LF	237.79	36.31	274.10
03.0115 120	Steel column, wide flange, W14x120	LF	370.96	56.29	427.25
03.0115 125	Steel column, wide flange, W14x176	LF	526.33	79.60	605.93
03.0115 130	Steel column, pipe, 3" diameter	LF	36.48	11.60	48.08
03.0115 135	Steel column, pipe, 4" diameter	LF	43.92	13.97	57.88
03.0115 140	Steel column, pipe, 6" diameter	LF	62.88	19.98	82.86
03.0115 145	Steel column, pipe, 8" diameter	LF	82.76	16.35	99.11
03.0115 150	Steel column, pipe, 10" diameter	LF	109.41	20.67	130.08
03.0115 155	Steel column, pipe, 12" diameter	LF	165.16	29.69	194.84
03.0115 160	Steel column, pipe, concrete filled, 3" diameter	LF	37.14	11.89	49.02
03.0115 165	Steel column, pipe, concrete filled, 4" diameter	LF	43.97	14.10	58.07
03.0115 170	Steel column, pipe, concrete filled, 6" diameter	LF	63.00	20.28	83.27
03.0115 175	Steel column, pipe, concrete filled, 8" diameter	LF	82.98	16.90	99.88
03.0115 180	Steel column, pipe, concrete filled, 10" diameter	LF	109.81	21.51	131.32
03.0115 190	Steel column, pipe, concrete filled, 12" diameter	LF	165.66	30.93	196.60
03.0115 195	Steel column, square tube, 3"	LF	38.39	9.73	48.12
03.0115 200	Steel column, square tube, 5"	LF	52.81	12.47	65.28
03.0115 205	Steel column, square tube, 6"	LF	60.34	13.89	74.23
03.0115 210	Steel column, square tube, 8"	LF	92.01	19.89	111.90
03.0115 215	Steel column, square tube, 10"	LF	133.12	27.69	160.81

3.0 - STRUCTURE

CSI#	Description	Unit	Material Cost	Install Cost	Total Cost
03.0115 220	Steel column, square tube, 12"	LF	183.34	37.22	220.57
03.0115 225	Steel column, square tube, concrete filled, 3"	LF	61.72	14.34	76.06
03.0115 230	Steel column, square tube, concrete filled, 6"	LF	61.64	14.88	76.52
03.0115 235	Steel column, square tube, concrete filled, 8"	LF	94.32	21.63	115.95
03.0115 240	Steel column, square tube, concrete filled, 10"	LF	136.73	30.42	167.15
03.0115 245	Steel column, square tube, concrete filled, 12"	LF	188.70	41.25	229.95
03.0120	**COLUMNS, WOOD**				
03.0120 100	Wood column, 4"x4" post	LF	2.86	8.51	11.37
03.0120 105	Wood column, 4"x6" post	LF	3.96	12.63	16.59
03.0120 110	Wood column, 4"x8" post	LF	36.97	17.71	54.68
03.0120 115	Wood column, 6"x6" post	LF	5.53	18.52	24.05
03.0120 120	Wood column, 6"x8" post	LF	38.82	19.48	58.30
03.0120 125	Wood column, 6"x10" post	LF	10.86	22.27	33.13
03.0120 130	Wood column, 6"x12" post	LF	12.37	26.69	39.07
03.0120 135	Wood column, 8"x8" post	LF	11.48	22.70	34.18
03.0120 140	Wood column, 8"x10" post	LF	13.44	28.06	41.51
03.0120 145	Wood column, 8"x12" post	LF	15.60	33.97	49.57
03.0120 150	Wood column, 12"x12" post	LF	21.78	50.87	72.65
03.0205	**BEAMS & GIRDERS, CAST IN PLACE**				
03.0205 100	Concrete beams & girders, cast in place, 12"x24"	LF	47.53	185.48	233.01
03.0205 110	Concrete beams & girders, cast in place, 12"x30"	LF	61.29	254.38	315.68
03.0205 115	Concrete beams & girders, cast in place, 12"x36"	LF	72.33	292.66	364.99
03.0205 120	Concrete beams & girders, cast in place, 18"x24"	LF	70.03	261.77	331.80
03.0205 125	Concrete beams & girders, cast in place, 18"x30"	LF	85.46	303.72	389.18
03.0205 130	Concrete beams & girders, cast in place, 18"x36"	LF	97.53	343.24	440.78
03.0210	**BEAMS & GIRDERS, PRECAST**				
03.0210 100	Concrete beams & girders, precast, rectangular, 12"x24"	LF	71.39	9.53	80.92
03.0210 105	Concrete beams & girders, precast, rectangular, 12"x30"	LF	89.72	11.97	101.69
03.0210 110	Concrete beams & girders, precast, rectangular, 12"x36"	LF	107.08	14.29	121.38
03.0210 115	Concrete beams & girders, precast, rectangular, 18"x24"	LF	107.08	14.29	121.38
03.0210 120	Concrete beams & girders, precast, rectangular, 18"x30"	LF	133.13	17.77	150.90
03.0210 125	Concrete beams & girders, precast, rectangular, 18"x36"	LF	161.12	21.50	182.62
03.0210 130	Concrete beams & girders, precast, inverted tee, 12"x24"	LF	83.27	11.91	95.17
03.0210 135	Concrete beams & girders, precast, inverted tee, 12"x30"	LF	104.65	14.96	119.61
03.0210 140	Concrete beams & girders, precast, inverted tee, 12"x36"	LF	124.89	17.86	142.75
03.0210 145	Concrete beams & girders, precast, inverted tee, 18x24"	LF	124.89	17.86	142.75
03.0210 150	Concrete beams & girders, precast, inverted tee, 18x30"	LF	155.28	22.20	177.48
03.0210 155	Concrete beams & girders, precast, inverted tee, 18x36"	LF	187.91	26.87	214.78
03.0215	**BEAMS & GIRDERS, STEEL**				
03.0215 100	Steel beams & girders, wide flange, W6x20	LF	68.54	13.97	82.50
03.0215 105	Steel beams & girders, wide flange, W8x24	LF	65.79	13.41	79.20
03.0215 110	Steel beams & girders, wide flange, W10x45	LF	123.36	25.14	148.50
03.0215 115	Steel beams & girders, wide flange, W12x72	LF	197.38	40.22	237.60
03.0215 120	Steel beams & girders, wide flange, W14x109	LF	298.81	60.89	359.71
03.0215 125	Steel beams & girders, wide flange, W14x176	LF	482.49	98.32	580.81
03.0220	**BEAMS & GIRDERS, WOOD**				
03.0220 100	Wood beams & girders, 4"x4"	LF	2.03	7.66	9.68
03.0220 105	Wood beams & girders, 4"x6"	LF	3.13	11.78	14.90
03.0220 110	Wood beams & girders, 4"x8"	LF	4.22	15.91	20.12
03.0220 115	Wood beams & girders, 6"x6"	LF	4.55	13.25	17.80
03.0220 120	Wood beams & girders, 6"x8"	LF	6.07	17.67	23.74
03.0220 125	Wood beams & girders, 6"x10"	LF	7.59	22.08	29.67
03.0220 130	Wood beams & girders, 6"x12"	LF	9.11	26.51	35.61
03.0220 135	Wood beams & girders, 8"x8"	LF	8.17	22.38	30.55
03.0220 140	Wood beams & girders, 8"x10"	LF	10.17	27.88	38.05
03.0220 145	Wood beams & girders, 8"x12"	LF	12.33	33.78	46.11
03.0220 150	Wood beams & girders, 12"x12"	LF	18.51	50.68	69.19
03.0220 155	Wood beams & girders, glu-lams, 3 1/8"x12"	LF	8.26	1.91	10.16

STRUCTURE - 3.0

CSI#	Description	Unit	Material Cost	Install Cost	Total Cost
03.0220 160	Wood beams & girders, glu-lams, 3 1/8"x18"	LF	17.13	3.95	21.09
03.0220 165	Wood beams & girders, glu-lams, 5 1/8"x12"	LF	13.54	3.13	16.67
03.0220 170	Wood beams & girders, glu-lams, 5 1/8"x18"	LF	20.30	4.69	24.99
03.0220 175	Wood beams & girders, glu-lams, 6 3/4"x12"	LF	17.85	4.13	21.98
03.0220 180	Wood beams & girders, glu-lams, 6 3/4"x12"	LF	26.70	6.17	32.87
03.0220 185	Wood beams & girders, glu-lams, 6 3/4"x18"	LF	40.19	9.29	49.47
03.0220 190	Wood beams & girders, glu-lams, 8 3/4"x12"	LF	23.13	5.35	28.48
03.0220 195	Wood beams & girders, glu-lams, 8 3/4"x18	LF	34.64	8.01	42.64
03.0220 200	Wood beams & girders, glu-lams, 8 3/4"x24	LF	46.27	10.69	56.96
03.0305	**FLOORS, CAST IN PLACE**				
03.0305 100	Floor, cast in place, slab, one way beams, with topping, 6" thick	SF	7.33	11.01	18.34
03.0305 105	Floor, cast in place, slab, one way beams, with topping, 8" thick	SF	9.74	14.62	24.37
03.0305 110	Floor, cast in place, slab, one way beams, with topping, 10" thick	SF	12.22	18.36	30.58
03.0305 115	Floor, cast in place, slab, two way beams, with topping, 6" thick	SF	8.75	13.17	21.92
03.0305 120	Floor, cast in place, slab, two way beams, with topping, 8" thick	SF	11.63	17.48	29.11
03.0305 125	Floor, cast in place, slab, two way beams, with topping, 10" thick	SF	14.59	21.94	36.53
03.0305 130	Floor, cast in place, slab, flat, with topping, 8" thick	SF	7.22	10.88	18.10
03.0305 135	Floor, cast in place, slab, flat, with topping, 12" thick	SF	9.89	25.85	35.74
03.0305 140	Floor, cast in place, slab, post tension with topping, 7" thick	SF	7.55	11.44	18.99
03.0305 145	Floor, cast in place, slab, with steel beam jacket topping, 6" thick	SF	7.09	10.73	17.83
03.0305 150	Floor, cast in place, slab, with steel beam jacket topping, 8" thick	SF	9.53	14.43	23.96
03.0305 155	Floor, cast in place, slab, on metal form, topping, 6" thick	SF	6.63	10.03	16.66
03.0305 160	Floor, cast in place, slab, on metal form, topping, 8" thick	SF	8.90	13.48	22.38
03.0305 165	Floor, cast in place, waffle slab, 30" square pan, 3" thick, 6"x12" joist	SF	9.20	13.90	23.10
03.0305 170	Floor, cast in place, waffle slab, 30" square pan, 3" thick, 6"x20" joist	SF	9.94	15.02	24.96
03.0305 175	Floor, cast in place, waffle slab, 20" square pan, 4-1/2" thick, 6"x12" joist	SF	10.69	16.17	26.86
03.0305 180	Floor, cast in place, waffle slab, 20" square pan, 4-1/2" thick, 6"x20" joist	SF	13.76	20.80	34.56
03.0310	**FLOORS, PRECAST**				
03.0310 100	Floor, precast concrete, single tees, with topping	SF	12.43	5.27	17.70
03.0310 105	Floor, precast concrete, double tees, with topping	SF	10.22	3.60	13.82
03.0310 110	Floor, precast concrete, plank, with topping, 6" solid	SF	6.94	2.87	9.81
03.0310 115	Floor, precast concrete, hollow plank, with topping, 4" thick	SF	6.37	2.32	8.69
03.0310 120	Floor, precast concrete, hollow plank, with topping, 6" thick	SF	6.73	2.59	9.32
03.0310 125	Floor, precast concrete, hollow plank, with topping, 8" thick	SF	7.19	2.87	10.06
03.0310 130	Floor, precast concrete, hollow plank, with topping, 10" thick	SF	7.64	3.24	10.88
03.0315	**FLOORS, CONCRETE ON STEEL JOISTS**				
03.0315 100	Floor, concrete on metal deck, open web steel joists, 40 PSF load	SF	13.18	4.39	17.57
03.0315 105	Floor, concrete on metal deck, open web steel joists, 75 PSF load	SF	15.02	4.80	19.82
03.0315 110	Floor, concrete on metal deck, open web steel joists, 125 PSF load	SF	17.82	5.49	23.31
03.0315 115	Floor, concrete slab, steel beam and deck, 40 PSF load	SF	24.97	7.22	32.19
03.0315 120	Floor, concrete slab, steel beam and deck, 75 PSF load	SF	30.36	8.42	38.79
03.0315 125	Floor, concrete slab, steel beam and deck, 125 PSF load	SF	33.48	9.17	42.65
03.0315 130	Floor, concrete slab, steel beam and deck, 200 PSF load	SF	42.22	11.22	53.44
03.0320	**FLOORS, WOOD**				
03.0320 100	Floor, wood, plywood, 2"x6" joists 16" OC	SF	2.73	3.25	5.98
03.0320 105	Floor, wood, plywood, 2"x8" joists 16" OC	SF	3.19	4.49	7.68
03.0320 110	Floor, wood, plywood, 2"x10" joists 16" OC	SF	3.46	4.63	8.09
03.0320 115	Floor, wood, plywood, 2"x12" joists 16" OC	SF	3.70	4.98	8.68
03.0320 120	Floor, wood, plywood, 2"x14" joists 16" OC	SF	4.02	5.73	9.75
03.0320 125	Add to above, 1" diagonal sheathing	SF	0.17	0.33	0.49
03.0320 130	Add to above, 2" tongue & groove sheathing	SF	0.97	0.25	1.22
03.0325	**ROOFS, METAL DECK ON STEEL JOISTS**				
03.0325 100	Roof, metal deck, open web steel joists, 20' span	SF	6.07	1.67	7.73
03.0325 105	Roof, metal deck, open web steel joists, 30' span	SF	8.72	2.18	10.90
03.0325 110	Roof, metal deck, open web steel joists, 40' span	SF	9.95	2.46	12.42
03.0325 115	Roof, metal deck, open web steel joists, 50 span	SF	12.97	3.11	16.08
03.0325 120	Roof, metal deck, open web steel joists, 60 span	SF	15.08	3.55	18.63
03.0325 125	Roof, metal deck, open web steel joists, 70' span	SF	16.32	3.83	20.15

3.0 - STRUCTURE

CSI#	Description	Unit	Material Cost	Install Cost	Total Cost
03.0330	**ROOFS, WOOD**				
03.0330 100	Roof, plywood, 2"x6" wood rafters 24" OC flat	SF	2.38	5.89	8.27
03.0330 105	Roof, plywood, 2"x6" wood rafters 24" OC 4 in 12 slope	SF	2.51	6.22	8.73
03.0330 110	Roof, plywood, 2"x6" wood rafters 24" OC 8 in 12 slope	SF	2.85	7.09	9.94
03.0330 115	Roof, plywood, 2"x6" wood rafters 24" OC 12 in 12 slope	SF	3.36	8.34	11.69
03.0330 120	Roof, plywood, 2"x8" wood rafters 24" OC flat	SF	2.71	7.56	10.27
03.0330 125	Roof, plywood, 2"x8" wood rafters 24" OC 4 in 12 slope	SF	2.85	7.97	10.83
03.0330 130	Roof, plywood, 2"x8" wood rafters 24" OC 8 in 12 slope	SF	3.25	9.09	12.34
03.0330 135	Roof, plywood, 2"x8" wood rafters 24" OC 12 in 12 slope	SF	3.82	10.69	14.51
03.0330 140	Roof, plywood, 2"x10" wood rafters 24" OC flat	SF	3.08	9.05	12.13
03.0330 145	Roof, plywood, 2"x10" wood rafters 24" OC 4 in 12 slope	SF	3.25	9.54	12.79
03.0330 150	Roof, plywood, 2"x10" wood rafters 24" OC 8 in 12 slope	SF	3.70	10.89	14.59
03.0330 155	Roof, plywood, 2"x10" wood rafters 24" OC 12 in 12 slope	SF	4.35	12.79	17.14
03.0330 160	Roof, plywood, 2"x12" wood rafters 24" OC flat	SF	3.38	10.53	13.91
03.0330 165	Roof, plywood, 2"x12" wood rafters 24" OC 4 in 12 slope	SF	3.57	11.11	14.67
03.0330 170	Roof, plywood, 2"x12" wood rafters 24" OC 8 in 12 slope	SF	4.05	12.66	16.71
03.0330 175	Roof, plywood, 2"x12" wood rafters 24" OC 12 in 12 slope	SF	4.78	14.89	19.66
03.0400	**FLOOR, TOPPING**				
03.0400 100	Monolithic topping, 1/16"	SF	0.08	1.51	1.59
03.0400 105	Monolithic topping, 3/16"	SF	0.32	1.59	1.92
03.0400 110	Monolithic topping, 1/2"	SF	0.64	1.68	2.32
03.0400 115	White cement	SF	21.62		21.62
03.0400 120	Felton sand	SF	72.48		72.48
03.0400 125	Wear course, monolithic rock, 3/8" thick	SF	0.43	1.46	1.89
03.0400 130	Wear course, Kal Man, 3/4" thick	SF	0.90	2.65	3.56
03.0400 135	Hardener, chemical	SF	0.10	0.30	0.41
03.0400 140	Hardener, granolithic	SF	0.01	0.30	0.31
03.0400 145	Hardener, iron base	SF	0.38	0.59	0.97
03.0500	**FIREPROOFING**				
03.0500 100	Fireproof, columns, sprayed fiber, 1-3/8", no finish	SYCA	13.26	26.08	39.34
03.0500 105	Fireproof, beams and girders, sprayed fiber, 1-3/8", no finish	SYCA	12.06	20.47	32.53
03.0500 110	Fireproof, metal deck, sprayed fiber, 1-/8", no finish	SYCA	9.77	16.72	26.48
03.0500 115	Fireproof, columns, 1 hour, furring, lath & plaster, painted	LF	42.67	71.52	114.19
03.0500 120	Fireproof, beams & girders, framing, lath & plaster, painted	LF	52.21	87.51	139.72
03.0500 125	Fireproof, deck, 1 hour, furring, lath & plaster, paint	SF	6.94	11.75	18.69
03.0500 130	Fireproof, columns, 1 hour, furring, gypsum board, paint	LF	24.93	38.30	63.22
03.0500 135	Fireproof, beams & girders, 1 hour, framing, gypsum board, paint	LF	31.16	47.87	79.03
03.0500 140	Fireproof, deck, 1 hour, furring, gypsum board, paint	SF	2.81	4.41	7.22
03.0500 145	Fireproof, columns, cementitious monokote	TON	150.06	224.57	374.62
03.0500 150	Fireproof, beams & girders, cementitious monokote	TON	150.06	224.57	374.62
03.0500 155	Fireproof, metal deck, cementitious monokote	SF	0.88	1.63	2.51

ENCLOSURE, VERTICAL - 4.1

CSI#	Description	Unit	Material Cost	Install Cost	Total Cost
	The costs in this section include materials, labor, equipment rental, supervision, and subcontractor overhead and profit. There are no allowances for the general contractor.				
	These costs are typical of those associated with the construction of a building.				
	Costs are complete, and represent normal conditions related to weather and soil. For cast in place concrete items, allowances for forms and finishing are included. Precast and tilt-up costs include equipment costs.				
	Curtain wall costs must be assembled to be complete, i.e.. the frame costs are separate from the glazing. Brick and concrete block costs represent standard running bond. For other bonding patterns the costs should be adjusted to account for additional installation time requirements.				
	The fenestration costs are a combination of framing and glazing.				
04.1105	**WALLS, CAST IN PLACE**				
04.1105 100	Concrete Wall, cast in place, cut & patch, 8' high, 6" thick	SF	8.33	29.85	38.18
04.1105 105	Concrete Wall, cast in place, cut & patch, 8' high, 8" thick	SF	10.01	30.59	40.60
04.1105 110	Concrete Wall, cast in place, cut & patch, 8' high, 10" thick	SF	11.96	31.67	43.62
04.1105 115	Concrete Wall, cast in place, cut & patch, 8' high, 12" thick	SF	14.81	34.00	48.81
04.1105 120	Concrete Wall, cast in place, cut & patch, 12' high, 6" thick	SF	8.51	30.59	39.11
04.1105 125	Concrete Wall, cast in place, cut & patch, 12' high, 8" thick	SF	10.18	31.34	41.52
04.1105 130	Concrete Wall, cast in place, cut & patch, 12' high, 10" thick	SF	12.13	32.41	44.55
04.1105 135	Concrete Wall, cast in place, cut & patch, 12' high, 12" thick	SF	14.46	33.76	48.22
04.1105 140	Concrete Wall, cast in place, cut & patch, 16' high, 6" thick	SF	9.52	34.05	43.57
04.1105 145	Concrete Wall, cast in place, cut & patch, 16' high, 8" thick	SF	11.12	34.76	45.88
04.1105 150	Concrete Wall, cast in place, cut & patch, 16' high, 10" thick	SF	13.15	35.85	49.00
04.1105 155	Concrete Wall, cast in place, cut & patch, 16' high, 12" thick	SF	15.39	37.18	52.57
04.1105 200	Finish Add, one side, concrete wall, cast in place, sandblast, light	SF	0.19	1.92	2.10
04.1105 205	Finish Add, one side, concrete wall, cast in place, sandblast, medium	SF	0.41	2.73	3.14
04.1105 210	Finish Add, one side, concrete wall, cast in place, sandblast, heavy	SF	0.73	3.31	4.04
04.1105 215	Finish Add, one side, concrete wall, cast in place, bush hammer, light	SF	0.73	1.93	2.66
04.1105 220	Finish Add, one side, concrete wall, cast in place, bush hammer, medium	SF	1.52	3.31	4.83
04.1105 225	Finish Add, one side, concrete wall, cast in place, bush hammer, heavy	SF	2.85	4.70	7.55
04.1105 230	Design mix Add, concrete wall, cast in place, Hi Early strength, 6" thick	SF	0.23		0.23
04.1105 235	Design mix Add, concrete wall, cast in place, Hi Early strength, 8" thick	SF	0.32		0.32
04.1105 240	Design mix Add, concrete wall, cast in place, Hi Early strength, 10" thick	SF	0.39		0.39
04.1105 245	Design mix Add, concrete wall, cast in place, Hi Early strength, 12" thick	SF	0.47		0.47
04.1105 250	Design mix Add, concrete wall, cast in place, lightweight aggregate, 6" thick	SF	0.99		0.99
04.1105 255	Design mix Add, concrete wall, cast in place, lightweight aggregate, 8" thick	SF	1.37		1.37
04.1105 260	Design mix Add, concrete wall, cast in place, lightweight aggregate, 10" thick	SF	1.70		1.70
04.1105 265	Design mix Add, concrete wall, cast in place, lightweight aggregate, 12" thick	SF	2.03		2.03
04.1105 270	Design mix Add, concrete wall, cast in place, granite aggregate, 6" thick	SF	0.25		0.25
04.1105 275	Design mix Add, concrete wall, cast in place, granite aggregate, 8" thick	SF	0.35		0.35
04.1105 280	Design mix Add, concrete wall, cast in place, granite aggregate, 10" thick	SF	0.44		0.44
04.1105 285	Design mix Add, concrete wall, cast in place, granite aggregate, 12" thick	SF	0.52		0.52
04.1110	**WALLS, TILT-UP CONCRETE PANEL**				
04.1110 100	Tilt-up, concrete wall, no pilasters, 6" thick	SF	8.03	8.46	16.48
04.1110 105	Tilt-up, concrete wall, no pilasters, 8" thick	SF	9.33	8.63	17.96
04.1110 110	Tilt-up, concrete wall, with pilasters, 6" thick	SF	8.51	9.83	18.34
04.1110 115	Tilt-up, concrete wall, with pilasters, 8" thick	SF	9.97	10.72	20.70
04.1115	**WALLS, PRECAST CONCRETE PANELS**				
04.1115 100	Precast concrete panel, single form	SF	40.29	11.95	52.24
04.1115 105	Precast concrete panel, double form	SF	53.88	11.95	65.83
04.1115 110	Precast concrete panel, single form, exposed aggregate	SF	56.09	11.95	68.04
04.1115 115	Precast concrete panel, single form, sandblast	SF	64.95	11.95	76.90
04.1115 120	Precast concrete panel, double form, exposed aggregate	SF	56.09	11.95	68.04
04.1115 125	Precast concrete panel, double form, sandblast	SF	71.62	11.95	83.57
04.1115 130	Precast concrete panel, single form, granite finish	SF	82.71	11.95	94.66

4.1 - ENCLOSURE, VERTICAL

CSI#	Description	Unit	Material Cost	Install Cost	Total Cost
04.1120	**WALLS, CONCRETE BLOCK**				
04.1120 100	Concrete Block, 8"X16", reinforced, 4" thick	SF	8.46	6.73	15.18
04.1120 110	Concrete Block, 8"X16", reinforced, 8" thick	SF	10.33	7.64	17.97
04.1120 115	Concrete Block, 8"X16", reinforced, 12" thick	SF	16.07	8.52	24.58
04.1120 120	Concrete Block, 8"X16", reinforced, filled 4" thick	SF	8.96	7.24	16.20
04.1120 130	Concrete Block, 8"X16", reinforced, filled, 8" thick	SF	12.15	7.81	19.95
04.1120 135	Concrete Block, 8"X16", reinforced, filled, 12" thick	SF	15.33	9.46	24.79
04.1120 140	Concrete Block, 8"X16", reinforced, grout lock, 8" thick	SF	12.81	6.09	18.90
04.1120 145	Concrete Block, 8"X16", reinforced, grout lock, 12" thick	SF	16.46	7.23	23.68
04.1120 150	Concrete Block, Slumpstone, reinforced, filled, 8"X4"X16"	SF	21.06	11.08	32.15
04.1120 155	Concrete Block, Slumpstone, reinforced, filled, 8"X8"X16"	SF	16.39	9.32	25.70
04.1120 157	Concrete Block, split face, reinforced, filled, 8"X4"X16"	SF	19.03	11.07	30.10
04.1120 160	Concrete Block, Glazed one side, reinforced, filled, 4"X8"X16"	SF	14.35	10.27	24.63
04.1120 165	Concrete Block, Glazed one side, reinforced, filled, 6"X8"X16"	SF	16.51	10.99	27.50
04.1120 170	Concrete Block, Glazed one side, reinforced, filled, 8"X8"X16"	SF	19.32	11.92	31.23
04.1120 175	Concrete Block, Glazed one side, reinforced, filled, 12"X8"X16"	SF	23.34	16.91	40.25
04.1120 180	Concrete Block, Glazed one side, reinforced, filled, 4"X4"X16"	SF	21.26	11.75	33.01
04.1120 185	Concrete Block, Glazed one side, reinforced, filled, 6"X4"X16"	SF	24.82	13.65	38.47
04.1120 190	Concrete Block, Glazed one side, reinforced, filled, 8"X4"X16"	SF	28.37	14.51	42.89
04.1120 195	Add for glazing both sides	SF	7.85		7.85
04.1120 200	Add for pilaster per square foot of pilaster	SF	34.39		34.39
04.1125	**WALLS, BRICK VENEER WITH BLOCK BACK-UP**				
04.1125 100	Brick veneer, 4" standard, 4"X8"X16" block back-up	SF	15.27	14.37	29.65
04.1125 110	Brick veneer, 4" standard, 8"X8"X16" block back-up	SF	17.15	15.29	32.43
04.1125 115	Brick veneer, 4" modular, 4"X8"X16" block back-up	SF	15.67	17.51	33.18
04.1125 120	Brick veneer, 4" modular, 6"X8"X16" block back-up	SF	17.03	18.01	35.04
04.1125 125	Brick veneer, 4" modular, 8"X8"X16" block back-up	SF	17.55	18.42	35.97
04.1125 130	Brick veneer, 4"X4"X12" Jumbo, 4"X8"X16" block back-up	SF	13.30	14.31	27.61
04.1125 140	Brick veneer, 4"X4"X12" Jumbo, 8"X8"X16" block back-up	SF	15.17	15.23	30.40
04.1125 145	Brick veneer, 6"X4"X12" Jumbo, 4"X8"X16" block back-up	SF	15.35	15.16	30.52
04.1125 155	Brick veneer, 6"X4"X12" Jumbo, 8"X8"X16" block back-up	SF	17.23	16.08	33.30
04.1125 160	Brick veneer, 8"X4"X12" Jumbo, 4"X8"X16" block back-up	SF	17.44	16.06	33.50
04.1125 165	Brick veneer, 8"X4"X12" Jumbo, 6"X8"X16" block back-up	SF	18.80	16.56	35.36
04.1125 170	Brick veneer, 8"X4"X12" Jumbo, 8"X8"X16" block back-up	SF	19.32	16.97	36.29
04.1125 175	Brick veneer, face brick modular, 4"X8"X16" block back-up	SF	16.38	18.33	34.70
04.1125 180	Brick veneer, face brick modular, 6"X8"X16" block back-up	SF	17.73	18.83	36.56
04.1125 185	Brick veneer, face brick modular, 8"X8"X16" block back-up	SF	18.25	19.24	37.49
04.1125 190	Brick veneer, Norman, 4"X8"X16" block back-up	SF	15.03	18.19	33.22
04.1125 195	Brick veneer, Norman, 6"X8"X16" block back-up	SF	16.39	18.69	35.07
04.1125 200	Brick veneer, Norman, 8"X8"X16" block back-up	SF	16.90	19.10	36.00
04.1125 205	Brick veneer, Roman, 4"X8"X16" block back-up	SF	16.08	23.02	39.10
04.1125 210	Brick veneer, Roman, 6"X8"X16" block back-up	SF	17.43	23.52	40.96
04.1125 215	Brick veneer, Roman, 8"X8"X16" block back-up	SF	17.95	23.94	41.89
04.1125 220	Brick veneer, Glazed, 4"X8"X16" block back-up	SF	20.64	25.17	45.81
04.1125 225	Brick veneer, Glazed, 6"X8"X16" block back-up	SF	21.99	25.67	47.66
04.1125 230	Brick veneer, Glazed, 8"X8"X16" block back-up	SF	22.51	26.08	48.59
04.1130	**WALLS, BRICK**				
04.1130 100	Brick wall, cavity, 10" thick	SF	13.47	19.68	33.15
04.1130 105	Brick wall, reinforced, 10" thick	SF	15.16	21.29	36.45
04.1130 110	Brick wall, reinforced, 13" thick	SF	18.27	23.28	41.55
04.1130 115	Brick wall, reinforced, 16" thick	SF	22.63	24.74	47.37
04.1130 120	Brick wall, reinforced, 20" thick	SF	25.35	31.20	56.55
04.1135	**WALLS, STONE VENEER WITH BLOCK BACK-UP**				
04.1135 100	Stone veneer, Granite, 4"X8"X16" block back-up	SF	36.84	38.11	74.95
04.1135 105	Stone veneer, Granite, 6"X8"X16" block back-up	SF	38.20	38.61	76.80
04.1135 110	Stone veneer, Granite, 8"X8"X16" block back-up	SF	38.72	39.02	77.73
04.1135 115	Stone veneer, Limestone, 4"X8"X16" block back-up	SF	29.21	41.93	71.14
04.1135 120	Stone veneer, Limestone, 6"X8"X16" block back-up	SF	30.57	42.43	73.00

ENCLOSURE, VERTICAL - 4.1

CSI#	Description	Unit	Material Cost	Install Cost	Total Cost
04.1135 125	Stone veneer, Limestone, 8"X8"X16" block back-up	SF	31.08	42.84	73.93
04.1135 130	Stone veneer, Marble, 4"X8"X16" block back-up	SF	38.72	37.65	76.37
04.1135 135	Stone veneer, Marble, 6"X8"X16" block back-up	SF	40.07	38.15	78.22
04.1135 140	Stone veneer, Marble, 8"X8"X16" block back-up	SF	40.59	38.56	79.15
04.1135 145	Stone veneer, Sandstone, 4"X8"X16" block back-up	SF	19.00	25.75	44.75
04.1135 155	Stone veneer, Sandstone, 8"X8"X16" block back-up	SF	20.87	26.67	47.54
04.1135 160	Stone veneer, Lava stone, 4"X8"X16" block back-up	SF	16.52	24.97	41.49
04.1135 165	Stone veneer, Lava stone, 6"X8"X16" block back-up	SF	17.88	25.47	43.35
04.1135 170	Stone veneer, Lava stone, 8"X8"X16" block back-up	SF	18.40	25.88	44.28
04.1135 175	Stone veneer, Arizona stone, 4"X8"X16" block back-up	SF	22.60	27.25	49.86
04.1135 185	Stone veneer, Arizona stone, 8"X8"X16" block back-up	SF	24.48	28.16	52.64
04.1135 190	Stone veneer, Rubble, 4"X8"X16" block back-up	SF	15.73	28.45	44.18
04.1135 195	Stone veneer, Rubble, 6"X8"X16" block back-up	SF	17.09	28.95	46.04
04.1135 200	Stone veneer, Rubble, 8"X8"X16" block back-up	SF	17.61	29.36	46.97
04.1140	**WALLS, BRICK VENEER ON STUD FRAME**				
04.1140 100	Brick veneer, 4" standard, sheathing, building paper, insulation, 2"X4" wood stud 16"OC	SF	8.50	11.98	20.48
04.1140 105	Brick veneer, 4" standard, sheathing, building paper, insulation, 2"X6" wood stud 16"OC	SF	9.05	12.96	22.01
04.1140 110	Brick veneer, 4" standard, sheathing, building paper, insulation, 16 ga 3-5/8 metal stud 16"OC	SF	9.74	14.24	23.98
04.1140 115	Brick veneer, 4" modular, sheathing, building paper, insulation, 2"X4" wood stud 16"OC	SF	8.90	15.11	24.01
04.1140 120	Brick veneer, 4" modular, sheathing, building paper, insulation, 2"X6" wood stud 16"OC	SF	9.45	16.10	25.55
04.1140 125	Brick veneer, 4" modular, sheathing, building paper, insulation, 16 ga 3-5/8 metal stud 16"OC	SF	10.15	17.37	27.52
04.1140 130	Brick veneer, 4"X4"X12" Jumbo, sheathing, building paper, insulation, 2"X4" wood stud 16"OC	SF	6.52	11.92	18.44
04.1140 135	Brick veneer, 4"X4"X12" Jumbo, sheathing, building paper, insulation, 2"X6" wood stud 16"OC	SF	7.07	12.90	19.97
04.1140 140	Brick veneer, 4"X4"X12" Jumbo, sheathing, building paper, insulation, 16 ga 3-5/8 metal stud 16"OC	SF	7.77	14.18	21.94
04.1140 160	Brick veneer, 8"X4"X12" Jumbo, sheathing, building paper, insulation, 2"X4" wood stud 16"OC	SF	10.67	13.66	24.33
04.1140 165	Brick veneer, 8"X4"X12" Jumbo, sheathing, building paper, insulation, 2"X6" wood stud 16"OC	SF	11.22	14.65	25.87
04.1140 170	Brick veneer, 8"X4"X12" Jumbo, sheathing, building paper, insulation, 16 ga 3-5/8 metal stud 16"OC	SF	11.92	15.92	27.84
04.1140 175	Brick veneer, face brick modular, sheathing, building paper, insulation, 2"X4" wood stud 16"OC	SF	9.60	15.93	25.53
04.1140 180	Brick veneer, face brick modular, sheathing, building paper, insulation, 2"X6" wood stud 16"OC	SF	10.15	16.92	27.07
04.1140 185	Brick veneer, face brick modular, sheathing, building paper, insulation, 16 ga 3-5/8 metal stud 16"OC	SF	10.85	18.19	29.04
04.1140 190	Brick veneer, Norman, sheathing, building paper, insulation, 2"X4" wood stud 16"OC	SF	8.26	15.79	24.05
04.1140 195	Brick veneer, Norman, sheathing, building paper, insulation, 2"X6" wood stud 16"OC	SF	8.80	16.78	25.58
04.1140 200	Brick veneer, Norman, sheathing, building paper, insulation, 16 ga 3-5/8 metal stud 16"OC	SF	9.50	18.05	27.55
04.1140 205	Brick veneer, Roman, sheathing, building paper, insulation, 2"X4" wood stud 16"OC	SF	9.30	20.63	29.93
04.1140 210	Brick veneer, Roman, sheathing, building paper, insulation, 2"X6" wood stud 16"OC	SF	9.85	21.61	31.46
04.1140 215	Brick veneer, Roman, sheathing, building paper, insulation, 16 ga 3-5/8 metal stud 16"OC	SF	10.55	22.89	33.44
04.1140 220	Brick veneer, Glazed, sheathing, building paper, insulation, 2"X4" wood stud 16"OC	SF	13.87	22.77	36.63
04.1140 225	Brick veneer, Glazed, sheathing, building paper, insulation, 2"X6" wood stud 16"OC	SF	14.41	23.76	38.17
04.1140 230	Brick veneer, Glazed, sheathing, building paper, insulation, 16 ga 3-5/8 metal stud 16"OC	SF	15.11	25.03	40.14

4.1 - ENCLOSURE, VERTICAL

CSI#	Description	Unit	Material Cost	Install Cost	Total Cost
04.1145	**WALLS, SIDING ON STUD FRAME**				
04.1145 100	Siding, Redwood Beveled, paint, sheathing, building paper, insulation, 2"X4" wood stud 16"OC	SF	5.55	7.73	13.28
04.1145 105	Siding, Redwood Beveled, paint, sheathing, building paper, insulation, 2"X6" wood stud 16"OC	SF	4.76	7.64	12.40
04.1145 110	Siding, Redwood Beveled, paint, sheathing, building paper, insulation, 16 ga 3-5/8 metal stud 16"OC	SF	6.80	9.99	16.78
04.1145 115	Siding, Cedar Beveled, paint, sheathing, building paper, insulation, 2"X4" wood stud 16"OC	SF	3.81	7.63	11.44
04.1145 120	Siding, Cedar Beveled, paint, sheathing, building paper, insulation, 2"X6" wood stud 16"OC	SF	3.63	7.58	11.21
04.1145 125	Siding, Cedar Beveled, paint, sheathing, building paper, insulation, 16 ga 3-5/8 metal stud 16"OC	SF	5.05	9.89	14.95
04.1145 130	Siding, board & batten, paint, sheathing, building paper, insulation, 2"X4" wood stud 16"OC	SF	7.82	7.17	14.99
04.1145 135	Siding, board & batten, paint, sheathing, building paper, insulation, 2"X6" wood stud 16"OC	SF	6.23	7.28	13.51
04.1145 140	Siding, board & batten, sheathing, paint, building paper, insulation, 16 ga 3-5/8 metal stud 16"OC	SF	9.07	9.43	18.50
04.1145 145	Siding, tongue & grove, paint, sheathing, building paper, insulation, 2"X4" wood stud 16"OC	SF	3.67	7.73	11.41
04.1145 150	Siding, tongue & grove, paint, sheathing, building paper, insulation, 2"X6" wood stud 16"OC	SF	4.22	8.72	12.94
04.1145 155	Siding, tongue & grove, paint, sheathing, building paper, insulation, 16 ga 3-5/8 metal stud 16"OC	SF	4.92	9.99	14.91
04.1145 160	Siding, Plywood texture 1-11 paint, sheathing, building paper, insulation, 2"X4" wood stud 16"OC	SF	3.77	7.73	11.50
04.1145 165	Siding, Plywood texture 1-11 paint, sheathing, building paper, insulation, 2"X6" wood stud 16"OC	SF	4.32	8.72	13.04
04.1145 170	Siding, Plywood texture 1-11 paint, sheathing, building paper, insulation, 16 ga 3-5/8 metal stud 16"OC	SF	5.02	9.99	15.01
04.1145 175	Siding, cedar shingles, stain, sheathing, building paper, insulation, 2"X4" wood stud 16"OC	SF	4.87	11.08	15.95
04.1145 180	Siding, cedar shingles, stain, sheathing, building paper, insulation, 2"X6" wood stud 16"OC	SF	5.42	12.06	17.49
04.1145 185	Siding, cedar shingles, stain, sheathing, building paper, insulation, 16 ga 3-5/8 metal stud 16"OC	SF	6.12	13.34	19.46
04.1145 190	Siding, hardboard, paint, sheathing, building paper, insulation, 2"X4" wood stud 16"OC	SF	2.79	8.34	11.13
04.1145 195	Siding, hardboard, paint, sheathing, building paper, insulation, 2"X6" wood stud 16"OC	SF	3.34	9.32	12.66
04.1145 200	Siding, hardboard, paint, sheathing, building paper, insulation, 16 ga 3-5/8 metal stud 16"OC	SF	4.04	10.60	14.64
04.1145 205	Siding, Glasweld, finish one side, sheathing, building paper, insulation, 2"X4" wood stud 16"OC	SF	8.26	10.06	18.32
04.1145 210	Siding, Glasweld, finish one side, sheathing, building paper, insulation, 2"X6" wood stud 16"OC	SF	8.81	11.05	19.85
04.1145 215	Siding, Glasweld, finish one side, sheathing, building paper, insulation, 16 ga 3-5/8 metal stud 16"OC	SF	9.51	12.32	21.83
04.1145 220	Siding, Glasweld, finish two sides, sheathing, building paper, insulation, 2"X4" wood stud 16"OC	SF	8.36	10.40	18.75
04.1145 225	Siding, Glasweld, finish two sides, sheathing, building paper, insulation, 2"X6" wood stud 16"OC	SF	8.90	11.38	20.29
04.1145 230	Siding, Glasweld, finish two sides, sheathing, building paper, insulation, 16 ga 3-5/8 metal stud 16"OC	SF	9.60	12.66	22.26
04.1150	**WALLS, STONE VENEER ON STUD FRAME**				
04.1150 100	Stone veneer, Granite, sheathing, building paper, insulation, 2"X4" wood stud 16"OC	SF	30.07	35.71	65.78
04.1150 105	Stone veneer, Granite, sheathing, building paper, insulation, 2"X6" wood stud 16"OC	SF	30.62	36.70	67.31
04.1150 110	Stone veneer, Granite, sheathing, building paper, insulation, 16 ga 3-5/8 metal stud 16"OC	SF	31.31	37.97	69.28
04.1150 115	Stone veneer, Limestone, sheathing, building paper, insulation, 2"X4" wood stud 16"OC	SF	22.44	39.53	61.97

ENCLOSURE, VERTICAL - 4.1

CSI#	Description	Unit	Material Cost	Install Cost	Total Cost
04.1150 120	Stone veneer, Limestone, sheathing, building paper, insulation, 2"X6" wood stud 16"OC	SF	22.99	40.52	63.50
04.1150 125	Stone veneer, Limestone, sheathing, building paper, insulation, 16 ga 3-5/8 metal stud 16"OC	SF	23.68	41.79	65.48
04.1150 145	Stone veneer, Sandstone, sheathing, building paper, insulation, 2"X4" wood stud 16"OC	SF	22.07	32.76	54.83
04.1150 150	Stone veneer, Sandstone, sheathing, building paper, insulation, 2"X6" wood stud 16"OC	SF	22.62	33.75	56.37
04.1150 155	Stone veneer, Sandstone, sheathing, building paper, insulation, 16 ga 3-5/8 metal stud 16"OC	SF	23.32	35.02	58.34
04.1150 160	Stone veneer, Lava stone, sheathing, building paper, insulation, 2"X4" wood stud 16"OC	SF	9.75	22.57	32.32
04.1150 165	Stone veneer, Lava stone, sheathing, building paper, insulation, 2"X6" wood stud 16"OC	SF	10.30	23.56	33.86
04.1150 170	Stone veneer, Lava stone, sheathing, building paper, insulation, 16 ga 3-5/8 metal stud 16"OC	SF	11.00	24.83	35.83
04.1150 175	Stone veneer, Arizona stone, sheathing, building paper, insulation, 2"X4" wood stud 16"OC	SF	15.83	24.85	40.68
04.1150 180	Stone veneer, Arizona stone, sheathing, building paper, insulation, 2"X6" wood stud 16"OC	SF	16.38	25.84	42.22
04.1150 185	Stone veneer, Arizona stone, sheathing, building paper, insulation, 16 ga 3-5/8 metal stud 16"OC	SF	17.08	27.11	44.19
04.1150 190	Stone veneer, Rubble, sheathing, building paper, insulation, 2"X4" wood stud 16"OC	SF	8.96	26.05	35.01
04.1150 195	Stone veneer, Rubble, sheathing, building paper, insulation, 2"X6" wood stud 16"OC	SF	9.51	27.04	36.55
04.1150 200	Stone veneer, Rubble, sheathing, building paper, insulation, 16 ga 3-5/8 metal stud 16"OC	SF	10.20	28.31	38.52
04.1155	**WALLS, STUCCO ON STUD FRAME**				
04.1155 100	Stucco, paint, sheathing, building paper, insulation, 2"X4" wood stud 16"OC	SF	9.89	16.21	26.10
04.1155 105	Stucco, paint, sheathing, building paper, insulation, 2"X6" wood stud 16"OC	SF	11.21	16.33	27.54
04.1155 110	Stucco, paint, sheathing, building paper, insulation, 16 ga 3-5/8 metal stud 16"OC	SF	14.52	17.93	32.45
04.1155 115	Stucco, paint, lath, building paper, insulation, 2"X4" wood stud 16"OC	SF	7.00	11.34	18.34
04.1155 120	Stucco, paint, lath, building paper, insulation, 2"X6" wood stud 16"OC	SF	9.47	13.02	22.49
04.1155 125	Stucco, paint, lath, building paper, insulation, 16 ga 3-5/8 metal stud 16"OC	SF	12.78	15.39	28.17
04.1160	**WALLS, METAL SIDING**				
04.1160 100	Metal Siding, corrugated aluminum, .032" thick	SF	3.87	3.61	7.48
04.1160 105	Metal Siding, corrugated aluminum, .032" thick, painted	SF	4.64	3.61	8.25
04.1160 110	Metal Siding, aluminum, simulated wood, insulated	SF	4.32	3.61	7.93
04.1160 115	Metal Siding, corrugated composition, 3/8" thick	SF	3.00	4.82	7.82
04.1160 120	Metal Siding, corrugated fiberglass, 8 oz.	SF	3.45	3.61	7.06
04.1160 125	Metal Siding, corrugated galvanized iron, 26 ga	SF	2.71	3.97	6.68
04.1160 130	Metal Siding, baked enamel	SF	39.40	4.38	43.78
04.1160 135	Metal Siding, porcelain	SF	43.78	4.62	48.39
04.1160 140	Metal Siding, insulated, baked enamel	SF	40.76	6.60	47.36
04.1160 145	Metal Siding, insulated, porcelain	SF	56.43	7.44	63.88
04.1165	**CURTAIN WALLS**				
04.1165 100	Curtain Wall, frame, anodized aluminum, bronze	SF	25.59		25.59
04.1165 105	Curtain Wall, frame, anodized aluminum, black	SF	29.47		29.47
04.1165 115	Curtain Wall, frame, aluminum, sloping section	SF	40.80		40.80
04.1165 120	Curtain Wall, frame, steel, painted	SF	24.84		24.84
04.1165 125	Curtain Wall, frame, steel, porcelain enamel	SF	26.82		26.82
04.1165 130	Curtain Wall, glazing, plate, 1/4" thick clear	SF	9.04		9.04
04.1165 135	Curtain Wall, glazing, plate, 1/4" thick tinted	SF	11.11		11.11
04.1165 140	Curtain Wall, glazing, plate, 1/4" solarcool reflective	SF	20.83		20.83
04.1165 145	Curtain Wall, glazing, plate, 1/4" veri-tran, reflective	SF	27.82		27.82
04.1165 150	Curtain Wall, glazing, tempered plate, 1/4" thick, clear	SF	16.64		16.64
04.1165 155	Curtain Wall, glazing, tempered plate, 1/4" thick, tinted	SF	19.49		19.49
04.1165 160	Curtain Wall, glazing, tempered plate, 1/4" solarcool, reflective	SF	31.32		31.32
04.1165 165	Curtain Wall, glazing, double glazed, 5/8" thick, clear	SF	16.64		16.64

4.1 - ENCLOSURE, VERTICAL

CSI#	Description	Unit	Material Cost	Install Cost	Total Cost
04.1165 170	Curtain Wall, glazing, double glazed, 5/8" thick, tinted	SF	18.08		18.08
04.1165 175	Curtain Wall, glazing, double glazed, 5/8" solarcool, reflective	SF	22.20		22.20
04.1165 180	Curtain Wall, glazing, double glazed, tempered, 5/8" thick, clear	SF	28.53		28.53
04.1165 185	Curtain Wall, glazing, double glazed, tempered, 5/8" thick, tinted	SF	30.61		30.61
04.1165 190	Curtain Wall, glazing, double glazed, tempered, 5/8" solarcool, reflective	SF	43.18		43.18
04.1165 195	Curtain Wall, glazing, double glazed, 1" thick, clear	SF	18.08		18.08
04.1165 200	Curtain Wall, glazing, double glazed, 1" thick, tinted	SF	19.49		19.49
04.1165 205	Curtain Wall, glazing, double glazed, 1" solarcool, reflective	SF	25.75		25.75
04.1165 210	Curtain Wall, glazing, double glazed, tempered, 1" thick, clear	SF	32.69		32.69
04.1165 215	Curtain Wall, glazing, double glazed, tempered, 1" thick, tinted	SF	35.49		35.49
04.1165 220	Curtain Wall, glazing, double glazed, tempered, 1" solarcool, reflective	SF	18.08		18.08
04.1165 225	Curtain Wall, glazing, spandrel	SF	19.49		19.49
04.1165 230	Curtain Wall, glazing, glasweld, 1/4" finished, one side	SF	34.14		34.14
04.1165 235	Curtain Wall, glazing, glasweld, 1" insulated, finished, one side, dual	SF	44.38		44.38
04.1165 240	Curtain Wall, glazing, mirawal, 5/16" finished, one side, dual	SF	34.79		34.79
04.1165 245	Curtain Wall, glazing, mirawal, 1" insulated, finished, one side, dual	SF	45.60		45.60
04.1165 250	Curtain Wall, glazing, aluca bond,	SF	50.08		50.08
04.1165 255	Curtain Wall, prefinished and insulated building panels with support, frame, aluca bond	SF	103.40		103.40
04.1165 260	Curtain Wall, Dryvit, with studs & 3" insulation	SF	33.30		33.30
04.1165 265	Curtain Wall, precast panel, fiberglass reinforced, 1 form, standard finish	SF	40.29		40.29
04.1165 270	Curtain Wall, precast panel, fiberglass reinforced, 2 form, standard finish	SF	53.88		53.88
04.1165 275	Curtain Wall, precast panel, fiberglass reinforced, 1 form, sandblast	SF	64.95		64.95
04.1165 280	Curtain Wall, precast panel, fiberglass reinforced, 2 form, sandblast	SF	71.62		71.62
04.1165 285	Curtain Wall, precast panel, fiberglass reinforced, 1 form, mo-sai finish	SF	56.09		56.09
04.1165 290	Curtain Wall, precast panel, fiberglass reinforced, 1 form, granite overlay	SF	82.71		82.71
04.1165 295	Curtain Wall, GFRC panel, fiberglass reinforced, 1 form, standard finish	SF	46.00		46.00
04.1165 300	Curtain Wall, GFRC panel, fiberglass reinforced, 2 form, standard finish	SF	53.83		53.83
04.1165 305	Curtain Wall, GFRC panel, fiberglass reinforced, 1 form, sandblast	SF	66.04		66.04
04.1165 310	Curtain Wall, GFRC panel, fiberglass reinforced, 2 form, sandblast	SF	70.20		70.20
04.1165 315	Curtain Wall, GFRC panel, fiberglass reinforced, 1 form, mo-sai finish	SF	53.88		53.88
04.1165 320	Curtain Wall, GFRC panel, fiberglass reinforced, 1 form, granite overlay	SF	82.84		82.84
04.1170	**WALLS, EXTERIOR COATING**				
04.1170 100	Exterior coating, concrete surface, paint, prime	SF	0.10	0.38	0.48
04.1170 105	Exterior coating, concrete surface, paint, prime + 1 finish	SF	0.18	0.72	0.91
04.1170 110	Exterior coating, concrete surface, paint, prime + 2 finish	SF	0.25	1.06	1.31
04.1170 115	Exterior coating, plaster surface, paint, prime	SF	0.10	0.38	0.48
04.1170 120	Exterior coating, plaster surface, paint, prime + 1 finish	SF	0.15	0.72	0.88
04.1170 125	Exterior coating, plaster surface, paint, prime + 2 finish	SF	0.25	1.06	1.31
04.1170 130	Exterior coating, masonry surface, paint, prime	SF	0.15	0.43	0.58
04.1170 135	Exterior coating, masonry surface, paint, prime + 1 finish	SF	0.15	0.76	0.91
04.1170 140	Exterior coating, masonry surface, paint, prime + 2 finish	SF	0.25	1.09	1.33
04.1170 145	Exterior coating, wood siding, paint, prime	SF	0.09	0.34	0.43
04.1170 150	Exterior coating, wood siding, paint, prime + 1 finish	SF	0.15	0.72	0.88
04.1170 155	Exterior coating, wood siding, paint, prime + 2 finish	SF	0.25	1.06	1.31
04.1170 160	Exterior coating, wood siding, stain	SF	0.10	0.34	0.44
04.1170 165	Exterior coating, wood siding, stain + 2 seal coats	SF	0.10	1.01	1.11
04.1170 170	Exterior coating, wood shingle, stain	SF	0.09	0.34	0.43
04.1170 175	Exterior coating, wood shingle, stain + 2 seal coats	SF	0.10	1.01	1.11
04.1170 180	Exterior coating, silicone seal, spray 1 coat	SF	0.27	0.34	0.61
04.1170 185	Exterior coating, silicone seal, spray 2 coats	SF	0.67	0.63	1.30
04.1170 190	Exterior coating, polyurethane, 1/8" thick	SF	1.74	3.37	5.10
04.1170 195	Exterior coating, sheet metal, paint, prime	SF	0.09	0.68	0.77
04.1170 200	Exterior coating, sheet metal, paint, prime + 1 finish	SF	0.15	1.14	1.29
04.1170 205	Exterior coating, sheet metal, paint, prime + 2 finish	SF	0.20	1.68	1.89
04.1170 210	Exterior coating, wood trim, paint, prime	SF	0.10	0.43	0.52
04.1170 215	Exterior coating, wood trim, paint, prime + 1 finish	SF	0.20	0.74	0.95
04.1170 220	Exterior coating, wood trim, paint, prime + 2 finish	SF	0.27	1.19	1.46

ENCLOSURE, VERTICAL - 4.1

CSI#	Description	Unit	Material Cost	Install Cost	Total Cost
04.1175	**INTERIOR SURFACE OF EXTERIOR WALLS**				
04.1175 100	Finish interior wall, gypsum board, on stud, painted, rubber base	SF	0.13	0.63	0.76
04.1175 105	Finish interior wall, gypsum board, on furring, painted, rubber base	SF	0.13	0.63	0.76
04.1175 110	Finish interior wall, lath and plaster, on stud, painted, rubber base	SF	0.13	0.63	0.76
04.1175 115	Finish interior wall, lath and plaster, on stud, painted, rubber base	SF	0.13	0.63	0.76
04.1175 120	Wall covering, ceramic tile, 4"x 4", mortar	SF	6.02	8.43	14.45
04.1175 125	Wall covering, ceramic tile, 4"x 4", mastic	SF	5.92	4.84	10.76
04.1175 130	Wall covering, ceramic tile, 6"x 4", mortar	SF	6.31	7.54	13.86
04.1175 135	Wall covering, ceramic tile, 6"x 4", mastic	SF	6.31	4.03	10.34
04.1175 140	Wall painting, concrete, prime	SF	0.04	0.34	0.39
04.1175 145	Wall painting, concrete, prime + 1 finished	SF	0.13	0.63	0.76
04.1175 150	Wall painting, concrete, prime + 2 finished	SF	0.23	1.01	1.24
04.1175 155	Concrete block painting, prime, brush	SF	0.04	0.38	0.42
04.1175 160	Concrete block painting, prime + 1 finished, brush	SF	0.15	0.72	0.88
04.1175 165	Concrete block painting, prime + 2 finished, brush	SF	0.25	1.07	1.32
04.1175 170	Concrete block painting, prime, roll	SF	0.04	0.25	0.30
04.1175 175	Concrete block painting, prime + 1 finished, roll	SF	0.15	0.48	0.63
04.1175 180	Concrete block painting, prime + 2 finished, roll	SF	0.25	0.72	0.96
04.1200	**FENESTRATION**				
04.1200 100	Window, steel frame, fixed, 1/4" glass, clear	SF	18.43	14.25	32.68
04.1200 105	Window, steel frame, fixed, 1/4" glass, tempered	SF	20.41	14.25	34.66
04.1200 110	Window, steel frame, fixed, security, 1/4" glass, wire	SF	42.30	14.25	56.55
04.1200 115	Window, steel frame, fixed, 1/4" glass, spandrel	SF	39.37	14.99	54.36
04.1200 120	Window, steel frame, fixed, 1/4" glass, obscure	SF	25.26	14.25	39.51
04.1200 125	Window, aluminum frame, fixed, 1/4" glass, clear	SF	18.11	11.06	29.17
04.1200 130	Window, aluminum frame, fixed, 1/4" glass, tempered	SF	20.09	11.06	31.16
04.1200 135	Window, aluminum frame, fixed, 1/4" glass, spandrel	SF	39.05	11.80	50.85
04.1200 140	Window, aluminum frame, fixed, 1/4" glass, obscure	SF	24.94	11.06	36.00
04.1200 145	Window, steel frame, fixed, 1/4" double glass, clear	SF	25.62	18.98	44.60
04.1200 150	Window, steel frame, fixed, 1/4" double glass, tinted	SF	27.89	18.98	46.87
04.1200 155	Window, steel frame, fixed, 1/4" double glass, tempered	SF	27.60	18.98	46.58
04.1200 160	Window, steel frame, fixed, security 1/4" double glass, wire	SF	49.49	18.98	68.47
04.1200 165	Window, steel frame, fixed, 1/4" double glass, spandrel	SF	46.56	19.72	66.28
04.1200 170	Window, steel frame, fixed, 1/4" double glass, obscure	SF	32.45	18.98	51.43
04.1200 175	Window, aluminum frame, fixed, 1/4" double glass, clear	SF	25.45	15.80	41.25
04.1200 180	Window, aluminum frame, fixed, 1/4" double glass, tinted	SF	27.72	15.80	43.52
04.1200 185	Window, aluminum frame, fixed, 1/4" double glass, tempered	SF	27.44	15.80	43.23
04.1200 190	Window, aluminum frame, fixed, 1/4" double glass, spandrel	SF	46.39	16.53	62.92
04.1200 195	Window, aluminum frame, fixed, 1/4" double glass, obscure	SF	32.28	15.80	48.08
04.1200 350	Window, metal, sliding, one vent, insulated glass, with screen, 2'0"X1'6"	EACH	72.84	38.10	110.94
04.1200 355	Window, metal, sliding, one vent, insulated glass, with screen, 2'0"X3'0"	EACH	95.94	38.10	134.04
04.1200 360	Window, metal, sliding, one vent, insulated glass, with screen, 3'0"X4'0"	EACH	135.67	45.67	181.34
04.1200 365	Window, metal, sliding, one vent, insulated glass, with screen, 4'0"X4'0"	EACH	151.28	53.36	204.64
04.1200 370	Window, metal, sliding, one vent, insulated glass, with screen, 5'0"X4'0"	EACH	170.06	53.36	223.41
04.1200 375	Window, metal, sliding, one vent, insulated glass, with screen, 6'0"X4'0"	EACH	185.69	60.94	246.63
04.1200 380	Window, metal, sliding, two vent, insulated glass, with screen, 8'0"X4'0"	EACH	278.51	76.17	354.69
04.1200 385	Window, metal, sliding, two vent, insulated glass, with screen, 10'0"X4'0"	EACH	318.50	91.33	409.83
04.1200 390	Window, metal, sliding, insulated glass, with screen, average	SF	11.74	2.66	14.40
04.1200 395	Window, metal, casement, one vent, insulated glass, with screen, 2'0"X2'6"	EACH	186.32	38.10	224.42
04.1200 400	Window, metal, casement, one vent, insulated glass, with screen, 2'6"X4'0"	EACH	249.00	38.10	287.09
04.1200 405	Window, metal, casement, one vent, insulated glass, with screen, 4'0"X5'0"	EACH	325.56	53.36	378.91
04.1200 410	Window, metal, casement, one vent, insulated glass, with screen, 6'0"X3'0"	EACH	300.55	53.36	353.91
04.1200 415	Window, metal, casement, two vent, insulated glass, with screen, 4'0"X5'0"	EACH	408.59	53.36	461.95
04.1200 420	Window, metal, casement, two vent, insulated glass, with screen, 6'0"X3'0"	EACH	380.46	53.36	433.82
04.1200 425	Window, metal, casement, insulated glass, with screen, average	SF	25.91	2.66	28.57
04.1200 430	Window, metal, awning, one vent, insulated glass, 2'0"X2'6"	EACH	186.32	38.10	224.42
04.1200 435	Window, metal, awning, one vent, insulated glass, 4'0"X5'0"	EACH	325.56	53.36	378.91
04.1200 440	Window, metal, awning, one vent, insulated glass, 6'0"X3'0"	EACH	300.55	53.36	353.91

4.1 - ENCLOSURE, VERTICAL

CSI#	Description	Unit	Material Cost	Install Cost	Total Cost
04.1200 445	Window, metal, awning, insulated glass, average	SF	25.91	2.66	28.57
04.1200 450	Window, vinyl, sliding, one vent, insulated glass, 2'6"X3'0"	EACH	263.09	39.56	302.66
04.1200 455	Window, vinyl, sliding, one vent, insulated glass, 3'0"X4'0"	EACH	305.16	47.66	352.82
04.1200 460	Window, vinyl, sliding, one vent, insulated glass, 5'0"X5'0"	EACH	406.44	62.93	469.37
04.1200 465	Window, vinyl, sliding, one vent, insulated glass, 6'0"X5'0"	EACH	458.47	79.12	537.59
04.1200 470	Window, vinyl, sliding, one vent, insulated glass, average	EACH	27.66	2.77	30.43
04.1200 475	Window, vinyl, casement, one vent, insulated glass, 2'0"X 2'6"	SF	407.90	39.56	447.46
04.1200 480	Window, vinyl, casement, one vent, insulated glass, 3'0"X 4'0"	EACH	436.25	47.66	483.91
04.1200 485	Window, vinyl, fixed, insulated glass, 3'0" X 3'0"	EACH	171.69	31.64	203.33
04.1200 490	Window, vinyl, fixed, insulated glass, 5'0" X 5'0"	EACH	327.09	50.35	377.45
04.1200 495	Window, vinyl, fixed, insulated glass, 10'0" X 5'0"	EACH	601.57	76.25	677.82
04.1200 500	Window, wood, double hung, insulated glass, 1'6"X2'6"	EACH	280.38	102.95	383.33
04.1200 505	Window, wood, double hung, insulated glass, 2'0"X2'6"	EACH	284.11	102.95	387.05
04.1200 510	Window, wood, double hung, insulated glass, 2'0"X3'0"	EACH	298.96	102.95	401.91
04.1200 515	Window, wood, double hung, insulated glass, 2'0"X3'6"	EACH	343.50	102.95	446.45
04.1200 520	Window, wood, double hung, insulated glass, 2'6"X4'0"	EACH	376.93	102.95	479.88
04.1200 525	Window, wood, double hung, insulated glass, 2'6"X4'6"	EACH	402.93	110.82	513.76
04.1200 530	Window, wood, double hung, insulated glass, 3'0"X4'0"	EACH	419.63	110.82	530.46
04.1200 535	Window, wood, double hung, insulated glass, 3'0"X4'6"	EACH	449.35	110.82	560.17
04.1200 540	Window, wood, double hung, insulated glass, 3'0"X5'0"	EACH	477.20	110.82	588.02
04.1200 545	Window, wood, double hung, insulated glass, 3'0"X6'0"	EACH	592.31	126.68	719.00
04.1200 550	Window, wood, double hung, insulated glass, 4'0"X4'6"	EACH	542.21	150.32	692.53
04.1200 555	Window, wood, double hung, insulated glass, 4'0"X6'0"	EACH	724.18	150.32	874.50
04.1200 560	Window, wood, double hung, insulated glass, Average	SF	41.45		41.45
04.1200 565	Window, wood, casement, insulated glass, 1'6"X2'0"	EACH	226.53	110.82	337.36
04.1200 570	Window, wood, casement, insulated glass, 2'0"X3'0"	EACH	291.52	118.70	410.22
04.1200 575	Window, wood, casement, insulated glass, 5'0"X4'0"	EACH	683.33	205.69	889.02
04.1200 580	Window, wood, casement, insulated glass, 8'0"X5'0"	EACH	1,245.91	221.55	1,467.47
04.1200 585	Window, wood, casement, insulated glass, 10'0"X5'0"	EACH	1,448.31	284.88	1,733.19
04.1200 590	Window, wood, fixed, insulated glass, 2'0"X2'0"	EACH	172.69	88.66	261.35
04.1200 595	Window, wood, fixed, insulated glass, 3'0"X3'6"	EACH	282.25	88.66	370.91
04.1200 600	Window, wood, fixed, insulated glass, 5'0"X6'0"	EACH	664.75	118.70	783.45
04.1200 605	Window, wood, fixed, insulated glass, average	SF	30.71		30.71
04.1300	**EXTERIOR DOORS**				
04.1300 100	Door, prehung, solid core, 3' X 7' X 1 3/8"	EACH	172.28	100.34	272.61
04.1300 105	Door, prehung, solid core, 3' X 7' X 1 3/4"	EACH	182.42	100.34	282.76
04.1300 107	Door, prehung, 9 lite, 3' X 7' X 1 3/4"	EACH	235.12	100.34	335.46
04.1300 110	Door, prehung, 12 lite, 3' X 7' X 1 3/4"	EACH	267.54	100.34	367.88
04.1300 115	Door, prehung, Dutch, 3' X 7' X 1 3/4"	EACH	389.80	150.51	540.31
04.1300 120	Door, prehung, 12 lite, 3'-6" X 9' X 1 3/4"	EACH	332.41	100.34	432.75
04.1300 130	Door, metal, panic hardware, single, 3' X 7'	EACH	467.99	250.84	718.83
04.1300 135	Door, metal, panic hardware, single, 4' X 7'	EACH	467.99	250.84	718.83
04.1300 140	Door, metal, panic hardware, double, 6' X 7'	EACH	983.46	376.27	1,359.72
04.1300 145	Door, metal and glass, panic hardware, single, 3' X 7'	EACH	467.99	250.84	718.83
04.1300 150	Door, metal and glass, panic hardware, single, 4' X 7'	EACH	467.99	250.84	718.83
04.1300 155	Door, metal and glass, panic hardware, double, 6' X 7'	EACH	983.46	376.27	1,359.72
04.1400	**STORE FRONT SYSTEMS**				
04.1400 100	Storefront, wall, aluminum & glass, stub wall to 9'	SF	24.72	11.23	35.95
04.1400 105	Storefront, wall, aluminum & glass, floor to 13'	SF	28.60	12.99	41.59
04.1400 110	Storefront, entrance, 3'x7', concealed closer, center pivot, narrow stile	EACH	959.73	576.22	1,535.95
04.1400 115	Storefront, entrance, 3'x7', concealed closer, center pivot, heavy duty	EACH	1,126.75	606.22	1,732.98
04.1400 120	Storefront, entrance, 3'x7', concealed closer, center pivot, 1/2" tempered	EACH	2,380.42	750.19	3,130.61
04.1400 125	Add to entrance cost, center stop	EACH	31.59	22.83	54.42
04.1400 130	Add to entrance cost, floor check	EACH	256.45	120.04	376.49
04.1400 135	Add to entrance cost, bronze anodized aluminum	EACH	188.85		188.85
04.1400 140	Add to entrance cost, black anodized aluminum	EACH	274.34		274.34
04.1400 145	Add to entrance cost, automatic opener	EACH	3,618.15	2,023.60	5,641.74

ENCLOSURE, VERTICAL - 4.1

CSI#	Description	Unit	Material Cost	Install Cost	Total Cost
04.1500	**SPECIAL DOORS**				
04.1500 100	Garage door, wood, spring balanced, 7' X 8'	EACH	234.95	445.65	680.60
04.1500 105	Garage door, wood, spring balanced, 7' X 16'	EACH	319.54	630.06	949.59
04.1500 110	Garage door, wood, track operated, 7' X 8'	EACH	244.39	461.01	705.40
04.1500 115	Garage door, wood, track operated, 7' X 16'	EACH	328.91	645.42	974.33
04.1500 120	Roll-up doors, chain operated, galvanized steel 20 GA, 10'x10'	EACH	985.73	2,180.58	3,166.31
04.1500 125	Roll-up doors, chain operated, galvanized steel 20 GA, 12'x12'	EACH	1,349.28	2,281.23	3,630.51
04.1500 130	Roll-up doors, chain operated, galvanized steel 20 GA, 14'x14'	EACH	1,686.55	2,281.23	3,967.78
04.1500 135	Roll-up grill, crank operated, aluminum, 8'x8'	EACH	1,384.96	1,140.63	2,525.59
04.1500 140	Roll-up grill, crank operated, aluminum, 10'x10'	EACH	1,607.36	1,509.67	3,117.03
04.1500 145	Overhead doors, sectional, steel, 8' X 8'	EACH	537.94	463.00	1,000.93
04.1500 150	Overhead doors, sectional, steel, 12'x12'	EACH	740.94	771.58	1,512.52
04.1500 155	Add to overhead door cost for motor	EACH	1,017.76	140.92	1,158.68
04.1500 160	Sliding door with track steel, 12'x14'	EACH	1,023.99	359.26	1,383.25
04.1500 165	Sliding door with track steel, 14'x16'	EACH	1,315.73	411.84	1,727.57
04.1500 170	Sliding door with track steel, 16'x20'	EACH	1,672.97	499.46	2,172.43
04.1500 175	Sliding door with track steel, industrial to 50' X 30'	SF	27.53	53.45	80.98
04.1500 177	Sliding door with track steel, industrial to 90' X 30'	SF	31.94	56.96	88.90
04.1500 180	Revolving door, aluminum, 7'	EACH	19,633.41	4,676.52	24,309.94
04.1500 185	Revolving door, stainless steel, satin finish, 7'	EACH	39,473.57	4,676.52	44,150.09
04.1500 190	Revolving door, stainless steel, mirror finish, 7'	EACH	47,740.28	6,963.15	54,703.43
04.1500 195	Revolving door, bronze, satin finish, 7'	EACH	38,964.59	6,830.25	45,794.84
04.1500 200	Revolving door, bronze, mirror finish, 7'	EACH	47,326.91	8,864.10	56,191.01

4.2 - ENCLOSURE HORIZONTAL

CSI#	Description	Unit	Material Cost	Install Cost	Total Cost
	The costs in this section include materials, labor, equipment rental, supervision, and subcontractor overhead and profit. There are no allowances for the general contractor.				
	These costs are typical of those associated with the construction of a building.				
	Costs are complete, and represent normal conditions related to weather.				
	Roof costs do not include the roof structure. These costs are found in section 3.0 Structure. The insulation costs are separate from the roofing material.				
04.2100	**ROOF & ROOF MATERIALS**				
04.2100 100	Roof cover, built-up, low rise, 3 ply	SQ	142.35	99.06	241.41
04.2100 105	Roof cover, built-up, low rise, 4 ply	SQ	165.31	108.06	273.37
04.2100 110	Roof cover, built-up, low rise, 5 ply	SQ	194.43	123.03	317.45
04.2100 115	Roof cover, built-up, high rise, 3 ply	SQ	142.35	222.88	365.23
04.2100 120	Roof cover, built-up, high rise, 4 ply	SQ	165.31	243.13	408.44
04.2100 125	Roof cover, built-up, high rise, 5 ply	SQ	194.43	276.80	471.23
04.2100 130	Roof cover, plastic, elastomeric membrane, 1/16"	SQ	235.73	318.19	553.92
04.2100 135	Roof cover, plastic, elastomeric membrane, 1/32"	SF	209.61	318.19	527.81
04.2100 140	Roof cover, plastic, elastomeric membrane, loose, trocal	SQ	223.95	113.04	336.98
04.2100 145	Roof cover, plastic, elastomeric membrane, neoprene	SQ	30.36	48.96	79.32
04.2100 150	Roof cover, bituthene, 1/16", selfseal	SQ	220.07	318.19	538.26
04.2100 155	Roof cover, silicone, 3 ply, rolled	SQ	298.74	390.16	688.91
04.2100 160	Roof cover, urethane foam, silicone cover, 1' thick	SQ	564.49	165.91	730.39
04.2100 165	Roof cover, acrylic, over existing cap sheet	SQ	266.07	73.34	339.41
04.2100 170	Roof cover, acrylic, over existing gravel	SQ	266.07	146.70	412.76
04.2100 175	Roof cover, acrylic, over existing deck	SQ	272.35	72.73	345.08
04.2100 180	Roof cover, acrylic, over existing corrugated metal	SQ	266.07	66.68	332.74
04.2100 185	Roof cover, arcylic, new	SQ	266.07	146.70	412.76
04.2100 190	Roof cover, shingles, composition asphalt, 240 #	SQ	78.81	97.20	176.00
04.2100 200	Roof cover, shingles, composition asphalt, 240 #, 'A'	SQ	101.51	102.09	203.60
04.2100 205	Roof cover, shingles, composition asphalt, 300 #	SQ	112.15	102.09	214.24
04.2100 210	Roof cover, shingles, composition asphalt, 325 #, 'A'	SQ	128.20	102.09	230.29
04.2100 215	Roof cover, shingles, composition asphalt, president	SQ	174.44	93.43	267.87
04.2100 220	Roof cover, shingles, valley roll	LF	0.54	1.51	2.05
04.2100 225	Roof cover, shingles, aluminum tab, 020"	SF	3.42	3.03	6.45
04.2100 230	Roof cover, shingles, aluminum tab, 030"	SF	3.98	3.03	7.01
04.2100 235	Roof cover, shingles, porcelain enamel, 18 GA	SQ	440.20	330.12	770.32
04.2100 240	Roof cover, shingles, fiberglass tabs, 300 #	SQ	129.09	129.77	258.86
04.2100 245	Roof cover, tile, clay, spanish, 2 piece	SQ	422.68	673.61	1,096.30
04.2100 250	Roof cover, tile, clay, flat bed	SQ	179.73	375.20	554.93
04.2100 255	Roof cover, tile, clay, glazed, interlock	SQ	293.07	407.36	700.43
04.2100 260	Roof cover, tile, clay, spanish	SQ	254.04	369.85	623.89
04.2100 265	Roof cover, tile, concrete, premium	SQ	367.95	399.94	767.89
04.2100 270	Roof cover, tile, concrete, flat	SQ	246.84	343.43	590.27
04.2100 275	Roof cover, tile, concrete, interlock	SQ	226.23	315.90	542.13
04.2100 280	Roof cover, tile, slate	SQ	701.66	257.99	959.65
04.2100 285	Roof cover, corrugated aluminum, 020"	SF	1.43	2.40	3.83
04.2100 290	Roof cover, corrugated aluminum, 032"	SF	2.39	2.79	5.17
04.2100 295	Roof cover, corrugated composition, 3/16", non-walk	SF	2.09	3.26	5.34
04.2100 300	Roof cover, corrugated composition, 3/8"	SF	2.52	3.61	6.13
04.2100 305	Roof cover, corrugated fiberglass, 8 oz	SF	3.46	3.26	6.72
04.2100 310	Roof cover, corrugated galvanized iron, 26 GA	SF	2.47	3.61	6.07
04.2100 315	Roof cover, copper clad stainless steel, 16 oz panels	SF	9.39	6.58	15.96
04.2100 320	Roof cover, copper clad stainless steel, 24 oz panels	SF	13.66	8.15	21.80
04.2100 325	Roof cover, 16 oz copper, standing seam, 16" panels	SF	8.75	6.58	15.32

ENCLOSURE HORIZONTAL - 4.2

CSI#	Description	Unit	Material Cost	Install Cost	Total Cost
04.2200	**RIGID INSULATION**				
04.2200 100	Rigid, insulation board, deck, mineral fiber, 1"	SF	0.77	0.29	1.06
04.2200 105	Rigid, insulation board, deck, mineral fiber, 1-1/2"	SF	1.11	0.38	1.48
04.2200 110	Rigid, insulation board, deck, mineral fiber, 2"	SF	1.49	0.38	1.87
04.2200 115	Rigid, insulation board, deck, mineral fiber, 6"	SF	4.59	0.68	5.27
04.2200 120	Rigid, insulation board, deck, fiberglass, 1-1/2"	SF	1.94	0.40	2.34
04.2200 125	insulation board, deck, fiberglass, 2"	SF	2.57	0.38	2.94
04.2200 130	Rigid, insulation board, deck, fiberglass, 3"	SF	3.39	0.40	3.79
04.2200 135	Rigid, insulation board, deck, firtex, 2"	SF	1.89	0.47	2.36
04.2200 140	Rigid, insulation board, deck, tectum, paint, 2"	SF	3.39	0.75	4.14
04.2200 145	Rigid, insulation board, deck, tectum, paint, 2-1/2"	SF	3.94	0.85	4.79
04.2200 150	Rigid, insulation board, deck, urethane, paint, 1-1/4" R9	SF	1.94	0.33	2.27
04.2200 155	Rigid, insulation board, deck, urethane, paint, 1-1/2" R11	SF	2.49	0.40	2.89
04.2200 160	Rigid, insulation board, deck, urethane, paint, 2" R14	SF	3.20	0.48	3.68
04.2200 165	Rigid, insulation board, deck, urethane, paint, 1" non-rated R7	SF	1.42	0.26	1.68
04.2200 170	Rigid, insulation board, deck, urethane, paint, 1-1/2" non-rated R11	SF	2.22	0.37	2.59
04.2200 175	Rigid, insulation board, deck, urethane, paint, 2" non-rated R14	SF	2.49	0.48	2.97
04.2200 180	Rigid, insulation board, deck, urethane, paint, 2-1/2" non-rated R19	SF	3.19	0.63	3.82
04.2200 185	Rigid, insulation board, deck, urethane, paint, 3" non-rated R25	SF	4.19	0.70	4.89
04.2200 190	Rigid, insulation board, deck, styrofoam, paint, 1-1/2"	SF	1.89	0.40	2.29
04.2200 195	Rigid, insulation board, deck, styrofoam, paint, 2"	SF	2.41	0.45	2.86
04.2200 200	insulation board, deck, styrofoam, paint, 3"	SF	3.65	0.48	4.13
04.2200 205	Cant strip, 3" fiber	LF	0.47	1.59	2.06
04.2200 210	Cant strip, 4" fiber	LF	0.59	1.59	2.18

4.3 SUPPORT ITEMS

CSI#	Description	Unit	Material Cost	Install Cost	Total Cost
	The costs in this section include materials, labor, equipment rental, supervision, and subcontractor overhead and profit. There are no allowances for the general contractor.				
	These costs are typical of those associated with the construction of a building.				
	Costs are complete, and represent normal conditions related to weather.				
	Where appropriate paint is included in the item cost.				
04.3100	**MISCELLANEOUS IRON**				
04.3100 100	Stairs, steel, concrete tread, 14 risers	FLT	4,959.46	637.01	5,596.47
04.3100 105	Stairs, steel, concrete tread, 18 risers	FLT	12,740.11	1,460.41	14,200.52
04.3100 110	Stairs, steel, steel tread, 18 risers	FLT	10,829.08	1,922.09	12,751.17
04.3100 115	Ladders, galvanized steel, 2-1/2"x3-3/8" bar, 3/4"rung, protective cage	RISER	274.59	70.37	344.96
04.3100 120	Railing, pipe 2 high, welded, with kick plate 1-1/2"	LF	68.89	17.15	86.04
04.3100 125	Railing, pipe 3 high, welded, with kick plate 1-1/2"	LF	88.60	17.15	105.75
04.3100 130	Railing, pipe 4 high, welded, with kick plate 1-1/2"	LF	108.30	17.15	125.45
04.3100 135	Railing, pipe 1-1/2", wall type	LF	32.23	17.15	49.38
04.3100 140	Railing, welded, flat bar and angle	LF	40.21	16.02	56.22
04.3100 145	Railing, welded, flat and square bar, ornamental	LF	48.32	17.34	65.66
04.3100 150	Railing, wrought iron, stock pattern	LF	170.08	18.19	188.27
04.3100 155	Railing, wrought iron, custom pattern	LF	376.48	38.60	415.08
04.3100 160	Railing, wrought iron, stair	LF	117.13	18.86	135.99
04.3100 165	Handrails, decorative, easy designs, brass/bronze, floor mounted	LF	379.02	36.46	415.48
04.3100 170	Handrails, decorative, easy designs, brass/bronze, wall mounted	LF	138.39	30.34	168.73
04.3100 175	Handrails, decorative, easy designs, stainless steel, floor mounted	LF	231.77	40.09	271.87
04.3100 180	Handrails, decorative, easy designs, stainless steel, wall mounted	LF	132.94	24.27	157.21
04.3100 185	Handrails, decorative, easy designs, aluminum, floor mounted	LF	49.87	20.26	70.13
04.3100 190	Handrails, decorative, easy designs, stainless steel, wall mounted	LF	83.54	32.18	115.72
04.3100 195	Fire escape, ladder and balcony	EACH	4,584.51	683.29	5,267.80
04.3100 200	Ornamental sight screen, aluminum	LF	58.26	12.38	70.64
04.3100 205	Ornamental sight screen, extruded metal	LF	44.80	12.38	57.19
04.3100 210	Ornamental sun screen, aluminum, manual	LF	58.62	6.15	64.77
04.3100 215	Ornamental sun screen, aluminum, motorized	LF	105.75	8.12	113.87
04.3100 220	Wrought iron gate, 6'x7'	EACH	2,696.01	353.77	3,049.78
04.3100 225	Expansion joint, 1-1/2", floor	LF	63.29	10.86	74.16
04.3100 230	Expansion joint, 1-1/2", wall	LF	44.65	13.23	57.88
04.3100 235	Expansion joint, 1-1/2", roof	LF	88.77	12.81	101.59
04.3100 240	Expansion joint, 4", floor	LF	122.88	21.80	144.69
04.3100 245	Expansion joint, 4", wall	LF	86.14	26.46	112.60
04.3100 250	Expansion joint, 4", roof	LF	130.34	24.54	154.88
04.3200	**SHEETMETAL**				
04.3200 100	Gutter, galvanized iron, facia, 5"	LF	1.86	4.48	6.34
04.3200 105	Gutter, aluminum, facia, 5"	LF	1.26	4.48	5.74
04.3200 110	Gutter, copper, facia, 5"	LF	9.07	4.64	13.70
04.3200 115	Downspout, galvanized iron, fabricated, 3"x4"	LF	2.64	3.87	6.51
04.3200 120	Downspout, aluminum, fabricated, 3"x4"	LF	1.95	3.87	5.82
04.3200 125	Downspout, copper, fabricated, 3"x4"	LF	10.07	4.10	14.17
04.3200 130	Downspout, add for over 2 stories, fabricated, 3"x4"	LF		1.60	1.60
04.3200 135	Gravel stop, galvanized iron, facia 10"	LF	2.19	3.87	6.05
04.3200 140	Gravel stop, aluminum, facia 10"	LF	3.06	3.87	6.93
04.3200 145	Gravel stop, copper, facia 10"	LF	5.76	4.17	9.93
04.3200 150	Sheet metal, fabricated, galvanized iron, 26 gauge	SF	1.42	6.14	7.56
04.3200 155	Sheet metal, fabricated, aluminum, .032	SF	2.43	6.14	8.56
04.3200 160	Sheet metal, fabricated, copper, 16 oz	SF	4.95	6.46	11.41
04.3200 165	Reglets, galvanized iron, 26 gauge	LF	1.05	2.35	3.41
04.3200 170	Reglets, aluminum, .032"	LF	1.31	2.35	3.66
04.3200 175	Reglets, copper, 16oz	LF	3.30	2.35	5.65

SUPPORT ITEMS - 4.3

CSI#	Description	Unit	Material Cost	Install Cost	Total Cost
04.3200 177	Add for neoprene gasket	LF	0.65	0.77	1.41
04.3200 180	Roof safe & cap, 4", aluminum, .032"	EA	13.55	12.89	26.44
04.3200 185	Plumber's flash cone, galvanized, 3"	EA	10.79	16.38	27.16
04.3200 190	Scupper, aluminum, .032"	EA	19.87	25.01	44.88
04.3200 195	Vent, frieze, galvanized, 4"x24"	EA	2.28	4.40	6.68
04.3200 197	Vent, block/brick, galvanized, 8"x16"	EA	19.03	33.35	52.39
04.3200 200	Vent, block/brick, galvanized, 12"x16"	EA	24.63	33.35	57.98
04.3200 205	Louvers, manual, galvanized	SF	11.94	19.70	31.65
04.3200 210	Screens, cooling tower, galvanized	SF	9.50	19.70	29.21
04.3200 212	Screens, bird	SF	1.91	2.28	4.19
04.3200 215	Gravity ventilator, galvanized 8"	EA	181.33	46.99	228.33
04.3200 220	Gravity ventilator, galvanized 36"	EA	770.13	95.49	865.62
04.3200 225	Gravity ventilator, galvanized 60"	EA	2,060.76	341.02	2,401.78
04.3200 230	Gravity ventilator, add for hand damper	EA	61.08		61.08
04.3200 235	Gravity ventilator, add for motor damper	EA	91.66		91.66
04.3200 240	Mushroom vent, 8", to 180 cubic feet per minute	EA	451.67	244.02	695.69
04.3200 245	Mushroom vent, 12", to 360 cubic feet per minute	EA	732.04	244.02	976.06
04.3200 250	Mushroom vent, 24", to 1200 cubic feet per minute	EA	1,619.94	399.37	2,019.31
04.3200 255	Expansion joints, dry wall, aluminum cover, 2"	LF	19.15	4.55	23.70
04.3200 260	Expansion joints, concrete wall, aluminum cover, 4"-6"	LF	40.34	27.29	67.63
04.3200 265	Expansion joints, roof, neoprene & aluminum, 2"	LF	27.27	6.07	33.33
04.3200 270	Roof hatches, frame/cover, galvanized, 3'x3'-6"	EA	552.12	425.82	977.94
04.3200 275	Roof hatches, frame/cover, galvanized, 3'x9'-6"	EA	1,391.61	548.50	1,940.12
04.3200 280	Smoke vent, automatic 160 degree, galvanized, 420#, 4'-8"x4'-8"	EA	1,464.47	472.73	1,937.20
04.3200 285	Smoke vent, automatic 160 degree, galvanized, 650#, 4'-8"x9'	EA	2,309.48	624.29	2,933.77
04.3200 290	Fire vent, automatic 160 degree, aluminum, 260#, 4'-8"x4'-8"	EA	1,560.26	397.86	1,958.12
04.3200 295	Fire vent, automatic 160 degree, aluminum, 380#, 4'-8"x9'-6"	EA	2,005.27	568.37	2,573.64
04.3200 300	Fire vent, automatic 160 degree, aluminum, 425#, 6'x9'-6"	EA	2,320.68	625.21	2,945.89
04.3300	**SKYLIGHTS**				
04.3300 100	Skylights, aluminum frame, plastic dome, 2'x2'	EA	155.53	98.67	254.20
04.3300 105	Skylights, aluminum frame, plastic dome, 4'x4'	EA	316.50	98.67	415.17
04.3300 110	Skylights, aluminum frame, plastic dome, 5'x5'	EA	598.88	98.67	697.55
04.3300 115	Skylights, aluminum frame, plastic dome, 3'x6'	EA	396.63	98.67	495.30
04.3300 120	Skylights, aluminum frame, plastic dome, 7'x7'	EA	1,547.02	133.16	1,680.18
04.3300 125	Skylights, aluminum frame, plastic dome, 8'x10'	EA	2,798.81	133.16	2,931.97
04.3300 130	Skylights, aluminum frame, pyramid dome, 2'x2'	EA	225.78	98.67	324.45
04.3300 135	Skylights, aluminum frame, pyramid dome, 3'x3'	EA	254.30	98.67	352.97
04.3300 140	Skylights, aluminum frame, pyramid dome, 4'x4'	EA	459.33	98.67	558.00
04.3300 145	Skylights, double glazed, aluminum frame 2'x2'	EA	272.76	112.16	384.93
04.3300 150	Skylights, double glazed, aluminum frame 3'x3'	EA	441.22	140.96	582.17
04.3300 155	Skylights, double glazed, aluminum frame 4'x4'	EA	680.88	168.24	849.12
04.3300 160	Skylights, double glazed, aluminum frame 6'x6'	EA	1,703.48	200.07	1,903.55
04.3300 165	Skylights, double glazed, aluminum frame 4'x8'	EA	1,623.34	200.07	1,823.41
04.3300 170	Skylights, fabricated, steel frame, laminated glass, 20' span	SF	100.06		100.06
04.3300 175	Skylights, fabricated, steel frame, laminated glass, 30' span	SF	123.37		123.37
04.3300 180	Skylights, fabricated, steel frame, laminated glass, 40' span	SF	144.44		144.44
04.3400	**INSULATION**				
04.3400 100	Acoustic insulation, sound board, vertical, 1/2"	SF	0.24	0.82	1.06
04.3400 105	Insulation, batt, wall and ceiling, mineral fiber, 2-1/2" R7	SF	0.27	0.44	0.71
04.3400 110	Insulation, batt, wall and ceiling, mineral fiber, 3" R11	SF	0.35	0.44	0.80
04.3400 115	Insulation, batt, wall and ceiling, mineral fiber, 3-1/2" R13	SF	0.38	0.44	0.82
04.3400 120	Insulation, batt, wall and ceiling, mineral fiber, 3-1/2", R13, fiberglass	SF	0.45	0.44	0.90
04.3400 125	Insulation, batt, wall and ceiling, mineral fiber, 6" R19	SF	0.56	0.44	1.01
04.3400 130	Insulation, batt, wall and ceiling, mineral fiber, 8-9" R30	SF	0.49	0.91	1.39
04.3400 135	Insulation, batt, wall and ceiling, mineral fiber, 11" blown fiber	SF	0.37	0.61	0.98
04.3400 140	Insulation, batt, wall and ceiling, mineral fiber, add for supporting Batts on wire	SF	0.01	0.23	0.24

4.3 SUPPORT ITEMS

CSI#	Description	Unit	Material Cost	Install Cost	Total Cost
04.3500	**CAULKING & SEALANTS**				
04.3500 100	Caulk, linseed base, 1/8"x1/8"	LF	0.02	1.11	1.13
04.3500 105	Caulk, linseed base, 1/4"x1/4"	LF	0.10	1.66	1.76
04.3500 110	Caulk, linseed base, 1/2"x1/2"	LF	0.23	2.21	2.44
04.3500 115	Caulk, linseed base, 3/4"x3/4"	LF	0.38	3.31	3.69
04.3500 120	Caulk, linseed base, 1"x1"	LF	1.01	4.42	5.44
04.3500 125	Caulk, butyl base, 1/8"x1/8"	LF	0.02	1.11	1.13
04.3500 135	Caulk, butyl base, 1/2"x1/2"	LF	0.40	2.21	2.61
04.3500 145	Caulk, butyl base, 1"x1"	LF	1.69	4.42	6.12
04.3500 150	Caulk, acrylic, 1/8"x1/8"	LF	0.06	1.11	1.18
04.3500 160	Caulk, acrylic, 1/2"x1/2"	LF	0.96	2.21	3.17
04.3500 170	Caulk, acrylic, 1"x1"	LF	3.48	4.42	7.90
04.3500 175	Caulk, polysulfide, 1/8"x1/8"	LF	0.34	1.11	1.46
04.3500 185	Caulk, polysulfide, 1/2"x1/2"	LF	1.11	2.21	3.32
04.3500 195	Caulk, polysulfide, 1"x1"	LF	4.56	4.42	8.98
04.3500 200	Caulk, silicone, 1/8"x1/8"	LF	0.10	1.11	1.21
04.3500 210	Caulk, silicone, 1/2"x1/2"	LF	1.49	2.21	3.70
04.3500 220	Caulk, silicone, 1"x1"	LF	5.90	4.42	10.32
04.3500 225	Caulk, mildew resistant, 1/8"x1/8"	LF	0.13	1.11	1.24
04.3500 235	Caulk, mildew resistant, 1/2"x1/2"	LF	2.31	2.21	4.52
04.3500 240	Caulk, mildew resistant, 3/4"x3/4"	LF	5.16	3.31	8.47
04.3500 245	Caulk, mildew resistant, 1"x1"	LF	7.72	4.42	12.14
04.3500 250	Caulk, elastomeric, for concrete	LF	2.71	1.84	4.55
04.3500 255	Caulk, acoustical, butyl rubber, 1/4"x1/2"	LF	0.53	2.49	3.02
04.3500 260	Sealants, self-leveling, polysulfide polymer, 1/4"x3/8"	LF	1.65	3.09	4.73
04.3500 265	Sealants, self-leveling, acrylic latex polymer, 1/4"x3/8"	LF	1.45	3.09	4.54
04.3500 270	Sealants, self-leveling, polyurethane, 1/4"x3/8"	LF	1.83	3.09	4.92

INTERNALS, VERTICAL - 5.1

CSI#	Description	Unit	Material Cost	Install Cost	Total Cost
	The costs in this section include materials, labor, equipment rental, supervision, and subcontractor overhead and profit. There are no allowances for the general contractor.				
	Costs are complete, walls are painted where applicable with allowances for rubber base and gypsum board or plaster on both sides.				
05.1105	**GYPSUM BOARD ON STUD**				
05.1105 100	Interior wall, non-rated, residential, 2"x4" wood stud 16"OC, 8' high, insulation, 5/8" gypsum board, both sides, painted, rubber base	SF	3.79	6.56	10.35
05.1105 120	Interior wall, non-rated, residential, 2"x6" wood stud 16"OC, 8' high, insulation, 5/8" gypsum board, both sides, painted, rubber base	SF	4.34	7.55	11.89
05.1105 135	Interior plumbing wall, water resistant, residential, 2"x6" wood stud 16"OC, 8' high, insulation, 5/8" gypsum board, one side, painted	SF	3.96	6.46	10.42
05.1105 150	Interior shearwall, non-rated, residential, 2"x4" wood stud 16"OC, 10' high, 3/8" structural plywood, insulation, 5/8" gypsum board, both sides painted, rubber base	SF	3.72	6.53	10.25
05.1105 165	Interior shearwall, 2 hour rated, residential, 2"x6" wood stud 16"OC, 10' high, 3/8" structural plywood, insulation, 5/8" gypsum board, both sides painted, rubber base	SF	5.31	8.83	14.14
05.1105 170	Interior wall, non-rated, commercial, 2"x4" wood stud 16"OC, 8' high, insulation, 5/8" gypsum board, both sides, painted, rubber base	SF	3.79	6.56	10.35
05.1105 175	Interior wall, 1 hour rated, commercial, 2"x4" wood stud 16"OC, 10' high, insulation, 5/8" gypsum board, both sides, painted, rubber base	SF	3.72	6.53	10.25
05.1105 180	Interior wall, 2 hour rated, commercial, 2"x4" wood stud 16"OC, 10' high, insulation, 5/8" gypsum board, both sides, painted, rubber base	SF	5.20	7.02	12.22
05.1105 190	Interior wall, non-rated, commercial, 2"x6" wood stud 16"OC, 8' high, insulation, 5/8" gypsum board, both sides, painted, rubber base	SF	4.34	7.55	11.89
05.1105 195	Interior wall, 1 hour rated, commercial, 2"x6" wood stud 16"OC, 10' high, insulation, 5/8" gypsum board, both sides, painted, rubber base	SF	4.18	7.76	11.95
05.1105 200	Interior wall, 2 hour rated, commercial, 2"x6" wood stud 16"OC, 10' high, insulation, 5/8" gypsum board, both sides, painted, rubber base	SF	5.16	9.04	14.20
05.1105 205	Interior plumbing wall, water resistant, commercial, 2"x6" wood stud 16"OC, 8' high, insulation, 5/8" water resistant gypsum board, one side, painted	SF	4.72	6.94	11.66
05.1105 220	Interior shearwall, non-rated, commercial, 2"x4" wood stud 16"OC, 3/8" structural plywood, insulation, 5/8" gypsum board, both sides, painted, rubber base	SF	3.79	6.56	10.35
05.1105 225	Interior shearwall, 2 hour rated, commercial, 2"x4" wood stud 16"OC, 3/8" structural plywood, insulation, 5/8" gypsum board, both sides, painted, rubber base	SF	4.76	7.84	12.60
05.1105 230	Interior shearwall, non-rated, commercial, 2"x6" wood stud 16"OC, 3/8" structural plywood, insulation, 5/8" gypsum board, both sides, painted, rubber base	SF	4.34	7.55	11.89
05.1105 235	Interior shearwall, 2 hour rated, commercial, 2"x6" wood stud 16"OC, 3/8" structural plywood, insulation, 5/8" gypsum board, both sides, painted, rubber base	SF	5.94	8.67	14.62
05.1105 240	Interior wall, non-rated, institutional, 2"x4" wood stud 16"OC, 8' high, insulation, 5/8" gypsum board, both sides, painted, rubber base	SF	3.79	6.56	10.35
05.1105 245	Interior wall, 1 hour rated, institutional, 2"x4" wood stud 16"OC, 10' high, insulation, 5/8" gypsum board, both sides, painted, rubber base	SF	3.72	6.53	10.25
05.1105 250	Interior wall, 2 hour rated, institutional, 2"x4" wood stud 16"OC, 10' high, insulation, 5/8" gypsum board, both sides, painted, rubber base	SF	5.33	7.65	12.98
05.1105 255	Interior plumbing wall, water resistant, institutional, 2"x4" wood stud 16"OC, 8' high, insulation, 5/8" gypsum board, one side, painted	SF	4.18	5.95	10.13
05.1105 260	Interior wall, non-rated, institutional, 2"x6" wood stud 16"OC, 8' high, insulation, 5/8" gypsum board, both sides, painted, rubber base	SF	4.34	7.55	11.89
05.1105 265	Interior wall, 1 hour rated, institutional, 2"x6" wood stud 16"OC, 10' high, insulation, 5/8" gypsum board, both sides, painted, rubber base	SF	4.18	7.76	11.95
05.1105 270	Interior wall, 2 hour rated, institutional, 2"x6" wood stud 16"OC, 10' high, insulation, 5/8" gypsum board, both sides, painted, rubber base	SF	5.79	8.89	14.68
05.1105 275	Interior plumbing wall, water resistant, institutional, 2"x6" wood stud 16"OC, 8' high, insulation, 5/8" gypsum board, one side, painted	SF	4.72	6.94	11.66
05.1105 290	Interior shearwall, non-rated, institutional, 2"x4" wood studs 16" OC, 3/8" structural plywood, 10' high, insulation, 5/8" gypsum board, both sides, painted, rubber base	SF	3.72	6.53	10.25
05.1105 295	Interior shearwall, 2 hour rated institutional, 2"x4" wood studs 16" OC, 3/8" structural plywood, 10' high, insulation, 5/8" gypsum board, both sides, painted, rubber base	SF	5.33	7.65	12.98

5.1 - INTERNALS, VERTICAL

CSI#	Description	Unit	Material Cost	Install Cost	Total Cost
05.1105 300	Interior shearwall, non-rated institutional, 2"x6" wood studs 16" OC, 3/8" structural plywood, 10' high, insulation, 5/8" gypsum board, both sides, painted, rubber base	SF	4.18	7.76	11.95
05.1105 305	Interior shearwall, 2 hour rated institutional, 2"x6" wood studs 16" OC, 3/8" structural plywood, 10' high, insulation, 5/8" gypsum board, both sides, painted, rubber base	SF	5.79	8.89	14.68
05.1105 310	Interior partitions, non-rated commercial, drywall studs 25 GA 3 5/8", insulation, 5/8" gypsum board, both sides, painted, rubber base	SF	3.86	6.46	10.32
05.1105 315	Interior partitions, 1 hour rated commercial, drywall studs 25 GA 3 5/8", insulation, 5/8" gypsum board, both sides, painted, rubber base	SF	3.86	6.46	10.32
05.1105 320	Interior partitions, 2 hour rated commercial, drywall studs 25 GA 3 5/8", insulation, 5/8" gypsum board, both sides, painted, rubber base	SF	5.46	7.59	13.05
05.1105 325	Interior plumbing wall, water resistant, commercial, drywall studs 25 GA 3 5/8", insulation, 5/8" gypsum board, one side, painted	SF	4.24	5.85	10.10
05.1105 330	Interior chase wall, 1 hour rated, commercial, drywall studs 25 GA 3 5/8", insulation, 5/8" gypsum board, one side, painted, rubber base	SF	3.73	5.83	9.56
05.1105 335	Interior chase wall, 2 hour rated, commercial, drywall studs 25 GA 3 5/8", insulation, 5/8" gypsum board, one side, painted, rubber base	SF	5.25	6.66	11.91
05.1105 375	Interior partitions, non-rated, institutional, drywall studs 25 GA 3 5/8", insulation, 5/8" gypsum board, both sides, painted, rubber base	SF	3.86	6.46	10.32
05.1105 380	Interior partitions, 1 hour rated, institutional, drywall studs 25 GA 3 5/8", insulation, 5/8" gypsum board, both sides, painted, rubber base	SF	3.86	6.46	10.32
05.1105 385	Interior partitions, 2 hour rated, institutional, drywall studs 25 GA 3 5/8", insulation, 5/8" gypsum board, both sides, painted, rubber base	SF	5.46	7.59	13.05
05.1105 390	Interior plumbing wall, water resistant, institutional, drywall studs 25 GA 3 5/8", insulation, 5/8" gypsum board, one side, painted	SF	5.01	6.33	11.34
05.1105 400	Interior chase wall, 2 hour rated, institutional, drywall studs 25 GA 3 5/8", insulation, 5/8" gypsum board, one side, painted, rubber base	SF	4.69	7.42	12.10
05.1105 440	Interior structural wall, non-rated, commercial, studs 14 GA 3 5/8", insulation, 5/8" gypsum board, both sides, painted, rubber base	SF	5.66	8.82	14.48
05.1105 450	Interior structural wall, 2 hour rated, commercial, studs 14 GA 3 5/8", insulation, 5/8" gypsum board, both sides, painted, rubber base	SF	7.27	9.95	17.21
05.1105 500	Interior shaft wall, 1 hour rated, commercial, studs CH 20 GA 4", insulation, 5/8" gypsum board, one side, 1/2" ashpalt core sheathing, one side, paint one side, rubber base	SF	6.23	6.48	12.71
05.1105 510	Interior structural wall, non-rated, institutional, studs 14 GA 3 5/8", insulation, 5/8" gypsum board, both sides, painted, rubber base	SF	5.66	8.82	14.48
05.1105 520	Interior structural wall, 2 hour rated, institutional, studs 14 GA 3 5/8", insulation, 5/8" gypsum board, both sides, painted, rubber base	SF	6.49	9.77	16.26
05.1105 575	Interior shaft wall, 1 hour rated, institutional, studs CH 25 GA 4", insulation, 5/8" gypsum board, one side, 1/2" asphalt core sheathing, one side, paint one side, rubber base	SF	5.41	5.97	11.37
05.1105 580	Column enclosure, non-rated, commercial, drywall studs 25 GA 1 5/8", 5/8" gypsum board, painted, rubber base	SF	3.56	5.41	8.97
05.1105 590	Column enclosure, 2 hour rated, commercial, drywall studs 25 GA 1 5/8", 5/8" gypsum board, painted, rubber base	SF	3.56	5.41	8.97
05.1110	**LATH AND PLASTER ON STUD**				
05.1110 100	Interior wall, non-rated, residential, 2"x4" wood stud 16"OC, 8' high, insulation, lath & plaster, both sides, painted, rubber base	SF	11.51	21.63	33.14
05.1110 120	Interior wall, non-rated, residential, 2"x6" wood stud 16"OC, 8' high, insulation, lath & plaster, both sides, painted, rubber base	SF	12.06	22.61	34.67
05.1110 135	Interior plumbing wall, residential, 2"x6" wood stud 16"OC, 8' high, insulation, lath & plaster, one side, painted	SF	11.93	21.98	33.91
05.1110 150	Interior shearwall, non-rated, residential, 2"x4" wood stud 16"OC, 10' high, 3/8" structural plywood, insulation, lath & plaster, both sides painted, rubber base	SF	11.44	21.60	33.04
05.1110 160	Interior shearwall, non-rated, residential, 2"x6" wood stud 16"OC, 10' high, 3/8" structural plywood, insulation, lath & plaster, both sides painted, rubber base	SF	12.06	22.61	34.67
05.1110 170	Interior wall, non-rated, commercial, 2"x4" wood stud 16"OC, 8' high, insulation, lath & plaster, both sides, painted, rubber base	SF	11.51	21.63	33.14
05.1110 175	Interior wall, 1 hour rated, commercial, 2"x4" wood stud 16"OC, 10' high, insulation, furring, lath & plaster, both sides, painted, rubber base	SF	11.44	21.60	33.04
05.1110 190	non-rated, commercial, 2"x6" wood stud 16"OC, 8' high, insulation, lath & plaster, both sides, painted, rubber base	SF	12.06	19.41	31.46
05.1110 195	Interior wall, 1 hour rated, commercial, 2"x6" wood stud 16"OC, 10' high, insulation, furring, lath & plaster, both sides, painted, rubber base	SF	13.45	24.43	37.89

INTERNALS, VERTICAL - 5.1

CSI#	Description	Unit	Material Cost	Install Cost	Total Cost
05.1110 205	Interior plumbing wall, commercial, 2"x6" wood stud 16"OC, 8' high, insulation, lath & plaster, one side, painted	SF	11.16	21.50	32.66
05.1110 220	Interior shearwall, non-rated, commercial, 2"x4" wood stud 16"OC, 3/8" structural plywood, insulation, lath & plaster, both sides, painted, rubber base	SF	11.51	21.63	33.14
05.1110 230	Interior shearwall, non-rated, commercial, 2"x6" wood stud 16"OC, 3/8" structural plywood, insulation, lath & plaster, both sides, painted, rubber base	SF	12.06	22.61	34.67
05.1110 240	Interior wall, non-rated, institutional, 2"x4" wood stud 16"OC, 8' high, insulation, lath & plaster, both sides, painted, rubber base	SF	11.51	21.63	33.14
05.1110 245	Interior wall, 1 hour rated, institutional, 2"x4" wood stud 16"OC, 10' high, insulation, furring, lath & plaster, both sides, painted, rubber base	SF	12.84	23.42	36.26
05.1110 255	Interior plumbing wall, institutional, 2"x4" wood stud 16"OC, 8' high, insulation, lath & plaster, one side, painted	SF	10.62	20.51	31.13
05.1110 260	Interior wall, non-rated, institutional, 2"x6" wood stud 16"OC, 8' high, insulation, lath & plaster, both sides, painted, rubber base	SF	12.06	22.61	34.67
05.1110 265	Interior wall, 1 hour rated, institutional, 2"x6" wood stud 16"OC, 10' high, insulation, furring, lath & plaster, both sides, painted, rubber base	SF	13.30	24.65	37.95
05.1110 275	Interior plumbing wall, institutional, 2"x6" wood stud 16"OC, 8' high, insulation, lath & plaster, one side, painted	SF	11.16	21.50	32.66
05.1110 290	Interior shearwall, non-rated, institutional, 2"x4" wood studs 16" OC, 3/8" structural plywood, 10' high, insulation, lath & plaster, both sides, painted, rubber base	SF	11.44	21.60	33.04
05.1110 300	Interior shearwall, non-rated institutional, 2"x6" wood studs 16" OC, 3/8" structural plywood, 10' high, insulation, lath & plaster, both sides, painted, rubber base	SF	11.90	22.83	34.73
05.1110 310	Interior partitions, non-rated commercial, drywall studs 25 GA 3 5/8", insulation, lath & plaster, both sides, painted, rubber base	SF	11.58	21.53	33.10
05.1110 315	Interior partitions, 1 hour rated commercial, drywall studs 25 GA 3 5/8", furring, lath & plaster, both sides, painted, rubber base	SF	12.97	23.35	36.32
05.1110 325	Interior plumbing wall, commercial, drywall studs 25 GA 3 5/8", insulation, lath & plaster, one side, painted	SF	11.45	20.89	32.34
05.1110 330	Interior chase wall, 1 hour rated, commercial, drywall studs 25 GA 3 5/8", insulation, furring, lath & plaster, one side, painted, rubber base	SF	10.68	20.41	31.10
05.1110 375	Interior partitions, non-rated, institutional, drywall studs 25 GA 3 5/8", insulation, lath & plaster, both sides, painted, rubber base	SF	11.58	21.53	33.10
05.1110 380	Interior partitions, 1 hour rated, institutional, drywall studs 25 GA 3 5/8", insulation, furring, lath & plaster, both sides, painted, rubber base	SF	12.97	23.35	36.32
05.1110 390	Interior plumbing wall, institutional, drywall studs 25 GA 3 5/8", insulation, lath & plaster, one side, painted	SF	10.68	20.41	31.10
05.1110 395	Interior chase wall, 1 hour rated, institutional, drywall studs 25 GA 3 5/8", insulation, furring, lath & plaster, one side, painted, rubber base	SF	12.84	22.71	35.56
05.1110 455	Interior structural wall, non-rated, commercial, studs 16 GA 3 5/8", insulation, lath & plaster, both sides, painted, rubber base	SF	12.68	23.89	36.56
05.1110 460	Interior structural wall, 1 hour rated, commercial, studs 16 GA 3 5/8", insulation, lath & plaster, both sides, painted, rubber base	SF	12.68	23.89	36.56
05.1110 500	Interior shaft wall, 1 hour rated, commercial, studs CH 20 GA 4", insulation, lath & plaster, one side, 1/2" ashpalt core sheathing, one side, paint one side, rubber base	SF	13.53	23.80	37.33
05.1110 510	Interior structural wall, non-rated, institutional, studs 14 GA 3 5/8", insulation, lath & plaster, both sides, painted, rubber base	SF	12.66	23.89	36.55
05.1110 545	Interior structural wall, 1 hour rated, institutional, studs 14 GA 3 5/8", insulation, lath & plaster, both sides, painted, rubber base	SF	12.66	23.89	36.55
05.1110 575	Interior shaft wall, 1 hour rated, institutional, studs CH 25 GA 4", insulation, lath & plaster, one side, 1/2" asphalt core sheathing, one side, paint one side, rubber base	SF	13.76	23.80	37.56
05.1110 580	Column furring, non-rated, commercial, drywall studs 25 GA 1 5/8", lath & plaster painted, rubber base	SF	12.67	22.48	35.15
05.1110 610	Column furring, non-rated, institutional, drywall studs 25 GA 1 5/8", lath & plaster, painted, rubber base	SF	12.67	22.48	35.15
05.1115	**CONCRETE BLOCK**				
05.1115 105	Concrete masonry unit, 4"x8"x16", #4 bar, 32" on center, both ways, unfilled	SF	8.46	6.73	15.18
05.1115 110	Concrete masonry unit, 6"x8"x16", #4 bar, 32" on center, both ways, unfilled	SF	9.81	7.23	17.04
05.1115 115	Concrete masonry unit, 8"x8"x16", #4 bar, 32" on center, both ways, unfilled	SF	10.33	7.64	17.97
05.1115 120	Concrete masonry unit, 12"x8"x16", #4 bar, 32" on center, both ways, unfilled	SF	16.07	8.52	24.58
05.1115 125	Concrete masonry unit, 4"x8"x16", #4 bar, 32" on center, both ways, filled	SF	8.96	7.24	16.20
05.1115 130	Concrete masonry unit, 6"x8"x16", #4 bar, 32" on center, both ways, filled	SF	11.31	7.39	18.70
05.1115 135	Concrete masonry unit, 8"x8"x16", #4 bar, 32" on center, both ways, filled	SF	12.15	7.81	19.95
05.1115 140	Concrete masonry unit, 12"x8"x16", #4 bar, 32" on center, both ways, filled	SF	15.33	9.46	24.79

5.1 - INTERNALS, VERTICAL

CSI#	Description	Unit	Material Cost	Install Cost	Total Cost
05.1115 145	Concrete masonry unit, 8"x4"x16", #4 bar, 32" on center, both ways, filled	SF	17.42	10.28	27.70
05.1115 150	Concrete masonry unit, 12"x4"x16", #4 bar, 32" on center, both ways, filled	SF	19.08	11.91	30.98
05.1115 155	Concrete masonry unit, grout lock, 6"x8"x24", #4 bar, filled	SF	12.11	5.01	17.13
05.1115 160	Concrete masonry unit, grout lock, 8"x8"x16", #4 bar, filled	SF	12.81	6.09	18.90
05.1115 165	Concrete masonry unit, grout lock, 12"x8"x16", #4 bar, filled	SF	16.46	7.23	23.68
05.1115 170	Concrete masonry unit, slumpstone, 8"x4"x16", #4 bar, filled	SF	21.06	11.08	32.15
05.1115 175	Concrete masonry unit, slumpstone, 8"x8"x16", #4 bar, filled	SF	16.39	9.32	25.70
05.1115 180	Concrete masonry unit, splitface, 8"x4"x16", #4 bar, filled	SF	19.03	11.07	30.10
05.1115 185	Concrete masonry unit, glazed 1 side, 4"x8"x16", reinforced, filled	SF	14.35	10.27	24.63
05.1115 190	Concrete masonry unit, glazed 1 side, 6"x8"x16", reinforced, filled	SF	16.51	10.99	27.50
05.1115 195	Concrete masonry unit, glazed 1 side, 8"x8"x16", reinforced, filled	SF	19.32	11.92	31.23
05.1115 200	Concrete masonry unit, glazed 1 side, 4"x4"x16", reinforced, filled	SF	21.26	11.75	33.01
05.1115 205	Concrete masonry unit, glazed 1 side, 6"x4"x16", reinforced, filled	SF	24.82	13.65	38.47
05.1115 210	Concrete masonry unit, glazed 1 side, 8"x4"x16", reinforced, filled	SF	28.37	14.51	42.89
05.1115 215	Concrete masonry unit, glazed 1 side, 12"x8"x16", reinforced, filled	SF	23.34	16.91	40.25
05.1115 220	Concrete masonry unit, add for glazing both sides,	SF	7.85		7.85
05.1115 225	Concrete masonry unit, screen block, 4"x12"x12"	SF	8.24	8.70	16.94
05.1115 250	add for pilasters	SF	34.39		34.39
05.1115 255	add for sill blocks	LF	3.71	6.64	10.35
05.1115 260	add for cutting blocks	LF	4.29	10.92	15.21
05.1115 265	add for bond beams	LF	8.15	14.77	22.92
05.1115 270	add for lintels, over openings	LF	15.89	24.97	40.87
05.1120	**STRUCTURAL TILE**				
05.1120 100	Structural tile, glazed 1 side, 2"x6"x12"	SF	7.17	12.66	19.83
05.1120 105	Structural tile, glazed 1 side, 4"x6"x12"	SF	10.55	14.46	25.01
05.1120 110	Structural tile, glazed 2 side, 4"x6"x12"	SF	12.25	16.67	28.92
05.1120 115	Structural tile, glazed 1 side, 6"x6"x12"	SF	10.14	15.26	25.40
05.1120 120	Structural tile, glazed 1 side, 3"x6"x12"	SF	7.49	14.69	22.18
05.1120 125	Structural tile, glazed 1 side, base	LF	10.18	14.69	24.87
05.1120 127	Structural tile, glazed 2 side, cap	SF	9.91	14.94	24.84
05.1120 130	Structural tile, clay backing, load bearing, 4"x12"x12"	SF	3.42	7.48	10.91
05.1120 135	Structural tile, clay backing, load bearing, 6"x12"x12"	SF	3.98	8.21	12.19
05.1120 140	Structural tile, clay backing, load bearing, 8"x12"x12"	SF	4.57	9.19	13.77
05.1120 145	Structural tile, clay backing, non-load bearing, 4"x12"x12"	SF	3.36	6.74	10.10
05.1120 150	Structural tile, clay backing, non-load bearing, 6"x12"x12"	SF	3.56	7.49	11.05
05.1120 155	Structural tile, clay backing, non-load bearing, 8"x12"x12"	SF	5.63	8.20	13.83
05.1125	**WALL FINISHES**				
05.1125 100	Wall finish, Granite, 3/4" thick	SF	27.69	28.01	55.71
05.1125 105	Wall finish, Granite, 1-1/4" thick	SF	28.38	31.38	59.76
05.1125 110	Wall finish, Limestone, 2" thick	SF	19.52	28.14	47.67
05.1125 115	Wall finish, Limestone, 3" thick	SF	20.75	35.20	55.96
05.1125 120	Wall finish, Marble, 7/8" thick	SF	26.47	27.73	54.20
05.1125 125	Wall finish, Marble, 1-1/4" thick	SF	30.26	30.92	61.18
05.1125 130	Wall finish, Travertine,	SF	23.05	27.22	50.27
05.1125 135	Wall finish, Travertine, base 9"x3/4"	SF	32.85	38.64	71.49
05.1125 145	Wall finish, Sandstone, 2" thick	SF	17.19	25.55	42.74
05.1125 147	Wall finish, Sandstone, 3" thick	SF	20.39	28.43	48.82
05.1125 155	Wall cover, paper hanging, normal conditions	SF		0.95	0.95
05.1125 160	Wall cover, wall paper, 36 SF/roll,	ROLL	34.13	28.07	62.20
05.1125 165	Wall cover, vinyl, 7oz, light	SF	1.83	0.62	2.45
05.1125 170	Wall cover, vinyl, 14oz, medium	SF	2.00	0.78	2.78
05.1125 175	Wall cover, vinyl, 22oz, heavy	SF	2.27	0.90	3.17
05.1125 180	Wall cover, vinyl, 14oz, aluminum back	SF	1.69	2.29	3.98
05.1125 185	Wall cover, linen, acrylic back	SF	1.42	1.22	2.65
05.1125 190	Wall cover, glass cloth	SF	1.55	0.84	2.39
05.1125 195	Wall cover, felt	SF	2.71	1.73	4.43
05.1125 200	Wall cover, cork sheating, 1/8"	SF	2.16	1.73	3.89
05.1125 205	Wall cover, flexible wood, veneer	SF	5.12	3.12	8.24

INTERNALS, VERTICAL - 5.1

CSI#	Description	Unit	Material Cost	Install Cost	Total Cost
05.1125 210	Laminated plastics, standard patterns & colors, w/o backing, adhesive 1/16"	SF	2.36	2.23	4.59
05.1125 230	Wall cover, hardboard, photo repro, w/o backing, plastic, 1/8", trim	SF	2.75	1.39	4.14
05.1125 235	Wall cover, hardboard, photo repro, w/o backing, plastic, 1/4", trim	SF	3.64	1.39	5.04
05.1125 240	Wall cover, pegboard, photo repro, w/o backing, plastic, 1/4", trim	SF	0.91	1.39	2.30
05.1125 245	Wall cover, pegboard, photo repro, w/o backing, plastic, 1/8", trim	SF	0.83	1.39	2.22
05.1125 250	Wall covering, tile, plastic tile, 4 1/4"x4 1/4"x.11"	SF	1.79	2.23	4.02
05.1125 255	Wall covering, tile, plastic tile, 4 1/4"x4 1/4"x.05"	SF	1.15	2.23	3.38
05.1125 260	Wall covering, tile, aluminum tile, 4 1/2"x4 1/2"	SF	8.02	2.67	10.69
05.1125 265	Wall covering, tile, copper/aluminum tile, 4 1/4"x4 1/4"	SF	8.02	2.67	10.69
05.1125 270	Wall covering, tile, stain steel tile, 4 1/4"x4 1/4"	SF	16.30	3.17	19.46
05.1125 275	Wainscote, galvanized sheet metal	SF	1.65	2.32	3.97
05.1125 280	Wall covering, ceramic tile, 1"x 1", mortar	SF	6.83	10.51	17.34
05.1125 285	Wall covering, ceramic tile, 1"x 1", mastic	SF	6.72	5.45	12.18
05.1125 290	Wall covering, ceramic tile, 2"x 1", mortar	SF	6.47	8.65	15.11
05.1125 295	Wall covering, ceramic tile, 2"x 1", mastic	SF	6.31	4.84	11.15
05.1125 300	Wall covering, ceramic tile, 4"x 4", mortar	SF	6.02	8.43	14.45
05.1125 305	Wall covering, ceramic tile, 4"x 4", mastic	SF	5.92	4.84	10.76
05.1125 310	Wall covering, ceramic tile, 6"x 3", mortar	SF	6.47	8.21	14.67
05.1125 315	Wall covering, ceramic tile, 6"x 3", mastic	SF	6.31	4.18	10.49
05.1125 320	Wall covering, ceramic tile, 6"x 4", mortar	SF	6.31	7.54	13.86
05.1125 325	Wall covering, ceramic tile, 6"x 4", mastic	SF	6.31	4.03	10.34
05.1125 330	Wall painting, concrete, prime	SF	0.10	0.38	0.48
05.1125 332	Wall painting, concrete, prime + 1 finished	SF	0.18	0.72	0.91
05.1125 335	Wall painting, concrete, prime + 2 finished	SF	0.25	1.06	1.31
05.1125 340	Wall painting, gypsum wall board, prime brush	SF	0.04	0.34	0.39
05.1125 345	Wall painting, gypsum wall board, prime + 1 finished, roll	SF	0.13	0.43	0.55
05.1125 350	Wall painting, gypsum wall board, prime + 2 finished, roll	SF	0.23	0.68	0.91
05.1125 355	Wall painting, plaster, prime, roll	SF	0.10	0.38	0.48
05.1125 360	Wall painting, plaster, prime + 1 finished, roll	SF	0.15	0.72	0.88
05.1125 365	Wall painting, plaster, prime + 2 finished, roll	SF	0.25	1.06	1.31
05.1125 370	Concrete block painting, prime, brush	SF	0.04	0.34	0.39
05.1125 375	Concrete block painting, prime + 1 finished, brush	SF	0.13	0.63	0.76
05.1125 380	Concrete block painting, prime + 2 finished, brush	SF	0.23	1.01	1.24
05.1125 385	Concrete block painting, prime, roll	SF	0.04	0.23	0.27
05.1125 390	Concrete block painting, prime + 1 finished, roll	SF	0.13	0.43	0.55
05.1125 395	Concrete block painting, prime + 2 finished, roll	SF	0.23	0.68	0.91
05.1125 400	Sheet metal painting, prime	SF	0.09	0.63	0.72
05.1125 405	Sheet metal painting, prime + 1 finished	SF	0.15	1.10	1.25
05.1125 410	Sheet metal painting, prime + 2 finished	SF	0.20	1.65	1.85
05.1125 415	Wood trim painting, prime	SF	0.10	0.38	0.48
05.1125 420	Wood trim painting, prime + 1 finished	SF	0.20	0.70	0.90
05.1125 425	Wood trim painting, prime + 2 finished	SF	0.27	1.06	1.33
05.1200	**DOORS, INCLUDING HARDWARE**				
05.1200 105	Door, hollow metal, stock, hardware, painted, door 18 GA, frame 16 GA, 3'0"x7'0", non-rated	EACH	740.69	382.48	1,123.17
05.1200 115	Door, hollow metal, stock, hardware, painted, door 18 GA, frame 16 GA, 4'0"x8'0", non-rated	EACH	808.05	455.22	1,263.26
05.1200 125	Door, hollow metal, stock, hardware, painted, door 18 GA, frame 16 GA, 3'0"x7'0", 1 hour rated	EACH	799.05	382.48	1,181.53
05.1200 135	Door, hollow metal, stock, hardware, painted, door 18 GA, frame 16 GA, 4'0"x8'0", 1 hour rated	EACH	866.41	455.22	1,321.62
05.1200 145	Door, hollow metal, stock, hardware, painted, door 18 GA, frame 16 GA, 3'0"x7'0", 3 hour rated	EACH	935.11	382.48	1,317.59
05.1200 155	Door, hollow metal, stock, hardware, painted, door 18 GA, frame 16 GA, 4'0"x8'0", 3 hour rated	EACH	866.41	455.22	1,321.62
05.1200 165	Door, hollow metal, custom, hardware, painted, door 18 GA, frame 16 GA, 3'0"x7'0", non-rated	EACH	882.57	382.48	1,265.05
05.1200 175	Door, hollow metal, custom, hardware, painted, door 18 GA, frame 16 GA, 4'0"x8'0", non-rated	EACH	1,005.20	455.22	1,460.41

5.1 - INTERNALS, VERTICAL

CSI#	Description	Unit	Material Cost	Install Cost	Total Cost
05.1200 185	Door, hollow metal, custom, hardware, painted, door 18 GA, frame 16 GA, 3'0"x7'0", 1 hour rated	EACH	947.71	409.85	1,357.57
05.1200 195	Door, hollow metal, custom, hardware, painted, door 18 GA, frame 16 GA, 4'0"x8'0", 1 hour rated	EACH	1,063.56	455.22	1,518.77
05.1200 205	Door, hollow metal, custom, hardware, painted, door 18 GA, frame 16 GA, 3'0"x7'0", 3 hour rated	EACH	1,076.99	382.48	1,459.47
05.1200 215	Door, hollow metal, custom, hardware, painted, door 18 GA, frame 16 GA, 4'0"x8'0", 3 hour rated	EACH	1,063.56	455.22	1,518.77
05.1200 220	Door, hollow metal, institutional, hardware, painted, door 16 GA, frame 14 GA, 3'0"x7'0", non-rated	EACH	988.80	430.53	1,419.33
05.1200 230	Door, hollow metal, institutional, hardware, painted, door 16 GA, frame 14 GA, 4'0"x7'0", non-rated	EACH	1,131.43	453.45	1,584.89
05.1200 235	Door, hollow metal, institutional, hardware, painted, door 16 GA, frame 14 GA, 3'0"x7'0", 1 hour rated	EACH	992.77	438.97	1,431.74
05.1200 245	Door, hollow metal, institutional, hardware, painted, door 16 GA, frame 14 GA, 4'0"x7'0", 1 hour rated	EACH	1,189.80	453.45	1,643.25
05.1200 250	Door, hollow metal, institutional, hardware, painted, door 16 GA, frame 14 GA, 3'0"x7'0", 3 hour rated	EACH	1,184.99	430.53	1,615.51
05.1200 260	Door, hollow metal, institutional, hardware, painted, door 16 GA, frame 14 GA, 4'0"x7'0", 3 hour rated	EACH	1,325.85	453.45	1,779.31
05.1200 265	Add for galvanizing	EACH	67.64		67.64
05.1200 267	Add for baked enamel	EACH	22.81		22.81
05.1200 270	Add for porcelain	EACH	187.49		187.49
05.1200 272	Add for 10" x 10" vision light	EACH	99.26		99.26
05.1200 274	Add for half glass opening	EACH	123.96		123.96
05.1200 275	Door, wood, commercial, door and frame, hardware, solid core, prehung, 3'0"x6'8", painted	EACH	145.73	100.34	246.07
05.1200 285	Door, wood, commercial, door and frame, hardware, solid core, prehung, 3'6"x6'8", painted	EACH	163.63	100.34	263.97
05.1200 290	Door, wood, commercial, door and frame, hardware, solid core, prehung, 3'0"x7'0", painted	EACH	163.60	110.37	273.97
05.1200 295	Door, wood, commercial, solid core, hardware, job hung, metal frame 16 GA, 2'8"x6'8", painted	EACH	607.75	503.12	1,110.86
05.1200 305	Door, wood, commercial, solid core, hardware, job hung, metal frame 16 GA, 3'0"x7'0", painted	EACH	589.46	512.77	1,102.23
05.1200 310	Door, wood, commercial, solid core, hardware, job hung, metal frame 16 GA, 3'6"x7'0", painted	EACH	632.71	521.21	1,153.92
05.1200 335	Door, wood, institutional, solid core, hardware, job hung, metal frame 14 GA, 3'0"x7'0", painted	EACH	862.46	551.16	1,413.62
05.1200 340	Door, wood, institutional, solid core, hardware, job hung, metal frame 14 GA, 3'6"x7'0", painted	EACH	876.14	551.16	1,427.30
05.1200 345	Add for formica clad	EACH	108.90		108.90
05.1200 350	Add for C label 1 hour	EACH	108.08		108.08
05.1200 355	Add for B label 1-1/2 hour	EACH	154.92		154.92
05.1200 360	Add for A label 3 hour	EACH	201.71		201.71
05.1200 365	Add for sound proof, STC 40	EACH	212.99	9.14	222.13
05.1200 370	Add for sound proof, STC 45	EACH	258.18	12.25	270.43
05.1200 375	Add for sound proof, STC 51	EACH	516.38	37.23	553.61
05.1300	**INTERNAL FENESTRATION**				
05.1300 100	Glazing, interior, clear, fixed, commercial, glass 1/4", steel frame, painted	SF	15.55	12.51	28.06
05.1300 105	Glazing, interior, tempered, fixed, commercial, glass 1/4", steel frame, painted	SF	17.33	12.51	29.84
05.1300 110	Glazing, interior, security, fixed, commercial, glass wire 1/4", steel frame, painted	SF	37.04	12.51	49.55
05.1300 115	Glazing, interior, spandrel, fixed, commercial, glass 1/4", steel frame, painted	SF	34.39	13.17	47.56
05.1300 120	Glazing, interior, obscure, fixed, commercial, glass 1/4", steel frame, painted	SF	21.69	12.51	34.20
05.1300 125	Sidelight, 2'x7', commercial, tempered glass 1/4", steel frame, painted	EACH	285.77	199.52	485.28
05.1300 130	Sidelight, 3'x7', commercial, tempered glass 1/4", steel frame, painted	EACH	428.65	299.27	727.93
05.1300 135	Glazing, interior, clear, fixed, commercial, glass 1/4", aluminun frame	SF	15.12	9.32	24.44
05.1300 140	Glazing, interior, tempered, fixed, commercial, glass 1/4", aluminum frame	SF	16.90	9.32	26.22
05.1300 145	Glazing, interior, spandrel, fixed, commercial, glass 1/4", aluminum frame	SF	33.96	9.98	43.94
05.1300 150	Glazing, interior, obscure, fixed, commercial, glass 1/4", aluminum frame	SF	21.26	9.32	30.58
05.1300 155	Sidelight, 2'x7', commercial, tempered glass 1/4", aluminum frame	EACH	281.30	154.89	436.19
05.1300 160	Sidelight, 3'x7', commercial, tempered glass 1/4", aluminum frame	EACH	421.95	232.33	654.28

INTERNALS, HORIZONTAL - 5.2

CSI#	Description	Unit	Material Cost	Install Cost	Total Cost
	The costs in this section include materials, labor, equipment rental, supervision, and subcontractor overhead and profit. There are no allowances for the general contractor. Costs are complete, the suspension systems and painting are part of the ceiling costs and appropriate adhesives and finishes are included in flooring costs.				
05.2100	**CEILING**				
05.2100 100	Acoustical ceiling, suspended, 2'x2'	SF	2.21	1.81	4.03
05.2100 105	Acoustical ceiling, suspended, 2'x4'	SF	1.95	1.31	3.26
05.2100 110	Acoustical ceiling, concealed spline, suspended, 1'x1'	SF	4.06	2.91	6.97
05.2100 115	Acoustical ceiling, concealed spline, suspended, 2'x2'	SF	6.36	2.74	9.10
05.2100 120	Acoustical ceiling, glue on, wood studs, 5/8" gypsum board, vinyl tile 1'x1', painted	SF	4.38	5.96	10.34
05.2100 125	Acoustical ceiling, glue on, metal studs, 5/8" gypsum board, vinyl tile 1'x1', painted	SF	3.91	5.39	9.30
05.2100 130	Gypsum board ceiling, wood frame, 5/8" gypsum board, painted	SF	1.97	4.47	6.44
05.2100 135	Gypsum board ceiling, metal frame, 5/8" gypsum board, painted	SF	1.57	3.93	5.51
05.2100 140	Acoustical ceiling, suspended, 2'x2', insulated	SF	2.50	2.17	4.67
05.2100 145	Acoustical ceiling, suspended, 2'x4', insulated	SF	2.27	1.69	3.95
05.2100 150	concealed spline, suspended, 1'x1', insulated	SF	4.25	3.23	7.48
05.2100 155	Acoustical ceiling, concealed spline, suspended, 2'x2', insulated	SF	6.31	3.07	9.38
05.2100 160	Acoustical ceiling, glue on, wood studs, 5/8" gypsum board, tile vinyl 1'x1' painted, insulated	SF	4.76	6.39	11.14
05.2100 165	glue on, metal studs, 5/8" gypsum board, tile vinyl 1'x1' painted, insulated	SF	4.28	5.82	10.10
05.2100 170	Gypsum board ceiling, wood frame, 5/8" gypsum board, painted, insulated	SF	2.35	4.90	7.24
05.2100 175	Gypsum board ceiling, framed, metal frame, 5/8" gypsum board, painted, insulated	SF	1.95	4.36	6.31
05.2200	**FLOOR, RESILIENT**				
05.2200 100	Resilient floor, tile, asphalt, 1/8", 'B'	SF	1.53	1.22	2.75
05.2200 105	Resilient floor, tile, asphalt, 1/8", 'C'	SF	1.70	1.22	2.92
05.2200 110	Resilient floor, tile, vinyl, grease resistant, 1/8"	SF	4.88	1.22	6.10
05.2200 115	Resilient floor, tile, cork, 5/16"	SF	6.96	1.22	8.18
05.2200 120	Resilient floor, tile, cork 3/16"	SF	4.63	1.22	5.85
05.2200 125	Resilient floor, tile, vinyl composition, standard 1/16"	SF	1.30	1.22	2.52
05.2200 130	Resilient floor, tile, vinyl composition, standard 3/32"	SF	1.70	1.22	2.92
05.2200 135	Resilient floor, tile, vinyl composition, standard 1/8"	SF	1.91	1.22	3.13
05.2200 140	Resilient floor, tile, vinyl composition, metallic 1/8"	SF	1.62	1.22	2.84
05.2200 145	Resilient floor, tile, vinyl solid, standard 1/16"	SF	4.54	1.22	5.76
05.2200 150	Resilient floor, tile, vinyl solid, heavy duty 1/8"	SF	6.71	1.22	7.93
05.2200 155	Resilient floor, tile, vinyl decorative, best 125"	SF	13.76	2.44	16.20
05.2200 160	Resilient floor, vinyl, 065", w/cove	SY	24.31	10.72	35.04
05.2200 165	Resilient floor, vinyl, 040", w/cove	SY	19.90	10.72	30.62
05.2200 170	Resilient floor, vinyl, metallic 065", w/cove	SY	30.81	10.72	41.54
05.2200 175	Resilient floor, vinyl, metallic 090", custom	SY	42.88	2.57	45.45
05.2200 180	Resilient floor, linoleum, standard grade, w/o cove	SY	26.30	10.72	37.02
05.2200 185	Resilient floor, polyvinylchloride, edge sealed, hot welded	SF	2.18	4.14	6.32
05.2200 190	Resilient base, top set, vinyl 2-1/2"	LF	0.70	1.01	1.71
05.2200 195	Resilient base, top set, vinyl 6"	LF	1.62	1.01	2.62
05.2200 200	Resilient base, top set, vinyl 4"	LF	1.11	1.01	2.12
05.2200 205	Resilient base, top set, rubber 2 1/2"	LF	1.11	1.01	2.12
05.2200 210	Resilient base, top set, rubber 6"	LF	2.24	1.01	3.25
05.2200 215	Resilient base, top set, rubber 4"	LF	1.67	1.41	3.09
05.2200 220	Resilient base, cove set, rubber 2-1/2"	LF	1.16	1.41	2.57
05.2200 225	Seamless floor, resilient, large area	SF	1.55	3.44	4.99
05.2200 230	Seamless floor, resilient, small area	SF	4.52	9.47	13.99
05.2200 235	Seamless floor, chemical resistant, 4" base	SF	1.90	4.14	6.04
05.2200 240	Elasto/poly cove base, large area	SF	1.55	3.44	4.99
05.2200 245	Elasto/poly cove base, small area	SF	1.83	4.00	5.83
05.2200 250	Magnesite floor, medium area	SF	3.55	4.00	7.55
05.2200 255	Magnesite floor, large area	SF	3.10	3.09	6.19
05.2200 260	Magnesite base medium area	LF	3.57	4.14	7.71
05.2200 265	Magnesite base, large area	LF	3.23	3.44	6.67
05.2200 270	Asphalt plank, 1/2"x12"x24"	SF	2.42	0.90	3.33

5.3 - FINISHES, SPECIAL

CSI#	Description	Unit	Material Cost	Install Cost	Total Cost
	The costs in this section include materials, labor, equipment rental, supervision, and subcontractor overhead and profit. There are no allowances for the general contractor. Costs are complete, floors are finished, walls are painted or appropriately finished all ancillary costs are included.				
05.3100	**TILE**				
05.3100 100	Tile, ceramic, floor, 1"x1", mortar set	SF	6.67	7.54	14.21
05.3100 105	Tile, ceramic, floor, 4"x4", mortar set	SF	6.51	7.44	13.95
05.3100 110	Tile, ceramic, wall, 1"x1", mortar set, base included	SF	7.94	11.52	19.46
05.3100 115	Tile, ceramic, wall, 4"x4", mortar set, base included	SF	7.13	9.44	16.57
05.3200	**TERRAZZO**				
05.3200 100	Terrazzo, floor, mud bonded, 2", non-slip abrasive	SF	7.57	2.25	9.82
05.3200 105	Terrazzo, floor, thin set, polyester, 3/8", polished	SF	7.41	14.00	21.41
05.3200 110	Terrazzo, floor, thin set, latex, 3/8", polished	SF	8.83	16.63	25.46
05.3200 115	Terrazzo, floor, thin set, epoxy chemical resistant, 3/8", polished	SF	7.31	13.71	21.02
05.3200 120	Divider strip, brass, 12 gauge adder	LF	4.46	0.50	4.96
05.3200 125	Divider strip, white metal, 12 gauge adder	LF	1.77	0.50	2.28
05.3300	**HARDWOOD FLOORS**				
05.3300 100	Hardwood floor, oak, common, finished	SF	5.68	8.25	13.93
05.3300 105	Hardwood floor, oak, select, finished	SF	5.78	8.53	14.31
05.3300 110	Hardwood floors, oak parquet 5/16", pre-finished	SF	5.42	6.03	11.45
05.3300 115	Hardwood floor, oak parquet 25/32", pre-finished	SF	5.84	6.21	12.04
05.3300 120	Hardwood floor, maple parquet 5/16", pre-finished	SF	9.06	6.48	15.53
05.3300 125	Hardwood floor, walnut parquet 5/16", pre-finished	SF	9.24	6.74	15.98
05.3300 130	Hardwood floor, teak parquet 5/16", pre-finished	SF	9.62	7.17	16.79
05.3300 135	Hardwood floor, gym, maple	SF	10.05	6.73	16.78
05.3300 140	Hardwood floor, gym, on steel springs	SF	11.59	9.81	21.40
05.3300 145	Hardwood floor, gym, on steel channels	SF	10.33	8.76	19.09
05.3400	**WALL COVERING**				
05.3400 100	Vinyl wall covering, light, 7 ounces	SF	1.83	0.62	2.45
05.3400 105	Vinyl wall covering, medium, 14 ounces	SF	2.00	0.78	2.78
05.3400 110	Vinyl wall covering, heavy, 22 ounces	SF	2.27	0.90	3.17
05.3400 115	Wall covering, plastic hardboard, 1/8"	SF	2.75	1.39	4.14
05.3400 120	Wall covering, plastic hardboard, 1/4"	SF	3.64	1.39	5.04
05.3500	**WOOD**				
05.3500 100	Paneling, birch or ash, select	SF	2.20	4.68	6.88
05.3500 105	Paneling, birch or ash, economy	SF	1.43	4.30	5.72
05.3500 110	Paneling, mahogany, African	SF	3.53	5.76	9.29
05.3500 115	Paneling, mahogany, Phillipine, select	SF	1.58	3.91	5.49
05.3500 120	Paneling, mahogany, ribbon cut, good	SF	1.64	3.91	5.55
05.3500 125	Paneling, mahogany, rotary cut, economy	SF	1.42	3.91	5.32
05.3500 130	Paneling, teak, select	SF	3.00	5.76	8.76
05.3500 135	Paneling, teak, finish V plank, good	SF	5.02	5.85	10.87
05.3500 140	Paneling, walnut, select	SF	7.78	5.85	13.64
05.3500 145	Paneling, walnut, domestic	SF	4.77	5.47	10.24
05.3500 150	Paneling, redwood, clear, all heart	SF	3.50	0.36	3.86
05.3500 155	Paneling, douglas fir, clear, vertical grain	SF	3.41	3.64	7.06
05.3500 160	Paneling, hardboard, printed finish	SF	1.03	4.26	5.30
05.3500 165	Base, 1-5/8", pre-finished, hardwood	LF	1.95	2.50	4.46
05.3500 170	Base, 2-1/4", softwood	LF	1.58	2.50	4.09
05.3500 175	Base, 3-1/4", softwood	LF	1.84	2.50	4.35
05.3500 180	Molding, chair rail	LF	1.70	3.90	5.59
05.3500 185	Molding, cove, 3/4"	LF	1.02	2.23	3.25
05.3500 190	Molding, cove, 1-3/4"	LF	1.21	2.23	3.44
05.3500 195	Molding, picture	LF	1.25	3.90	5.14
05.3500 200	Trim, apron, 1-3/8"	LF	1.17	2.89	4.06
05.3500 205	Trim, jamb, 2" pine w/stop	LF	2.79	2.37	5.16
05.3500 210	Trim, jamb, 4-5/8" pine	LF	4.13	10.15	14.28

INTERIORS - 5.4

CSI#	Description	Unit	Material Cost	Install Cost	Total Cost
	The costs in this section include materials, labor, equipment rental, supervision, and subcontractor overhead and profit. There are no allowances for the general contractor. Costs are complete, as expected under standard construction conditions. Certain adders are noted among the costs.				
05.4100	**CABINETS & TOPS**				
05.4100 100	Cabinets and laminated plastic tops, multi-unit, hardwood, economy, base	LF	46.02	22.79	68.81
05.4100 105	Cabinets and laminated plastic tops, multi-unit, hardwood, economy, wall	LF	29.71	40.26	69.97
05.4100 110	Cabinets and laminated plastic tops, multi-unit, hardwood, economy, full height	LF	114.71	51.74	166.46
05.4100 115	Cabinets and laminated plastic tops, multi-unit, hardwood, custom, base	LF	66.22	22.79	89.01
05.4100 120	Cabinets and laminated plastic tops, multi-unit, hardwood, custom, wall	LF	47.99	40.26	88.26
05.4100 125	Cabinets and laminated plastic tops, multi-unit, hardwood, custom, full height	LF	115.95	51.74	167.70
05.4100 130	Cabinets and laminated plastic tops, premium, birch institutional, base	LF	132.87	44.64	177.52
05.4100 135	Cabinets and laminated plastic tops, premium, birch institutional, wall	LF	111.49	78.31	189.79
05.4100 140	Cabinets and laminated plastic tops, premium, birch institutional, full height	LF	188.47	56.36	244.83
05.4100 143	Add for ash	LF	12.13		12.13
05.4100 144	Add for walnut	LF	66.07		66.07
05.4100 145	Add for overlay	LF	8.12		8.12
05.4100 146	Add for flush overlay	LF	12.13		12.13
05.4100 147	Add for edge banding	LF	8.12		8.12
05.4100 148	Add for pre-finished exterior	LF	6.21		6.21
05.4100 149	Add for pre-finished interior	LF	13.33		13.33
05.4100 150	Add for standard hardware	LF	5.52		5.52
05.4100 151	Add for institutional hardware	LF	8.12		8.12
05.4100 152	Add for roller guides	SET	6.21		6.21
05.4100 153	Add for roller guides, full suspension	SET	30.88		30.88
05.4100 154	Cabinets and laminated plastic tops, units, premium, open with one shelf	LF	83.06		83.06
05.4100 155	Cabinets and laminated plastic tops, units, premium, with door and one shelf	LF	102.98		102.98
05.4100 160	Cabinets and laminated plastic tops, units, premium, with door, 1 shelf & 1 drawer	LF	142.96		142.96
05.4100 165	Cabinets and laminated plastic tops, units, premium, sink	LF	91.42		91.42
05.4100 170	Cabinets and laminated plastic tops, units, premium, 4 drawers	LF	164.59		164.59
05.4100 175	Cabinets and laminated plastic tops, units, premium, add for apron (knee space)	LF	23.22		23.22
05.4100 180	Cabinets and laminated plastic tops, units, premium, add for backsplash	LF	13.33		13.33
05.4100 185	Cabinets and laminated plastic tops, metal, base, commercial, with door, shelf	LF	132.03	23.76	155.79
05.4100 190	Cabinets and laminated plastic tops, metal, wall, commercial, with door, 2 shelves	LF	97.23	28.53	125.76
05.4100 195	Cabinets and laminated plastic tops, metal, full, commercial, with door, 5 shelves	LF	229.11	37.28	266.39
05.4100 200	Cabinets and laminated plastic tops, metal, full, library, shelving	LF	85.04	20.84	105.88
05.4100 205	Cabinets and laminated plastic tops, metal, full, wardrobe, 4' wide	EA	555.78	62.95	618.73
05.4100 210	Cabinets and laminated plastic tops, wood, formica faced, base, w/door, 1 shelf, school	LF	252.92	39.50	292.42
05.4100 215	Cabinets and laminated plastic tops, wood, formica faced, wall, w/door, 1 shelves, school	LF	183.64	47.43	231.07
05.4100 220	Cabinets and laminated plastic tops, wood, formica faced, full, w/doors, school	LF	300.15	48.80	348.95
05.4100 225	Cabinets and laminated plastic tops, wood, formica faced, add for laboratory cabinets, school	LF	59.30	27.26	86.57
05.4100 230	Cabinets and laminated plastic tops, wood and metal, w/door, 1 shelf, 1 drawer, hospital	LF	191.77	79.91	271.69
05.4100 235	Cabinets and laminated plastic tops, wood and metal, w/door, 2 shelves, hospital	LF	112.39	53.24	165.64
05.4100 240	Cabinets and laminated plastic tops, wood and metal, w/door, full, hospital	LF	231.89	41.59	273.48
05.4100 245	Cabinets and laminated plastic tops, furniture grade, base, laboratory	LF	268.54	140.40	408.94
05.4100 250	Cabinets and laminated plastic tops, furniture grade, wall, laboratory	LF	127.92	67.81	195.73
05.4100 255	Cabinets and laminated plastic tops, furniture grade, wardrobe, laboratory	LF	287.22	58.10	345.32
05.4100 260	Cabinets and laminated plastic tops, furniture grade, lab island, laboratory	LF	383.99	169.44	553.43
05.4100 265	Cabinets and laminated plastic tops, furniture grade, fume hood, laboratory	LF	530.78	377.56	908.34
05.4100 270	Cabinets and laminated plastic tops, furniture grade, fume hood, stainless steel, laboratory	LF	708.76	363.08	1,071.84
05.4100 275	Cabinets and laminated plastic tops, furniture grade, add for premium quality, laboratory	LF	30.57	26.14	56.71
05.4100 280	Laminated plastic & simulated marble tops, multi-residential	LF	36.32		36.32
05.4100 285	Laminated plastic & simulated marble tops, small projects	LF	41.93		41.93
05.4100 290	Laminated plastic & simulated marble tops, custom jobs	LF	55.98		55.98

5.4 - INTERIORS

CSI#	Description	Unit	Material Cost	Install Cost	Total Cost
05.4100 295	Laminated plastic & simulated marble tops, vanity top, cultered marble, no bowl	LF	50.13		50.13
05.4100 300	Laminated plastic & simulated marble tops vanity top, add to vanity for molded bowl	EA	46.48		46.48
05.4100 305	Laminated plastic & simulated marble tops, vanity top, add to vanity for molded clam shell	EA	81.34		81.34
05.4100 310	Laminated plastic & simulated marble tops, acid proof tops	LF	112.45		112.45
05.4100 315	Laminated plastic & simulated marble tops, komar/simulated marble, molded section	SF	20.98		20.98
05.4300	**CARPET**				
05.4300 100	Carpet, 30 oz polyester, with 50 oz pad	SY	27.50	7.86	35.36
05.4300 105	Carpet, 35 oz polyester, with 50 oz pad	SY	30.53	7.86	38.40
05.4300 110	Carpet, 40 oz polyester, with 50 oz pad	SY	32.54	7.86	40.40
05.4300 115	Carpet, 50 oz polyester, with 50 oz pad	SY	38.19	7.86	46.05
05.4300 120	Carpet, 20 oz nylon shag, with 50 oz pad	SY	15.99	7.86	23.85
05.4300 125	Carpet, 24 oz nylon shag, with 50 oz pad	SY	18.22	7.86	26.08
05.4300 130	Carpet, 24 oz nylon shag, with 50 oz pad	SY	15.99	7.86	23.85
05.4300 135	Carpet, 20 oz nylon filament, with 50 oz pad	SY	15.99	7.86	23.85
05.4300 140	Carpet, 24 oz nylon filament, with 50 oz pad	SY	22.10	7.86	29.96
05.4300 145	Carpet, 15 oz nylon level loop, with rubber back	SY	18.22	6.74	24.95
05.4300 150	Carpet, 21 oz nylon level loop, with rubber back	SY	22.10	6.74	28.83
05.4300 155	Carpet, 28 oz nylon level loop, with rubber back	SY	30.14	6.74	36.88
05.4300 160	Carpet, 15 oz nylon level loop, with 50 oz pad	SY	19.09	7.86	26.95
05.4300 165	Carpet, 21 oz nylon level loop, with 50 oz pad	SY	23.19	7.86	31.05
05.4300 170	Carpet, 28 oz nylon level loop, with 50 oz pad	SY	31.76	7.86	39.62
05.4300 175	Carpet, 48 oz nylon level loop, with 50 oz pad	SY	38.41	7.86	46.27
05.4300 180	Carpet, 21 oz antron, anti-static, with 50 oz pad	SY	30.14	7.86	38.00
05.4300 185	Carpet, wool commercial, with 50 oz pad	SY	64.51	7.86	72.37
05.4300 190	Carpet, exterior, premium, without pad	SY	29.77	9.00	38.76
05.4300 195	Carpet pad, 40 oz, jute	SY	3.98	1.67	5.66
05.4300 200	Carpet pad, 50 oz, jute/hair	SY	5.58	1.67	7.25
05.4300 205	Carpet pad, 50 oz, hair	SY	7.00	1.67	8.67
05.4300 210	Carpet pad, 72 oz, rubber waffle	SY	5.10	1.67	6.77
05.4300 215	Carpet pad, 100 oz, rubber waffle	SY	9.64	1.67	11.31
05.4300 220	Carpet pad, 72 oz, rubber slab	SY	6.44	1.67	8.11
05.4300 225	Carpet pad, 88 oz, rubber slab	SY	9.64	1.67	11.31
05.4300 230	Carpet pad, urethane, 15#	SY	3.29	1.67	4.97
05.4300 235	Carpet pad, urethane, 2#	SY	5.29	1.67	6.96
05.4300 240	Carpet pad, urethane, 25#	SY	6.21	1.67	7.88
05.4300 245	Carpet pad, urethane, rebound	SY	6.78	1.67	8.46
05.4300 250	Carpet average, housing	SY	24.58		24.58
05.4300 255	Carpet average, commercial	SY	34.79		34.79
05.4300 260	Carpet average, school	SY	37.94		37.94
05.4300 265	Carpet average, hotel/motel/theatre	SY	41.13		41.13
05.4300 270	Carpet average, custom housing	SY	60.13		60.13
05.4300 275	Floor mat, rubber, 1/4" recessed	SF	7.91	3.30	11.21
05.4300 280	Floor mat, rubber, 1/2" recessed	SF	12.50	4.15	16.65
05.4300 285	Floor mat, vinyl, 1/4" recessed	SF	11.63	3.30	14.93
05.4300 290	Floor mat, vinyl, 1/2" recessed	SF	15.39	4.15	19.55

SPECIALTIES - 6.0

CSI#	Description	Unit	Material Cost	Install Cost	Total Cost
	The costs of Specialties in this section include material, installation, and subcontractor overhead and profit. There are no allowances for general contractor markup and profit.				
	The items in this section are typical of those found in offices, hotel meeting rooms and buildings that are designed for high concentrations of public use.				
	Costs represent standard grade materials and normal installation. Adjustments should be made for economy quality or custom and heavy duty materials and commensurate installation.				
06.0100	**CHALK & TACK BOARDS**				
06.0100 110	Tackboard, 1/4" vinyl cork, 1/4" hardboard, without frame	SF	25.82	3.54	29.36
06.0100 115	Tackboard, vinyl and fiberboard 1/2", without frame, installed	SF	14.57	2.13	16.69
06.0100 120	Tackboard, 1/8", burlap back, without frame, installed	SF	14.39	1.51	15.91
06.0100 125	Tackboard, 1/8", cork unbacked 1/4", without frame, installed	SF	7.11	0.71	7.82
06.0100 128	Tackboard, cork or vinyl, with trim, installed	SF	15.69	2.74	18.43
06.0100 130	Aluminum map, display rail, deluxe	LF	6.34	1.14	7.48
06.0100 135	Aluminum chalk tray	LF	11.96	2.63	14.59
06.0100 140	Aluminum chalk board frame with trim	LF	5.25	0.91	6.15
06.0100 145	Chalkboard, slate 3/8", without frame, installed	SF	27.07	5.72	32.79
06.0100 150	Chalkboard, hardboard, tempered 1/4", without frame, installed	SF	15.84	1.95	17.79
06.0100 155	Chalkboard, metal 24 GA 1/2", without trim, installed	SF	17.95	2.63	20.59
06.0100 160	Chalkboard, reversible, roll, 4'x8'	EACH	1,018.05	499.51	1,517.56
06.0100 165	Chalkboard, horizontal sliding, installed	SF	63.99	17.49	81.47
06.0100 170	Chalkboard, vertical sliding, installed	SF	98.47	26.76	125.23
06.0100 180	Chalkboard, without trim, average, installed	SF	10.35	2.84	13.19
06.0100 185	Chalkboard, with trim, map, rail, installed	SF	24.68	5.65	30.33
06.0200	**TOILET PARTITIONS**				
06.0200 100	Toilet partition, baked enamel, floor mounted	EACH	714.44	140.86	855.31
06.0200 105	Toilet partition, baked enamel, ceiling mounted	EACH	754.17	164.32	918.49
06.0200 110	Toilet partition, porcelin enamel, floor mounted	EACH	1,055.53	172.19	1,227.72
06.0200 115	Toilet partition, porcelin enamel, ceiling mounted	EACH	1,111.33	200.87	1,312.20
06.0200 120	Toilet partition, laminated plastic, floor mounted	EACH	923.86	172.19	1,096.05
06.0200 125	Toilet partition, laminated plastic, ceiling mounted	EACH	938.62	200.87	1,139.49
06.0200 130	Toilet partition, stainless steel, floor mounted	EACH	2,423.32	172.19	2,595.51
06.0200 135	Toilet partition, stainless steel, ceiling mounted	EACH	2,250.24	200.87	2,451.11
06.0200 140	Add for best quality	EACH	162.45		162.45
06.0200 145	Urinal screen, baked enamel, wall mounted	EACH	257.83	48.44	306.28
06.0200 150	Urinal screen, porcelin enamel, wall mounted	EACH	485.19	59.20	544.39
06.0200 155	Urinal screen, laminated plastic, wall mounted	EACH	357.70	59.20	416.90
06.0200 160	Urinal screen, stainless steel, wall mounted	EACH	656.97	59.20	716.17
06.0200 165	Add for best quality	EACH	45.17	23.27	68.44
06.0200 170	Sight screen, 3'x7', baked enamel, floor mounted	EACH	406.10	101.64	507.74
06.0200 175	Sight screen, laminated plastic, floor mounted	EACH	539.63	124.27	663.89
06.0200 180	Dressing cubicle, baked enamel, floor mounted	EACH	1,320.20	113.13	1,433.34
06.0200 185	Dressing cubicle, porcelin enamel, floor mounted	EACH	1,954.67	138.31	2,092.99
06.0200 190	Dressing cubicle, stainless steel, floor mounted	EACH	2,446.19	138.31	2,584.50
06.0200 195	Dressing cubicle, laminated plastic, floor mounted	EACH	1,716.30	138.31	1,854.62
06.0300	**DEMOUNTABLE PARTITIONS**				
06.0300 100	Relocatable partitions, 5/8"	SF	9.62	2.87	12.49
06.0300 105	Accordian partitions, vinyl, 8', STC 36, economy	SF	21.70	6.41	28.10
06.0300 110	Accordian partitions, vinyl, 30'x17', STC 43, good	SF	28.28	8.39	36.68
06.0300 115	Accordian partitions, vinyl, large, open, STC 45, bottom	SF	40.91	12.04	52.95
06.0300 120	Accordian partitions, wide slat, birch/ash, prefinished	SF	21.70	6.41	28.10
06.0300 125	Folding partitions, vinyl, metal panel, STC 52	SF	67.84	26.94	94.78
06.0300 130	Folding partitions, vinyl, wide panel, STC 40	SF	49.92	19.77	69.69
06.0300 135	Add for laminated plastic	SF	4.17	1.33	5.49
06.0300 140	Add for wood veneer	SF	5.45	1.66	7.11
06.0300 145	Folding partitions, wide, side coil, single, crank (operable walls)	SF	31.60	12.70	44.30
06.0300 150	Folding partitions, motor, under 30'x9'	EACH	1,944.06	777.91	2,721.97

6.0 SPECIALTIES

CSI#	Description	Unit	Material Cost	Install Cost	Total Cost
06.0400	**TOILET ACCESSORIES**				
06.0400 100	Toilet accessories, paper towel and waste combination, stainless steel/lam, 14"x24", recessed	EACH	516.66	36.15	552.81
06.0400 105	Toilet accessories, paper towel and waste combination, stainless steel, 17"x54", semi-recessed	EACH	779.00	36.15	815.15
06.0400 110	Toilet accessories, paper towel and waste combination, stainless steel/lam, 12"x72", semi-recessed	EACH	900.85	36.15	937.00
06.0400 115	Toilet accessories, paper towel and waste combination, stainless steel, 17"x54", semi-recessed	EACH	1,298.32	36.15	1,334.47
06.0400 120	Toilet accessories, paper towel dispenser, stainless steel, 12"x17", surface	EACH	127.18	36.15	163.33
06.0400 125	Toilet accessories, paper towel dispenser, stainless steel, 12"x15", recessed	EACH	246.46	36.15	282.61
06.0400 130	Toilet accessories, paper towel dispenser, stainless steel, 14"x26", recessed	EACH	368.28	36.15	404.42
06.0400 140	Toilet accessories, paper towel dispenser, stainless steel trim, 17"x28"	EACH	877.06	36.15	913.21
06.0400 150	Toilet accessories, paper towel and soap combo, laminated	EACH	556.44	36.15	592.59
06.0400 160	Toilet accessories, towel, soap and mirror, laminated	EACH	839.95	36.15	876.09
06.0400 165	Toilet accessories, waste receptacles, stainless steel, 3 GAL, recessed	EACH	380.87	36.15	417.02
06.0400 170	Toilet accessories, waste receptacles, stainless steel, 10 GAL, semi-recessed	EACH	577.51	36.15	613.66
06.0400 200	Toilet accessories, toilet seat cover dispensers, stainless steel, surface	EACH	79.49	36.15	115.64
06.0400 210	Toilet accessories, toilet seat cover dispensers, laminated, recessed	EACH	348.75	36.15	384.90
06.0400 220	Toilet accessories, toilet paper dispensers, aluminum, single, surface	EACH	28.18	36.15	64.33
06.0400 225	Toilet accessories, toilet paper dispensers, stainless steel, single, surface	EACH	51.29	36.15	87.44
06.0400 245	Toilet accessories, toilet paper dispenser, stainless steel, double, recessed	EACH	87.18	36.15	123.33
06.0400 250	Toilet accessories, feminine napkin dispenser, stainless steel, surface	EACH	769.22	36.15	805.37
06.0400 265	Toilet accessories, feminine napkin dispensers/dispoasl, stainless steel	EACH	1,161.58	36.15	1,197.73
06.0400 285	Toilet accessories, feminine napkin dispoasl, stainless steel, recessed, large	EACH	325.68	36.15	361.83
06.0400 295	Toilet accessories, soap dispensers, plastic, liquid, surface, stainless steel lid	EACH	48.72	36.15	84.87
06.0400 310	Toilet accessories, soap dispensers, stainless steel, liquid, recessed	EACH	166.64	36.15	202.79
06.0400 325	Toilet accessories, soap dispensers, stainless steel, liquid, recessed, shelf	EACH	325.68	36.15	361.83
06.0400 345	Toilet accessories, facial tissue dispensers, stainless steel, surface	EACH	41.04	36.15	77.19
06.0400 360	Toilet accessories, facial tissue dispensers, stainless steel, recessed, with shelf	EACH	230.78	36.15	266.93
06.0400 370	Toilet accessories, grab bars, stainless steel, 1 1/2"x24", exposed mounted	EACH	82.06	36.15	118.21
06.0400 380	Toilet accessories, grab bars, swing away, exposed floor mounted	EACH	910.27	36.15	946.42
06.0400 405	Toilet accessories, miscellaneous, towel bar, stainless steel, 24"	EACH	76.92	36.15	113.07
06.0400 415	Toilet accessories, miscellaneous, shower rod, stainless steel, 1"x6'	EACH	53.83	36.15	89.97
06.0400 435	Toilet accessories, miscellaneous, mirror, stainless steel frame, tilt	EACH	206.16	36.15	242.30
06.0400 445	Toilet accessories, miscellaneous, mirror, stainless steel frame, 16"x24", shelf	EACH	147.76	36.15	183.91
06.0400 455	Toilet accessories, miscellaneous, medicine cabinet and mirror, baked enamel, recessed	EACH	226.05	36.15	262.20
06.0400 460	Toilet accessories, miscellaneous, medicine cabinet and mirror, stainless steel, recessed	EACH	656.05	36.15	692.20
06.0400 515	Toilet accessories, miscellaneous, shelf, stainless steel, 24"	EACH	118.82	36.15	154.97
06.0400 530	Toilet accessories, miscellaneous, towel shelf, stainless steel, 24"	EACH	156.21	36.15	192.36
06.0400 545	Toilet accessories, miscellaneous, electric hand dryer, surface, 40 second cycle	EACH	2,061.71	36.15	2,097.86

EQUIPMENT - 7.0

CSI#	Description	Unit	Material Cost	Install Cost	Total Cost
	The equipment costs, in this section, include material, installation, and subcontractor overhead and profit. There are no allowances for general contractor markup and profit.				
	Costs represent standard grade materials and normal installation. Adjustments should be made for economy quality or custom and heavy duty materials and commensurate installation.				
07.0100	**BANK EQUIPMENT**				
07.0100 100	Vault door for safekeeping, 78"x44"x3 1/2", steel, class 1	EACH	37,673.09	1,157.02	38,830.11
07.0100 105	Vault door for safekeeping, 78"x44"x3 1/2", steel, class 2	EACH	42,534.11	1,157.02	43,691.14
07.0100 110	Vault door for safekeeping, 78"x44"x3 1/2", steel, class 3	EACH	51,040.96	1,157.02	52,197.99
07.0100 115	Vault door for safekeeping, 78"x44"x3 1/2", steel, 6 hour	EACH	28,432.50	1,157.02	29,589.52
07.0100 120	Vaults for record keeping only, 78"x33", 1 hour, single	EACH	5,004.99	598.30	5,603.28
07.0100 125	Vaults for record keeping only, 84"x51 1/8", 2 hour, single	EACH	6,099.89	598.30	6,698.18
07.0100 130	Vaults for record keeping only, 84"x51 1/8", 4 hour, single	EACH	6,375.97	586.99	6,962.96
07.0100 135	Vaults for record keeping only, 84"x51 1/8", 6 hour, single	EACH	7,400.08	594.53	7,994.61
07.0100 140	Modular vaults, class 1, 15 minute	EACH	35,395.97	9,753.65	45,149.62
07.0100 145	Modular vaults, class 2, 1 hour	EACH	58,993.25	9,752.71	68,745.97
07.0100 150	Modular vaults, class 3, 2 hour	EACH	82,590.52	9,757.42	92,347.94
07.0100 155	Teller Window, drive-up, manual	EACH	11,628.88	1,281.83	12,910.70
07.0100 160	Teller Window, drive-up, motorized	EACH	13,096.99	1,281.83	14,378.82
07.0100 165	Teller Window, walk-up, 1 teller	EACH	7,994.69	1,025.44	9,020.14
07.0100 170	Teller Window, walk-up, 2 teller	EACH	8,832.72	1,025.44	9,858.17
07.0100 175	Night deposit, bag/envelope, illuminated	EACH	6,583.64	1,061.86	7,645.50
07.0100 180	Nigth deposit, envelope	EACH	2,720.71	293.01	3,013.73
07.0100 185	Night deposit, bag, flush mounted	EACH	3,324.80	481.56	3,806.35
07.0100 190	Teller counter, modular component	LF	397.33	97.05	494.38
07.0100 195	Check desk, round 48", 4 person	EACH	2,911.91	587.65	3,499.56
07.0100 200	Check desk, square 48", 4 person	EACH	2,454.06	602.34	3,056.41
07.0100 205	Check desk, 72"x24", 4 person	EACH	2,916.62	364.73	3,281.35
07.0100 210	Check desk, 72"x36", 4 person	EACH	3,624.54	477.16	4,101.70
07.0100 215	Safe deposit, 42 opening, 2"x5"	EACH	3,884.08	380.23	4,264.31
07.0100 220	Safe deposit, 30 opening, 2"x5"	EACH	2,774.99	283.99	3,058.98
07.0100 225	Safe deposit, 18 opening, 5"x5"	EACH	2,005.74	236.63	2,242.37
07.0100 230	Safe deposit, base, 32"x24"x3"	EACH	259.55	86.09	345.64
07.0100 235	Safe deposit, canopy top	EACH	66.65	47.36	114.01
07.0100 240	Safe deposit, 9 open, 5"x10 3/8"	EACH	1,510.20	273.44	1,783.63
07.0100 245	Safe deposit, 15 open, 3"x10 3/8"	EACH	1,746.21	270.98	2,017.18
07.0100 250	Safe deposit, 1 section, 3 open, 5"x10"	EACH	1,533.82	492.77	2,026.59
07.0100 255	Deposit, 1 section, 3 open, 10"x10 3/8"	EACH	1,533.82	501.34	2,035.16
07.0100 260	Safe, floor, 10"x24", minimum security	EACH	633.15	127.06	760.21
07.0100 265	Safe, wall, minimum security	EACH	3,315.82	905.16	4,220.98
07.0100 270	Safe, cabinet, medium security, full door	EACH	4,778.00	1,304.38	6,082.38
07.0100 275	Book drop, minimum security	EACH	1,185.77	323.72	1,509.49
07.0100 280	Night depository	EACH	1,934.16	528.02	2,462.18
07.0200	**EDUCATIONAL EQUIPMENT**				
07.0200 100	Equipment, educational, wardrobes, teacher, 40"x78"x26 1/4"	EACH	1,279.66	70.10	1,349.76
07.0200 105	Equipment, educational, wardrobes, student, 40"x78"x26 1/4"	EACH	838.58	52.57	891.15
07.0200 110	Equipment, educational, seating, pedestal, folding arm	EACH	141.45	25.27	166.71
07.0200 115	Equipment, educational, seating, horizontal 2 section, 5 chair	EACH	152.03	27.12	179.15
07.0200 120	Equipment, educational, tables, fixed pedestal, 48"x16"	EACH	589.65	105.23	694.88
07.0200 125	Equipment, educational, tables, fixed pedestal, chair, 48"x16"	EACH	812.33	144.98	957.31
07.0200 130	Equipment, educational, slide screen, pull, 70"x70", ceiling	SF	6.88	1.49	8.36
07.0200 135	Equipment, educational, slide screen, electric, ceiling	SF	59.06	6.12	65.18
07.0200 140	Equipment, educational, drafting furniture, table, steel base, 60"x37 1/2"	EACH	657.53		657.53
07.0200 145	Equipment, educational, drafting furniture, table, hardwood, 60"x37 1/2"	EACH	1,356.90		1,356.90
07.0200 150	Equipment, educational, drafting furniture, table, 2 station, 10 drawer, flex	EACH	918.89		918.89
07.0200 155	Equipment, educational, drafting furniture, table, 1 station, 6 drawer, flex	EACH	734.25		734.25
07.0200 160	Equipment, educational, drafting furniture, desk, metal frame, mechanical drawing	EACH	985.33		985.33

7.0 EQUIPMENT

CSI#	Description	Unit	Material Cost	Install Cost	Total Cost
07.0200 165	Equipment, educational, drafting furniture, desk, wood, mechanical drawing	EACH	492.62		492.62
07.0200 170	Equipment, educational, drafting furniture, tracing table, pedestal, 24"x22"	EACH	879.56		879.56
07.0200 175	Equipment, educational, file cabinet, steel, 10 drawer	EACH	2,389.03		2,389.03
07.0200 180	Equipment, educational, file cabinet, wood, 10 drawer	EACH	1,857.06		1,857.06
07.0200 185	Equipment, educational, files, modular, 8 tube, 48"	EACH	145.37		145.37
07.0200 190	Equipment, educational, files, 26 binder, 48 3/4"	EACH	2,651.80		2,651.80
07.0200 195	Equipment, educational, audio visual, video tape recorder	EACH	5,258.64	448.37	5,707.01
07.0200 200	Equipment, educational, audio visual, camera	EACH	2,102.70	480.35	2,583.04
07.0200 205	Equipment, educational, audio visual, monitor	EACH	931.49	453.65	1,385.14
07.0200 210	Equipment, educational, audio visual, tape recorder	EACH	581.10		581.10
07.0200 215	Equipment, educational, audio visual, head set	EACH	56.11		56.11
07.0200 220	Equipment, educational, audio visual, projector, movie, 8MM	EACH	686.78		686.78
07.0200 225	Equipment, educational, audio visual, projector, slide, carousel	EACH	528.27		528.27
07.0200 230	Equipment, educational, study carrel, plastic laminated wood, 48"x30"x54", 1 station	EACH	1,003.98	106.71	1,110.70
07.0200 235	Equipment, educational, study carrel, plastic laminated wood, 73"x30"x47", 2 station	EACH	1,757.27	160.12	1,917.39
07.0200 240	Equipment, educational, study carrel, plastic laminated wood, 66"x66"x47", 4 station	EACH	2,079.90	240.13	2,320.03
07.0200 245	Equipment, educational, audio-visual center, mobile, with control panels, 10 listening stations with earphones, folding table, electric, 4'x8'	EACH	1,447.42		1,447.42
07.0200 250	Equipment, educational, audio-visual center, stack chairs	EACH	154.58		154.58
07.0200 255	Equipment, educational, accessories for carrels, rear projection module with light	EACH	351.24		351.24
07.0200 260	Equipment, educational, accessories for carrels, power column, study carrel	EACH	70.21		70.21
07.0300	**FOOD SERVICE EQUIPMENT**				
07.0300 100	Food service equipment, oven, single, self clean	EACH	1,323.03		1,323.03
07.0300 105	Food service equipment, oven, double, self clean	EACH	1,541.35		1,541.35
07.0300 110	Food service equipment, oven, microwave, built-in	EACH	2,452.35		2,452.35
07.0300 115	Food service equipment, cook top	EACH	890.96		890.96
07.0300 120	Food service equipment, range & oven, drop-in	EACH	1,276.24		1,276.24
07.0300 125	Food service equipment, range hood with microwave	EACH	906.50		906.50
07.0300 130	Food service equipment, hood	EACH	155.91		155.91
07.0300 140	Food service equipment, electric grill	EACH	913.22		913.22
07.0300 145	Food service equipment, refrigerator, 12 CF	EACH	933.25		933.25
07.0300 150	Food service equipment, refrigerator, 18 CF	EACH	1,218.35		1,218.35
07.0300 155	Food service equipment, refrigerator with icemaker	EACH	1,173.81		1,173.81
07.0300 160	Food service equipment, freezer, 16 CF	EACH	861.96		861.96
07.0300 165	Food service equipment, dishwasher, built-in	EACH	741.67		741.67
07.0300 170	Food service equipment, garbage disposal	EACH	218.29		218.29
07.0300 175	Food service equipment, trash compactor	EACH	652.61		652.61
07.0300 180	Food service equipment, washer	EACH	815.22		815.22
07.0300 185	Food service equipment, dryer, electric	EACH	648.14		648.14
07.0300 190	Food service equipment, dryer, gas	EACH	1,176.06		1,176.06
07.0300 195	Food service equipment, tables, counters with sinks, shelves, racks, economy	LF	401.70	90.10	491.81
07.0300 200	Food service equipment, tables, counters with sinks, shelves, racks, best	LF	647.23	137.51	784.75
07.0300 205	Food service equipment, service center, open storage	LF	985.93	203.89	1,189.82
07.0300 210	Food service equipment, service center, closed storage	LF	985.93	260.80	1,246.73
07.0300 215	Food service equipment, cook's table, with sink, 6'	EACH	5,657.99	616.55	6,274.54
07.0300 220	Food service equipment, vegetable prep table, with sink, 12'	EACH	5,939.35	792.19	6,731.54
07.0300 225	Food service equipment, average cost	SF	505.06		505.06
07.0300 230	Food service equipment, table and counter, stainless steel, rolled edge, 6" splash	LF	394.43	122.89	517.32
07.0300 240	Food service equipment, serving fixture, stainless steel, economy	LF	338.12	163.80	501.92
07.0300 245	Food service equipment, serving fixture, all stainless steel, best	LF	479.05	163.80	642.85
07.0300 250	Food service equipment, adders for basic built-up fixtures, shelves, stainless steel, 85 square feet, base	LF	78.33		78.33
07.0300 255	Food service equipment, adders for basic built-up fixtures, shelves, galvanized iron, 375 square feet, base	LF	42.73		42.73
07.0300 260	Food service equipment, adders for basic built-up fixtures, angle or pipe stretchers	LF	26.68		26.68
07.0300 265	Food service equipment, adders for basic built-up fixtures, tray slide, stainless steel	LF	90.84		90.84

EQUIPMENT - 7.0

CSI#	Description	Unit	Material Cost	Install Cost	Total Cost
07.0300 270	Food service equipment, adders for basic built-up fixtures, display shelf, sneeze guard	LF	110.43		110.43
07.0300 275	Food service equipment, adders for basic built-up fixtures, each additional shelf	LF	65.93		65.93
07.0300 280	Food service equipment, adders for basic built-up fixtures, plastic laminate on plywood	LF	55.32		55.32
07.0300 285	Food service equipment, adders for basic built-up fixtures, stainless steel facing, 18 GA, 3' high	LF	82.03		82.03
07.0300 290	Food service equipment, adders for basic built-up fixtures, stainless steel facing, 22 GA, 3' high	LF	52.06		52.06
07.0300 295	Food service equipment, adders for basic built-up fixtures, stainless steel dirty dish table, 12" splash	LF	41.00		41.00
07.0300 300	Food service equipment, adders for basic built-up fixtures, slop cutter, 4" square, with sump	LF	78.12		78.12
07.0300 305	Food service equipment, shop installed accessories, ventilating grill, 24"x12"	EACH	115.72		115.72
07.0300 310	Food service equipment, shop installed accessories, vegetable sink	EACH	1,177.84		1,177.84
07.0300 315	Food service equipment, disposer, with stainless steel cone, 1 HP	EACH	1,913.99	552.95	2,466.94
07.0300 320	Food service equipment, disposer, with stainless steel cone, 1-1/2 HP	EACH	2,626.39	552.95	3,179.34
07.0300 325	Food service equipment, disposer, with stainless steel cone, 3 HP	EACH	3,996.65	552.95	4,549.60
07.0300 330	Food service equipment, disposer, with stainless steel cone, 3 HP	EACH	3,996.65	552.95	4,549.60
07.0300 335	Food service equipment, water heater, electric, sink mounted	EACH	1,116.45	552.95	1,669.40
07.0300 340	Food service equipment, pot & pan rack, 6', table mounted	EACH	696.51		696.51
07.0300 345	Food service equipment, glass rack dispenser, self level	EACH	1,015.02	294.88	1,309.89
07.0300 350	Food service equipment, cup & plate dispenser, heater	EACH	773.32	184.29	957.61
07.0300 355	Food service equipment, hot food well, electric, 12"x20"	EACH	628.28	276.47	904.76
07.0300 360	Food service equipment, drop-in water cooler	EACH	1,643.34	184.29	1,827.63
07.0300 370	Food service equipment, refrigerator, undercounter, without compressor	EACH	406.95		406.95
07.0300 375	Food service equipment, refrigerator, undercounter, without compressor, add for each door	EACH	481.54		481.54
07.0300 380	Food service equipment, refrigerator, undercounter, without compressor, add for each drawer	EACH	425.25		425.25
07.0300 385	Food service equipment, refrigerator, undercounter, without compressor, add for each 16"x16" coil	EACH	276.68		276.68
07.0300 390	Food service equipment, refrigerator, undercounter, without compressor, add for each remote compressor, 1/4 HP	EACH	1,063.30	1,751.04	2,814.34
07.0300 395	Food service equipment, refrigerator, undercounter, without compressor, soft drink dispenser, 4 spout, 1 remote	EACH	10,730.32	3,686.25	14,416.56
07.0400	**GYMNASIUM & PLAYGROUND EQUIPMENT**				
07.0400 100	Field equipment, basketball, post and steel backstop, single	EACH	1,086.24	454.22	1,540.46
07.0400 105	Field equipment, basketball, post and steel backstop, double	EACH	1,525.18	487.21	2,012.38
07.0400 110	Field equipment, basketball, add for Fiberglass backstop	EACH	65.79		65.79
07.0400 115	Field equipment, basketball backstop, 34x10, with hood	EACH	3,280.83	1,359.35	4,640.18
07.0400 120	Field equipment, basketball backstop, 60x15, with hood	EACH	9,184.19	2,385.77	11,569.96
07.0400 125	Field equipment, football goal, single post, two	SET	3,910.67	1,109.67	5,020.34
07.0400 130	Field equipment, soccer goal, two	SET	3,039.42	1,442.59	4,482.01
07.0400 135	Field equipment, tennis post, two	SET	384.04	233.03	617.06
07.0400 140	Field equipment, tennis net, nylon	EACH	331.32		331.32
07.0400 145	Field equipment, tennis net, metal	EACH	702.22		702.22
07.0400 150	Field equipment, volley ball post, two	SET	329.13	244.67	573.80
07.0400 155	Field equipment, tether ball post	EACH	137.99	128.18	266.17
07.0400 160	Field equipment, add for ground sock, tether ball	EACH	105.63	128.18	233.81
07.0400 165	Field equipment, court striping, tennis, basketball	EACH	344.61		344.61
07.0400 170	Field equipment, court striping, volleyball	EACH	206.75		206.75
07.0400 175	Field equipment, swings, 10' high, 4 seats	SET	1,422.03	757.36	2,179.39
07.0400 180	Field equipment, swings, 10' high, 6 seats	SET	1,744.60	903.02	2,647.62
07.0400 185	Field equipment, horizontal ladder, 8'x16'	EACH	917.32	466.05	1,383.37
07.0400 190	Field equipment, horizontal bar, single, 6 1/2'	EACH	270.98	221.38	492.36
07.0400 195	Field equipment, bleachers, on concrete riser, benches only, fiberglass	SEAT	32.69		32.69
07.0400 200	Field equipment, bleachers, on concrete riser, benches only, wood	SEAT	61.95		61.95
07.0400 205	Field equipment, bleachers, on concrete riser, benches only, aluminum	SEAT	68.88		68.88
07.0400 210	Field equipment, bleachers, on concrete riser, benches only, fiberglass, with back	SEAT	82.68		82.68

7.0 EQUIPMENT

CSI#	Description	Unit	Material Cost	Install Cost	Total Cost
07.0400 215	Field equipment, bleachers, on concrete riser, individual seats	SEAT	117.10		117.10
07.0400 220	Field equipment, bleachers, on concrete riser, benches, portable, 16 rows, 500 minimum	SEAT	65.43		65.43
07.0400 225	Field equipment, outdoor equipment, miscellaneous, benches, wood slats, 6'	EACH	808.11	43.59	851.71
07.0400 230	Field equipment, outdoor equipment, miscellaneous, benches, pre-cast concrete, 6'	EACH	1,818.27	65.39	1,883.67
07.0400 235	Field equipment, outdoor equipment, miscellaneous, picnic table and benches, 6'	EACH	1,817.25	87.19	1,904.44
07.0400 240	Field equipment, outdoor equipment, miscellaneous, picnic table and benches, 6'	EACH	1,401.79	87.19	1,488.97
07.0400 245	Field equipment, outdoor equipment, miscellaneous, bicycle rack, galvanized iron, 10', 2 side	EACH	1,609.51	87.19	1,696.70
07.0400 250	Field equipment, outdoor equipment, miscellaneous, bicycle rack, pre-cast concrete, single	EACH	228.29		228.29
07.0400 255	Field equipment, athletic field, synthetic surface, uniturf, embossed, running, 3/8"	SF	10.10		10.10
07.0400 260	Field equipment, athletic field, synthetic surface, turf, 3 layer	SF	15.21		15.21
07.0400 265	Field equipment, athletic field, synthetic surface, turf, 2 layer	SF	13.30		13.30
07.0400 270	Field equipment, athletic field, synthetic surface, running track, volcanic cinder, 7"	SF	2.62		2.62
07.0400 275	Field equipment, athletic field, synthetic surface, running track, bitum/cork, 2"	SF	4.17		4.17
07.0400 280	Field equipment, athletic field, synthetic surface, running track, AC, 1/4 mile, 55 MSF, 1/4" synth	EACH	638,393.89		638,393.89
07.0400 285	Field equipment, athletic field, synthetic surface, running track, 1/4 mile, 55 MSF, 2" cinder, 6" curb	EACH	179,667.71		179,667.71
07.0400 290	Field equipment, athletic field, synthetic surface, running track, bitum/cork, 1/4 mile, 55 MSF,	EACH	229,363.08		229,363.08
07.0400 295	Field equipment, athletic field, synthetic surface, tennis court, AC base, all weather	SF	4.74		4.74
07.0400 300	Field equipment, score board	EACH	19,113.53		19,113.53
07.0400 305	Field equipment, basketball backstops, wall, out-rigger, fixed	EACH	818.90	788.18	1,607.08
07.0400 310	Field equipment, basketball backstops, wall, out-rigger, swing	EACH	2,510.48	1,505.39	4,015.87
07.0400 315	Field equipment, basketball backstops, ceiling, swing up, manual	EACH	5,048.90	2,016.40	7,065.30
07.0400 320	Field equipment, basketball backstops, add for glass, fan backstop	EACH	1,146.49		1,146.49
07.0400 325	Field equipment, basketball backstops, add for glass, rectangular backs	EACH	1,369.24		1,369.24
07.0400 330	Field equipment, basketball backstops, add for power operation	EACH	910.59		910.59
07.0400 335	Field equipment, gym walls, padded	SF	10.78		10.78
07.0400 340	Field equipment, gym floors, sub-floor or base not included, synthetic gym floor, 3/16"	SF	8.88		8.88
07.0400 345	Field equipment, gym floor, sub-floor or base not included, synthetic gym floor, 3/8"	SF	9.75		9.75
07.0400 350	Field equipment, gym floor, sub-floor or base not included, maple, wood, spring	SF	25.66		25.66
07.0400 355	Field equipment, gym floor, sub-floor or base not included, rubber cusion maple	SF	14.36		14.36
07.0400 360	Field equipment, gym floor, sub-floor or base not included, maple over sleepers	SF	20.62		20.62
07.0400 365	Field equipment, gym floor, add for apparatus inserts	EACH	68.77		68.77
07.0400 370	Field equipment, gym seating, bleachers, telescoping, manual	SEAT	84.04	19.02	103.06
07.0400 375	Field equipment, gym seating, bleachers, portable, hydraulic	SEAT	85.43		85.43
07.0400 380	Field equipment, score boards basketball, economy	EACH	3,182.05	774.27	3,956.32
07.0400 385	Field equipment, score boards basketball, good	EACH	3,196.96	1,419.50	4,616.46
07.0400 390	Field equipment, score boards basketball, best	EACH	4,758.33	2,258.36	7,016.70
07.0500	**INDUSTRIAL EQUIPMENT**				
07.0500 100	Industrial equipment, service station, 3 island, 4 tanks	EACH	739,643.95		739,643.95
07.0500 105	Industrial equipment, service station, remodel to self-serve, average	EACH	211,326.83		211,326.83
07.0500 110	Industrial equipment, service station, air compressor, 2 HP, with receiver	EACH	3,719.33	366.82	4,086.15
07.0500 115	Industrial equipment, service station, air compressor, 3 HP, with receiver	EACH	3,909.52	366.82	4,276.34
07.0500 120	Industrial equipment, service station, gas pump, full size, 1/2 HP	EACH	2,747.22	183.37	2,930.59
07.0500 125	Industrial equipment, service station, gas pump, submerge turbine, 1/3 HP	EACH	1,299.60	92.05	1,391.65
07.0500 130	Industrial equipment, service station, gas pump, submerge turbine, 3/4 HP	EACH	1,504.63	92.05	1,596.68
07.0500 135	Industrial equipment, service station, gas dispenser, computing, single hose	EACH	4,617.47	94.34	4,711.81
07.0500 140	Industrial equipment, service station, add for vapor recovery	EACH	950.93		950.93
07.0500 145	Industrial equipment, service station, gas dispenser, computing, dual hose	EACH	9,234.97	94.34	9,329.30
07.0500 147	Industrial equipment, service station, fill boxes, 12", cast iron	EACH	72.88	84.69	157.57
07.0500 150	service station, air/water, bibbs, underground reels	EACH	496.58	75.50	572.08
07.0500 155	Industrial equipment, service station, add for electric thermal unit	EACH	200.77		200.77
07.0500 160	Industrial equipment, service station, hoist, 1 post, 8000#, semi-hydraulic	EACH	4,194.84	820.73	5,015.57

EQUIPMENT - 7.0

CSI#	Description	Unit	Material Cost	Install Cost	Total Cost
07.0500 165	Industrial equipment, service station, hoist, 2 post, 8000#, semi-hydraulic	EACH	6,915.81	1,760.87	8,676.67
07.0500 170	Industrial equipment, service station, cash box & pedestal stand	EACH	200.77	91.70	292.47
07.0500 175	Industrial equipment, service station, tire changer, air operated	EACH	2,019.68	113.20	2,132.88
07.0500 180	Industrial equipment, service station, lube, oil and air reels/rem pump	EACH	2,430.21	849.15	3,279.36
07.0500 185	Industrial equipment, service station, exhaust fume system, underground	STA.	1,315.50		1,315.50
07.0500 190	Industrial equipment, service station, island metal furring for 3 islands	1 STA	1,267.90	177.96	1,445.87
07.0500 195	Industrial equipment, service station, dynamometer	EACH	79,531.56	10,329.94	89,861.50
07.0500 200	Industrial equipment, shop miscellaneous, sawdust collector	EACH	14,624.93	1,056.38	15,681.32
07.0500 205	Industrial equipment, shop miscellaneous, sawdust collector, large capacity	EACH	19,384.61	1,056.38	20,440.99
07.0500 210	Industrial equipment, shop miscellaneous, paint spray booths	SF	15.04		15.04
07.0500 215	Industrial equipment, shop miscellaneous, paint spray booth car	EACH	65,897.62	2,877.96	68,775.57
07.0500 220	Industrial equipment, shop miscellaneous, paint spray booth, 60' bus	EACH	284,041.47	65,999.66	350,041.13
07.0600	**PARKING LOT EQUIPMENT**				
07.0600 100	Parking lot equipment, automatic gate, automatic arm, 8'	EACH	4,235.00	177.96	4,412.96
07.0600 105	Parking lot equipment, traffic detector	EACH	1,355.19	222.46	1,577.65
07.0600 110	Parking lot equipment, ticket dispenser, control unit	EACH	5,048.15	148.30	5,196.45
07.0600 115	Parking lot equipment, gate operator card/coin	EACH	1,355.19	510.18	1,865.37
07.0700	**MATERIAL HANDLING EQUIPMENT**				
07.0700 100	Dock bumper, 10"x4 1/2"x2'	EA	104.82	63.32	168.14
07.0700 105	Scissors lift, 56", 2000#, 4'x4' deck	EA	9,991.57		9,991.57
07.0700 110	Dock leveler, heavy duty, hydraulic	EA	8,669.87	822.77	9,492.64
07.0700 115	Dock leveler, accordian door, 7'6"x8'	EA	1,351.90	850.62	2,202.51
07.0800	**LABORATORY EQUIPMENT**				
07.0800 100	Laboratory equipment, instructor's table, includes sinks, fixtures and acid resistant tops, 12'x36"x36"	EACH	3,451.89	301.01	3,752.91
07.0800 105	Laboratory equipment, student table, for 8 students, 15'x5'x36"	EACH	7,281.42	301.01	7,582.43
07.0800 110	Laboratory equipment, steel cabinet, base	LF	297.97	173.70	471.67
07.0800 115	Laboratory equipment, wood cabinet, base	LF	245.74	173.70	419.44
07.0800 120	Laboratory equipment, plastic laminated cabinet, base	LF	222.92	173.70	396.62
07.0800 125	Laboratory equipment, steel cabinet, knee space, drawer unit	LF	190.89	110.37	301.26
07.0800 130	Laboratory equipment, wood cabinet, knee space, drawer unit	LF	159.01	110.26	269.27
07.0800 135	Laboratory equipment, plastic laminated cabinet, knee space, drawer unit	LF	143.12	110.16	253.28
07.0800 140	Laboratory equipment, steel cabinet, wall hung	LF	218.97	84.37	303.34
07.0800 145	Laboratory equipment, wood cabinet, wall hung	LF	228.07	84.37	312.44
07.0800 150	Laboratory equipment, plastic laminated cabinet, wall hung	LF	144.00	84.37	228.37
07.0800 155	Laboratory equipment, plastic laminated cabinet top	LF	81.81		81.81
07.0800 160	Laboratory equipment, epoxy resin, acid resistant cabinet top	LF	190.89		190.89
07.0800 165	Laboratory equipment, stainless steel cabinet top	LF	274.95		274.95
07.0800 170	Laboratory equipment, stainless steel cabinet top	LF	109.04		109.04
07.0800 175	Laboratory equipment, centrifuge, high speed, portable	EACH	2,794.96		2,794.96
07.0800 180	Laboratory equipment, centrifuge, ultra high speed, explosion proof	EACH	12,725.04		12,725.04
07.0800 185	Laboratory equipment, centrifuge, ultra high speed, refer	EACH	20,223.76		20,223.76
07.0800 190	Reagent rack, metal upright, wall, 1 tier	EACH	109.04	49.95	158.99
07.0800 195	Reagent rack, metal upright center, 1 tier	EACH	134.05	49.95	184.01
07.0800 200	Laboratory stoarage, wardrobe	LF	305.55	70.83	376.38
07.0800 205	Laboratory equipment miscellaneous, fume hood	LF	1,363.39	367.90	1,731.29
07.0800 210	Laboratory equipment miscellaneous, add for stainless steel	LF	681.68		681.68
07.0800 215	Laboratory equipment miscellaneous, water distiller, 10 GPM	EACH	8,606.29	414.35	9,020.64
07.0800 220	Laboratory equipment miscellaneous, water tank, stainless steel, 50 GAL	EACH	9,392.63	414.35	9,806.98
07.0800 225	Laboratory equipment miscellaneous, washer	EACH	11,190.64	828.66	12,019.30
07.0800 230	Laboratory equipment miscellaneous, portable water distiller, 18 LPH	EACH	1,442.09	150.51	1,592.60
07.0800 235	Laundry equipment, unloading washer, 60"x44", 400#	EACH	117,513.40	17,125.81	134,639.21
07.0800 240	Laundry equipment, extractor, 200#, with accessories	EACH	94,636.15	13,968.94	108,605.09
07.0800 245	Laundry equipment, washer/extractor, 600#	EACH	171,591.23	25,032.25	196,623.48
07.0800 250	Laundry equipment, dryer, gas or steam, 200/400#	EACH	77,552.87	11,314.80	88,867.67
07.0800 255	Laundry equipment, flat ironer, 6 roller	EACH	174,511.24	25,445.71	199,956.95
07.0800 260	Laundry equipment, ironer/folder, 2 lane	EACH	70,057.30	10,336.96	80,394.26
07.0800 265	Laundry equipment, folder, 3 lane	EACH	44,095.25	6,425.67	50,520.92

7.0 EQUIPMENT

CSI#	Description	Unit	Material Cost	Install Cost	Total Cost
07.0900	**LIBRARY EQUIPMENT**				
07.0900 100	Library equipment, study carrels, hardwood, 36"x24"x29", 2 face	EACH	508.15	113.38	621.53
07.0900 105	Library equipment, study carrels, hardwood, individual lights	EACH	90.64	68.08	158.72
07.0900 110	Library equipment, study carrels, hardwood, power post recept	EACH	25.30	68.08	93.38
07.0900 115	Library equipment, table, 60"x36"x39"	EACH	1,621.27		1,621.27
07.0900 120	Library equipment, table, 48" round	EACH	1,468.81		1,468.81
07.0900 125	Library equipment, chairs, wood	EACH	274.03		274.03
07.0900 130	Library equipment, card catalog cabinet, wood, 60 tray	EACH	1,239.89		1,239.89
07.0900 135	Library equipment, card catalog cabinet, wood, 30 tray	EACH	713.31		713.31
07.0900 140	Library equipment, charging counters, hardwood	LF	180.65	68.08	248.73
07.0900 145	Library equipment, charging counter top	SF	16.05	11.34	27.39
07.0900 150	Library equipment, shelving, metal, bracket, 3 tier, 3x3	EACH	209.66	62.36	272.02
07.0900 155	Library equipment, shelving, metal, bracket, 5 tier, 3x3	EACH	225.83	68.08	293.91
07.1000	**HOSPITAL EQUIPMENT**				
07.1000 100	Nurses monitoring equipment, CCU/ICU, cardioscope, 1 channel, bed side	EACH	5,357.98		5,357.98
07.1000 105	Nurses monitoring equipment, CCU/ICU, BP monitor, bed side, with readout	EACH	8,226.54		8,226.54
07.1000 110	Nurses monitoring equipment, CCU/ICU, CCU bed & nurse display, 8 bed	EACH	167,394.86		167,394.86
07.1000 115	Nurses monitoring equipment, CCU/ICU, telemetry, wireless, 8 bed	EACH	457,832.93		457,832.93
07.1000 120	Nursing acute care equipment, modular wall unit, 1 bed core, prewire	EACH	6,087.81	780.34	6,868.15
07.1000 125	Nursing acute care equipment, modular wall unit, 24" nurse treatment	EACH	807.71	180.29	988.01
07.1000 130	Nursing acute care equipment, mod wall unit, 24" storage center	EACH	874.45	198.31	1,072.77
07.1000 135	Nursing acute care equipment, patient communications & convenience unit	EACH	1,887.26		1,887.26
07.1000 140	Nursing acute care equipment, cubicle track & curtain, to 8' high	LF	14.77	4.40	19.18
07.1000 145	Nursing acute care equipment, mirror, medicine cabinet, stainless steel, resess, light	EACH	581.77	131.35	713.11
07.1000 150	Nursing station/core equipment, nurse chart desk	EACH	1,327.47		1,327.47
07.1000 155	Nursing station/core equipment, nurse call, 40 station, two way	EACH	25,020.25	6,644.36	31,664.61
07.1000 160	Nursing station/core equipment, nourishment station, 84" x 80", stainless steel	EACH	22,692.77	401.35	23,094.12
07.1000 165	Nursing station/core equipment, medical prep cab, 72" x 80", stainless steel, lock	EACH	10,473.53	401.35	10,874.88
07.1000 170	Nursing station/core equipment, IV prep center, 60" x 80", stainless steel	EACH	9,018.89	401.35	9,420.24
07.1000 175	Nursing station/core equipment, sani-prep maintenance station, stainless steel	EACH	6,545.95	1,738.36	8,284.30
07.1000 180	Nursing station/core equipment, multi-use tote carts, average cost	EACH	1,838.43		1,838.43
07.1000 185	Nursing station/core equipment, modular storage systems	LF	418.03		418.03
07.1000 190	Corridors, hospital corner guard, plastic, 8'	EACH	163.65	48.87	212.52
07.1000 195	Corridors, hospital corner guard, plastic, 4'	EACH	93.52	36.42	129.94
07.1000 200	Corridors, hospital corner guard, stainless steel, 8'	EACH	394.97	156.66	551.62
07.1000 205	Corridors, hospital corner guard, stainless steel, 4'	EACH	197.48	84.75	282.22
07.1000 210	Surgery tables & accessories, surgery table, electric, stationary	EACH	27,461.50	2,558.61	30,020.12
07.1000 215	Surgery tables & accessories, surgery service island, complete	EACH	9,941.34	2,082.01	12,023.35
07.1000 220	Surgery tables & accessories, surgical monitor, conduct casters	EACH	10,362.57		10,362.57
07.1000 225	Surgery tables & accessories, conductivity meter, current leakage detector	EACH	1,147.00	289.73	1,436.73
07.1000 230	Surgery tables & accessories, surgical clock, stainless steel, auto reset	EACH	1,385.44	297.46	1,682.90
07.1000 235	Obstetrical & nursery equipment, delivery table	EACH	7,953.01	2,008.76	9,961.77
07.1000 240	Obstetrical & nursery equipment, incubator, isolation servo-care	EACH	7,354.00		7,354.00
07.1000 245	Obstetrical & nursery equipment, incubator, warming	EACH	2,406.76		2,406.76
07.1000 250	Surgical lighting, surgery light, surface mounted, 3 arm	EACH	31,056.66	2,993.79	34,050.45
07.1000 255	Surgical lighting, surgery light, surface mounted, 2 arm	EACH	21,568.29	2,756.57	24,324.86
07.1000 260	Surgical lighting, surgery light, surface mounted, 1 arm	EACH	15,045.63	1,999.90	17,045.53
07.1000 265	Surgical lighting, add for auxiliary light head	EACH	7,827.21	831.12	8,658.33
07.1000 270	Surgical lighting, surgical light intensity control, 600W	EACH	869.70	623.34	1,493.04
07.1000 275	Surgical lighting, surgical light intensity control, 300W	EACH	652.22	401.35	1,053.57
07.1000 280	Scrub & clean room equipment, scrub station, stainless steel, 3 bay, base mounted	EACH	11,335.84	1,722.61	13,058.44
07.1000 285	Scrub & clean room equipment, scrub station, stainless steel, 1 bay, base mounted	EACH	8,097.02	1,129.58	9,226.60
07.1000 290	Scrub & clean room equipment, solution warm cab, stainless steel, 24"X30"X74"	EACH	8,372.67	282.39	8,655.06
07.1000 295	Scrub & clean room equipment, surgery storage console, stainless steel, 12'	EACH	5,576.10	200.68	5,776.78
07.1000 300	Scrub & clean room equipment, suture/drug cab, stainless steel, 36"X18"X81"	EACH	1,844.27	150.51	1,994.78
07.1000 305	Scrub & clean room equipment, instrument cab, stainless steel, 48"X18"X60"	EACH	2,501.06	150.51	2,651.57
07.1000 310	Scrub & clean room equipment, sterilizer, 16"X16"X26"	EACH	25,861.87	501.69	26,363.56

EQUIPMENT - 7.0

CSI#	Description	Unit	Material Cost	Install Cost	Total Cost
07.1000 315	Cardiac emergency equipment, monitor/resuscitation unit, with DC defibrillator	EACH	20,262.41		20,262.41
07.1000 320	Cardiac emergency equipment, mobile emergency utility, crash cart	EACH	11,052.14		11,052.14
07.1000 325	Cardiac emergency equipment, DC defibrillator	EACH	5,894.27		5,894.27
07.1000 330	Cardiac emergency equipment, external cardiac compressor	EACH	6,631.35		6,631.35
07.1000 335	Emergency accessories, treatment cabinet, mobile unit	EACH	2,026.24		2,026.24
07.1000 340	Emergency accessories, treatment cabinet, with suction compressor	EACH	3,592.08		3,592.08
07.1000 345	Emergency accessories, exam lights, 22", ceiling mounted	EACH	3,946.18	701.26	4,647.44
07.1000 350	Emergency accessories, plaster sink with trap & fittings	EACH	1,578.86	571.17	2,150.03
07.1000 355	Emergency accessories, splint cabinet with plaster bins	EACH	1,376.44	257.51	1,633.95
07.1000 360	Sterilizers, recessed, with automatic doors, sterilizer, 24"x36"x36", 1 door, steam	EACH	104,661.25	28,172.89	132,834.14
07.1000 365	Sterilizers, recessed, with automatic doors, sterilizer, 24"x36"x48", 1 door, steam	EACH	106,667.59	28,172.89	134,840.48
07.1000 370	Sterilizers, recessed, with automatic doors, sterilizer, gas, 24"x36"x60", 1 door	EACH	125,409.44	22,755.93	148,165.37
07.1000 375	Sterilizers, recessed, with automatic doors, sterilizer, gas, 24"x36"x60", pass-thru	EACH	131,540.81	22,755.93	154,296.74
07.1000 380	Sterilizers, recessed, with automatic doors, sterilizer, steam, 24"x36"x48", pass-thru	EACH	133,932.41	28,172.89	162,105.30
07.1000 385	Sterilizers, recessed, with automatic doors, sterilizer, steam, 24"x36"x60", pass-thru	EACH	136,293.63	28,172.89	164,466.52
07.1000 390	Sterilizers, recessed, with automatic doors, sterilizer, gas/cyro, 24"x36"x60", pass-thru	EACH	112,407.55	24,423.29	136,830.83
07.1000 395	Sterilizers, recessed, with automatic doors, sterilizer, gas/cyro, 24"x36"x48", 1 door	EACH	95,666.04	11,228.82	106,894.86
07.1000 400	Sterilizers, recessed, with automatic doors, gas, aerator, 24"x36"x60"	EACH	11,958.24	3,037.27	14,995.50
07.1000 405	Sterilizers, recessed, with automatic doors, gas, aerator, 24"x36"x48"	EACH	6,457.44	1,628.22	8,085.66
07.1000 410	Sterilizers, accessories, loading car & carriage, large	EACH	7,892.41		7,892.41
07.1000 415	Sterilizers, accessories, loading car & carriage, medium	EACH	6,457.44		6,457.44
07.1000 420	Sterilizers, accessories, steam general, 10 BHP/110, PSIG/208V, 3PH	EACH	5,680.11	1,228.99	6,909.11
07.1000 425	Instrument & utensil washer/sterilizer, instrument cleaner, sonic, 12x11x24, 1 comp	EACH	18,535.26	5,733.57	24,268.83
07.1000 430	Instrument & utensil washer/sterilizer, instrument cleaner, sonic, 12x11x24, 2 comp	EACH	25,003.56	5,733.57	30,737.13
07.1000 435	Instrument & utensil washer/sterilizer, glass wash, steam, 24x20x24, pass-thru	EACH	39,331.77	8,274.44	47,606.20
07.1000 440	Instrument & utensil washer/sterilizer, utensil wash, 27x27x60, free stand	EACH	16,741.53	3,265.05	20,006.58
07.1000 445	Instrument & utensil washer/sterilizer, utensil wash, conveyor, Pass-thru	EACH	56,801.64	12,166.25	68,967.89
07.1000 450	Instrument & utensil washer/sterilizer, drying Oven, 25"x25"x50", steam	EACH	13,871.55	2,771.71	16,643.26
07.1000 455	Instrument & utensil washer/sterilizer, charging Oven, 5', stainless steel, 2 sink	EACH	2,630.79	684.96	3,315.75
07.1000 460	Hospital cart wash, pit mounted, cart wash, 92"x72"x98", pass-thru, pit	EACH	109,011.40	23,744.86	132,756.25
07.1000 465	Hospital cart wash, pit mounted, cart wash, 92"x72"x196", auto, 2 stage	EACH	273,275.24	28,702.58	301,977.82
07.1000 470	Central pharmacy equipment, pharmacy unit w/basic modules	LF	671.68		671.68
07.1000 475	Central pharmacy equipment, medication refer, 115 CF, stainless steel	EAcH	4,655.77	333.43	4,989.20
07.1000 480	Central pharmacy equipment, water purification mod, rev Osm, 6 LPM	EACH	3,846.03	832.10	4,678.13
07.1000 485	Central pharmacy equipment, laminated flow hood, 36" work area	EACH	8,744.75	1,587.94	10,332.69
07.1000 490	Central pharmacy equipment, hi-density storage, caster/track	EACH	3,643.59	260.93	3,904.52
07.1000 495	Central laboratory equipment, lab work counter, with base units	LF	591.70		591.70
07.1000 500	Central laboratory equipment, lab sterilizer, 16"x26", 208V, 3PH	EACH	7,287.32	834.99	8,122.31
07.1000 505	Central laboratory equipment, liquid nitrogen refer, 30 CF, 190 DEG	EACH	32,388.13	869.79	33,257.92
07.1000 510	Central laboratory equipment, water distribution, steam, 10 GPH, wall	EACH	17,937.31	1,188.70	19,126.01
07.1000 515	Central laboratory equipment, water tank, stainless steel, 50 GAL, wall, mounted	EACH	10,762.41	1,066.94	11,829.36
07.1000 520	Central laboratory equipment, specimen pass-thru box, stainless steel	EACH	222.66	140.72	363.39
07.1000 525	X-ray equipment, x-ray, ceiling mounted, telescoping	EACH	43,724.00	5,183.71	48,907.72
07.1000 530	X-ray equipment, x-ray, ceiling wall mounted, chest	EACH	12,955.21	2,559.86	15,515.07
07.1000 535	X-ray equipment, mobile x-ray unit	EACH	54,156.20		54,156.20
07.1000 540	X-ray equipment, x-ray control unit	EACH	14,331.67	2,831.85	17,163.52
07.1000 545	X-ray equipment, multix table	EACH	27,327.50	5,399.70	32,727.20
07.1000 550	X-ray processing equipment, auto film processor, complete	EACH	43,027.93		43,027.93
07.1000 555	X-ray processing equipment, developer tank, 10 G, stainless steel, 2 comp, mix valve	EACH	2,325.85		2,325.85
07.1000 560	X-ray processing equipment, x-ray pass box, 2 comp, RO-IN frame	EACH	1,943.21		1,943.21
07.1000 565	X-ray processing equipment, x-ray film loading bin	EACH	789.45		789.45
07.1000 570	X-ray processing equipment, revolving door, 36"x80", safe hinge	EACH	4,007.96		4,007.96
07.1000 575	X-ray viewing equipment, x-ray film illuminating, wet, drip tray	EACH	386.61		386.61
07.1000 580	X-ray viewing equipment, x-ray film illluminating, 1 panel 14x17	EACH	308.64		308.64
07.1000 585	X-ray viewing equipment, x-ray film illluminating, 2 panel 30x18	EACH	587.01		587.01

7.0 EQUIPMENT

CSI#	Description	Unit	Material Cost	Install Cost	Total Cost
07.1000 590	X-ray viewing equipment, x-ray film illluminating, multibank, 4/4	EACH	2,797.51		2,797.51
07.1000 595	X-ray viewing equipment, x-ray film illluminating, multibank, 6/6	EACH	3,724.59		3,724.59
07.1000 600	X-ray viewing equipment, x-ray shield, view glass, deluxe	EACH	1,275.28		1,275.28
07.1000 605	Nuclear equipment, gamma camera, 10" view, complete	EACH	481,450.31		481,450.31
07.1000 610	Nuclear equipment, gamma camera, 15" view, complete	EACH	542,178.18		542,178.18
07.1000 615	Ultra sound equipment, ultra sound unit complete	EACH	280,663.82		280,663.82
07.1000 620	Hydrotherapy units, hubbard tank, 400 gallon, twin eject	EACH	32,792.95		32,792.95
07.1000 625	Hydrotherapy units, treatment/wade tank, 1000 gallon	EACH	40,080.32		40,080.32
07.1000 630	Hydrotherapy units, whirlpool, stainless steel, 85 gallon, leg & hip	EACH	6,194.21		6,194.21
07.1000 635	Hydrotherapy units, whirlpool, stainless steel, 80 gallon, arm, leg & hip	EACH	5,344.00		5,344.00
07.1000 640	Hydrotherapy units, whirlpool, stainless steel, 80 gallon, arm	EACH	3,947.31		3,947.31
07.1000 645	Hydrotherapy units, moisture heat therapy unit, table	EACH	5,167.69		5,167.69
07.1000 650	Hydrotherapy units, mobile paraffin bath	EACH	1,893.70		1,893.70
07.1000 655	Hydrotherapy units, mobile sitz bath	EACH	516.36		516.36
07.1000 660	Inhalation therapy equipment, ventilator, complete	EACH	17,559.95		17,559.95
07.1000 665	Inhalation therapy equipment, suction unit, stainless steel cabinet	EACH	1,618.42		1,618.42
07.1000 670	Inhalation therapy equipment, IPPB inhaler	EACH	2,134.68		2,134.68
07.1000 675	Inhalation therapy equipment, air volume tester, lung	EACH	8,607.79		8,607.79
07.1000 680	Physical therapy equipment, exercise unit, complete	EACH	8,798.38		8,798.38
07.1000 685	Physical therapy equipment, exercise chair	EACH	2,582.21		2,582.21
07.1000 690	Physical therapy equipment, treadmill, motorized, 5 speed	EACH	5,597.92		5,597.92
07.1000 695	Physical therapy equipment, treadmill, adjustable angle	EACH	1,205.10		1,205.10
07.1000 700	Physical therapy equipment, rowing machine	EACH	1,119.55		1,119.55
07.1000 705	Physical therapy equipment, rehabilitation loom	EACH	1,807.29		1,807.29
07.1100	**DENTAL EQUIPMENT**				
07.1100 100	Dental equipment, dental chair, deluxe, with lift	EACH	11,428.98		11,428.98
07.1100 105	Dental equipment, dental chair, standard, tilt	EACH	6,651.68		6,651.68
07.1100 110	Dental equipment, instrument unit, 4 port, tray, chair mounted	EACH	7,465.40		7,465.40
07.1100 115	Dental equipment, assistant instrument unit, chair mounted	EACH	5,684.04		5,684.04
07.1100 117	Dental equipment, mobile instrument unit, cabinet	EACH	8,546.38		8,546.38
07.1100 120	Dental equipment, mobile assistant's instrument unit, cabinet	EACH	7,343.92		7,343.92
07.1100 125	Dental equipment, instrument unit, cabinet, wall mounted	EACH	21,372.13		21,372.13
07.1100 130	Dental equipment, dental light, chair or unit mounted	EACH	1,939.21		1,939.21
07.1100 135	Dental equipment, dental light, ceiling mounted	EACH	1,939.21		1,939.21
07.1100 140	Dental equipment, dental x-ray, wall mounted	EACH	15,821.55		15,821.55
07.1100 141	Dental equipment, dental x-ray, wall mounted, extra remote HD	EACH	25,011.72		25,011.72
07.1100 142	Dental equipment, dental sterilizer, chemiclave	EACH	2,228.70		2,228.70
07.1100 143	Dental equipment, dental sterilizer, vibraclean	EACH	1,359.43		1,359.43
07.1100 144	Dental equipment, dental compressor with dryer	EACH	4,033.04		4,033.04
07.1100 145	Dental equipment, laboratory, dust collector, pedestal	EACH	2,825.81	495.78	3,321.58
07.1100 150	Dental equipment, laboratory, waxing unit, 3 compartment	EACH	111.91		111.91
07.1100 155	Dental equipment, laboratory, pneumatic curing unit	EACH	239.84		239.84
07.1100 160	Dental equipment, laboratory, double pneumatic press	EACH	1,679.27		1,679.27
07.1100 165	Dental equipment, laboratory, curing tank assembly	EACH	1,305.61	224.46	1,530.07
07.1100 170	Dental equipment, laboratory, boilout assembly	EACH	1,072.86	184.39	1,257.25
07.1100 175	Dental equipment, laboratory, plaster bin, 4 compartment, 300#	EACH	744.86	127.61	872.47
07.1100 180	Dental equipment, laboratory furniture, dental lab tech bench, 5 drawer	EACH	1,415.49	216.13	1,631.62
07.1100 185	Dental equipment, laboratory furniture, dental lab, 2 door cabinet, 36"x24"x36"	EACH	1,603.14	278.14	1,881.28
07.1100 190	Dental equipment, laboratory furniture, dental lab, 1 door cabinet, 24"x24"x36"	EACH	1,246.86	216.37	1,463.24
07.1100 195	Dental equipment, laboratory furniture, dental lab, 1 door cabinet, corner	EACH	1,959.43	339.94	2,299.37
07.1200	**STAGE EQUIPMENT**				
07.1200 100	Stage equipment, lighting instruments, footlights, disappearing, reflect	PER 5	1,028.22		1,028.22
07.1200 105	Stage equipment, lighting instruments, add for motorized	PER 5	1,769.94		1,769.94
07.1200 110	Stage equipment, lighting instruments, border lights, 1 row, reflect, 3 color	PER 8	859.67		859.67
07.1200 115	Stage equipment, lighting instruments, border lights, 1 row, roundel, 3 color	PER 8	1,196.81		1,196.81
07.1200 120	Stage equipment, lighting instruments, border lights, 2 row, roundel, 4 color	PER 8	3,034.25		3,034.25
07.1200 123	Stage equipment, lighting instruments, ball, mirrored, rotating, 30"	EACH	2,697.33	1,605.88	4,303.21
07.1200 125	Stage equipment, lighting instruments, spotlight, follow, carbon arc	EACH	7,417.24		7,417.24

EQUIPMENT - 7.0

CSI#	Description	Unit	Material Cost	Install Cost	Total Cost
07.1200 130	Stage equipment, lighting instruments, spotlight, follow, quartz halogen	EACH	2,359.98		2,359.98
07.1200 135	Stage equipment, lighting instruments, quartz spot, ellip, iod lamp, 3000W	EACH	2,191.38		2,191.38
07.1200 140	Stage equipment, lighting instruments, quartz spot, ellip, iod lamp, 1500W	EACH	758.54		758.54
07.1200 145	Stage equipment, lighting instruments, quartz spot, ellip, iod lamp, 500W	EACH	471.95		471.95
07.1200 150	Stage equipment, lighting instruments, quartz spot, fresnl, iod lamp, 1000W	EACH	539.33		539.33
07.1200 155	Stage equipment, lighting instruments, quartz spot, fresnl, iod lamp, 500W	EACH	320.22		320.22
07.1200 160	Stage equipment, lighting instruments, color wheel, motorized, 20"	EACH	320.22		320.22
07.1200 165	Stage equipment, lighting instruments, quartz beam proj, iod lamp, 1500W	EACH	522.45		522.45
07.1200 170	Stage equipment, lighting instruments, floodlight, utility, incandescent, 1000W	EACH	286.50		286.50
07.1200 175	Stage equipment, lighting instruments, floodlight, scoop, iod lamp, 18", 750W	EACH	320.22		320.22
07.1200 180	Stage equipment, lighting and control accessories, light tower, 4'x6'x15'	EACH	8,428.63		8,428.63
07.1200 185	Stage equipment, lighting and control accessories, light stand, cast iron base, 24"	EACH	387.69		387.69
07.1200 190	Stage equipment, lighting and control accessories, plug strip	LF	57.25		57.25
07.1200 195	Stage equipment, lighting and control accessories, floor pocket, plug, 4 outlet, 100A	EACH	269.64		269.64
07.1200 200	Stage equipment, lighting and control accessories, floor pocket, plug, 2 outlet, 50A	EACH	235.96		235.96
07.1200 205	Stage equipment, lighting and control accessories, wall pocket, surf plug, 3 outlet, 50A	EACH	84.25		84.25
07.1200 210	Stage equipment, lighting and control accessories, wall pocket, surf plug, 1 outlet, 50A	EACH	185.36		185.36
07.1200 215	Stage equipment, lighting and control accessories, wall pocket, flush, 2 outlet, 100A	EACH	326.96		326.96
07.1200 220	Stage equipment, lighting and control accessories, wall pocket, flush, 4 outlet, 50A	EACH	286.50		286.50
07.1200 225	Stage equipment, lighting and control accessories, catwalk, wood, metal pipe mount rail	LF	23.45	101.57	125.02
07.1200 230	Stage equipment, rigging curtains and drops, acoustic cloud, adjust, wood frame	SF	4.36	8.38	12.75
07.1200 235	Stage equipment, rigging curtains and drops, curtain track, heavy duty, straight	LF	17.37	21.58	38.95
07.1200 240	Stage equipment, rigging curtains and drops, curtain track, medium duty, straight	LF	10.93	21.58	32.51
07.1200 245	Stage equipment, rigging curtains and drops, curtain track, heavy duty, curved	LF	69.72	32.42	102.15
07.1200 250	Stage equipment, rigging curtains and drops, curtain track, medium duty, curved	LF	17.98	32.42	50.40
07.1200 255	Stage equipment, rigging curtains and drops, add for electric driven heavy duty	EACH	1,547.59	279.58	1,827.17
07.1200 260	Stage equipment, rigging curtains and drops, add for electric driven medium duty	EACH	1,486.89	279.58	1,766.47
07.1200 265	Stage equipment, rigging curtains and drops, T-bar rig, 4 loft blk, 55x35	SET	53,943.60		53,943.60
07.1200 270	Stage equipment, rigging curtains and drops, rig, wire guard, 4 loft blk 45x30	SET	47,200.70		47,200.70
07.1200 275	Stage equipment, rigging curtains and drops, T-bar rig, 4 loft blk caster mounted	SET	57,315.11		57,315.11
07.1200 280	Stage equipment, rigging curtains and drops, add for electric control	LUMP	53,943.60		53,943.60
07.1200 285	Stage equipment, rigging curtains and drops, curtain, asbestos, straight, lift, 23'x47'	LUMP	75,858.21		75,858.21
07.1200 290	Stage equipment, rigging curtains and drops, curtain, asbestos, trip, 23'x47'	LUMP	72,486.75		72,486.75
07.1200 295	Stage equipment, rigging curtains and drops, curtain, main, velour, heavy	SY	26.08		26.08
07.1200 300	Stage equipment, rigging curtains and drops, curtain, main, velour, medium	SY	18.30		18.30
07.1200 305	Stage equipment, rigging curtains and drops, curtain, light weight, cyclorama	SY	13.74		13.74
07.1200 310	Stage equipment, rigging curtains and drops, drops, velour, heavy, 6'x40'	EACH	2,160.73		2,160.73
07.1200 315	Stage equipment, rigging curtains and drops, drops, velour, medium, 6'x40'	EACH	2,022.86		2,022.86
07.1200 320	Stage equipment, rigging curtains and drops, wings, velour, heavy, 10'x30'	EACH	1,517.11		1,517.11
07.1200 325	Stage equipment, rigging curtains and drops, wings, velour, medium, 10'x30'	EACH	1,415.94		1,415.94
07.1200 330	Stage equipment, rigging curtains and drops, scissors lift, not including any structural work, 5'x10', 12' lift	LUMP	20,228.80		20,228.80
07.1200 335	Stage equipment, television studio lighting, including lamps, studio, 8000 SF, very well equipped	LUMP	212,129.53	17,184.64	229,314.17
07.1200 340	Stage equipment, television studio lighting, including lamps, studio, 6, 60'x72', well equipped	LUMP	659,181.91	16,059.07	675,240.98
07.1200 345	Stage equipment, television studio lighting, including lamps, studio, medium equipped	LUMP	156,113.15	4,496.52	160,609.67
07.1200 350	Stage equipment, television studio lighting, including lamps, studio, 2, 20'x30', medium equipped	LUMP	121,383.79	3,211.86	124,595.66
07.1200 355	Stage equipment, television studio lighting, including lamps, studio, port, minimum equipped, no control	LUMP	4,383.20	642.33	5,025.54

7.0 EQUIPMENT

CSI#	Description	Unit	Material Cost	Install Cost	Total Cost
07.1300	**GARBAGE COMPACTORS**				
07.1300 100	Garbage compactor, 1-1/2 CY	EACH	15,760.34	1,129.58	16,889.92
07.1300 105	Garbage compactor, 2 CY	EACH	23,400.03	1,129.58	24,529.61
07.1300 110	Garbage compactor, 2-1/2" CY	EACH	28,709.01	1,355.49	30,064.50

SPECIAL CONSTRUCTION - 8.0

CSI#	Description	Unit	Material Cost	Install Cost	Total Cost
	The special construction costs in this section include material, installation, and subcontractor overhead and profit. There are no allowances for general contractor markup and profit.				
	The items in this section are associated with computer rooms, X-Ray rooms and swimming pools. The treatment of these items is not extensive, in this publication, but the costs can be used for non detailed estimating.				
	Costs represent standard grade materials and normal installation. Adjustments should be made for economy quality or custom and heavy duty materials and commensurate installation.				
08.0100	**RAISED FLOORS**				
08.0100 100	Pedestal floors, vinyl tile, gridless	SF	22.61		22.61
08.0100 105	Pedestal floors, vinyl tile, grid	SF	27.15		27.15
08.0100 110	Pedestal floors, perma kleen, grid	SF	29.59		29.59
08.0100 115	Pedestal floors, carpeted system	SF	30.56		30.56
08.0100 120	Pedestal floors, ramps	SF	32.28		32.28
08.0100 125	Add for cutouts	EACH	256.29		256.29
08.0100 130	Add for floor grills	EACH	113.95		113.95
08.0100 140	Add for sheetmetal trim and casing	LF	126.47		126.47
08.0100 145	Add for seismic bracing	SF	0.07		0.07
08.0100 150	Add for CO2 fire system (smoke detector)	SF	7.24		7.24
08.0100 155	Add for automatic fire alarm	SF	0.48		0.48
08.0200	**X-RAY ROOM CONSTRUCTION**				
08.0200 100	Lead lined lath 2#	SF	10.25		10.25
08.0200 105	Lead lined lath 4#	SF	15.16		15.16
08.0200 110	Lead lined lath 6#	SF	18.16		18.16
08.0200 115	Lead lined lath 8#	SF	23.41		23.41
08.0200 120	Lead glass windows with lead frames	SF	380.79		380.79
08.0200 125	Lead lined doors to 4#	SF	39.67		39.67
08.0200 130	Lead lined door frames	EACH	519.09		519.09
08.0300	**POOLS**				
08.0300 100	Swimming pool, residential	SF	73.74		73.74
08.0300 105	Swimming pool, multiple residential	SF	83.27		83.27
08.0300 110	Swimming pool, community	SF	88.94		88.94
08.0300 115	Swimming pool, hotel/resort	SF	95.86		95.86
08.0300 120	Swimming pool, school 42'x75'	SF	98.23		98.23
08.0300 125	Swimming pool, school 42'x165'	SF	101.98		101.98
08.0300 130	Swimming pool, school 30'x30'	SF	106.29		106.29
08.0300 135	Pool deck concrete	SF	7.94		7.94
08.0300 140	Pool cool concrete	SF	11.65		11.65

9.0 CONVEYING

CSI#	Description	Unit	Material Cost	Install Cost	Total Cost
	The costs in this section include materials, labor, equipment rental, supervision, and subcontractor overhead and profit. There are no allowances for the general contractor.				
	Costs are complete, stairs are painted or finished where appropriate and ancillary installation and equipment costs are included where expected.				
09.0100	**STAIRS**				
09.0100 100	Wood stair, straight, wood rail, 36" wide	RISER	136.96	140.61	277.57
09.0100 105	Wood stair, switch back, wood rail, 36" wide	RISER	147.97	165.18	313.15
09.0100 110	Wood stair, circular, wood rail, 36" wide	RISER	174.52	140.61	315.13
09.0100 115	Wood stair, circular, wood rail, 36" wide	RISER	174.52	140.61	315.13
09.0100 120	Concrete stair, straight, wood rail, 36" wide	RISER	710.64	120.20	830.84
09.0100 125	Concrete stair, straight, wood rail, 48" wide	RISER	710.64	120.20	830.84
09.0100 130	Concrete stair, straight, wood rail, 72" wide	RISER	710.64	120.20	830.84
09.0100 135	Steel stair, concrete tread, wrought iron rail, 44" wide	RISER	524.31	47.99	572.30
09.0100 140	Steel stair, steel pan, concrete tread, wrought iron rail, 44" wide	RISER	877.86	119.63	997.49
09.0100 145	Steel stair, steel pan, concrete tread, wrought iron rail, 48" wide	RISER	877.86	119.63	997.49
09.0100 150	Steel stair, cast iron, circular, wrought iron rail, 32" wide	RISER	658.59	59.97	718.56
09.0100 155	Steel stair, steel Riser, wrought iron rail, 36" wide	RISER	524.31	47.99	572.30
09.0205	**ELEVATORS, HYDRAULIC**				
09.0205 100	Elevator, hydraulic, 125'min, 2000 lb, 5'x6'cab, auto exit, 2 stop	EACH	77,611.96		77,611.96
09.0205 105	Elevator, hydraulic, 125'min, 2000 lb, 5'x6'cab, auto exit, 4 stop	EACH	101,342.05		101,342.05
09.0205 110	Elevator, hydraulic, 125'min, 2000 lb, 5'x6'cab, auto exit, 5 stop	EACH	116,851.45		116,851.45
09.0205 115	Elevator, hydraulic, 125'min, 2000 lb, 5'x6'cab, auto exit, additional stops	EACH	15,721.81		15,721.81
09.0205 120	Elevator, hydraulic, 125'min, 2500 lb, 5'x7'cab, auto exit, 2 stop	EACH	79,867.60		79,867.60
09.0205 125	Elevator, hydraulic, 125'min, 2500 lb, 5'x7'cab, auto exit, 3 stop	EACH	92,055.58		92,055.58
09.0205 130	Elevator, hydraulic, 125'min, 2500 lb, 5'x7'cab, auto exit, 4 stop	EACH	104,286.58		104,286.58
09.0205 135	Elevator, hydraulic, 125'min, 2500 lb, 5'x7'cab, auto exit, 5 stop	EACH	120,253.68		120,253.68
09.0205 140	Elevator, hydraulic, 125'min, 2500 lb, 5'x7'cab, auto exit, additional stops	EACH	16,178.05		16,178.05
09.0205 145	Elevator, hydraulic, 125'min, 4000 lb, 6'x8'cab, auto exit, 3 stop	EACH	107,667.36		107,667.36
09.0205 150	Elevator, hydraulic, 125'min, 20000 lb, 10'x16'cab, auto exit, 3 stop	EACH	204,567.98		204,567.98
09.0210	**ELEVATORS, ELECTRIC GEAR**				
09.0210 100	Elevator, electric gear, 350'min, 3500 lb, 5'x8'cab, auto exit, 10 stop	EACH	268,149.82		268,149.82
09.0210 105	Elevator, electric gear, 350'min, 3500 lb, 5'x8'cab, auto exit, 15 stop	EACH	319,717.12		319,717.12
09.0210 110	Elevator, electric gear, 350'min, 3500 lb, 5'x8'cab, auto exit, additional stops	EACH	10,313.45		10,313.45
09.0210 115	Elevator, electric gear, 350'min, 4000 lb, 6'x8'cab, auto exit, 5 stop	EACH	344,535.52		344,535.52
09.0210 120	Elevator, electric gear, 200'min, 4000 lb, 6'x8'cab, auto exit, 3 stop	EACH	301,468.58		301,468.58
09.0210 125	Elevator, electric gear, 350'min, 4500 lb, 6'x9'cab, auto exit, 10 stop	EACH	307,074.82		307,074.82
09.0210 130	Elevator, electric gear, 350'min, 4500 lb, 6'x9'cab, auto exit, 15 stop	EACH	377,938.27		377,938.27
09.0210 135	Elevator, electric gear, 350'min, 4500 lb, 6'x9'cab, auto exit, additional stops	EACH	15,353.71		15,353.71
09.0210 140	Elevator, electric gear, 350'min, 6000 lb, 8'x10'cab, auto exit, 3 stop	EACH	323,002.06		323,002.06
09.0215	**ELEVATORS, ELECTRIC GEARLESS**				
09.0215 100	Elevator, electric gearless, 500'min, 3500 lb, 6'x9'cab, auto exit, 10 stop	EACH	377,111.48		377,111.48
09.0215 105	Elevator, electric gearless, 500'min, 3500 lb, 6'x9'cab, auto exit, 15 stop	EACH	450,100.85		450,100.85
09.0215 110	Elevator, electric gearless, 500'min, 3500 lb, 6'x9'cab, auto exit, additional stops	EACH	18,247.30		18,247.30
09.0215 115	Elevator, electric gearless, 700'min, 4500 lb, 6'x9'cab, auto exit, 10 stop	EACH	401,331.61		401,331.61
09.0215 120	Elevator, electric gearless, 700'min, 4500 lb, 6'x9'cab, auto exit, 15 stop	EACH	463,581.53		463,581.53
09.0215 125	Elevator, electric gearless, 700'min, 4500 lb, 6'x9'cab, auto exit, 20 stop	EACH	525,831.52		525,831.52
09.0215 130	Elevator, electric gearless, 700'min, 4500 lb, 6'x9'cab, auto exit, additional stops	EACH	12,747.41		12,747.41
09.0215 135	Elevator, electric gearless, 1200'min, 4500 lb, 6'x9'cab, auto exit, 20 stops	EACH	748,834.41		748,834.41
09.0215 140	Elevator, electric gearless, 1200'min, 4500 lb, 6'x9'cab, auto exit, additional stops	EACH	14,105.76		14,105.76
09.0215 145	Elevator, freight, electric gearless, 100'min, 4500 lb, 6'x9'cab, manual door, 2 stop	EACH	159,342.86		159,342.86
09.0215 150	Elevator, freight, electric gearless, 100'min, 4500 lb, manual door, additional stops	EACH	13,703.43		13,703.43
09.0220	**HOISTS AND CRANES**				
09.0220 100	Air hoist, 30'lift, rail, 2000 lb,	EACH	11,411.64	755.68	12,167.32
09.0220 105	Air hoist, 30'lift, rail, 500 lb,	EACH	12,425.93	755.68	13,181.61
09.0220 110	Electric hoist, 30'lift, rail, 1000 lb,	EACH	7,426.58	755.68	8,182.26
09.0220 115	Electric hoist, 30'lift, rail, 500 lb,	EACH	6,448.44	755.68	7,204.12

CONVEYING - 9.0

CSI#	Description	Unit	Material Cost	Install Cost	Total Cost
09.0220 120	Crane, hydraulic, portable, 2000 lb,	EACH	4,067.56		4,067.56
09.0220 125	Crane, hydraulic, gantry, 4000 lb,	EACH	4,227.59		4,227.59
09.0220 130	Crane, monorail, overhead, 200#/LF, manual	LF	30.08	29.69	59.77
09.0220 135	Crane, monorail, overhead, 100#/LF, manual	LF	12.36	24.68	37.04
09.0220 140	bridge Crane, 1 ton,	EACH	10,622.82		10,622.82
09.0220 150	Bridge crane, 2 ton	EACH	21,245.69		21,245.69
09.0220 155	Bridge crane, 3 ton	EACH	31,868.57		31,868.57
09.0225	**MANLIFT**				
09.0225 100	manlift, 2 stop	EACH	25,561.58		25,561.58
09.0225 105	manlift, 3 stop	EACH	28,060.95		28,060.95
09.0225 110	manlift, 4 stop	EACH	30,560.35		30,560.35
09.0300	**DUMB-WAITERS**				
09.0300 100	Dumbwaiter, manual, 200#, 2 stop	EACH	11,275.50		11,275.50
09.0300 105	Dumbwaiter, manual, 200#, additional stops	EACH	1,931.28		1,931.28
09.0300 110	Dumbwaiter, electric, 300#, 2 stops	EACH	28,046.78		28,046.78
09.0300 115	Dumbwaiter, electric, 300#, additional stops	EACH	3,890.94		3,890.94
09.0400	**ESCALATORS**				
09.0400 100	Escalator, 12' floor to floor, 24" wide	FLOOR	183,901.11		183,901.11
09.0400 105	Escalator, 12' floor to floor, 32" wide	FLOOR	187,327.87		187,327.87
09.0400 110	Escalator, 12' floor to floor, 36" wide	FLOOR	192,129.67		192,129.67
09.0400 115	Escalator, 12' floor to floor, 40" wide	FLOOR	192,711.89		192,711.89
09.0400 120	Escalator, 12' floor to floor, 44" wide	FLOOR	196,975.94		196,975.94
09.0400 125	Escalator, 12' floor to floor, 48" wide	FLOOR	205,326.51		205,326.51
09.0400 130	Escalator, 12' floor to floor, add for baked enamel sides	FLOOR	7,651.91		7,651.91
09.0400 135	Escalator, 12' floor to floor, add for glass sides	FLOOR	7,881.45		7,881.45
09.0400 140	Escalator, 12' floor to floor, add for stainless steel sides	FLOOR	10,508.64		10,508.64
09.0500	**PNEUMATIC SYSTEMS**				
09.0500 100	Pneumatic tube systems, Twin 3", 2 station	LUMP	28,344.97		28,344.97
09.0500 105	Pneumatic tube systems, Twin 3", add for additional station	STATION	11,815.11		11,815.11
09.0500 110	Pneumatic tube systems, Twin 4", 2 station	LUMP	29,253.82		29,253.82
09.0500 115	Pnuematic tube systems, Twin 4", add for additional station	STATION	12,269.55		12,269.55

10.1 PLUMBING & FIRE PROTECTION

CSI#	Description	Unit	Material Cost	Install Cost	Total Cost
	The equipment costs in this section include material, installation, and subcontractor overhead and profit. There are no allowances for general contractor markup and profit.				
	Costs represent standard grade materials and normal installation. Adjustments should be made for economy quality or custom and heavy duty materials and commensurate installation.				
10.1100	**EQUIPMENT**				
10.1100 100	Water heater, commercial, electric, 6 gallon, 17 GPH	EACH	719.42	242.92	962.34
10.1100 105	Water heater, commercial, electric, 50 gallon, 100 GPH	EACH	3,985.46	605.00	4,590.46
10.1100 110	Water heater, commercial, gas, 50 gallon, 100 GPH	EACH	2,041.28	651.55	2,692.83
10.1100 115	Water heater, commercial, gas, 85 gallon, 168 GPH	EACH	6,906.05	698.08	7,604.13
10.1100 120	Interceptor grease, cast iron, 10 gallons per minute, 20#	EACH	1,321.83	279.24	1,601.07
10.1100 125	Interceptor grease, cast iron, 50 gallons per minute, 100#	EACH	4,408.01	1,116.93	5,524.95
10.1100 130	Pump, circulating, in line, flanged, iron body, 3/4" to 1-1/2"	EACH	709.69	176.86	886.55
10.1100 135	Pumps, circulating, in line, flanged, iron body, 3/4" to 2-1/2"	EACH	2,261.08	260.62	2,521.70
10.1100 140	Pump, sewage ejector, w/tank & fittings, single, 1/2 horsepower, 3"	EACH	2,884.11	288.54	3,172.65
10.1100 145	Pump, sewage ejector, w/tank & fittings, single, 2 horsepower, 4"	EACH	2,884.11	288.54	3,172.65
10.1100 150	Pump, sewage ejector, w/tank & fittings, duplex, 2 horsepower	EACH	20,603.60	1,982.56	22,586.15
10.1100 155	Pump, sewage ejector, w/tank & fittings, duplex, 5 horsepower	EACH	24,713.43	2,820.24	27,533.67
10.1100 160	Pump, sump, electric, w/iron guard accessories, 1/3 horsepower, 3'deep, 1/2" outlet	EACH	2,177.01	232.69	2,409.71
10.1100 165	Pump, sump, electric, w/iron guard accessories, 1/2 horsepower, 6'deep, 2" outlet	EACH	2,743.69	325.78	3,069.47
10.1100 170	Pressure booster system, 2 pump, 100 GPM, 50 PSI	EACH	25,386.00	2,915.04	28,301.04
10.1100 175	Pressure booster system, 3 pump, 300 GPM, 100 PSI	EACH	39,436.78	3,886.72	43,323.50
10.1100 180	Septic tank, steel, 200 gallon, buried	EACH	1,160.21	328.92	1,489.13
10.1100 185	Septic tank, steel, 500 gallon, buried	EACH	1,654.02	548.19	2,202.20
10.1100 190	Septic tank, steel, 1000 gallon, buried	EACH	3,642.85	785.74	4,428.59
10.1100 195	Septic tank, steel, 10,000 gallon, buried	EACH	24,869.43	2,996.76	27,866.20
10.1100 200	Compressor air, simplex, 5 horsepower, reciprocating, tank mounted	EACH	9,538.62	1,943.36	11,481.98
10.1100 205	Compressor air, simplex, 10 horsepower, reciprocating, tank mounted	EACH	17,292.26	2,429.20	19,721.46
10.1100 210	Tank, water, glass lined, 200 gallon, ASME	EACH	3,247.72	833.05	4,080.77
10.1100 215	Tank, water, glass lined, 400 gallon, ASME	EACH	4,426.22	1,331.01	5,757.23
10.1100 220	Water softener, w/brine tank, 25 GPM, start-up	EACH	4,050.48	744.62	4,795.10
10.1100 225	Water softeners, w/brine tank, 50 GPM, start-up	EACH	7,382.36	1,349.63	8,731.99
10.1205	**FIXTURES, ECONOMY GRADE**				
10.1205 100	Fixture, economy grade, bath tub, steel, w/shower	EACH	464.81	325.78	790.59
10.1205 105	Fixture, economy grade, bath tub, steel, w/o shower	EACH	397.72	279.24	676.96
10.1205 110	Fixture, economy grade, tub, fiberglass, integral walls	EACH	911.55	325.78	1,237.33
10.1205 115	Fixture, economy grade, bidet, floor mounted	EACH	631.38	186.16	817.55
10.1205 120	Fixture, economy grade, lavatory, steel, wall hung	EACH	260.37	186.16	446.54
10.1205 125	Fixture, economy grade, lavatory, steel, vanity mounted	EACH	203.52	186.16	389.68
10.1205 130	Fixture, economy grade, service sink	EACH	481.39	232.69	714.09
10.1205 135	Fixture, economy grade, shower & drain receptor, 32" square	EACH	374.77	279.24	654.01
10.1205 140	Fixture, economy grade, shower cabinet, w/door, 32" square	EACH	797.05	418.85	1,215.91
10.1205 145	Fixture, economy grade, sink, porcelain on steel, counter, single	EACH	177.49	139.62	317.11
10.1205 150	Fixture, economy grade, sink, porcelain on cast iron, counter, single	EACH	252.46	139.62	392.07
10.1205 155	Fixture, economy grade, sink, stainless steel, counter, single	EACH	264.34	139.62	403.96
10.1205 160	Fixture, economy grade, sink, porcelain on steel, counter, double	EACH	195.97	139.62	335.59
10.1205 165	Fixture, economy grade, sink, porcelain on cast iron, counter, double	EACH	268.26	139.62	407.88
10.1205 170	Fixture, economy grade, sink, stainless steel, counter double	EACH	302.97	139.62	442.59
10.1205 175	Fixture, economy grade, sink, bar	EACH	256.87	139.62	396.49
10.1205 180	Fixture, economy grade, sink, floor mounted	EACH	189.31	93.09	282.40
10.1205 185	Fixture, economy grade, urinal, floor, w/flush valve	EACH	571.78	186.16	757.94
10.1205 190	Fixture, economy grade, urinal, wall, w/flush valve, carrier	EACH	603.98	186.16	790.14
10.1205 195	Fixture, economy grade, water closet, floor, w/tank	EACH	587.88	186.16	774.04
10.1205 200	Fixture, economy grade, water closet, wall, w/flush valve, carrier	EACH	623.41	186.16	809.57

PLUMBING & FIRE PROTECTION - 10.1

CSI#	Description	Unit	Material Cost	Install Cost	Total Cost
10.1210	**FIXTURES, STANDARD GRADE**				
10.1210 100	Fixture, standard grade, bathtub, cast iron, enamel, with shower	EACH	1,549.87	325.78	1,875.65
10.1210 105	Fixture, standard grade, lavatory, wall hung	EACH	737.57	186.16	923.74
10.1210 110	Fixture, standard grade, lavatory, vanity mounted	EACH	525.73	186.16	711.90
10.1210 115	Fixture, standard grade, service sink	EACH	1,297.16	232.69	1,529.85
10.1210 120	Fixture, standard grade, shower and drain receptor, 32" square	EACH	953.16	279.24	1,232.40
10.1210 125	Fixture, standard grade, sink, stainless steel, counter top with trim, single	EACH	996.01	186.16	1,182.17
10.1210 130	Fixture, standard grade, sink, stainless steel, counter top with trim, double	EACH	1,121.25	186.16	1,307.41
10.1210 135	Fixture, standard grade, bar sink	EACH	652.07	186.16	838.23
10.1210 140	Fixture, standard grade, urinal, wall mounted	EACH	1,372.26	232.69	1,604.95
10.1210 145	Fixture, standard grade, water closet with tank and accessories	EACH	978.89	232.69	1,211.58
10.1215	**FIXTURES, INSTITUTIONAL GRADE**				
10.1215 100	Fixture, institutional, hospital, lavatory, wall hung	EACH	964.67	232.69	1,197.36
10.1215 105	Fixture, institutional, hospital, lavatory, counter top with trim	EACH	761.08	232.69	993.77
10.1215 110	Fixture, institutional, hospital, service sink	EACH	1,537.25	232.69	1,769.94
10.1215 115	Fixture, institutional, hospital, sink, clinic	EACH	1,613.23	232.69	1,845.92
10.1215 120	Fixture, institutional, hospital, sink, stainless steel, single	EACH	1,024.31	186.16	1,210.47
10.1215 125	Fixture, institutional, hospital, sink, stainless steel, double	EACH	1,403.41	186.16	1,589.57
10.1215 130	Fixture, institutional, hospital, sitz bath	EACH	2,102.48	325.78	2,428.26
10.1215 135	Fixture, institutional, hospital, urinal, wall hung	EACH	1,372.26	232.69	1,604.95
10.1215 140	Fixture, institutional, hospital, water closet, floor mounted	EACH	1,001.37	232.69	1,234.07
10.1215 145	Fixture, institutional, hospital, water closet, wall mounted	EACH	978.89	232.69	1,211.58
10.1215 150	Fixture, institutional, hospital, water closet, floor with bed pan	EACH	1,161.20	279.24	1,440.43
10.1215 155	Fixture, institutional, hospital, water closet, wall with bed pan	EACH	2,104.80	279.24	2,384.04
10.1215 160	Fixture, institutional, hospital, shower cabinet, corner with door	EACH	5,468.48	465.39	5,933.86
10.1215 165	Fixture, institutional, hospital, shower cabinet, 36" square, with door	EACH	1,784.95	465.39	2,250.34
10.1215 170	Fixture, institutional, hospital, shower and drain receptor, 36" square	EACH	1,077.89	325.78	1,403.67
10.1215 175	Fixture, institutional, jail, water closet, in floor, stainless steel	EACH	3,287.44	279.24	3,566.68
10.1215 180	Fixture, institutional, jail, water closet/lavatory, in floor, stainless steel	EACH	4,546.42	325.78	4,872.20
10.1215 185	Fixture, institutional, jail, shower, wall unit, stainless steel, floor drain	EACH	1,248.26	465.39	1,713.65
10.1305	**ROUGH-INS**				
10.1305 100	Rough-in for fixtures, piping and valves, fixture to 5' beyond building perimeter, tract housing	FIX	386.97	209.80	596.77
10.1305 105	Rough-in for fixtures, piping and valves, fixture to 5' beyond building perimeter, custom housing	FIX	740.82	389.60	1,130.41
10.1305 110	Rough-in for fixtures, piping and valves, fixture to 5' beyond building perimeter, apartment building	FIX	536.27	291.14	827.40
10.1305 115	Rough-in for fixtures, piping and valves, fixture to 5' beyond building perimeter, industrial building	FIX	988.13	500.94	1,489.06
10.1305 120	Rough-in for fixtures, piping and valves, fixture to 5' beyond building perimeter, commercial building	FIX	1,211.37	543.73	1,755.10
10.1305 125	Rough-in for fixtures, piping and valves, fixture to 5' beyond building perimeter, institutional structures	FIX	2,706.69	1,862.41	4,569.09
10.1305 130	Rough-in for fixtures, piping and valves, fixture to 5' beyond building perimeter, schools	FIX	1,956.03	1,344.37	3,300.39
10.1305 135	Rough-in for fixtures, piping and valves, fixture to 5' beyond building perimeter, 1 or 2 story hospital	FIX	3,290.05	2,256.28	5,546.33
10.1305 140	Rough-in for fixtures, piping and valves, fixture to 5' beyond building perimeter, high rise hospital	FIX	3,809.04	2,611.66	6,420.70
10.1305 145	Rough-in for fixtures, piping and valves, fixture to 5' beyond building perimeter, high rise office building	FIX	2,609.15	1,806.74	4,415.89
10.1305 150	Rough-in at fixtures, bath tub, fittings and valving, fixture to waste line, vent riser and water runs	EA	381.11	325.78	706.89
10.1305 155	Rough-in at fixtures, fountains and coolers, fittings and valving, fixture to waste line, vent riser and water runs	EA	222.58	209.43	432.01
10.1305 160	Rough-in at fixtures, lavatory, fittings and valving, fixture to waste line, vent riser and water runs	EA	314.42	302.51	616.93
10.1305 165	Rough-in at fixtures, shower, fittings and valving, fixture to waste line, vent riser and water runs	EA	349.99	325.78	675.77

10.1 PLUMBING & FIRE PROTECTION

CSI#	Description	Unit	Material Cost	Install Cost	Total Cost
10.1305 170	Rough-in at fixtures, sinks, fittings and valving, fixture to waste line, vent riser and water runs	EA	315.58	302.51	618.09
10.1305 175	Rough-in at fixtures, urinal, fittings and valving, fixture to waste line, vent riser and water runs	EA	246.91	302.51	549.42
10.1305 180	Rough-in at fixtures, washing machine, fittings and valving, fixture to waste line, vent riser and water runs	EA	329.34	279.24	608.58
10.1305 185	Rough-in at fixtures, water closet, fittings and valving, fixture to waste line, vent riser and water runs	EA	397.04	390.93	787.97
10.1305 190	Rough-in at fixtures, wash fountain, fittings and valving, fixture to waste line, vent riser and water runs	EA	718.31	465.39	1,183.70
10.1310	**PIPE, CAST IRON**				
10.1310 100	Cast iron pipe, soil, service weight, single hub, 2"	LF	11.33	22.34	33.67
10.1310 105	Cast iron pipe, soil, service weight, single hub, 3"	LF	13.90	23.73	37.63
10.1310 110	Cast iron pipe, soil, service weight, single hub, 4"	LF	18.49	25.60	44.09
10.1310 115	Cast iron pipe, soil, service weight, single hub, 5"	LF	26.19	28.86	55.05
10.1310 120	Cast iron pipe, soil, service weight, single hub, 6"	LF	32.12	33.05	65.17
10.1310 125	Cast iron pipe, soil, service weight, single hub, 8"	LF	61.95	36.30	98.26
10.1310 130	Cast iron pipe, soil, service weight, single hub, 10"	LF	81.33	42.03	123.36
10.1310 135	Cast iron pipe, soil, service weight, hubless, 1-1/2"	LF	11.60	20.65	32.25
10.1310 140	Cast iron pipe, soil, service weight, hubless, 2"	LF	11.96	20.65	32.61
10.1310 145	Cast iron pipe, soil, service weight, hubless, 3"	LF	13.51	22.59	36.10
10.1310 150	Cast iron pipe, soil, service weight, hubless, 4"	LF	18.06	23.77	41.83
10.1310 155	Cast iron pipe, soil, service weight, hubless, 5"	LF	28.12	27.27	55.39
10.1310 160	Cast iron pipe, soil, service weight, hubless, 6"	LF	33.14	31.31	64.46
10.1310 165	Cast iron pipe, soil, service weight, hubless, 8"	LF	59.50	35.37	94.87
10.1310 170	Cast iron pipe, soil, service weight, hubless, 10"	LF	85.65	40.12	125.77
10.1310 175	Cast iron pipe, soil, extra heavy, single hub, 2"	LF	13.10	22.34	35.44
10.1310 180	Cast iron pipe, soil, extra heavy, single hub, 3"	LF	16.17	23.73	39.90
10.1310 185	Cast iron pipe, soil, extra heavy, single hub, 4"	LF	21.38	25.60	46.99
10.1310 190	Cast iron pipe, soil, extra heavy, single hub, 5"	LF	28.28	28.86	57.14
10.1310 195	Cast iron pipe, soil, extra heavy, single hub, 6"	LF	37.13	33.05	70.18
10.1310 200	Cast iron pipe, soil, extra heavy, single hub, 8"	LF	71.55	36.30	107.85
10.1310 205	Cast iron pipe, soil, extra heavy, single hub, 10"	LF	93.92	42.03	135.94
10.1310 210	Cast iron pipe, soil, extra heavy, single hub, 12"	LF	135.26	47.47	182.73
10.1310 215	Cast pipe, soil, dur iron, 2"	LF	60.72	21.41	82.13
10.1310 220	Cast iron pipe, soil, dur iron, 3"	LF	76.07	23.73	99.81
10.1310 225	Cast iron pipe, soil, dur iron, 4"	LF	106.14	25.23	131.37
10.1310 230	Cast iron pipe, soil, dur iron, 6"	LF	168.49	32.59	201.07
10.1310 235	Cast iron pipe, soil, dur iron, 8"	LF	317.76	35.25	353.00
10.1315	**PIPE, COPPER**				
10.1315 100	Copper pipe, underground, "K", soft, w/trenching, 1/2" coils	LF	6.65	7.68	14.32
10.1315 105	Copper pipe, underground, "K", soft, w/trenching, 3/4" coils	LF	11.81	8.84	20.65
10.1315 110	Copper pipe, underground, "K", soft, w/trenching, 1" coils	LF	15.87	8.84	24.72
10.1315 115	Copper pipe, underground, "K", soft, w/trenching, 1-1/4" coils	LF	20.80	15.81	36.62
10.1315 120	Copper pipe, underground, "K", hard, straight, 1-1/2"	LF	26.20	16.30	42.50
10.1315 125	Copper pipe, underground, "K", hard, straight, 2"	LF	40.48	19.54	60.02
10.1315 130	Copper pipe, underground, "K", hard, straight, 2-1/2"	LF	57.71	19.54	77.25
10.1315 135	Copper pipe, underground, "K", hard, straight, 3"	LF	80.38	20.02	100.39
10.1315 140	Copper pipe, underground, "K", hard, straight, 4"	LF	132.92	22.29	155.20
10.1315 145	Copper pipe, underground, "K", hard, straight, 5"	LF	292.54	29.32	321.86
10.1315 155	Copper pipe, underground, "K", hard, straight, 6"	LF	387.26	32.11	419.37
10.1315 160	Copper pipe, in building, "L", w/fittings & supports, 1/2"	LF	5.31	9.61	14.91
10.1315 165	Copper pipe, in building, "L", w/fittings & supports, 3/4"	LF	8.25	11.68	19.94
10.1315 170	Copper pipe, in building, "L", w/fittings & supports, 1"	LF	11.94	13.87	25.81
10.1315 175	Copper pipe, in building, "L", w/fittings & supports, 1-1/4"	LF	17.10	14.98	32.07
10.1315 180	Copper pipe, in building, "L", w/fittings & supports, 1-1/2"	LF	21.32	15.92	37.24
10.1315 185	Copper pipe, in building, "L", w/fittings & supports, 2"	LF	26.44	19.20	45.64
10.1315 190	Copper pipe, in building, "L", w/fittings & supports, 2-1/2"	LF	46.61	21.28	67.89
10.1315 195	Copper pipe, in building, "L", w/fittings & supports, 3"	LF	64.12	24.44	88.56

PLUMBING & FIRE PROTECTION - 10.1

CSI#	Description	Unit	Material Cost	Install Cost	Total Cost
10.1315 200	Copper pipe, in building, "L", w/fittings & supports, 4"	LF	107.68	29.78	137.46
10.1315 205	Copper pipe, in building, "L", w/fittings & supports, 5"	LF	256.74	37.32	294.07
10.1315 210	Copper pipe, in building, "L", w/fittings & supports, 6"	LF	302.66	45.69	348.36
10.1315 215	Copper pipe, in building, "M", w/fittings & supports, 1/2"	LF	4.37	9.61	13.98
10.1315 220	Copper pipe, in building, "M", w/fittings & supports, 3/4"	LF	6.85	11.68	18.54
10.1315 225	Copper pipe, in building, "M", w/fittings & supports, 1"	LF	10.23	13.87	24.10
10.1315 230	Copper pipe, in building, "M", w/fittings & supports, 1-1/4"	LF	15.37	14.98	30.35
10.1315 235	Copper pipe, in building, "M", w/fittings & supports, 1-1/2"	LF	20.45	15.92	36.37
10.1315 240	Copper pipe, in building, "M", w/fittings & supports, 2"	LF	32.21	19.20	51.41
10.1315 245	Copper pipe, in building, "M", w/fittings & supports, 3"	LF	57.64	24.44	82.08
10.1315 250	Copper pipe, in building, "M", w/fittings & supports, 4"	LF	103.56	29.78	133.34
10.1315 255	Copper pipe, in building, "M", w/fittings & supports, 5"	LF	245.47	38.16	283.63
10.1315 260	Copper pipe, in building, "M", w/fittings & supports, 6"	LF	315.30	37.06	352.36
10.1315 265	Copper pipe, "DWV", drainage tube, 2"	LF	41.11	19.20	60.31
10.1315 270	Copper pipe, "DWV", drainage tube, 3"	LF	55.08	24.44	79.52
10.1320	**PIPE, PVC**				
10.1320 100	PVC pipe, schedule 40, in building, w/fittings & supports, 1/2"	LF	1.63	6.75	8.38
10.1320 105	PVC pipe, schedule 40, in building, w/fittings & supports, 3/4"	LF	1.90	7.54	9.45
10.1320 110	PVC pipe, schedule 40, in building, w/fittings & supports, 1"	LF	2.51	7.54	10.05
10.1325	**PIPE, POLYPROPYLENE**				
10.1325 100	Pipe, polypropylene, acid waste, 2"	LF	25.21	6.13	31.35
10.1325 105	Pipe, polypropylene, acid waste, 3"	LF	34.78	9.21	43.99
10.1325 115	Pipe, polypropylene, acid waste, 4"	LF	46.75	11.08	57.82
10.1330	**PIPE, PLASTIC**				
10.1330 100	Plastic, "DWV", ABS, pipe, 1-1/2"	LF	9.92	14.89	24.82
10.1330 105	Plastic, "DWV", ABS, pipe, 2"	LF	12.84	19.36	32.20
10.1330 110	Plastic, "DWV", ABS, pipe, 3"	LF	19.21	23.37	42.58
10.1330 115	Plastic, "DWV", ABS, pipe, 4"	LF	32.66	25.87	58.53
10.1330 120	Plastic, "DWV", ABS, pipe, 6"	LF	123.46	26.53	149.99
10.1335	**PIPE, PYREX UNIT COST**				
10.1335 100	Pyrex, glass, pipe, 1"	LF	28.18	25.36	53.54
10.1335 105	Pyrex, glass, pipe, 1-1/2"	LF	39.84	27.91	67.75
10.1335 110	Pyrex, glass, pipe, 2"	LF	48.85	30.46	79.32
10.1335 115	Pyrex, glass, pipe, 3"	LF	53.58	39.24	92.82
10.1335 120	Pyrex, glass, pipe, 4"	LF	93.38	53.14	146.52
10.1340	**PIPE, STEEL UNIT COST**				
10.1340 100	Steel pipe, black, weld, schedule 40, A-120, screwed, 3/4"	LF	2.75	11.31	14.06
10.1340 105	Steel pipe, black, weld, schedule 40, A-120, screwed, 1"	LF	3.56	13.63	17.19
10.1340 110	Steel pipe, black, weld, schedule 40, A-120, screwed, 2"	LF	7.41	23.17	30.59
10.1340 115	Steel pipe, black, weld, schedule 40, A-120, screwed, 3"	LF	15.37	32.48	47.85
10.1340 120	Steel pipe, galvanized, weld, schedule 40, A-120, screwed, 3/4"	LF	3.47	11.31	14.78
10.1340 125	Steel pipe, galvanized, weld, schedule 40, A-120, screwed, 1-1/2"	LF	6.44	17.03	23.47
10.1400	**MISCELLANEOUS PLUMBING SPECIALTIES**				
10.1400 100	Access door, painted steel, 8"x8"	EACH	85.29	37.24	122.53
10.1400 105	Access door, painted steel, 12"x12"	EACH	95.38	37.24	132.62
10.1400 110	Access door, painted steel, 18"x18"	EACH	136.30	46.54	182.84
10.1400 115	Access door, painted steel, 24"x24"	EACH	194.26	46.54	240.80
10.1400 120	Access door, painted steel, 36"x36"	EACH	378.23	65.16	443.39
10.1400 125	Cleanout, cast iron, floor & wall, 2" & 3"	EACH	178.15	190.83	368.97
10.1400 130	Cleanout, cast iron, floor & wall, 4"	EACH	238.75	197.32	436.07
10.1400 135	Cleanout, cast iron, to grade, 4"	EACH	291.22	102.39	393.62
10.1400 140	Cleanout, cast iron, to grade, 6"	EACH	476.62	116.35	592.96
10.1400 145	Cleanout, cast iron, wall, 2"	EACH	40.40	93.09	133.48
10.1400 150	Drains, cast iron, area, w/5"x5" strainer	EACH	120.17	139.62	259.79
10.1400 155	Drains, cast iron, floor, 2"-4", with trap	EACH	142.66	139.62	282.27
10.1400 160	Drains, cast iron, roof, 2"-4", aluminum dome	EACH	191.37	176.86	368.23

10.1 PLUMBING & FIRE PROTECTION

CSI#	Description	Unit	Material Cost	Install Cost	Total Cost
10.1500	**MEDICAL GASES, ACCESSORIES**				
10.1500 100	Manifold, 4 cylinder	EACH	10,631.83	595.70	11,227.53
10.1500 105	Manifold, 6 cylinder	EACH	11,179.82	837.70	12,017.51
10.1500 110	Zone valve, w/box 3 @ 1/2"	EACH	1,037.24	183.54	1,220.78
10.1500 115	Zone valve, w/box 5 @ 1/2"	EACH	1,899.69	255.07	2,154.75
10.1500 120	Shut-off valve, w/o box 1"	EACH	197.30	41.89	239.19
10.1500 122	Shut-off valve, w/o box 1-1/4"	EACH	288.41	55.85	344.25
10.1500 125	Gas outlet, wall	EACH	136.95	57.69	194.64
10.1500 130	Alarm, line press, local, 1 gas	EACH	3,739.74	71.52	3,811.25
10.1500 135	Alarm, line press, local, 2 gas	EACH	4,634.42	93.09	4,727.51
10.1500 140	Alarm, line press, local, 3 gas & liquid	EACH	5,806.68	139.62	5,946.29
10.1500 145	Alarm, line press, local, 5 gas	EACH	6,850.06	236.34	7,086.40
10.1500 150	Alarm, master 15 signal	EACH	5,044.73	304.86	5,349.59
10.1500 155	Vacuum pump, duplex, w/all related accessories, 30 CFM, 5 HP	EACH	35,790.61	2,792.32	38,582.92
10.1500 160	Vacuum pump, duplex, w/all related accessories, 210 CFM, 15 HP	EACH	45,979.44	3,723.09	49,702.53
10.1500 165	Air compressor, duplex, medical, 2 @ 3 HP	EACH	19,517.32	1,489.25	21,006.57
10.1500 170	Air compressor, duplex, medical, 2 @ 10 HP	EACH	78,988.73	2,792.32	81,781.05
10.1515	**MEDICAL GAS PIPING**				
10.1515 100	Gas pipe, medical, copper, 'L', 1/2"	LF	13.18	19.21	32.39
10.1515 105	Gas pipe, medical, copper, 'L', 3/4"	LF	17.90	23.34	41.25
10.1515 110	Gas pipe, medical, copper, 'L', 1"	LF	23.41	27.72	51.13
10.1515 115	Gas pipe, medical, copper, 'L', 1-1/4"	LF	31.09	29.93	61.02
10.1515 120	Stainless steel pipe, non-spooled,'304', schedule 10, seamless, 1/2"	LF	11.38	10.09	21.47
10.1515 125	Stainless steel pipe, non-spooled,'304', schedule 10, seamless, 3/4"	LF	12.96	12.49	25.45
10.1515 130	Stainless steel pipe, non-spooled,'304', schedule 10, seamless, 1"	LF	18.08	15.04	33.12
10.1515 140	Stainless steel pipe, non-spooled,'304', schedule 10, seamless, 1-1/4"	LF	22.23	16.24	38.48
10.1515 145	Stainless steel pipe, '304', schedule 80, seamless, 1/2"	LF	17.82	12.10	29.92
10.1515 150	Stainless steel pipe, '304', schedule 80, seamless, 3/4"	LF	22.47	14.98	37.45
10.1515 155	Stainless steel pipe, '304', schedule 80, seamless, 1"	LF	29.52	18.02	47.55
10.1515 160	Stainless steel pipe, '304', schedule 80, seamless, 1-1/4"	LF	37.47	19.48	56.96
10.1515 165	Stainless steel pipe, '316', extra low carbon, schedule 10, seamless, 1/2"	LF	13.01	10.97	23.98
10.1515 170	Stainless steel pipe, '316', extra low carbon, schedule 10, seamless, 3/4"	LF	15.04	13.58	28.61
10.1515 175	Stainless steel pipe, '316', extra low carbon, schedule 10, seamless, 1"	LF	21.45	16.35	37.80
10.1515 180	Stainless steel pipe, '316', extra low carbon, schedule 10, seamless, 1-1/4"	LF	25.66	17.66	43.32
10.1515 190	Stainless steel pipe, '316', extra low carbon, schedule 40, seamless, 1/2"	LF	18.42	11.50	29.92
10.1515 195	Stainless steel pipe, '316', extra low carbon, schedule 40, seamless, 3/4"	LF	21.19	14.25	35.44
10.1515 200	Stainless steel pipe, '316', extra low carbon, schedule 40, seamless, 1"	LF	28.60	17.17	45.78
10.1515 205	Stainless steel pipe, '316', extra low carbon, schedule 40, seamless, 1-1/4"	LF	34.35	18.55	52.89
10.1515 210	Stainless steel pipe, '316', extra low carbon, schedule 80, seamless, 1/2"	LF	23.41	12.10	35.51
10.1515 215	Stainless steel pipe, '316', extra low carbon, schedule 80, seamless, 3/4"	LF	27.29	14.98	42.26
10.1515 220	Stainless steel pipe, '316', extra low carbon, schedule 80, seamless, 1"	LF	35.85	18.02	53.87
10.1515 225	Stainless steel pipe, '316', extra low carbon, schedule 80, seamless, 1-1/4"	LF	48.11	19.48	67.59
10.1600	**FEES, PERMITS, STERILIZATION**				
10.1600 105	Water meter fee, 3/4" connection	EACH	1,506.27		1,506.27
10.1600 110	Water meter fee, 1" connection	EACH	2,317.87		2,317.87
10.1600 115	Water meter fee, 1-1/2" connection	EACH	4,464.41		4,464.41
10.1600 120	Water meter fee, 2" connection	EACH	6,916.43		6,916.43
10.1600 125	Water meter fee, 3" connection	EACH	12,725.06		12,725.06
10.1600 130	Water meter fee, 4" connection	EACH	20,190.10		20,190.10
10.1600 135	Water meter fee, 6" connection	EACH	40,247.38		40,247.38
10.1600 140	Sewer fee, average, connection	FIX	12,623.93		12,623.93
10.1600 145	Sewer fee, average, connection, no plant or line charge	FIX	346.67		346.67
10.1600 150	Sterilization, testing & cleaning, per fixture	FIX	94.55		94.55
10.1700	**FIRE PROTECTION SYSTEMS**				
10.1700 100	Fire protection, concealed system, wet, normal hazard, 1 to 5,000 SF	SF	3.82	2.96	6.77
10.1700 105	Fire protection, concealed system, wet, normal hazard, 6,000 to 15,000 SF	SF	3.35	2.19	5.54
10.1700 110	Fire protection, concealed system, wet, normal hazard, over 15,000 SF	SF	3.15	2.06	5.21
10.1700 115	Fire protection, concealed system, wet, high hazard, 1 to 5,000 SF	SF	4.96	3.40	8.36

PLUMBING & FIRE PROTECTION - 10.1

CSI#	Description	Unit	Material Cost	Install Cost	Total Cost
10.1700 120	Fire protection, concealed system, wet, high hazard, 6,000 to 15,000 SF	SF	4.35	2.52	6.87
10.1700 125	Fire protection, concealed system, wet, high hazard, over 15,000 SF	SF	4.10	2.37	6.46
10.1700 130	Fire protection, concealed system, wet, light hazard, 1 to 5,000 SF	SF	3.24	2.96	6.20
10.1700 135	Fire protection, concealed system, wet, light hazard, 6,000 to 15,000 SF	SF	2.84	2.19	5.04
10.1700 140	Fire protection, concealed system, wet, light hazard, over 15,000 SF	SF	2.68	2.06	4.74
10.1700 145	Fire protection, exposed system, wet, normal hazard, 1 to 5,000 SF	SF	2.38	2.73	5.11
10.1700 150	Fire protection, exposed system, wet, normal hazard, 6,000 to 15,000 SF	SF	1.98	1.98	3.97
10.1700 155	Fire protection, exposed system, wet, normal hazard, over 15,000 SF	SF	1.90	1.88	3.78
10.1700 160	Fire protection, exposed system, wet, high hazard, 1 to 5,000 SF	SF	3.10	3.14	6.23
10.1700 165	Fire protection, exposed system, wet, high hazard, 6,000 to 15,000 SF	SF	2.58	2.28	4.86
10.1700 170	Fire protection, exposed system, wet, high hazard, over 15,000 SF	SF	2.47	2.16	4.64
10.1700 175	Fire protection, exposed system, wet, light hazard, 1 to 5,000 SF	SF	2.03	2.73	4.75
10.1700 180	Fire protection, exposed system, wet, light hazard, 6,000 to 15,000 SF	SF	1.69	1.98	3.67
10.1700 185	Fire protection, exposed system, wet, light hazard, over 15,000 SF	SF	1.62	1.88	3.50
10.1700 190	Fire protection, halon system, 2000 SF, 8' ceiling & 1" raised floor	SF	26.55		26.55
10.1700 195	Fire protection, halon system, 4000 SF, 8' ceiling & 1" raised floor	SF	19.67		19.67
10.1700 200	Fire protection, halon system, 10,000 SF, 8' ceiling & 1" raised floor	SF	16.19		16.19
10.1700 205	Standpipe, dry, 6" diameter, hook up, connection, pumper 6"	EACH	1,934.64	554.75	2,489.39
10.1700 210	Standpipe, dry, 6" diameter, hook up, connection, pumper 4"	EACH	1,842.82	540.21	2,383.04
10.1700 215	Fire pump, electric, 2500 GPM, @ 40 psi	EACH	20,621.28	4,700.70	25,321.98
10.1700 220	Fire pump, electric, 750 GPM, @ 100 psi	EACH	30,036.19	5,640.84	35,677.03
10.1700 225	Fire pump, diesel, 500 GPM, @ 100 psi	EACH	57,499.16	5,223.00	62,722.16
10.1700 230	Fire pump, diesel, 1000 GPM, @ 150 psi	EACH	93,727.77	6,894.36	100,622.13

10.2 HEAT, VENT & AIR CONDITIONING

CSI#	Description	Unit	Material Cost	Install Cost	Total Cost
	The costs in this section include material, installation, and subcontractor overhead and profit. There are no allowances for general contractor markup and profit.				
	Costs represent standard grade materials and normal installation. Adjustments should be made for economy quality or custom and heavy duty materials and commensurate installation.				
10.2105	**FURNACES**				
10.2105 100	Furnace, up flow, 80 MBTU, gas fired	EACH	816.94	446.88	1,263.81
10.2105 105	Furnace, up flow, 120 MBTU, gas fired	EACH	962.81	604.58	1,567.39
10.2105 110	Furnace, horizontal flow, 80 MBTU, gas fired	EACH	946.74	552.05	1,498.79
10.2105 115	Furnace, horizontal flow, 100 MBTU, gas fired	EACH	1,039.95	624.30	1,664.25
10.2105 120	Unit heater, suspended, 75 MBTU, gas fired	EACH	1,188.22	498.42	1,686.64
10.2105 125	Unit heater, suspended, 175 MBTU, gas fired	EACH	1,900.17	799.53	2,699.70
10.2105 130	Duct heater, indoor, 100 MBTU, gas fired	EACH	1,617.61	481.95	2,099.56
10.2105 135	Duct heater, indoor, 175 MBTU, gas fired	EACH	2,039.03	576.69	2,615.72
10.2105 140	Duct heater, roof, 125 MBTU, gas fired	EACH	2,662.35	539.67	3,202.02
10.2105 145	Duct heater, roof, 225 MBTU, gas fired	EACH	3,351.31	862.40	4,213.71
10.2105 150	Electric furnace, unit heater, 10 MBH, 3 KW, 240 V	EACH	546.52	230.03	776.56
10.2105 155	Electric furnace, unit heater, 34 MBH, 10 KW, 240 V	EACH	806.97	344.98	1,151.95
10.2105 160	Electric furnace, duct heater, 3.4 MBH, 1 KW, no enclosure	EACH	216.22	94.79	311.01
10.2105 165	Electric furnace, duct heater, 18 MBH, 5 KW, no enclosure	EACH	323.34	131.69	455.02
10.2105 170	Electric furnace, baseboard heater, hot water, 1030 BTU	LF	37.93	14.05	51.99
10.2105 175	Electric furnace, baseboard heater, hot water, 2 row	LF	54.26	18.06	72.32
10.2105 180	Electric furnace, baseboard heater, 240/160 V, aluminum finish	LF	55.12	12.62	67.74
10.2105 185	Electric furnace, baseboard heater, 240/140 V	LF	34.41	12.62	47.03
10.2110	**BOILERS**				
10.2110 100	Boiler, steel tube, gas fired, 670 MBH, 150# steam, pump 200 GPM at 95 TDH, expansion tank, air separator, pipe, valves	EACH	27,124.59	1,442.70	28,567.29
10.2110 105	Boiler, steel tube, gas fired, 2700 MBH, 150# steam, pump 108 GPM at 60 TDH, expansion tank, air separator, pipe, valves	EACH	55,378.04	2,047.71	57,425.75
10.2110 110	Boiler, steel tube, gas fired, 6700 MBH, 150# steam, pump 250 GPM at 90 TDH, expansion tank, air separator, pipe, valves	EACH	99,101.44	4,653.86	103,755.30
10.2110 115	Boiler, steel tube, gas fired, 3350 MBH, 15# steam, 30" water, pump 108 GPM at 60 TDH, expansion tank, air separator, pipe, valves	EACH	77,117.07	4,095.40	81,212.47
10.2110 120	Boiler, steel tube, gas fired, 670 MBH, 15# steam, 30" water, pump 200 GPM at 95 TDH, expansion tank, air separator, pipe, valves	EACH	102,253.63	4,653.86	106,907.49
10.2115	**CHILLERS**				
10.2115 100	Chiller, reciprocating, air cool, 20 TON	EACH	25,176.97	4,691.10	29,868.07
10.2115 105	Chiller, reciprocating, air cool, 50 TON	EACH	52,982.63	6,701.56	59,684.19
10.2115 110	Chiller, reciprocating, air cool, 150 TON	EACH	116,987.01	13,403.12	130,390.14
10.2115 115	Chiller, centrifugal, water cool, 60 TON	EACH	61,930.68	5,584.63	67,515.31
10.2115 120	Chiller, centrifugal, water cool, 150 TON	EACH	116,007.00	12,844.66	128,851.66
10.2115 125	Chiller, centrifugal, water cool, 400 TON	EACH	235,987.08	27,225.09	263,212.17
10.2115 130	Chiller, absorption, hot water, 100 TON	EACH	160,561.82	9,493.87	170,055.69
10.2115 135	Chiller, absorption, hot water, 200 TON	EACH	294,364.40	16,753.90	311,118.30
10.2115 140	Chiller, absorption, hot water 650 TON	EACH	835,100.24	29,598.56	864,698.80
10.2120	**COOLING TOWERS**				
10.2120 100	Cooling tower, 20 TON, compressor chiller	EACH	9,367.16	1,116.92	10,484.08
10.2120 105	Cooling tower, 60 TON, compressor chiller	EACH	28,099.28	2,233.86	30,333.13
10.2120 110	Cooling tower, 150 TON, compressor chiller	EACH	60,209.87	4,132.64	64,342.51
10.2120 115	Cooling tower, 400 TON, compressor chiller	EACH	134,014.44	8,376.95	142,391.39
10.2120 120	Cooling tower, 200 TON, absorption chiller	EACH	41,491.91	5,249.56	46,741.47
10.2120 125	Cooling tower, 350 TON, absorption chiller	EACH	51,345.88	8,153.57	59,499.45
10.2120 130	Cooling tower, 1000 TON, absorption chiller	EACH	138,792.18	17,200.68	155,992.86
10.2125	**AIR CONDITIONERS**				
10.2125 100	Air conditioner, 8 TON, roof mounted, D-X	EACH	14,702.51	1,116.93	15,819.44
10.2125 105	Air conditioner, 15 TON, roof mounted, D-X	EACH	23,324.16	1,489.25	24,813.41
10.2125 110	Air conditioner, 2 TON, thru-wall, D-X	EACH	2,428.07	558.47	2,986.54
10.2125 115	Air conditioner, 4 TON, thru-wall, D-X	EACH	4,826.92	744.62	5,571.54

HEAT, VENT & AIR CONDITIONING - 10.2

CSI#	Description	Unit	Material Cost	Install Cost	Total Cost
10.2130	**SPLIT SYSTEMS**				
10.2130 100	Split system, air conditioner, residential, 2 TON	EACH	3,368.53	418.85	3,787.39
10.2130 105	Split system, air conditioner, residential, 3 TON	EACH	4,267.12	442.13	4,709.25
10.2130 110	Split system, air conditioner, residential, 4 TON	EACH	5,647.27	651.55	6,298.81
10.2130 115	Split system, air conditioner, residential, 5 TON	EACH	6,570.12	791.17	7,361.28
10.2130 120	Split system, air conditioner, commercial, 10 TON	EACH	11,644.03	1,559.55	13,203.58
10.2130 125	Split system, air conditioner, commercial, 20 TON	EACH	19,629.14	1,535.78	21,164.92
10.2130 130	Split system, air conditioner, commercial, 30 TON	EACH	29,488.10	2,978.48	32,466.57
10.2130 135	Split system, air conditioner, commercial, 50 TON	EACH	42,504.87	3,630.01	46,134.88
10.2130 140	Split system, air conditioner, commercial, 70 TON	EACH	52,893.31	4,653.86	57,547.18
10.2135	**COMPUTER ROOM AIR CONDITIONING**				
10.2135 100	Computer room, air conditioning unit, 5 TON	EACH	19,029.74	2,978.48	22,008.22
10.2135 105	Computer room, air conditioning unit, 10 TON	EACH	34,812.29	5,956.94	40,769.23
10.2135 110	Computer room, air conditioning unit, 20 TON	EACH	56,743.51	8,935.42	65,678.93
10.2140	**ROOF MOUNTED UNITS**				
10.2140 100	Roof mounted, air-conditioner, gas heat, DX cooling, 3 TON cooling, 100 MBH heating	EACH	7,656.50	1,303.08	8,959.59
10.2140 105	Roof mounted, air-conditioner, gas heat, DX cooling, 4 TON cooling, 140 MBH heating	EACH	10,258.78	1,396.16	11,654.93
10.2140 110	Roof mounted, air-conditioner, gas heat, DX cooling, 5 TON cooling, 140 MBH heating	EACH	10,932.48	1,582.32	12,514.80
10.2140 115	Roof mounted, air-conditioner, gas heat, DX cooling, 8 TON cooling, 200 MBH heating	EACH	17,516.48	1,861.54	19,378.02
10.2140 120	Roof mounted, air-conditioner, gas heat, DX cooling, 10 TON cooling, 360 MBH heating	EACH	21,130.08	2,420.02	23,550.10
10.2140 125	Roof mounted, air-conditioner, gas heat, DX cooling, 20 TON cooling, 400 MBH heating	EACH	48,999.10	3,723.09	52,722.19
10.2140 130	Roof mounted, air-conditioner, gas heat, DX cooling, 40 TON cooling, 760 MBH heating	EACH	88,807.75	7,911.56	96,719.31
10.2140 135	Roof mounted, air-conditioner, gas heat, DX cooling, 60 TON cooling, 1000 MBH heating	EACH	125,555.43	10,238.49	135,793.93
10.2145	**HEAT PUMPS**				
10.2145 100	Heat pump, 2 TON, 26000 BTU, thru-wall	EACH	2,376.57	744.62	3,121.19
10.2145 105	Heat pump, 4 TON, 52000 BTU, thru-wall	EACH	4,840.25	1,116.93	5,957.19
10.2145 110	Heat pump, 10 TON, 120000 BTU, roof, duct	EACH	17,763.32	2,420.02	20,183.33
10.2145 115	Heat pump, 20 TON, 240000 BTU, roof, duct	EACH	33,685.59	3,723.09	37,408.67
10.2145 120	Heat pump, 2 TON, 26000 BTU, plenum	EACH	3,981.02	651.55	4,632.57
10.2145 125	Heat pump, 4 TON, 52000 BTU, plenum	EACH	6,737.40	744.62	7,482.02
10.2150	**HYDRONIC SYSTEMS**				
10.2150 100	Hydronic unit, 50 TON, 300 MBH	EACH	40,074.02	1,943.36	42,017.38
10.2150 105	Hydronic unit, 100 TON, 600 MBH	EACH	71,516.74	2,915.04	74,431.78
10.2150 110	Hydronic unit, 150 TON, 900 MBH	EACH	91,247.25	3,643.80	94,891.05
10.2150 115	Hydronic unit, 200 TON, 1200 MBH	EACH	110,974.07	3,886.72	114,860.79
10.2155	**INFRA-RED AND RADIANT SYSTEMS**				
10.2155 100	Infra-red heater, 30,000 BTU, gas	EACH	1,212.40	272.58	1,484.98
10.2155 105	Infra-red heater, 60,000 BTU, gas	EACH	1,923.73	327.64	2,251.37
10.2155 110	Radiant heat panel, 24x24, 357 W	EACH	383.00	93.09	476.08
10.2155 115	Radiant heat panel, 24x48, 500 W	EACH	427.12	93.09	520.21
10.2160	**HUMIDIFIERS**				
10.2160 100	Humidifier, steam, 50#/hr	EACH	1,064.52	744.62	1,809.14
10.2160 105	Humidifier, steam, 300#/hr	EACH	2,913.88	1,745.20	4,659.07
10.2160 110	Humidifier, electric, 50#/hr	EACH	5,270.78	558.47	5,829.25
10.2160 115	Humidifier, electric, 150#/hr	EACH	9,154.52	1,116.93	10,271.46
10.2165	**PUMPS**				
10.2165 100	Pump, condensate, duplex, 40 GPM	EACH	7,670.08	1,861.54	9,531.63
10.2165 105	Pump, condensate, duplex, 75 GPM	EACH	8,994.61	2,792.32	11,786.93
10.2165 110	Pump, condensate, duplex, 120 GPM	EACH	11,745.24	3,443.86	15,189.10
10.2165 115	Pump, water service, 20 GPM, 50' head	EACH	1,416.98	418.85	1,835.83
10.2165 120	Pump, water service, 60 GPM, 50' head	EACH	2,016.24	930.77	2,947.01

10.2 HEAT, VENT & AIR CONDITIONING

CSI#	Description	Unit	Material Cost	Install Cost	Total Cost
10.2165 125	Pump, water service, 200 GPM, 50' head	EACH	3,409.55	1,582.32	4,991.87
10.2165 130	Pump, water service, 400 GPM, 50' head	EACH	4,995.25	2,326.93	7,322.18
10.2165 135	Pump, water service, 2000 GPM, 50' head	EACH	15,232.38	4,095.40	19,327.78
10.2165 140	Pump, water service, 4000 GPM, 50' head	EACH	20,375.81	5,119.25	25,495.06
10.2165 145	Pump, fuel oil, 1 GPM, 1/4 horse power	EACH	1,205.80	605.00	1,810.80
10.2165 150	Pump, fuel oil, 5 GPM, 1/4 HP	EACH	1,394.85	930.77	2,325.62
10.2165 155	Pump, fuel oil, 10 GPM, 1/4 HP	EACH	1,544.29	1,023.86	2,568.15
10.2170	**AIR HANDLERS**				
10.2170 100	Air handler, 2000 CFM, 5 TON	EACH	7,641.64	956.90	8,598.53
10.2170 105	Air handler, 5200 CFM, 12 TON	EACH	21,504.69	1,611.61	23,116.31
10.2170 110	Air handler, 15,000 CFM, 30 TON	EACH	66,117.23	3,021.75	69,138.98
10.2170 115	Air handler, 28,000 CFM, 75 TON	EACH	114,769.73	6,547.13	121,316.86
10.2170 120	Air handler, 2500 CFM, 8 TON	EACH	5,648.22	805.81	6,454.02
10.2170 125	Air handler, 4000 CFM, 10 TON	EACH	9,658.51	906.53	10,565.03
10.2170 130	Air handler, 6250 CFM, 15 TON	EACH	13,218.97	1,208.71	14,427.68
10.2170 140	Air handler, 15,000 CFM, 30 TON	EACH	28,470.58	2,215.96	30,686.54
10.2175	**COILS**				
10.2175 100	Coil, chill water, 22,000 CFM, 44 SF	EACH	15,434.49	1,638.17	17,072.66
10.2175 105	Coil, chill water, 28,000 CFM, 56 SF	EACH	19,908.86	2,084.93	21,993.79
10.2175 110	Coil, chill water, 32,000 CFM, 64 SF	EACH	22,114.10	2,233.86	24,347.95
10.2175 115	Coil, hot water, 16,000 CFM, 16 SF	EACH	2,548.60	595.70	3,144.30
10.2175 120	Coil, hot water, 22,000 CFM, 22 SF	EACH	3,029.69	819.09	3,848.78
10.2175 125	Coil, hot water, 28,000 CFM, 28 SF	EACH	4,401.22	1,042.47	5,443.68
10.2180	**FANS**				
10.2180 100	Fan, supply, low pressure, 10,000 CFM, 1-1/2" utility set	EACH	5,204.93	1,393.92	6,598.85
10.2180 105	Fan, supply, low pressure, 20,000 CFM, 1-1/2" utility set	EACH	8,880.88	2,130.09	11,010.97
10.2180 110	Fan, supply, low pressure, 40,000 CFM, 1-1/2" utility set	EACH	15,158.17	3,516.07	18,674.24
10.2180 115	Fan, supply, high pressure, 10,000 CFM, 3-1/2" utility set	EACH	7,467.01	1,472.25	8,939.25
10.2180 120	Fan, supply, high pressure, 30,000 CFM, 3-1/2" utility set	EACH	13,177.89	2,764.31	15,942.20
10.2180 125	Fan, supply, high pressure, 40,000 CFM, 3-1/2" utility set	EACH	18,935.79	3,680.62	22,616.41
10.2180 130	Fan, exhaust, wall, 100 CFM	EACH	113.52	84.87	198.39
10.2180 135	Fan, exhaust, wall, 300 CFM	EACH	193.22	234.88	428.10
10.2180 140	Fan, exhaust, roof, 1/2"SP, 1600 CFM	EACH	1,525.76	509.04	2,034.80
10.2180 145	Fan, exhaust, roof, 1/2"SP, 6000 CFM	EACH	5,847.49	1,912.12	7,759.61
10.2180 150	Fan, exhaust, vibration mount, 3/4"SP, 2000 CFM	EACH	2,231.59	580.81	2,812.39
10.2180 155	Fan, exhaust, vibration mount, 3/4"SP, 10,000 CFM	EACH	4,134.71	1,089.83	5,224.55
10.2180 160	Fan, return, vibration mount, 5 HP, 17,500 CFM	EACH	9,806.14	1,158.34	10,964.48
10.2180 165	Fan, return, vibration mount, 10 HP, 31,000 CFM	EACH	19,999.31	2,014.50	22,013.82
10.2180 170	Fan, centrifugal in-line, 4,000 CFM	EACH	4,093.43	595.15	4,688.59
10.2180 175	Fan, centrifugal in-line, 10,000 CFM	EACH	7,392.47	1,007.02	8,399.50
10.2180 180	Fan, centrifugal in-line, 20,000 CFM	EACH	12,010.22	1,228.09	13,238.31
10.2180 185	Fan coil unit, duct mount, 2 pipe, 1 coil, 600 CFM	EACH	1,567.18	248.93	1,816.12
10.2180 190	Fan coil unit, duct mount, 2 pipe, 1 coil, 1000 CFM	EACH	2,284.54	373.33	2,657.87
10.2180 195	Fan coil unit, w/cabinets, 2 pipe, 1 coil, 600 CFM	EACH	1,652.17	308.16	1,960.33
10.2180 200	Fan coil unit, w/cabinets, 2 pipe, 1 coil, 1500 CFM	EACH	3,403.52	497.84	3,901.36
10.2185	**TANKS**				
10.2185 100	Expansion tank, chilled water, 44 GAL	EACH	1,446.98	190.60	1,637.57
10.2185 105	Expansion tank, ASME code, 88 GAL	EACH	4,378.04	833.22	5,211.26
10.2185 110	Expansion tank, ASME code, 250 GAL	EACH	8,436.56	1,647.00	10,083.56
10.2200	**CONTROLS**				
10.2200 100	Controls, pneumatic, air conditioning unit, to 10 Tons	LUMP	1,280.84	171.55	1,452.39
10.2200 105	Controls, pneumatic, air conditioning unit, to 20 Tons	LUMP	2,159.21	168.94	2,328.14
10.2200 110	Controls, pneumatic, air conditioning unit, to 30 Tons	LUMP	3,243.69	327.63	3,571.32
10.2200 115	Controls, pneumatic, boiler, to 1,000 MBH	LUMP	11,677.25	469.11	12,146.36
10.2200 120	Controls, pneumatic, boiler, to 4,000 MBH	LUMP	22,962.25	1,146.71	24,108.96
10.2200 125	Controls, pneumatic, boiler, to 10,000 MBH	LUMP	36,988.24	1,863.41	38,851.65
10.2200 130	Controls, pneumatic, chiller, to 50 Ton	LUMP	17,340.59	1,563.70	18,904.29

HEAT, VENT & AIR CONDITIONING - 10.2

CSI#	Description	Unit	Material Cost	Install Cost	Total Cost
10.2200 135	Controls, pneumatic, chiller, to 300 Ton	LUMP	48,008.36	5,808.02	53,816.38
10.2200 140	Controls, pneumatic, chiller, to 600 Ton	LUMP	77,397.65	7,260.02	84,657.67
10.2200 145	Controls, pneumatic, cooling tower, to 100 Ton	LUMP	11,239.16	938.22	12,177.37
10.2200 150	Controls, pneumatic, cooling tower, to 300 Ton	LUMP	28,323.31	1,876.44	30,199.75
10.2200 155	Controls, pneumatic, cooling tower, to 600 Ton	LUMP	48,554.78	2,680.62	51,235.41
10.2200 160	Controls, pneumatic, exhaust fan	LUMP	806.36		806.36
10.2200 165	Controls, pneumatic, variable air volume box	LUMP	580.63		580.63
10.2200 170	Controls, pneumatic, variable air volume box with reheat coil	LUMP	645.17		645.17
10.2200 175	Controls, pneumatic, air compressor, 1/2 HP	LUMP	28,816.20	2,792.32	31,608.52
10.2200 180	Controls, pneumatic, air compressor, 1-1/2 HP	LUMP	19,982.61		19,982.61
10.2200 185	Controls, pneumatic, air compressor, 5 HP	LUMP	38,566.98		38,566.98
10.2300	**DUCTWORK**				
10.2300 100	Duct rectangular, galvanized iron, with supports, to 10000 LBS	POUND	7.24	6.23	13.47
10.2300 105	Duct rectangular, galvanized iron, with supports, to 20000 LBS	POUND	6.73	5.24	11.97
10.2300 110	Duct rectangular, galvanized iron, with supports, to 40000 LBS	POUND	5.98	4.60	10.58
10.2300 115	Duct flexible, insulated, with clamps, 4"	LF	4.39	6.72	11.11
10.2300 120	Duct flexible, insulated, with clamps, 6"	LF	5.92	8.31	14.23
10.2300 125	Duct flexible, insulated, with clamps, 8"	LF	8.43	12.00	20.44
10.2300 130	Duct flexible, insulated, with clamps, 10"	LF	8.80	15.79	24.59
10.2300 135	Duct flexible, insulated, with clamps, 12"	LF	10.80	22.19	32.98
10.2300 140	Duct flexible, insulated, with clamps, 14"	LF	11.92	27.23	39.15
10.2300 145	Duct flexible, insulated, with clamps, 16"	LF	15.18	29.15	44.34
10.2300 150	Duct flexible, insulated, with clamps, 18"	LF	18.50	35.33	53.83
10.2300 155	Duct flexible, insulated, with clamps, 20"	LF	19.80	41.65	61.45
10.2300 160	Fire dampers, insulated, UL rated, fusible link to 1 SF	EACH	219.47	110.95	330.42
10.2300 165	Fire dampers, insulated, UL rated, fusible link to 2 SF	EACH	289.75	137.04	426.79
10.2300 170	Fire dampers, insulated, UL rated, fusible link to 4 SF	EACH	370.69	202.31	573.00
10.2300 175	Fire dampers, insulated, UL rated, fusible link to 6 SF	EACH	704.39	234.88	939.27
10.2300 180	Fire dampers, insulated, UL rated, fusible link to 10 SF	EACH	760.87	339.34	1,100.21
10.2300 185	Fire dampers, insulated, UL rated, fusible link to 16 SF	EACH	939.70	411.09	1,350.79
10.2300 190	Fire dampers, insulated, UL rated, fusible link to 25 SF	EACH	1,144.34	576.69	1,721.03
10.2300 195	Air outlets, supply, to 12" louver face	EACH	85.44	84.87	170.31
10.2300 200	Air outlets, supply, to 18" louver face	EACH	111.46	97.86	209.32
10.2300 205	Air outlets, supply, to 24" louver face	EACH	140.31	110.95	251.25
10.2300 210	Air outlets, exhaust, to 12" louver face	EACH	64.83	65.27	130.10
10.2300 215	Air outlets, exhaust, to 18" louver face	EACH	90.87	84.87	175.74
10.2300 220	Air outlets, exhaust, to 24" louver face	EACH	125.77	104.44	230.22
10.2300 225	Duct insulation, wrap with vapor barrier, 1" thick	SFCA	1.87	2.01	3.88
10.2300 230	Duct insulation, lining with vapor barrier, 1" thick	SFCA	1.87	2.01	3.88
10.2300 235	Duct insulation, lining with vapor barrier, 2" thick	SFCA	1.87	2.01	3.88
10.2300 240	Plenum insulation, rigid, 2" thick	SFCA	3.09	4.63	7.73
10.2400	**PIPING**				
10.2400 100	Steel pipe, seamless, A-53, schedule 40, welded, 2"	LF	19.40	23.37	42.77
10.2400 105	Steel pipe, seamless, A-53, schedule 40, welded, 3"	LF	26.63	33.83	60.46
10.2400 110	Steel pipe, seamless, A-53, schedule 40, welded, 4"	LF	37.84	43.52	81.36
10.2400 115	Steel pipe, seamless, A-53, schedule 40, welded, 6"	LF	47.96	53.42	101.38
10.2400 120	Valve, gate, globe & check, iron body, flange, 125#, bolt & gasket, 2"	EACH	442.36	171.45	613.81
10.2400 125	Valve, gate, globe & check, iron body, flange, 125#, bolt & gasket, 2-1/2"	EACH	483.55	204.78	688.33
10.2400 130	Valve, gate, globe & check, iron body, flange, 125#, bolt & gasket, 3"	EACH	551.77	238.11	789.88
10.2400 135	Valve, gate, globe & check, iron body, flange, 125#, bolt & gasket, 5"-6"	EACH	1,403.97	461.55	1,865.52
10.2400 140	Valve, gate, globe & check, iron body, flange, 125#, bolt & gasket, 8"	EACH	2,772.98	680.18	3,453.16
10.2400 145	Steam traps, cast iron, screwed, 3/4", bucket	EACH	425.73	55.01	480.74
10.2400 150	Steam traps, cast iron, screwed, 1", bucket	EACH	846.56	69.62	916.18
10.2400 155	Butterfly valve, iron body, nylon, 2", wafer body	EACH	167.80	128.59	296.39
10.2400 160	Butterfly valve, iron body, nylon, 3", wafer body	EACH	177.01	153.59	330.60
10.2400 165	Butterfly valve, iron body, nylon, 6", wafer body	EACH	418.33	346.16	764.49
10.2400 170	Butterfly valve, iron body, nylon, 8", wafer, chain	EACH	625.28	510.13	1,135.41
10.2400 175	Butterfly valve, iron body, nylon, 2", lug type	EACH	199.96	128.59	328.55

10.2 HEAT, VENT & AIR CONDITIONING

CSI#	Description	Unit	Material Cost	Install Cost	Total Cost
10.2400 180	Butterfly valve, iron body, nylon, 3", lug type	EACH	227.58	178.58	406.16
10.2400 185	Butterfly valve, iron body, nylon, 6", lug type	EACH	485.06	346.16	831.22
10.2400 190	Butterfly valve, iron body, nylon, 8", lug, chain	EACH	673.50	510.13	1,183.63
10.2400 195	Valve, balancing, circuit setter, 1/2"	EACH	89.93	31.09	121.02
10.2400 200	Valve, balancing, circuit setter, 3/4"	EACH	109.72	37.24	146.96
10.2400 205	Valve, balancing, circuit setter, 1 1/2"	EACH	168.62	81.73	250.35
10.2400 210	Valve, balancing, circuit setter, 2"	EACH	277.30	138.46	415.77

ELECTRICAL - 11.0

CSI#	Description	Unit	Material Cost	Install Cost	Total Cost
	The costs in this section include material, installation, and subcontractor overhead and profit. There are no allowances for general contractor markup and profit.				
	Costs represent standard grade materials and normal installation. Adjustments should be made for economy quality or custom and heavy duty materials and commensurate installation.				
11.0100	**SWITCHGEAR**				
11.0100 100	Substation, 150 KVA, 1200 AMP, with transformer, breakers, pad, grounding	EACH	58,343.80	7,573.53	65,917.34
11.0100 105	Substation, 1000 KVA, 3000 AMP, with transformer, breakers, pad, grounding	EACH	127,395.79	20,007.12	147,402.92
11.0100 110	Service switchboard, 600 A, 3PH, 277/480V MCMB, breakers, bus, meter, enclosure	EACH	11,086.85	2,506.39	13,593.24
11.0100 115	Service switchboard, 1200 A, 3PH, 277/480V MCMB, breakers, bus, meter, enclosure	EACH	17,788.10	3,211.61	20,999.71
11.0100 120	Service switchboard, 2000 A, 3PH, 277/480V MCMB, breakers, bus, meter, enclosure	EACH	26,745.42	5,444.18	32,189.60
11.0100 125	Distribution switchboard, 600 A, 3PH, 120/208 V, enclosure, breakers	EACH	3,145.31	1,039.39	4,184.70
11.0100 130	Distribution switchboard, 800 A, 3PH, 120/208 V, enclosure, breakers	EACH	3,759.09	1,095.96	4,855.05
11.0100 135	Distribution switchboard, 1600 A, 3PH, 120/208 V, enclosure, breakers	EACH	5,953.58	2,283.89	8,237.47
11.0100 140	Distribution switchboard, 400 A, 3PH, 277/480 V, enclosure, breakers	EACH	2,731.23	1,039.39	3,770.63
11.0100 145	Distribution switchboard, 800 A, 3PH, 277/480 V, enclosure, breakers	EACH	3,575.76	1,114.82	4,690.58
11.0100 150	Distribution switchboard, 1200 A, 3PH, 277/480 V, enclosure, breakers	EACH	3,366.53	1,303.37	4,669.90
11.0100 151	Transformer, 480/120 V, 3 phase, 500 KVA	EACH	30,774.19	4,468.40	35,242.59
11.0100 152	Transformer, 480/120 V, 3 phase, 750 KVA	EACH	50,008.00	5,362.08	55,370.08
11.0100 153	Transformer, 480/120 V, 3 phase, 1,000 KVA	EACH	60,327.12	5,808.92	66,136.04
11.0100 155	Panelboards, bolt-on breakers, 277/480V, 3PH, 4W, 100 A, 24 circuits	EACH	3,802.33	518.95	4,321.28
11.0100 160	Panelboards, bolt-on breakers, 277/480V, 3PH, 4W, 225 A, 42 circuits	EACH	4,027.50	1,182.92	5,210.42
11.0100 165	Panelboards, bolt-on breakers, 277/480V, 3PH, 4W, 400 A, 42 circuits	EACH	5,234.46	1,683.38	6,917.84
11.0100 205	Panelboards, bolt-on breakers, 120/208 V, 3PH, 4W, 100 A, 30 circuits	EACH	1,234.16	499.25	1,733.42
11.0100 210	Panelboards, bolt-on breakers, 120/208 V, 3PH, 4W, 225 A, 42 circuits	EACH	2,804.75	1,272.29	4,077.04
11.0100 215	Panelboards, bolt-on breakers, 120/208 V, 3PH, 4W, 400 A, 42 circuits	EACH	4,485.06	1,951.48	6,436.54
11.0100 220	Transformer, 480/120 V, 3 phase, 45 KVA	EACH	4,335.21	1,787.36	6,122.57
11.0100 225	10.1100 185	EACH	6,533.36	2,010.78	8,544.14
11.0100 230	10.1100 190	EACH	8,693.96	2,234.20	10,928.16
11.0200	**EMERGENCY SYSTEMS**				
11.0200 100	Emergency generator to 30 KW	EACH	35,639.70	3,656.94	39,296.64
11.0200 105	Emergency generator to 100 KW	EACH	60,761.46	7,313.88	68,075.34
11.0200 110	Emergency generator to 400 KW	EACH	176,208.20	12,799.28	189,007.49
11.0200 115	Emergency generator to 750 KW	EACH	382,797.22	18,284.69	401,081.91
11.0200 120	Emergency generator to 1,000 KW	EACH	440,094.68	21,941.63	462,036.31
11.0200 125	Automatic transfer switch 30 A	EACH	8,374.68	365.69	8,740.37
11.0200 130	Automatic transfer switch 100 A	EACH	9,263.55	731.39	9,994.94
11.0200 135	Automatic transfer switch 400 A	EACH	17,990.43	1,462.78	19,453.21
11.0200 140	Automatic transfer switch 1,000 A	EACH	56,618.55	2,377.01	58,995.56
11.0200 145	Automatic transfer switch 2,000 A	EACH	73,211.33	3,290.98	76,502.31
11.0200 150	Exit sign, incandescent, battery, surface mounted	EACH	133.65	68.57	202.22
11.0200 155	Exit sign, incandescent, battery, recessed	EACH	122.13	114.28	236.41
11.0200 160	Exit sign, fluorescent, battery, surface mounted	EACH	133.65	68.57	202.22
11.0200 165	Exit sign, fluorescent, battery, recessed	EACH	134.05	114.28	248.33
11.0300	**FEEDER CONDUIT & WIRE**				
11.0300 100	Conduit and wire, in slab, 60 A, 3 CU wire 600 V THHN #6, 1 CU soft #10, in 1" PVC	LF	3.42	4.67	8.09
11.0300 105	Conduit and wire, in slab, 100 A, 4 CU wire 600 V THHN #2, 1 CU soft #8, in 1" PVC	LF	5.29	5.18	10.47
11.0300 110	Conduit and wire, in slab, 225 A, 4 CU wire 600 V THHN #4/0, 1 CU soft #1, in 2" PVC	LF	13.40	8.13	21.53
11.0300 115	Conduit and wire, in slab, 300 A, 4 CU wire 600 V THHN 350 MCM, 1 CU soft #4, in 3" PVC	LF	11.22	12.04	23.26
11.0300 120	Conduit and wire, in slab, 400 A, 4 CU wire 600 V THHN 500 MCM, 1 CU soft #2, in 3" PVC	LF	12.47	12.33	24.80
11.0300 125	Conduit and wire, in slab, 600 A, 8 CU wire 600 V THHN 350 MCM, 1 CU soft #1, in 2" PVC	LF	8.60	7.15	15.75
11.0300 130	Conduit and wire, in slab, 800 A, 6 CU wire 600 V THHN 500 MCM, 2 CU soft #1/0, in 3" PVC	LF	14.23	12.73	26.97

11.0 ELECTRICAL

CSI#	Description	Unit	Material Cost	Install Cost	Total Cost
11.0300 135	Conduit and wire, in slab, 1000 A, 12 CU wire 600 V THHN 400 MCM, 3 bare CU soft #2/0, in 3" PVC	LF	15.10	12.96	28.06
11.0300 140	Conduit and wire, in slab, 1200 A, 16 CU wire 600 V THHN 350 MCM, 4 bare CU soft #3/0, in 3" PVC	LF	18.15	13.49	31.64
11.0300 145	Conduit and wire, in slab, 1600 A, 20 CU wire 600 V THHN 400 MCM, 5 bare CU soft #4/0, in 3" PVC	LF	20.21	13.76	33.97
11.0300 150	Conduit and wire, in slab, 2000 A, 18 CU wire 600 V THHN 400 MCM, 6 bare CU soft 250 MCM, in 3" PVC	LF	26.83	14.66	41.48
11.0300 155	Conduit and wire, 60 A, 3 CU wire 600 V THHN #6, 1 bare CU soft #10, in 1" EMT	LF	3.25	6.43	9.68
11.0300 160	Conduit and wire, 100 A, 4 CU wire 600 V THHN #2, 1 bare CU soft #8, in 1" EMT	LF	5.20	6.87	12.07
11.0300 165	Conduit and wire, 225 A, 4 CU wire 600 V THHN #4/0, 1 bare CU soft #1, in 2" EMT	LF	8.77	10.93	19.69
11.0300 170	Conduit and wire, 300 A, 4 CU wire 600 V THHN 350 MCM, 1 bare CU soft #4, in 3" EMT	LF	15.40	14.40	29.80
11.0300 175	Conduit and wire, 400 A, 4 CU wire 600 V THHN 500 MCM, 1 bare CU soft #2, in 3" EMT	LF	16.53	14.69	31.22
11.0300 180	Conduit and wire, 600 A, 8 CU wire 600 V THHN 350 MCM, 1 bare CU soft #1, in 2" EMT	LF	8.77	10.93	19.69
11.0300 185	Conduit and wire, 800 A, 6 CU wire 600 V THHN 500 MCM, 2 bare CU soft #1/0, in 3" EMT	LF	18.46	15.01	33.48
11.0300 190	Conduit and wire, 1000 A, 12 CU wire 600 V THHN 400 MCM, 3 bare CU soft #2/0, in 3" EMT	LF	19.43	15.19	34.63
11.0300 195	Conduit and wire, 1200 A, 16 CU wire 600 V THHN 350 MCM, 4 bare CU soft #3/0, in 3" EMT	LF	20.91	15.40	36.32
11.0300 200	Conduit and wire, 1600 A, 20 CU wire 600 V THHN 400 MCM, 5 bare CU soft #4/0, in 3" EMT	LF	22.56	15.62	38.18
11.0300 205	Conduit and wire, 2000 A, 18 CU wire 600 V THHN 400 MCM, 6 bare CU soft 250 MCM, in 3" EMT	LF	24.85	15.83	40.68
11.0300 210	Conduit and wire, 60 A, 3 CU wire 600 V THHN #6, 1 bare CU soft #10, in 1" GRS	LF	13.81	8.01	21.82
11.0300 215	Conduit and wire, 100 A, 4 CU wire 600 V THHN #2, 1 bare CU soft #8, in 1" GRS	LF	5.56	6.08	11.63
11.0300 220	Conduit and wire, 225 A, 4 CU wire 600 V THHN #4/0 1 bare CU soft #1, in 2" GRS	LF	18.58	12.04	30.62
11.0300 225	Conduit and wire, 300 A, 4 CU wire 600 V THHN 350 MCM 1 bare CU soft #4, in 3" GRS	LF	21.18	14.40	35.58
11.0300 230	Conduit and wire, 400 A, 4 CU wire 600 V THHN 500 MCM 1 bare CU soft #2, in 3" GRS	LF	22.31	14.69	37.00
11.0300 235	Conduit and wire, 600 A, 8 CU wire 600 V THHN 350 MCM 2 bare CU soft #1, in 2" GRS	LF	13.56	11.26	24.82
11.0300 240	Conduit and wire, 800 A, 6 CU wire 600 V THHN 500 MCM, 2 bare CU soft #1/0, in 3" GRS	LF	24.24	15.01	39.25
11.0300 245	Conduit and wire, 1000 A, 12 CU wire 600 V THHN 400 MCM, 3 bare CU soft #2/0, in 3" GRS	LF	25.21	15.19	40.40
11.0300 250	Conduit and wire, 1200 A, 16 CU wire 600 V THHN 350 MCM, 4 bare CU soft #3/0, in 3" GRS	LF	26.69	15.40	42.09
11.0300 255	Conduit and wire, 1600 A, 20 CU wire 600 V THHN 400 MCM, 5 bare CU soft #4/0, in 3" GRS	LF	28.34	15.62	43.96
11.0300 260	2000 A, 18 CU wire 600 V THHN 400 MCM, 6 bare CU soft 250 MCM, in 3" GRS	LF	37.56	16.54	54.10
11.0300 265	Cable trays ladder type 12", with ells, tees, 4" drop cover, unistrut hangers	LF	19.42	12.51	31.93
11.0300 270	Cable trays ladder type 24", with ells, tees, 4" drop cover, unistrut hangers	LF	21.26	14.30	35.56
11.0400	**FIXTURES**				
11.0400 100	Fixture, incandescent, commercial grade, surface mounted, with junction box, wire, conduit, 100 watt	EACH	139.11	121.73	260.84
11.0400 105	Fixture, incandescent, commercial grade, recessed downlight, with junction box, wire, conduit, 100 watt	EACH	178.92	158.29	337.21
11.0400 110	Fixture, incandescent, commercial grade, surface mounted, with junction box, wire, conduit, 200 watt	EACH	165.66	144.59	310.25
11.0400 115	Fixture, incandescent, commercial grade, recessed downlight, with junction box, wire, conduit, 200 watt	EACH	218.82	181.15	399.97
11.0400 120	Fixture, fluorescent, commercial grade, recessed, with junction box, wire, conduit, 2' x 2'	EACH	180.02	122.21	302.24
11.0400 125	Fixture, fluorescent, commercial grade, recessed, with junction box, wire, conduit, 2' x 4'	EACH	225.64	127.80	353.43
11.0400 130	Fixture, fluorescent strip, surface mounted, with junction box, wire, conduit, 4', 2 lamp	EACH	94.70	133.38	228.08

ELECTRICAL - 11.0

CSI#	Description	Unit	Material Cost	Install Cost	Total Cost
11.0400 135	Fixture, fluorescent strip, surface mounted, with junction box, wire, conduit, 8', 4 lamp	EACH	265.44	211.58	477.02
11.0400 140	Fixture, fluorescent strip, surface mounted, with junction box, wire, conduit, 4', 4 lamp	EACH	164.07	144.55	308.62
11.0400 145	Fixture, mercury vapor, hi-bay, with junction box, wire, conduit, 250 watt	EACH	341.10	249.72	590.82
11.0400 150	Fixture, mercury vapor, hi-bay, with junction box, wire, conduit, 400 watt	EACH	458.66	272.57	731.23
11.0400 155	Fixture, mercury vapor, hi-bay, with junction box, wire, conduit, 1000 watt	EACH	458.66	272.57	731.23
11.0400 160	Fixture, metal halide, hi-bay, with junction box, wire, conduit, 250 watt	EACH	330.63	275.25	605.89
11.0400 165	Fixture, metal halide, hi-bay, with junction box, wire, conduit, 400 watt	EACH	701.69	310.82	1,012.51
11.0400 170	Fixture, high pressure sodium, hi-bay, with junction box, wire, conduit, 250 watt	EACH	580.46	249.72	830.18
11.0400 175	Fixture, high pressure sodium, hi-bay, with junction box, wire, conduit, 400 watt	EACH	625.79	318.28	944.07
11.0400 180	Fixture, high pressure sodium, hi-bay, with junction box, wire, conduit, 1000 watt	EACH	824.44	351.22	1,175.65
11.0400 185	Perimeter security lights, industrial, 400 watt HPS wallpack, 40' on center	LF	17.62	9.86	27.47
11.0400 190	Perimeter security lights, school, 250 watt HPS wallpack, 30' on center	LF	20.92	10.95	31.87
11.0500	**SIGNAL & COMMUNICATIONS**				
11.0500 100	Telephone system, warehouse, 10,000 SF	SYS	3,778.53	2,692.49	6,471.02
11.0500 105	Telephone system, hospital, 220,000 SF	SYS	27,055.24	55,507.20	82,562.44
11.0500 110	Telephone system, office, 250,000 SF	SYS	50,886.83	54,395.37	105,282.21
11.0500 115	Telephone system, school, 80,000 SF	SYS	7,198.14	9,376.62	16,574.77
11.0500 120	Telephone system, library, 10,000 SF	SYS	5,818.63	4,043.12	9,861.75
11.0500 125	Fire alarm system, warehouse, 10,000 SF	SYS	8,929.97	7,298.65	16,228.62
11.0500 130	Fire alarm system, hospital, 220,000 SF	SYS	128,366.50	145,239.09	273,605.58
11.0500 135	Fire alarm system, office, 250,000 SF	SYS	104,253.05	118,518.99	222,772.05
11.0500 140	Fire alarm system, school, 80,000 SF	SYS	46,163.67	47,918.27	94,081.95
11.0500 145	Fire alarm system, library, 10,000 SF	SYS	14,333.53	9,709.63	24,043.15
11.0500 150	Public address/intercom system, warehouse, 10,000 SF	SYS	5,950.74	2,685.92	8,636.66
11.0500 155	Public address/intercom system, hospital, 220,000 SF	SYS	45,663.83	35,326.65	80,990.48
11.0500 160	Public address/intercom system, office, 250,000 SF	SYS	50,422.22	39,371.96	89,794.19
11.0500 165	Public address/intercom system, school, 80,000 SF	SYS	17,661.76	13,532.57	31,194.33
11.0500 175	Public address/intercom system, library, 10,000 SF	SYS	9,190.90	4,106.65	13,297.55
11.0500 180	Security system, warehouse, 10,000 SF	SYS	4,302.59	2,574.19	6,876.78
11.0500 185	Security system, hospital, 220,000 SF	SYS	38,002.18	20,617.62	58,619.80
11.0500 190	Security system, office, 250,000 SF	SYS	62,519.68	30,372.16	92,891.84
11.0500 195	Security system, school, 80,000 SF	SYS	22,041.42	14,385.89	36,427.31
11.0500 200	Security system, library, 10,000 SF	SYS	3,962.39	2,490.94	6,453.33
11.0600	**DEVICES**				
11.0600 100	Receptacle, duplex, commercial, standard grade	EACH	21.99	84.90	106.89
11.0600 105	Receptacle, duplex, hospital grade	EACH	50.82	102.77	153.60
11.0600 110	Receptacle, locking type, tamperproof	EACH	33.00	84.90	117.90
11.0600 115	60 A, welding, heavy duty	EACH	32.43	113.50	145.93
11.0600 120	Receptacle, 100 A, welding, heavy duty	EACH	432.16	195.49	627.66
11.0600 125	Switch, commercial, single pole	EACH	106.80	45.71	152.51
11.0600 130	Switch, commercial, double pole	EACH	147.25	45.71	192.96
11.0600 135	Switch, commercial, three way toggle	EACH	172.14	100.57	272.72

Commercial Square Foot Building Costs

This Cost Manual is ideal for evaluating "Trade-offs" on materials and design. There are 65 square foot tables. Each has its own detailed description with the construction parameters clearly stated. The functional assemblies section can be used for unusual situations.

Table of Contents

1.0 Apartment, 2-3 story 317	34.0 Laundromat 323
2.0 Apartment, 4-7 story 317	35.0 Library 323
3.0 Apartment, 8-30 story 317	36.0 Manufacturing, heavy 324
4.0 Auditorium 317	37.0 Manufacturing, light 324
5.0 Auto Sales/Showroom 317	38.0 Medical office 324
6.0 Bank 318	39.0 Motel 324
7.0 Bowling alley 318	40.0 Multiple residence 324
8.0 Car wash 318	41.0 Multiple residence, elderly 325
9.0 Club, country 318	42.0 Office, 2-3 story 325
10.0 Club, social 318	43.0 Office, 4-7 story 325
11.0 Convenience market 319	44.0 Office, 8-30 story 325
12.0 Courthouse 319	45.0 Post office 325
13.0 Day care center 319	46.0 Prison 326
14.0 Dispensary 319	47.0 Restaurant 326
15.0 Dormitory 319	48.0 Restaurant, fast food 326
16.0 Fire station 320	49.0 Rink, hockey 326
17.0 Fraternal building 320	50.0 School, elementary 326
18.0 Garage, mini-lube 320	51.0 School, secondary 326
19.0 Garage, parking 320	52.0 School, vocational 327
20.0 Garage, repair 320	53.0 Shopping center, strip 327
21.0 Garage service station 321	54.0 Store, department 327
22.0 Garage underground parking 321	55.0 Store, discount 327
23.0 Government building 321	56.0 Store, retail 328
24.0 Gymnasium 321	57.0 Supermarket 328
25.0 Handball/Racquetball club 321	58.0 Surgical center 328
26.0 Hangar, aircraft 322	59.0 Swimming pool, enclosed 328
27.0 Healthclub 322	60.0 Terminal, airport 328
28.0 Hospital, convalescent 322	61.0 Terminal bus 329
29.0 Hospital, general 322	62.0 Theater, movie 329
30.0 Hotel 4-7 story 322	63.0 Veterinary hospital 329
31.0 Hotel 8-30 story 323	64.0 Warehouse 329
32.0 Indoor tennis club 323	65.0 Warehouse, self-storage 329
33.0 Jail 323	

Commercial Square Foot Building Costs

1.0 APARTMENT, 2-3 STORY

Building Parameters: 2 Story, 10 Ft Story Height, 15,000 Square Feet

Exterior	Zone 0, 1	Zone 2	Zone 3	Zone 4
Wood siding on stud frame	163.07	165.02	168.24	171.00
Brick veneer on stud frame	167.32	169.27	172.49	175.25
Stucco on stud frame	162.38	164.34	167.55	170.31
Brick, concrete block, back-up	172.49	174.44	177.66	180.41
Decorative concrete block	169.16	171.11	174.33	177.08

2.0 APARTMENT, 4-7 STORY

Building Parameters: 6 Story, 11 Ft Story Height, 65,000 Square Feet

Exterior	Zone 0, 1	Zone 2	Zone 3	Zone 4
Decorative concrete block, steel frame	191.21	195.00	201.31	206.71
Brick, concrete block back-up steel frame	194.08	197.87	204.19	209.58
Brick, concrete block back-up, reinforced concrete frame	170.65	174.44	180.76	186.16
Precast panels, steel frame	199.02	202.81	209.12	214.52
Precast panels, reinforced concrete frame	164.57	168.36	174.67	180.07

3.0 APARTMENT, 8-30 Story

Building Parameters: 15 Story, 11 Ft Story Height, 175,000 Square Feet

Exterior	Zone 0, 1	Zone 2	Zone 3	Zone 4
Decorative concrete block, steel frame	242.66	248.40	258.16	266.43
Decorative concrete block, reinforced concrete frame	199.02	204.76	214.52	222.79
Brick, concrete block back-up, steel frame	245.53	251.27	261.03	269.30
Precast panels, steel frame	239.44	245.18	254.94	263.21
Precast panels, reinforced concrete frame	182.71	188.45	198.21	206.48

4.0 AUDITORIUM

Building Parameters: 1 Story, 35 Ft Story Height, 25,000 Square Feet

Exterior	Zone 0, 1	Zone 2	Zone 3	Zone 4
Precast panels, steel frame	354.86	360.25	369.44	377.25
Brick, concrete block back-up, steel frame	345.21	350.61	371.28	367.60
Decorative concrete block, steel frame	339.24	344.63	353.82	361.63
Stone veneer, block back-up, steel frame	389.88	395.28	404.47	412.28
Tilt-up panels, steel frame	305.01	310.41	319.60	327.41

5.0 AUTO SALES/SHOWROOM

Building Parameters: 1 Story, 14 Ft Story Height, 25,000 Square Feet

Exterior	Zone 0, 1	Zone 2	Zone 3	Zone 4
Brick veneer on stud frame	168.01	169.85	172.72	175.25
Stucco on stud frame	164.80	166.63	169.50	172.03
Decorative concrete block, steel frame	206.14	207.98	210.85	213.37
Brick, concrete block back-up, steel frame	205.22	207.06	209.93	212.45
Precast panels, steel frame	207.75	209.58	212.45	214.98

Costs include General Contractor's Overhead and Profit.
See page VI for seismic zones. See page VII-X for Cities Index.

Commercial Square Foot Building Costs

6.0 BANK
Building Parameters: 1 Story, 14 Ft Story Height, 4,000 Square Feet

Exterior	Zone 0, 1	Zone 2	Zone 3	Zone 4
Brick veneer on stud frame	277.80	279.18	281.47	283.42
Stone veneer, block back-up, steel frame	356.12	357.50	359.79	361.75
Precast panels, steel frame	366.45	367.83	370.13	372.08
Tilt-up panels, steel frame	334.18	335.56	337.86	339.81
Wood siding on stud frame	274.93	276.30	278.60	280.55

7.0 BOWLING ALLEY
Building Parameters: 1 Story, 14 Ft Story Height, 20,000 Square Feet

Exterior	Zone 0, 1	Zone 2	Zone 3	Zone 4
Tilt-up panels, steel frame	205.45	208.66	214.18	218.77
Concrete block, steel roof frame	210.39	213.60	219.11	223.71
Decorative concrete block, steel frame	211.08	214.29	219.80	224.40
Stucco on stud frame	162.96	166.17	171.69	176.28
Wood siding on stud frame	163.53	166.75	172.26	176.85

8.0 CAR WASH
Building Parameters: 1 Story, 12 Ft Story Height, 2,500 Square Feet

Exterior	Zone 0, 1	Zone 2	Zone 3	Zone 4
Metal siding on steel frame	144.93	146.88	150.10	152.85
Insulated metal panel, steel frame	185.47	187.42	190.63	193.39
Brick, concrete block back-up	160.89	162.84	166.06	168.81
Concrete block, steel roof frame	154.57	156.53	159.74	162.50
Precast panels, steel frame	286.41	179.27	182.48	185.24

9.0 CLUB, COUNTRY
Building Parameters: 1 Story, 14 Ft Story Height, 40,000 Square Feet

Exterior	Zone 0, 1	Zone 2	Zone 3	Zone 4
Wood siding on stud frame	299.85	293.07	294.68	299.85
Stucco on stud frame	299.27	292.50	294.10	299.27
Brick, concrete block back-up	320.86	314.09	315.69	320.86
Decorative concrete block	317.88	311.10	312.71	317.88
Insulated metal panel, steel frame	286.41	279.64	281.24	286.41

10.0 CLUB, SOCIAL
Building Parameters: 1 Story, 12 Ft Story Height, 20,000 Square Feet

Exterior	Zone 0, 1	Zone 2	Zone 3	Zone 4
Decorative concrete block, steel frame	229.11	231.63	235.88	239.44
Stone veneer, block back-up, steel frame	245.76	248.28	252.53	256.09
Decorative concrete block	186.39	188.91	193.16	196.72
Wood siding on stud frame	186.73	189.26	193.51	197.07
Brick veneer on stud frame	189.26	191.78	196.03	199.59

Costs include General Contractor's Overhead and Profit.
See page VI for seismic zones. See page VII-X for Cities Index.

Commercial Square Foot Building Costs

11.0 CONVENIENCE MARKET

Building Parameters: 1 Story, 12 Ft Story Height, 5,000 Square Feet

Exterior	Zone 0, 1	Zone 2	Zone 3	Zone 4
Wood siding on stud frame	172.95	174.56	177.43	179.72
Decorative concrete block	181.91	183.51	186.39	188.68
Stucco on stud frame	172.03	173.64	176.51	178.81
Insulated metal panel, steel frame	180.18	181.79	184.66	186.96
Brick veneer on stud frame	179.15	180.76	183.63	185.93

12.0 COURTHOUSE

Building Parameters: 2 Story, 12 Ft Story Height, 40,000 Square Feet

Exterior	Zone 0, 1	Zone 2	Zone 3	Zone 4
Precast panels, steel frame	324.42	329.13	337.28	344.18
Brick, concrete block back-up	313.74	318.45	326.60	333.49
Stone veneer, block back-up, steel frame	329.71	334.41	342.57	349.46
Decorative concrete block, steel frame	311.68	316.38	324.54	331.43
Curtain wall, metal and glass	284.46	289.17	297.32	304.21

13.0 DAY CARE CENTER

Building Parameters: 1 Story, 10 Ft Story Height, 6,000 Square Feet

Exterior	Zone 0, 1	Zone 2	Zone 3	Zone 4
Decorative concrete block	196.03	197.64	200.28	202.46
Brick, concrete block back-up, steel frame	217.16	218.77	221.41	223.59
Wood siding on stud frame	190.86	192.47	195.11	197.29
Brick veneer on stud frame	194.42	196.03	198.67	200.85
Stucco on stud frame	190.29	191.90	194.54	196.72

14.0 DISPENSARY

Building Parameters: 2 Story, 10 Ft Story Height, 10,000 Square Feet

Exterior	Zone 0, 1	Zone 2	Zone 3	Zone 4
Decorative concrete block, steel frame	308.69	311.91	317.30	322.01
Brick, concrete block back-up, steel frame	312.48	315.69	321.09	325.80
Wood siding on stud frame	265.97	269.18	274.58	279.29
Brick veneer on stud frame	270.91	274.12	279.52	284.23
Stucco on stud frame	265.28	268.50	273.89	278.60

15.0 DORMITORY

Building Parameters: 3 Story, 10 Ft Story Height, 30,000 Square Feet

Exterior	Zone 0, 1	Zone 2	Zone 3	Zone 4
Brick, concrete block back-up	165.37	166.75	168.93	170.88
Decorative concrete block	162.73	164.11	166.29	168.24
Brick, concrete block back-up, steel frame	205.45	206.83	209.01	210.96
Decorative concrete block, steel frame	202.81	204.19	206.37	208.32
Precast panels, steel frame	209.81	211.19	213.37	215.32

Costs include General Contractor's Overhead and Profit.
See page VI for seismic zones. See page VII-X for Cities Index.

Commercial Square Foot Building Costs

16.0 FIRE STATION
Building Parameters: 2 Story, 14 Ft Story Height, 9,000 Square Feet

Exterior	Zone 0, 1	Zone 2	Zone 3	Zone 4
Tilt-up panels, steel frame	269.87	275.27	284.34	292.15
Insulated metal panel, steel frame	291.12	296.52	305.59	290.43
Brick, concrete block back-up	234.04	239.44	248.51	256.32
Decorative concrete block	229.56	234.96	244.03	251.84
Brick veneer on stud frame	226.92	232.32	241.39	249.20

17.0 FRATERNAL BUILDING
Building Parameters: 1 Story, 12 Ft Story Height, 20,000 Square Feet

Exterior	Zone 0, 1	Zone 2	Zone 3	Zone 4
Decorative concrete block, steel frame	226.12	230.02	236.57	242.20
Stone veneer, block back-up, steel frame	246.33	250.24	256.78	262.41
Brick, concrete block back-up	204.76	208.66	215.21	220.84
Decorative concrete block	178.81	182.71	189.26	194.88
Insulated metal panel, steel frame	171.23	175.13	181.68	187.30

18.0 GARAGE, MINI-LUBE
Building Parameters: 1 Story, 14 Ft Story Height, 1,500 Square Feet

Exterior	Zone 0, 1	Zone 2	Zone 3	Zone 4
Metal siding on steel frame	211.88	213.83	217.05	219.80
Insulated metal panel, steel frame	268.84	270.79	274.01	276.76
Concrete block, steel roof frame	248.63	250.58	253.80	256.55
Stucco on stud frame	227.27	229.22	232.44	235.19
Decorative concrete block	242.20	244.15	247.37	250.12

19.0 GARAGE, PARKING
Building Parameters 4 Story, 10 Ft Story Height, 185,000 Square Feet

Exterior	Zone 0, 1	Zone 2	Zone 3	Zone 4
Precast panels, steel frame	97.04	102.09	110.59	117.83
Precast panels, reinforced concrete frame	61.32	66.38	74.88	82.11
Brick, concrete block back-up, reinforced concrete frame	64.77	69.82	78.32	85.56
Reinforced concrete, cast in place	63.85	68.90	77.40	84.64
Decorative concrete block, reinforced concrete frame	57.31	62.36	70.86	78.09

20.0 GARAGE, REPAIR
Building Parameters: 1 Story, 14 Ft Story Height, 8,000 Square Feet

Exterior	Zone 0, 1	Zone 2	Zone 3	Zone 4
Wood siding on stud frame	116.22	117.25	118.97	120.35
Stucco on stud frame	115.53	116.56	118.29	119.66
Metal siding on steel frame	113.12	114.15	115.87	117.25
Reinforced concrete, cast in place	132.98	134.02	135.74	137.12
Concrete block, steel roof frame	142.06	143.09	144.81	146.19

Costs include General Contractor's Overhead and Profit.
See page VI for seismic zones. See page VII-X for Cities Index.

Commercial Square Foot Building Costs

21.0 GARAGE SERVICE STATION
Building Parameters: 1 Story, 10 Ft Story Height, 1,500 Square Feet

Exterior	Zone 0, 1	Zone 2	Zone 3	Zone 4
Metal siding on steel frame	157.45	159.40	162.61	165.37
Insulated metal panel, steel frame	198.10	200.05	203.27	206.02
Concrete block, steel roof frame	170.54	172.49	175.71	178.46
Wood siding on stud frame	152.62	154.57	157.79	160.55
Brick veneer on stud frame	159.40	161.35	164.57	167.32

22.0 GARAGE UNDERGROUND PARKING
Building Parameter: 2 Story, 10 Ft Story Height, 90,000 Square Feet

Exterior	Zone 0, 1	Zone 2	Zone 3	Zone 4
Reinforced concrete, cast in place (Two levels below grade)	90.84	93.59	98.30	102.32
Reinforced concrete, cast in place (Three levels below grade)	85.79	88.54	93.25	97.27
Reinforced concrete, cast in place (Four levels below grade)	67.41	70.17	74.88	78.89

23.0 GOVERNMENT BUILDING
Building Parameters: 2 Story, 12 Ft Story Height, 25,000 Square Feet

Exterior	Zone 0, 1	Zone 2	Zone 3	Zone 4
Brick veneer on stud frame	249.66	251.15	253.57	255.63
Brick, concrete block back-up	282.51	284.00	286.41	288.48
Stone veneer block back-up, steel frame	302.60	304.10	306.51	308.57
Decorative concrete block, steel frame	279.86	281.36	283.77	285.84
Decorative concrete block, reinforced concrete frame	255.52	257.01	259.42	261.49

24.0 GYMNASIUM
Building Parameters: 1 Story, 35 Ft Story Height, 30,000 Square Feet

Exterior	Zone 0, 1	Zone 2	Zone 3	Zone 4
Brick, concrete block back-up, steel frame	275.27	278.60	284.23	289.05
Decorative concrete block, steel frame	268.95	272.29	277.91	282.74
Insulated metal panel, steel frame	258.62	261.95	267.58	272.40
Reinforced concrete block, steel roof frame	267.69	271.02	276.65	281.47
Precast panels, steel frame	285.61	288.94	294.56	299.39

25.0 HANDBALL/RACQUETBALL CLUB
Building Parameters: 2 Story, 12 Ft Story Height, 30,000 Square Feet

Exterior	Zone 0, 1	Zone 2	Zone 3	Zone 4
Precast panels, steel frame	245.30	249.66	257.13	263.44
Brick, concrete block back-up	240.25	244.61	252.07	258.39
Tilt-up panels, steel frame	231.29	241.51	248.97	255.29
Decorative concrete block, steel frame	237.14	235.65	243.12	249.43
Insulated metal panel, steel frame	250.24	254.60	262.06	268.38

Costs include General Contractor's Overhead and Profit.
See page VI for seismic zones. See page VII-X for Cities Index.

Commercial Square Foot Building Costs

26.0 HANGAR, AIRCRAFT
Building Parameters: 1 Story, 30 Ft Story Height, 50,000 Square Feet

Exterior	Zone 0, 1	Zone 2	Zone 3	Zone 4
Metal siding on steel frame	152.28	154.92	159.51	163.30
Insulated metal panel, steel frame	173.29	175.93	180.53	184.32
Concrete block, steel roof frame	161.46	164.11	168.70	172.49
Tilt-up panels, steel frame	156.99	159.63	164.22	168.01
Precast panels, steel frame	169.04	171.69	176.28	180.07

27.0 HEALTH CLUB
Building Parameters: 2 Story, 12 Ft Story Height. 27,000 Square Feet

Exterior	Zone 0, 1	Zone 2	Zone 3	Zone 4
Brick, concrete block back-up	306.62	312.02	321.09	328.90
Precast panels, steel frame	311.91	317.30	326.37	334.18
Decorative concrete block, steel frame	303.29	308.69	317.76	325.57
Insulated metal panel, steel frame	317.07	322.47	331.54	339.35
Tilt-up panels, steel frame	297.09	302.49	311.56	319.37

28.0 HOSPITAL, CONVALESCENT
Building Parameters: 2 Story, 10 Ft Story Height, 28.000 Square Feet

Exterior	Zone 0, 1	Zone 2	Zone 3	Zone 4
Wood siding on stud frame	234.62	236.11	238.64	240.82
Brick veneer on stud frame	237.49	238.98	241.51	243.69
Stucco on stud frame	234.16	235.65	238.18	240.36
Brick, concrete block back-up	236.11	237.60	240.13	242.31
Decorative concrete block, steel frame	264.94	266.43	268.95	271.14

29.0 HOSPITAL GENERAL
Building Parameters: 4 Story, 15 Ft Story Height, 140,000 Square Feet

Exterior	Zone 0, 1	Zone 2	Zone 3	Zone 4
Precast panels, steel frame	527.57	536.07	550.31	562.49
Precast panels, reinforced concrete frame	443.97	452.47	466.71	478.88
Brick, concrete block back-up, steel frame	518.16	526.66	540.90	553.07
Brick, concrete block back-up, reinforced concrete frame	440.99	449.48	463.72	475.90
Curtain wall, metal and glass	530.79	539.29	553.53	565.70

30.0 HOTEL 4-7 STORY
Building Parameters: 5 Story, 10 Ft Story Height 100,000 Square Feet

Exterior	Zone 0, 1	Zone 2	Zone 3	Zone 4
Brick, concrete block back-up, steel frame	271.02	276.88	286.64	295.02
Brick, concrete block back-up, reinforced concrete frame	257.01	262.87	272.63	281.01
Precast panels, steel frame	273.89	279.75	289.51	297.89
Precast panels, reinforced concrete frame	259.88	265.74	275.50	283.88
Curtain wall, metal and glass	285.38	291.23	301.00	309.38

Costs include General Contractor's Overhead and Profit.
See page VI for seismic zones. See page VII-X for Cities Index.

Commercial Square Foot Building Costs

31.0 HOTEL 8-30 STORY
Building Parameters: 15 Story, 10 Ft Story Height, 470,000 Square Feet

Exterior	Zone 0, 1	Zone 2	Zone 3	Zone 4
Brick, concrete block back-up, steel frame	260.46	268.73	282.74	294.56
Precast panels, steel frame	262.41	270.68	284.69	296.52
Precast panels, reinforced concrete frame	238.29	246.56	260.57	272.40
Curtain wall, metal and glass	269.64	277.91	291.92	303.75
Decorative concrete block, steel frame	259.19	267.46	281.47	293.30

32.0 INDOOR TENNIS CLUB
Building Parameters: 1 Story, 24 Ft Story Height, 23,000 Square Feet

Exterior	Zone 0, 1	Zone 2	Zone 3	Zone 4
Tilt-up panels, steel frame	176.28	178.58	182.60	185.93
Metal siding on steel frame	167.55	169.85	173.87	177.20
Insulated metal panel, steel frame	167.32	169.62	173.64	176.97
Concrete block, steel roof frame	184.66	186.96	190.98	194.31
Precast panels, steel frame	198.79	201.08	205.10	208.43

33.0 JAIL
Building Parameters: 2 Story, 12 Ft Story Height, 20,000 Square Feet

Exterior	Zone 0, 1	Zone 2	Zone 3	Zone 4
Reinforced concrete, cast in place	376.33	383.91	396.66	407.57
Precast panels, reinforced concrete frame	319.03	326.60	339.35	350.26
Brick, concrete block back-up, steel frame	383.68	391.26	404.01	414.92
Brick, concrete block back-up, reinforced concrete frame	313.51	321.09	333.84	344.75
Decorative concrete block, steel frame	382.07	378.17	390.91	401.82

34.0 LAUNDROMAT
Building Parameters: 1 Story, 12 Ft Story, 8,000 Square Feet

Exterior	Zone 0, 1	Zone 2	Zone 3	Zone 4
Wood siding on stud frame	167.90	169.39	172.03	174.21
Stucco on stud frame	167.21	168.70	171.34	173.52
Decorative concrete block	176.39	177.89	180.53	182.71
Precast panels, steel frame	195.11	196.61	199.25	201.43
Insulated metal panel, steel frame	169.39	170.88	173.52	175.71

35.0 LIBRARY
Building Parameters: 2 Story, 14 Ft Story Height, 15,000 Square Feet

Exterior	Zone 0, 1	Zone 2	Zone 3	Zone 4
Decorative concrete block, steel frame	303.75	309.26	318.45	326.26
Brick, concrete block back-up, steel frame	308.35	313.86	323.04	330.85
Precast panels, steel frame	315.81	321.32	330.51	338.32
Tilt-up panels, steel frame	295.02	300.54	309.72	317.53
Stone veneer, block back-up, steel frame	342.68	348.19	357.38	365.19

Costs include General Contractor's Overhead and Profit.
See page VI for seismic zones. See page VII-X for Cities Index.

Commercial Square Foot Building Costs

36.0 MANUFACTURING, HEAVY
Building Parameters: 1 Story, 20 Ft Story Height, 40,000 Square Feet

Exterior	Zone 0, 1	Zone 2	Zone 3	Zone 4
Concrete block, steel roof frame	229.11	236.91	250.24	261.61
Reinforced concrete, cast in place	236.00	243.81	257.13	268.50
Reinforced concrete block, steel roof frame	229.11	236.91	250.24	261.61
Precast panels, reinforced concrete frame	236.68	244.49	257.82	269.18
Decorative concrete block, steel frame	229.68	237.49	250.81	262.18

37.0 MANUFACTURING, LIGHT
Building Parameters: 1 Story, 12 Ft Story Height, 35,000 Square Feet

Exterior	Zone 0, 1	Zone 2	Zone 3	Zone 4
Precast panels, steel frame	170.31	173.87	179.84	184.89
Tilt-up panels, steel frame	162.61	166.17	172.14	177.20
Metal siding on steel frame	159.63	163.19	169.16	174.21
Insulated metal panel, steel frame	172.95	176.51	182.48	187.53
Concrete block, steel roof frame	165.48	169.04	175.02	180.07

38.0 MEDICAL OFFICE
Building Parameters: 2 Story, 10 Ft Story Height, 8,000 Square Feet

Exterior	Zone 0, 1	Zone 2	Zone 3	Zone 4
Brick veneer on stud frame	349.57	351.41	354.51	357.15
Stucco on stud frame	343.60	345.44	348.54	351.18
Brick, concrete block back-up	355.77	357.61	360.71	363.35
Precast panels, steel frame	382.65	384.48	387.58	390.23
Curtain wall, metal and glass	416.64	418.48	421.58	424.22

39.0 MOTEL
Building Parameters: 3 Story, 9 Ft Story Height, 46,000 Square Feet

Exterior	Zone 0, 1	Zone 2	Zone 3	Zone 4
Wood siding on stud frame	202.00	203.73	206.60	209.12
Brick veneer on stud frame	204.30	206.02	208.89	211.42
Stucco on stud frame	201.66	203.38	206.25	208.78
Brick, concrete block back-up	207.17	208.89	211.76	214.29
Decorative concrete block, steel frame	242.43	244.15	247.02	249.55

40.0 MULTIPLE RESIDENCE
Building Parameters: 2 Story, 9 Ft Story Height, 7,000 Square Feet

Exterior	Zone 0, 1	Zone 2	Zone 3	Zone 4
Wood siding on stud frame	189.83	191.21	193.62	195.57
Brick veneer on stud frame	195.00	196.38	198.79	200.74
Decorative concrete block	197.41	198.79	201.20	203.15
Stucco on stud frame	189.03	190.40	192.82	194.77
Decorative concrete block, steel frame	212.45	213.83	216.24	218.20

Costs include General Contractor's Overhead and Profit.
See page VI for seismic zones. See page VII-X for Cities Index.

Commercial Square Foot Building Costs

41.0 MULTIPLE RESIDENCE, ELDERLY
Building Parameters: 3 Story, 9 Ft Story Height, 12,000 Square Feet

Exterior	Zone 0, 1	Zone 2	Zone 3	Zone 4
Brick, concrete block back-up	194.19	195.46	197.75	199.59
Decorative concrete block	190.40	191.67	193.96	195.80
Brick, concrete block back-up, steel frame	243.12	244.38	246.68	248.51
Decorative concrete block, steel frame	239.33	240.59	242.89	244.72
Precast panels, steel frame	249.20	250.47	252.76	254.60

42.0 OFFICE, 2-3 STORY
Building Parameters: 3 Story, 12 Ft Story Height, 23,000 Square Feet

Exterior	Zone 0, 1	Zone 2	Zone 3	Zone 4
Wood siding on stud frame	193.51	195.34	198.56	201.31
Brick veneer on stud frame	197.87	199.71	202.92	205.68
Stucco on stud frame	192.82	194.65	197.87	200.63
Decorative concrete block	201.08	202.92	206.14	208.89
Brick, concrete block back-up, steel frame	242.20	244.03	247.25	250.01

43.0 OFFICE, 4-7 STORY
Building Parameters: 6 Story, 12 Ft Story Height, 64,000 Square Feet

Exterior	Zone 0, 1	Zone 2	Zone 3	Zone 4
Decorative concrete block, steel frame	261.03	266.89	276.76	285.15
Brick, concrete block back-up, steel frame	263.90	269.76	279.64	288.02
Precast panels, steel frame	268.61	274.47	284.34	291.69
Curtain wall, metal and glass	292.84	298.70	308.57	316.96
Decorative concrete block, reinforced concrete frame	242.54	248.40	258.27	266.66

44.0 OFFICE, 8-30 STORY
Building Parameters: 20 Story, 12 Ft Story Height, 135,000 Square Feet

Exterior	Zone 0, 1	Zone 2	Zone 3	Zone 4
Precast panels, steel frame	313.97	321.78	335.22	346.59
Precast panels, reinforced concrete frame	257.93	265.74	279.18	290.54
Brick, concrete block back-up, steel frame	308.12	315.92	329.36	340.73
Brick, concrete block back-up, reinforced concrete frame	252.07	259.88	273.32	284.69
Curtain wall, metal and glass	344.63	352.44	365.88	377.25

45.0 POST OFFICE
Building Parameters: 1 Story, 18 Ft Story Height, 13,000 Square Feet

Exterior	Zone 0, 1	Zone 2	Zone 3	Zone 4
Brick, concrete block back-up, steel frame	286.64	290.20	296.40	301.57
Decorative concrete block, steel frame	282.39	285.95	292.15	297.32
Precast panels, steel frame	293.53	297.09	303.29	308.46
Decorative concrete block, steel frame	282.39	285.95	292.15	297.32
Tilt-up panels, steel frame	274.24	277.80	284.00	289.17

Costs include General Contractor's Overhead and Profit.
See page VI for seismic zones. See page VII-X for Cities Index.

Commercial Square Foot Building Costs

46.0 PRISON
Building Parameters: 2 Story, 12 Ft Story Height, 40,000 Square Feet

Exterior	Zone 0, 1	Zone 2	Zone 3	Zone 4
Reinforced concrete, cast in place	415.26	422.84	435.59	446.50
Precast panels, reinforced concrete frame	357.73	365.31	378.05	388.96
Brick, concrete block back-up, steel frame	425.25	432.83	445.58	456.49
Brick, concrete block back-up, reinforced concrete frame	353.71	361.29	374.03	384.94
Decorative concrete block, steel frame	408.03	415.61	428.35	439.26

47.0 RESTAURANT
Building Parameters: 1 Story, 12 Ft Story Height, 5,000 Square Feet

Exterior	Zone 0, 1	Zone 2	Zone 3	Zone 4
Wood siding on stud frame	283.65	285.26	288.02	337.51
Brick veneer on stud frame	289.51	291.12	293.88	331.20
Brick, concrete block back-up, steel frame	299.96	301.57	304.33	346.82
Decorative concrete block, steel frame	295.48	297.09	299.85	342.57
Stone veneer, block back-up, steel frame	333.61	335.22	337.97	360.48

48.0 RESTAURANT, FAST FOOD
Building Parameters: 1 Story, 10 Ft Story Height, 3,000 Square Feet

Exterior	Zone 0, 1	Zone 2	Zone 3	Zone 4
Brick veneer on stud frame	330.85	332.46	335.22	223.82
Stucco on stud frame	324.54	326.15	328.90	236.46
Brick, concrete block back-up, steel frame	340.16	341.76	344.52	221.76
Decorative concrete block steel frame	335.91	337.51	340.27	231.29
Insulated metal panel, steel frame	353.82	355.43	358.19	243.58

49.0 RINK, HOCKEY
Building Parameters: 1 Story, 24 Ft Story Height, 30,000 Square Feet

Exterior	Zone 0, 1	Zone 2	Zone 3	Zone 4
Tilt-up panels, steel frame	214.18	216.47	220.49	223.82
Brick, concrete block back-up	226.81	229.11	233.12	236.46
Insulated metal panel, steel frame	212.11	214.41	218.43	221.76
Concrete block, steel roof frame	221.64	223.94	227.96	231.29
Precast panels, steel frame	233.93	236.23	240.25	243.58

50.0 SCHOOL, ELEMENTARY
Building Parameters: 1 Story, 14 Ft Story Height, 43,000 Square Feet

Exterior	Zone 0, 1	Zone 2	Zone 3	Zone 4
Brick veneer on stud frame	286.64	290.43	290.43	302.26
Stucco on stud frame	327.98	331.77	296.29	343.60
Brick, concrete block back-up	334.18	337.97	306.74	349.80
Decorative concrete block, steel frame	332.12	335.91	302.26	347.74
Tilt-up panels, steel frame	328.33	332.12	340.39	343.95

Costs include General Contractor's Overhead and Profit.
See page VI for seismic zones. See page VII-X for Cities Index.

Commercial Square Foot Building Costs

51.0 SCHOOL, SECONDARY
Building Parameters: 2 Story, 14 Ft Story Height, 100,000 Square Feet

Exterior	Zone 0, 1	Zone 2	Zone 3	Zone 4
Brick, concrete block back-up, steel frame	338.09	347.39	363.12	376.45
Precast panels, steel frame	336.71	346.01	361.75	375.07
Stone veneer, block back-up, steel frame	384.94	394.25	409.98	423.30
Decorative concrete block, steel frame	337.40	346.70	362.43	375.76
Insulated metal panel, steel frame	339.35	348.65	364.39	377.71

52.0 SCHOOL, VOCATIONAL
Building Parameters: 2 Story, 14 Ft Story Height, 50,000 Square Feet

Exterior	Zone 0, 1	Zone 2	Zone 3	Zone 4
Decorative concrete block, steel frame	366.11	376.90	395.16	410.78
Brick, concrete block back-up, steel frame	366.91	377.71	395.97	411.59
Tilt-up panels, steel frame	354.28	365.08	383.34	398.95
Concrete block, steel roof frame	358.30	369.10	387.35	402.97
Precast panels, steel frame	365.08	375.87	394.13	409.75

53.0 SHOPPING CENTER, STRIP
Building Parameters: 1 Story, 10 Ft Story Height, 6,000 Square Feet

Exterior	Zone 0, 1	Zone 2	Zone 3	Zone 4
Brick, concrete block back-up, steel frame	237.60	240.70	245.87	250.35
Wood siding on stud frame	202.46	205.56	210.73	215.21
Brick veneer on stud frame	205.91	209.01	214.18	218.66
Decorative concrete block, steel frame	235.08	238.18	243.35	247.82
Stucco on stud frame	202.00	205.10	210.27	214.75

54.0 STORE, DEPARTMENT
Building Parameters: 2 Story, 16 Ft Story Height, 150,000 Square Feet

Exterior	Zone 0, 1	Zone 2	Zone 3	Zone 4
Brick, concrete block back-up, steel frame	199.82	204.53	212.57	219.34
Decorative concrete block, steel frame	198.21	202.92	210.96	217.74
Precast panels steel frame	202.46	207.17	215.21	221.99
Stone veneer, block back-up, steel frame	211.99	216.70	224.74	231.52
Tilt-up panels, steel frame	195.11	199.82	207.86	214.64

55.0 STORE, DISCOUNT
Building Parameters: 1 Story, 18 Ft Story Height, 80,000 Square Feet

Exterior	Zone 0, 1	Zone 2	Zone 3	Zone 4
Decorative concrete block, steel frame	167.44	170.19	174.67	178.46
Precast panels, steel frame	171.80	174.56	179.04	182.83
Brick, concrete block back-up, steel frame	169.04	171.80	176.28	180.07
Tilt-up panels, steel frame	164.22	166.98	171.46	175.25
Insulated metal panel, steel frame	170.77	173.52	178.00	181.79

Costs include General Contractor's Overhead and Profit.
See page VI for seismic zones. See page VII-X for Cities Index.

Commercial Square Foot Building Costs

56.0 STORE, RETAIL
Building Parameters: 1 Story, 14 Ft Story Height, 35,000 Square Feet

Exterior	Zone 0, 1	Zone 2	Zone 3	Zone 4
Brick, concrete block back-up, steel frame	175.13	177.77	182.25	186.16
Precast panels, steel frame	178.69	181.33	185.81	189.72
Decorative concrete block, steel frame	172.95	175.59	180.07	183.97
Tilt-up panels, steel frame	168.70	171.34	175.82	179.72
Stucco on stud frame	146.31	148.95	153.43	157.33

57.0 SUPERMARKET
Building Parameters: 1 Story, 20 Ft Story Height, 20,000 Square Feet

Exterior	Zone 0, 1	Zone 2	Zone 3	Zone 4
Brick, concrete block back-up	172.03	174.21	177.89	181.10
Decorative concrete block	169.50	171.69	175.36	178.58
Tilt-up panels, steel frame	168.24	170.42	174.10	177.31
Insulated metal panel, steel frame	179.84	182.02	185.70	188.91
Precast panels, steel frame	176.05	178.23	181.91	185.12

58.0 SURGICAL CENTER
Building Parameters: 2 Story, 14 Ft Story Height, 10,000 Square Feet

Exterior	Zone 0, 1	Zone 2	Zone 3	Zone 4
Decorative concrete block, steel frame	475.67	480.38	488.18	494.85
Brick, concrete block back-up, steel frame	481.64	486.35	494.16	500.82
Precast panels, steel frame	491.51	496.22	504.03	510.69
Stone veneer, block back-up, steel frame	526.66	531.36	539.17	545.83
Tilt-up panels, steel frame	464.30	469.01	476.82	483.48

59.0 Swimming POOL, ENCLOSED
Building Parameters: 1 Story, 24 Ft Story Height, 20,000 Square Feet

Exterior	Zone 0, 1	Zone 2	Zone 3	Zone 4
Brick, concrete block backup, steel frame	311.56	313.63	317.19	320.29
Decorative concrete block, steel frame	306.39	308.46	312.02	315.12
Insulated metal panel, steel frame	458.79	460.85	464.41	467.51
Reinforced concrete block, steel roof frame	305.47	307.54	311.10	314.20
Precast panels, steel frame	319.83	321.90	325.46	328.56

60.0 TERMINAL, AIRPORT
Building Parameters: 3 Story, 16 Ft Story Height, 140,000 Square Feet

Exterior	Zone 0, 1	Zone 2	Zone 3	Zone 4
Decorative concrete block, steel frame	352.90	358.64	368.29	376.45
Precast panels, steel frame	351.98	357.73	367.37	375.53
Brick, concrete block back-up, steel frame	353.71	359.45	369.10	377.25
Insulated metal panel, steel frame	355.20	360.94	370.59	378.74
Curtain wall, metal and glass	369.21	374.95	384.60	392.75

Costs include General Contractor's Overhead and Profit.
See page VI for seismic zones. See page VII-X for Cities Index.

Commercial Square Foot Building Costs

61.0 Terminal BUS
Building Parameters: 1 Story, 14 Ft Story Height, 15,000 Square Feet

Exterior	Zone 0, 1	Zone 2	Zone 3	Zone 4
Brick, concrete block back-up steel frame	236.91	240.36	245.99	250.81
Decorative concrete block, steel frame	233.93	237.37	243.00	247.82
Tilt-up panels, steel frame	228.07	231.52	237.14	241.97
Wood siding on stud frame	183.05	186.50	192.13	196.95
Brick veneer on stud frame	186.96	190.40	196.03	200.85

62.0 THEATER, MOVIE
Building Parameters: 1 Story, 20 Ft Story Height, 16,000 Square Feet

Exterior	Zone 0, 1	Zone 2	Zone 3	Zone 4
Decorative concrete block, steel frame	270.91	273.09	276.88	280.09
Precast panels, steel frame	284.00	286.18	289.97	293.19
Brick, Concrete block back-up, steel frame	275.85	278.03	281.82	285.03
Tilt-up panels, steel frame	261.38	263.56	267.35	270.56
Insulated metal panel, steel frame	291.81	293.99	297.78	301.00

63.0 VETERINARY HOSPITAL
Building Parameters: 1 Story, 12 Ft Story Height, 6,000 Square Feet

Exterior	Zone 0, 1	Zone 2	Zone 3	Zone 4
Brick veneer on stud frame	324.54	326.60	330.16	333.27
Wood siding on stud frame	318.57	320.63	324.19	327.29
Brick, concrete block back-up, steel frame	352.33	354.40	357.96	361.06
Decorative concrete block, steel frame	347.62	349.69	353.25	356.35
Insulated metal panel, steel frame	367.26	369.33	372.89	375.99

64.0 WAREHOUSE
Building Parameters: 1 Story, 24 Ft Story Height, 45,000 Square Feet

Exterior	Zone 0, 1	Zone 2	Zone 3	Zone 4
Concrete block, steel roof frame	143.32	146.31	151.36	155.61
Insulated metal panel, steel frame	158.25	161.24	166.29	170.54
Tilt-up panels, steel frame	137.69	140.68	145.73	149.98
Precast panels, steel frame	152.97	155.95	161.01	165.25
Metal siding on steel frame	131.72	134.71	139.76	144.01

65.0 WAREHOUSE, SELF STORAGE
Building Parameters: 1 Story, 12 Ft Story Height, 33,000 Square Feet

Exterior	Zone 0, 1	Zone 2	Zone 3	Zone 4
Concrete block, steel roof frame	175.71	177.77	181.33	184.43
Insulated metal panel, steel frame	184.89	186.96	190.52	193.62
Tilt-up panels, steel frame	172.26	174.33	177.89	180.99
Precast panels, steel frame	181.68	183.74	187.30	190.40
Metal siding on steel frame	168.58	170.65	174.21	177.31

Costs include General Contractor's Overhead and Profit.
See page VI for seismic zones. See page VII-X for Cities Index.

Index

A

Accordion partitions 126
Aftercoolers 8
Aggregate 46
Air compressors 6
    portable 7
Air conditioners 185
Air floor scrabblers 14
Air hoists 158
Air hoses 8
Air manifolds 8
Air vibrators 12
Alarms 174
Altars 134
Arks 133
Asphalt cutter 9
Auger holes 32
Augers 16
Awnings 132

B

Backhoe loaders 16
Backhoes 16
Baggage carousels 159
Barrel pumps 22
Baseball backstops 140
Basketball backstop 140
Beams 52, 76
    steel 69
Bell footings 33
Bench, cast iron 50
Bench, wood 50
Berths
    open 51
Berths, covered 51
Bidets 162
Bike locks 47
Bits
    carbide 9
Bitumals 84
Bleachers 131, 141
Blinds 150, 152
Board
    chalk 125
    tack 125
Boilers 183
Bookcases 153
Bowstring trusses 81
Box scrapers 15
Breakers
    paving 8
Brick 67
Brick veneer 65
Brick walls 65
Brooms
    rotary 16
Buckets
    concrete 13

C

Cabinets 80, 150
Cable trays 230
Camera 135
Canopies 132
Canopy framing 73
Capacitors 219
Carbide bits 9
Carpets 151
Cast iron pipes 35
Cathedral chairs 134
Cedar facia 79
Ceiling
    removal 27
Cement 46
Centrifugal pumps 17
Centrifuges 143
Chalk board 125
Chemical toilets 3
Chillers 184
Chisels 8
Chutes 129
Circuit breakers 216
Clay diggers 9
Colored glass 135
Columns, travertine 67
Communion rails 134
Compactors, vibratory 10
Compressors
    air, portable 7
    electric 7
Compressors, air 6
Concrete
    floor grinders 14
    lightweight 53
Concrete buckets 13
Concrete conveyors 4
Concrete floor grinders 14
Concrete hardeners 47
Concrete saws 13
Concrete troweling machines 13
Concrete walls 52
Confessionals 134
Conveyor belts 14
Conveyors
    concrete 4
Cook tops 136
Cooling towers 185
Coping 90
Corrugated roofing 88
Covered berths 51
Cranes
    crawler 4
    handy 4
    tower 3
    truck 4
    truck, hydraulic 4
Crawler cranes 4
Crawler tractors 15
Curbs 46
Curtains 152
Cutter
    asphalt 9

D

Deck, timber 77
Demolition hammers 11
Dental equipment 146
Dental laboratory furniture 147
Derricks
    guy 4
    stiff leg 4
Diamond core bits 12
Diaphragm pumps 18
Directories 129
Disc harrows 15
Dishwashers 136
Domes, observation 136
Doors
    refrigerator 101
    revolving 101
    wood 80
Downspouts 90
Draftstops 117
Draperies 152
Dredging 51
Dressing cubicles 126
Drill
    rock 9
Drill steel 9
Drinking fountains 164
Dryers 136
Ducts 193
Dumbwaiters 157

Index

E

Edging 50
Electric compressors 7
Electric grills 136
Elevators 157
Emergency generators 207
Epoxy injection 64
Escalators 158
Excavators 15
    mini 15
Extinguishers 130

F

Facet glass 135
Facia
    cedar 79
    redwood 79
Fans 189
Fees, sewer 2
Fees, water meter 2
Fence
    chain link, removal 26
Fences
    temporary 3
Fertilizer 50
File cabinets 153
Fill boxes 141
Fills 54
Fire hose cabinets 131
Fireplaces 68
Fireproofing 87
Fire proofing treatment 78
Fire protection systems 201
Fixtures 209
Flag poles 129
Flanges
    cast iron 197
    steel 197
Flashings 91
Flex couplings 229
Floor decking 71
Floor grinders
    concrete 14
Floor sheathing 77
Fluorescent fixtures 238
Fluorescent step lights 45
Folding partitions 127
Footlights 148
Forklifts
    rental 4
Foundations 52, 55
    institutional 63
Foundations, accessories 62
Frames 96
Freezers 136
Fuel storage tanks 22
Furring 77

G

Garbage compactors 149
Garbage disposals 136
Gin poles 4
Girders 52
    steel 69
Glass
    colored 135
    facet 135
    stained 135
Glazing, 107
Gradings 30
Granite 67
Gravel stops 91
Grinders 149
Ground covers 49
Guardrail
    removal 26
Gunite 54
Gutters 90
Guy derricks 4

H

Hammers
    demolition 11
Handy cranes 4
Hoods 136
Hoses
    air 8
HVAC 182
Hydraulic truck cranes 4

I

Incandescent fixtures 238
Insulation 200
Insulation board deck 85
Insulation, piping 171
Irrigation pipes 49

J

Jackhammer 9
Jack posts 76
Jack studs 76
Jitterbugs 14

K

Kick plates 106
Kneelers 134

L

Latchsets 105
Lathes 149
Lavatories 162
Lecterns 133
Letters 130
Library shelving 143
Lighting
    fluorescent 45
    parking lot 45
Lights, football field 45
Lightweight concrete 53
Lime bulk 46
Limestone 67
Loaders
    backhoe 16
    skid steer 16
Lockers 131
Locksets 105

Index

M

Mail chutes 131
Mail conveyors 159
Manholes
  removal 26
Manifolds 174
  air 8
Marble 67
Masonry wall finishes 68
Mini excavators 15
Mixers 12
Monitor 135
Mortuary refrigerator 147
Mowers
  rotary 15
Mulch 50

N

Nails 82

O

Office trailers 3
Open berths 51
Ovens 136

P

Panelboards 206
Panic devices 106
Paper holders 118
Pargeting 68
Parking lot lighting 45
Pavement
  removal 26
Paving breakers 8
Permits 2, 3
Pews 134
Phone enclosures 132
Piers 32
Piles 31
  sheet 33
Pilot lights 218
Pipe jacking 34
Pipes
  cast iron 35
  irrigation 49
Plaques 130
Plaster 114
Plastic roofing 88
Playgrounds 45
Poles
  gin 4
  wood 44
Portable air compressors 7
Power pole, temporary 3
Prison equipment 147
Public address/intercoms 244
Pulpits 133
Pumps
  barrel 22
  centrifugal 17
  diaphragm 18
  sump 18
  trash 17

R

Rafters 75
Railings
  temporary 3
Rammers 10
Readymix 60
Redwood facia 79
Redwood trim 79
Reel banks 141
Refrigerator
  mortuary 147
Refrigerator doors 101
Refrigerators 136
Reglets 91
Relocatable partitions 126
Revolving doors 101
Robe hooks 129
Rock drill 9
Rollers, static 11
Roof
  removal 27
Roof decking 71
Roof drain 173
Roofing
  corrugated 88
  plastic 88
Rotary hammer drills 12
Rotary mowers 15
Roto tillers 15
Rowing machines 146
Running tracks 140

S

Sandblasting 62
Sandstone 67
Saws
  concrete 13
Scaffolds 23
Scales 131
Scissor lift 142
Scissors lift 148
Score boards 141
Scrabblers
  floor, air 14
Seating 135
Seeding 50
Septic tanks 162
Sewer fees 2
Shades 150
Shakes 87
Sheet piles 33
Shelves 131
Shelving
  library 143
Shingles 87
Shower drain 173
Shrubs 50
Siding 79
Signals
  traffic 45
Skid steer loaders 16
Soccer goals 140
Sodding 50
Space heaters 22
Spades 9
Spandrel 110
Spotlights 148
Sprinkler heads 49
Stacks 184
Stage equipment 147
Stained glass 135
Stair, circular 81
Stair, straight 81
Standpipes 202
Steps, concrete 61
Stiff leg derrick 4
Stonework 67
Storefront systems 109
Striping 47
Stripping 77
Stucco 114
Studs 112
Study carrel 135, 143
Submersibles 17
Sump pumps 18
Switchboards 215

Index

Switchboards, main 206
Switches 44, 174

T

Tack board 125
Tanks
    fuel storage 22
Telephone cables 244
Telephone, temporary 3
Telescopes 136
Teller windows 133
Temporary power pole 3
Tennis courts 45
Tennis nets 140
Tennis posts 140
Thermostats 193
Tile, clay 87
Tile, concrete 87
Toilet accessories 81
Toilet partitions 67, 125
Toilets, chemical 3
Towel dispensers 127
Tower cranes 3
Tractors
    crawler 15
    wheel 15
Traffic signals 45
Trailers, office 3
Transformers 44, 207
Trash compactor 136
Trash pumps 17
Travertine 67
Treadmills 146
Trees 50
Trenchers 14
Trim
    redwood 79
Trollies 149
Troweling machines
    concrete 13
Truck cranes 4
    hydraulic 4
Trucks
    rental 4
Trusses
    heavy steel 69
    light steel 69
Turnstiles 130
Turntables 149

U

Urinals 163
Urinal screens 125

V

Vacuum systems 147
Vaults
    bank 133
Veneer, brick 65
Vibrators, air 12
Video tape recorder 135
Voltmeters 218

W

Wainscot, neoprene 119
Walls
    brick 65
    concrete 52
Wardrobes 135
Washer 136
Waste receptacles 127
Water closets 163
Water coolers 164
Water heater 161
Water meter fees 2
Waterproof, deck 84
Water softeners 162
Water stop 62
Weatherstrip 107
Welders 20
Wharfs 51
Wheelbarrows 14
Wheel tractors 15
Whip blast 124
Windows
    bank teller 133
Windows, aluminum 103
Windows, vinyl 102
Wiremolds 231
Wood poles 44

X

X-rays 145

Z

Zone valves 174
Zonolite 64

Printed in the USA
CPSIA information can be obtained
at www.ICGtesting.com
LVHW081123261023
762202LV00012B/336